Power-to-X in Regional Energy Systems

Power-to-X in Regional Energy Systems discusses the role of these technologies in achieving a carbon-neutral economy and the impact on the energy markets, with implications for electricity, gas, hydrogen, and ancillary services. It focuses on the challenges and benefits of implementing PtX technologies in regional-scale applications.

Emphasizing the role of PtX technologies as enablers of sector coupling, the book provides a comprehensive understanding of how these technologies integrate and interact with the industry, transportation, and residential sectors. It describes the significance of PtX, optimal planning, and cost-effective operation of PtX technologies across different sectors and the impact of PtX devices on energy markets. The book considers investing in PtX technologies and contributing to the transition to a sustainable economy.

The book will interest professionals and policymakers working in various energy sectors. Researchers and academics in electrical engineering, power systems, renewable energy, and energy economics will also find the content useful.

Power-to-X in Regional Energy Systems

Planning, Operation, Control, and Market Perspectives

Edited by Amjad Anvari-Moghaddam,
Sina Ghaemi, Shi You, and Frede Blaabjerg

CRC Press
Taylor & Francis Group
Boca Raton London New York

CRC Press is an imprint of the
Taylor & Francis Group, an **informa** business

Designed cover image: iStockphoto

MATLAB® and Simulink® are trademarks of The MathWorks, Inc. and are used with permission. The MathWorks does not warrant the accuracy of the text or exercises in this book. This book's use or discussion of MATLAB® or Simulink® software or related products does not constitute endorsement or sponsorship by The MathWorks of a particular pedagogical approach or particular use of the MATLAB® and Simulink® software.

First edition published 2025
by CRC Press
2385 NW Executive Center Drive, Suite 320, Boca Raton FL 33431

and by CRC Press
4 Park Square, Milton Park, Abingdon, Oxon, OX14 4RN

CRC Press is an imprint of Taylor & Francis Group, LLC

ISBN: 978-1-032-71941-2 (hbk)
ISBN: 978-1-032-71942-9 (pbk)
ISBN: 978-1-032-71943-6 (ebk)

DOI: 10.1201/9781032719436

Typeset in Times
by Apex CoVantage, LLC

Contents

Editors' Bios

Amjad Anvari-Moghaddam (S'10–M'14–SM'17) is a full professor and leader of the Intelligent Energy Systems and Flexible Markets (iGRIDS) Research Group at the Department of Energy (AAU Energy), Aalborg University, where he is also acting as the vice-leader of Power Electronic Control, Reliability, and System Optimization (PESYS) and the coordinator of Integrated Energy Systems Laboratory (IES-Lab). His research interests include planning, control, and operation management of microgrids, renewable/hybrid power systems, and integrated energy systems with appropriate market mechanisms. He has (co)authored more than 300 technical articles, 9 books, and 17 book chapters in the field. Dr. Anvari-Moghaddam is the editor-in-chief of the *Academia Engineering* journal and serves as the associate editor of several leading journals, such as the *IEEE Transactions on Power Systems, IEEE Systems Journal, IEEE Open Access Journal of Power and Energy*, and *IEEE Power Engineering Letters*. He is the chair of IEEE Denmark, a member of IEC SC/8B Working Group (WG3, WG6, and WG8), and as well as a technical committee member of several IEEE PES/IES/PELS and CIGRE WGs. He also serves as an alternate member of the Council for Energy Efficient Transition at the Danish Ministry of Climate, Energy, and Utilities; a member of Reference Group 5, "Climate, Energy, and Mobility" for Horizon Europe (HE) at the Danish Agency for Education and Research (UFS); and a board member of Energy Mission at Aalborg University. He was the recipient of the 2020 DUO–India Fellowship Award, the DANIDA Research Fellowship grant from the Ministry of Foreign Affairs of Denmark in 2018 and 2021, the IEEE-CS Outstanding Leadership Award 2018 (Halifax, Nova Scotia, Canada), and the 2017 IEEE-CS Outstanding Service Award (Exeter, UK).

Sina Ghaemi is a postdoctoral researcher in the Energy Department at Aalborg University. His primary research areas include renewable energy, the hydrogen economy, energy markets, and integrated energy systems. He has extensive experience in addressing the challenges and opportunities associated with carbon-neutral energy transitions in regional areas, through the development of market-oriented and interdisciplinary solutions.

Shi You is a senior researcher with the Department of Wind and Energy Systems, Technical University of Denmark (DTU). He received his BSc and MSc degrees in electrical engineering from Harbin Institute of Technology, China, in 2004 and Chalmers Institute of Technology, Sweden, in 2007, respectively. After he received his PhD in electrical engineering from DTU in 2010, Shi has been working on developing theoretical and practical smart grid and integrated energy solutions that aim at enabling a future energy system with an extremely high share of renewables. He has worked on energy storage, power-to-heat, power-to-gas, integrated planning and operation, demand response, advanced control and automation, and market-based energy management, etc. His expertise in this field was developed through deep involvement in 30+ national and international R&D projects and close collaboration with related industrial stakeholders. Shi has also published more than 150 peer-reviewed articles.

Frede Blaabjerg (S'86–M'88–SM'97–F'03) was with ABB–Scandia, Randers, Denmark, from 1987 to 1988. He became an assistant professor in 1992, an associate professor in 1996, and a full professor of power electronics and drives in 1998 at AAU Energy. In 2017, he became a Villum Investigator. He is honoris causa at University Politehnica Timisoara (UPT), Romania, in 2017, and at Tallinn Technical University (TTU) in 2018. He has published more than 600 journal papers in the fields of power electronics and its applications. He is the coauthor of eight monographs and the editor of 14 books in power electronics and its applications. He has received 38 IEEE Prize Paper Awards, the IEEE PELS Distinguished Service Award in 2009, the EPE-PEMC Council Award in 2010, the IEEE William E. Newell Power Electronics Award in 2014, the Villum Kann Rasmussen Research Award in 2014, the Global Energy Prize in 2019, and the 2020 IEEE Edison Medal. He was the editor-in-chief of the *IEEE Transactions on Power Electronics* from 2006 to 2012. Between 2019 and 2020, he served as the president of IEEE Power Electronics Society. He has been vice-president of the Danish Academy of Technical Sciences.

Contributors

Mehdi Abapour
Energy Systems Research Institute (ESRI)
Smart Energy Systems Lab
Iran

Abdullah A. Almehizia
Future Energy Technologies Institute
King Abdulaziz City for Science and Technology
Saudi Arabia

Ali Aminlou
Energy Systems Research Institute (ESRI)
University of Tabriz
Iran

Shiva Ansaripour
Energy Modelling and Sustainable Energy
   System (METSAP) Research Lab
University of Tehran
Iran

Arash Asrari
Department of Electrical and Computer
   Engineering
Purdue University Northwest
USA

Hamed Barfan
Faculty of Electrical Engineering
Shahid Behestri University
Iran

Nicola Bianco
Department of Industrial Engineering
University of Naples Federico II
Italy

Houhe Chen
Department of Electrical Engineering
Northeast Electric Power University
China

Rahman Dashti
Clinical Laboratory Center of Power System
   Protection
Persian Gulf University
Iran

Marialaura Di Somma
Department of Industrial Engineering
University of Naples Federico II
Italy

M. Edwin
Department of Mechanical Engineering
University College of Engineering
India

M. C. Eniyan
Department of Mechanical Engineering
University College of Engineering
India

Danial Esmaeili Aliabadi
Helmholtz Centre for Environmental
   Research–UFZ
Germany

Samuel Raafat Fahim
Electrical Power and Machines Department
Ain Shams University
Egypt

Khashayar Fardnia
Energy Modeling and Sustainable Energy
   Systems (METSAP) Lab
University of Tehran
Iran

Asghar Akbari Foroud
Electrical and Computer Engineering
   Faculty
Semnan University
Iran

Ahmad Nazari Gazik
Energy Modeling and Sustainable Energy
   Systems (METSAP) Lab
University of Tehran
Iran

Gevork B. Gharehpetian
Department of Electrical Engineering
Amirkabir University of Technology
Iran

Reza Gharibi
Clinical Laboratory Center of Power System
    Protection
Persian Gulf University
Iran

Mohammad Hasan Ghodusinejad
Energy Modeling and Sustainable Energy
    Systems (METSAP) Lab
University of Tehran
Iran

Giorgio Graditi
Italian National Agency for New Technologies
Energy and Sustainable Economic
    Development (ENEA)
Italy

Sarmad Hanif
Electric Power Research Institute (EPRI)
USA

Hany M. Hasanien
Electrical Power and Machines Department
Ain Shams University
Egypt

Seyed Mohsen Hashemi
Power Systems Operation and Planning
    Research Department
Niroo Research Institute (NRI)
Iran

Mohammad Mohsen Hayati
Energy Systems Research Institute (ESRI)
Smart Energy Systems Lab
Iran

Chunjun Huang
Delft University of Technology
The Netherlands

Farkhondeh Jabari
Power Systems Operations and Planning
    Research Department
Niroo Research Institute (NRI)
Iran

Tao Jiang
Department of Electrical Engineering
Northeast Electric Power University
China

Matthias Jordan
Helmholtz Centre for Environmental
    Research–UFZ
Germany

Mohammad Karrabi
Electrical and Computer Engineering Faculty
Semnan University
Iran

Reza Khalili
Department of Electrical and Computer
    Engineering
Stony Brook University
USA

Adil Khurram
University of California–San Diego
USA

Fangxing Li
Department of Electrical Engineering and
    Computer Science
The University of Tennessee
USA

Guoqing Li
Department of Electrical Engineering and
    Computer Science
The University of Tennessee
USA

Xue Li
Department of Electrical Engineering
Northeast Electric Power University
China

Abbas Marini
Power Systems Operation and Planning
    Research Department
Niroo Research Institute (NRI)
Iran

Amir Meydani
Department of Electrical Engineering
Amirkabir University of Technology
Iran

Andreas Meurer
German Aerospace Center–DLR
Institute of Networked Energy Systems
Germany

Ehsan Naderi
Department of Electrical Engineering
Arkansas State University
USA

Mehrdad Setayesh Nazar
Faculty of Electrical Engineering
Shahid Beheshti University
Iran

Nawaf Nazir
Pacific Northwest National Laboratory (PNNL)
USA

Mahshid Noorollahi
Faculty of Economics
University of Tehran
Iran

Younes Nourolahi
Energy Modelling and Sustainable Energy
    System (METSAP) Research Lab
University of Tehran
Iran

Arman Oshnoei
Department of Energy
Aalborg University
Denmark

Christina Papadimitriou
Department of Electrical Engineering
Eindhoven University of Technology
The Netherlands

Zhihua Qu
Department of Electrical and Computer
    Engineering
University of Central Florida
USA

Towfiq Rahman
Department of Electrical and Computer
    Engineering
University of Central Florida
USA

Ashkan Safari
Faculty of Electrical and Computer
    Engineering
University of Tabriz
Iran

Miadreza Shafie-khah
School of Technology and
    Innovations
University of Vaasa
Finland

Hossein Shahinzadeh
Department of Electrical Engineering
Amirkabir University of Technology
Iran

Mohammad Reza Sheibani
Power Systems Operation and Planning
    Research Department
Niroo Research Institute (NRI)
Iran

Behrooz Vahidi
Electrical Engineering Department
Amirkabir University of Technology
Iran

Niklas Wulff
German Aerospace Center–DLR
Institute of Networked Energy Systems
Germany

Shi You
Technical University of Denmark
Denmark

Hossein Yousefi
Energy Modeling and Sustainable Energy
    Systems (METSAP) Lab
University of Tehran
Iran

Mehdi Zeraati
Power Systems Operation and Planning
    Research Department
Niroo Research Institute (NRI)
Iran

Rufeng Zhang
Department of Electrical Engineering
Northeast Electric Power University
China

Yi Zheng
Inner Mongolia University of Technology
China

Preface

The intensifying impact of global warming and climate change has heightened the urgency for adopting more environmentally friendly energy systems. Currently, nearly two-thirds of the world's carbon emissions originate from activities tied to these energy systems. As a result, transitioning to greener energy sectors has become a priority for nations committed to the Paris Agreement, which aims to limit global temperature increases to below 2°C. Achieving this ambitious goal requires the implementation of sustainable energy solutions, particularly to decarbonize hard-to-abate sectors in regions where energy generation and consumption are most concentrated.

Innovative technologies, such as power-to-X (PtX), offer a promising pathway for this transition by converting renewable power into usable energy carriers, facilitating the decarbonization process. These advancements not only enable flexible sector integration across regional landscapes but also provide a critical solution for accelerating the shift to a carbon-neutral economy. However, the effective deployment of PtX technologies in energy-intensive regions to meet environmental targets requires a comprehensive analysis from technical, economic, and regulatory perspectives.

This book serves as a key resource, guiding readers through the transformative potential of PtX technologies and their pivotal role in shaping carbon-neutral economies at the regional level. It explores the diverse applications of PtX technologies, which convert surplus renewable energy into versatile energy carriers, such as hydrogen, synthetic fuels, and chemicals. Through these technologies, PtX emerges as a cornerstone in the effort to promote carbon-neutral practices across regions, industries, and communities.

Whether the readers are industry leaders, policymakers, researchers, or entrepreneurs, this book is designed to provide valuable insights into the future of sustainable energy. It offers a roadmap to a future where carbon neutrality is not a distant goal but a tangible reality. The central focus of the book is to explore the potential of PtX technologies in decarbonizing hard-to-abate sectors in regional contexts. To achieve this, the book is structured around four key themes: concepts (Chapters 1–5), planning (Chapters 6–9), operation (Chapters 10–17), and market (Chapters 18–21).

In Chapter 1, PtX systems are introduced as vital for converting renewable energy into valuable products, like synthetic fuels, chemicals, and ammonia. The chapter explores key technologies, such as water electrolyzers, carbon capture, and hydrogen storage, while also examining the techno-economic, thermodynamic, and environmental benefits of PtX. It highlights the integration of hybrid renewable energy systems (HRESs) to improve efficiency and flexibility across energy sectors. The chapter concludes by emphasizing PtX's transformative potential in the global energy transition and identifies key areas for future research and development.

The focus of Chapter 2 is on the accelerating adoption of renewable energy sources (RESs), driven by declining costs and the need to replace environmentally harmful fossil fuels amid rising energy demands. The chapter also explores the criteria for selecting appropriate PtX technologies based on local features, examines international initiatives, and addresses the challenges of implementation.

In Chapter 3, the focus shifts to the need for alternative energy sources due to population growth and the depletion of fossil fuels. The chapter highlights the importance of optimizing the synergy between hydrogen carriers and RESs to enhance sustainability and reduce waste. It also emphasizes the benefits of hybrid renewable energy systems (HRESs) for resource diversity and system reliability, exploring PtX pathways and identifying gaps and future research opportunities in this area.

The integration of sustainable energy sources with traditional power systems in developing countries is explored in Chapter 4, addressing environmental concerns and the energy crisis. The chapter reviews advances in carbon capture, methane production, and green hydrogenation, focusing on how CO_2 captured from thermal plants can be used to generate methane for power, heat, and water production. Additionally, it discusses hydrogen production via electrolysis using renewable energy, offering an eco-friendly approach to meeting methane demand. The chapter concludes with an overview of current technologies, challenges, and future research directions in green gasification and decarbonization.

In Chapter 5, the PtX concept is presented as a crucial solution for decarbonizing the economy by addressing the intermittency of RESs. The chapter highlights three main energy storage forms: power-to-gas (PtG) for hydrogen gas; power-to-liquid (PtL) for synthetic carbon-neutral fuels, like methane and methanol; and power-to-heat (PtH). It discusses the synergy between these PtX technologies and their potential integration across various sectors, including transportation and residential applications, emphasizing that optimizing this synergy can accelerate the adoption of PtX solutions.

The role of PtX technologies in energy hubs (EH) is examined as a vital component of the transition to a sustainable energy sector in Chapter 6, where it details the modeling of EHs and their key components, such as energy storage systems and demand-side management, while highlighting technologies like power-to-green hydrogen, power-to-methane, and power-to-ammonia. Additionally, the chapter discusses PtX's impact on electricity markets, emphasizing its potential to stabilize operations and foster a balanced energy landscape, ultimately calling for collaborative efforts to fully leverage PtX for global decarbonization.

Chapter 7 discusses the necessity for new decision support tools and frameworks to assist engineers and researchers in navigating emerging complexities, such as the increasing integration of renewables along with flexible loads, such as electric vehicles (EVs). The chapter also reviews recent advancements, ranging from traditional model-based optimization to modern data-driven methods, including machine learning and reinforcement learning. Finally, it outlines the current challenges in grid management, evaluates existing methods, and identifies open research problems that need to be addressed for the future of grid operations.

The focus of Chapter 8 is on integrated energy systems (IES) and the vital role of PtX technologies in enhancing resilience and sustainability in modern energy systems. The chapter defines *IES* as a complex network where diverse energy sources and infrastructures collaborate. It highlights PtX technologies' potential to revolutionize energy storage, address grid disruptions, and improve flexibility, thereby aiding decarbonization, facilitating synthetic fuel generation, and enabling electrification in underserved sectors. The chapter also underscores the importance of aligning supply with regional demand and enhancing energy self-sufficiency, influenced by factors such as economic viability and policy support. Through case studies, the chapter illustrates the effectiveness of PtX technologies and predicts their crucial role in addressing future energy and environmental challenges.

Chapter 9 elaborates on the critical need for defossilizing transportation and heat provision, as RESs comprise over 50% of the electricity mix. The chapter discusses the potential of biofuels as substitutes for synthetic fuels derived from renewable sources, highlighting trade-offs, synergies, and competition within energy planning. It offers a systematic terminology for different fuel types and provides an overview of synthetic fuel production routes and biofuel supply options. Additionally, two case studies are presented, illustrating the impact of synthetic and biofuels on energy system transformation, followed by key research questions and future research perspectives.

In Chapter 10, the integration of hydrogen production, electricity generation, and artificial intelligence (AI) technologies within smart grids is examined as a key strategy for developing sustainable and resilient energy systems. The chapter utilizes simulations and machine learning models to analyze and predict hydrogen production, electricity consumption, and state-of-charge (SoC) levels in hydrogen-enabled energy systems, focusing on hydrogen generation from wind and solar energy through alkaline technology.

The complexity of current power systems, driven by increased integration of renewable energy, fluctuating production, distributed generation, and rising demand, is addressed in Chapter 11, highlighting the role of intelligent techniques in secure grid management. This chapter also involves analyzing demand curves and consumer behavior using AI algorithms to predict unexpected patterns. Finally, it focuses on AI-based DSM applications for real-time response to consumption patterns, demand forecasting, and implementing demand response strategies in power-to-X systems, ultimately enhancing the performance, sustainability, and economic viability of these energy systems.

In Chapter 12, the importance of renewable energy in modern power systems is emphasized, particularly for achieving carbon neutrality, which requires efficient monitoring and control through information and communication technology (ICT). This chapter examines how false data injection attacks (FDIAs) can target power-to-X components by injecting harmful information into load centers, leading to operational issues like system congestion and potential cascading outages.

The power-to-mobility aspects of energy systems are explored in Chapter 13, where the increasing integration of EVs into the distribution grid and the advancements in EV technologies, such as vehicle-to-grid (V2G) systems, enable their participation in grid ancillary services. The chapter also elaborates on multi-layer time-dependent optimization models to maximize the utilization of EVs for ancillary services.

Chapter 14 focuses on the power-to-gas (P2G) facility as a key element in integrating power and gas networks. The chapter discusses recent advancements in P2G technologies and the decreasing costs of renewable energy, which have facilitated the global deployment of P2G facilities. It presents a model for operating integrated gas and power networks that incorporates P2G facilities, energy storage systems, renewable energy resources, and various power generators, including combined heat and power (CHP) units.

In Chapter 15, the focus is on achieving a climate-neutral energy system through an integrated decarbonization approach across sectors like electricity, heating, cooling, transport, and industry. It highlights sector coupling's role in enhancing system flexibility and efficiency, while addressing the challenges posed by interdependencies among energy carriers and PtX technologies. The chapter presents an analytical framework for optimizing the operation of sector coupling and PtX technologies within a multi-carrier energy system (MCES) with multiple sustainability objectives, supported by two illustrative case studies of varying complexity.

Enhancing the flexibility of energy systems through the integration of RES and PtX systems is detailed in Chapter 16, where various types of PtX systems commonly employed to improve this flexibility are explored. The role of PtX systems as complementary infrastructures to other solutions, such as energy storage and demand management tools, is examined. Finally, strategies for operational and expansion planning of modern power systems incorporating PtX technologies are discussed.

In Chapter 17, a foundational introduction to the engineering aspects of modeling, control, and operation of grid-connected electrolyzer plants is provided, with a focus on the commercialized alkaline electrolyzer (AEL) technology. The critical role of electrolysis in renewables-dominated power systems is emphasized, particularly regarding energy storage, grid balancing, and sector coupling. Various modeling methods are examined to accurately represent the dynamic behavior of AEL systems, alongside control strategies for temperature stabilization, power regulation, and dynamic frequency support to enhance grid stability and flexibility. Additionally, operational aspects are explored, including optimization techniques for maximizing economic benefits through participation in power markets.

The integration of RESs into the electricity grid is examined in Chapter 18, highlighting the challenges and opportunities in ancillary services due to variable resources, like wind and solar power. The chapter discusses the PtX concept as a solution for converting surplus renewable electricity into carbon-neutral fuels, facilitating grid balancing, peak shaving, and reducing carbon emissions. It emphasizes the role of dynamic energy pricing and demand-side management in optimizing PtX operations and explores the potential for cross-sector coupling to enhance synergies among electricity, heating, and transportation sectors. The chapter also contributes to the discourse on sustainable energy transitions and innovative market designs for maximizing renewable energy resources.

In Chapter 19, the focus is on P2G as an emerging energy storage technology for renewable energy, emphasizing the need for efficient bidding strategies to enhance profitability. A bi-level strategic bidding model is proposed, incorporating a carbon emission trading scheme (CETS), locational marginal pricing (LMP), and wind power uncertainty, in which the upper-level model aims to maximize P2G profits, while the lower-level model handles market clearing through a stochastic low-carbon economic dispatch (ED) process.

Chapter 20 highlights P2G technology as a means to enhance flexibility in electricity systems characterized by high renewable energy integration. It proposes a coordinated bidding strategy for wind farms and P2G facilities, employing a cooperative game approach to optimize profits. The framework includes wind power participation in day-ahead, real-time, and reserve markets, while P2G facilities engage in day-ahead and reserve markets to capitalize on arbitrage opportunities. Additionally, the chapter discusses how wind farms can strategically purchase reserve capacity to mitigate high imbalance penalties while supplying energy to P2G facilities in exchange for reserve capacity.

Finally, Chapter 21 examines the significant structural changes in energy markets over the past two decades, focusing on two key aspects: the emergence of peer-to-peer energy exchange mechanisms and the power-to-X (PtX) concept. It discusses how the intersection of these two concepts has transformed energy markets, highlighting their combined impact on market dynamics and operations.

As we conclude this preface, we want to take a moment to express our sincere appreciation to all the contributors who poured their hearts and minds into this project. Each chapter reflects their dedication, expertise, and passion for advancing sustainable energy solutions. We truly believe that the diverse insights and innovative ideas shared in this book can spark meaningful conversations and inspire action toward a greener future. We hope that you, dear readers, enjoy the journey through these pages as much as we have enjoyed bringing this work to life. May it serve as a guiding light in your pursuit of knowledge and progress in the exciting world of power-to-X technologies and their potential to transform our energy landscape. Happy reading!

Editors
Amjad Anvari-Moghaddam
Sina Ghaemi
Shi You
Frede Blaabjerg

1 Power-to-X
*Pioneering the Future
of Sustainable Energy*

*Amir Meydani, Hossein Shahinzadeh,
Gevork B. Gharehpetian, Mohammad Mohsen Hayati,
and Mehdi Abapour*

1.1 INTRODUCTION

Global energy consumption has significantly increased, from 94,847 TWh in 1985 to 178,899 TWh in 2022. Similarly, electricity generation rose from 9,753.72 TWh in 1985 to 28,660.98 TWh in 2022. Fossil fuels, including natural gas (NG), coal, and oil, remain the primary sources of energy, contributing over 80% of the global energy demand and more than 60% of electricity generation. In 2022, fossil fuels accounted for 137,236.66 TWh of global energy consumption, up from 73,917.16 TWh in 1985. Major industrialized nations, such as the USA, China, Russia, Germany, and India, heavily depend on fossil fuels for their energy needs, which exacerbates global environmental concerns [1, 2]. The rise in fossil fuel use correlates with increasing greenhouse gas (GHG) emissions, a key driver of global warming. Since 1950, GHG emissions have surged from 16.52 billion tons to 53.85 billion tons in 2022. The United States, China, Russia, India, and the EU are among the highest emitters [3]. Carbon dioxide (CO_2) emissions, primarily from the combustion of fossil fuels, represent 93.8% of global GHG emissions, making it the dominant contributor to air pollution [4]. Between 1950 and 2022, global CO_2 emissions from coal, oil, and gas combustion rose substantially, and cumulative CO_2 emissions increased by 650% during the same period. The rising CO_2 emissions have had a direct impact on global temperatures, with Earth's average temperature increasing by 1.45°C by 2023 compared to the late 19th century. This rise brings the planet closer to the 1.5°C threshold set by the Paris Agreement. Since the 1980s, each decade has been warmer than the previous one, emphasizing the urgent need for action to curb GHG emissions and prevent further climate change [5]. Renewable energy sources (RES) are increasingly viewed as a viable alternative to fossil fuels. In 2023, global renewable capacity reached 3,870 GW, with 8,914 TWh of electricity generated from low-carbon sources, accounting for 30.24% of global electricity production [6]. According to the IEA's Renewables 2023 report, renewable capacity is expected to grow significantly, with projections indicating an additional 3,700 GW of capacity by 2028. This expansion would bring total renewable capacity to 7,338.7 GW, covering 41.6% of global electricity generation. China's RE capacity is set to lead this growth, followed by substantial increases in the United States and the EU [7]. To support this transition, international organizations, such as the Green Climate Fund and UN–Energy, are actively promoting collaborations and initiatives to accelerate the shift toward RE, reflecting a global commitment to sustainable energy solutions.

Integrating RE into today's electricity systems poses several challenges, primarily due to the unpredictable and intermittent nature of RE generation. This variability, in the short and long term, complicates efforts to maintain a constant power balance between supply and demand. The limited dispatchability of RE sources makes it difficult to align generation with consumption needs, leading to power quality (PQ) issues, such as voltage fluctuations, harmonics, and reduced reliability,

DOI: 10.1201/9781032719436-1

particularly during blackouts. Also, RE integration can cause voltage congestion at the distribution level, further destabilizing the power supply and potentially leading to inefficiencies and interruptions. The rising penetration of RE also complicates the optimal power flow within the grid, causing an increased voltage profile, constraints on grid capacity, and instability at lower grid levels. These issues are exacerbated by limited transmission capacity and long transmission distances, which increase inefficiencies and costs. Also, the restricted control and monitoring capabilities of RE systems hinder effective grid management. Grid stability, already at risk from RE integration, suffers due to reduced reactive power support, decreased fault detection, and a decline in grid inertia, which can lead to difficulty maintaining grid frequency and increased instability [8, 9]. In addition, the rapid growth of electric vehicles (EV)—with the global number of EVs surging from 20,000 in 2010 to 7.2 million in 2019—presents new issues for the grid [10]. Increased demand during peak charging periods raises the risk of overloading local distribution networks and exacerbates fluctuations in electricity consumption. Many of these challenges can be addressed by employing energy storage systems (ESS); they mitigate short-term power generation imbalances and support long-term generation operations by storing excess energy during high production periods and releasing it during low production times. ESS enhances the dispatchability of RE, stabilizes voltage levels, reduces flickering, and provides backup power during outages, ultimately improving PQ. ESS helps alleviate grid capacity constraints, maintain optimal voltage levels, and stabilize local grids. ESS also plays a crucial role in improving grid stability by providing reactive power support and simulating the inertia response, which helps maintain grid frequency [11, 12]. ESS can also resolve issues arising from EV integration by supplying energy during peak demand periods and storing it during off-peak times, reducing the strain on the grid. It can provide local power to fast-charging stations, thereby preventing grid destabilization [13]. The global energy storage capacity has seen remarkable growth in recent years. In 2023, the total installed energy storage capacity was estimated to reach around 99 GWh, a significant jump from previous years. This represents a 34% increase from 2022. Specifically, utility-scale projects were a key driver of this expansion, with regions like China, the United States, and Europe contributing substantially to the overall growth. In 2023, the United States alone added 25.5 GWh of storage capacity, while China led with 39 GWh. By 2030, projections indicate that global energy storage could exceed 1,000 GWh, further emphasizing the rapid expansion of this sector [14].

Hydrogen (H_2) is a highly promising energy carrier due to its versatility, ease of transport, and environmental benefits, especially when produced using RESs. It possesses the highest energy density per unit of weight, among other alternatives, but its energy density per unit of volume remains low [15]. Until 2010, hydrogen had a minimal role in the energy economy due to the high energy requirements for production and the reliance on fossil fuels. Additionally, a lack of infrastructure for hydrogen storage and delivery limited its use. At that time, hydrogen production was largely reliant on NG (48%), petroleum (30%), and coal (18%), with only 4% being derived from electrolysis. However, falling RE costs since the 2010s have spurred renewed interest in water electrolysis as a method for producing hydrogen. To distinguish between hydrogen production methods, various color designations have emerged. Green hydrogen is produced from RE, blue hydrogen from NG using carbon capture and sequestration (CCS), gray hydrogen from NG without CCS, black hydrogen from bituminous coal, and brown hydrogen from lignite. Red hydrogen is generated using nuclear power [16, 17]. While hydrogen is recognized for its benefits in power generation, it is also a key element in power-to-X (PtX) systems, which utilize hydrogen to create synthetic products, such as fuels, chemicals, heat, and power. This approach allows for sustainable energy storage, efficient distribution, and the creation of carbon-neutral fuels. PtX involves synthesizing hydrocarbon-based renewable fuels by combining renewable hydrogen with CO_2 captured from the air or industrial processes [18]. Also, nitrogen (N_2) can be used to produce synthetic ammonia (NH_3). At the heart of PtX is the power-to-hydrogen (PtH) process, which generates hydrogen through water electrolysis utilizing RE. This technology has global applications, offering high energy density by weight, making hydrogen suitable for stationary energy generation or transportation. When used in fuel cells or combustion systems, hydrogen produces only water as a by-product, further enhancing its environmental appeal. Hydrogen also enables the creation of alternative fuels through processes such as power-to-methanol (PtCH$_4$) and power-to-ammonia (PtA). PtM synthesizes methane (CH$_4$) by combining electrolytic hydrogen with CO$_2$, and though it supports carbon capture, burning methanol (CH$_3$OH) still emits CO$_2$. In contrast, PtA combines hydrogen with nitrogen from the air to produce ammonia, which releases no carbon during energy production but produces water and nitrogen compounds. Since PtA relies solely on water and air, it offers geographical adaptability similar to PtH, and its use in energy applications is increasingly widespread [19].

The efficient and continuous operation of PtX systems is highly dependent on effectively managing the variability of RESs. Demand-side management (DSM) plays a critical role in many PtX models by adjusting production processes in response to the fluctuating availability and cost of power and raw materials. Whether process units within PtX systems function in a steady-state or dynamic condition determines how the fluctuating power inputs are managed [20]. While some units can maintain steady operation through the use of electricity storage, others must operate dynamically in the absence of such storage, particularly electrolyzers and other key components. Storage tanks for raw materials or intermediates, such as hydrogen and carbon dioxide, become essential when production is variable but downstream synthesis processes need to remain consistent. To reduce reliance on expensive storage, another approach involves variable operation of downstream conversion processes. In cases where all units can operate flexibly, storage may not be crucial; however, it is vital to evaluate the flexibility of these operations, including parameters like start-up and shutdown times, ramp rates, operating ranges, and settling times. If certain process units are unable to adjust quickly to load changes, small-scale electricity storage might still be needed to ensure stability. Despite the benefits of flexible operation, it can lead to a reduced capacity factor and may negatively impact efficiency, product quality, and the longevity of process units. Also, this method of operation poses issues in terms of integrating heat and mass flows on a larger scale, which requires careful consideration to avoid inefficiencies and operational bottlenecks.

Koji Hashimoto introduced the concept of power-to-X (PtX) in 1994 as a method for producing artificial methane, with pilot facilities constructed at Tohoku University in 1995 and 2003 [21, 22]. Since then, many industrial and demonstration projects have been initiated to manufacture renewable fuels and chemicals, with Germany and France leading in methanation applications. One notable example is the Audi e-gas [23, 24] facility in Germany, which uses wind-powered alkaline electrolyzers to produce hydrogen, which is then combined with CO$_2$ from biogas plants to synthesize methane. The facility produces 1,000 t of synthetic methane annually, with a system efficiency of 54%. In Germany, the Carbon2Chem® [25] project captures 2 t of CO$_2$ daily from steel production, converting it into valuable chemicals like methanol. Powered by RE employing alkaline water electrolyzers (AWE), the project demonstrates the potential to reduce greenhouse gas emissions while fostering a circular carbon economy. Involving industrial partners such as ThyssenKrupp and BASF, Carbon2Chem® is a major step toward scalable CO$_2$ conversion for sustainable chemical feedstock. The STORE&GO [26] project captures around 1.2 t of CO$_2$ per day, converting it into synthetic methane using hydrogen produced from polymer electrolyte membrane (PEM) electrolyzers. This European initiative showcases the viability of large-scale power-to-gas (PtG) technology by integrating advanced methanation reactors, and it contributes to the EU's decarbonization and energy transition goals by enhancing cross-border cooperation in RE. The SOLETAIR [27] project, launched in 2018, successfully produced 6.2 kg of oil and wax per day over 300 hr, also generating gas as a by-product. PtX technologies are increasingly recognized for their role in sector coupling, a critical element of the EU's pathway toward energy decarbonization. The EU aims to increase electrolyzer capacity from 0.1 GW today to 40 GW by 2030, although scaling up presents challenges, such as cost reductions and improvements in storage and transportation infrastructure [28]. Significant progress has been made, with a 40% decrease in the cost of AWE since 2015, making environmentally friendly hydrogen more competitive with NG. Projections suggest hydrogen prices will range between €0.68 and €1.36/kg by 2050. The cost of green hydrogen varies globally due to differences in climate and electricity prices. By 2030, the EU expects electrolyzer costs to fall by €200–320/kW. While Japan leads in hydrogen technology investment, the EU remains the largest contributor to power and storage technologies, investing nearly €380 million in 2021 [29, 30].

The purpose of this chapter is to provide a comprehensive examination of PtX technologies and their pivotal role in advancing sustainable energy systems. By exploring the fundamental principles and mechanisms of various PtX pathways, conducting in-depth system-level analyses, and investigating optimization strategies, this chapter aims to equip readers with a thorough understanding of how PtX can effectively convert and store RE into valuable products. Also, it seeks to highlight the economic, thermodynamic, and environmental benefits of PtX, while addressing the challenges and future directions essential for its widespread adoption. Through this detailed exploration, the chapter underscores the transformative potential of PtX technologies in achieving a cleaner, more resilient, and sustainable energy future.

1.2   PtX PATHWAYS: PRINCIPLES AND MECHANISMS

There is increasing interest, particularly in Europe, in advancing research and development (R&D) for PtX systems, driven by the EU's commitment to reducing GHG emissions and increasing the share of RESs in energy production. PtX systems utilize surplus electricity generated by RESs to produce energy carriers in the form of gases, liquids, or chemicals. The core of renewable PtX technology is the production of hydrogen through water splitting electrolysis (WE). With a high heating value (HHV) of 142 MJ/kg, hydrogen serves as an efficient energy carrier, especially in fuel cells (FC). However, its low density of 0.0813 g/L in normal conditions presents challenges for storage and transportation, requiring careful consideration of technical and economic factors. The "X" in *PtX* refers to the final product formed when hydrogen reacts with either CO_2 or nitrogen. CO_2 can be sourced from stationary point sources, from biogenic processes, or even directly from the atmosphere. PtX technologies are seen as a promising solution for reducing CO_2 emissions and integrating RE into the broader energy industry, promoting both sustainability and decarbonization goals.

1.2.1   WATER ELECTROLYZER TECHNOLOGY

Water electrolysis is the process by which electricity is transformed into a chemical energy source that may be stored. Electricity is employed to undergo electrochemical decomposition of water (H_2O) into hydrogen and oxygen (O_2) in the presence of an electrocatalyst (equation 1):

$$2H_2O \rightarrow 2H_2 + O_2 \Delta H^0_{298.15} = -285.8\,kj.mol^{-1} \tag{1}$$

The process occurs within an electrolysis cell, comprising two electrodes—an anode and a cathode—that are physically and electrically isolated by a diaphragm or membrane. Applying voltage to the electrodes causes an exchange of ions between them, which then triggers a redox reaction. At the anode, oxidation occurs to produce elementary oxygen, whereas at the cathode, a reduction event takes place to produce elementary hydrogen. There are three main technologies used for electrolysis: (1) alkaline water electrolyzers (AWE); (2) proton-exchange membrane (PEM), also known as polymer electrolyte membrane water electrolyzers (PEMWE); and (3) high-temperature, specifically solid oxide water electrolyzers (SOWE). The key distinction among these types is the type of electrolyte that facilitates the ion exchange and simultaneously separates the two electrochemical partial reactions occurring at the anode and cathode.

1.2.1.1   Alkaline Water Electrolyzers (AWE)

AWE is the most advanced electrolysis technology, utilizing an intensely concentrated alkaline solution, typically potassium hydroxide (KOH), at 20–30% by weight. This solution is supplied to the cathode, where water molecules are split into hydrogen and hydroxide ions (OH^-). Other

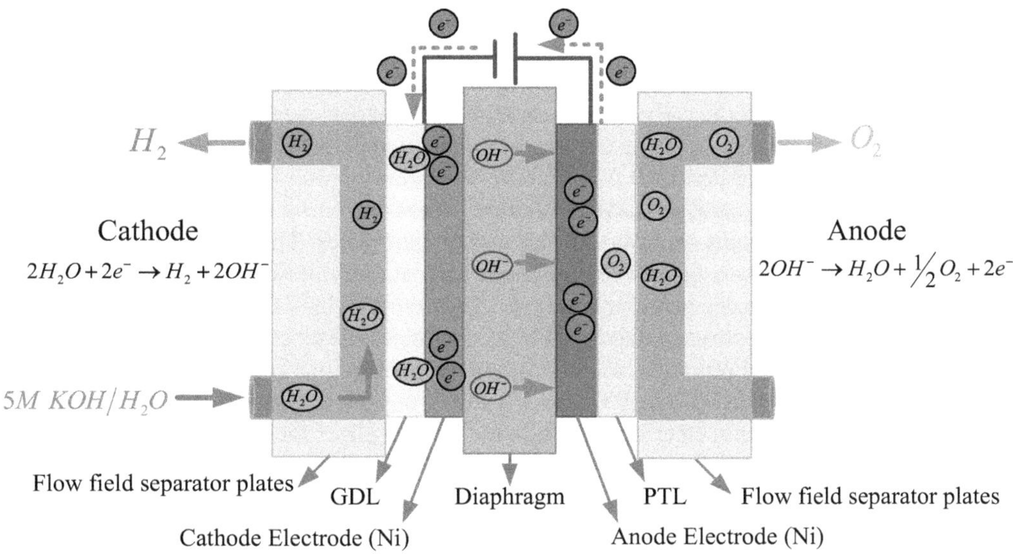

FIGURE 1.1 AWE process for hydrogen and oxygen production.

alkaline electrolytes, such as sodium hydroxide (NaOH), sodium chloride (NaCl), and zinc hydroxide (ZnOH), have also been explored. Hydroxide ions undergo oxidation at the anode, producing oxygen and water. AWE systems come in two main commercial versions: the traditional AWE, which uses a KOH-based alkaline solution with electrodes separated by a diaphragm, allowing OH⁻ ions to pass, and an enhanced AWE, featuring porous electrodes divided by a gas-tight barrier. The zero-gap technology in enhanced AWE offers superior voltage efficiency over traditional systems. Operational parameters for AWE typically range between 40 and 90°C, pressures of 10 to 30 bar, and cell voltages of 1.80 to 2.4 V, with efficiency between 62% and 82% [31]. AWEs have several advantages, including lower investment costs, due to the absence of expensive noble metals like platinum or iridium. Their efficiency rates, typically 63–70%, surpass those of PEMWEs, which achieve 56–60%. However, AWEs struggle with power fluctuations from RESs, like wind power. Additionally, AWEs have lower current densities and working pressures and face significant maintenance costs due to the corrosive nature of alkaline electrolytes. Efforts are underway to improve current density, reduce costs, and introduce non-noble catalysts like cobalt nanosheets and NiMo alloys to enhance performance. AWE systems operate by connecting multiple cells into a stack, which can be arranged in series or parallel to meet output needs. The cells use non-noble metal electrodes and alkaline electrolytes, typically KOH. To function, AWE systems require auxiliary units, including pumps for electrolyte circulation, a power supply to convert AC to DC electricity, and sensors and valves to control gas flow, pressure, and temperature. Water purification systems ensure feedwater purity, and cooling systems regulate operational temperatures. AWEs generate hydrogen and oxygen gases at low pressures, necessitating external compressors to meet high-pressure application requirements. Gas separation and purification units ensure hydrogen and oxygen meet purity standards, critical for safe operation and handling of hydrogen's flammability [32–34]. Figure 1.1 depicts the AWE process, wherein water is dissociated into hydrogen and oxygen via electrochemical reactions at the cathode and anode, aided by the migration of hydroxide ions over a diaphragm under the influence of an external electrical current.

1.2.1.2 Polymer Electrolyte Membrane Water Electrolyzers (PEMWE)

PEMWE technology was first introduced in the 1950s and further refined by General Electric in the 1960s [35]. In recent decades, it has seen significant advancements. PEMWEs use a polymeric membrane, such as Nafion™ or fumapem, to separate the anode, where water is supplied and oxygen is formed, from the cathode, where hydrogen is collected. The primary advantages of PEMWE over AWE include faster cold starts, greater flexibility, and better compatibility with dynamic and intermittent energy sources. Due to the acidic nature of the PEM, noble metal catalysts, such as iridium at the anode and platinum at the cathode, are used, but these expensive materials have been a major cost barrier. PEMWE operates at temperatures between 20 and 150°C and pressures of up to 200 bar, with current densities of up to 4.0 A/cm² and efficiencies ranging from 67% to 82%. PEMWE cells have a sandwich structure consisting of multiple layers, including a polymer electrolyte membrane, electrodes, porous transport layers (PTL), and bipolar plates. The membrane, which is only 20–300 μm thick, acts as a barrier between the anode and cathode, allowing protons to pass through while blocking other molecules and contaminants. The combination of membrane and electrodes, known as the membrane electrode assembly (MEA), facilitates the electrochemical process. One of the main challenges in PEMWE is the use of costly noble metals, like iridium and platinum, which are highly efficient but expensive. The PTLs evenly distribute reactants across the catalyst layer and remove gases produced during electrolysis, while the bipolar plates facilitate the flow of reactants, electrons, and heat and physically separate the cells in a stack [31].

PEMWE systems connect multiple cells in series to form a stack, which can also be connected in parallel. These systems rely on auxiliary components, like pumps, sensors, and power converters. To maintain efficiency, feedwater is pre-treated to remove contaminants, and excessive water at the anode side prevents membrane dehydration and aids in cooling. Cooling systems, such as refrigeration cycles, help reduce steam output, while purification processes ensure hydrogen and oxygen meet purity standards. Pressure/temperature swing adsorption (PSA/TSA) methods dehydrate hydrogen, achieving high purity levels. Power electronics, including AC/DC and DC/DC converters, allow flexible operation of the electrolyzer and maintain power quality when connected to the grid [36]. Hydrogen pressurization is essential for various applications, such as pipeline injection, chemical synthesis, underground storage, and refueling hydrogen cars. This is achieved either by pressurizing within the PEMWE, using an electrochemical compressor, or with a mechanical compressor. High-pressure hydrogen generation can increase efficiency but also poses challenges, such as reducing faradaic efficiency and safety risks due to hydrogen crossover to the anode. Solutions include using thicker membranes or adding a third electrode to convert crossed hydrogen into water. Electrochemical compression can prevent hydrogen crossover but may reduce compression efficiency. Figure 1.2 illustrates the PEMWE process, wherein water molecules are dissociated into hydrogen and oxygen, with protons (H⁻) traversing the membrane, leading to hydrogen gas production at the cathode and oxygen gas creation at the anode.

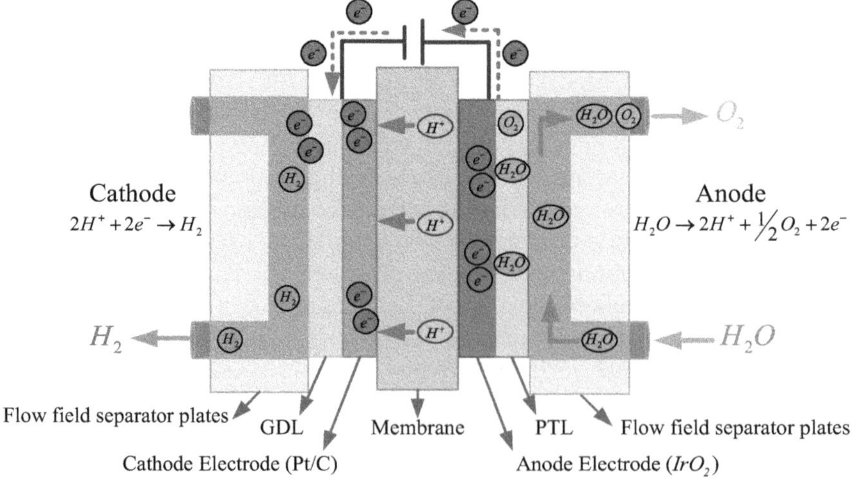

FIGURE 1.2   PEMWE process for hydrogen and oxygen production.

1.2.1.3 Solid Oxide Water Electrolyzers (SOWE)

SOWE is an emerging technology that utilizes a ceramic substance capable of conducting oxygen ions (O^{2-}) to separate the anode and cathode. SOWE is particularly suitable for PtX applications and does not require noble materials, while also showing low sensitivity to contaminants in the feed stream. Additionally, SOWEs are reversible, meaning, they can function as solid oxide FCs to generate electricity. Operating at temperatures between 800°C and 1,000°C, SOWEs achieve higher efficiency (74–81%, based on the LHV) compared to AWE and PEMWE systems. The high temperature enhances ionic conductivity and accelerates the kinetics of electrochemical processes at the electrode surfaces, reducing overall power consumption by supplementing it with thermal energy. A solid electrolyte made of zirconium oxide (ZrO_2) doped with 8% yttrium oxide (Y_2O_3) is commonly used, with SOWE cells achieving voltage efficiencies exceeding 80% and cell voltages between 0.7 and 1.5 V, depending on current densities [37]. The SOWE process is illustrated in Figure 1.3. Hydrogen and oxygen are produced at the cathode and anode, respectively, as water is divided into hydrogen and oxygen at high temperatures. The oxygen ions (O^{2-}) migrate through a solid oxide membrane.

SOWE cells can be arranged in series and parallel configurations to form a stack, enabling desired output levels. The system operates at high temperatures (700–1,000°C) and relies on auxiliary units, including pumps for gas circulation and steam supply, as well as power supply units that convert AC to DC electricity. Sensors and valves are employed to monitor gas flow, pressure, and temperature, while water purification units ensure the steam is free from impurities. High-temperature heat exchangers are incorporated to recover and utilize waste heat, improving overall efficiency. To meet high-pressure hydrogen application requirements, mechanical compressors are used to pressurize the hydrogen produced. Gas separation and purification processes are essential to maintain the purity standards of the hydrogen and oxygen gases, ensuring safe and effective system operation at high temperatures. However, the high operating temperatures of SOWEs pose issues, including material degradation and short-term stability. Fluctuations in power supply, common in RESs, exacerbate these issues by causing temperature variations that lead to heat dissipation and the formation of micro-fractures in the membrane material. This ultimately shortens the lifespan of the electrolyzer, presenting a significant obstacle for long-term SOWE stability and reliability. Table 1.1 offers a summary of the main features of the electrolyzers [38, 39].

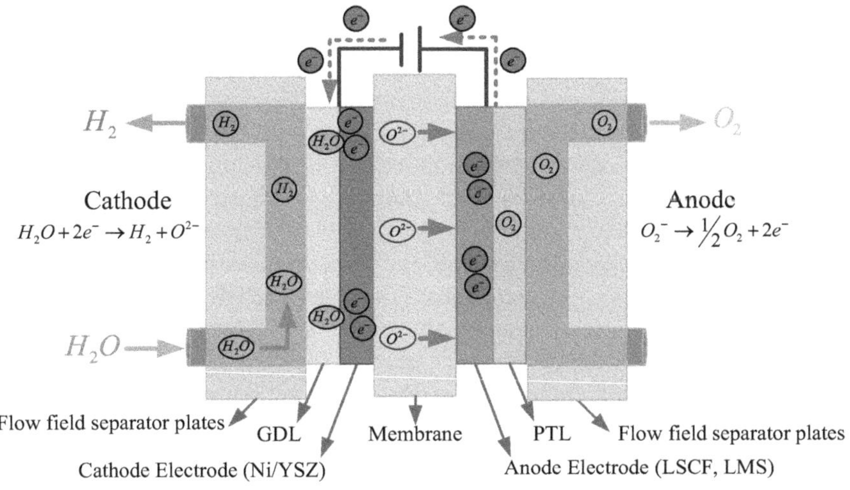

FIGURE 1.3 SOWE process for high-temperature water splitting.

1.2.2 CARBON DIOXIDE (CO_2) SUPPLY/CAPTURE

1.2.2.1 CO_2 Sources

The sustainability of PtX systems largely depends on the source of CO_2 used as a raw material. CO_2 can be sourced from various industrial sources, such as power plants, petrochemical refineries, iron and steel manufacturing facilities, and cement companies, which collectively emit substantial quantities of CO_2 annually (e.g., cement and iron and steel industries each emit approximately 2.5–2.6 GT of CO_2 globally). Biomass-based processes, such as fermentation, combustion, and gasification, or direct capture from the atmosphere (known as direct air capture, or DAC), also provide alternative sources of CO_2. Each source has varying CO_2 concentrations, which are typically not in a pure state, necessitating separation, purification, and upgrading procedures to make the CO_2 suitable for PtX processes. Industrial CO_2 capture commonly utilizes techniques like absorption, adsorption, membrane separations, and cryogenic fractionation. Among these, absorption using chemical solvents like monoethanolamine (MEA) is a widely adopted method for capturing CO_2 from flue gases [40]. Despite being effective, MEA has limitations, such as high energy consumption for solvent regeneration, limited CO_2 capture capacity, and issues like solvent degradation and equipment corrosion. To address these limitations, technologies such as adsorption, membrane separation, and cryogenic fractionation are being explored to improve CO_2 capture efficiency and reduce energy requirements. Incorporating CO_2 capture technologies into power plants that utilize coal, oil, or NG can significantly impact plant efficiency and costs, with electricity production costs increasing by 30–75% for NG plants and 40–85% for coal-fired plants. Also, transporting captured CO_2 from industrial sites to PtX facilities poses additional challenges. Transportation costs are influenced by factors such as proximity to the CO_2 source, separation technology, and terrain of the pipeline route, with estimates ranging from €1.9 to €5.2 per ton of CO_2 in the EU and €3.1 per ton in the United States. Hence, for PtX facilities to be sustainable and economically viable, it is beneficial for them to be located close to CO_2 sources to minimize transportation costs [41].

Biomass-derived CO_2 offers an alternative source perceived as having a net-zero carbon footprint, as the CO_2 released during combustion or gasification was previously absorbed during biomass growth. Processes like fermentation for biogas or bioethanol production yield relatively pure CO_2, which can be captured at a low additional cost. Integrating PtX technologies with biomass gasification can enhance overall efficiency by utilizing oxygen generated during electrolysis as a gasifying agent. However, challenges such as tar production during gasification and high purification costs remain significant hurdles [42]. Other industrial operations, like ammonia synthesis and oil refineries, are also major CO_2 sources. In ammonia production, CO_2 is produced during the steam reforming of NG, and using green hydrogen instead can significantly reduce emissions. Similarly, refineries and iron and steel manufacturing facilities emit large amounts of CO_2, which can be captured using advanced techniques like absorption with efficiencies up to 90% [43]. DAC is an emerging technology for capturing CO_2 directly from the atmosphere, potentially reducing the need for transportation from industrial point sources. DAC involves extracting CO_2 from ambient air using strong bases, such as sodium hydroxide (NaOH) or potassium hydroxide (KOH), and then purifying it to a usable form [44]. However, the low concentration of CO_2 in the atmosphere, approximately 415 ppm, presents a challenge that necessitates substantial energy inputs, with current levelized costs ranging between €85 and €209 per ton of CO_2 captured. Although amine-based solvents have the potential to reduce energy requirements, they face issues such as high energy consumption for solvent regeneration and corrosion. Despite these issues, CO_2 capture and utilization technologies remain crucial for the sustainability of PtX systems. Table 1.2 summarizes key CO_2 sources for PtX systems, outlining the CO_2 concentrations, suitable capture technologies, energy requirements, and notable case studies.

TABLE 1.1
Electrolyzers Comparison

Electrolyzer Type	AWE	PEMWE	SOWE
Electrolyte	NaOH/KOH (liquid)	Solid polymer membrane	Yttria (Y_2O_3)–stabilized zirconia (YSZ)
Anode	Ni, Ni–Co alloys	RuO_2, IrO_2	Lanthanum strontium manganite (LSM)/YSZ composite
Anode reaction	$2OH^-\rightarrow H_2O + 0.5O_2 + 2e^-$	$2H_2O\rightarrow O_2 + 2H^+ + 2e^-$	$O^{2-}\rightarrow 0.5O_2 + 2e^-$
Cathode	Ni, Ni–Mo alloys	Pt, Pt–Pd	Ni/YSZ
Cathode reaction	$2H_2O + 2e^-\rightarrow H_2 + 2OH^-$	$2H^+ + 2e^-\rightarrow H_2$	$H_2O + 2e^-\rightarrow H_2 + O^{2-}$
Charge carrier	OH^-	H^+, H_3O^+	O^{2-}
Working fluid	High-concentration solution	Distilled water	Steam
Cell temperature (°C)	40–90	50–90	500–1,000
Cell area (m²)	<4	<0.3	<0.01
Cell voltage	1.80–2.4	1.4–2.2	0.7–2.0
Operating pressure (bar)	2–10	15–30	<30
Hydrogen purity	>99.5	99.9	99.9
Specific electricity consumption (kWh/Nm³)	3.8–5.9	3.6–5.6	2.5–3.5
Degradation rate (μV/h)	<3	<14	<10
Cold start-up time (min)	60	<20	600
Voltage efficiency (%HHV)	62–82	67–84	~90
Current density (A/cm²)	0.13–0.5	0.6–4.0	0.3–2.0
Lower partial load range	10–40	0–10	>30
Lifetime stack (operating hours)	<90,000	<90,000	<10,000
Lifetime balance of plant (years)	20–30	10–20	N/A
Production rate	760	210	<40
Capital cost (€/kWel)	620–1,200	1,390–2,320	1,350–3,250
Technology maturity	Commercial	Commercial	R&D
Benefits	Long-term stability; well-established; non-noble materials; low capital cost; large stack size	Design simplicity; dynamic operation; high current density; rapid response; compact system	Non-noble materials; reversible operation as FC; high energy efficiency; low capital cost
Disadvantages	Slow dynamics; low current density; gas permeation; corrosive electrolyte	Low durability; noble materials; high membrane cost; acidic environment	Brittle ceramics; bulky design; sealing issues; unstable electrodes

1.2.2.2 CO_2 Capturing Technologies

CO_2 capture technologies in PtX systems are classified based on the position of the carbon capture process relative to combustion: pre-combustion, post-combustion, and oxy-fuel combustion [45]. Pre-combustion capture involves converting carbon-based fuels into syngas (a mixture of hydrogen and carbon monoxide) before combustion. The carbon monoxide is shifted to CO_2 and hydrogen

TABLE 1.2

Comparison of CO_2 Sources for PtX Systems

CO_2 Source Type	Description	CO_2 Concentration	Capture Technologies	Advantages	Challenges	Examples/Case Studies
Industrial point sources	Large stationary sources emitting CO_2 as a by-product of industrial processes. Includes power plants, cement production, iron and steel manufacturing, oil refineries, and chemical industries.	**High** (Typically, 5–30% by volume)	• Absorption (chemical solvents) • Adsorption • Membrane separation • Cryogenic fractionation	• High CO_2 concentration simplifying capture • Established technologies • Potential for large-scale capture	• Energy-intensive capture processes • High capital and operational costs • CO_2 purity possibly requiring further processing	• **Boundary Dam Power Station (Canada):** Captures CO_2 from coal-fired power plant. • **Petra Nova Project (USA):** Suspended; captured CO_2 from flue gas using MEA solvent.
Biomass processes	CO_2 emissions from biomass combustion, gasification, and fermentation processes. Biomass is considered carbon-neutral, as CO_2 released equals CO_2 absorbed during growth.	**Medium to high** (Up to ~15% for combustion) **High** (Nearly pure for fermentation)	• Absorption • Adsorption • Membrane separation	• Carbon-neutral or negative emissions • Fermentation producing high-purity CO_2 • Utilizes renewable resources	• Collection and transportation of biomass • Variability in biomass composition • Tar formation in gasification affecting efficiency	• **Drax Power Station (UK):** Transitioned from coal to biomass to reduce carbon footprint. • **Biogas plants in EU:** Integration with PtX technologies in anaerobic digestion processes.
Direct air capture (DAC)	Capturing CO_2 directly from the atmosphere using chemical processes. DAC can be located anywhere, reducing the need for CO_2 transport infrastructure.	**Very low** (~0.0415% or 415 ppm)	• Chemical absorption using strong bases (e.g., NaOH, KOH) • Solid sorbents • Amines in liquid solvents	• Unlimited CO_2 availability • Flexibility in plant location • Supports negative emissions when combined with storage	• High energy consumption due to low CO_2 concentration • High operational costs • Requires heat and electricity inputs • Technology still developing cost efficiencies	• **Climeworks (Switzerland):** Operates DAC plants capturing CO_2 from ambient air. • **Carbon Engineering (Canada):** Developing large-scale DAC facilities.

(Continued)

TABLE 1.2 (Continued)

Industrial processes with pure CO$_2$ streams	Certain industrial processes emit nearly pure CO$_2$ streams as by-products, such as ethanol fermentation, ammonia production, and natural gas processing.	**High** (Typically, >95% by volume)	• Compression and dehydration (minimal processing required)	• Low-cost capture due to high purity • Lower energy requirements for processing	• Limited availability based on industrial activity • Geographic limitations if not co-located with PtX facilities	• **Ethanol plants:** CO$_2$ captured during fermentation can be used for PtX. • **Natural gas processing plants:** Removal of CO$_2$ from raw gas streams.
Fossil fuel combustion (mobile sources)	CO$_2$ emissions from vehicles and mobile machinery using gasoline, diesel, or natural gas.	**Low** (Typically, ~10–15% in exhaust gases)	• Not typically captured due to dispersion • Research into mobile capture technologies	• Potential for capture at centralized facilities (e.g., shipping ports) • Large total emissions contributing to CO$_2$ availability	• Technologically challenging to capture dispersed emissions • Economic feasibility low • Requires development of new capture technologies	• **Research stage:** Mobile carbon capture technologies are being explored but are not yet commercially viable.
Geothermal sources	CO$_2$ emissions released naturally from geothermal reservoirs during energy extraction processes.	**Variable** (Can be high in certain locations)	• Gas separation and purification techniques	• Utilizes naturally occurring CO$_2$ • Can integrate with geothermal energy production	• Geographic limitations • CO$_2$ composition possibly including other gases requiring separation • Environmental regulations on geothermal emissions	• **Hellisheiði Power Plant (Iceland):** CarbFix project captures and mineralizes CO$_2$ from geothermal emissions.

using the water–gas shift reaction, and CO_2 is captured from syngas before combustion. This method is typically implemented in integrated gasification combined cycle (IGCC) facilities, although high costs and technical challenges have led to the discontinuation of some projects, such as the Kemper County Energy Facility. Post-combustion capture, on the other hand, extracts CO_2 from flue gases after combustion. It is the most common method used in existing power plants and industrial processes due to its compatibility with retrofitting existing infrastructure. The Boundary Dam Power Station implemented post-combustion capture, successfully reducing CO_2 emissions by around 90%. Oxy-fuel combustion replaces air with pure oxygen for combustion, producing flue gas primarily composed of CO_2 and water vapor, which simplifies the CO_2 capture process. The OXY-CFB-300 project demonstrated this approach using a circulating fluidized-bed boiler to enhance CO_2 capture efficiency and reduce nitrogen oxide emissions. Several CO_2 capture technologies are employed in PtX systems, each with distinct advantages and limitations [46]. Chemical absorption using solvents like MEA is the most mature technology but suffers from high energy consumption for solvent regeneration, corrosion issues, and limited CO_2 capture capacity. Alternative solvents, such as piperazine-enhanced amines and ammonia, are being explored to address these limitations. Chemical adsorption using solid sorbents, such as calcium oxide (CaO), is effective for high-temperature CO_2 capture but requires energy-intensive regeneration. PSA utilizes pressure changes to separate CO_2 based on its preferential adsorption on solid adsorbents and is commonly used in hydrogen production facilities. Membrane separation uses selective membranes to separate CO_2 from gas mixtures, while cryogenic separation condenses CO_2 at low temperatures, making it suitable for high-purity applications, but often too energy-intensive for large-scale use.

Innovative technologies such as molten carbonate fuel cells (MCFC) and biological capture using microalgae are also being explored. MCFCs can capture CO_2 from flue gases while simultaneously generating electricity, providing dual benefits of power generation and CO_2 concentration for capture. Biological capture involves using microalgae to absorb CO_2 through photosynthesis, producing biomass that can be converted into biofuels. While this is promising, issues in optimizing growth conditions and scaling up for industrial applications remain. Selecting the most suitable CO_2 capture technology depends on various factors, such as the specific application's technical, economic, and environmental requirements. Technical considerations include CO_2 concentration, pressure, and temperature in the gas stream. Economic factors involve capital and operational costs and potential revenue from captured CO_2 utilization. Environmental impact considerations focus on energy efficiency and potential emissions. By integrating various CO_2 capture technologies, PtX systems have the capability to enhance efficiency and sustainability, contributing to the global effort to reduce GHG emissions. Table 1.3 provides a detailed comparison of several CO_2 capture systems, outlining their descriptions, advantages, problems, and real-world case studies that illustrate their applicability in diverse industrial contexts.

1.2.3 Hydrogen (H_2) Storage and Supply

During periods of insufficient renewable electricity supply, it is essential to consider PtX processes to manage fluctuations in RE availability. PtX plants, like modern chemical plants, operate optimally under steady-state conditions, making energy storage solutions crucial to maintain uninterrupted operation. An adequately sized hydrogen storage system can help balance energy deficits by storing surplus H_2 produced during high RE availability for use during low-availability periods. Given that H_2 has a low density of 0.084 kg/m³ under standard conditions, large storage capacities are necessary to accommodate the high volumetric flow rates [47]. Salt caverns are commonly used for this purpose, as they can store up to 1 million cubic meters at depths ranging from several hundred to over a thousand meters and operate safely under pressures ranging from 60 to 180 bar, aligning with geostatic pressures at those depths. Alternatively, surplus H_2 can be injected into NG grids, which in Germany have a pipeline network of around 500,000 km and a storage capacity of up to 200,000 GWh [48, 49]. However, current regulations limit its proportion to 5–10% by volume

TABLE 1.3

Overview of CO_2 Capturing Technologies

Technology	Description	Advantages	Challenges	Applications	Technology Readiness Level (TRL)	Examples/Case Studies
Chemical absorption	Uses chemical solvents (e.g., monoethanolamine, MEA) to absorb CO_2 from gas streams. CO_2 is then released by heating the solvent for regeneration.	• High CO_2 capture efficiency • Well-established technology • Suitable for low CO_2 concentrations	• High energy requirement for solvent regeneration • Solvent degradation and corrosion issues • Large equipment size	• Post-combustion capture in power plants and industrial processes	TRL 9 (commercially deployed)	• **Boundary Dam (Canada):** Coal-fired power plant with post-combustion capture using amine solvents. • **Petra Nova (USA):** Used MEA absorption (suspended in 2020).
Physical absorption	Uses physical solvents (e.g., Selexol, Rectisol) to absorb CO_2 under high pressure and low temperature. CO_2 is released by reducing pressure.	• Lower energy requirement for regeneration • Suitable for high-pressure gas streams	• Less effective at low CO_2 partial pressures • Solvent costs can be high	• Pre-combustion capture in IGCC plants • Natural gas processing	TRL 9	• **Great Plains Synfuels Plant (USA):** Uses physical solvents for CO_2 capture from gasification processes.
Chemical adsorption with solid sorbents	Utilizes solid materials (e.g., calcium oxide) that chemically react with CO_2 to form compounds; CO_2 is released upon heating.	• High CO_2 selectivity • Potential for lower energy consumption	• Sorbent degradation over cycles • High-temperature operation required for regeneration • Heat management challenges	• Integration with high-temperature processes, like cement production and gasification	TRL 5–7 (pilot to demonstration)	• **Research projects:** Focus on calcium looping for CO_2 capture in cement plants.

(Continued)

TABLE 1.3 (Continued)

Physical adsorption (PSA/VPSA)	Employs solid adsorbents (e.g., zeolites, activated carbon) that physically adsorb CO_2 at high pressure or vacuum; CO_2 is released by reducing pressure or applying vacuum.	• Lower energy consumption compared to absorption • Good for high CO_2 concentrations	• Less effective for dilute gas streams • Requires multiple adsorption beds • Adsorbent degradation over time	• Hydrogen purification • CO_2 capture from high-pressure gas streams	TRL 8–9	• **Hydrogen production facilities:** Use PSA for gas purification.
Membrane separation	Uses semi-permeable membranes that selectively allow CO_2 to pass through while retaining other gases, based on differences in permeability or solubility.	• Compact and modular • Low energy consumption • No chemical solvents require	• Lower CO_2 purity in a single stage • Susceptible to contaminants • Membrane durability	• Natural gas sweetening • Biogas upgrading • Potential for flue gas CO_2 capture	TRL 7–8	• **Air Products PRISM®Membranes:** Used in industrial gas separation. • **Research at NREL:** On advanced membrane materials.
Cryogenic separation	Involves cooling gas streams to low temperatures to condense and separate CO_2 from other gases based on differences in boiling points.	• Produces high-purity CO_2 • No chemical solvents • Suitable for high CO_2 concentrations	• Energy-intensive due to refrigeration requirements • Not economical for low CO_2 concentrations • Complex equipment	• CO_2 capture from gas streams with high CO_2 concentrations • Liquefaction of CO_2 for transport	TRL 6–7	• **Air Liquide's Cryocap™:** Combines cryogenic separation with other methods for CO_2 capture.
Oxy-fuel combustion	Burns fuel in pure oxygen instead of air, producing flue gas that is mainly CO_2 and water vapor; after condensation, a high-purity CO_2 stream is obtained.	• Simplifies CO_2 capture due to high CO_2 concentration • Reduces NO^- emissions • Potential for integration with PtX systems using oxygen from electrolysis	• Requires air separation units for oxygen supply • High capital cost • Combustion system modifications needed	• Power plants • Cement kilns • Steel production	TRL 6–7	• **Schwarze Pumpe Pilot Plant (Germany):** Oxy-fuel combustion in a lignite-fired boiler. • **Callide Oxyfuel Project (Australia):** Demonstration plant.

(Continued)

TABLE 1.3 (Continued)

Pre-combustion capture	Converts fuel into a mixture of H2 and CO_2 (e.g., via gasification); CO_2 is captured before combustion, and H2 is used as fuel or feedstock.	• CO_2 capture at higher pressures, improving efficiency • Produces hydrogen for multiple uses	• Complex process requiring gasification • High capital costs • Limited to new builds or major retrofits	• IGCC power plants • Hydrogen production facilities	TRL 7–8	• **Edwardsport IGCC Plant (USA):** Pre-combustion capture demonstration. • **Kemper County Energy Facility (USA):** Project converted to natural gas without CO_2 capture.
Direct air capture (DAC)	Captures CO_2 directly from ambient air using chemical sorbents (solid or liquid); CO_2 is released for use or storage upon sorbent regeneration.	• Unlimited CO_2 supply • Flexibility in plant location • Can achieve negative emissions when combined with storage	• Very low atmospheric CO_2 concentration requires significant energy input • High operational costs • Technology still scaling up	• Production of synthetic fuels in PtX systems • Negative emissions technologies	TRL 6–7	• **Climeworks (Switzerland):** Operates DAC plants capturing CO_2 from air. • **Carbon Engineering (Canada):** Developing large-scale DAC facilities.
Molten carbonate fuel cells (MCFCs)	Uses MCFCs to capture CO_2 from flue gas while generating electricity; CO_2 is concentrated in the anode side and can be separated.	• Dual function of power generation and CO_2 capture • High CO_2 capture efficiency	• High-temperature operation • Fuel cell degradation over time • High capital costs	• Integration with power plants and industrial processes • Potential for capturing CO_2 from dilute gas streams	TRL 5–6	• **FuelCell Energy (USA):** Demonstrated MCFCs for CO_2 capture and power generation.
Biological capture (microalgae)	Utilizes microalgae to absorb CO_2 during photosynthesis; microalgae biomass can be harvested for biofuels or other products.	• Converts CO_2 into valuable biomass • Can treat flue gases with low CO_2 concentrations • Renewable and sustainable process	• Requires large land areas • Sensitivity to environmental conditions • Possibility of harvesting and processing biomass to be complex	• CO_2 capture from flue gases • Biofuel production • Wastewater treatment	TRL 4–5	• **AlgaePARC (Netherlands):** Research facility studying microalgae cultivation for CO_2 capture and biomass production.

due to concerns about compatibility with existing infrastructure. Any significant increase would require modifications to pipelines, compressors, turbines, and other industrial equipment to ensure safety and efficiency. Hydrogen production is primarily dominated by fossil fuel–based processes, such as SMR and coal gasification, which account for over 95% of global production. These methods are cost-effective but energy-intensive and generate significant CO_2 emissions unless combined with CCS technologies. In contrast, green hydrogen, produced through water electrolysis powered by RE, results in zero GHG emissions and is considered the most environmentally friendly option. Additional production methods include biochemical and thermochemical processes, which use biomass, and various electrolysis techniques, such as photo-electrolysis, photolysis, and thermolysis. Hydrogen produced through different methods is classified based on its carbon intensity: gray hydrogen (no CCS), blue hydrogen (with CCS), and green hydrogen (using RE for electrolysis). Despite its abundance in compounds like water and hydrocarbons, hydrogen is rarely found in its pure form on Earth due to its lightness and the planet's gravitational forces, making its extraction from these compounds a challenging process. The environmental impact, cost, and efficiency of hydrogen production vary depending on the method employed, with mature technologies generally offering higher efficiencies and lower costs compared to those still in the R&D phase [50].

Current global efforts focus on expanding green hydrogen production to support a sustainable energy transition. For instance, the EU aims to achieve an annual green hydrogen production of 10 Mt (million tons) by 2030, which will require a substantial increase in electrolyzer capacity and technological advancements [50]. To meet this target, significant cost reductions in electrolyzer technologies are anticipated. CAPEX for electrolyzer technologies, such as AWEs, PEMWEs, and SOWEs, is expected to decrease significantly (from nearly €750/kW to around €400/kW) by 2030 due to improvements in system design, materials, and supportive government policies. Emerging technologies such as anion exchange membrane (AEM) electrolyzers also show potential for competitive hydrogen production, enhancing the overall resilience and stability of energy systems by enabling efficient energy storage and utilization. Currently, hydrogen is utilized primarily in oil refining, ammonia, and methanol production, while its use in the transportation sector is relatively modest. In Europe, hydrogen consumption is highest in Germany and the Netherlands, where it is used predominantly in the refining, ammonia, and methanol industries. As interest and investment in hydrogen technologies grow, hydrogen is expected to play a key role in various sectors, facilitating the shift toward cleaner energy systems and contributing to global decarbonization efforts. The methods for supplying and transporting hydrogen, along with their descriptions, advantages, and limitations, are encapsulated in Tables 1.4 and 1.5, offering a thorough comparison of the principal technologies employed in PtX processes.

1.2.4 Nitrogen (N_2) Supply

In the PtX systems, sourcing high-purity nitrogen is crucial, especially for processes like PtA. The air separation industry employs various technologies for N2 supply, primarily classified into cryogenic and non-cryogenic processes [51]. Cryogenic processes, such as cryogenic distillation (CRD), leverage the distinct boiling points of nitrogen (−196°C) and oxygen (−183°C) to separate these gases from air. Achieving ultra-high nitrogen purity above 99.9% requires operating at temperatures slightly below the boiling point of oxygen, using a series of distillation columns working at pressures between 5 bar and 1.2–1.5 bar. CRD offers high energy efficiency, consuming around 0.11 MWh per ton of N2, but has limited dynamic flexibility, requiring lengthy start-up times and a minimum operating load of 60% [52, 53]. Large-scale air separation units (ASU) using CRD are the industry standard for producing substantial quantities of purified N2, often exceeding thousands of tons per day. ASUs are divided into warm and cold regions, with each region serving distinct roles: the cold region (below −170°C) contains the heat exchanger network, rectification units, and storage, which are critical for optimizing liquefaction efficiency, while the warm region includes equipment for compressing, pre-cooling, drying, and purifying incoming air. Non-cryogenic processes include PSA and membrane-based air

TABLE 1.4
Comparison of Various Hydrogen Supply Methods

Supply Method	Description	Energy Source	Benefits	Challenges	Technology Maturity
Steam methane reforming (SMR)	Hydrogen is produced by reacting methane with steam at high temperatures, resulting in hydrogen and CO_2.	Natural gas	• Low-cost hydrogen production • Established technology	• High CO_2 emissions (gray hydrogen) • Relies on fossil fuels • Requires carbon capture for emissions reduction (blue hydrogen)	Commercially mature
Coal gasification	Coal is converted into syngas (a mixture of hydrogen, carbon monoxide, and CO_2) through high-temperature reactions with oxygen and steam.	Coal	• Utilizes abundant coal resources • Established technology	• Highest CO_2 emissions • Environmental concerns • Requires carbon capture for emissions reduction	Commercially mature
Water electrolysis (green hydrogen)	Water is split into hydrogen and oxygen using electricity from renewable energy sources through electrolysis processes (e.g., PEM, AWE, SOE).	Renewable energy (wind, solar, hydro)	• Zero CO_2 emissions • Supports renewable energy integration • Enhances energy storage capabilities	• Higher production costs • Dependent on availability of renewable electricity • Requires investment in infrastructure	Commercial and expanding
Methane pyrolysis (turquoise hydrogen)	Methane is thermally decomposed into hydrogen and solid carbon without emitting CO_2, potentially using renewable electricity or heat.	Natural gas	• Lower CO_2 emissions compared to SMR • Solid carbon by-product potentially stored or used	• Emerging technology • High energy input required • Need for market development for solid carbon	Pilot and demonstration projects
Biomass gasification	Biomass is thermochemically converted into syngas, followed by hydrogen extraction; can be carbon-neutral if biomass is sustainably sourced.	Biomass	• Renewable feedstock • Potential for carbon-neutral or negative emissions • Utilizes waste materials	• Feedstock availability and logistics • Technology complexity • Higher costs compared to fossil methods	Demonstration stage
Photocatalytic water splitting	Water is split into hydrogen and oxygen using sunlight and specialized photocatalysts, directly harnessing solar energy for hydrogen production.	Solar energy	• Zero CO_2 emissions • Direct use of abundant solar energy • Simple process concept	• Very low efficiency currently • Technology in research phase • Challenges in scaling up	Experimental

TABLE 1.5

Comparison of Various Hydrogen Transport Methods

Transport Method	Description	Advantages	Challenges	Suitable Distance	Technology Maturity
Pipeline transport	Hydrogen is transported through pipelines, either dedicated hydrogen pipelines or blended with natural gas in existing pipelines.	• Efficient for large volumes • Established technology	• High capital cost for new pipelines • Material compatibility issues (hydrogen embrittlement) • Leakage concerns	Short to long distances	Commercially mature
Compressed gas transport	Hydrogen is compressed to high pressures (200–700 bar) and stored in cylinders or tube trailers for transportation by road or rail.	• Flexible and scalable • Suitable for small to medium quantities	• Low energy density per volume • High transportation cost per unit of hydrogen • Safety concerns at high pressures	Short distances (local distribution)	Commercially mature
Liquefied hydrogen transport	Hydrogen is cooled to cryogenic temperatures (−253°C) to become liquid and transported in insulated tanks via trucks, ships, or rail.	• Higher energy density than compressed gas • Suitable for bulk transport	• Energy-intensive liquefaction process • Boil-off losses during transport • High infrastructure costs	Medium to long distances	Commercially mature
Liquid organic hydrogen carriers (LOHC)	Hydrogen is chemically bound to a liquid organic carrier (e.g., dibenzyltoluene), transported under ambient conditions, and released at the destination through dehydrogenation.	• Transport under ambient conditions • Uses existing fuel infrastructure • Safe and non-flammable during transport	• Energy required for hydrogenation/dehydrogenation • Catalyst costs • Lower overall efficiency due to conversion steps	Long distances (international trade)	Demonstration stage
Ammonia as a hydrogen carrier	Hydrogen is converted into ammonia via the Haber–Bosch (HB) process, transported as a liquid under mild pressure or refrigeration, and reconverted to hydrogen at the destination.	• High hydrogen density • Established transport and storage infrastructure for ammonia	• Energy required for synthesis and cracking • Toxicity and handling risks of ammonia • Need for efficient catalysts for reconversion	Long distances (international trade)	Demonstration stage
Metal hydrides	Hydrogen is stored by forming metal hydrides with certain alloys and transported in solid form; hydrogen is released by heating at the destination.	• High volumetric hydrogen density • Safe storage form (non-flammable) • Reversible storage	• Heavy weight of storage materials • Slow kinetics of hydrogen absorption/desorption • High cost of hydride materials	Short to medium distances	Research and development

separation [52]. PSA operates cyclically at ambient temperature using O_2-selective carbon molecular sieve adsorbents, alternating between pressures of 6–8 bar for adsorption and 1 bar for desorption. PSA is suitable for small to medium production scales, up to several hundred tons of N_2 per day. It offers excellent dynamic flexibility with start-up times of less than an hour and a lower operating limit of around 30%. However, PSA has limitations, such as a lower nitrogen purity (99.8%), higher energy usage (0.22–0.31 MWh per ton), and decreased cost-effectiveness at larger production scales. Scaling up PSA systems for larger production targets becomes increasingly inefficient compared to CRD, as the process's complexity and energy demand increase disproportionately with size.

Membrane-based air separation technology employs O_2-permeable materials, such as polysulfone hollow fibers or carbon molecular sieves, to separate nitrogen from air. This process operates at ambient temperature and is driven by pressure differences between the feed and permeate sides and is suitable for small-scale production. Membrane systems provide greater dynamic flexibility and simpler operation due to their continuous nature and lack of moving parts. However, achieving purities above 99.5% is challenging, and energy consumption can reach up to 0.65 MWh per ton of N_2 for higher purities [51]. Additionally, membrane systems do not remove argon, resulting in nitrogen products with argon content similar to atmospheric air, necessitating further separation steps if lower argon levels are required. This limitation restricts the application of membrane-based separation, where ultra-high nitrogen purity or low argon content is essential. Dynamic flexibility and DSM strategies can improve the economic viability of ASUs by mitigating RE curtailment. For cryogenic units like ASUs, DSM strategies include maintaining cold conditions to reduce start-up times and employing operational adjustments, such as modulating rotating speed, using variable inlet guide vanes, and optimizing blade geometry in the centrifugal compressor train to enhance operating efficiency. In contrast, non-cryogenic units, like PSA and membrane systems, with their faster start-up times and lower minimum operating loads (down to 30%), are more adaptable to variable renewable energy inputs. This adaptability makes them well-suited for integration with PtX systems, where dynamic flexibility is essential for balancing intermittent renewable energy supplies. The methods for supplying nitrogen, along with their descriptions, advantages, and limitations, are encapsulated in Tables 1.6, offering a thorough comparison of the principal technologies employed in PtX processes.

1.2.5 PtX System Infrastructure

The infrastructure necessary for power-to-X (PtX) systems is complex, involving the transportation and storage of electricity, CO_2, water, hydrogen, and various products. Efficient transportation is critical since electricity generation and subsequent conversion processes often occur at different locations. Although some existing infrastructure can be repurposed with minimal modifications, new infrastructure is often required, depending on the energy source or product. PtX systems can leverage existing electrical grids to transmit electricity from generation sites to PtX facilities. However, significant upgrades to transmission and distribution lines may be necessary to handle increased loads and ensure grid stability, particularly with the integration of large-scale renewable energy sources. Similarly, the existing natural gas pipeline network can be adapted to transport synthetic methane produced through PtX processes. Minor adjustments, such as compression and blending compatibility, can enable these networks to accommodate novel fuel blends, like oxymethylene ether diesel (OME1-blends). In some cases, developing new infrastructure is unavoidable. Hydrogen pipelines, for instance, require specialized materials, such as stainless steel or composites, to prevent hydrogen-induced embrittlement and leakage due to hydrogen's high diffusivity and unique properties. Additionally, when energy generation, electrolysis, and synthesis processes are consolidated at a single PtX facility, these systems can operate independently, reducing grid strain and improving resilience.

Various transportation methods can be employed to move PtX products. Trucks provide flexible options for transporting liquid fuels, H_2, and CO_2. Hydrogen can be transported in high-pressure tanks for gaseous H_2, cryogenic tanks for liquefied H_2, or as a LOHC, which chemically binds hydrogen to a liquid carrier for safer and more efficient transport under ambient conditions. Pipelines are

TABLE 1.6

Comparison of Various Nitrogen Supply Methods

Supply Method	Description	Advantages	Disadvantages	Production Scale	Nitrogen Purity	Energy Usage	Dynamic Flexibility	Technology Maturity
Cryogenic distillation (CRD)	Separates air components by cooling air to cryogenic temperatures and distilling nitrogen from oxygen and other gases based on boiling points.	• Produces high-purity N_2 (>99.9%) • Economical at large scales • Low energy consumption per unit	• High capital cost • Limited dynamic flexibility (slow start-up times) • Minimum load ~60%	Large-scale production (>1000 t/day)	>99.9%	~0.11 MWh/ ton N_2	Limited (start-up times in hours)	Commercially mature
Pressure swing adsorption (PSA)	Uses adsorbent materials that preferentially adsorb oxygen over nitrogen; pressure is cycled to adsorb and desorb gases, allowing nitrogen to be collected.	• Modular and compact • Faster start-up times (<1 hr) • Good dynamic flexibility (operates down to ~30% load)	• Lower nitrogen purity (~99.8%) • Higher energy consumption than CRD • Less economical at large scales	Small to medium-scale production (Up to several hundred tons/day)	Up to ~99.8%	0.22– 0.31 MWh/ ton N_2	Good (start-up times <1 hr)	Commercially mature
Membrane separation	Uses selective membranes that allow faster permeation of oxygen over nitrogen, effectively separating nitrogen from air under pressure differences.	• Simple operation • Continuous process • Excellent dynamic flexibility • Low maintenance	• Limited nitrogen purity (up to ~99.5%) • High energy consumption at higher purities • Not economical at large scales	Small-scale production (Up to a few hundred tons/day)	Up to ~99.5%	Up to 0.65 MWh/ ton N_2	Excellent (rapid response)	Commercially mature
Vacuum pressure swing adsorption (VPSA)	Similar to PSA but incorporates vacuum during the desorption phase to improve efficiency and reduce energy consumption.	• Improved energy efficiency over PSA • Good dynamic flexibility • Faster cycle times	• Complexity of vacuum equipment • Limited nitrogen purity compared to CRD • Capital cost	Small to medium-scale production	Up to ~99.8%	Slightly lower than PSA	Good	Commercially mature
Fractional distillation (small-scale cryogenic units)	Similar to CRD but designed for smaller capacities; uses cryogenic cooling to separate nitrogen from air.	• Produces high-purity N_2 • Suitable for medium scales, where high purity is needed	• Higher energy consumption than large-scale CRD • Higher per-unit costs • Less flexibility than PSA or membranes	Medium-scale production (Hundreds of tons/ day)	>99.9%	Higher than large-scale CRD	Limited	Commercially mature
Electrochemical separation	Uses electrochemical cells to separate nitrogen from air by selectively transporting ions through membranes under an electric field.	• Potential for high purity • Compact and modular design	•Emerging technology •Energy-intensive •Not yet commercially viable	Small-scale production	Variable	High	Limited	Research and development

typically preferred for transporting gaseous products like synthetic methane and hydrogen, but they must be designed to withstand the challenges associated with hydrogen, such as embrittlement and leakage. PtX storage solutions vary depending on the state and scale of the product. Gaseous energy carriers can be stored in on-site tanks for small-scale, short-term storage or in large-scale geological storage, such as salt caverns, for long-term, high-density energy storage over several months. Liquid hydrogen tanks are suitable for smaller quantities of hydrogen but require energy-intensive liquefaction and advanced insulation. Metal hydride storage offers a compact solution where metal hydrides absorb and release hydrogen, making it ideal for applications requiring high-density storage in limited space. The distribution of PtX products to end users also requires specific infrastructure, such as charging stations for electric vehicles and fueling stations for liquid fuels, hydrogen, and SNG. These fueling stations must be adapted to handle the specific properties of each fuel type. Compressor stations play a critical role in maintaining gas pressure in pipelines to enable the efficient long-distance transportation of gases like hydrogen and synthetic methane. However, hydrogen's lower volumetric energy density may necessitate more frequent compression or higher compression ratios, impacting energy consumption and operational costs.

Several projects showcase how existing infrastructure can be integrated or adapted for PtX applications. The Linde Ammonia Concept utilizes cryogenic ASUs to produce nitrogen for large-scale ammonia synthesis, demonstrating how existing systems can support PtX processes. The Boundary Dam Power Station in Canada captures CO_2 from a coal-fired power plant for reuse, showcasing the retrofitting capabilities of conventional facilities. Climeworks's DAC facilities in Switzerland integrate DAC technologies into PtX systems, highlighting the potential of new technologies for sustainable energy solutions. Similarly, the Drax Power Station in the UK transitioned from coal to biomass to reduce its carbon footprint, demonstrating the adaptation of existing facilities for sustainable PtX operations. Overall, the infrastructure required for PtX systems encompasses a diverse range of transportation and storage solutions designed to manage electricity, CO_2, water, hydrogen, and various products. While existing infrastructure can be repurposed to some extent, the unique properties of hydrogen and specific requirements of PtX processes often necessitate new, specialized infrastructure. Ensuring compatibility, safety, and efficiency in transportation and storage is essential for the successful integration and scalability of PtX systems. Strategic placement of PtX facilities near CO_2 sources and renewable energy generation sites can further enhance economic viability and reduce logistical challenges, supporting the sustainable implementation of PtX technologies.

1.2.6 PtX Pathways

PtX technologies are innovative processes that convert surplus renewable energy into chemical products, fuels, or other energy carriers, thereby enabling the storage of renewable energy and facilitating its integration into various sectors. These pathways play a vital role in decarbonization efforts by utilizing captured CO_2 for valuable products. One key PtX pathway is hydrogen generation and catalytic conversion (HGCC), which produces hydrogen through water electrolysis powered by RESs. The generated hydrogen is then combined with captured CO_2 or CO in catalytic reactors to synthesize hydrocarbon-based fuels and chemicals like methanol or synthetic gasoline. HGCC benefits from using existing fuel infrastructure and helps reduce GHG emissions; however, it faces challenges such as high energy requirements for electrolysis and the cost of electrolyzers and catalysts [54]. An example is the H_2FUTURE project in Austria, which uses PEMWE to produce green hydrogen for steel manufacturing, reducing carbon emissions. Another pathway is electrochemical CO_2 conversion and catalysis (ECCC), which uses electrochemical methods and specialized catalysts to directly convert CO_2 into valuable compounds, such as CO, formic acid, or ethanol. This process eliminates the need for intermediate hydrogen production and offers a direct approach to CO_2 utilization. The main benefits include high selectivity and modular, scalable systems, but challenges such as catalyst durability, low current densities, and slow reaction rates hinder large-scale applications [55]. The Twelve project, formerly known as Opus 12, exemplifies

ECCC by developing an electrochemical system that converts CO_2 into chemicals like ethylene and ethanol using renewable electricity. Direct electrochemical synthesis (DES) is another PtX pathway that focuses on converting basic feedstocks into chemicals or fuels via electrochemical processes within a single reactor, thus eliminating the need for pure hydrogen production. This simplification reduces infrastructure requirements and can potentially result in compact and energy-efficient systems. Despite these benefits, DES struggles with low energy efficiency, slow reaction rates, and the stability of electrocatalysts. Current research is exploring the DES of ammonia and other chemicals, but commercial-scale implementations are still under development.

Co-electrolysis of CO_2 and H_2O (CoSOE) enables the simultaneous electroreduction of these molecules to generate syngas, a mixture primarily composed of CO and H_2, with minor quantities of CH_4, CO_2, and H_2O, depending on the thermodynamic conditions. Syngas can then be used as a precursor in the Sabatier chemical reactor to produce a high concentration of CH_4. CoSOE technology is versatile and can be used to generate a variety of fuels, including methanol, ethanol, dimethyl ether (DME), and methane, for both direct and indirect fuel applications. Also, CO_2 electrolysis systems can convert CO_2 into CO using a range of cathodic catalysts, such as silver, platinum, copper, gold, zinc, and palladium. The resulting CO can then be used in subsequent fuel generation processes through catalytic reactions. Depending on the choice of catalyst and operating conditions, CO_2 can be efficiently converted into CO using specialized electrochemical cells, such as modified solid oxide cells, or emerging technologies, like molten carbonate electrolyzers. This process supports the production of low-carbon fuels, such as CH_4, ethylene (C_2H_4), and CH_3OH, thereby contributing to the development of sustainable energy systems and advancing the utilization of renewable energy in fuel production.

Bio-catalytic conversion (BCC) utilizes biological catalysts, such as enzymes or microorganisms, to convert CO_2 into fuels or chemicals. Techniques like microbial electrolysis cells (MEC) use microbes to produce hydrogen or methane from organic substrates, offering the dual benefit of waste treatment and fuel production. The main advantages include the utilization of waste streams and operation under mild conditions; however, low reaction rates, scaling up biological systems, and maintaining microbial activity over time present significant challenges. A notable example is the development of MECs at Penn State University [56] that produce hydrogen gas from wastewater, demonstrating the benefits of integrating wastewater treatment with hydrogen production. High-temperature CO_2 conversion (HTCC) employs high temperatures and catalysts to convert CO_2 and water into syngas—a mixture of hydrogen and carbon monoxide—which can be further processed into synthetic fuels. This pathway often uses solar thermal energy to achieve the necessary temperatures, enhancing overall efficiency. While HTCC offers high efficiency due to elevated operating temperatures, it also faces issues, such as material durability and the high energy requirements to maintain these temperatures. Sunfire [57] GmbH in Germany has developed a high-temperature co-electrolysis plant using SOWEs to produce syngas from CO_2 and water, which can then be converted into synthetic fuels. Lastly, plasma-enhanced conversion (PEC) uses plasma, an ionized gas, to activate and dissociate CO_2 molecules, facilitating their conversion into useful products, like carbon monoxide, hydrocarbons, or oxygenates. Plasma reactors can operate at lower temperatures compared to conventional thermal processes and offer potential for high conversion rates. However, issues such as the energy efficiency of plasma generation, scalability, and cost-effectiveness remain [58]. Research is ongoing in plasma-assisted CO_2 conversion, but industrial-scale examples are still under development. Each of these PtX pathways offers unique methods for converting renewable energy and captured CO_2 into valuable products, contributing to sustainable energy systems and reducing carbon emissions. Despite significant progress, improving energy efficiency, catalyst performance, scalability, and cost reduction remain critical challenges for the widespread adoption of these technologies. Continued research and collaboration between academia and industry are essential to overcoming these hurdles and realizing the full potential of PtX pathways. Table 1.7 offers a detailed comparison of PtX pathways, emphasizing their principal technologies, benefits, drawbacks, and case studies.

TABLE 1.7

Comparison of Various PtX Pathways

Pathway	Description	Technologies	Benefits	Issues (Challenges)	Case Study
Hydrogen generation and catalytic conversion (HGCC)	Involves producing hydrogen via water electrolysis and combining it with CO_2 and/or CO or N_2 in chemical reactors to synthesize hydrocarbon-based fuels and chemicals.	• **Water electrolysis:** PEMWE, AWE, SOWE • **Catalytic reactors:** Methanol synthesis, Fischer–Tropsch, Sabatier reactors • **Catalysts:** $Cu/ZnO/Al2O_3$ for methanol synthesis • **Reactors:** Fixed-bed, fluidized-bed, microreactors	• Produces environmentally friendly synthetic fuels • Enables storage and utilization of renewable energy • Integrates with existing fuel infrastructure	• High energy requirements for electrolysis • Cost of electrolyzers and catalysts • Need for CO_2 capture infrastructure	**H2FUTURE Project (Austria):** Uses PEMWE powered by RE to produce green hydrogen for steel manufacturing, reducing carbon emissions.
Electrochemical CO_2 conversion and catalysis (ECCC)	Uses electrochemical methods and catalysts to convert CO_2 directly into valuable compounds, like CO, formic acid, or hydrocarbons, which can be used as fuels or chemical feedstocks.	• **Electrochemical CO_2 reduction cells** • **Catalysts:** Metals like silver, gold, copper, etc. • **Specialized electrolyzers** designed for CO_2 reduction	• Direct conversion of CO_2 using renewable electricity • Potential for high selectivity to desired products • Modular and scalable	• Catalyst durability and efficiency • Low current densities and slow reaction rates • Scaling up for commercial applications	**Twelve (formerly Opus 12):** Developed an electrochemical system that converts CO_2 into products like ethylene and ethanol using renewable electricity.
Direct electrochemical synthesis (DES)	Directly converts feedstocks into chemicals or fuels via electrochemical processes in a single reactor, often avoiding the need for pure H_2 production.	• **Electrochemical reactors for direct synthesis** • **Electrocatalysts** for target reactions (e.g., N_2 reduction to NH_3) • **Solid-state electrochemical cells**	• Simplifies the process by eliminating intermediate steps • Reduces infrastructure for H_2 handling • Potentially compact and energy-efficient systems	• Low energy efficiency and slow reaction rates • Challenges in achieving high selectivity and yields • Electrocatalyst stability and durability	—

(Continued)

TABLE 1.7 (Continued)

Bio-catalytic conversion (BCC)	Utilizes biological catalysts, like microorganisms or enzymes, to convert CO_2 into fuels or chemicals, sometimes using sunlight or other energy sources.	• **Microbial electrolysis cells (MECs)** • **Enzymatic catalysis** • **Microbial fuel cells**	• Can utilize waste streams or wastewater • Combines waste treatment with fuel production • Operates under mild conditions	• Low reaction rates and efficiencies • Scaling up biological systems • Maintaining microbial activity over time	**Penn State University MECs:** Developed MECs that produce hydrogen gas from wastewater using microbes, demonstrating benefits of wastewater treatment and hydrogen production.
High-temperature CO_2 conversion (HTCC)	Uses high temperatures and catalysts to convert CO_2 and H_2O into syngas or other chemicals, often utilizing solar thermal energy.	• **High-temperature co-electrolysis using SOWEs** • **Thermochemical cycles** • **Solar thermal reactors**	• High efficiencies due to operating at elevated temperatures • Syngas possibly converted into various fuels • Utilizes abundant solar thermal energy	• Material durability at high temperatures • High energy requirements to maintain temperatures • Scaling up technologies for commercial use	**Sunfire GmbH (Germany):** Developed a high-temperature co-electrolysis plant using SOECs to produce syngas from CO_2 and water, which can then be converted into synthetic fuels.
Plasma-enhanced conversion (PEC)	Uses plasma (ionized gas) to activate and dissociate CO_2 molecules, facilitating their conversion into useful products, like CO, hydrocarbons, or oxygenates.	• **Plasma reactors:** Dielectric barrier discharge, microwave plasma • **Plasma–catalyst systems**	• Plasma possibly activating CO_2 at lower bulk temperatures • Potential for high conversion rates • Can be powered by renewable electricity	• Energy efficiency of plasma generation • Scalability and cost-effectiveness • Integration with catalysts for selectivity	–

1.2.7 Overview of PtX Conversion, Products, and Applications

PtG involves the generation of hydrogen, methane, and syngas (a mixture of H_2 and carbon monoxide, CO) from renewable or excess electricity [59]. The process begins with electrolysis, which splits water into hydrogen and oxygen using electricity. The produced hydrogen can be utilized directly in FCs for transportation, industrial processes, or power generation, thereby reducing carbon emissions in these sectors. Methane is synthesized through the methanation process, where hydrogen combines with CO_2 to produce SNG. SNG can be injected into existing NG networks for heating, power generation, or use as a transportation fuel. This approach leverages existing infrastructure, minimizing the need for significant new investments. Syngas is produced via processes such as the reverse water–gas shift (rWGS), dry reforming of methane, or co-electrolysis of CO_2 and water. The H_2 ratio in syngas can be adjusted depending on its intended application. Syngas serves as a crucial intermediary in the production of chemicals and fuels, particularly within the chemical industry. PtL focuses on converting renewable electricity into liquid fuels, with methanol being a primary output [60]. Methanol can be produced through the direct hydrogenation of CO_2 and hydrogen or by synthesizing syngas, which is then converted into methanol. Methanol is versatile; it can be used as a fuel for internal combustion engines (ICE), thereby reducing transportation emissions, or as a feedstock for producing chemicals and synthetic fuels. Its compatibility with existing fuel infrastructures allows for a gradual transition to greener energy without necessitating immediate extensive infrastructure changes. PtChem involves synthesizing chemical products by combining hydrogen with carbon-based feedstocks. This process yields essential chemicals, such as ethylene and propylene, which are fundamental in manufacturing plastics, resins, and other industrial materials. Additionally, processes like Fischer–Tropsch synthesis can produce polyoxymethylene and other polymers. PtChem significantly reduces the carbon footprint of chemical production by utilizing renewable hydrogen and CO_2, fostering a circular economy and promoting sustainable manufacturing practices. PtHeat converts excess electricity into thermal energy for applications such as residential heating, industrial processes, and district heating networks. Technologies like heat pumps and electrode boilers are employed to transform electricity into heat, enabling the use of renewable energy for heating purposes and reducing reliance on fossil fuels. In both residential and industrial settings, PtHeat enhances energy efficiency and lowers carbon emissions, supporting the transition to more sustainable heating practices and cleaner industrial operations [61, 62]. Figure 1.4 demonstrates that PtX includes different conversion processes and infrastructure that facilitate the conversion of excess renewable electricity into various products, such as gases, liquids, chemicals, and heat, for use in numerous sectors.

Governments and enterprises are increasingly recognizing the significant economic potential of renewable PtX technologies, particularly within the fossil fuel and related industries, which are leading global decarbonization efforts. As of late 2023, over 200 PtX plants are operational worldwide, encompassing both large-scale facilities and smaller pilot projects aimed at assessing the scalability and commercial viability of these technologies [63]. Initial adoption of PtX technologies has been observed at the small scale, typically within pilot projects and limited commercial applications. However, widespread household-level adoption remains in the emerging stages. PtX technologies are expected to expand to commercial and industrial applications, and eventually to large-scale hubs, thereby supporting national economies and contributing significantly to energy transition goals. The growth trajectory of PtX mirrors the global adoption of solar PV technology, which began with rooftop installations and progressed to utility-scale PV farms. However, many large-scale PV projects have faced delays due to grid congestion. PtX offers a potential solution by utilizing excess renewable electricity that might otherwise contribute to grid instability, thereby enhancing grid flexibility and resilience without needing extensive new infrastructure. A key issue and benefit for the PtX economy is identifying initial specialized applications that require minimal new infrastructure while offering strong commercial prospects. Critical factors for success include the scalability of technologies, early market opportunities for high-value products, effective risk

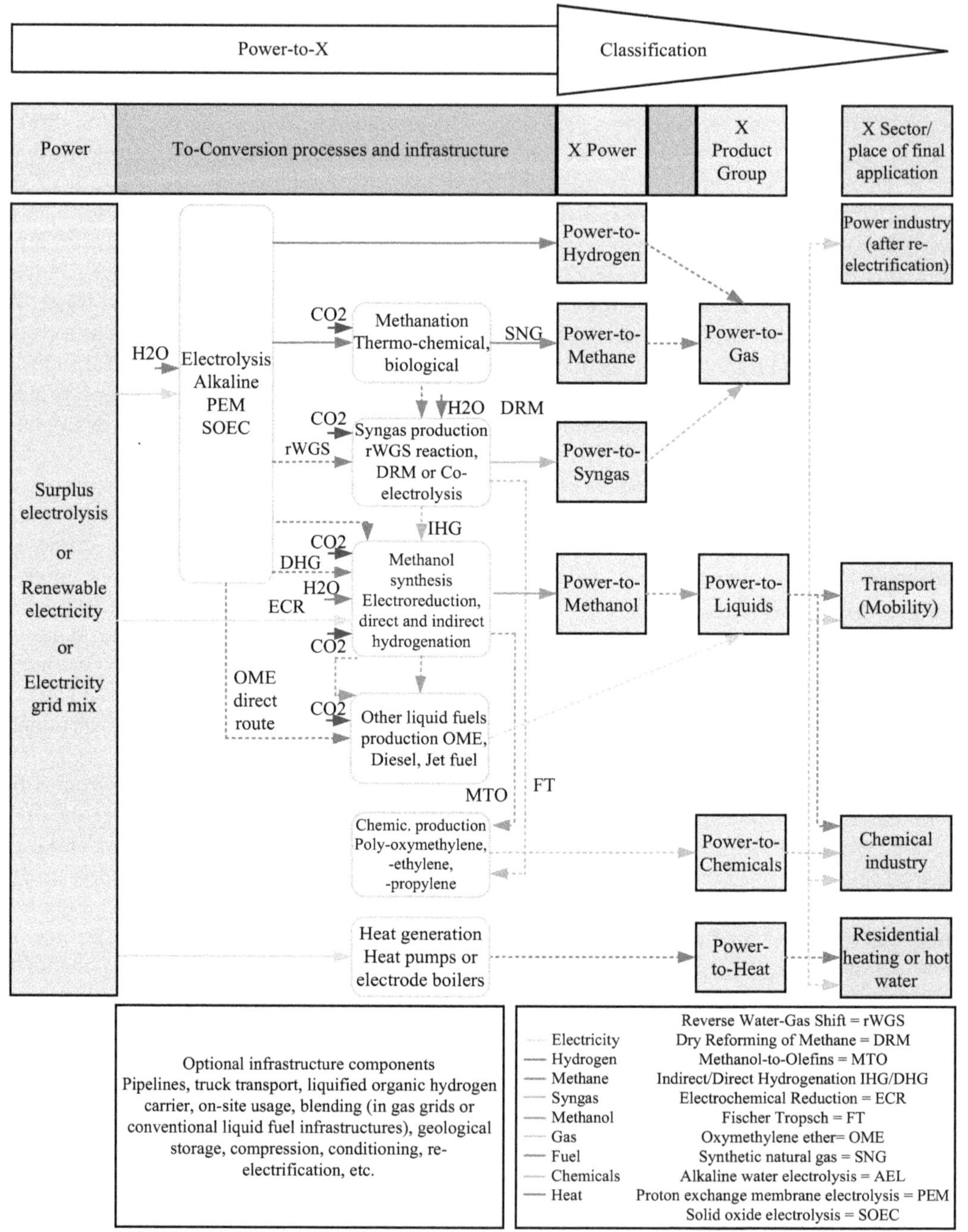

FIGURE 1.4   Overview of PtX conversion pathways and infrastructure.

management, and broad social acceptance. Based on current trends, a prospective PtX roadmap can be structured into three phases:

1. ***Small-scale/household application (kW capacity).*** The first phase focuses on deploying PtX technologies at the micro level, suitable for individual households or small communities. These decentralized systems include hydrogen storage units, residential fuel cells, and solar panels. They enable localized production and use of hydrogen for applications such

as home energy systems. An example of early-stage deployment is the integration of residential FCs that convert hydrogen into electricity and heat, enhancing energy efficiency at the household level [64].

2. *Utility-scale application (MW capacity).* In the second phase, PtX technologies are scaled up to meet the energy and material needs of larger communities or municipalities. PtX solutions are integrated into public infrastructure and industrial processes, converting captured CO_2 into valuable products and powering mass transit systems with hydrogen fuel cells. The production of synthetic fuels like ammonia and methanol using renewable energy and electrochemical processes helps reduce carbon emissions. A notable example is the Advanced Clean Energy Storage (ACES) project in Utah [65], USA, which focuses on large-scale hydrogen production and storage to meet regional energy demands.

3. *Economy-scale (hundreds of MW—GW capacity).* The final phase involves the creation of large multi-product PtX centers, leading to substantial industrial scaling of PtX technologies. These centers aim to produce significant quantities of hydrogen, syngas, CO, and other chemicals using renewable energy sources, contributing to a circular and sustainable economy. A prominent example is the UK's Gigastack initiative [66], which leverages offshore wind energy to produce green hydrogen for large-scale industrial processes. This phase represents a major transition toward integrating PtX technologies into the broader energy and industrial landscape, aligning with long-term sustainability and climate goals.

1.3 PtX SYSTEM–LEVEL ANALYSIS

The PtX system–level analysis is a comprehensive methodology that assesses the sustainability, efficacy, and feasibility of transforming RE into a variety of chemical fuels and raw materials. It consists of three essential components: techno-economic analysis (TEA), thermodynamic analysis (THA), and life cycle assessment (LCA). TEA evaluates factors such as capital expenditure, operational costs, and return on investment in order to determine the economic viability, cost-effectiveness, and market potential of PtX technologies. THA evaluates the thermodynamic feasibility and energy efficiency of the processes in question to guarantee that the energy conversion pathways are optimized for optimal performance. LCA assesses the environmental impact of PtX systems throughout their entire life cycle, from the extraction of raw materials to their dispersal at the end of life, encompassing emissions, resource consumption, and ecological footprint. Collectively, these analyses offer a comprehensive perspective on PtX technologies, which assists in the transition to a low-carbon economy and informs decision-making for sustainable energy solutions.

1.3.1 PtX Techno-Economic Analysis (TEA)

1.3.1.1 PtX TEA Methods

Conducting a TEA of PtX is crucial for the advancement and successful implementation of PtX techniques. Nevertheless, the absence of a universally recognized approach for TEA persists due to the lack of uniformity and comparability across various research investigations. Also, when considering future alternative energy systems to achieve decarbonization objectives, conducting TEA may be challenging due to limited data availability, insufficiently precise process descriptions, and unpredictable regulatory changes. Comparing research on TEA can be challenging due to the use of many markers and presumed input values when reporting results. Differences between studies might arise from variations in space and time, as well as changes in governmental rules. The key technical performance measure utilized in TEA studies of PtX is plant efficiency. The various methods and equations employed in the TEA of PtX systems are summarized in Table 1.8, where key metrics are detailed alongside their respective equations and symbol definitions.

TABLE 1.8

Overview of Key Methods and Equations in TEA of PtX Systems

Method/Metric	Equation	Symbols	Explanation
Levelized cost of energy (LCOE)	$$LCOE = \dfrac{\sum_{t=0}^{TH} \dfrac{I_t + M_t + F_t + C_t + T_t}{(1+r)^t}}{\sum_{t=0}^{T} \dfrac{E_t}{(1+r)^t}}$$ $$LCOE = \dfrac{\sum_{t=1}^{TH} \dfrac{CAPEX_t + OPEX_t + FUEL_t}{(1+r)^t}}{\sum_{t=1}^{TH} \dfrac{E_t}{(1+r)^t}}$$	I_t: (investment expenditure); M_t: (maintenance cost); $FUEL_t\ and\ F_t$: (fuel cost); C_t: (carbon cost); T_t: (taxes); r: (discount rate); E_t: (electricity generation); TH: (time horizon)	Evaluates the total cost per unit of stored energy over the system's lifespan.
Capital expenditure (CAPEX)	$$CAPEX = \dfrac{FCI}{0.85}$$ $$FCI = 1.18 \times \sum_i C_{BM} + 0.5 \times \sum_i C_{BM}^{0}$$	FCI: (fixed capital investment); C_{BM}: (installed cost of base module); C_{BM}^{0}: (installed cost of base module, at initial time or a reference module)	Represents initial investment costs, calculated using FCI.
Operating expenditure (OPEX)	$$OPEX = C_M + C_{OL} + C_{RM} + C_{WT} + C_R$$	C_M: (maintenance cost); (labor cost); C_{RM}: (raw material cost); C_{WT}: (waste treatment cost); C_R: (repair cost)	Ongoing operational costs, including labor, materials, and waste treatment.
Cost of manufacturing product (COP)	$$C_{COP,x} = \dfrac{OPEX + C_{dep} + C_{dep}^{byp}}{P_x}$$	C_{dep}: (depreciation cost); C_{dep}^{byp}: (by-product revenue); P_x: (production capacity)	Total cost of manufacturing, accounting for depreciation and by-product revenue.

(Continued)

TABLE 1.8 (Continued)

	$$C_{dep} = FCI \times \frac{i \times (1+i)^n}{(1+i)^n - 1}$$		
Payback time (τ)	$$\tau = \frac{FCI}{C_{dep}^{byp} + C_{rev}^{x} - OPEX}$$	τ: (payback time); C_{dep}^{byp}: (revenue from production); C_{rev}^{x}: (X revenue)	The time required to recover initial investment through generated revenue.
Levelized cost of the product (LCOP)	$$LCOP = \frac{COP}{TH} \sum_{t=1}^{TH} (1 + IR)^t$$	IR: (inflation rate)	Comprehensive cost per unit of product over the entire life cycle.
Net present value (NPV)	$$NPV = \sum_{t=1}^{TH} \frac{C_{F,t}}{(1+i)^t} - FCI$$ $$NPV = E_x (LHV) \times LCOP - OPEX - E_{el}.LCOE$$	$C_{F,t}$: (cash flow in year t); i: (discount rate); FCI: (fixed capital investment); E_x: (quantity of electric energy provided by renewables); LHV: (lower heating value of the energy product); E_{el}: (the energy value of the produced X)	Profitability assessment using the time value of money and cash flows. A positive NPV indicates profit.

1.3.1.2 Key Determinants in PtX TEA

PtX systems, which convert electricity into various fuels and chemicals, have significant potential to support the transition toward a sustainable energy future. Recent evaluations have identified several key factors influencing the techno-economic feasibility of PtX technologies. One of the most critical determinants is the cost of electricity, as lower electricity prices directly reduce production costs and shorten payback periods, improving the economic viability of renewable fuels. The declining costs and increasing availability of wind and solar power render these energy sources particularly beneficial, enhancing the financial attractiveness of PtX systems. However, the economic performance of PtX is highly sensitive to fluctuations in electricity prices, and the reliance on renewable electricity also introduces challenges related to the variability of supply, which must be considered in techno-economic assessments [67]. The choice of electrolyzer technology is another crucial factor. AWEs are the most commonly employed for hydrogen production due to their cost-effectiveness and technological maturity, resulting in lower capital and operational costs. However, newer technologies, such as PEMWEs and SOWEs, offer advantages under specific conditions. PEMWEs are particularly suited for flexible operation when integrated with variable RES, thus rendering them a key technology for systems reliant on intermittent renewable power. SOWEs, known for their high efficiency and lower energy requirements, present significant cost savings in large-scale operations, particularly where waste heat integration is possible [68, 69]. The economies of scale play a critical role in reducing costs in PtX systems; for instance, scaling up electrolyzer stacks for green hydrogen production can significantly reduce capital expenditures, improving competitiveness with conventional fossil fuel technologies. However, achieving these benefits depends on substantial upfront investment and large-scale deployment, which may not be feasible everywhere [70].

The cost of CO_2 capture also significantly affects the economic feasibility of PtX systems. DAC methods are considerably more expensive than capturing CO_2 from concentrated sources like flue gases, impacting the overall cost structure, particularly in hydrocarbon-based fuel production, where CO_2 is a key input. While DAC offers more flexibility in site selection, its high energy consumption and cost remain barriers to widespread adoption, especially in the near term. In contrast, capturing CO_2 from industrial flue gases is currently more economically feasible, though availability and concentration of CO_2 can vary by industry [71, 72]. Renewable methane has emerged as one of the most economically feasible PtX products, particularly given recent increases in natural gas prices. However, the economic viability of other PtX products, such as methanol, DME, and gasoline, depends on a range of factors, including hydrogen production and storage costs, which account for a significant portion of overall system expenses [73, 74]. Ammonia remains less competitive due to its higher production costs, though improvements in hydrogen production efficiency may enhance its prospects [75]. The efficiency of the overall system and the affordability of the technology are central concerns across all PtX pathways. The performance of electrolysis units and CO_2 capture technologies directly affects both system efficiency and cost-effectiveness. Implementing effective heat management strategies, such as recovering and utilizing excess heat from PtG processes, can improve system efficiency. However, traditional reactor designs, such as fixed-bed reactors, face operational challenges, particularly at scale, which may limit their economic viability [76, 77]. PtL products, such as methanol and synthetic fuels for aviation and shipping, are highly valued for their energy density and potential to replace fossil fuels in transportation. Nonetheless, their production costs remain high, and the absence of flexible process designs continues to present barriers to widespread adoption [78]. PtG products, like renewable methane, are nearing economic feasibility, while PtChem products, such as formic acid and ammonia, hold promise for future competitiveness, depending on the source and cost of CO_2 [79].

Geographical location and infrastructure availability also play critical roles in determining the feasibility of PtX systems. Proximity to RESs such as wind or solar farms, combined with the availability of electricity transmission infrastructure, influences both cost and system efficiency. Transportation and storage infrastructure for CO_2, hydrogen, and final PtX products pose additional logistical issues, particularly for remote or offshore locations [80]. Hydrogen storage technologies, such as compressed gas or liquefied hydrogen, significantly impact both technical feasibility and overall system costs. The choice of storage technology must balance capital costs, operational complexity, and efficiency losses associated with each option [81]. PtX technology also offers broader benefits by enabling the conversion of renewable electricity into gases, chemicals, and liquid fuels, providing a valuable solution for storing excess RE and reducing carbon emissions across multiple sectors. However, the intermittent nature of RES presents a challenge, as continuous PtX operation requires a reliable backup power source [8]. To maintain the environmental benefits of PtX systems, this backup should ideally come from low-carbon or renewable sources. Market demand and price competitiveness for PtX products are other critical determinants of techno-economic feasibility. The demand for green hydrogen, synthetic fuels, and renewable chemicals is influenced by regulatory frameworks and market conditions, including the price volatility of fossil-based alternatives [82]. As such, PtX products' economic attractiveness may fluctuate with changes in fossil fuel prices. Additionally, clear standards and certifications for green hydrogen and other PtX products will be essential for market acceptance and large-scale deployment [83].

While most TEA studies provide system-wide evaluations, they often overlook key process parameters, such as temperature, pressure, catalyst selection, and CO_2 purity, which can significantly affect system performance and costs. More detailed analysis of these variables is necessary to explore new catalysts that can improve efficiency, lower costs, and reduce environmental impacts. The dynamic behavior of PtX systems under varying operational conditions is also a critical area for further research, particularly as flexible operation becomes more important for integration with RES. The source of CO_2—whether from biogas or DAC—strongly influences both economic and environmental outcomes for PtX systems. DAC offers flexibility in site selection, but its high energy demand makes it less economically viable in the near term compared to CO_2 capture from industrial sources. Advancements in CCS technologies will be crucial for integrating PtX into industrial operations, particularly in hard-to-abate sectors, such as cement and steel. Government policies and incentives play a pivotal role in determining the scale and speed of PtX technology deployment. Regulatory support will be especially important in sectors like transportation, where PtG and PtL pathways show great potential for reducing CO_2 emissions. Policymakers and investors must carefully evaluate the levels of subsidies and support mechanisms needed to accelerate PtX adoption and achieve decarbonization goals. Additionally, the role of carbon pricing and emissions trading systems is critical, as higher carbon prices can improve the economic viability of PtX by incentivizing emissions reductions and making renewable alternatives more competitive with fossil fuels.

1.3.2 PtX Thermodynamic Analysis (THA)

1.3.2.1 PtX THA Methods

THA is vital for assessing the performance and efficiency of PtX processes. These processes transform excess RESs into different types of energy carriers and chemicals. It examines the conversion of energy and the efficiency of various processes, offering important insights into their practicality and improvement. The key objective of THA is to evaluate the efficacy of energy conversion processes and to identify the potential for performance improvement. This pertains to the investigation of the conversion of electrical energy into chemical energy or other forms of stored energy, including liquid fuels, methane, or hydrogen, in the context of PtX. The analysis is instrumental in determining the extent to which the PtX process complies with the laws of thermodynamics and the principles of energy conservation. Energy and exergy analysis, process efficiency, and energy conversion efficiency are core concepts in THA. According to the first law of thermodynamics, which is also referred to as the law of energy conservation, energy cannot be generated or destroyed; rather, it can only be converted from one form to another. This law is employed in PtX processes to determine the efficiency of energy conversion by comparing the input electrical energy to the output chemical or stored energy; for instance, the efficiency of electrolysis for hydrogen production is contingent upon the ratio of the chemical energy contained in hydrogen to the electrical energy supplied. The degree of disorder in a system is measured by the concept of entropy, which is introduced by the second law of thermodynamics and addresses the direction of energy transformations. The second law is employed in PtX processes to assess the irreversible process and losses that are linked to energy conversions. Greater energy losses and a decrease in overall efficacy are indicative of increased entropy generation. Table 1.9 provides a summary of key thermodynamic metrics and their respective equations.

TABLE 1.9

Key Methods and Equations in THA of PtX Systems

Method/Metric	Equation	Symbols	Explanation
Energy efficiency (η)	$$\eta = \frac{\sum_x \left(\dot{m}_X \times LHV_X \left(or\ HHV_X \right) \right)}{\sum \dot{W}_{p,in}}$$	$\dot{m}_X$: (mass flow rate of product X); LHV_X / HHV_X: (lower/higher heating value; $\dot{W}_{p,in}$: (power input from RES)	Assesses the ratio of useful energy output to energy input in the PtX process.
Exergy efficiency (ε)	$$\varepsilon = \frac{\dot{E}_{flow,X}}{\dot{E}_{heat} + \dot{W}_{p,in} + \sum \dot{E}_{flow,in} - \sum \dot{E}_{flow,out}}$$	$\dot{E}_{flow,X}$: (exergy rate of product X); $\dot{E}_{heat}$: (exergy rate of heat input); $\dot{E}_{flow,in}$: (exergy rate of inflows); $\dot{E}_{flow,out}$: (exergy rate of outflows)	Evaluates the quality and quantity of energy conversions by considering irreversibilities in the system.
Carnot efficiency (η_{Carnot})	$$\eta_{Carnot} = 1 - \frac{T_C}{T_H}$$	T_C: (temperature of the cold reservoir, heat sink); T_H: (temperature of the hot reservoir, heat source)	Establishes the theoretical maximum efficiency based on temperature difference between heat source and sink.
Carbon conversion rate (CCR)	$$CCR = \frac{\dot{m}_X}{\left(\dot{m}_{CO} + \dot{m}_{CO2} \right)_{in}}$$	$\dot{m}_X$: (mass flow rate of product X); $\dot{m}_{CO}$: (mass flow rate of carbon monoxide); $\dot{m}_{CO_2}$: (mass flow rate of carbon dioxide)	Measures the efficiency of carbon-containing feedstocks conversion into desired products.

There are various applications and benefits of THA. It enables the identification of inefficiencies and opportunities for development in PtX processes. Engineers enhance performance by optimizing system design and operational parameters by comprehending the thermodynamic limitations and energy losses. The practical feasibility of implementing PtX technologies on a large scale is determined by evaluating the thermodynamic performance. It guides decision-making and investment by providing insights into the energy requirements and prospective benefits. The design and scaling up of PtX systems are contingent upon the availability of precise thermodynamic data. It ensures that the systems operate within their optimal efficiency range and satisfy the intended performance criteria. Yet THA is frequently difficult due to the complex multi-step reactions and energy conversions that are frequently present in PtX processes. Detailed knowledge of reaction kinetics, heat transfer, and material properties is necessary for accurate modeling. Also, the accuracy of the analysis may be compromised by the scarcity of reliable thermodynamic data for new or emergent PtX technologies. In order to establish precise thermodynamic parameters, it is frequently necessary to conduct experimental validation and pilot-scale studies.

1.3.2.2 Key Determinants in PtX THA

The THA of PtX systems provides critical insights into the efficiency and feasibility of various technologies and processes. Several software tools are used for such analyses, including MATLAB, UniSim® Design R491 [84], Engineering Equation Solver [85], and ASPEN Hysys™ [86]. However, Aspen Plus is often preferred due to its comprehensive modeling capabilities and extensive use in the chemical industry, particularly for process simulation [87, 88]. While most studies employ steady-state (SS) analysis [89, 90], a smaller number incorporate dynamic (D) modeling [91]. Dynamic analysis is particularly valuable for capturing transient behaviors when PtX systems are integrated with variable RESs. Some studies combine both methodologies (SS + D), offering a more holistic understanding of system performance under different operational conditions [27, 86, 92]. Among the electrolyzer technologies, SOWEs—particularly in steam and co-electrolysis modes— are recognized for their high efficiency due to superior thermal integration. SOWEs' high operating temperatures allow for efficient waste heat utilization, leading to notable improvements in overall system efficiency. Co-electrolysis is often highlighted for its enhanced electrical efficiency, especially for producing hydrocarbon-based fuels, due to the simultaneous production of hydrogen and CO [86, 93, 94]. However, the benefits of co-electrolysis can vary depending on the specific product synthesized and operational conditions. For example, in processes where exothermic reactions (e.g., methanation) generate steam internally, co-electrolysis may offer limited additional benefits, as the required steam can be produced without external energy input. Thus, its applicability should be evaluated case by case.

Catalyst selection is another crucial factor in PtX systems, as advanced catalysts can reduce activation energy, improve reaction rates, and lower operating temperatures, enhancing thermodynamic performance. For example, nickel- and iron-based catalysts are commonly used in methanation to maximize system efficiency. However, it is vital to note that the effectiveness of catalyst materials relies on the specific reactions involved and operational conditions. Also, reaction kinetics are key in improving thermodynamic performance, as faster reaction rates can reduce residence times in reactors, making the overall process more efficient [94, 95]. Energy demand for CO_2 capture also heavily influences the thermodynamic performance of PtX systems. Membrane-integrated systems tend to be more thermodynamically efficient compared to chemical absorption or physical adsorption, as they avoid the significant reboiler duties typically required in solvent-based systems. Yet the choice of CO_2 capture technology should be tailored to factors such as CO_2 concentration in the gas stream and operational costs. For example, in certain industrial settings with low-concentration CO_2 streams, solvent-based methods may still be more practical despite their higher energy consumption. In addition, carbon efficiency, or the percentage of captured CO_2 converted into valuable products, is an essential determinant of system performance. Higher carbon efficiency reduces waste and improves the overall viability of PtX systems [88, 92].

When considering product routes in PtX processes, methane synthesis is often favored for its compatibility with existing gas infrastructure and versatility in applications such as heating, transportation, and power generation. Methanation, an exothermic reaction, offers additional opportunities for thermal integration, improving overall system efficiency [88]. However, other PtX products, such as methanol and ammonia, offer alternative pathways that may be more suitable in specific industrial contexts, though they typically require more complex downstream processes and thermal management. Each product route should be analyzed individually to determine the optimal balance between thermodynamic performance and system complexity [92, 96]. Thermal integration is fundamental to optimizing PtX systems' efficiency. Techniques such as recycling process streams, pressurizing reaction units, and pinch analysis help identify opportunities to minimize energy losses [86, 95]. For instance, pinch analysis can optimize the integration of exothermic and endothermic processes, ensuring that heat generated during methanation is effectively used in electrolysis or other energy-intensive processes. Also, waste heat recovery through heat exchangers and advanced heat storage technologies, such as thermocline storage or phase-change materials (PCM), can further improve energy efficiency by reducing reliance on external energy inputs [97, 98]. Exergy analysis complements traditional thermodynamic analysis by focusing on the quality of energy utilized within the system and identifying areas where exergy destruction occurs. This deeper level of analysis can help pinpoint inefficiencies, such as heat losses or suboptimal CO_2 capture, that reduce the system's potential for useful work. Systems with high exergy efficiency tend to have better overall thermodynamic performance, as they minimize energy losses across all stages of the process [85, 90].

System flexibility is another critical aspect, particularly in PtX systems integrated with intermittent RESs. These systems must maintain high thermodynamic performance under varying load conditions, which is crucial for large-scale PtX deployment [84–86]. Part load efficiency is especially important in systems subject to fluctuating energy inputs from RESs like wind and solar. Flexible systems that can operate efficiently across a wide range of loads are better suited for long-term, large-scale implementation. Process intensification is another strategy to improve thermodynamic performance [95]. Optimizing reactor designs and reducing the size of system components—such as through the use of microchannel reactors or modular systems—can enhance heat and mass transfer efficiency, resulting in more compact and energy-efficient designs. These innovations help reduce thermal gradients and energy losses, further improving overall system performance. Downstream processes, such as fuel purification, compression, or liquefaction, also play an important role in thermodynamic analysis. For example, the energy-intensive step of hydrogen liquefaction significantly affects overall system efficiency [99, 100]. Comprehensive thermodynamic analysis must account for these downstream processes to offer a complete picture of the PtX system's performance. By incorporating thermodynamic strategies like thermal integration, catalyst optimization, and exergy analysis, PtX systems can significantly improve their efficiency and viability. This holistic approach is vital for scaling up PtX systems to support the transition toward a sustainable energy future. Yet continued development of flexible systems, coupled with advancements in process intensification, will be key to ensuring that PtX technologies can meet the demands of large-scale implementation in a variety of contexts.

1.3.3 PtX Life Cycle Assessment (LCA)

1.3.3.1 PtX LCA Method

LCA is a commonly employed approach for conducting environmental evaluations. The ISO 14040 and ISO 14044 standards provide a description of the structural and methodological features [101, 102]. An LCA study is organized into four separate phases: aim and scope definition, life cycle inventory (LCI) analysis, life cycle impact assessment (LCIA), and interpretation. Each stage is essential for guaranteeing a thorough and precise assessment of the environmental effects of a product or process.

The objective and scope definition is the initial phase of an LCA study. This phase entails the establishment of the assessment's objective, the specification of the intended application, the rationale behind the study, the target audience, and the manner in which the results will be utilized. For example, the objective may be to evaluate the environmental impacts of various RESs or to determine the most sustainable hydrogen production method. Also, the scope establishes the study's boundaries and level of detail. Key elements include the functional unit, which is a quantified description of the primary function of the system being studied, serving as a reference for all subsequent analysis (e.g., producing 1 kg of hydrogen). It encompasses the processes and phases of the life cycle that are determined by the system boundaries, such as the extraction, production, use, and disposal of raw materials. This phase also delineates the impact categories to be evaluated, as well as the assumptions and limitations (including global warming potential, acidification, and resource depletion). The second phase, LCI analysis, involves the collection and quantification of data for all inputs and outputs of the system. This comprises the generation of waste; the use of basic materials; the emission of pollutants to the air, water, and soil; and the consumption of energy during the life cycle stages that are specified in the scope. The objective is to create an exhaustive inventory of all interactions between the system and the environment. Data sourcing can involve either primary data, which are direct measurements or information obtained from specific processes under study or secondary data, derived from literature, databases, or industry reports, particularly when primary data is unavailable. It is imperative to verify the data's accuracy, reliability, and relevance by conducting consistency checks and cross-referencing data sources. The LCIA converts the LCI data into potential environmental impacts in the third step. This requires numerous procedures. Initially, inventory data is classified into a variety of environmental impact categories (e.g., all emissions that contribute to global warming are grouped together). The subsequent step is characterization, which involves quantifying the classified data by converting LCI results into common units within each impact category. This is achieved by utilizing specific models and characterization factors to determine the magnitude of impacts, such as converting greenhouse gas emissions to CO_2 equivalents. Normalization and weighting are optional procedures that further interpret the results by comparing them to reference values (normalization) or designating relative importance to various impact categories (weighting). The significance of the impacts in a broader context is facilitated by these steps. The final phase is interpretation, during which the results of the LCI and LCIA are analyzed to draw conclusions, make suggestions, and establish a foundation for decision-making. This entails the identification of significant issues by identifying the primary contributors to environmental impacts and comprehending the underlying reasons. The evaluation evaluates the results' robustness and completeness, which includes analyses of uncertainty and sensitivity to comprehend the impact of assumptions and data variability on the results. After conducting the analysis, conclusions are derived, including the identification of the most environmentally impactful stages or processes, as well as recommendations for improvement. This may entail recommending modifications to process design, material selection, or other strategies to mitigate environmental effects. Table 1.10 illustrates the most often employed LCIA indicators.

1.3.3.2 Key Determinants in PtX LCA

PtX systems, particularly those powered by RESs, such as wind and surplus electricity, generally demonstrate a lower environmental footprint compared to systems dependent on grid mixes dominated by fossil fuels. This is evident in key metrics such as GWP, one of the 18 environmental indicators commonly used in LCA. However, it is important to recognize that the environmental performance of PtX systems is highly context-dependent, influenced by factors such as local energy mix, system design, and technological choices [103, 104]. The source of CO_2 is also a critical factor in determining environmental results. Renewable fuel generation systems typically show substantial environmental advantages when CO_2 is sourced from renewable methods, such as biomass or DAC, rather than from fossil-based processes [105]. Additionally, the choice of CO_2 capture technology can significantly influence LCA outcomes. While physical adsorption techniques often

TABLE 1.10
LCIA Indicators

1	Global warming potential (GWP)	Measures the potential impact on global climate change due to greenhouse gas emissions, expressed in CO_2 equivalents	13	Freshwater ecotoxicity (FEC)	Assesses the potential toxic effects of substances on freshwater ecosystems
2	Fossil depletion (FD)	Evaluates the reduction in available fossil fuel resources due to extraction and use, affecting future energy availability	14	Marine ecotoxicity (MEC)	Evaluates the potential toxic effects of substances on marine ecosystems
3	Water depletion (WD)	Assesses the reduction in freshwater availability due to consumption in processes, impacting ecosystems and human water supply	15	Terrestrial ecotoxicity (TE)	Measures the potential toxic effects of substances on terrestrial ecosystems
4	Particulate matter formation (PMF)	Measures the potential human health impacts from inhalable particles formed by emissions of pollutants	16	Resource depletion (RD)	Evaluates the depletion of natural resources (both renewable and non-renewable), affecting future availability
5	Human toxicity (HT)	Evaluates the potential adverse effects on human health from exposure to toxic substances released during a product's life cycle	17	Metal depletion (MD)	Measures the reduction in metal resources due to extraction and use
6	Ozone depletion (OD)	Measures the potential impact on the stratospheric ozone layer, which protects the Earth from harmful ultraviolet radiation	18	Total material requirement (TMR)	Assesses the total amount of materials required throughout the life cycle of a product, reflecting resource efficiency

(Continued)

TABLE 1.10 (Continued)

7	Acidification potential (AP)	Assesses the potential for emissions to cause acid rain, which can damage ecosystems, soil, and water bodies	19	Raw material input (RMI)	Measures the input of raw materials necessary for the production of goods, impacting resource availability
8	Terrestrial acidification (TA)	Specific assessment of acidification effects on terrestrial ecosystems, impacting soil quality and plant life	20	Ionizing radiation (IR)	Evaluates the potential health effects from exposure to ionizing radiation, typically from nuclear energy or medical sources
9	Freshwater eutrophication (FE)	Evaluates the potential for nutrient enrichment in freshwater environments, causing algal blooms and hypoxia	21	Land occupation (LO)	Assesses the amount of land occupied by processes or products, affecting land use and ecosystem services
10	Marine eutrophication (ME)	Assesses the potential for nutrient enrichment in marine environments, causing algal blooms and hypoxia	22	Urban and occupation (UO)	Measures the impact of urban development and land use changes on natural habitats and biodiversity
11	Terrestrial eutrophication (TEU)	Measures the potential for nutrient enrichment in terrestrial environments, leading to excessive plant growth and soil nutrient imbalance	23	Natural and transformation (NT)	Evaluates the changes in natural landscapes due to human activities, impacting biodiversity and ecosystem functions
12	Photochemical oxidant formation (POF)	Measures the formation of ground-level ozone (smog), which can harm human health and ecosystems	24	Primary energy (PE)	Assesses the total primary energy consumption throughout the life cycle of a product, reflecting energy efficiency

exhibit lower energy consumption and simpler operational processes compared to chemical absorption methods, the optimal choice may vary depending on the specific system, CO_2 concentration, and technological advancements [106].

The electricity source powering PtX systems is a key determinant of their environmental impact. Hydropower, along with other RESs, such as wind and solar, offers substantial environmental benefits, particularly when considering the electricity emission factor—the amount of CO_2 emitted per kWh. However, the temporal and spatial variability of these RESs can affect LCA results [105]. PtX systems located in regions with high RES penetration are likely to have significantly lower environmental impacts than those relying on fossil fuel–dominated grids. Advances in energy storage and grid management may help mitigate the variability in RES, improving the overall performance of PtX systems in such regions [107]. Despite the generally lower environmental impact of PtX systems, certain pathways, such as PtMethane, can exhibit environmental disadvantages in some impact categories. In specific contexts, these systems may perform worse than conventional NG production in areas such as metal depletion, water depletion, and toxicity (marine, terrestrial, and human). However, these outcomes are highly dependent on system design, operational conditions, and the type of energy used. For instance, if powered entirely by RES, PtMethane systems may offer significant benefits despite potential drawbacks in other categories [108]. Water consumption is a key concern, particularly for electrolyzers, which can contribute to water scarcity and land acidification in regions where water is a limited resource [109]. To address this issue, water conservation measures, such as recirculation and incorporating a water stress index, help improve the sustainability of PtX systems, especially in water-scarce regions. Also, the use of gray water or recycled water in electrolyzers presents a promising option, although its practical implementation may depend on regional infrastructure, water quality, and safety standards [110]. The infrastructure required for PtX systems, including pipeline transport and storage facilities, generally has a lower environmental impact compared to the core conversion processes. Yet end-of-life (EoL) considerations are critical for a complete LCA. Disposal and recycling of materials such as electrolyzer components, pipelines, and storage facilities can introduce environmental burdens if not managed properly. The recyclability of materials like steel, nickel, and rare earth elements plays a crucial role in determining long-term sustainability. Infrastructure decommissioning, particularly in large-scale PtX projects, must also be factored into LCA assessments to avoid underestimating future environmental impacts [111, 112].

The supply chain for materials used in PtX systems is another important consideration. The extraction, processing, and transportation of raw materials, such as metals for electrolyzers, can contribute to the overall environmental burden, particularly in categories like resource depletion and toxicity. This is especially relevant for rare earth elements, where supply chain impacts may vary depending on sourcing practices. In future LCAs, incorporating improvements in responsible mining and recycling practices could help mitigate these effects [113, 114]. Also, the transportation of PtX products (e.g., hydrogen or synthetic fuels) over long distances can contribute to environmental impacts. Optimizing local production and distribution systems could further reduce these effects, though the feasibility of such measures will depend on regional infrastructure and market demand [107]. The choice of functional units in LCA studies can also significantly affect the outcomes. In PtX systems, the functional unit could be based on energy produced (e.g., per kWh of fuel) or CO_2 captured (e.g., per ton of CO_2). The choice of functional unit should align with the primary objective of the PtX system and be consistent across studies to ensure comparability. Also, allocation methods in multi-functional PtX systems—where multiple products, such as fuel, heat, and chemicals, are generated—must be carefully considered to avoid skewed results [108]. Clearly defining system boundaries—whether cradle-to-grave or cradle-to-gate—is vital to avoid underestimating environmental impacts [115]. Material utilization plays a significant role in PtX systems, particularly for those relying on renewable electricity sources. While renewable systems reduce operational emissions, they often require more materials and higher primary energy demand compared to fossil-based technologies [106]. Materials encompassing steel, nickel, and polytetrafluoroethylene, employed in electrolyzer components, contribute to environmental issues, such as ozone depletion, acidification, and human toxicity [116]. Applying circular economy principles, such as recycling and reusing materials, can significantly reduce the overall environmental footprint. Designing PtX systems with closed-loop processes—where waste heat and CO_2 are reused—can also enhance environmental performance, though practical implementation may vary depending on technological and regional constraints [117].

Moreover, LCAs of PtX systems reveal several methodological gaps and inconsistencies. A key issue is the lack of transparency regarding methodological choices, such as how multi-functionality is handled and what data is employed for LCI. Many LCAs focus predominantly on GWP, which, while important, may overlook other crucial indicators, such as material use and resource depletion. Broadening the focus to include these indicators can provide a more comprehensive picture of PtX systems' environmental impacts. Additionally, while renewable electricity sources reduce operational emissions, they often involve higher material consumption, which could negatively affect the overall environmental profile if not properly accounted for. EoL considerations remain critical to a full LCA. The disposal and recycling of electrolyzers, pipelines, and other PtX infrastructure should be included in cradle-to-grave LCAs to avoid underestimating long-term impacts. Improvements in the LCA of PtX systems should focus on advancing low-carbon electricity generation and integrating carbon capture technologies more effectively. This could significantly reduce the environmental footprint of PtX products compared to conventional technologies. To improve the consistency, scope, and transparency of LCA studies, detailed explanations of the technological and methodological decisions involved should be provided. Integrating LCA with TEA can offer a fuller picture of PtX systems' sustainability by balancing environmental performance with economic viability. Together, these improvements will support the large-scale implementation of PtX systems in a sustainable energy future.

Figure 1.5 provides a visual representation of the key determinants categorized by their level of importance across TEA, THA, and LCA analyses, offering a comprehensive overview of the critical factors influencing PtX systems. However, it is important to note that the significance of key determinants in the PtX systems is dynamic and influenced by multiple factors. Their interdependency means that changes in one determinant, such as government policies, can directly impact others, like electricity costs. Also, the importance of each determinant can vary based on regional contexts, such as resource availability or infrastructure; for instance, in water-scarce regions, water usage in electrolyzers may become a high-importance factor. Market conditions and regulatory landscapes are also continually evolving, altering the relevance of determinants over time. While categorizing determinants offers a useful framework, their significance is not fixed and must be continually reassessed based on the specific context and market dynamics to ensure an accurate evaluation of PtX systems' feasibility.

1.4 OPTIMIZATION OF PTX SYSTEMS

PtX technologies represent a transformative solution to the global energy crisis by converting renewable electricity into various energy carriers and compounds. The diverse applications and advantages of PtX systems are critical to advancing sustainable energy solutions and enabling the transition to an LC economy. The key benefits of PtX include enhancing hybrid renewable energy systems (HRESs), facilitating sector coupling, enabling flexible operations, and contributing to grid stability. Also, PtX plays a pivotal role in the circular economy and supports the re-electrification of hydrogen, thus improving energy security and independence while reducing GHGs. PtX technologies are applied across multiple sectors, including energy, transportation (EVs, aviation, and maritime), manufacturing, refining, agriculture, and heating. They are also utilized in the chemical industry, mining, waste management, remote and off-grid energy solutions, and offshore applications. As PtX technologies continue to evolve, they hold significant potential for RE surplus storage, job creation, and economic development.

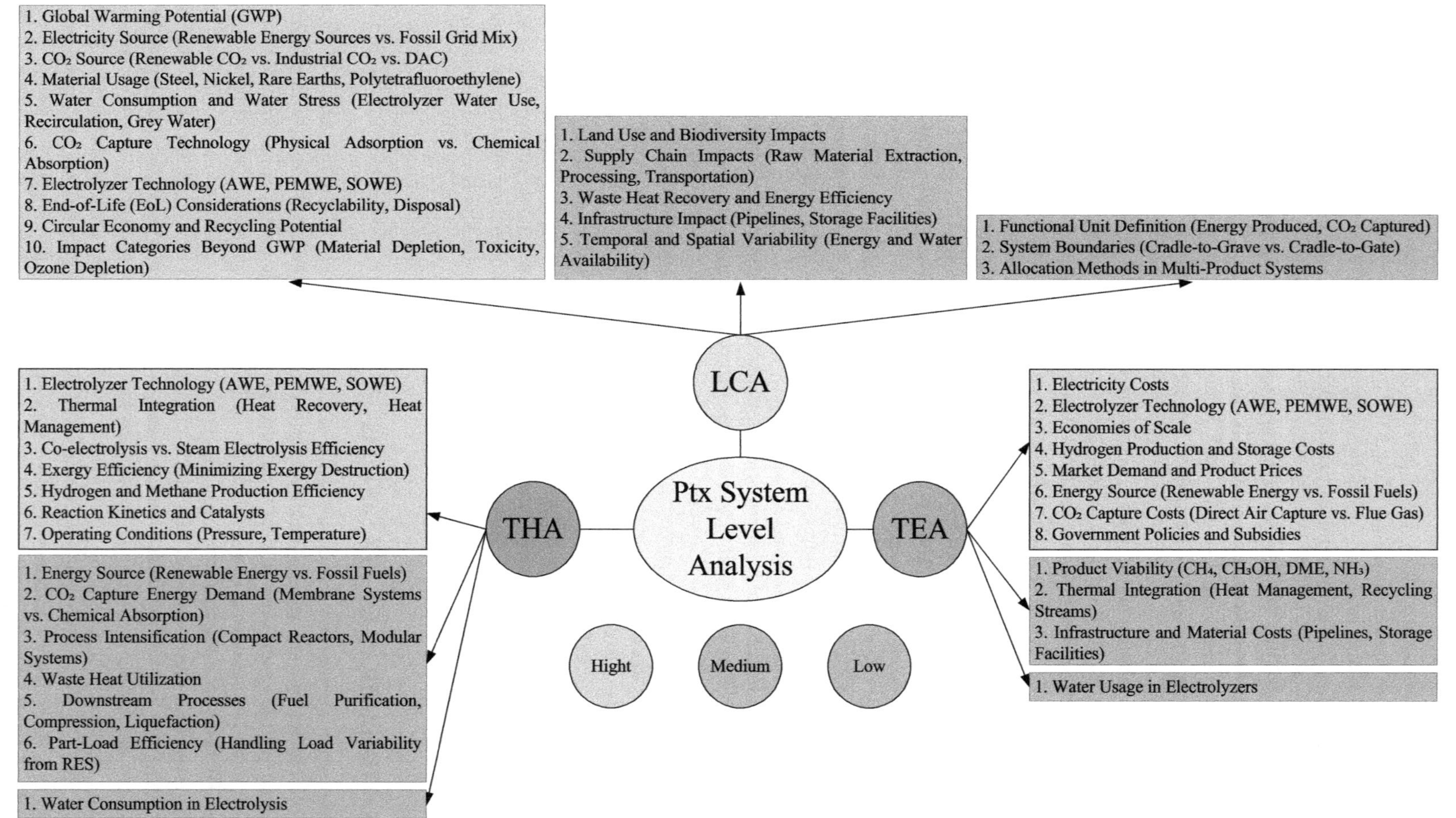

FIGURE 1.5 Schematic diagram of the key determinants of PtX system–level analysis.

1.4.1 HYBRID RENEWABLE ENERGY SYSTEMS (HRESS)

In PtH technology, the process of electrolysis is utilized to split water into hydrogen and oxygen utilizing electricity, which plays a crucial role in HRESs. This process offers significant environmental benefits by enabling the integration of various energy applications, including electric power systems (EPS), natural gas systems (NGS), district heating systems (DHS), transportation, and the production of synthetic fuels and chemicals, while avoiding carbon emissions during conversion. Hydrogen and its derivatives are versatile and can be integrated into other industrial processes, such as carbon derivative production or the treatment of effluent gases from biogas plants. Also, the heat generated during electrolysis, typically ranging from 200 to 400°C, can be repurposed for district heating, thus enhancing overall energy efficiency. Among the primary electrolyzer types, PEMWE and AWE are most suitable for small-scale applications due to their rapid start-up and shutdown capabilities, making them ideal for HRESs. In contrast, SOWE offers greater efficiency but is typically used only when waste heat is available, requiring high operating temperatures. Electrolysis produces hydrogen that is carbon-free and of high purity, positioning it as an environmentally sustainable alternative.

Hydrogen storage remains a significant challenge for the scalability of HRESs. Salt caverns offer a cost-effective, large-scale storage solution with high-quality capabilities but require proximity to HRESs. Pressurized H_2 storage, while commonly used, is less efficient, is expensive, and faces technical issues, such as material embrittlement. FCs, powered by H_2, CH_3OH, or NH_3, are being explored for emission reduction in energy applications, though they currently come with high costs and relatively low efficiency (~40%). Experimental projects in South Africa and Kenya are showcasing the use of FCs, fueled by CH_4 or CH_3OH, to supply electricity to rural areas [118–120]. A novel hybrid FC system combining a low-temperature polymer electrolyte membrane fuel cell (LT PEM-FC) with a methanol steam reformer (MSR) and a methanation (MET) reactor achieved a power density of 245.2 mW/cm², compared to 268.8 mW/cm² using pure H_2. The system reduces carbon monoxide to prevent FC catalyst poisoning and provides in situ H_2 production, addressing both storage and transportation issues and promoting FC commercialization [121]. Also, a comparison of two integrated power systems for RE storage in Italy—one using PV panels and batteries and another with a PV–H_2 system—revealed that the H_2-based system, while environmentally beneficial, was more costly due to the niche status of FCs and electrolyzers. In contrast, the battery-based system was more cost-effective, benefiting from mature technology and market availability, despite higher environmental burdens from battery production [122]. Finally, an optimization method for standalone HRES that integrates wind, PV, H_2 storage, and batteries, using a bi-level mixed-integer nonlinear programming (MINLP) and economic model predictive control (EMPC), demonstrated significant cost reductions and enhanced system reliability. Genetic algorithms (GA) optimized system scalability, while a rolling horizon MILP handled energy management, resulting in improved flexibility, energy balance, and operational efficiency compared to rule-based energy management systems (EMS) [123].

Ammonia's high energy density and carbon-free nature make it particularly appealing for heavy transportation, positioning PtA as an emerging technology within HRESs. NH_3 stands out from other chemical fuels, such as alcohols, formic acid, and hydrides, due to its ability to be employed in FCs for carbon-free transportation. However, conventional NH_3 production through the HB process is a major contributor to global CO_2 emissions, accounting for about 1% of global anthropogenic emissions. This has driven interest in developing alternative, lower-carbon methods for NH_3 synthesis, which can significantly reduce its carbon footprint by adopting electrical alternatives and optimizing industrial processes. One study [124] explored an integrated system comprising a gas turbine cycle, Rankine cycle, two organic Rankine cycles, ejector-based cooling, and subsystems for H_2 and NH_3 production, drying, and hot water generation. This system achieved energy and exergetic efficiencies of 62.18% and 58.37%, respectively, producing multiple outputs, such as power, H_2, and NH_3. Another study [125] examined NH_3 as a hydrogen storage medium and proposed on-site H_2 production at refueling stations via NH_3 decomposition. The system, which involves separating H_2 from an H_2–N2 mixture, improved H_2 recovery to over 95% using a PSA-to-membrane subsystem, increasing system efficiency from 59.1% to 85.37%, while cutting costs by 22% to €4.31 per kg. NH_3's role in long-term, seasonal energy storage, combined with H_2's use for short-term energy balancing in HRESs, was shown to enhance system performance and reduce costs across various regions, including Spain [126], Texas [127], and New York City [128], and remote locations, like Mahaka, Hawaii, and the Northwest Arctic Borough, Alaska [129]. The utilization of multiple energy carriers, such as NH_3, CH_3OH, and H_2, was found to optimize logistics and reduce overall costs from a supply chain perspective. For instance, producing CH_3OH near CO_2 sources and NH_3 in other locations has been identified as a cost-effective strategy. Also, research in South Korea examined transportation energy systems utilizing a variety of fuels for different vehicle types, further improving economic viability and system efficiency through spatial synergies [130–132].

HRESs are primarily concerned with energy storage for the EPS, with power-to-power (PtP) systems being a significant component of PtH applications. H_2 is generated through electrolysis in PtP systems during periods of surplus energy generation and is subsequently utilized in FCs to generate electricity when required. H_2 storage and a DC bus are typical components of this configuration. The PtP approach prioritizes the storage of energy surplus, ensuring a consistent power supply and contributing to grid stability. PtP system studies can vary in complexity, from long-term planning to day-ahead and real-time operational planning. Various optimization methods have been explored to improve the design and operation of these systems. For instance, a decentralized energy management system (DEMS) utilizing game theory (GT) was developed in [133] for polygeneration microgrids (MG), where MG devices are modeled as interactive agents. The system uses Nash equilibrium to optimize power control strategies among agents, leading to reduced power losses and costs. Another study [134] compared DEMS based on GT with a fuzzy cognitive map–based DEMS, showing that the GT method reduces power losses and costs over a 20-year period, highlighting its operational and financial advantages. Teaching learning-based optimization (TLBO), a parameter-free algorithm, was employed to optimize a hybrid system of PVs, wind turbines (WT), FCs, and a diesel generator (DG) in Pakistan. The system was designed to lower total annual cost (TAC), including capital, maintenance, and fuel expenses, while meeting household energy demands. TLBO successfully identified the optimal system components, achieving the lowest TAC. Lastly, a comparison of particle swarm optimization (PSO) and GA for a grid-connected H_2–wind–solar hybrid system demonstrated that GA is more effective in minimizing costs for residential applications. Combined heat and power (CHP) technologies were integrated into the system, and thermal recovery further improved efficiency and cost savings, with GA outperforming PSO in cost reduction [135].

$PtCH_4$ involves converting RE into methane or hydrogen through electrolysis, followed by methanation processes using technologies such as the Sabatier reactor and fixed-bed technologies. This process generates both SNG and usable heat, supporting DHSs at temperatures between 200 and 400°C. While $PtCH_4$ offers substantial environmental benefits, it comes with higher installation costs compared to conventional energy systems. Studies emphasize the importance of reactor design, system optimization, and optimal sizing and performance assessment of $PtCH_4$ units under various operational conditions. One study integrates battery storage with a PtG system connected to a large PV facility to optimize H_2 production through a mixed-integer linear programming (MILP) approach. The system uses battery storage to support electrolyzer operation during periods of low renewable generation, though H_2 production costs are higher [136]. Another study optimizes a hybrid off-grid system integrating methanation, battery storage, and PV arrays employing PSO and sequential quadratic programming (SQP), achieving a 17.77% cost reduction [137]. In stand-alone $PtCH_4$ systems, increasing electrolyzer capacity is often more cost-effective than expanding battery operations. In the day-ahead market, a hybrid system combining gas-fired units (GFU), PtG technology, and wind power units, managed through robust optimization, reduces NG dependency and enhances resilience against electricity price fluctuations, with a nearly 1% increase in system profit due to PtG technology [138]. Another energy storage system integrates compressed

CO_2, H_2 generation, and methanation for SNG production, achieving a nominal point efficiency of 45.08% and storage efficiencies ranging from 35.06% to 63.93%. This system shows high potential, especially in regions with mining infrastructure, and uses Aspen Plus software to model thermodynamics, parameters, and energy fluxes [139]. Finally, a day-ahead dispatch model for integrated port energy systems (IPES), utilizing PEM electrolysis, methanation, and multi-energy storage systems, demonstrated improved economic and environmental performance through efficient CO_2 reduction and energy cost savings. The model uses second-order cone programming (SOCP) to resolve nonlinearities in power flow and energy conversion, ensuring computational efficiency and accuracy [140].

1.4.2 FLEXIBLE OPERATIONS

PtX systems are vital for the transition to a decarbonized and sustainable energy landscape. These systems convert surplus RE into a variety of storable and usable energy carriers, including liquid fuels, compounds, synthetic CH_4, and H_2. The efficient and reliable energy conversion processes are contingent upon the flexible operation of PtX systems, which is essential for optimizing their integration with intermittent RESs.

The flexible operation of PtX systems is essential for adapting to the variability of RESs. Dynamic load management, the ability to adjust production rates in real time based on renewable electricity availability, is crucial for aligning hydrogen production and other conversion processes with the fluctuating output of RESs. This adaptability allows PtX systems to operate efficiently, minimize curtailment, and optimize RE use, thereby contributing to overall grid stability by balancing supply and demand. Short start-up and shutdown durations are vital for operational flexibility, enabling PtX systems to quickly respond to changes in power supply and demand, while preventing energy waste during periods of low renewable generation. Technologies such as PEMWEs excel in providing fast response times, allowing the systems to capitalize on brief periods of RE surplus. However, frequent load changes and start–stop cycles can expedite material degradation, posing an issue to the long-term durability and performance of PtX systems. Addressing this issue requires advancements in protective coatings and materials to prevent performance deficits, reduce maintenance costs, and extend the systems' lifespan. Effective thermal management is also critical, as it ensures PtX systems operate within optimal temperature ranges, preventing overloads and damage from combustion and ensuring consistent performance and safety. Advanced cooling technologies and real-time temperature monitoring are particularly vital for high-temperature systems, such as SOWEs. The use of sophisticated control systems and algorithms for real-time performance regulation further enhances the flexibility and efficiency of PtX systems, enabling stable operations and optimizing production processes in response to external conditions. Implementing these systems, along with real-time monitoring technologies, is essential to optimize the integration of RESs and preserve grid stability.

Capacity planning, which includes determining the appropriate size and configuration of PtX systems, as well as infrastructure planning, is essential for meeting future energy demands. Comprehensive modeling and forecasting inform these strategic investments, ensuring seamless integration with existing energy infrastructure. PtX systems also play a vital role in providing grid stability and ancillary services, such as demand response and frequency regulation, by adjusting their operations in real time. This capability is particularly important, given the growing penetration of intermittent RESs, and helps maintain grid reliability. The economic feasibility of PtX systems is a critical consideration, as the costs associated with advanced control systems, durable materials, and thermal management technologies impact their widespread adoption. Reducing capital and operational costs through technological innovations and economies of scale, along with favorable regulatory frameworks and economic incentives, can enhance the financial viability of PtX investments. Scalability is another issue, as PtX technologies must be able to expand effectively to meet growing hydrogen and RE demand. Overcoming technical challenges, such as maintaining performance and durability at a larger scale, as well as economic challenges, like reducing costs through mass production and infrastructure development (storage facilities, pipelines, and distribution networks), will be essential for PtX systems to contribute significantly to the global energy transition. Finally, the deployment and operation of PtX systems depend on robust safety protocols, adherence to regulatory standards, and continuous monitoring to ensure the safe management and storage of hydrogen and other energy carriers.

1.4.3 HYDROGEN RE-ELECTRIFICATION

Hydrogen re-electrification, often referred to as power-to-hydrogen-to-power (PtHtP), is the process of converting hydrogen produced from RESs back into electrical energy. This mechanism functions as a means to store and distribute RE, mitigating its inherent unpredictability. The conversion process primarily relies on the utilization of gas turbines and FCs.

1.4.3.1 Gas Turbine Method for Hydrogen Re-Electrification

The gas turbine method for hydrogen re-electrification represents a transformative approach to power generation, utilizing hydrogen as a clean fuel. This method involves using hydrogen to power gas turbines, efficiently converting chemical energy into electrical energy while significantly reducing carbon emissions. The process encompasses several critical components working together to achieve this conversion with minimal environmental impact. A gas turbine comprises three primary components: a compressor, a combustion chamber, and a turbine [141]. The compressor draws in ambient air and compresses it to a high pressure using a series of stationary and rotating blades. In hydrogen applications, the compressor typically does not require modifications, as it handles only air. Instead, adjustments are necessary in the fuel delivery and combustion systems to accommodate hydrogen's unique properties. In the combustion chamber, the compressed air is mixed with hydrogen and ignited, significantly increasing the temperature and volume of the gases. Due to hydrogen's higher flame speed and wider flammability range compared to natural gas (NG), modifications to burner design and materials are essential for stable and effective combustion. Challenges such as flame flashback and increased nitrogen oxides (NO_X) emissions must be addressed through advanced combustion techniques and materials that can withstand hydrogen's combustion dynamics. The high-temperature, high-pressure gases produced in the combustion chamber expand through the turbine, causing the blades to rotate. This rotational motion drives the compressor and, in power generation applications, turns a generator to produce electricity. Turbine blades and cooling systems must be designed to handle the specific thermal and mechanical stresses associated with hydrogen combustion. In a combined cycle system, residual heat from the turbine's exhaust gases can be harnessed to generate additional electricity through a heat recovery steam generator (HRSG), which produces steam to power a steam turbine, thereby increasing overall efficiency.

Historically, gas turbines have relied on traditional fuels, such as NG and propane. However, hydrogen is gaining attention as a fuel due to its potential to reduce carbon emissions. Hydrogen-powered gas turbines offer a reliable and flexible power solution, particularly advantageous when RE generation is limited. Companies like Siemens and General Electric are actively developing gas turbines capable of operating on hydrogen. While hydrogen combustion offers environmental benefits, it also presents issues. Hydrogen has a lower volumetric energy density than methane, requiring a larger volume to produce the same energy output. This necessitates adjustments to the fuel handling and combustion systems to manage higher fuel flow rates. Also, hydrogen's higher flame speed and shorter ignition delay increase the risk of NO_X emissions and flame flashback. Safety concerns also arise from hydrogen's nearly invisible flame, high flammability, and ability to permeate materials, requiring advancements in flame detection, sealing systems, and overall plant design. Ensuring gas turbines are hydrogen-ready is crucial for future transitions, considering their long operational lifespan. Presently, the high cost of green hydrogen, approximately three to ten times higher than that of natural gas, limits its widespread use as a primary fuel for power generation.

A proposed solution is blending hydrogen with NG to reduce costs and CO_2 emissions. Yet this approach requires careful analysis of the blend's heat input, as hydrogen contributes less energy per unit volume compared to methane. While emissions reductions are a favorable outcome, the economic viability of hydrogen blends remains challenging due to the high production costs associated with green hydrogen. Therefore, initial blends may include smaller quantities of hydrogen to balance economic and environmental objectives.

Vattenfall's Magnum Power Station in the Netherlands, in collaboration with Mitsubishi Power and Equinor, is undergoing a transformation to convert one of its gas turbines to operate solely on hydrogen by 2025. This project emphasizes the technical modifications required, such as advanced burner designs and materials capable of handling hydrogen's unique combustion characteristics. Preliminary results have shown substantial CO_2 emissions reductions, validating the feasibility and scalability of the concept. Similarly, the Long Ridge Energy Terminal in Ohio, USA, is partnering with GE Gas Power and New Fortress Energy to blend hydrogen with natural gas in a GE 7HA.02 gas turbine, with plans for a full transition to 100% hydrogen. Early operations have successfully implemented a 20% hydrogen blend, offering a blueprint for integrating hydrogen into power generation systems. Meanwhile, the HYFLEXPOWER project in France, led by Engie Solutions and Siemens Energy, aims to operate an industrial gas turbine using hydrogen blends, with the ultimate objective of transitioning to 100% hydrogen. Hydrogen is produced on-site through electrolysis powered by renewable energy sources, and the project has successfully utilized a 30% hydrogen blend, achieving significant emissions reductions and establishing a model for future industrial applications.

1.4.3.2 Fuel Cell (FC) Method for Hydrogen Re-Electrification

The FC method for hydrogen re-electrification is an advanced technique that utilizes hydrogen as a fuel to generate electrical power in a clean and efficient manner. This process involves the electrochemical conversion of hydrogen into electrical energy, bypassing combustion and its associated emissions. Fuel cells offer exceptional efficiency and produce no carbon emissions, making them a promising technology for sustainable power generation. As shown in Figure 1.6, the primary components of fuel cells include an anode, an electrolyte, and a cathode. A catalyst, such as platinum, facilitates the introduction of hydrogen gas at the anode, where it is split into protons and electrons. The most common electrolyte is the proton-exchange membrane (PEM), which allows protons to pass through while blocking electrons. Electricity is generated as electrons flow through an external circuit. At the cathode, oxygen reacts with protons and electrons to form water and release heat. Direct current (DC) electricity is generated, which can then be converted to alternating current (AC) using an inverter for various applications. Fuel cells directly convert the chemical energy of hydrogen into electricity, with electrical efficiencies typically ranging from 40% to 60%. In combined heat and power (CHP) applications, overall efficiencies can reach up to 80–90% when utilizing both electricity and heat. This efficiency is significantly higher than that of hydrogen-based internal combustion engines (ICE), which achieve efficiencies of about 25% [142]. During operation, hydrogen and oxygen are supplied to the anodes and cathodes, eliminating CO_2 emissions and producing only water, electricity, and heat without combustion.

FC systems are categorized based on the type of electrolyte membrane they utilize. Solid oxide fuel cells (SOFC) operate at high temperatures (800–1,000°C) and can use a variety of fuels, including hydrogen and hydrocarbons; however, they face challenges, such as durability issues and slow start-up times, making them suitable primarily for stationary power generation. Molten carbonate fuel cells (MCFC) are also high-temperature fuel cells (600–700°C) that can utilize CO-rich gases and are employed in industrial applications, but they require complex thermal management. Alkaline fuel cells (AFC) use a liquid potassium hydroxide electrolyte and offer cost-effective and highly efficient performance, but their sensitivity to CO_2 limits their use mainly to controlled environments, like space applications. Phosphoric acid fuel cells (PAFC), which use liquid phosphoric acid as the electrolyte, are more tolerant to fuel impurities and are commonly used in stationary

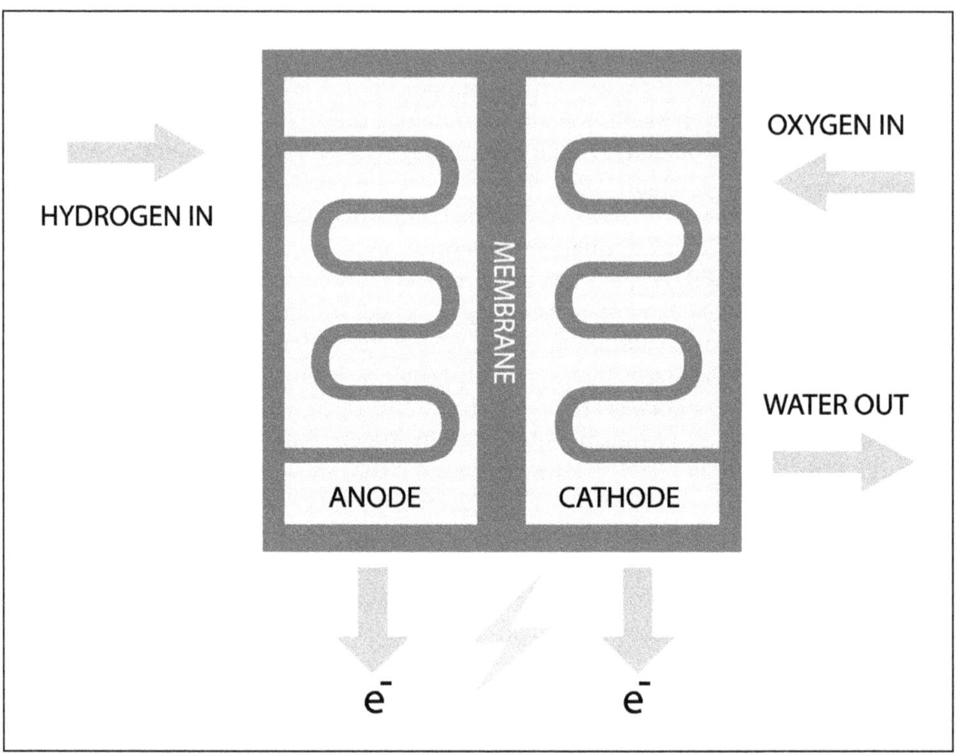

FIGURE 1.6 A schematic of a fuel cell.

power generation, especially in cogeneration systems [143]. Proton-exchange membrane fuel cells (PEM-FC) employ a solid polymer electrolyte and operate at lower temperatures (60–80°C). PEM-FCs have gained popularity due to their durability, high power density, and fast start-up times, making them ideal for both mobility and stationary applications. Direct methanol fuel cells (DMFC), a subtype of PEM-FC, use methanol directly as fuel and are designed for small-scale and portable applications but encounter challenges, such as methanol crossover and lower efficiency. The modular design of PEM-FCs allows flexibility in their applications [144]. Cells can be connected in series or parallel to meet different power requirements. Each cell produces an operating voltage of approximately 0.6–0.8 V (volts) under load. PEM-FCs are suitable for electric vehicles (20–250 kW), residential power (100 W–5 kW), and larger stationary applications (100 kW–2 MW) due to their versatility [145].

The mobility sector is a primary application for PEM-FCs. As of recent data, the global fleet of fuel cell electric vehicles (FCEV) exceeded 50,000 units by the end of 2022, with significant deployments in countries like Japan, the United States, South Korea, and China [146]. FCEVs are particularly utilized in urban areas, where air quality is a concern. Despite their superior efficiency compared to ICE vehicles (ICEV), FCEVs face higher operational costs, primarily due to the price of hydrogen, rendering them less competitive than battery electric vehicles (BEVs) and ICEVs. In stationary applications, PEM-FCs are used as backup power sources and in micro-cogeneration systems for residential and small commercial applications, especially in Japan and Europe. These systems can enhance overall energy efficiency by utilizing the heat generated during hydrogen re-electrification. While overall efficiencies in CHP systems can reach up to 80–90%, achieving such high efficiency requires optimal utilization of both electricity and heat. PEM-FCs face challenges, such as the high cost of catalysts, typically platinum, and sensitivity to fuel impurities. Effective thermal management is essential to maintain efficiency and extend the lifespan of the cells. The International Organization for Standardization (ISO) has established hydrogen quality standards (ISO 14687) to mitigate issues related to fuel impurities. Additionally, cogeneration and trigeneration systems offer solutions for heat recovery, further improving the overall efficiency of these systems. Various models, including steady-state and dynamic models, have been developed to optimize the cost-efficiency and performance of PEM-FCs. These models simulate performance, assess life cycle impacts, and provide TEA for optimizing operating conditions under various scenarios. Such modeling is crucial for enhancing performance, reducing environmental impacts, and managing costs.

FuelCell Energy's SureSource power plants utilize proprietary carbonate fuel cell technology and operate globally, demonstrating the feasibility of fuel cells in generating continuous, reliable power with minimal environmental impact. When powered by hydrogen, these plants produce electricity, heat, and water with high efficiency and zero emissions. The HAEOLUS project [147] in Norway integrates a 45 MW wind farm with a 100 kW PEM-FC system, combining RESs with fuel cells to generate clean and consistent power, showcasing the potential of integrating RESs with hydrogen production and fuel cell power generation. Similarly, Toyota's Tri-Gen Facility in California, USA, located at the Port of Long Beach, generates hydrogen from biogas derived from agricultural waste to power fuel cells that produce electricity, heat, and hydrogen for vehicle refueling, creating a sustainable, circular energy system. Additionally, Bloom Energy Servers, developed by Bloom Energy, provide reliable, on-site power using solid oxide fuel cell (SOFC) technology. Hundreds of these servers have been deployed across the United States and internationally, offering a flexible and scalable solution for sustainable energy by operating on either natural gas or hydrogen. When fueled with hydrogen, they produce electricity with zero carbon emissions. The detailed overview of the various methods for PtHtP, including their efficiency, emissions, applications, and challenges, is summarized in Table 1.11.

1.5 CONCLUSION

The investigation of PtX technologies underscores their revolutionary capacity to promote sustainable energy solutions across various industries. Our comprehensive analysis reveals substantial findings that highlight the efficiency and long-term potential of PtX systems. Specifically, the assessment of PtX pathways demonstrates the vast capacity of water electrolyzer technology in efficiently generating hydrogen, a fundamental element for PtX applications. The effective capture and utilization of CO_2 not only provide a feasible approach to reducing GHG emissions but also contribute to the production of valuable synthetic fuels and chemicals. Advancements in H_2 storage and delivery systems have made significant strides in overcoming safety and efficiency challenges, thereby establishing H_2 as a viable energy carrier. Additionally, the incorporation of nitrogen supply systems has proven indispensable for the sustainable production of ammonia and other N2-derived products. The existing PtX infrastructure supports these developments, enabling seamless and effective system operations. Overall, the diverse PtX pathways illustrate a wide range of conversion processes and end products, highlighting the flexibility and extensive advantages of PtX technology. Our system-level analysis confirms the economic feasibility, thermodynamic effectiveness, and environmental sustainability of PtX technologies. The TEA indicates that PtX systems can achieve cost-effectiveness by optimizing crucial economic factors. THA emphasizes the significant energy savings achievable in PtX processes, confirming their practicality and scalability. LCA verifies the considerable environmental benefits of PtX, such as notable reductions in carbon footprints and resource consumption, positioning PtX as a crucial technology for sustainable development. Optimization methodologies have demonstrated that integrating HRESs with PtX technology enhances system resilience and efficiency. Sector coupling effectively optimizes PtX technologies across multiple energy sectors by facilitating the integration of power, heat, and mobility applications. Furthermore, PtX systems exhibit flexibility, allowing them to adjust to varying energy inputs and demands, ensuring constant and efficient performance. The process of hydrogen

TABLE 1.11
PtHtP Methods Overview

Method	Description	Advantages	Challenges	Applications	Efficiency	TRL	Examples/Case Studies
Gas turbine combustion	Uses hydrogen as a fuel in gas turbines to generate electricity through combustion. Modifications to existing turbines allow for hydrogen or hydrogen–natural gas blends to be used as fuel.	• High power output • Fast ramp-up times • Utilizes existing infrastructure • Reduces CO_2 emissions when using hydrogen	• NO_x emissions due to high flame temperatures • Flame flashback risks • Material compatibility with hydrogen • High cost of green hydrogen	• Utility-scale power generation • Peak load balancing • Integration with RESs	**35–45%** (simple cycle) **55–60%** (combined cycle)	TRL 7–9 (system prototype to commercial deployment)	• **Vattenfall's Magnum Power Station (Netherlands)** • **Long Ridge Energy Terminal (USA)** • **HYFLEXPOWER project (France)**
Fuel cells (general)	Electrochemical devices that convert hydrogen directly into electricity and heat without combustion. Types include PEM-FC, SOFC, MCFC, AFC, and PAFC.	• High electrical efficiency • Zero emissions at point of use • Quiet operation • Scalability and modularity	• High capital costs • Durability and lifespan concerns • Sensitivity to fuel impurities • Infrastructure for hydrogen supply	• Backup and primary power for buildings • Distributed generation • Transportation (vehicles, buses, forklifts) • Portable power devices	**40–60%** (electrical efficiency) Up to **80–90%** with CHP	TRL 7–9 (system prototype to commercial deployment)	• **FuelCell Energy's SureSource Plants** • **Bloom Energy Servers (USA and internationally)** • **Toyota's Tri-Gen Facility (USA)**
Proton-exchange membrane fuel cells (PEM-FC)	Use a solid polymer electrolyte membrane to conduct protons from anode to cathode. Operate at low temperatures (~60–80°C) and have quick start-up times.	• High power density • Fast start-up and response times • Suitable for variable loads • Ideal for transportation applications	• High cost of platinum catalysts • Sensitive to impurities (CO, sulfur) • Thermal and water management required	• Transportation (cars, buses, trucks) • Backup power systems • Residential and commercial CHP systems	**40–60%** (electrical efficiency) Up to **80–90%** with CHP	TRL 8–9 (first-of-a-kind commercial systems)	• **Toyota Mirai fuel cell vehicle** • **Hyundai NEXO fuel cell vehicle** • **Micro-CHP units in Japan (e.g., ENE-FARM program)**
Solid oxide fuel cells (SOFC)	High-temperature fuel cells (operating at 600–1,000°C) using a solid ceramic electrolyte. Can internally reform hydrocarbon fuels and have high-quality heat output suitable for CHP applications.	• High electrical efficiency • Fuel flexibility (hydrogen, natural gas, biogas) • High-temperature waste heat for CHP • Less sensitive to fuel impurities	• Long start-up times • Material degradation at high temperatures • Thermal cycling reducing lifespan • Higher capital costs	• Distributed power generation • Industrial CHP systems • Data centers and hospitals requiring reliable power	**45–65%** (electrical efficiency) Up to **85%** with CHP	TRL 7–8 (demonstration to early commercial)	• **Bloom Energy Servers** • **FuelCell Energy installations**

(Continued)

TABLE 1.11 (Continued)

Molten carbonate fuel cells (MCFC)	High-temperature fuel cells (operating at ~650°C) using a molten carbonate electrolyte. Suitable for large-scale stationary power generation with fuel flexibility.	• High efficiency for large-scale applications • Can use CO-rich fuels • High-temperature waste heat usable for CHP • Tolerant to some fuel impurities	• High operating temperatures leading to material challenges • Corrosion of components • Complex thermal management • Longer start-up times	• Utility-scale power plants • Industrial facilities • Combined heat and power applications	**45–55%** (electrical efficiency) Up to **85%** with CHP	TRL 7–8 (demonstration to early commercial)	• **FuelCell Energy's SureSource plants** • **Gyeonggi Green Energy Fuel Cell Park (South Korea)**
Hydrogen internal combustion engines (H2-ICE)	Modified internal combustion engines running on hydrogen instead of conventional fuels. Can be based on spark–ignition or compression–ignition designs.	• Utilizes existing engine technology • Quick refueling times • Lower emissions than fossil fuels (when NO_X controlled) • Can be retrofitted from existing engines	• Lower efficiency compared to fuel cells • NO_X emissions unless controlled • Limited infrastructure for hydrogen fueling • Engine durability concerns with hydrogen combustion	• Heavy-duty vehicles (trucks, buses) • Off-road equipment • Backup power generators	**25–40%** (thermal efficiency)	TRL 6–7 (technology demonstrated in relevant environment)	• **BMW Hydrogen 7 prototype vehicles** • **Research projects on H2-ICE buses and trucks**
Hydrogen reciprocating engines	Similar to H2-ICE, but specifically designed for stationary power generation using hydrogen in reciprocating engine generators.	• Mature technology base • Fast start-up times • Can utilize existing gas engine designs • Lower capital costs compared to fuel cells	• Lower efficiency than fuel cells • NO_X emissions requiring after-treatment • Frequent maintenance possibly needed • Hydrogen embrittlement of engine components	• Backup and emergency power • Remote power generation • Integration with renewable hydrogen production	**30–40%** (electrical efficiency)	TRL 7–8 (demonstration to early commercial)	• **INNIO Jenbacher hydrogen engines** • **Caterpillar hydrogen engine projects**
Hydrogen micro-gas turbines	Small-scale gas turbines modified to burn hydrogen, suitable for distributed generation and CHP applications.	• Compact size • Capable of rapid load changes • Integration with CHP systems • Lower emissions than larger turbines when designed properly	• Technical challenges with hydrogen combustion (flashback, NO_X) • Material compatibility • Lower efficiency compared to larger turbines and fuel cells	• Distributed generation • Commercial and industrial CHP • Backup power systems	**25–35%** (electrical efficiency) Up to **80%** with CHP	TRL 6–7 (technology demonstrated in relevant environment)	• **Capstone Turbine Corporation demonstrations** • **Micro-gas turbine projects in Europe and Japan**

re-electrification, involving the use of gas turbines and fuel cells, effectively converts stored H_2 back into electricity, significantly improving H_2's versatility as an energy carrier.

Despite these promising advancements, several challenges remain. Reducing the costs associated with H_2 production, storage, and fuel cell technologies is essential for broader economic competitiveness. Developing a resilient infrastructure for H_2 transport and storage, including pipelines, refueling stations, and large-scale storage facilities, is critical for mainstream adoption. Effective policies and regulations, such as incentives, subsidies, and standards, are necessary to support the development and deployment of PtX technologies. Integration with RESs requires ongoing research into hybrid systems, advanced grid management, and sophisticated energy storage solutions to optimize collaboration between PtX technologies and RES. Public knowledge and acceptance are vital for gaining social endorsement of H_2 and PtX technologies; therefore, outreach and education initiatives should be prioritized to build trust and understanding of their benefits and safety. Additionally, the advancement of sophisticated modeling and simulation tools will aid in optimizing PtX systems by forecasting performance, evaluating environmental impacts, and guiding decision-making processes. In conclusion, the findings from this extensive investigation confirm the transformative potential of PtX technologies in the sustainable energy sector. PtX pathways efficiently utilize RES, demonstrating their ability to reduce carbon emissions and decarbonize multiple sectors. With their economic feasibility, high efficiency, and environmental benefits, PtX technologies are positioned as a crucial element of future energy systems. Enhancing integration, flexibility, and strategic operation of PtX systems is essential for improving their performance and resilience. The successful demonstration of H_2 re-electrification highlights the adaptability of H_2 as an energy transporter, further cementing its significance in future energy solutions. To fully unlock the potential of PtX technologies, continued research and development in key areas are imperative, ensuring that PtX can effectively contribute to a cleaner, more sustainable energy future.

REFERENCES

[1] Ritchie H, Rosado P, Roser M (2024) *Energy production and consumption.* Our World in Data. https://ourworldindata.org/energy-production-consumption.

[2] Ritchie H, Rosado P, Roser M (2024) *Electricity mix.* Our World in Data. https://ourworldindata.org/electricity-mix.

[3] Our World in Data (2024) *Greenhouse gas emissions.* Our World in Data. https://ourworldindata.org/grapher/total-ghg-emissions?tab=chart.

[4] Ritchie H, Rosado P, Roser M (2024) *CO$_2$ emissions by fuel.* Our World in Data. https://ourworldindata.org/emissions-by-fuel.

[5] *Average temperature anomaly.* Our World in Data. https://ourworldindata.org/grapher/temperature-anomaly.

[6] IEA (2023) Electricity: Renewables 2023 – Analysis. www.iea.org/reports/renewables-2023/electricity.

[7] IEA (2024) *Renewable energy progress tracker: Data tools.* IEA. www.iea.org/data-and-statistics/data-tools/renewable-energy-progress-tracker.

[8] Khalid M (2024) Smart grids and renewable energy systems: Perspectives and grid integration challenges. *Energy Strategy Reviews* 51:101299. doi: 10.1016/j.esr.2024.101299.

[9] Khan KA, Quamar MM, Al-Qahtani FH, et al. (2023) Smart grid infrastructure and renewable energy deployment: A conceptual review of Saudi Arabia. *Energy Strategy Reviews* 50:101247. doi: 10.1016/j.esr.2023.101247.

[10] Meydani A, Shahinzadeh H, Nafisi H, et al. (2024) Optimum energy management strategies for the hybrid sources-powered electric vehicle settings. In *2024 28th International Electrical Power Distribution Conference (EPDC).* doi: 10.1109/epdc62178.2024.10571762.

[11] Mitali J, Dhinakaran S, Mohamad AA (2022) Energy storage systems: A review. *Energy Storage and Saving* 1:166–216. doi: 10.1016/j.enss.2022.07.002.

[12] Nadeem F, Hussain SMS, Tiwari PK, et al. (2019) Comparative review of energy storage systems, their roles, and impacts on future power systems. *IEEE Access* 7:4555–4585. doi: 10.1109/access.2018.2888497.

[13] Hasan MK, Mahmud M, Habib AKMA, et al. (2021) Review of electric vehicle energy storage and management system: Standards, issues, and challenges. *Journal of Energy Storage* 41:102940. doi: 10.1016/j.est.2021.102940.

[14] Colthorpe A (2023) World's energy storage capacity forecast to exceed a terawatt-hour by 2030. *Energy-Storage News.* www.energy-storage.news/worlds-energy-storage-capacity-forecast-to-exceed-a-terawatt-hour-by-2030/.

[15] Ishaq H, Dincer I, Crawford C (2022) A review on hydrogen production and utilization: Challenges and opportunities. *International Journal of Hydrogen Energy* 47:26238–26264. doi: 10.1016/j.ijhydene.2021.11.149.

[16] Dash SK, Chakraborty S, Elangovan D (2023) A brief review of hydrogen production methods and their challenges. *Energies* 16:1141. doi: 10.3390/en16031141.

[17] Zainal BS, Ker PJ, Mohamed H, et al. (2024) Recent advancement and assessment of green hydrogen production technologies. *Renewable and Sustainable Energy Reviews* 189:113941. doi: 10.1016/j.rser.2023.113941.

[18] https://orsted.com/en/what-we-do/renewable-energy-solutions/power-to-x.

[19] Daiyan R, MacGill I, Amal R (2020) Opportunities and challenges for renewable power-to-X. *ACS Energy Letters* 5(12):3843–3847. doi: 10.1021/acsenergylett.0c02249.

[20] Burre J, Bongartz D, Brée L, et al. (2019) Power-to-X: Between electricity storage, e-production, and demand side management. *Chemie Ingenieur Technik* 92(1–2):74–84. doi: 10.1002/cite.201900102.

[21] Hashimoto K (1994) Metastable metals for "green" materials for global atmosphere conservation and abundant energy supply. *Materials Science and Engineering A* 179–180:27–30. doi: 10.1016/0921-5093(94)90158-9.

[22] Hashimoto K, Kumagai N, Izumiya K, et al. (2014) The production of renewable energy in the form of methane using electrolytic hydrogen generation. *Energy Sustainability and Society* 4. doi: 10.1186/s13705-014-0017-5.

[23] *Power-to-gas plant > g-tron > technology > audi malta.* Audi.de. www.audi.com.mt/mt/web/en/models/layer/technology/g-tron/power-to-gas-plant.html.

[24] Autoblog (2013) Audi opens renewable energy E-gas plant in Germany. *Autoblog.* www.autoblog.com/2013/07/08/audi-opens-renewable-energy-e-gas-plant-in-germany/.

[25] www.thyssenkrupp-steel.com/en/company/sustainability/carbon2chem/carbon2chem.html.

[26] Schlautmann R, Al-Breiki M, Zauner A, et al. (2021) Renewable power-to-gas: A technical and economic evaluation of three demo sites within the STORE&GO project. *Chemie Ingenieur Technik* 93:568–579. doi: 10.1002/cite.202000187.

[27] Vázquez FV, Koponen J, Ruuskanen V, et al. (2018) Power-to-X technology using renewable electricity and carbon dioxide from ambient air: SOLETAIR proof-of-concept and improved process concept. *Journal of CO$_2$ Utilization* 28:235–246. doi: 10.1016/j.jcou.2018.09.026.

[28] www.spglobal.com/commodityinsights/en/market-insights/latest-news/electric-power/070721-hydrogen-fever-in-eu-puts-2024-target-of-6-gw-electrolyzer-capacity-in-reach.

[29] Sparkedadmin (2024) Hydrogen insights 2021. *Hydrogen Council.* https://hydrogencouncil.com/en/hydrogen-insights-2021/.

[30] IEA, *Energy technology RD&D budgets data explorer: Data tools.* IEA. www.iea.org/data-and-statistics/data-tools/energy-technology-rdd-budgets-data-explorer.

[31] Carmo M, Fritz DL, Mergel J, et al. (2013) A comprehensive review on PEM water electrolysis. *International Journal of Hydrogen Energy* 38:4901–4934. doi: 10.1016/j.ijhydene.2013.01.151.

[32] Hnát J, Plevova M, Tufa RA, et al. (2019) Development and testing of a novel catalyst-coated membrane with platinum-free catalysts for alkaline water electrolysis. *International Journal of Hydrogen Energy* 44:17493–17504. doi: 10.1016/j.ijhydene.2019.05.054.

[33] Kuroda Y, Nishimoto T, Mitsushima S (2019) Self-repairing hybrid nanosheet anode catalysts for alkaline water electrolysis connected with fluctuating renewable energy. *Electrochimica Acta* 323:134812. doi: 10.1016/j.electacta.2019.134812.

[34] Peng Y, Jiang K, Hill W, et al. (2019) Large-scale, low-cost, and high-efficiency water-splitting system for clean H$_2$ generation. *ACS Applied Materials & Interfaces* 11:3971–3977. doi: 10.1021/acsami.8b19251.

[35] Kumar SS, Himabindu V (2019) Hydrogen production by PEM water electrolysis – A review. *Materials Science for Energy Technologies* 2:442–454. doi: 10.1016/j.mset.2019.03.002.

[36] Tjarks G, Gibelhaus A, Lanzerath F, et al. (2018) Energetically-optimal PEM electrolyzer pressure in power-to-gas plants. *Applied Energy* 218:192–198. doi: 10.1016/j.apenergy.2018.02.155.

[37] Laguna-Bercero MA (2012) Recent advances in high temperature electrolysis using solid oxide fuel cells: A review. *Journal of Power Sources* 203:4–16. doi: 10.1016/j.jpowsour.2011.12.019.

[38] Chi J, Yu H (2018) Water electrolysis based on renewable energy for hydrogen production. *Chinese Journal of Catalysis* 39:390–394. doi: 10.1016/s1872-2067(17)62949-8.

[39] Dincer I, Acar C (2015) Review and evaluation of hydrogen production methods for better sustainability. *International Journal of Hydrogen Energy* 40:11094–11111. doi: 10.1016/j.ijhydene.2014.12.035.

[40] Lockwood T (2017) A comparative review of next-generation carbon capture technologies for coal-fired power plant. *Energy Procedia* 114:2658–2670. doi: 10.1016/j.egypro.2017.03.1850.

[41] Simonsen KR, Hansen DS, Pedersen S (2024) Challenges in CO_2 transportation: Trends and perspectives. *Renewable and Sustainable Energy Reviews* 191:114149. doi: 10.1016/j.rser.2023.114149.

[42] Santos RGD, Alencar AC (2020) Biomass-derived syngas production via gasification process and its catalytic conversion into fuels by Fischer Tropsch synthesis: A review. *International Journal of Hydrogen Energy* 45:18114–18132. doi: 10.1016/j.ijhydene.2019.07.133.

[43] Yapicioglu A, Dincer I (2019) A review on clean ammonia as a potential fuel for power generators. *Renewable and Sustainable Energy Reviews* 103:96–108. doi: 10.1016/j.rser.2018.12.023.

[44] Keith DW, Holmes G, St Angelo D, et al. (2018) A process for capturing CO_2 from the atmosphere. *Joule* 2:1573–1594. doi: 10.1016/j.joule.2018.05.006.

[45] Lyons M, Durrant P, Kochhar K (2021) *Reaching zero with renewables: Capturing carbon.* International Renewable Energy Agency.

[46] IEA (2021) *About CCUS: Analysis.* IEA. www.iea.org/reports/about-ccus.

[47] The Engineering Toolbox (2024) Hydrogen: Density and specific weight vs. temperature and pressure. www.engineeringtoolbox.com/hydrogen-H2-density-specific-weight-temperature-pressure-d_2044.html.

[48] https://pgjonline.com/magazine/2020/august-2020-vol-247-no-8/features/germany-plans-ambitious-hydrogen-pipeline-network.

[49] https://pgjonline.com/magazine/2024/january-2024-vol-251-no-1/features/spotlight-on-germany-nation-s-pipelines-nearing-major-changes.

[50] IRENA (2020) Global renewables outlook: Energy transformation 2050. www.irena.org/publications/2020/Apr/Global-Renewables-Outlook-2020.

[51] Smith AR, Klosek J (2001) A review of air separation technologies and their integration with energy conversion processes. *Fuel Processing Technology* 70:115–134. doi: 10.1016/s0378-3820(01)00131-x.

[52] Böcker N, Grahl M, Tota A, et al. (2013) Nitrogen. In *Ullmann's encyclopedia of industrial chemistry*, pp. 1–27. doi: 10.1002/14356007.a17_457.pub2.

[53] Miller J, Luyben WL, Belanger P, et al. (2007) Improving agility of cryogenic air separation plants. *Industrial & Engineering Chemistry Research* 47:394–404. doi: 10.1021/ie070975t.

[54] Varvoutis G, Lampropoulos A, Mandela E, et al. (2022) Recent advances on CO_2 mitigation technologies: On the role of hydrogenation route via green H_2. *Energies* 15(13):4790. doi: 10.3390/en15134790.

[55] Stephens IE, Chan K, Bagger A, et al. (2022) 2022 Roadmap on low temperature electrochemical CO_2 reduction. *Journal of Physics Energy* 4(4):042003. doi: 10.1088/2515-7655/ac7823.

[56] Penn State Hydrogen Energy (H_2E) Center – Energy 2100. https://energy2100.psu.edu/reports/penn-state-hydrogen-energy-h2e-center/.

[57] Sunfire—Progress within Kopernikus P2X research project: High-temperature electrolyzer successfully commissioned. *Sunfire.* www.sunfire.de/en/news/detail/progress-within-kopernikus-p2x-research-project-high-temperature-electrolyzer-successfully-commissioned.

[58] Mierczyński P, Mierczynska-Vasilev A, Szynkowska-Jóźwik MI, et al. (2023) Plasma-assisted catalysis for CH_4 and CO_2 conversion. *Catalysis Communications* 180:106709. doi: 10.1016/j.catcom.2023.106709.

[59] Eveloy V, Gebreegziabher T (2018) A review of projected power-to-gas deployment scenarios. *Energies* 11:1824. doi: 10.3390/en11071824.

[60] Chen C, Yang A (2021b) Power-to-methanol: The role of process flexibility in the integration of variable renewable energy into chemical production. *Energy Conversion and Management* 228:113673. doi: 10.1016/j.enconman.2020.113673.

[61] Sillman J, Hynynen K, Dyukov I, et al. (2023) Emission reduction targets and electrification of the Finnish energy system with low-carbon Power-to-X technologies: Potentials, barriers, and innovations – A Delphi survey. *Technological Forecasting and Social Change* 193:122587. doi: 10.1016/j.techfore.2023.122587.

[62] Connolly D, Lund H, Mathiesen BV, et al. (2014) Heat roadmap Europe: Combining district heating with heat savings to decarbonise the EU energy system. *Energy Policy* 65:475–489. doi: 10.1016/j.enpol.2013.10.035.

[63] Chehade Z, Mansilla C, Lucchese P, et al. (2019) Review and analysis of demonstration projects on power-to-X pathways in the world. *International Journal of Hydrogen Energy* 44:27637–27655. doi: 10.1016/j.ijhydene.2019.08.260.

[64] Simader G, Vidovic P (2022) Success factors for demonstration projects of small-scale stationary fuel cells in residential buildings. *E3S Web of Conferences* 334:04007. doi: 10.1051/e3sconf/202233404007.

[65] Safari A, Hayati MM, Nazari-Heris M (2024) Hydrogen-combined smart electrical power systems: An overview of united states projects. In *Green Energy and Technology*, pp. 321–340.

[66] https://assets.publishing.service.gov.uk/media/61f957ca8fa8f5388b582d82/Phase_2_Report_-_ITM_-_Gigastack.pdf.

[67] Parra D, Valverde L, Pino FJ, et al. (2018) A review on the role, cost and value of hydrogen energy systems for deep decarbonisation. *Renewable and Sustainable Energy Reviews* 101:279–294. doi: 10.1016/j.rser.2018.11.010.

[68] Pratschner S, Hammerschmid M, Müller S, et al. (2024) Off-grid vs. grid-based: Techno-economic assessment of a power-to-liquid plant combining solid-oxide electrolysis and Fischer-Tropsch synthesis. *Chemical Engineering Journal* 481:148413. doi: 10.1016/j.cej.2023.148413.

[69] Salomone F, Giglio E, Ferrero D, et al. (2019) Techno-economic modelling of a Power-to-Gas system based on SOEC electrolysis and CO_2 methanation in a RES-based electric grid. *Chemical Engineering Journal* 377:120233. doi: 10.1016/j.cej.2018.10.170.

[70] Böhm H, Zauner A, Rosenfeld DC, et al. (2020) Projecting cost development for future large-scale power-to-gas implementations by scaling effects. *Applied Energy* 264:114780. doi: 10.1016/j.apenergy.2020.114780.

[71] Leeson D, Mac Dowell N, Shah N, et al. (2017) A techno-economic analysis and systematic review of carbon capture and storage (CCS) applied to the iron and steel, cement, oil refining and pulp and paper industries, as well as other high purity sources. *International Journal of Greenhouse Gas Control* 61:71–84. doi: 10.1016/j.ijggc.2017.03.020.

[72] Mac Dowell N, Fennell PS, Shah N, et al. (2017) The role of CO_2 capture and utilization in mitigating climate change. *Nature Climate Change* 7(4):243–249. doi: 10.1038/nclimate3231.

[73] Al-Breiki M, Bicer Y (2023) Techno-economic evaluation of a power-to-methane plant: Levelized cost of methane, financial performance metrics, and sensitivity analysis. *Chemical Engineering Journal* 471:144725. doi: 10.1016/j.cej.2023.144725.

[74] Michailos S, McCord S, Sick V, et al. (2019) Dimethyl ether synthesis via captured CO_2 hydrogenation within the power to liquids concept: A techno-economic assessment. *Energy Conversion and Management* 184:262–276. doi: 10.1016/j.enconman.2019.01.046.

[75] Zhang H, Wang L, Van Herle J, et al. (2020) Techno-economic comparison of green ammonia production processes. *Applied Energy* 259:114135. doi: 10.1016/j.apenergy.2019.114135.

[76] Skov IR, Schneider N, Schweiger G, et al. (2024) Power-to-X in Denmark: An analysis of strengths, weaknesses, opportunities, and threats. *Sustainable Energy Research* 9:9. doi: 10.1186/s40807-024-00129-9.

[77] Bram MV, Liniger J, Majidabad SS, et al. (2024) Challenges in Power-to-X: A perspective of the configuration and control process for E-methanol production. *International Journal of Hydrogen Energy* 76:315–325. doi: 10.1016/j.ijhydene.2024.05.273.

[78] Eyberg V, Dieterich V, Bastek S, et al. (2024) Techno-economic assessment and comparison of Fischer–Tropsch and Methanol-to-Jet processes to produce sustainable aviation fuel via Power-to-Liquid. *Energy Conversion and Management* 315:118728. doi: 10.1016/j.enconman.2024.118728.

[79] Cuevas-Castillo GA, Michailos S, Akram M, et al. (2024) Techno economic and life cycle assessment of olefin production through CO_2 hydrogenation within the power-to-X concept. *Journal of Cleaner Production* 469:143143. doi: 10.1016/j.jclepro.2024.143143.

[80] Pfennig M, Böttger D, Häckner B, et al. (2023) Global GIS-based potential analysis and cost assessment of Power-to-X fuels in 2050. *Applied Energy* 347:121289. doi: 10.1016/j.apenergy.2023.121289.

[81] Gorre J, Ruoss F, Karjunen H, et al. (2020) Cost benefits of optimizing hydrogen storage and methanation capacities for Power-to-Gas plants in dynamic operation. *Applied Energy* 257:113967. doi: 10.1016/j.apenergy.2019.113967.

[82] Ball M, Weeda M (2015) The hydrogen economy: Vision or reality? *International Journal of Hydrogen Energy* 40(25):7903–7919. doi: 10.1016/j.ijhydene.2015.04.032.

[83] International Renewable Energy Agency (IRENA) (2020) *Green hydrogen: A guide to policy making*. Abu Dhabi: IRENA. www.irena.org/publications/2020/Nov/Green-hydrogen.

[84] Fahr S, Schiedeck M, Schwarzhuber J, et al. (2023) Design and thermodynamic analysis of a large-scale ammonia reactor for increased load flexibility. *Chemical Engineering Journal* 471:144612. doi: 10.1016/j.cej.2023.144612.

[85] Hjeij D, Biçer Y, Koç M (2022) Thermodynamic analysis of a multigeneration system using solid oxide cells for renewable power-to-X conversion. *International Journal of Hydrogen Energy* 48(32):12056–12071. doi: 10.1016/j.ijhydene.2022.09.024.

[86] Ancona MA, Bianchi M, Branchini L, et al. (2020) Numerical prediction of off-design performance for a Power-to-Gas system coupled with renewables. *Energy Conversion and Management* 210:112702. doi: 10.1016/j.enconman.2020.112702.

[87] Koytsoumpa EI, Karellas S, Kakaras E (2020) Modelling of methanol production via combined gasification and power to fuel. *Renewable Energy* 158:598–611. doi: 10.1016/j.renene.2020.05.169.

[88] Jeanmonod G, Wang L, Diethelm S, et al. (2019) Trade-off designs of power-to-methane systems via solid-oxide electrolyzer and the application to biogas upgrading. *Applied Energy* 247:572–581. doi: 10.1016/j.apenergy.2019.04.055.

[89] Samavati M, Martin A, Nemanova V, et al. (2018) Integration of solid oxide electrolyser, entrained gasification, and Fischer-Tropsch process for synthetic diesel production: Thermodynamic analysis. *International Journal of Hydrogen Energy* 43(10):4785–4803. doi: 10.1016/j.ijhydene.2018.01.138.

[90] Wang L, Düll J, Maréchal F, et al. (2019) Trade-off designs and comparative exergy evaluation of solid-oxide electrolyzer based power-to-methane plants. *International Journal of Hydrogen Energy* 44(19):9529–9543. doi: 10.1016/j.ijhydene.2018.11.151.

[91] Gu C, Tang C, Xiang Y, et al. (2019) Power-to-gas management using robust optimisation in integrated energy systems. *Applied Energy* 236:681–689. doi: 10.1016/j.apenergy.2018.12.028.

[92] Bos MJ, Kersten SRA, Brilman DWF (2020) Wind power to methanol: Renewable methanol production using electricity, electrolysis of water and CO_2 air capture. *Applied Energy* 264:114672. doi: 10.1016/j.apenergy.2020.114672.

[93] Luo Y, Wu X-Y, Shi Y, et al. (2018) Exergy analysis of an integrated solid oxide electrolysis cell-methanation reactor for renewable energy storage. *Applied Energy* 215:371–383. doi: 10.1016/j.apenergy.2018.02.022.

[94] Wang L, Rao M, Diethelm S, et al. (2019) Power-to-methane via co-electrolysis of H_2O and CO_2: The effects of pressurized operation and internal methanation. *Applied Energy* 250:1432–1445. doi: 10.1016/j.apenergy.2019.05.098.

[95] Ince AC, Colpan CO, Keles A, et al. (2022) Scaling and performance assessment of power-to-methane system based on an operation scenario. *Fuel* 332:126182. doi: 10.1016/j.fuel.2022.126182.

[96] Marchese M, Giglio E, Santarelli M, et al. (2020) Energy performance of Power-to-Liquid applications integrating biogas upgrading, reverse water gas shift, solid oxide electrolysis and Fischer-Tropsch technologies. *Energy Conversion and Management X* 6:100041. doi: 10.1016/j.ecmx.2020.100041.

[97] Schwarzmayr P, Birkelbach F, Walter H, et al. (2024) Exergy efficiency and thermocline degradation of a packed bed thermal energy storage in partial cycle operation: An experimental study. *Applied Energy* 360:122895. doi: 10.1016/j.apenergy.2024.122895.

[98] Ortega-Fernández I, Rodríguez-Aseguinolaza J (2019) Thermal energy storage for waste heat recovery in the steelworks: The case study of the REslag project. *Applied Energy* 237:708–719. doi: 10.1016/j.apenergy.2019.01.007.

[99] Chen J (2022) Hydrogen production in microchannel methanol steam reforming reactors. *Basrah Journal of Sciences* 40(1):83–106. doi: 10.29072/basjs.20220105.

[100] Kolb G, Keller S, O'Connell M, et al. (2013) Microchannel fuel processors as a hydrogen source for fuel cells in distributed energy supply systems. *Energy & Fuels* 27(8):4395–4402. doi: 10.1021/ef302039x.

[101] International Organization for Standardization (ISO). *ISO 14040: Environmental management—Life cycle assessment—Principles and framework*, 2006.

[102] International Organization for Standardization (ISO). *ISO 14044: Environmental management—Life cycle assessment—Requirements and guidelines*, 2006.

[103] Watanabe MDB, Hu X, Ballal V, et al. (2023) Climate change mitigation potentials of on grid-connected Power-to-X fuels and advanced biofuels for the European maritime transport. *Energy Conversion and Management X* 20:100418. doi: 10.1016/j.ecmx.2023.100418.

[104] Rojas-Michaga MF, Michailos S, Cardozo E, et al. (2023) Sustainable aviation fuel (SAF) production through power-to-liquid (PtL): A combined techno-economic and life cycle assessment. *Energy Conversion and Management* 292:117427. doi: 10.1016/j.enconman.2023.117427.

[105] Navajas A, Mendiara T, Gandía LM, et al. (2022) Life cycle assessment of power-to-methane systems with CO_2 supplied by the chemical looping combustion of biomass. *Energy Conversion and Management* 267:115866. doi: 10.1016/j.enconman.2022.115866.

[106] Rigamonti L, Brivio E (2022) Life cycle assessment of methanol production by a carbon capture and utilization technology applied to steel mill gases. *International Journal of Greenhouse Gas Control* 115:103616. doi: 10.1016/j.ijggc.2022.103616.

[107] Blanco H, Nijs W, Ruf J, et al. (2018) Potential for hydrogen and Power-to-Liquid in a low-carbon EU energy system using cost optimization. *Applied Energy* 232:617–639. doi: 10.1016/j.apenergy.2018.09.216.

[108] Blanco H, Codina V, Laurent A, et al. (2019) Life cycle assessment integration into energy system models: An application for Power-to-Methane in the EU. *Applied Energy* 259:114160. doi: 10.1016/j. apenergy.2019.114160.

[109] Boyce J, Sacchi R, Goetheer E, et al. (2024) A prospective life cycle assessment of global ammonia decarbonisation scenarios. *Heliyon* e27547. doi: 10.1016/j.heliyon.2024.e27547.

[110] PtX Hub (2021) PtX sustainability: Dimensions and concerns. *PtX Hub.* https://ptx-hub.org/wp-content/ uploads/2022/05/PtX-Hub-PtX.Sustainability-Dimensions-and-Concerns-Scoping-Paper.pdf.

[111] MissionGreenFuels. Guideline: Safe and faster PtX. https://brandogsikring.dk/files/Pdf/FogU/Guideline-Safe-and-Faster-PtX.pdf.

[112] Mendoza JMF, Ibarra D (2023) Technology-enabled circular business models for the hybridisation of wind farms: Integrated wind and solar energy, power-to-gas and power-to-liquid systems. *Sustainable Production and Consumption* 36:308–327. doi: 10.1016/j.spc.2023.01.011.

[113] Schreiber A, Peschel A, Hentschel B, et al. (2020) Life cycle assessment of Power-to-Syngas: Comparing high temperature co-electrolysis and steam methane reforming. *Frontiers in Energy Research* 8. doi: 10.3389/fenrg.2020.533850.

[114] Schreiber A, Troy S, Weiske S, et al. (2024) Comparative well-to-tank life cycle assessment of methanol, dimethyl ether, and oxymethylene dimethyl ethers via the power-to-liquid pathway. *Journal of CO_2 Utilization* 82:102743. doi: 10.1016/j.jcou.2024.102743.

[115] Ouda M, Hank C, Nestler F, et al. (2019) *Power-to-methanol: Techno-economical and ecological insights.* Springer eBooks, pp. 380–409. doi: 10.1007/978-3-662-58006-6_17.

[116] Ajeeb W, Baptista P, Neto RC (2024) Life cycle analysis of hydrogen production by different alkaline electrolyser technologies sourced with renewable energy. *Energy Conversion and Management* 316:118840. doi: 10.1016/j.enconman.2024.118840.

[117] Buffo G, Marocco P, Ferrero D, et al. (2019) *Power-to-X and power-to-power routes.* Elsevier eBooks, pp. 529–557. doi: 10.1016/B978-0-12-814853-2.00015-1.

[118] Valera-Medina A, Xiao H, Owen-Jones M, et al. (2018) Ammonia for power. *Progress in Energy and Combustion Science* 69:63–102. doi: 10.1016/j.pecs.2018.07.001.

[119] Ikäheimo J, Lindroos TJ, Kiviluoma J (2023) Impact of climate and geological storage potential on feasibility of hydrogen fuels. *Applied Energy* 342:121093. doi: 10.1016/j.apenergy.2023.121093.

[120] Mukelabai MD, Wijayantha UKG, Blanchard RE (2022) Renewable hydrogen economy outlook in Africa. *Renewable and Sustainable Energy Reviews* 167:112705. doi: 10.1016/j.rser.2022.112705.

[121] Xing S, Zhao C, Ban S, et al. (2021) A hybrid fuel cell system integrated with methanol steam reformer and methanation reactor. *International Journal of Hydrogen Energy* 46:2565–2576. doi: 10.1016/j. ijhydene.2020.10.107.

[122] Belmonte N, Girgenti V, Florian P, et al. (2016) A comparison of energy storage from renewable sources through batteries and fuel cells: A case study in Turin, Italy. *International Journal of Hydrogen Energy* 41:21427–21438. doi: 10.1016/j.ijhydene.2016.07.260.

[123] Rullo P, Braccia L, Luppi P, et al. (2019) Integration of sizing and energy management based on economic predictive control for standalone hybrid renewable energy systems. *Renewable Energy* 140:436–451. doi: 10.1016/j.renene.2019.03.074.

[124] Tukenmez N, Yilmaz F, Ozturk M (2020) A thermal performance evaluation of a new integrated gas turbine-based multigeneration plant with hydrogen and ammonia production. *International Journal of Hydrogen Energy.* doi: 10.1016/j.ijhydene.2020.11.054.

[125] Lin L, Tian Y, Su W, et al. (2020) Techno-economic analysis and comprehensive optimization of anonsitehydrogen refuelling station system using ammonia: Hybrid hydrogen purification with both high H_2 purity and high recovery. *Sustainable Energy & Fuels* 4:3006–3017. doi: 10.1039/c9se01111k.

[126] Bødal EF, Mallapragada D, Botterud A, et al. (2020) Decarbonization synergies from joint planning of electricity and hydrogen production: A Texas case study. *International Journal of Hydrogen Energy* 45:32899–32915. doi: 10.1016/j.ijhydene.2020.09.127.

[127] Sánchez A, Zhang Q, Martín M, et al. (2022) Towards a new renewable power system using energy storage: An economic and social analysis. *Energy Conversion and Management* 252:115056. doi: 10.1016/j. enconman.2021.115056.

[128] Demirhan CD, Tso WW, Powell JB, et al. (2020) A multiscale energy systems engineering approach for renewable power generation and storage optimization. *Industrial & Engineering Chemistry Research* 59:7706–7721. doi: 10.1021/acs.iecr.0c00436.

[129] Palys MJ, Mitrai I, Daoutidis P (2021) Renewable hydrogen and ammonia for combined heat and power systems in remote locations: Optimal design and scheduling. *Optimal Control Applications and Methods* 44:719–738. doi: 10.1002/oca.2793.

[130] Tso WW, Demirhan CD, Lee S, et al. (2019) Energy carrier supply chain optimization: A Texas case study. In *Computer-aided chemical engineering*, pp. 1–6.

[131] Sánchez A, Martín M, Zhang Q (2021) Optimal design of sustainable power-to-fuels supply chains for seasonal energy storage. *Energy* 234:121300. doi: 10.1016/j.energy.2021.121300.

[132] Han S, Kim J (2019) A multi-period MILP model for the investment and design planning of a national-level complex renewable energy supply system. *Renewable Energy* 141:736–750. doi: 10.1016/j.renene.2019.04.017.

[133] Karavas C-S, Arvanitis K, Papadakis G (2017) A game theory approach to multi-agent decentralized energy management of autonomous polygeneration microgrids. *Energies* 10:1756. doi: 10.3390/en10111756.

[134] Khan A, Javaid N (2019) Optimum sizing of PV-WT-FC-DG hybrid energy system using teaching learning-based optimization. In *2019 International Conference on Frontiers of Information Technology (FIT)*, pp. 127–1275. IEEE. doi: 10.1109/fit47737.2019.00033.

[135] Maleki A, Hafeznia H, Rosen MA, et al. (2017) Optimization of a grid-connected hybrid solar-wind-hydrogen CHP system for residential applications by efficient metaheuristic approaches. *Applied Thermal Engineering* 123:1263–1277. doi: 10.1016/j.applthermaleng.2017.05.100.

[136] Gillessen B, Heinrichs HU, Stenzel P, et al. (2017) Hybridization strategies of power-to-gas systems and battery storage using renewable energy. *International Journal of Hydrogen Energy* 42:13554–13567. doi: 10.1016/j.ijhydene.2017.03.163.

[137] Xu X, Hu W, Cao D, et al. (2021) Enhanced design of an offgrid PV-battery-methanation hybrid energy system for power/gas supply. *Renewable Energy* 167:440–456. doi: 10.1016/j.renene.2020.11.101.

[138] Hayati MM, Safari A, Nazari-Heris M, et al. (2024) Hydrogen-incorporated sector coupled smart grids: A systematic review & future concepts. In *Green hydrogen in power systems*. doi: 10.1007/978-3-031-52429-5_2.

[139] Skorek-Osikowska A, Bartela Ł, Katla D, et al. (2021) Thermodynamic assessment of the novel concept of the energy storage system using compressed carbon dioxide, methanation and hydrogen generator. *Fuel* 304:120764. doi: 10.1016/j.fuel.2021.120764.

[140] Wang X, Huang W, Wei W, et al. (2022) Day-ahead optimal economic dispatching of integrated port energy systems considering hydrogen. *IEEE Transactions on Industry Applications* 58:2619–2629. doi: 10.1109/tia.2021.3095830.

[141] Goldmeer J (2019) *Power to gas: Hydrogen for power generation*. GE Power, General Electric Company.

[142] Mekhilef S, Saidur R, Safari A (2011) Comparative study of different fuel cell technologies. *Renewable and Sustainable Energy Reviews* 16(1):981–989. doi: 10.1016/j.rser.2011.09.020.

[143] Fan L, Tu Z, Chan SH (2021) Recent development of hydrogen and fuel cell technologies: A review. *Energy Reports* 7:8421–8446. doi: 10.1016/j.egyr.2021.08.003.

[144] Sharaf OZ, Orhan MF (2014) An overview of fuel cell technology: Fundamentals and applications. *Renewable and Sustainable Energy Reviews* 32:810–853. doi: 10.1016/j.rser.2014.01.012.

[145] Wang Y, Seo B, Wang B, et al. (2020) Fundamentals, materials, and machine learning of polymer electrolyte membrane fuel cell technology. *Energy and AI* 1:100014. doi: 10.1016/j.egyai.2020.100014.

[146] Samsun R, Rex M, Antoni L, et al. (2022) Deployment of fuel cell vehicles and hydrogen refueling station infrastructure: A global overview and perspectives. *Energies* 15(14):4975. doi: 10.3390/en15144975.

[147] Abdelghany MB, Shehzad MF, Mariani V, et al. (2022) Two-stage model predictive control for a hydrogen-based storage system paired to a wind farm towards green hydrogen production for fuel cell electric vehicles. *International Journal of Hydrogen Energy* 47(75):32202–32222. doi: 10.1016/j.ijhydene.2022.07.136.

2 Defining "X," Solving Global Problems by Local Solutions

Abdullah A. Almehizia

2.1 INTRODUCTION

This chapter analyzes the relationship between renewable energy input and its deriving process (X). As power-to-X technologies provide a flexibility advantage, optimizing process selection and operation is crucial to achieving higher efficiencies technically and economically. Typically, power-to-X technologies are implemented locally within a region, necessitating an evaluation of implementation options based on geographic location and available resources. The typical challenges of renewable energy integration with the grid and the opportunity of power-to-X in alleviating such challenges are presented.

Furthermore, describing the demand-side management (DSM) as the core part of the power-to-X process and the various applications of DSM in conjunction with power-to-X. Energy storage systems are also introduced, including typical and emerging technologies and their integration with power-to-X.

This leads to the localized power-to-X and its potential solutions to the identified challenges with renewable integration. Two main cases are introduced: water desalination plants and industrial facilities.

This chapter also analyzes related case studies in-depth and summarizes global initiatives, demonstrations, and commercial projects promoting power-to-X. Finally, the challenges associated with the broad implementation of power-to-X technology are discussed from various perspectives.

2.2 RENEWABLE ENERGY INTEGRATION: CHALLENGES AND OPPORTUNITIES

2.2.1 THE GLOBAL CHALLENGES OF GRID INTEGRATION WITH RENEWABLE ENERGY SOURCES AND THE OPPORTUNITY OF POWER-TO-X

2.2.1.1 Variability and Intermittency

Renewable energy is being widely used for power generation globally. The traditional way of generating power has raised environmental concerns within society. Renewable energy systems, including solar, hydro, wind, bioenergy, and geothermal, have unique operating parameters and energy conversion efficiency, making them site- and condition-dependent [1].

With all the advantages that renewable energy sources provide, they still suffer from unpredictability, which can affect the reliability of the power supply—imposing a challenge for high renewable sources penetration into the electricity grid. The variability and uncertainty of solar radiation and wind speed pose severe challenges for grid operators. *Variability* is defined as the continuous fluctuation of power generation based on the availability of primary fuel sources (solar radiation, wind). *Uncertainty* is related to the magnitude and timing of the renewable generation output being less predictable than conventional power generation systems [2]. These issues must be dealt with to maintain system balance. Thus, the system operators must ensure sufficient resources are available to accommodate significant ramps up or down in the wind and solar power generation to maintain system balance.

The mismatch between renewable generation and load demand can be characterized as producing more power during low load demand, which would require curtailment or storage elements. On

DOI: 10.1201/9781032719436-2

the other hand, the opposite could happen, experiencing low renewable power resources when high demand is required. In this situation, storage elements would need to be brought up, and conventional generators would need to ramp up. They could go beyond their maximum ramp capability, causing severe damage [3].

This issue is more attributed to wind power generation, as solar power coincides with the load [3]. Also, since long-term accurate wind speed forecasts are not yet possible, power production from wind turbines can suffer from low or extremely high wind speeds, which is problematic for power dispatching.

2.2.1.2 Grid Stability and Frequency Control

Another aspect of renewable energy integration is the grid stability and frequency control challenge. Traditional power plants provide inertia to the grid, which helps maintain frequency stability. Renewable sources, especially wind and solar, typically do not provide inertia, leading to challenges in maintaining the stability of the grid. The disturbance in the grid can lead to faster frequency changes, increasing the risk of instability and blackouts. With less inertia in the system, the grid becomes more vulnerable to slight imbalances [4].

2.2.1.3 Grid Infrastructure Modernization

The existing grid infrastructure is typically designed for large-scale centralized power generation, whereas renewable energy sources are often decentralized and distributed. The integration of renewable energy would require significant modernization of the grid and handling of bidirectional power flows. The transformation of the electrical grid would involve capacity expansion and extending the reach of the grid for dispersed renewables, such as offshore wind resources. In addition, intelligent monitoring and control systems are needed for the reliable operation of the grid [5].

One of the leading technologies proposed to handle these issues is power-to-X, which refers to converting surplus electricity from renewable energy resources into energy carriers or products that can be easily stored. Also, power-to-X technologies can help manage grid stability by acting as flexible loads that can be ramped up or down, depending on the grid's needs [6]. In addition, power-to-X technologies can alleviate the pressure on electrical grid infrastructure by converting electricity into other forms of energy that do not rely on the electricity grid. For instance, converting electricity to hydrogen or synthetic gas reduces the need for extensive transmission infrastructure. Thus, by diversifying the types of energy carriers, power-to-X reduces the demand on the electrical grid and can lead to more efficient use of existing infrastructure, delaying or reducing the need for costly grid upgrades [7].

From the definition of power-to-X, renewable energy is a core component of the technology. The other component is the X component, which includes the energy carrier or product produced through relevant conversion processes [8]. The structure diagram is shown in Figure 2.1.

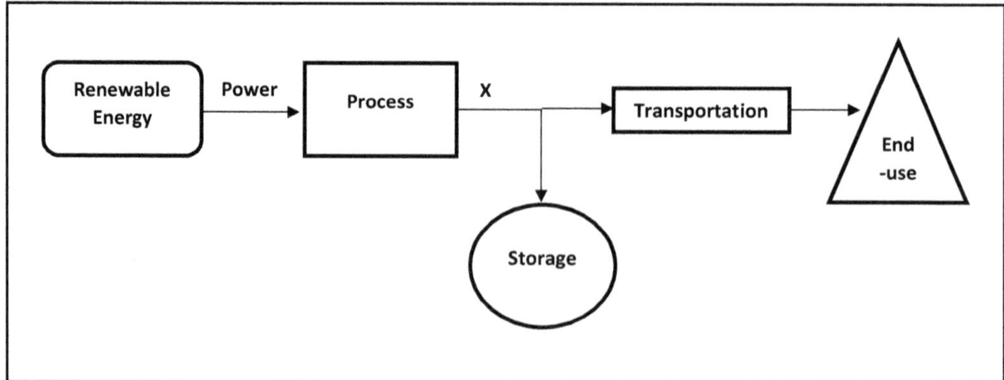

FIGURE 2.1 Basic components of the power-to-X process.

2.2.2 Leveraging Renewables for Efficient Power-to-X Processes

Power-to-X has the potential to accelerate the energy transition toward a sustainable energy future. It is advancing the coupling with other energy sectors, including transportation, industry, and buildings, through various energy domains, such as heating/cooling and gasification. The technology could increase the flexibility of the power system as it acts as a buffer, offers storage for surplus energy from renewable resources, and reduces carbon by displacing fossil fuel energy sources in other sectors [9].

Globally, the renewable energy capacity reported in 2024 is around 4,704 GW, dominated by solar PV at 1,949.4 GW, followed by hydropower at 1,432 GW, and wind turbine technology at 1,125 GW, constituting approximately 96% of total capacity [10].

Different countries have different renewable energy mix targets, depending on the availability of the source and applicability of the technology. Some countries have abundant solar radiation, others have more wind potential, and some are blessed with water resources for hydropower.

The characteristics of renewable energy resources differ in their availability and intermittency. Solar energy is typically more predictable than energy from wind turbines, as the wind is less predictable than solar radiation. Solar power coincides more with the load. Therefore, achieving high power-to-X efficiencies requires careful consideration when selecting renewable energy resources and identifying the sector to be coupled with. Geographic locations have different potential for solar radiation, wind speed, water resources, and various sectors. It is thus crucial for the countries adopting the power-to-X concept to investigate their potential and available resources. For example, geothermal technology has the potential to provide several other benefits in addition to steam driving the turbines, including direct use of heat for different purposes and other extracted valuable gases, which would increase the overall efficiency of the system [11].

As for the coupled sector, the surplus energy from renewable resources powering the processes is typically intermittent and cannot be controlled as a conventional thermal generator. Thus, the process to be coupled must have operational flexibility and can withstand fluctuations in the energy supply.

One of the leading power-to-X applications is the production of hydrogen from electricity generated by renewable energy, mostly from solar PV and wind turbine systems, through water electrolysis, referred to as green hydrogen. Water electrolysis splits the water atoms into hydrogen and oxygen, with electricity as the prime energy source. Electrolysis cells comprise an anode and a cathode separated by an electrolyte [13]. This electrochemical process involves reduction–oxidation (redox) chemical reactions.

The popularity of hydrogen comes from the various pathways it enables. Hydrogen can be used directly as a fuel or can be transformed into other products, such as ammonia, methane, and liquid fuels.

TABLE 2.1

Characteristics of Renewable Energy Technologies

Renewable Energy Technology	Resource Availability	Efficiency	Affordability	Scalability
Solar photovoltaic	Abundant	15–22%	₵4.9/kWh	Very high
Concentrated solar power	Abundant	15–30%	₵11.8/kWh	Medium
Wind turbines	Coastal and elevated areas	25–45%	₵3.3/kWh	Very high
Hydro	Requires a reliable water source	90%	₵6.1/kWh	Very high
Geothermal	Location-specific	10–20%	₵5.6/kWh	Low
Biomass	Depends on regional agricultural and forestry activities	25–40%	₵6.1/kWh	Medium

Source: Data from [12].

2.3 DEMAND-SIDE MANAGEMENT (DSM) AND ITS SYNERGY WITH POWER-TO-X

2.3.1 IMPORTANCE OF DSM IN ENHANCING POWER-TO-X EFFICIENCY

DSM is regarded as a function of the smart grid (SG) framework, and it is defined as a set of measures that can be carried out to improve the energy system on the load side. These measures include demand response (DR), energy efficiency installation, and distributed generator control (DGs). DSM intends to manipulate the electricity demand from the consumers, resulting in a smoother load profile [14]. Implementation of DSM is through economic incentives that encourage electricity consumers to alter their energy consumption [15].

The techniques that can be used to realize DSM are shown in Figure 2.2. *Load shifting* is the process of shifting time-dependent loads from peak hours to off-peak hours. Some applications of this technique include storage space heating and water heating, which can be used instead of the conventional electric water heater. It is the most effective load management technique commonly used in distribution networks. *Valley filling* aims to reduce the difference between the peak and valley load demands, resulting in a smoother load curve and, thus, fewer peak units needed. *Peak clipping*, as the name suggests, is concerned with reducing the peak demand by directly controlling the load demand, which can lower the operating cost of the generation units. *Strategic conservation* is focused on optimizing the load profile by implementing demand reduction methods at the consumer's end. *Strategic load growth* deals with more considerable demand than is usually handled by valley filling by optimizing the daily response. It can involve an increased market share of loads that non-conventional energy sources and energy storage can serve, as well as balancing the growing load demand with the construction of the infrastructure needed to serve the load. The *flexible load shape* concept considers the load shape forecast, incorporating all DSM techniques. Then,

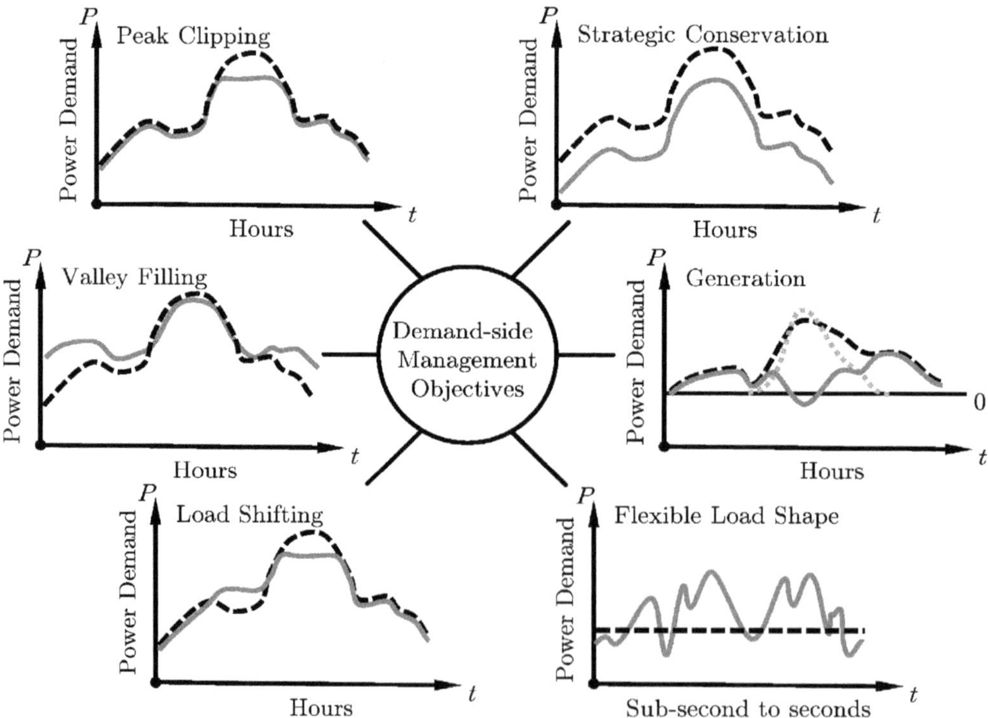

FIGURE 2.2 Load shape management.

Source: [12].

consumers with flexible loads willing to let the utilities control their load during high-demand periods are identified. In turn, they would receive compensation [14, 16].

A balance between generation and demand has to be met at all times. The introduction of intermittent renewable energy sources can jeopardize this balance. Implementing power-to-X applying DSM techniques could potentially be the answer for stretching the grid capabilities, as the load can be utilized to provide an additional degree of freedom for system operators and, as a result, provides a less-expensive option to supply the load rather than build new power generation systems or expand the transmission and distribution networks, yielding a higher efficiency (utilization) of the renewable energy resources.

2.3.2 Strategies for Implementing DSM in Conjunction with Power-to-X

Demand response is the DSM technique for implementing power-to-X and achieving load shape management, as shown in Figure 2.2. Demand response requires *controllable loads*, defined as loads whose demand can be deferred to a later time interval, such as hydrogen electrolyzers, given that a total amount of energy must be supplied by the end of a predefined period. Thus, power balance constraint is no longer an issue, as the operator is not bonded to specific power demands while adhering to technical constraints, such as minimum and maximum power input. This demand flexibility would be beneficial to accommodate substantial penetration of renewable energy sources, as it virtually eliminates the variability and intermittency issues related to renewable sources, adhering to technical constraints of the deferred load. The definition of *controllable load* is also stretched to account for power demand, which can be stored [17, 18].

2.4 ENERGY STORAGE SOLUTIONS FOR POWER-TO-X APPLICATIONS

2.4.1 Overview of Energy Storage Technologies and Power-to-X Processes

Many energy storage technologies and processes have been investigated in the literature to implement power-to-X. Energy storage systems can be seen in Figure 2.3. With power-to-X, the process (X) determines the storage type.

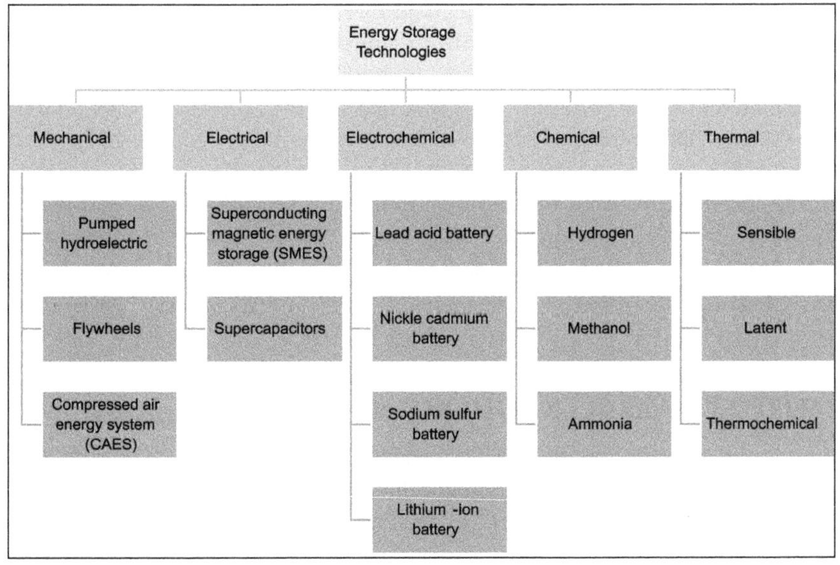

FIGURE 2.3   Types of energy storage technologies.
Source: [19, 20].

2.4.1.1 Power-to-Hydrogen with a Hydrogen Tank

Electrolyzers utilize electricity to split water into hydrogen and oxygen. Hydrogen, a versatile and clean energy carrier, can be a fuel for power generation or an industrial feedstock. It can be stored and transported in liquid or gas form and used in combustion or fuel cells to produce heat and electricity. Consequently, hydrogen can be crucial for the seasonal storage of renewable electricity and for decarbonizing various sectors, including mobility, industry, and the gas grid. As an energy carrier, hydrogen offers multiple pathways to its final applications [8].

Three leading technologies are currently available for electrolyzers: alkaline electrolyzer (AEC); proton electrolyte membrane electrolyzer (PEMEC), shortened as PEM; and solid oxide electrolyzer (SOEC), with AEC being the most mature technology. AEC and PEMEC typically operate at 50–80°C, and efficiencies reach 85%. AEC has the lowest capital cost and operates up to 90,000 hr. However, AEC suffers from limited dynamic flexibility and a minimum operation of 10–40% of total capacity, with start-up time ranging up to several hours. PEMEC has a higher capital cost than AEC and achieves better operational flexibility, a lower operating range of 0–10%, response times of a few milliseconds, and start-up times of less than 20 min, making it suitable for incorporating renewable energy. PEMEC can operate at a pressure of 200 bars. Thus, it is ideal for gaseous hydrogen storage. The operating lifetime is in the range of 20,000 to 60,000 hr. SOEC is still in the early stages of development and is limited to demonstration projects. It operates at temperatures ranging from 650 to 1,000°C, with lifetimes of less than 10,000 hr, with a lower operating bound of 30% capacity and a start-up time of up to an hour [13].

Hydrogen can be stored as compressed gas, cryogenic liquid, or metal hydrides, with compression considered the most developed technology for hydrogen storage. Cryogenic liquid storage involves cooling to −253°C, with low-pressure compression approximately at 4 bar, allowing for an increased energy density compared to compressed gas while avoiding safety concerns associated with high pressure.

Hydrogen has many pathways producing value-added chemicals. Two main approaches are considered: CO_2 hydrogenation and CO_2 electrochemical reduction. The chemicals, liquids, and gases produced include formic acid, methanol, methane, and alkane, as shown in Figure 2.4 [21].

2.4.1.2 Power to Heat with Thermal Storage

Power-to-heat systems use boilers, heat pumps, and thermal storage to produce heat from surplus renewable electricity. Centralized or decentralized approaches can be used to build power-to-heat systems. Regarding central heating systems, heat is produced by large-scale electric boilers or heat pumps and distributed to many buildings via a network of pipes. *District heating system* is another name for these systems. Decentralized systems use electric boilers or heat pumps to heat individual buildings or homes [8].

Three leading technologies used for thermal storage are sensible, latent, and thermochemical. *Sensible* heat storage involves storing energy in a temperature difference in solids, such as sand, or liquids, such as molten salt. *Latent* heat storage stores the energy through phase-change materials, such as salts, metals, and organics. *Thermochemical* heat storage stores the energy in chemical bounds [22].

2.5 LOCALIZED POWER-TO-X SOLUTIONS

Most countries are attracted to the green hydrogen approach as an implementation of power-to-X. However, depending on the geographic location, other approaches could be beneficial. This section defines and models localized power-to-X [23, 24]. One of the significant aspects of adopting the concept of power-to-X is to select the most appropriate (X) technology, which can be summarized as follows.

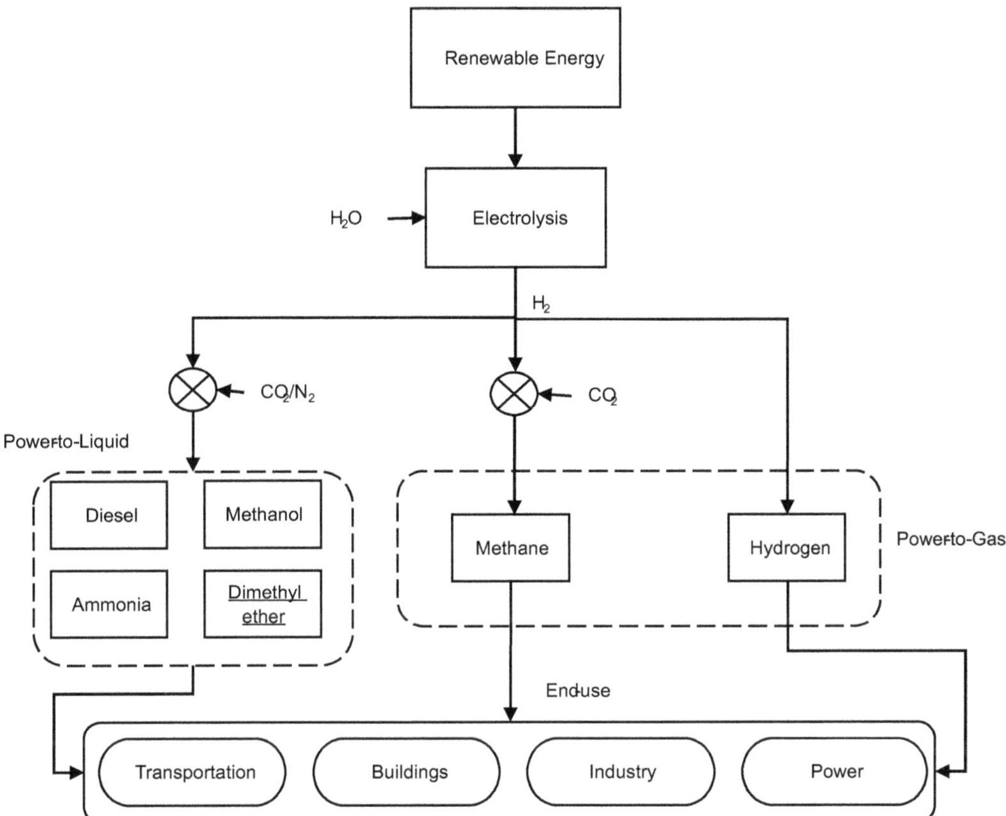

FIGURE 2.4 Hydrogen pathways.

2.5.1 Features of a Localized Solution

- **Nature of the load demand.** Every region is unique in terms of its electrical load demand. Some regions have high industrial facilities which can benefit from implementing power-to-X. The desalination industry in Saudi Arabia is a clear example of this.
- **Renewable energy resource type and availability.** It involves identifying available resources in the selected region, as different regions have different resource availability. For example, coastal areas are known to have high wind energy potentials.
- **Ability to preserve natural resources.** Countries with valuable fossil resources can preserve their resources by electrifying potential sectors.
- **Cost-effectiveness.** Investing in low-cost renewable energy, engaging in contracts such as power purchase agreements with renewable energy providers, optimizing plant size, and establishing partnerships and collaborations with various stakeholders can make the project affordable and competitive.
- **Maturity of the technology.** Careful selection of the process (X) is a significant factor for the project's success. For example, technological readiness for electrolyzers differs between AEC, PEM, and SOEC, and the supply chain of these technologies would impact the project's feasibility and profitability.
- **Applicability and scalability.** In some cases, the feasibility of full-scale implementation of power-to-X can be limited. Small-scale pilot implementations can reduce some of the technological and financial burdens.

- **Infrastructure readiness.** Assessment of the existing energy infrastructure available. If natural gas pipelines or energy grids exist, integrating power-to-X products (e.g., hydrogen, methane) into these networks reduces distribution challenges and costs. Areas with natural gas infrastructure could produce synthetic methane for injection into gas pipelines.
- **Market demand for the produced products.** Depending on the location, the needed product and available markets would differ. Agricultural regions can benefit from ammonia production as it can be used for fertilizers or as a carbon-free fuel. Regions with strong transportation or aviation sectors may focus on producing synthetic fuels. Securing offtake agreements with industries that commit to purchasing power-to-X products solidifies the project's business case.
- **Availability of policy and regulations.** Established regulations, subsidies, and incentives for the selected process (X) can boost and expedite the implementation of the power-to-X projects.

Localized power-to-X involves the selection of the renewable resource (power) and the process (X) that would exploit the previous points to the limits. An example of a localized power-to-X is a system of renewable solar energy (CSP/PV) and desalination plants. The example can be related to regions where most potable water comes from high-electricity-demand desalination plants and have enormous solar energy resources, such as in the Middle East.

Also, for regions having industries with heavy electricity demand, such as the aluminum industry in Iceland, some of the industries could be a part of a power-to-X system where renewable energy resources can supply the factory, and the products of the facility can be utilized as a means of storage rather than using the typical electrochemical storage system.

2.5.2 Water Desalination Plant with Water Storage (Power-to-Water)

There are 150 countries where water desalination technology is used, and more than 300 million people worldwide depend on desalinated water for their daily needs [25]. Power-to-water applies to countries that heavily rely on desalinated potable water. Water desalination technology can be categorized into three fundamental processes: thermal, phase change, and membrane. The *thermal* process mimics the natural water cycle of evaporation and condensation, in which feedwater is heated to boiling temperature. The pollutants in the water, including salt and minerals, remain in the base water, as they are heavy elements. The steam is then cooled and condensed, producing water with low salinity levels. *Membrane* or *single-phase* processes separate salts without the need for phase transition, and they have the advantage of lower energy requirements than thermal processes. The thermal process plant includes multi-stage flash (MSF) and multiple-effect distillation (MED), and the reverse osmosis (RO) plant is an example of a membrane process plant [26, 27].

The specific energy consumption per unit volume models the desalination plants as electrical power demand. One unit volume of desalinated water would require a certain amount of electrical energy. This electrical energy is the electrical demand used to construct the load profile for the desalination plant. Desalination plants are usually built with an on-site storage tank. This tank can be utilized to store water when the production of renewable sources is higher than the demand, and the water stored can then be used when renewable sources cannot provide enough power. Therefore, the storage tank acts as a buffer, which constitutes the concept of value storage, where instead of storing electrical energy, the value of that energy is stored in a different form (water in this case).

$$E_{desal,e} = \kappa E_{desal,w} \tag{2.1}$$

$$P_{desal,e} = \kappa P_{desal,w} \tag{2.2}$$

Where $E_{desal,w}$ is the volume of desalinated water (m³), $E_{desal,e}$ is the electrical energy required to produce $E_{desal,w}$ (MWh). $P_{desal,w}$ is the water demand (in m³/h), and $P_{desal,e}$ is the corresponding power demand (in MW). This transformation is done by the specific energy requirement κ, represented in kWh/m³.

$$P_{re}(t) + \kappa W_{tk}(t) = P_{desal,e}(t) + P_{curt}(t) \tag{2.3}$$

Equation (2.3) represents the power balance constraint, where $P_{re}(t)$ is the forecasted hourly power supplied from the renewables, $W_{tk}(t)$ is the hourly water supplied from the storage tank and multiplied by κ to translate into electrical terms, and $P_{dump}(t)$ is the hourly power curtailed when the storage tank reaches its maximum capacity. Thus, the hourly demand is decoupled from the available renewable power as they do not need to be equal, since the water storage tank will supply or store the difference. The reader is advised to read [23, 24] for more details.

2.5.3 Industrial Facilities with a Storage Warehouse (Power-to-Product)

Industrial factories are also considered a load that can utilize the value storage concept. Different factories have various processes, and the industrial facilities of interest produce materialized products, such as steel or aluminum plants, petrochemical plants, and many others. These facilities usually include a storage warehouse within the plant's premises to store their products. This storage facility can be used to realize the concept of value storage. The assumption made with industrial facilities is that they have a range of minimum and maximum power demand, which translates to the highest and lowest product production. A distinction is made between water desalination plants and industrial facilities, in which the electricity demand is provided as an energy demand of a longer time interval than an hour. In other words, the system operator can construct the appropriate load profile under the condition that a certain amount of electrical energy will be supplied within a predefined period and power supply range. With this approach, the hinges of renewable energy sources' unpredictability and intermittency can be alleviated, as the system operator is no longer required to supply a specific demand every hour. Instead, the system operator will serve the industrial facility demand based on the availability of resources (i.e., solar radiation and wind speed). Thus, this type of load is referred to here as a generation following load.

$$E_{fact,\,e} = \delta E_{fact,\,ton} \tag{2.4}$$

Here, $E_{fact,ton}$ (in tons) is the number of products (in tons) representing the demand of the factory. The electrical counterpart is the electrical energy needed to produce $E_{fact,ton}$, which is $E_{fact,e}$ (given in MWh). This conversion is done similarly to the desalination plant by utilizing a specific energy factor, δ (in kWh/ton). The total energy required for a year, $E_{fact,e}^{year}$, is uniformly divided within 12 months, $E_{fact,e}^{month}$. As such, the system operator must serve the same amount of energy each month, considering that different months of the year have different days. $\mathcal{H}$ is the set number of hours in each month, and $P_{fact,e}$, in MW, is the factory electrical power demand.

$$E_{fact,e}^{month} = \frac{E_{fact,e}^{year}}{12} \tag{2.5}$$

$$E_{fact,e}^{month} = \sum_{h=1}^{H \in H} P_{fact,\,e}(h) = E_{fact,e}^{month} \tag{2.6}$$

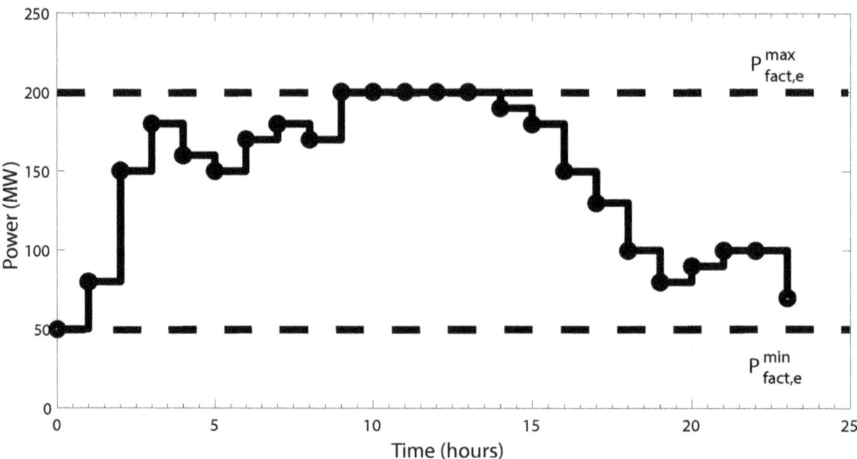

FIGURE 2.5 Daily tracking model for factory load demand.

The scheduling process to construct the load profile ($\boldsymbol{P}_{fact,e}^{day}$) for the industrial load is formulated as a least squares optimization problem following the approach in [23].

$$minimize\ \frac{1}{2}\left\|\left(\boldsymbol{P}_{fact,e}^{day} - \boldsymbol{P}_{re}^{day}\right)\right\|_2^2 \tag{2.7}$$

$$subject\ to\ \boldsymbol{P}_{fact,e}^{min} \leq \boldsymbol{P}_{fact,e}^{day} \leq \boldsymbol{P}_{fact,e}^{max} \tag{2.8}$$

$$-\boldsymbol{P}_{fact,e}^{dn} \leq \Delta \boldsymbol{P}_{fact,e}^{day} \leq \boldsymbol{P}_{fact,e}^{up} \tag{2.9}$$

$$\sum \boldsymbol{P}_{fact,e}^{day} = E_{fact,e}^{day} \tag{2.10}$$

Where $\boldsymbol{P}_{re}^{day}$ and $\boldsymbol{P}_{fact,e}^{day}$ are the daily vectors of forecasted renewable power output and factory power demand, respectively (MW). $\boldsymbol{P}_{fact,e}^{day}$ is shown in Figure 2.5. Note that this figure is only an illustration, and each day is expected to be different, depending on the availability of renewable energy on that particular day. $\boldsymbol{P}_{fact,e}^{min}$ and $\boldsymbol{P}_{fact,e}^{max}$ are the minimum and maximum factory power limits (MW) vectors. $\boldsymbol{P}_{fact,e}^{dn}$ and $\boldsymbol{P}_{fact,e}^{up}$ are the factory ramp-down and ramp-up rates (MW/h). The annual vector for the factory's power demand is given as $\boldsymbol{P}_{fact,e}$:

$$\boldsymbol{P}_{fact,e} = \left[\boldsymbol{P}_{fact,e}^{(1)} \cdots \boldsymbol{P}_{fact,e}^{(365)}\right] \tag{2.11}$$

where $\boldsymbol{P}_{fact,e}^{(1)}$ and $\boldsymbol{P}_{fact,e}^{(365)}$ represent the power demand vector for the first and last day of the year, comprising 24-hourly data points.

The primary objective is to construct a load profile for the factory, which ideally follows the renewable power output. The constraints imposed are to account for the process of the industrial plant. Constraint (8) is related to the range of power supply that the plant can withstand to stay online. Constraint (9) is to account for the process that can only tolerate a specific number of ramps each hour. Constraint (10) ensures that the daily energy required is satisfied. The reader is advised to read [23, 24] for more details.

2.6 INTERNATIONAL AND NATIONAL POWER-TO-X INITIATIVES

Different technologies to realize power-to-X have been investigated, with most of the projects and initiatives focused on power-to-hydrogen, in what is referred to as green hydrogen.

TABLE 2.2
Electrolyzer Capacity Targets

Country	Electrolyzer Capacity Target by 2030
France	6.5 GW
Netherlands	3–4 GW
Germany	5 GW
Portugal	2–2.5 GW
Spain	4 GW

2.6.1 National Strategies

Many European countries have developed national strategies related to hydrogen production. The European Union's hydrogen strategy aims to increase electrolyzer capacity to reach 40 GW by 2030 for the production of green hydrogen. About 50% of this capacity would be from the frontier countries, including France, Germany, Spain, Portugal, and the Netherlands [28].

2.6.2 International Power-to-X Hub

The International Power-to-X Hub is a center of expertise, collaboration, and innovation in power-to-X. Implemented by the Deutsche Gesellschaft für Internationale Zusammenarbeit (GIZ) GmbH and financed by the International Climate Initiative and regarded as a contribution to the German Hydrogen Strategy of 2020, the hub delivers policy and regulation advice to stakeholders, developing power-to-X solutions and accelerating sustainable aviation and shipping [29].

2.6.3 Mission Innovation

Mission Innovation is a global initiative aimed at stimulating action and investment in the research, development, and demonstration of clean energy solutions. It comprises seven missions investigating various pathways toward net-zero emissions. One of the prominent missions is the clean hydrogen mission, which involves countries such as Australia, Chile, the United Kingdom, the United States of America, and the European Union. The mission's primary goal is to reach a 2 USD/kg production cost for hydrogen through research, development, and pilot demonstration projects [30].

2.6.4 Clean Hydrogen Initiative

The World Economic Forum (WEF) leads the Accelerating Clean Hydrogen Initiative, and Accenture is a knowledge partner. The initiative focuses on developing roadmaps for the acceleration of clean hydrogen and realizing the role of clean hydrogen in the energy transition journey [31].

2.6.5 Kopernikus Project

The German Federal Ministry of Education and Research (BMBF) is funding the Kopernikus Project, a research initiative aiming to advance the energy transition in Germany. The initiative comprises four sub-projects: the ENSURE, P2X, SynErgie, and the Ariadne projects. The P2X sub-project focuses on the research, development, and demonstration of power-to-X technologies from the fundamentals up to an industrial scale [32].

2.6.6 NEOM GREEN HYDROGEN

A joint venture between NEOM, ACWA Power, and Air Products formed the NEOM Green Hydrogen Company (NGHC). Aiming to build the world's largest green hydrogen plant to produce green ammonia by 2026 in the Kingdom of Saudi Arabia, the plant is expected to be powered by 4 GW of solar energy capable of producing 600 t/day of carbon-free hydrogen by the year 2026 in the form of green ammonia [33].

2.7 SUCCESS STORIES OF POWER-TO-X IMPLEMENTATION

In this section, a selection of power-to-X projects is identified and reported. The readers are advised to check [34–36] for more comprehensive details about global projects.

TABLE 2.3
Power-to-X Notable Projects

Country	**Germany**
Project	Refhyne [37]
Renewable source	Wind energy
Technology type	Power-to-hydrogen
Electrolyzer capacity and technology	10 MW PEM (one of the world's largest hydrogen electrolyzers)
Hydrogen production capacity	1,300 t/year
Usage	In refineries, for desulfurization of conventional fuels
Remarks	• Demonstrated the applicability of integration with existing industrial processes without significant infrastructure changes.
Country	**Denmark**
Project	HyBalance [38]
Renewable source	Wind energy
Technology type	Power-to-hydrogen
Electrolyzer capacity and technology	1.2 MW PEM
Hydrogen production capacity	180 t/year
Usage	Transportation and industrial sectors
Remarks	• Demonstrate the complete value chain from hydrogen production to end users, including hydrogen stations and Fuel Cell Electric Vehicles (FCEVs) in circulation. • Capable of supplying a fleet of more than 1000 FCEVs and could contribute up to 0.5% of the transport sector GHG reduction targets in Denmark.
Country	**Netherland**
Project	Nouryon and Gasunie Green Hydrogen Project [39]
Renewable source	Wind energy
Technology type	Power-to-hydrogen
Electrolyzer capacity and technology	20 MW PEM
Hydrogen production capacity	3,000 t/year
Usage	Utilization of hydrogen in chemical plants
Remarks	• Reducing CO_2 emissions by 27,000 t/year.
Country	**United Kingdom**
Project	BIGH2IT [40]
Renewable source	Wind energy
Technology type	Power-to-hydrogen
Electrolyzer capacity and technology	1.5 MW PEM

(Continued)

TABLE 2.3 (Continued)

Hydrogen production capacity	50 t/year
Usage	Heating and fuel cells
Country	**France**
Project	Jupiter 1000 [41]
Renewable source	Wind and solar energy
Technology type	Power-to-gas
Electrolyzer capacity and technology	0.5 MW Alkaline and 0.5 MW PEM
Hydrogen production capacity	Not specified
Usage	Demonstrate the feasibility of developing power-to-gas on an industrial scale; hydrogen and methane injected into the gas network • The first industrial demonstrator of power-to-gas in the country. • The project includes methanation and carbon capture.
Country	**Austria**
Project	H2Future [42]
Renewable source	Hydropower
Technology type	Power-to-hydrogen
Electrolyzer capacity and technology	6 MW PEM
Hydrogen production capacity	945 t/year
Usage	Providing hydrogen to a steel plant.
Notable impact	• Demonstrated the possibility of decarbonizing the steel industry
Country	**Spain**
Project	Iberdrola–Puertollano I [43]
Renewable source	Solar PV
Technology type	Power-to-ammonia
Electrolyzer capacity and technology	20 MW PEM
Hydrogen production capacity	3,000 t/year
Usage	Fertilization
Investment	€150 million
Notable impact	• Prevent emissions of 48,000 t CO_2/year. • Creation of 1,000 jobs.
Country	**Japan**
Project	Fukushima Hydrogen Energy Research Field (FH2R) [44]
Renewable source	Solar energy
Technology type	Power-to-hydrogen
Electrolyzer capacity and technology	10 MW ALK
Hydrogen production capacity	945 t/year
Usage	System balance
Country	**China**
Project	Shell China–Zhangjiakou [45]
Renewable source	Onshore wind
Technology type	Power-to-hydrogen
Electrolyzer capacity and technology	20 MW ALK
Hydrogen production capacity	3,400 t/year
Usage	Public and commercial transportation

Notes: ALK: alkaline electrolysis; PEM: proton-exchange membrane electrolysis; SOEC: solid oxide electrolysis cells.

2.8 CASE STUDIES

The following case studies demonstrate detailed local power-to-X projects that demonstrate the use of local resources and industry potential—the motivations behind these projects, their design and implementation, and their contributions to global challenges.

Project	H2RES Project [46]
Country	Denmark
Motivations	• Denmark's high wind energy potential often results in excess electricity production, which can destabilize the grid if not managed effectively. • There is a need to reduce carbon emissions in the transportation sector, particularly in maritime applications, which are challenging to electrify.
Rationale of defining X	"X" is defined as hydrogen produced using local wind energy. Hydrogen is chosen for its versatility in storing energy and its potential as a clean fuel for maritime transportation.
Exploiting local resources	• Denmark's extensive wind power capacity is a critical local resource, and the maritime industry is a significant part of the local economy. • Taking advantage of the country's established wind energy infrastructure and its expertise in renewable energy technologies.
Methods of design and implementation	• The H2RES project involves a 2 MW electrolyzer that converts wind-generated electricity into green hydrogen. • The hydrogen is stored and used to fuel ships, helping decarbonize the maritime sector. • The project is a collaboration between Ørsted, a leading energy company in Denmark, and other local stakeholders, with support from the Danish government.
Achievements	• Successfully demonstrated the feasibility of using surplus wind energy for hydrogen production and its application in maritime transport. • Established a model for integrating power-to-X technologies with existing renewable energy infrastructure. • Contributed to reducing carbon emissions in Denmark's maritime sector.
Obstacles	• The high cost of electrolyzer technology and the infrastructure needed for hydrogen storage and distribution. • Regulatory challenges related to hydrogen production and use in maritime applications. • The need to scale up production and infrastructure to meet the growing demand for green hydrogen.
Contribution to solving global problems	• The H2RES project contributes to global decarbonization efforts by providing a solution for sectors like maritime transport that are difficult to electrify. • It offers a model for other coastal regions with abundant wind resources to follow, helping reduce global carbon emissions and advancing the use of green hydrogen.
Project	Western Green Energy Hub [47]
Country	Australia
Motivations	• Australia has vast areas of land with high solar and wind energy potential but faces challenges in transmitting this energy across its large, sparsely populated areas. • The need to diversify the economy and reduce reliance on fossil fuel.
Rationale of defining X	"X" is defined as green hydrogen and ammonia produced using local solar and wind energy. These can be exported as clean energy carriers or used locally in various industrial applications.
Exploiting local resources and industry	• The project capitalizes on Australia's vast solar and wind resources, particularly in the western region. • It aims to transform the region into a global hub for green hydrogen and ammonia production, leveraging Australia's existing export infrastructure and experience in large-scale resource projects.

(Continued)

(Continued)

Methods of design and implementation	• The project is one of the world's largest proposed renewable energy projects, with plans for 50 GW of combined solar and wind capacity. • This renewable energy will power electrolyzers to produce green hydrogen, which will be converted into ammonia for easier storage and export. • The project involves a consortium of global and local companies, with strong support from the Australian government.
Achievements	• Positioned Australia as a potential leader in the global green hydrogen and ammonia markets. • Providing a large-scale solution for storing and exporting renewable energy. • Contribution to job creation and economic diversification in Australia.
Obstacles	• The scale of the project presents significant technical, financial, and regulatory challenges. • Building the necessary infrastructure for hydrogen production, storage, and export requires substantial investment. • Environmental concerns related to land use and water consumption for electrolysis.
Contribution to solving global problems	• The project addresses global energy security and climate change by producing large quantities of green hydrogen and ammonia, which can replace fossil fuels in various applications. • It sets a precedent for other regions with abundant renewable resources to contribute to the global energy transition.
Project	Haru Oni [48]
Country	Chile
Motivations	• Chile's abundant renewable energy resources, particularly in wind-rich Patagonia, present an opportunity to produce clean fuels. • The need to decarbonize the transportation sector locally and for export markets, particularly in hard-to-electrify sectors like aviation.
Rationale of defining X	"X" is synthetic fuels (e-fuels) produced from green hydrogen and captured CO_2, which can be used as a carbon-neutral alternative to traditional fossil fuels.
Exploiting local resources and industry	• The project exploits Patagonia's strong and consistent winds, making it an ideal location for wind-powered hydrogen production. • Chile's experience exporting natural resources is leveraged to create an export market for synthetic fuels.
Methods of design and implementation	• The project involves the construction of a pilot plant that uses wind energy to power electrolyzers, producing green hydrogen. • Hydrogen is combined with CO_2 captured from the air to produce synthetic methanol, which is further processed into e-fuels. • The project is supported by a consortium of international companies, including Siemens Energy and Porsche, with backing from the Chilean government.
Achievements	• Successfully demonstrated the production of e-fuels using local renewable resources. • Positioned Chile as a pioneer in the production of carbon-neutral synthetic fuels. • Contributed to the global effort to decarbonize the transport sector, particularly aviation.
Obstacles	• The high cost of producing e-fuels, particularly at the pilot stage. • The need for significant investment in infrastructure and scaling up production. • Regulatory challenges related to the certification and use of synthetic fuels.
Contribution to solving global problems	• The project contributes to the global reduction of carbon emissions by providing an alternative to fossil fuels in the transportation sector that is difficult to electrify. • Serving as a model for other countries with abundant renewable resources to produce and export synthetic fuels, helping reduce global reliance on fossil fuels.

2.9 CHALLENGES AND OPPORTUNITIES [8, 20]

2.9.1 IDENTIFYING CHALLENGES IN IMPLEMENTING LOCALIZED POWER-TO-X SOLUTIONS

With all the potential power-to-X has, it still suffers from serious challenges, which are classified in the following sections.

2.9.1.1 Technical

From a technical point of view, obstacles manifest themselves in several ways, ranging from discrete elements to their incorporation into more extensive systems. Many of the power-to-X technologies suffer from low efficiency of the processes. Most power-to-X technologies are still in their infancy, and it will take significant advancements in their performance before they can be widely implemented and integrated. Furthermore, it is essential to emphasize that the widespread use of power-to-X technologies only makes sense when a surplus of inexpensive, carbon-free electricity is available—indicating that technologies that harvest renewable energy resources have already been widely used. It is also evident that certain nations lack the renewable capacity to generate such enormous amounts of excess electricity, which restricts the usefulness of power-to-X technology as a temporary fix.

It is still necessary to construct the essential infrastructure for energy conversion, transporting, processing, and storing produced products. Lastly, the use of synthetic fuels made of carbon compounds, such as synthetic methane or methanol, still results in net positive carbon dioxide emissions and, therefore, does not reduce the emissions of the sectors in which these fuels are used unless direct air (carbon) capture technologies are implemented.

2.9.1.2 Economic

The cost of technologies along the power-to-X chain remains high. It is important to note that significant cost savings would require the combination of technological and manufacturing advances and a production scale-up. Also, it remains unclear if the services these technologies may provide are worth the ongoing and substantial investment required for their development and deployment. Furthermore, it is critical to begin creating applications and markets for secondary fuels and by-products as early as possible. Indeed, the first step in the power-to-hydrogen cycle is water electrolysis, which generates a large amount of oxygen but has very few markets or applications for this by-product. Similarly, building proper market arrangements that allow power-to-gas plants to supply a wide range of auxiliary services to electricity or gas networks is critical. It is also worth noting that some power-to-X technologies may face stiff direct competition from fossil fuel–based alternatives, such as steam methane reformers equipped with carbon capture equipment, thereby undermining their business case.

The challenges investors and financiers face in power-to-X projects are linked to the market's early stages. Uncertainty about prices, rules, and standards impacts long-term offtake agreements. The inability of equipment vendors to provide precise delivery timelines and technical assurances, as well as the possibility of technology obsolescence, enlarges these issues. Potential investors have worries about the shipping and logistics of power-to-X products.

2.9.1.3 Regulatory

Several institutional and regulatory concerns are also expected to surface. The lack of standardizations and frameworks covering the whole power-to-X value chain, including production, transportation, and utilization, could hinder the wide adaptation of the technology. Furthermore, the early stages of power-to-X adaptation require significant provision of subsidies and incentives. Tax policies should be designed to favor power-to-X products against fossil fuel–based alternatives. Another dimension of the regulations is the environmental and safety aspect, which entails safety codes and regulations and addresses the environmental impact of power-to-X processes. The sector coupling regulations must also be established, coordinating with various regulatory bodies, such as electricity, gas, and water. On a higher level, international trade policies for the import and export of power-to-X products have to be developed to ensure the quality and sustainability of the products produced.

2.10 CONCLUSIONS

Power-to-X represents an alternative to conventional battery energy storage through an electrochemical process where surplus renewable energy is utilized to produce valuable products, referred to as *value storage*. The unique characteristics and potential of the location are a primary concern for the implementation of power-to-X projects. Careful consideration of selecting renewable energy resources and the process (*X*) is the pinnacle for a successful power-to-X project. Different dimensions exist in implementing power-to-X solutions, including economics, technical, and regulatory aspects. Many pilot and commercial projects are currently operational to produce clean hydrogen that is converted into different valuable products.

REFERENCES

[1] T.-Z. Ang, M. Salem, M. Kamarol, H. S. Das, M. A. Nazari, and N. Prabaharan, "A comprehensive study of renewable energy sources: Classifications, challenges and suggestions," *Energy Strategy Reviews*, vol. 43, p. 100939, 2022.

[2] NERC, "Accommodating high levels of variable generation," 2009.

[3] M. M. Lori Bird and Debra Lew, "Integrating variable renewable energy: Challenges and solutions," 2013.

[4] J. Song, X. Zhou, Z. Zhou, Y. Wang, Y. Wang, and X. Wang, "Review of low inertia in power systems caused by high proportion of renewable energy grid integration," *Energies*, vol. 16, p. 6042, 2023.

[5] M. Krishnan, C. Bradley, H. Tai, T. Devesa, S. Smit, and D. Pacthod, *The Hard Stuff: Navigating the Physical Realities of the Energy Transition*. McKinsey Global Institute, 2024.

[6] V. A. Martinez Lopez, H. Ziar, J. W. Haverkort, M. Zeman, and O. Isabella, "Dynamic operation of water electrolyzers: A review for applications in photovoltaic systems integration," *Renewable and Sustainable Energy Reviews*, vol. 182, p. 113407, 2023.

[7] R. Daiyan, I. MacGill, and R. Amal, "Opportunities and challenges for renewable Power-to-X," *ACS Energy Letters*, vol. 5, pp. 3843–3847, 2020.

[8] C. Doczekal, *Smart Strategies for the Transition in Coal Intensive Regions, Fact Sheet: Power-to-X.* Güssing Energy Technologies, 2019.

[9] IRENA, *Innovation Landscape for a Renewable-Powered Future: Solutions to Integrate Variable Renewables*. Abu Dhabi: International Renewable Energy Agency, 2019.

[10] IEA, "Renewable Energy Progress Tracker," 2024. www.iea.org/data-and-statistics/data-tools/renewable-energy-progress-tracker.

[11] J. H. Ingmar Budach and M. Wunsch, *A Geothermal Approach to Power-to-X*. Berlin: International PtX Hub, 2023.

[12] IRENA, "Renewable power generation costs in 2022," Abu Dhabi, 2023.

[13] M. J. Palys and P. Daoutidis, "Power-to-X: A review and perspective," *Computers & Chemical Engineering*, vol. 165, p. 107948, 2022.

[14] T. Logenthiran, D. Srinivasan, and T. Z. Shun, "Demand side management in smart grid using heuristic optimization," *IEEE Transactions on Smart Grid*, vol. 3, pp. 1244–1252, 2012.

[15] P. Palensky and D. Dietrich, "Demand side management: Demand response, intelligent energy systems, and smart loads," *IEEE Transactions on Industrial Informatics*, vol. 7, pp. 381–388, 2011.

[16] D. Leherbauer and P. Hehenberger, "Physics-based modeling and parameter tracing for industrial demand-side management applications: A novel approach," *Sustainability*, vol. 16, p. 1995, 2024.

[17] J. Aghaei and M.-I. Alizadeh, "Demand response in smart electricity grids equipped with renewable energy sources: A review," *Renewable and Sustainable Energy Reviews*, vol. 18, pp. 64–72, 2013.

[18] P. Siano, "Demand response and smart grids—A survey," *Renewable and Sustainable Energy Reviews*, vol. 30, pp. 461–478, 2014.

[19] F. Rahman, S. Rehman, and M. A. Abdul-Majeed, "Overview of energy storage systems for storing electricity from renewable energy sources in Saudi Arabia," *Renewable and Sustainable Energy Reviews*, vol. 16, pp. 274–283, 2012.

[20] M. Münster, D. Sneum, R. Bramstoft, F. Bühler, B. Elmegaard, S. Giannelos, X. Zhang, G. Strbac, M. Berger, D. Radu, and D. Elsaesser, *Sector Coupling: Concepts, State-of-the-Art and Perspectives*, 2020.

[21] B. Rego de Vasconcelos and J.-M. Lavoie, "Recent advances in power-to-X technology for the production of fuels and chemicals," *Frontiers in Chemistry*, vol. 7, 2019.

[22] G. Alva, Y. Lin, and G. Fang, "An overview of thermal energy storage systems," *Energy*, vol. 144, pp. 341–378, 2018.

[23] A. A. Almehizia, H. M. K. Al-Masri, and M. Ehsani, "Integration of renewable energy sources by load shifting and utilizing value storage," *IEEE Transactions on Smart Grid*, vol. 10, pp. 4974–4984, 2019.

[24] A. A. Almehizia, H. M. K. Al-Masri, and M. Ehsani, "Feasibility study of sustainable energy sources in a fossil fuel rich country," *IEEE Transactions on Industry Applications*, vol. 55, pp. 4433–4440, 2019.

[25] (2017, 2024). "Desalination by the numbers," http://idadesal.org/desalination-101/desalination-by-the-numbers/.

[26] T. Mezher, H. Fath, Z. Abbas, and A. Khaled, "Techno-economic assessment and environmental impacts of desalination technologies," *Desalination*, vol. 266, pp. 263–273, 2011.

[27] B. Valdez Salas and M. Schorr Wiener, "Desalination, trends and technologies," *Desalination and Water Treatment*, vol. 42, pp. 347–348, 2012.

[28] IRENA, "Green hydrogen: A guide to policy making," Abu Dhabi, 2020.

[29] I. P. Hub, "International PtX hub flyer," 2024.

[30] M. E, "Clean hydrogen mission action plan," 2022.

[31] World Economic Forum. "Accelerating clean hydrogen initiative," https://initiatives.weforum.org/accelerating-clean-hydrogen-initiative/home.

[32] "Kopernikus-project: P2X," www.kopernikus-projekte.de/en/projects/p2x.

[33] "NEOM green hydrogen company," 2024. www.neom.com/en-us/newsroom/neom-green-hydrogen-investment.

[34] Z. Chehade, C. Mansilla, P. Lucchese, S. Hilliard, and J. Proost, "Review and analysis of demonstration projects on power-to-X pathways in the world," *International Journal of Hydrogen Energy*, vol. 44, pp. 27637–27655, 2019.

[35] C. Wulf, J. Linßen, and P. Zapp, "Review of power-to-gas projects in Europe," *Energy Procedia*, vol. 155, pp. 367–378, 2018.

[36] C. Wulf, P. Zapp, and A. Schreiber, "Review of power-to-X demonstration projects in Europe," *Frontiers in Energy Research*, vol. 8, 2020–September 25, 2020.

[37] "REFHYNE clean refinery hydrogen for Europe," www.refhyne.eu/.

[38] "HyBalance project," https://hybalance.eu/.

[39] "Nouryon and Gasunie green hydrogen project," www.nouryon.com/news-and-events/news-overview/2019/nouryon-and-gasunie-study-scale-up-of-green-hydrogen-project-to-meet-aviation-fuels-demand/.

[40] "Big hit," www.bighit.eu/.

[41] "Jupiter 1000," https://mcphy.com/en/achievements/power-to-gas-en/jupiter-1000/.

[42] "H$_2$FUTURE," www.h2future-project.eu/.

[43] "Puertollano green hydrogen plant," www.iberdrola.com/about-us/what-we-do/green-hydrogen/puertollano-green-hydrogen-plant.

[44] "New Energy and Industrial Technology Development Organization (NEDO): Fukushima Hydrogen Energy Research Field (FH2R)," www.nedo.go.jp/english/news/AA5en_100422.html.

[45] "Shell China—Zhangjiakou," www.shell.com/news-and-insights/newsroom/news-and-media-releases/2022/shell-starts-up-hydrogen-electrolyser-in-china-with-20mw-product.html#vanity-aHR0cHM6Ly93d3cuc2hlbGwuY29tL21lZGlhL25ld3MtYW5kLW1lZGlhLXJlbGVhc2VzLzIwMjIvc2hlbGwtc3RhcnRzLXVwLWh5ZHJvZ2VuLWVsZWN0cm9seXNlci1pbi1jaGluYS13aXRoLTIwbXctcHJvZHVjdC5odG1s.

[46] "H2RES: State-of-the-art-green hydrogen production," 2022, https://stateofgreen.com/en/solutions/h2res-green-hydrogen-production/.

[47] "Western green energy hub sets new heights for green energy projects worldwide," 2024, www.blackridgeresearch.com/project-profiles/western-green-hydrogen-energy-hub-australia.

[48] "Haru Oni: eFuel plant of the future," 2022, www.siemens-energy.com/global/en/home/stories/haru-oni.html.

3 Power-to-X and the Art of Hybrid Systems

M. Edwin and M. C. Eniyan

3.1 INTRODUCTION

The growth in urbanization, industrialization, population, and conventional fuel depletion threatened global energy security. Moreover, these advancements will increase global warming gas emissions by 37.55 $GtCO_2$ per year in 2023, which is 1.1% greater than in 2022 [1]. It is anticipated that global CO_2 emission will peak at around 39 $GtCO_2$ by 2025 [1]. Coal, oil, and natural gas are the prime contributors to harmful global warming gas emissions [2]. Further, the global temperature increasing profile shows that the mean global temperature can be expected to reach more than 4°C by the end of the 21st century, leading to adverse effects on living things [3]. Therefore, the Paris Agreement states that an effort must be made to limit the mean global temperature rise to below 1.5°C [4]. The Intergovernmental Panel on Climate Change (IPCC) report also states that necessary policies must be implemented to lower the global temperature rise below 1.5°C above pre-industrial levels, and global warming gas emissions must cease by 2050 [5]. Henceforth, to address and maintain low carbon emissions, the modern energy system has to be designed based on the following factors: low cost, higher efficiency, reliability, and lower carbon emission [6, 7]. In this connection, renewable energy sources (RES), namely, solar, wind, hydro, and biomass, are thought to be the primary energy source choices to fulfill the growing energy demand and reduce carbon emissions. Currently, RES contributes 25% of the global electricity demand and is targeted to increase the share to two-third of world electricity generation by 2040 [6]. The inconsistent nature of RES is a significant drawback, since it can cause transient electricity shortage, destabilization in the grid network, and excess energy wastage. Therefore, energy storage techniques are crucial for maintaining a stable supply chain in the grid network and storing the excess energy produced by RES. In recent years, a novel energy storage technique utilized to stabilize the grid network and avoid excess energy wastage is the power-to-X (P2X) concept.

Power-to-X (P2X) is a cutting-edge technology that transforms excess electrical energy, especially from RES, into various energy or chemical compounds (product X). The reported potential P2X technologies are power-to-X products, such as heat, methane, hydrogen, chemicals, and liquids. Besides the power industry, P2X technologies enable flexible energy conversion and storage, decarbonization, improvement of energy security, facilitation of RES integration into large-scale energy systems, and replacement of fossil fuels.

In typical P2X technology, hydrogen is initially produced through an electrolysis process powered by renewable resources [2]. In recent years, several nations' future energy schemes focus on hydrogen as an energy carrier for various technological processes, such as fuel cell, hydrocracking, hydrodesulfurization, fertilizer production, rocket fuel, transportation, and power supply sectors, because of the decrease in renewable energy production cost observed, and further costs for generating hydrogen via electrolysis [8, 9]. The hydrogen production cost through electrolysis is predicted to reach €1/kg of H_2 by 2030–2040 [10]. Even though the transportation and storage of H_2 are difficult due to its small molecular size and relatively low energy density, the P2X concept provides an alternative solution for the storage and transportation of H_2 by converting H_2 into value-added products using microbial synthesis, ammonia, and methane [11–13]. The conversion methods are called power-to-protein, power-to-chemical, and power-to-methane, which are nothing but P2X.

DOI: 10.1201/9781032719436-3

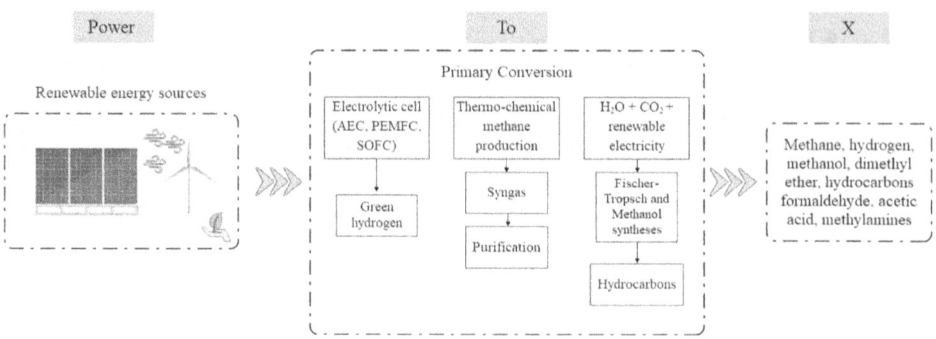

FIGURE 3.1   General layout of power-to-X concept with multiple *X*s.

The general layout of the power-to-X concept with multiple *X*s is illustrated in Figure 3.1. Therefore, the implementation of P2X technologies can address some of the current energy challenges and climatic change.

3.2 P2X INTEGRATION IN HYBRID RENEWABLE ENERGY SYSTEMS (HRESs)

3.2.1 OVERVIEW OF HYBRID RENEWABLE ENERGY SYSTEMS (HRESs)

Even RES has numerous merits based on ecological growth due to the diverse nature of RES, which limits the constant energy supply. This limitation can be reduced by integrating two or more RESs to meet the energy demand [14, 15]. The integration of more than one RES to form a multi-energy renewable system is called a hybrid renewable energy system (HRES). The HRES is a more resilient energy system that has advantages by enhancing system reliability, stability, energy efficiency, energy security, cost savings, and environmental benefits over sole RES [16]. The general layout of HRES is shown in Figure 3.2. Standalone solar PV–wind energy system, solar PV–wind–battery, solar–wind–biomass–grid, solar–hydro–grid, solar–wind–fuel cell, and so on are some of the HRES combinations listed from the reported literature [17–19]. In general, HRESs are categorized into two distinct groups. In the first group, the HRES operates in parallel with the conventional power grid (on-grid HRES), whereas in the second group, the HRES operates in standalone mode (off-grid HRES) [20, 21]. The criteria for the combination of RES for HRES depend on the potential availability of renewable resources and energy demand in the location where HRES will be installed. The energy demand for the locality to install HRES is crucial to determining the need for an energy storage system. Moreover, the HRES has to be evaluated and optimized regarding technical, economic, environmental, and social accessibility. HRES's reliability and scalability provide advantages in the P2X concept to produce valuable, sustainable products.

3.2.2 BENEFITS OF INTEGRATING POWER-TO-X IN HRESs

Generally, HRES can be implemented to improve the energy supply, meet the energy demand of the locality, and reduce the cost of a sustainable energy system with one energy source. However, excess energy generated in HRES was often dumped, but by combining the P2X technology with HRES, the surplus power can be utilized and converted into useful X products, as mentioned earlier. Thus, integrating power-to-X (P2X) in HRES may enhance the power system's versatility, efficiency, and sustainability. The benefits of integrating P2X in HRES are listed as follows:

1. *Enhanced energy storage.* P2X technologies, such as power-to-hydrogen (P2H$_2$), power-to-methane, and power-to-chemicals, convert surplus electricity into other forms of energy, like hydrogen, methane, ammonia, formic acid, and methanol [22]. P2X acts as an energy

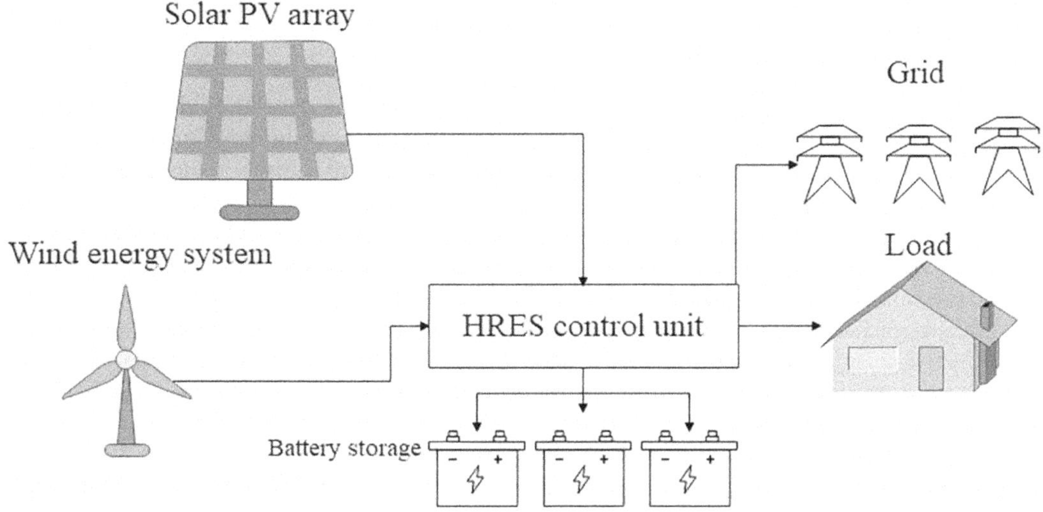

FIGURE 3.2 General hybrid solar PV–wind–battery storage energy system.

carrier and allows stored energy to be used in various sectors, including power, transportation, and even air-conditioning requirements, and it improves the overall energy system efficiency. Moreover, P2X addresses seasonal fluctuations in renewable energy supply and demand by enabling long-term storage.

2. *Decarbonization of sectors.* The integration of P2X with HRES leads to lower carbon emissions in the transportation and processing industry sectors and further helps achieve the reduced carbon emission and global temperature goals set in the Paris Agreement [3]. In the transportation sector, P2X technology promotes the usage of synthetic fuels and hydrogen, lowering the dependence on fossil fuels and reducing carbon emissions [23].

3. *Enhancing reliability and efficiency.* P2X technologies boost the overall efficiency of HRES by enabling excess renewable energy to transform valuable products. P2X integration makes it possible to connect the energy sectors of mobility, heat, gas, and electricity, improving renewable energy use in various contexts [24].

4. *Energy security.* P2X can improve energy security by lowering their reliance on imported fossil fuels by locally generating hydrogen and synthetic fuels [25]. Further, by utilizing local RES, P2X promotes energy independence and lowers exposure to volatility in the global energy market.

5. *Environmental benefits.* Compared to traditional fossil fuels, P2X considerably lowers greenhouse gas emissions by facilitating the generation of clean chemicals and fuels [22].

3.3 CHALLENGES WITH P2X IN RENEWABLE ENERGY DENSITY

The quantity of energy contained in a system or area of space per unit volume is referred to as its *energy density* [26]. To match the production of conventional power plants, large surface areas are required to catch significant amounts of energy, necessitating massive installations. Further, in order to capture sufficient energy, for example, wind energy, wind turbines need large amounts of land or offshore space, which presents problems for land usage and environmental effect if the land is agricultural land [27]. Similar effect was observed in installing a solar power plant. Achieving a high energy density is essential for the practicality of renewable energy sources. The following are some of the difficulties with energy density in RES:

1. *Intermittency and reliability.* Because solar and wind powers depend on the presence of both sunlight and wind, they are, by nature, sporadic. Meeting continuous energy demands becomes difficult due to this intermittency, which also has an impact on the constancy of energy supply [28].
2. *Efficiency losses.* Efficiency losses occur during the transformation of renewable energy into useable forms (heat, electricity). In comparison to fossil fuels, solar panels often have conversion efficiency of 30% or less, resulting in large losses.
3. *Geographical and environmental constraints.* Geographical and climatic factors influence effective solar and wind energy via less solar intensity and wind velocity, leading to a lower potential energy density, which makes them less viable as sources of energy in specific location [25].

Overcoming the challenges of low energy density in renewable energy involves a multifaceted approach, combining technological innovation, improved storage solutions, efficient energy conversion, and supportive policies. P2X technology integrated with HRES provides a positive solution to the low-energy-density nature of RES [29]. The excess energy generated from HRES is converted into other useful forms of energy, such as hydrogen, methane, and liquid fuels, which can be stored for later usage without notable losses. This lessens the need for costly battery storage, which is typically more expensive for long-term storage. Therefore, power-to-X technologies provide significant advantages by enhancing energy storage, supporting grid stability, integrating renewable energy across multiple sectors, and facilitating the production of green fuels and chemicals [30]. These benefits contribute to the decarbonization of the energy system, reduction of fossil fuel dependency, and overall environmental and economic sustainability.

3.4 POWER-TO-X TECHNOLOGIES

P2X plays a vital role in decarbonizing the energy system and efficiently integrating renewable energy sources through the transformation of surplus electrical power typically derived from RES into other kinds of fuels or chemicals.

3.4.1 POWER-TO-HYDROGEN (P2H$_2$)

The majority of P2X chains need hydrogen to continue the production of products. Thus, power-to-hydrogen utilizes the electrochemical process of splitting water into hydrogen and oxygen with a surplus electricity generated from RES as an energy input [31]. The general layout for power-to-hydrogen technology is depicted in Figure 3.3. In order to create hydrogen, an electrolyzer uses an electrolytic cell with energy input and water to initiate the electrochemical decomposition process [32]. The electrochemical decomposition of water to produce hydrogen is given in equation (3.1).

$$2H_2O \rightarrow 2H_2 + O_2 \tag{3.1}$$

The most suitable electrolytic cell designs reported in the literature for hydrogen production via electrolysis are the alkaline electrolytic cell (AEC), proton-exchange membrane electrolytic cell (PEMEC), and solid oxide electrolytic cell (SOEC) [33]. Moreover, PEM-FC and SOEC are dominating research and large-scale application, while the AEC seems to have presently the largest commercial application. PEMEC and AEC are the most practical electrolyzers for P2H$_2$ small-scale system applications. Since SOEC requires high temperatures, its optimum efficiency can only be

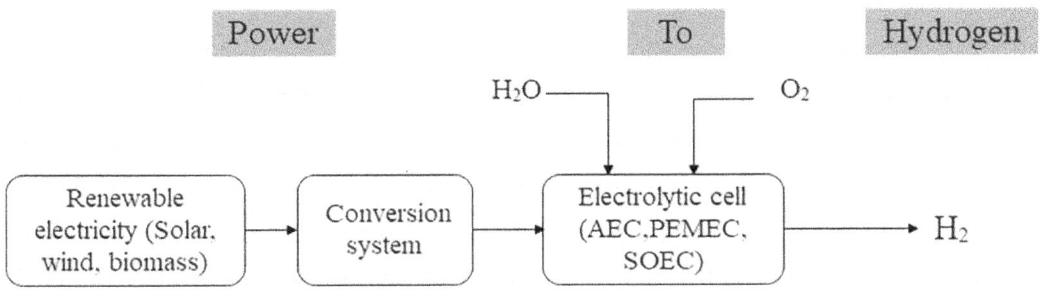

FIGURE 3.3 Layout of the power-to-hydrogen system.

TABLE 3.1
Different Processes of Hydrogen Production

Hydrogen Production Process	Energy Consumption, in kWh/m³H$_2$	Efficiency, in %	Initial Investment, in USD/kWH$_2$	References
Alkaline electrolytic cell (AEC)	3.8–8.2	60–79	2,100–5,700	[35]
Proton-exchange membrane electrolytic cell (PEMEC)	4.4–7.1	62–82	3,100–6,600	[36]
Solid oxide electrolytic cell (SOEC)	3.7	Up to 100	5,200	[37]

realized best if a waste heat source is located close to the power system. Of these three choices, PEM-FC and AEC have quick start-up and shutdown times that are appropriate for HRES [34]. The different processes of hydrogen production and their performance are tabulated in Table 3.1.

Electrolyzers account for the majority of P2H$_2$ facility capital expenditures, whereas electricity accounts for the majority of running costs. According to reported literatures, there is a significant variation in the investment prices of PEMEC, which range from 3,100 to 6,600 USD per kWH$_2$, and AEC, which range from 2,100 to 5,700 USD per kWH$_2$ [35, 36]. Additionally, it was found that, in the long run, alkaline electrolyzers are more sophisticated and economical than PEM-FC. P2H$_2$ pilots are now operating at a fairly limited scale (in MW). Before P2H$_2$ can capture an even greater portion of the energy market, a number of issues need to be resolved. These include the availability of inexpensive renewable energy, technological innovation to improve the efficiency and reduce the conversion losses, the requirement for low electrolyzer capital costs, and so on. Hence, numerous investigations are committed to improve P2H$_2$'s overall system effectiveness.

3.4.2 POWER-TO-SYNTHETIC NATURAL GAS (P2SNG)

Power-to-synthetic natural gas (P2SNG), also known as power-to-gas (P2G), is defined as the process of converting excess energy from RES to gaseous fuels, such as methane or hydrogen [22]. The term *P2G* refers to a set of procedures that include power-to-syngas, P2H$_2$, and P2SNG [22]. A typical schematic representation of P2SNG is shown in Figure 3.4.

Power-to-methane (P2M) is the process of converting hydrogen to methane by reacting it with carbon dioxide to produce a gas mixture that contains both methane and water vapor. Methanation, or methane production, can be carried out in two distinct operations, namely, thermochemical method and biochemical method. Methane generation in the thermochemical method proceeds with a chemical catalyst, and similarly, a bio-catalyst is used in the biochemical methane production method [38]. The biomethane generation efficiency in the thermochemical method is slightly lower than that of the biological method. Efficiency ranges for biological methanation are reported as being between 75% and 98%, and efficiency levels from thermochemical methanation are typically

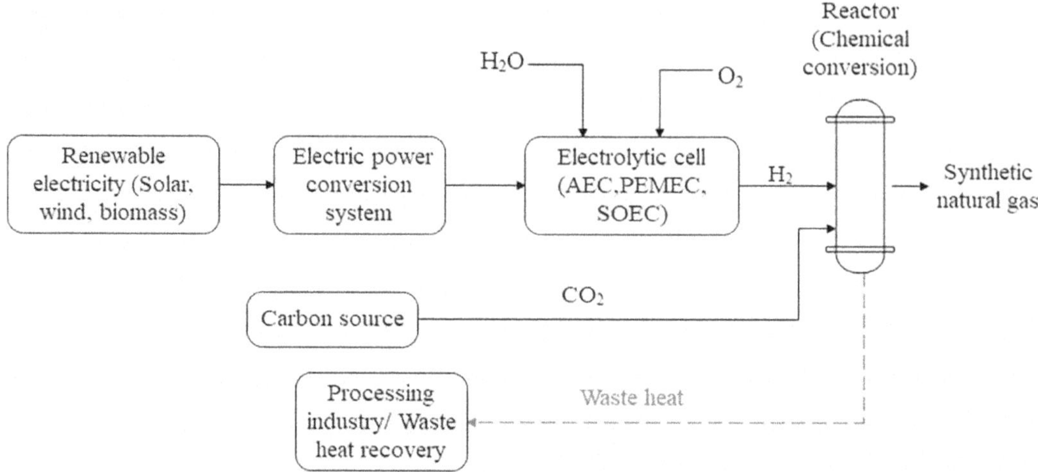

FIGURE 3.4 Schematic representation of power-to-synthetic natural gas system.

between 70% and 85% [39]. The methane synthesis reaction is generally known as the Sabatier reaction [40], as shown in equation (3.2)

$$CO_2 + 4H_2 \rightleftharpoons CH_4 + 2H_2O \tag{3.2}$$

Reactor size, type, optimization, and system analysis for both planning and operation have all been the subject of numerous research [41]. In a study performed by Xiao Xu et al. [42] on P2M technology comprising of off-grid solar PV–battery–methanation hybrid energy system, a bi-level programming was used to optimize the system cost, with maximum reliability on HRES as goal. They found that the proposed HRES can supply continuous electricity and gas to the locality [42]. Similarly, in another study, an HRES consists of PV–battery–AEC to produce hydrogen, and system cost optimization was done by adopting the mixed integer linear programming method. The results concluded that the battery could increase the operation of AEC, and also, battery cost is comparatively higher than AEC cost [43]. In conclusion, the thermochemical methane production method is a highly investigated technology, and its technology readiness level (TRL) is noted to be 5–7 [44]. Moreover, the biochemical method is less-developed than the others. It should be mentioned that heat is produced during the thermochemical method, which might be utilized rather than released.

Power-to-syngas is a form of P2SNG [45]. Syngas is mainly comprised of hydrogen, carbon monoxide, and carbon dioxide. In addition to the standard method of producing syngas through steam reforming, other viable techniques include co-electrolysis, dry reforming of methane, and the reverse water–gas shift reaction [2, 46]. Often, syngas is utilized as an intermediate product for further processing of the chemical and/or the transportation industries.

3.4.3 POWER-TO-LIQUID FUEL (P2L)

The transport sector contributes significantly to greenhouse gas emissions worldwide, as well as for the increasing demand for transportation day by day. The power-to-liquid (P2L) concept provides an alternative fuel to transport and reduce CO_2 emissions. Power-to-liquid systems use renewable electricity, water, and CO_2 to produce liquid fuels [47]. Such a system is shown in Figure 3.5. The P2L system produces synthetic fuels, like methanol, dimethyl ether, and hydrocarbons (alcohols, paraffins, and olefins), in two pathways, namely, Fischer–Tropsch and methanol syntheses. In the Fischer–Tropsch synthesis, initially through water–gas shift reaction, CO_2 is transformed into CO,

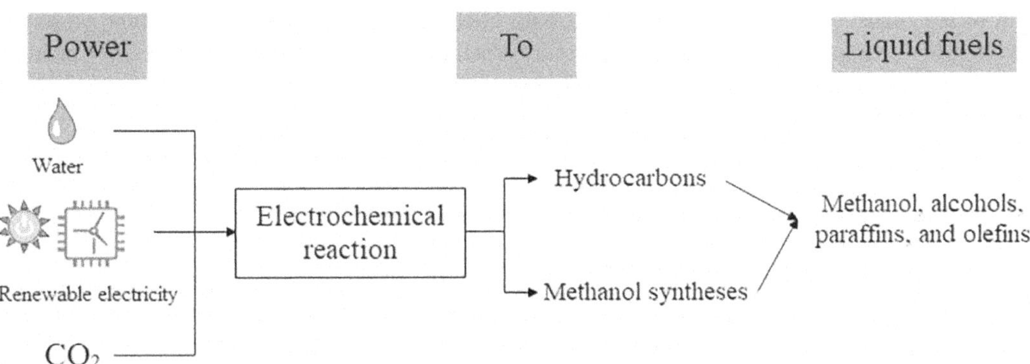

FIGURE 3.5   Layout of the P2L system.

then by water electrolysis, H_2 is produced [48]. After that, H_2 reacts with CO to generate hydrocarbons, as shown in equations (3.3) to (3.5).

$$\text{Alcohols: } nCO + 2nH_2 \leftrightarrow C_nH_{2n} + 2OH + (n-1)H_2O \tag{3.3}$$

$$\text{Olefins: } nCO + 2nH_2 \leftrightarrow C_nH_{2n} + nH_2O \tag{3.4}$$

$$\text{Paraffins: } nCO + (2n+1)H_2 \leftrightarrow C_nH_{(2n+2)}OH + nH_2O \tag{3.5}$$

Methanol is a necessary chemical for the production of formaldehyde, acetic acid, methylamines, dimethyl ether, and so on [49]. Therefore, methanol plays an important role in chemical industries. Methanol is used as a marine fuel and in direct methanol fuel cell for stationary applications and transportation. The installed manufacturing methanol capacity worldwide has increased to an average annual rate of 10% [50].

3.5 INNOVATIONS, STATE OF THE ART, AND FUTURE DIRECTIONS IN POWER-TO-X INTEGRATION

The power-to-X (PtX) technology has gained much popularity since it converts renewable electricity into easily stored and transported chemicals and fuels. Recent investigations into the P2X concept mainly focus on incorporating HRES efficiently and economically. These innovations play a crucial role in enhancing the flexibility and sustainability of energy systems by linking the electricity sector with other sectors, like transport, heating, and industry. By fostering technological innovation, reducing reliance on imported fossil fuels, and generating new companies and jobs in the renewable energy sector, power-to-X technologies also aid in economic development.

Power-to-X technologies are essential for the energy transition because they facilitate the decarbonization of a number of industries where reducing carbon emissions would be challenging otherwise. These technologies will become more and more important in the journey toward a sustainable and carbon-neutral future as long as we keep innovating and optimizing them.

Power-to-X technology that converts renewable electricity to gas, chemicals, and liquid fuels could be a key component for the transition to a sustainable energy regime [51]. Even though P2X technologies provide lots of merits in sustainable energy systems, several bottlenecks need to be addressed to explore the potential of P2X. The major bottlenecks in P2X include high capital and operational costs, energy efficiency and conversion losses, system integration, and policy barriers. To overcome these obstacles will take sustained technology innovation, large infrastructure investments, and appropriate legislative frameworks.

In recent times, several studies about P2X with HRES have been carried out to improve process efficiency and to minimize the production cost for further development. Several assumptions, including system configurations, operating strategies, CO_2 sources for X syntheses, as well as spatiotemporal scale for decarbonization targets, have been investigated. Recently, analyses were done on the feasibility of present and future scenarios of a decarbonized world. It can be observed that a significant issue related to this technology is on the improvement of water electrolysis, both in terms of efficiency and cost.

It is anticipated that in the near future, the amount of electricity generated by RES would rise quickly. Also, energy demand is increasing due to the electrification of vehicles. However, because of their innate unreliability, they were unable to consistently provide the demand for electricity. Therefore, a backup power source is required to guarantee an uninterrupted functioning of P2X systems. It is important to note that P2X has been considered as a technique for storing excess electricity from renewable energy sources for sector decarbonization. It is recommended that low-carbon electricity sources provide this backup. H_2 must generally originate from GHG-free sources for P2X to be considered an environmentally acceptable process [52].

Several sorts of methods or algorithms can be used for long-term analysis of HRES-incorporated P2X. The P2X model is preset when using the current tools, and features cannot be changed, because input and output values do not accurately reflect transient events. Depending on the implemented technique, using programming platforms like MATLAB, Julia, or Python allows access to more defined component levels [53]. The electrolytic cell model uses linear or nonlinear equations to provide additional information on the voltage and current using mixed integer linear programming (MILP). By using heuristic and meta-heuristic algorithms, one can obtain a deeper understanding of the component's dynamics. Through validation analyses using an experimental validation, a more precise model, including the component's performance in transient and steady-state, physical, and thermochemical characteristics, can be reached [54]. To determine the P2X value in HRES, a few gaps need to be filled. Further analysis in the long-term planning to identify components and future scenarios would result from the P2X combining HRES to generate energy carriers, like gas, heat, and chemicals.

3.5.1 FUTURE RESEARCH FIELDS AND RECOMMENDATIONS

Numerous avenues of investigation in HRESs with P2X principles remain unexplored, and numerous domains of inquiry require examination. Due to the lack of sustainability and environmental effect indices in the literature at this time, the technical and financial advantages of P2X in HRES, including the hydrogen carrier, have not been thoroughly studied. In order to identify the advantages of new P2X technology in HRES and for improvements to be inevitably introduced in the future sustainability, further component investigations are advised in optimal configuration analysis. It is now hard to uncover the fundamental interactions of a multi-energy system with P2X in HRES since the electrolytic cell is frequently connected to a fuel cell to provide only the electrical system, or occasionally in a methanator to generate synthetic natural gas.

The real use of hydrogen and its advantages to the environment are further limited by hydrogen storage capability. To make the technology competitive in the market, the right innovative studies have to be carried out. Additionally, research should be done to assess the effects of hydrogen storage and transport as they stand right now. Opportunities for economic progress for developing nations with abundant renewable energy sources could be jeopardized if technology, financial resources, and political clout are not dispersed equally. Ultimately, given that the current cost of green hydrogen is too high, concessions will need to be made.

Also, gaps in HRES real-time operation should be further examined in order to comprehend the underlying interactions. There has not been much research done on a more precise model of P2X components on the system and uncertainties associated with other elements, like political incentives, the carbon trade, and social acceptance. Real-time stochastic power management techniques would reveal this important point, greatly enhancing understanding of the overall system performance. Moreover, no research has been done on the distributed coordination of several HRES using P2X.

Finally, a thorough investigation of HRESs, including P2X, is advised in accordance with sustainable development. The present investigation aims to provide an overview of the long- and short-term impacts of P2X vector exploitation by shedding light on key elements and routes.

REFERENCES

[1] P. Friedlingstein, M. O'Sullivan, M. W. Jones, R. M. Andrew, D. C. E. Bakker, J. Hauck, P. Landschützer, C. Le Quéré, I. T. Luijkx, G. P. Peters, W. Peters, J. Pongratz, C. Schwingshackl, S. Sitch, J. G. Canadell, P. Ciais, R. B. Jackson, S. R. Alin, P. Anthoni, L. Barbero, N. R. Bates, J. Zeng, and B. Zheng, *Global Carbon Budget 2023*, 2023. doi: 10.5194/essd-2023-409.

[2] A. C. Ince, C. O. Colpan, A. Hagen, and M. F. Serincan, "Modeling and simulation of Power-to-X systems: A review," *Fuel*, vol. 304, p. 121354, Nov. 2021. doi: 10.1016/j.fuel.2021.121354.

[3] J. Rogelj, M. Den Elzen, N. Höhne, T. Fransen, H. Fekete, H. Winkler, R. Schaeffer, F. Sha, K. Riahi, and M. Meinshausen, "Paris Agreement climate proposals need a boost to keep warming well below 2°C," *Nature*, vol. 534, no. 7609, pp. 631–639, Jun. 2016. doi: 10.1038/nature18307.

[4] EIA, "EIA energy outlook 2020," *Annu. Energy Outlook 2020 with Proj. to 2050*, vol. 58, no. 12, pp. 7250–7257, 2020.

[5] M. Allen, M. Babiker, Y. Chen, and H. C. de Coninck, "IPCC SR15: Summary for policymakers," in *IPCC Special Report Global Warming of 1.5°C*, Intergovernmental Panel on Climate Change, 2018.

[6] K. Mohn, "The gravity of status quo: A review of IEA's World Energy Outlook," *Econ. Energy Environ. Policy*, vol. 9, no. 1, Jan. 2020. doi: 10.5547/2160-5890.8.2.kmoh.

[7] J. Ikäheimo, J. Kiviluoma, R. Weiss, and H. Holttinen, "Power-to-ammonia in future North European 100% renewable power and heat system," *Int. J. Hydrogen Energy*, vol. 43, no. 36, pp. 17295–17308, Sep. 2018. doi: 10.1016/j.ijhydene.2018.06.121.

[8] Federal Ministry of Economic Affairs & Energy, "The national hydrogen strategy," *Berlin*, pp. 1–32, 2020. [Online]. www.bmwi.de.

[9] W. Logroño, S. Kleinsteuber, J. Kretzschmar, F. Harnisch, J. De Vrieze, and M. Nikolausz, "The microbiology of Power-To-X applications," *FEMS Microbiol. Rev.*, vol. 47, no. 2, pp. 1–23, 2023. doi: 10.1093/femsre/fuad013.

[10] A. van Wijk, E. van der Roest, and J. Boere, *Solar power to the people*, November 2018. doi: 10.1525/9780520938403-012.

[11] A. C. Ince, C. Ozgur Colpan, A. Keles, M. F. Serincan, and U. Pasaogullari, "Scaling and performance assessment of power-to-methane system based on an operation scenario," *Fuel*, vol. 332, p. 126182, Jan. 2023. doi: 10.1016/j.fuel.2022.126182.

[12] Z. Yu, J. Lin, F. Liu, J. Li, Y. Zhao, Y. Song, Y. Song, and X. Zhang, "Optimal sizing and pricing of grid-connected renewable power to ammonia systems considering the limited flexibility of ammonia synthesis," *IEEE Trans. Power Syst.*, vol. 39, no. 2, pp. 3631–3648, 2024. doi: 10.1109/TPWRS.2023.3279130.

[13] S. W. S. Chan, H. Marami, L. L. Tayo, E. Fog, T. A. Andrade, M. Ambye-Jensen, M. Birkved, and B. Khoshnevisan, "Environmental impacts of a novel biorefinery platform integrated with power-to-protein technology to decrease dependencies on soybean imports," *Sci. Total Environ.*, vol. 907, p. 167943, 2024. doi: 10.1016/j.scitotenv.2023.167943.

[14] M. Edwin and S. Joseph Sekhar, "Techno-economic studies on hybrid energy based cooling system for milk preservation in isolated regions," *Energy Convers. Manag.*, vol. 86, pp. 1023–1030, 2014. doi: 10.1016/j.enconman.2014.06.075.

[15] M. Edwin and S. Joseph Sekhar, "Techno-economic evaluation of milk chilling unit retrofitted with hybrid renewable energy system in coastal province," *Energy*, vol. 151, pp. 66–78, 2018. doi: 10.1016/j.energy.2018.03.050.

[16] M. Edwin and S. J. Sekhar, "Hybrid thermal energy based cooling system for a remote seashore villages," *Adv. Mater. Res.*, vol. 984–985, pp. 719–724, 2014. doi: 10.4028/www.scientific.net/AMR.984-985.719.

[17] M. Edwin and S. J. Sekhar, "Thermal performance of milk chilling units in remote villages working with the combination of biomass, biogas and solar energies," *Energy*, vol. 91, pp. 842–851, 2015. doi: 10.1016/j.energy.2015.08.103.

[18] M. Edwin and S. Joseph Sekhar, "Thermo-economic assessment of hybrid renewable energy based cooling system for food preservation in hilly terrain," *Renew. Energy*, vol. 87, pp. 493–500, 2016. doi: 10.1016/j.renene.2015.10.056.

[19] E. Mohan, S. N. Mohan, and J. S. Santhappan, "Hybrid energy-based chilling system for food preservation in remote areas," *Intech*, vol. 11, no. tourism, p. 13, 2016 [Online]. www.intechopen.com/books/advanced-biometric-technologies/liveness-detection-in-biometrics.

[20] M. S. Nair and S. J. S. M. Edwin, *Resource Assessment and Implementation of Hybrid Renewable Energy Systems for Food Preservation in Agro-Tropical Areas: A Techno-Economic Approach M*, 2020. doi: 10.1007/978-3-030-34021-6_1.

[21] M. Edwin, M. S. Nair, and S. J. Sekhar, "Techno-economic modeling of stand-alone and hybrid renewable energy systems for thermal applications in isolated areas," *Renew. Energy Syst. Model. Optim. Control*, pp. 279–308, 2021. doi: 10.1016/B978-0-12-820004-9.00013-9.

[22] I. Sorrenti, T. B. Harild Rasmussen, S. You, and Q. Wu, "The role of power-to-X in hybrid renewable energy systems: A comprehensive review," *Renew. Sustain. Energy Rev.*, vol. 165, no. March, p. 112380, 2022. doi: 10.1016/j.rser.2022.112380.

[23] Y. Wu, Y. Zhang, C. Xia, A. Chinnathambi, O. Nasif, B. Gavurová, M. Sekar, A. Anderson, N. T. L. Chi, and A. Pugazhendhi, "Assessing the effects of ammonia (NH_3) as the secondary fuel on the combustion and emission characteristics with nano-additives," *Fuel*, vol. 336, p. 126831, 2023. doi: 10.1016/j.fuel.2022.126831.

[24] Y. Han, Y. Liao, X. Ma, and X. Guo, "Simulation study of a novel methanol production process based on an off-grid wind/solar/oxy-fuel power generation system," *Energy Convers. Manag.*, vol. 314, p. 118672, 2024. doi: 10.1016/j.enconman.2024.118672.

[25] R. Daiyan, I. MacGill, and R. Amal, "Opportunities and challenges for renewable power-to-X," *ACS Energy Lett.*, vol. 5, no. 12, pp. 3843–3847, Dec. 2020. doi: 10.1021/acsenergylett.0c02249.

[26] T.-Z. Ang, M. Salem, M. Kamarol, H. S. Das, M. A. Nazari, and N. Prabaharan, "A comprehensive study of renewable energy sources: Classifications, challenges and suggestions," *Energy Strateg. Rev.*, vol. 43, p. 100939, Sep. 2022. doi: 10.1016/j.esr.2022.100939.

[27] E. A. Nanaki and G. A. Xydis, "Deployment of renewable energy systems: Barriers, challenges, and opportunities," in *Advances in Renewable Energies and Power Technologies*, Elsevier, 2018, pp. 207–229. doi: 10.1016/B978-0-12-813185-5.00005-X.

[28] M. A. Basit, S. Dilshad, R. Badar, and S. M. Sami ur Rehman, "Limitations, challenges, and solution approaches in grid-connected renewable energy systems," *Int. J. Energy Res.*, vol. 44, no. 6, pp. 4132–4162, May 2020. doi: 10.1002/er.5033.

[29] M. J. Palys and P. Daoutidis, "Power-to-X: A review and perspective," *Comput. Chem. Eng.*, vol. 165, p. 107948, Sep. 2022. doi: 10.1016/j.compchemeng.2022.107948.

[30] C. Schnuelle, J. Thoeming, T. Wassermann, P. Thier, A. von Gleich, and S. Goessling-Reisemann, "Socio-technical-economic assessment of power-to-X: Potentials and limitations for an integration into the German energy system," *Energy Res. Soc. Sci.*, vol. 51, pp. 187–197, May 2019. doi: 10.1016/j.erss.2019.01.017.

[31] M. A. Hannan, S. M. Abu, A. Q. Al-Shetwi, M. Mansor, M. N. M. Ansari, K. M. Muttaqi, and Z. Y. Dong, "Hydrogen energy storage integrated battery and supercapacitor based hybrid power system: A statistical analysis towards future research directions," *Int. J. Hydrogen Energy*, vol. 47, no. 93, pp. 39523–39548, 2022. doi: 10.1016/j.ijhydene.2022.09.099.

[32] Q. Hassan, S. Algburi, A. Z. Sameen, H. M. Salman, and M. Jaszczur, "A review of hybrid renewable energy systems: Solar and wind-powered solutions: Challenges, opportunities, and policy implications," *Results Eng.*, vol. 20, no. November, p. 101621, 2023. doi: 10.1016/j.rineng.2023.101621.

[33] S. Y. Wong and C. Li, "Techno-economic analysis of optimal hybrid renewable energy systems – A case study for a campus microgrid," *Energy Rep.*, vol. 9, no. September, pp. 134–138, 2023. doi: 10.1016/j.egyr.2023.09.153.

[34] M. Genovese, A. Schlüter, E. Scionti, F. Piraino, O. Corigliano, and P. Fragiacomo, "Power-to-hydrogen and hydrogen-to-X energy systems for the industry of the future in Europe," *Int. J. Hydrogen Energy*, vol. 48, no. 44, pp. 16545–16568, 2023. doi: 10.1016/j.ijhydene.2023.01.194.

[35] A. R. Dahiru, A. Vuokila, and M. Huuhtanen, "Recent development in Power-to-X: Part I—A review on techno-economic analysis," *J. Energy Storage*, vol. 56, no. PA, p. 105861, 2022. doi: 10.1016/j.est.2022.105861.

[36] H. Zhu, and H. Zhang, "Integration of proton exchange membrane fuel cell with air gap membrane distillation for sustainable electricity and freshwater cogeneration: Performance, influential mechanism, multi-objective optimization and future perspective," *Renewable and Sustainable Energy Reviews*, vol. 199, p. 114523, 2024. doi: 10.1016/j.rser.2024.114523.

[37] H. van't Noordende, F. van Berkel, and M. Stodolny, *Next level solid oxide electrolysis*. Institute for Sustainable Process Technology (ISPT), 2023.

[38] M. Götz, J. Lefebvre, F. Mörs, A. M. Koch, F. Graf, S. Bajohr, R. Reimert, and T. Kolb, "Renewable power-to-gas: A technological and economic review," *Renew. Energy*, vol. 85, pp. 1371–1390, 2016. doi: 10.1016/j.renene.2015.07.066.

[39] T. T. Q. Vo, A. Xia, F. Rogan, D. M. Wall, and J. D. Murphy, "Sustainability assessment of large-scale storage technologies for surplus electricity using group multi-criteria decision analysis," *Clean Technol. Environ. Policy*, vol. 19, no. 3, pp. 689–703, 2017. doi: 10.1007/s10098-016-1250-8.

[40] P. Styring, S. McCord, and S. Rackley, "Chapter 17: Carbon dioxide utilization," in S. Rackley, G. Andrews, D. Clery, R. De Richter, G. Dowson, P. Knops, W. Li, S. Mccord, T. Ming, A. Sewel, P. Styring, and M. B. T.-N. E. T. for C. C. M. Tyka, Eds. *Negative Emissions Technologies for Climate Change Mitigation*, Elsevier, 2023, pp. 391–413. doi: 10.1016/B978-0-12-819663-2.00005-8.

[41] X. Li, X. Yang, G. Y. Zhou, S. Mu, and J. Lemmon, "Sustainable energy ecosystem based on power to X technology," in *International Conference on Applied Energy*, Mar. 2019. doi: 10.46855/energy-proceedings-2896.

[42] X. Xu, W. Hu, D. Cao, W. Liu, Q. Huang, Y. Hu, and Z. Chen, "Enhanced design of an offgrid PV-battery-methanation hybrid energy system for power/gas supply," *Renew. Energy*, vol. 167, pp. 440–456, 2021. doi: 10.1016/j.renene.2020.11.101.

[43] B. Gillessen, H. U. Heinrichs, P. Stenzel, and J. Linssen, "Hybridization strategies of power-to-gas systems and battery storage using renewable energy," *Int. J. Hydrogen Energy*, vol. 42, no. 19, pp. 13554–13567, 2017. doi: 10.1016/j.ijhydene.2017.03.163.

[44] J. de Bucy, O. Lacroix, and L. Jammes, "The potential of power-to-gas Technology review and economic potential assessment," France. [Online]. http://inis.iaea.org/search/search.aspx?orig_q=RN:51056921.

[45] B. R. de Vasconcelos and J. M. Lavoie, "Recent advances in power-to-X technology for the production of fuels and chemicals," *Front. Chem.*, vol. 7, no. June, pp. 1–24, 2019. doi: 10.3389/fchem.2019.00392.

[46] J. C. Koj, C. Wulf, and P. Zapp, "Environmental impacts of power-to-X systems—A review of technological and methodological choices in life cycle assessments," *Renew. Sustain. Energy Rev.*, vol. 112, no. June, pp. 865–879, 2019. doi: 10.1016/j.rser.2019.06.029.

[47] M. Fasihi, D. Bogdanov, and C. Breyer, "Economics of global gas-to-liquids (GtL) fuels trading based on hybrid Pv-wind power plants," in *Proceedings of the ISES Solar World Congress 2015*, 2016, pp. 1–20. doi: 10.18086/swc.2015.09.03.

[48] D. C. de Oliveira, E. E. S. Lora, O. J. Venturini, D. M. Y. Maya, and M. Garcia-Pérez, "Gas cleaning systems for integrating biomass gasification with Fischer-Tropsch synthesis—A review of impurity removal processes and their sequences," *Renew. Sustain. Energy Rev.*, vol. 172, p. 113047, 2023. doi: 10.1016/j.rser.2022.113047.

[49] N. Onishi, G. Laurenczy, M. Beller, and Y. Himeda, "Recent progress for reversible homogeneous catalytic hydrogen storage in formic acid and in methanol," *Coord. Chem. Rev.*, vol. 373, pp. 317–332, 2018. doi: 10.1016/j.ccr.2017.11.021.

[50] M. Pérez-Fortes, J. C. Schöneberger, A. Boulamanti, and E. Tzimas, "Methanol synthesis using captured CO_2 as raw material: Techno-economic and environmental assessment," *Appl. Energy*, vol. 161, pp. 718–732, 2016. doi: 10.1016/j.apenergy.2015.07.067.

[51] J. Liu, X. Duan, Z. Yuan, Q. Liu, and Q. Tang, "Experimental study on the performance, combustion and emission characteristics of a high compression ratio heavy-duty spark-ignition engine fuelled with liquefied methane gas and hydrogen blend," *Appl. Therm. Eng.*, vol. 124, pp. 585–594, 2017. doi: 10.1016/j.applthermaleng.2017.06.067.

[52] H. Blanco, W. Nijs, J. Ruf, and A. Faaij, "Potential for hydrogen and power-to-liquid in a low-carbon EU energy system using cost optimization," *Appl. Energy*, vol. 232, pp. 617–639, Dec. 2018. doi: 10.1016/j.apenergy.2018.09.216.

[53] H. Aki, I. Sugimoto, T. Sugai, M. Toda, M. Kobayashi, and M. Ishida, "Optimal operation of a photovoltaic generation-powered hydrogen production system at a hydrogen refueling station," *Int. J. Hydrogen Energy*, vol. 43, no. 32, pp. 14892–14904, 2018. doi: 10.1016/j.ijhydene.2018.06.077.

[54] C. Varela, M. Mostafa, and E. Zondervan, "Modeling alkaline water electrolysis for power-to-x applications: A scheduling approach," *Int. J. Hydrogen Energy*, vol. 46, no. 14, pp. 9303–9313, 2021. doi: 10.1016/j.ijhydene.2020.12.111.

4 A Comprehensive Review on Green Hydrogen, CCSU, and Methane Production Technologies

Mohammad Karrabi, Farkhondeh Jabari,
and Asghar Akbari Foroud

4.1 URGENT NEED FOR GREEN HYDROGENATION AND CARBON CAPTURE

In recent years, green hydrogen and carbon reduction have attracted worldwide attention due to the limited resources of petroleum products, huge volume of natural gas consumption, climate change, and global warming [1]. For this purpose, hydrogen can be utilized as a carbon-free alternative in energy systems, fuel cell vehicles, and natural gas production facilities. Meanwhile, renewable energy sources (RES), such as solar, wind, and hydro, can be utilized for water separation into H_2 and O_2. It should be noted that hydrogen-air combustion can chemically be formulated as (4.1).

$$H_{2(g)} + 0.5 \times \left(O_{2(g)} + 3.76\ N_{2(g)}\right) \rightarrow H_2O_{(g)} + 1.88 \times N_{2(g)} \tag{4.1}$$

It is obvious from chemical balance (4.1) that H_2 can be used in conventional thermal power plants aiming to mitigate the emission of pollutant gases and reduce carbon credit. Furthermore, the consumption of hydrogen in combined heat and power (CHP) generation systems reduces CH_4 demand and congestion of natural gas pipelines. Moreover, a part of natural gas demand can be procured by integrating H_2 and CO_2 captured from hydrocarbon combustion, as fulfilled in (4.2) [2].

$$CO_{2(g)} + 4H_{2(g)} \rightarrow CH_{4(g)} + 2H_2O_{(g)} \tag{4.2}$$

Some of the pros and cons of hydrogen are discussed in what follows.

Since hydrogen has a high energy density, it can be used to fuel energy-intensive industrial processes that are difficult to electrify. H_2 can be stored for a long time and with very low losses in suitable tanks as a clean and nature-friendly energy. Compared to renewable electricity connected to distribution grids, it can be transported over long distances due to greater flexibility. H_2 fuel cell vehicles in road transportation sector and H_2-fueled internal combustion engines in maritime and air industries reduce land, sea, and air pollutant substances, significantly. Penetration of green hydrogen in energy portfolio of low-income countries may increase energy security and reduce natural gas import. Production of hydrogen from RES causes higher cost saving due to no need to purchase electricity from local power distribution network for driving electrolyzer. Power-to-gas energy storage technology makes it possible to produce H_2 from surplus electrical power output of RES while improving the frequency stability of power distribution networks. H_2 can integrate with district heating and cooling systems, natural gas networks, fuel cells and hydrogen boilers, as well as for residential heating systems. In a Haber–Bosch process, H_2 and N_2 are mixed at high

DOI: 10.1201/9781032719436-4

temperature and pressure to produce ammonia, which can fuel a combined cycle power plant with no carbon footprints. The green hydrogen can be considered as a sustainable energy resource both in technology development and political prospects, which are used in fuel cell vehicles, electrocatalysts, photocatalysts, and solar hydrogen panels.

Transition to green hydrogen plays a vital role in air quality improvement by its applications in various heating and cooling industries and the transportation sector while improving efficiency, flexibility, security, and resiliency of energy systems. In solar hydrogen panels, photocatalytic materials with power semiconductors are used as an advanced technology and have the practical and scalable capabilities for hydrogen production. Another outstanding property of these materials is the direct conversion of solar energy into hydrogen, which has the least energy loss and optimizes the overall efficiency of the system. The green hydrogenation process is considered as the best type of hydrogen from an environmental viewpoint, which can be distinguished from other hydrogen production approaches. In contrast to green hydrogen, we can refer to blue hydrogen and gray hydrogen, all of which have carbon emissions because of using fossil fuel–based prime movers in electrolyzing process. Blue hydrogen, which is based on carbon absorption and storage technology, emits less carbon compared to gray one. However, hydrogen production technologies are progressing, aiming to have a zero-emission community. In terms of production cost, green hydrogen has higher cost than the gray one [3]. However, the wide development of hydrogen-based infrastructure is associated with many challenges, which can point to high investment cost in carbon-free hydrogenation methods. So far, hydrogen is produced in various industries, from raw materials of fossil fuels, coal, and natural gas. Moreover, several companies and industries have made efforts to produce green hydrogen using renewable energies, such as wind and solar [4].

The commercialization of hydrogen production is faced with some limitations, such as slow growth of infrastructure and high capital investment cost. Reducing the costs of green hydrogen production and welcoming it as zero emission has close competition with the production of blue hydrogen, which is caused by fossil fuels [4]. To mitigate CO_2, NO_x, and SO_x pollutants, some countries, such as China, the United States, the European Union, and India, have planned for green hydrogen development by promoting and expanding electric vehicles and completing the electrification of the transportation sector [5]. It is easier and more possible to continue and increase electricity supply to domestic and industrial centers, but in cases such as air, marine, and heavy cargo vessels, because of the high capacity of electric storage, it is faced with some limitations in electricity supply and storage. Therefore, converting energy into different forms is a solution for solving this issue. One of these forms is power-to-fuel (PtF), which has recently attracted attention. PtH_2 is introduced as a cost-effective type of power to fuel conversation procedures. By using RES, hydrogen can be used in the electricity industry, to be converted into natural gas during the methanation process at off-peak electricity consumption periods and be used by thermal power plants for generating electricity in peak load time intervals [6]. About 80% of global trade is carried out via sea transportation. Therefore, a huge volume of pollutant substances is emitted from fossil fuels. According to an International Maritime Organization (IMO) report, the emission of greenhouse gases caused by ships has reached 1,076 Mt (million tons) in 2018. Thus, the use of hydrogen as a green fuel in this industry will greatly contribute to decarbonization. It is expected that carbon emissions of the marine industry can be reduced to almost 70% by 2050. These policies and regulations, applied at regional, national, and international levels, are necessary, and it is expected that a huge change will occur [7].

4.2 HYDROGEN PRODUCTION, STORAGE, AND APPLICATIONS

In this section, hydrogen production, purification, compression, storage, and applications are introduced. To produce green hydrogen, renewable energies, such as wind, solar, hydroelectric, tidal, biogas, geothermal, biomass, etc., are usually used as clean and sustainable resources. Several key criteria should be considered in the design, development, and deployment of hydrogen production, storage, and transportation facilities [8]:

- Investment, operation, and maintenance costs
- Payback period, break-even point, and life cycle
- Demand- and generation-side energy management implementation capability
  - o Demand response schemes
  - o Energy storage technologies
- Economic, energetic, exergetic, and environmental (4E) efficiencies
- Degradation of system equipment
- Storage infrastructure of gaseous or liquefied hydrogen

In an electrolyzer unit, anode and cathode electrodes, as well as an electrolyte, are exposed to an electric field using a DC power supply, as modeled by equation (4.3). The energy efficiency of the pure water electrolyzing process is defined as energy content of produced hydrogen and consumed electrical power. It should be noted that H_2 energy content is determined based on its lower heating value.

$$H_2O_{(l)} + \text{Electricity} \rightarrow H_{2(g)} + 0.5O_{2(g)} \tag{4.3}$$

4.2.1 HYDROGEN PRODUCTION APPROACHES

As illustrated in Figure 4.1, there are three types of electrolyzers:

- Proton-exchange membrane (PEM)
- Alkaline water electrolysis (AWE)
- Solid oxide electrolysis cell (SOEC)
- Mechanical
  - o Reciprocating
  - o Diaphragm
  - o Centrifugal
  - o Liquid compressors
- Non-mechanical
  - o Cryogenic
  - o Metal hydride
  - o Electrochemical
  - o Absorption

The maximum hydrogen compression capacity of these methods is less than or equal to 1,000 Nm^3/h. Centrifugal and reciprocating compressors are more suitable options for large-scale hydrogen production and compression capacities. For low flows and high pressures, reciprocating compressors are appropriate to create high pressures in one step. Meanwhile, centrifugal compressors are more convenient with higher flow rates, but they have lower single-stage pressure than the absorption type. Its discharge pressure, which depends on gas velocity and density, is low because H_2 density is almost one-eighth of natural gas [9]. Both discharge pressure and flow rate of H_2 during compression after purification should be taken into account for selecting a compressor. To reach higher discharge pressures, compression should be carried out in several stages, requiring more simultaneous cooling and relatively higher operating costs. Hence, cost estimation and techno-economic analysis are important factors that should be considered in the operation and planning stages. Due to low density and molecular weight of hydrogen, its compression is faced with some challenges from safety and protection prospects. Hydrogen is expanded at its inversion temperature

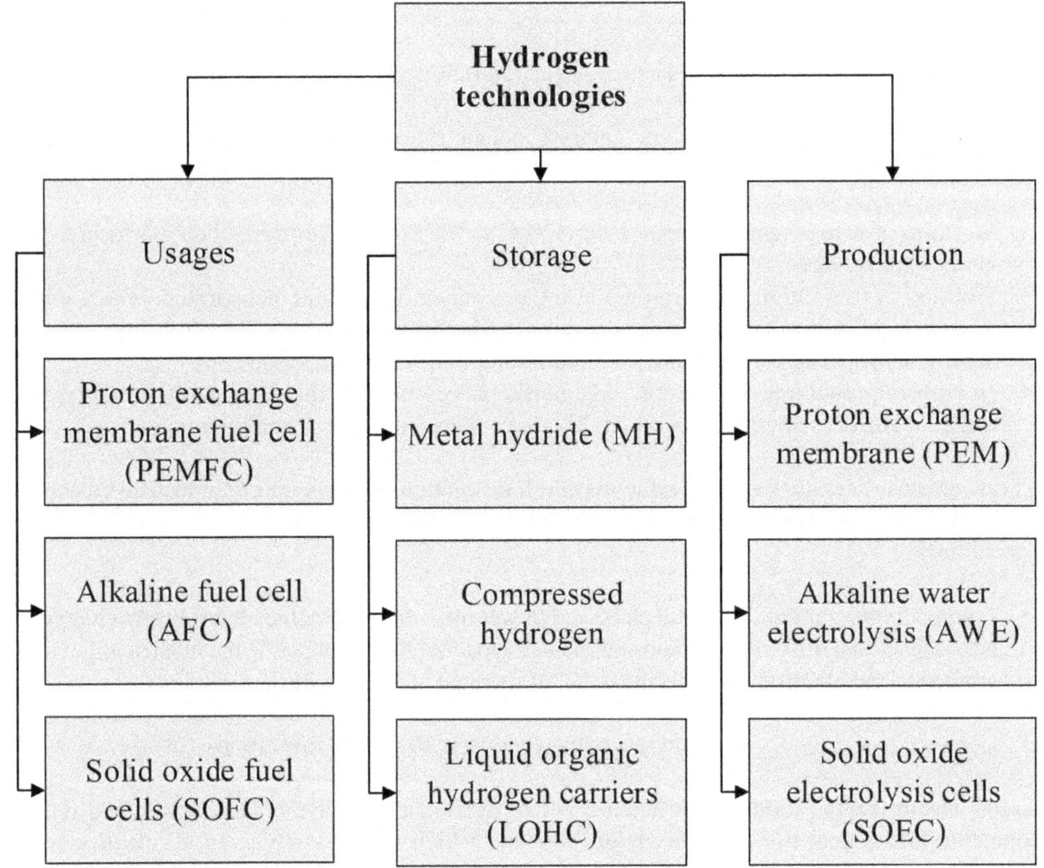

FIGURE 4.1 Electrolytic hydrogen production, storage, and applications.

and heated up instead of cooling. It has become more complicated and may cause several limitations, such as hydrogen fragility and maximum discharge temperature. The total investment cost of PEM electrolyzer is greater than €2,000/kW, which is higher than that of the alkaline type. The operation temperature of the PEM electrolyzer changes within 20–100°C. The purity degree of H_2 will be larger than 99%, with no need for additional purification processes. Their cold-start time is less than several minutes, which makes them compatible with alkaline ones. One main advantage of PEM electrolyzers is the high purity of H_2 and its compact design in combination with RES, which makes it suitable for industrial applications. Using specific catalysts, cheaper materials can be used to reduce investment costs. Lower operation temperature (<85°C), higher energy efficiency, greater current density (>3 A/cm^3), high purity of H_2, fast response, and lower operation and maintenance costs as a result of smaller size can be stated as the other advantages of PEM electrolyzers [10]. However, it should be mentioned that due to easy accessibility and scalability and cheaper AWE types, it is still more practical in the water electrolysis industry [9]. Applying an external voltage source to anode and cathode electrodes, H_2O is oxidized in the anode, as chemically formulated by equation (4.4). In the cathode side, hydrogen protons are coupled with electrons to produce $H_{2(g)}$ as chemical equilibrium (4.5).

$$H_2O \rightarrow 2H^+ + 0.5O_2 + 2e^- \tag{4.4}$$

$$2H^+ + 2e^- \rightarrow H_2 \tag{4.5}$$

Therefore, we have:

$$H_2O_{(l)} \rightarrow H_{2(g)} + 0.5O_{2(g)} \qquad (4.6)$$

The disadvantages of the PEM-type catalyst include the need of an expensive membrane and catalyst [8]. Some features of PEM can be expressed as follows:

- Working at a high temperature of 60–120°C in PEMs will improve their performance under high voltages.
- Working at these high temperatures also causes them to become dehydrated, which will require more water supply for its operation at high temperatures. Therefore, the use of materials including Nafion mixed with titanium oxide solves this issue.
- At ambient condition, H_2 has low volumetric energy density that is recently pressurized using mechanical compressors which have low energy efficiency and high noise.

Alkaline electrolyzers are based on aqueous alkaline solutions and applied to industrial customers. Their electricity consumption widely varies from 10 kW to several MW. Some of their challenges are as follows:

- *Limited loading point.* The total electrical power utilized by alkaline electrolyzers changes between 10 and 40% of their nominal power capacity. For example, if the electrical power applied to the alkaline electrolyzer is 5% of its rated value, it will be turned off.
- Time delays in operation due to transient response time from several seconds to minutes.
- Long cold-start (up to 1 hr) and restarting (between 30 and 60 min) times.

Alkaline electrolyzers produce significant flows of hydrogen and have relatively safe and reliable equipment, with a long life cycle of about 30 years, which requires replacing all electrodes and membranes every 8 years. Their maximum efficiency and operating temperature reach up to 80% and 80°C. Its chemical process in anode and cathode electrodes is given by equations (4.7) and (4.8):

$$4OH^- \rightarrow O_2 + 2H_2O + 4e^- \qquad (4.7)$$

$$4H^+ + 4e^- \rightarrow 2H_2 \qquad (4.8)$$

AWEs can operate at high and low pressures of up to 30 bar, working at pressures that cause the simultaneous compression of hydrogen and eliminate recompression after production. They have some advantage of eliminating the compression stage, but they also reduce product purity due to the higher permeability of membrane compared to gas pressure [10]. The SOEC electrolysis method is very popular due to its high efficiency. This electrolysis process takes place at high temperatures (about 1,000°C). In this method, special waterproof and heat-resistant materials are used. This waterproof feature makes it possible to minimize the loss of water vapor for hydrogen production at high working temperatures. This method is used in high pressure and temperature and makes it more economic and efficient than the other two methods. This technology can be used in fuel cell mode to produce synthesis gas. In large-scale applications, the degradation and brittleness of ceramic materials due to high operating temperature and pressure are the two major problems of this method. Although this method has high efficiency and durability, it has been researched on a laboratory scale and has not yet been commercialized. The high operating temperature and infeasibility of this technology in the microgrid and in large scale enforce higher thermal shock resistances [10]. Equations (4.9) and (4.10) occur at the anode and cathode, respectively:

$$O^{2-} \rightarrow 0.5O_2 + 2e^- \qquad (4.9)$$

$$H_2O + 2e^- \rightarrow H_2 + 4O^{2-} \qquad (4.10)$$

The waste heat of the industrial processes can be recovered as an auxiliary heat reservoir for SOEC. The features of this method are as follows:

- Increase in the efficiency of the electrolysis process, especially in long-term processes that have high currents, and an attempt at the reduction of the further deterioration of the system.
- Improvement and development of electrochemical materials for chemical stabilization in terms of high levels of conductivity at high working temperatures.
- Improvement of electrodes and electrolytic materials for greater conductivity during oxidation and a reduction in the coefficient of thermal expansion, for better compatibility and stability of all system components under hot environmental conditions.

4.2.2 Hydrogen Storage

The green hydrogen can be stored after production and can be transported and then stored at the consumption side. Hydrogen transportation and storage are necessary because some pollutant industrial plants are far away from the consumption sides. Moreover, the intermittent nature of renewable energy sources may limit hydrogen production. Therefore, paying attention to the hydrogen supply chain for continuous procurement of demand is important [9]. In remote and off-grid areas, storing hydrogen and providing energy services are necessary and vital. In these regions, electricity demand is provided by renewable energy sources. Three hydrogen storage technologies include compressed gaseous hydrogen (CH_2), metal hydride (MH), and liquid organic hydrogen carriers (LOHC). The simplest, most common, and most mature way of storing hydrogen is in the form of compressed gas (CH_2), and it is stored at carbon fiber tanks that have a long life cycle, simple design, the ability to refuel and release it, and a high storage pressure of up to 350 bar [10]. Due to its commercial availability and lower cost compared to liquid hydrogen, hydrogen is mainly stored as a compressed gas. But storing hydrogen as a gas has some problems: due to volumetric density of hydrogen gas, storage requires more space and is expensive, so solid or liquid storage can resolve this problem. Due to the lower volumetric density of hydrogen gas and the higher storage pressure, safety issues must be considered in design and development. For example, hydrogen gas storage in fuel cell electrolyzers should be planned according to operating standards and at low working pressures [10]. Due to its utilization in the transportation sector, hydrogen gas is stored at pressures of 200 to 750 bar, and its volumetric energy density increases to 4,276 MJ/m^3 [11]. It is not possible to store hydrogen in the form of compressed gas if space and location are limited, so in this condition, hydrogen can be stored in the liquid phase. By cooling hydrogen to low temperatures of −253°C, it can be turned into a liquid. This cooling process increases energy demand by 20 to 50% of lower heating value (LHV). With very strong insulation in liquid storage tanks, liquid evaporation and energy losses can be reduced, but due to energy losses, this technology is not very useful for medium- and long-term storage [11]. The LOHC includes n-heterocycles and cycloalkanes. In long-term storage, land and sea transportation systems, as well as power-to-X-to-power (P2X2P) applications, it has been taken into account. Liquid ethylene glycol, dimethyl ether, N-ethylcarbazole, dibenzyltoluene, toluene, and biphenyl are some of these storage carriers. When storing liquid H_2, about 43% of hydrogen is lost, which enforces more advanced technologies and infrastructures to be used for reducing hydrogen losses and improving storage efficiency [9]. The advantages of LOHC are summarized as follows:

- Easy transportation
- Low cost due to lack of high-pressure tanks
- Need for lower insulation
- Possibility of hydrogen storage in ambient conditions

Three types of electrolyzers are shown in Figure 4.2.

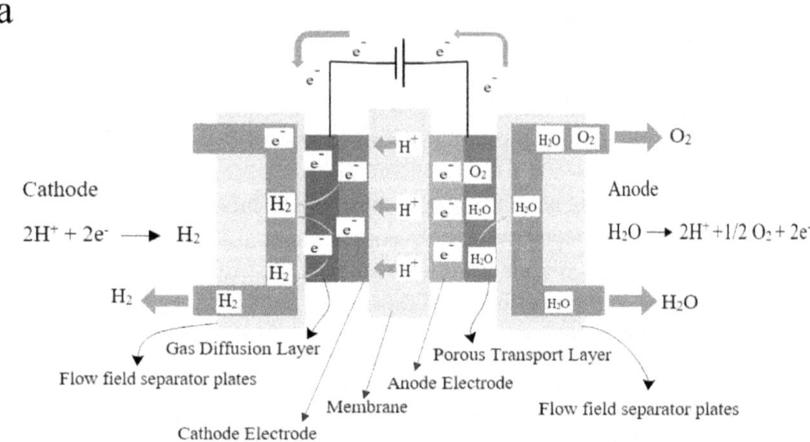

Solid Oxide Electrolysis

$$Anode: \quad O2^- \longrightarrow 1/2\ O_2 + 2e^-$$

$$Cathode: \quad H_2O + 2e^- \longrightarrow H_2 + O^{2-}$$

FIGURE 4.2 Structure of (a) PEM, (b) alkaline, and (c) solid oxide electrolyzers.

Metal hydrides (MH) have both stationary and portable capabilities for H_2 storage in various industries and transportation systems. For temperatures below 100°C, especially in the transportation sector, they have attracted wide attention. In cases of non-portable applications, it can be used at high and low temperature levels.

Under operating temperatures less than 100°C, this technology can be used and stored for domestic scales. Metal hydrides can also be operated at moderate temperatures and capacities using long-term thermal energy storages for discharging H_2. Compared to liquefied and compressed gas H_2, this technology has several advantages, including:

- Low working pressure
- Higher safety
- No welding
- Simple and convenient operation
- Easier hydrogen absorption using magnesium-based hydrides

Solid metal hydride storages have advantages compared to other ones, that is, liquid and gas storages, such as:

- Higher density of hydrogen storage
- Higher safety
- Easier transportation, in accordance with ambient conditions

Various characteristics must be considered for hydrogen storage and technology selection, including:

- Costs
- Durability
- Efficiency
- Safety

Each of these issues may be different according to its type and use case. In all technologies, safety should be prioritized. For small scales, grid-connected, and off-grid operating modes, CH_2 gas is desired because of its simple structure and scalability. But it should be noted that due to spillage and release of hydrogen gas, as well as the risk of explosion and fire, safety issues must be considered. In contrast, the use of metal hydrides has less safety requirements due to low pressure and cryogenic working temperature. It has simpler operation and transportation, and H_2 leakage is as low as possible, and it has also been successfully used in microgrids. Liquid storage technology is a suitable option for off-grid users due to storage in little space and small footprint [9].

4.2.3 Hydrogen Use Cases

Fuel cells are a type of electrochemical converter that produces electricity and heat directly and from hydrogen without carbon emissions. Fuel cells consist of three important components in their internal structure:

- Anode
- Cathode
- Electrolyte

Free electrons are released by a chemical reaction in the hydrogen anode. These free electrons flow through an electrolyte toward the cathode, and a reduction reaction occurs in the cathode, in which electrons and cations reduce oxygen into water [10]. Fuel cells have many advantages, including:

- High energy and power density
- Versatility
- Modular design
- Mechanical stability
- Thermal stability
- Very low noise and vibration
- Low maintenance cost
- Long charge-and-discharge cycle (around ±20,000 length of 15 years)
- Easy installation and transportation

High investment cost in large scales are introduced as a main challenge of fuel cells. Based on electrolyte type, fuel cells are divided into three categories:

- Proton-exchange membrane fuel cell (PEM FC)
- Alkaline fuel cell (AFC)
- Solid oxide fuel cell (SOFC)

The PEM FC has a working temperature of 50 to 100°C and output power less than 250 kW. Its electrical efficiency is limited to 60%. Two chemical reactions in the anode and the cathode are modeled as equations (4.11) and (4.12), respectively:

$$H_2 \rightarrow 2H^+ + 2e^- \tag{4.11}$$

$$0.5O_2 + 2H^+ + 2e^- \rightarrow H_2O \tag{4.12}$$

AFC operates at a temperature of 90 to 100°C. Its output power is 10–100 kW, with electrical efficiency between 60 and 70%. The AFC chemical process is given by equations (4.13) and (4.14) [10]:

$$\text{Anode: } H_2 + 2OH^- \rightarrow H_2O + 2e^- \tag{4.13}$$

$$\text{Cathode: } 0.5O_2 + H_2O + 2e^- \rightarrow 2OH^- \tag{4.14}$$

The SOFC has an operating temperature between 650 and 1,000°C, and its output power reaches 3,000 kW. It has an electrical efficiency less than 60%, with the following chemical reactions in the anode and the cathode, respectively:

$$\text{Anode: } H_2O + O^{2-} \rightarrow H_2O + 2e^- \tag{4.15}$$

$$\text{Cathode: } 0.5O_2 + 2e^- \rightarrow O^{2-} \tag{4.16}$$

4.3 CARBON CAPTURE, STORAGE, AND UTILIZATION (CCSU)

According to the research report of the International Energy Agency (IEA), the emission of greenhouse gases and the emission of CO_2 in the world have reached more than 34 billion tons, which has severely changed weather conditions, and the result is severe environmental pollution as well as the melting of glaciers, polarization, and rising seas and oceans. Therefore, according to the 2015 Paris Agreement, countries are forced to reduce CO_2 emissions and limit the increase in global temperature.

The use of technologies and innovations related to carbon absorption and storage is very vital and important. In CCS technology, carbon is absorbed and permanently stored in suitable tanks, while in CCU technology, carbon is absorbed as a raw material for the preparation of gases, such as

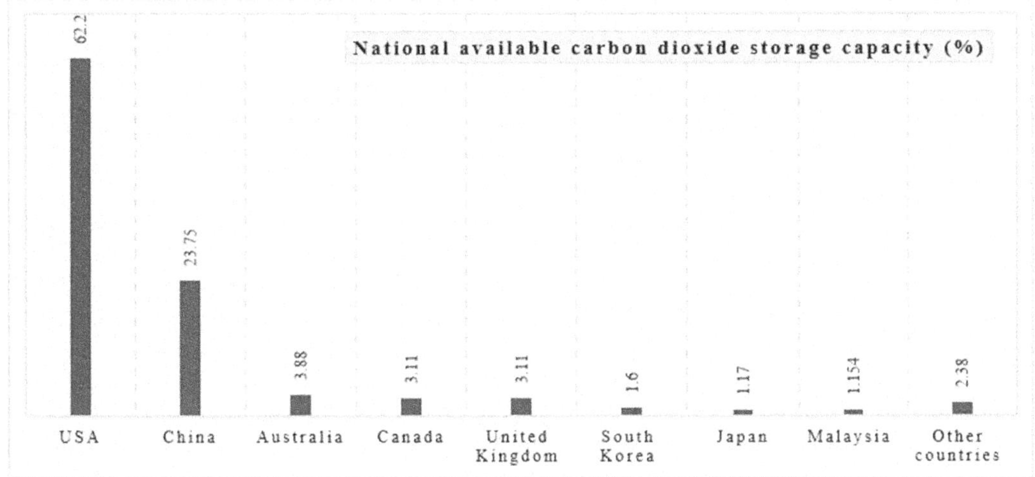

FIGURE 4.3 The international available CO_2 storage capacity all around the world.

synthetic natural gas (SNG), liquid fuels, dimethyl ether (DME), urea, ethylene, and formic acid. In these processes, power-to-gas (PTG) and power-to-liquid (PTL) have wide applications, and special attention has been paid to them [12].

In fact, the purpose of PTG and PTL processes is to convert renewable energies with intermittent production into safe and stable chemical energies. This energy conversion process is divided into three parts:

- Absorption of CO_2 from the exhaust gases of various industries, such as gas power plants and steel and cement plants
- Production of H_2 through the process of water electrolysis with a device and its combination with renewable energies
- Production of hydrocarbons with added value through the hydrogenation of CO_2 by direct and indirect methods

The difference between the two is that in the PTL method, synthesizing methods are used to convert CO_2 into hydrocarbons, which have a high volumetric density and are used for fuel production and applications such as gasoline, diesel, and jet fuel. But in the power-to-gas (PTG) method with the methanation process, CO_2 is converted into synthetic natural gas (SNG), which can be injected into portable fuel tankers or connected to gas pipelines [12]. Figure 4.3 represents the international available CO_2 storage capacity all around the world. For example, USA, China, and Australia developed carbon dioxide storage projects with volumes of 8,062, 3,077, and 502 Gt, or 62.2, 23.75, and 3.88% of total CO_2 storage capacity in the world. In these projects, carbon dioxide is mainly absorbed based on petroleum refining, chemical production, and natural gas and hydrogen processing. In general, the CCSU process consists of the following stages [13]:

- Identifying the resource of CO_2
- Extracting and absorbing CO_2
- Refining CO_2
- Compressing CO_2
- Transporting and storing CO_2

4.3.1 CARBON CAPTURE

Carbon capture, sequestration, and storage is a process to absorb CO_2, and it can be done from industrial processes and fossil fuel power plants. This CO_2 absorption process leads to a huge reduction in CO_2 emissions and prevents atmospheric pollution. CCS facilities installed in power plants can absorb about 90% of CO_2 produced. So far, there are about 51 CCS centers around the world. Three main steps are defined for CCS in every industry or power plant, and they include capture, transportation, and storage. The CO_2 absorption stage itself includes three stages: carbon capture after combustion (this method is mostly used in power plants), carbon capture before combustion (this method is mostly used in various industrial centers), and oxygen combustion systems [14].

The solubility, density, reactivity, thermal capacity, and pressure sensitivity of a CO_2 absorption process are introduced as some prominent features and characteristics which should be considered in design and planning strategies. Gasification before combustion and extraction of CO_2 from the fuel input to the combustion chamber usually takes place in the combined cycle power plants. The input fuel gases are a combination of H_2, CO, and CO_2 absorbed in the carbon dioxide absorption process, and a clean hydrogen fuel is injected into the combustion chamber. In this cycle, CO_2 has a high concentration ratio, and its absorption process is easier, with low energy consumption and higher investment cost [13].

The technology of carbon capture before combustion is widely used in industrial centers. In this method, fuel is entered as gas and CO_2 is extracted from it. The extraction of CO_2 from fossil fuels, gas, or biomass before combustion produces synthesis gas. *Synthesis gas* is a combination of CO, H_2, and a very small amount of CO_2, which is used as a green fuel for the combustion chamber and turbine. This technology is used in high pressure without pollutants, such as NO_x and SO_x. Before the synthesized gas is injected into the combustion chamber, it needs to be purified and cleaned from contaminated gases. In centers where pre-combustion carbon capture systems are installed, CO is absorbed through the reaction of water and gas according to the following formula [14]:

$$CO + H_2O \leftrightarrow H_2 + CO_2 \tag{4.17}$$

Carbon capture in the post-combustion method is absorbed from the exhaust. Many different industries use the pre-combustion method to absorb CO_2. The carbon capture, storage, and utilization (CCSU) in fossil fuel thermal power plants (gas power plants and combined cycle) can be achieved from the exhaust outlet, where CO_2 is absorbed from flue gases and can be stored [14]. According to IEA's report, greenhouse gases are emitted by heavy industries, such as steel, cement, and chemical industries. About 60% of energy consumption in these three industries and about 70% of emissions are created by these three industries [15]. The decarbonization process takes place from the exit chimney of the industrial centers and the exhaust gases. The exhaust gases pressure is usually low, causing higher energy utilization and greater CO_2 absorption investment cost [13]. In a power plant or industrial center, absorption units such as desulfurization, nitrogen removal, and dust removal facilities are placed after the purification systems. This method is more economical compared to other methods, but it requires the installation of new equipment. This method has disadvantages, such as the low efficiency of CO_2 capture and the difficulty of connecting to power plants. The main obstacle to using this technology is the low pressure of CO_2 in the exhaust chimney. Therefore, this technology requires research and development for its evolution and for reaching new generation and in increasing its efficiency. The carbon capture methods in this technology are as follows [14]:

1. Absorption solvent–based methods
2. Adsorption–physical separation
3. Membrane separation
4. Chemical looping combustion and calcium looping process
5. Cryogenic methods

The membrane-based installation used to capture CO_2 in a power plant consumes about 162 kWh of energy to absorb 1 t of CO_2 from the chimney, and this energy consumption causes the efficiency of the power plant to decrease from 38 to 33%. This reduction in efficiency should be compensated by the government and supporting policies and the growth and development of CCS and CCU technologies and be an incentive for the use of CO_2 capture systems [16]. Aqueous mineral carbonation is used to convert CO_2 extracted from flue gases into carbonates with different rocks, such as dunnite, and minerals, such as antigorite, olivine, and lizardite, and it greatly reduces CO_2 emissions. This mineral carbonation is a mature and complete process for absorbing and storing carbon. This carbonation process can be done in one-step, two-step, and acid dissolution processes [17]. Mineral waste can be used to capture CO_2, which is a useful method in terms of environmental issues and CO_2 reduction, as well as with the growing concerns of environmental issues; it is thus possible to produce a carbonated product by using this mineral waste, and its reaction with CO_2 has made it useful for the environment and humans. Therefore, with this process in the mining industry, it is possible to both reduce the pollution of carbon emissions and industrial waste and help clean the Earth. And this is a circular economy approach that contributes greatly to the long-term sustainability of economic and environmental issues [15]. In oil and gas refining facilities, excess CO_2 can be used to produce chemical fertilizers and urea. The production of CO_2 in this process is temporary, and it is immediately used for oil recovery. Although this method is a temporary production and storage, it can be used to reduce CO_2 emissions. It should be noted that the possibility of CO_2 leakage in this type of facility is high due to its temporary nature [17]. Oxygen combustion, or oxy-fuel, is a method in which fuel is burned inside pure oxygen and CO_2 is extracted from it. Separation is done with high density and temperature higher than ambient conditions. This process is used in energy systems with gas turbines. This absorbed CO_2 is compressed and liquefied and then transported to a storage location through transportation systems, such as ships or trains, or through pipelines. It must be said that gas and oil wells (extracted and unused), coal beds, and saltwater aquifers are the best places for long-term storage of CO_2. From the point of view of energy production and energy efficiency, oxygen fuel combustion is considered a promising method [14].

Direct capture of CO_2 from the atmosphere is one of the methods of CO_2 capture. According to IEA's report, by 2070, 10 GT of greenhouse gases should be absorbed from the air, and 90% of these greenhouse gases should be stored underground and 10% converted into CO_2-based fuels. This capture process is carried out by chemical adsorbents, such as calcium, potassium, and sodium hydroxides. So far, about 27 CCSU facilities have been installed in the world to absorb 40 Mt (million tons) of CO_2 [17]. Among the advantages of this technology, we can mention the relatively simple carbon capture technology, cost reduction due to reducing the dimensions and size of the steam boiler, the possibility of integrating it with existing technologies, and reducing the mass flow rate of exhaust gases. And it has disadvantages, such as high working temperature and the need for special materials, reduced efficiency, and high investment cost. In addition to the aforementioned methods, other methods for CO_2 capture and O_2 increase and air optimization can be mentioned. Afforestation or reforestation, biochar method, soil reclamation, and advanced weathering techniques are some of these techniques [14]. Trees and plants use CO_2 to carry out photosynthesis and help absorb CO_2 in the world. Therefore, planting trees and plants around the world is an urgent need and should be given special attention [17]. With all these explanations, operating costs, storage capacity, technical and technical maturity, and economic issues and environmental impacts may vary according to the technology used. With the research conducted, and according to IEA's report, new energies, especially biomass for electricity generation, have more potential and effects on reducing environmental pollution, especially carbon emissions [16]. By using ionic liquids, CO_2 absorption efficiency can be improved, and thus, the energy required for solvent regeneration can be reduced. In fact, these are special solvents that have less impurities and are not prone to degradation. By adding nanoparticles such as SiO_2 or carbon nanotubes to the solvent, the performance of CO_2 absorption from energy efficiency and economic points of view will be improved [18].

4.3.2 CARBON UTILIZATION

CO_2 has been used both as a green fuel and in chemicals in recent years. Due to the lack of environmental pollution, it can be considered a green fuel. Its emissions are very low compared to fossil fuels. In general, CO_2 consumption is done in two ways, direct and indirect. The methods of direct use of CO_2 are natural mineral carbonation and oil recovery. Among the direct uses of CO_2, we can mention carbonated soft drinks, fire extinguishers, and foaming agents. *Direct use* means the direct and unmediated consumption of CO_2 as a raw material in an industrial process without turning it into another product, such as the use of CO_2 in iron and steel manufacturing, in alumina industries, in the recovery of CO_2 rare earth metals from acid mine drainage, in advanced oil recovery, and in concrete as a building material. The method of *indirect consumption* of CO_2 is actually its conversion into fine chemicals and fuels in a carbon cycle. By consuming CO_2 in this method, CO_2 can be captured and stored again. But with a smaller amount at this stage, the fuel is converted into green energy by photocatalytic, electrochemical, thermochemical, or biological approach [14].

Direct carbonation has more advantages than indirect carbonation and is more practical. The carbonation of minerals is a mature and advanced method and has the ability of permanent storage and the possibility of storage in any place, even in the vicinity of CO_2 [17]. In thermochemical conversion, the required energy is provided in the presence of a reducing agent to convert CO_2 into chemicals and fuel. In this method, it is difficult to achieve high efficiency, and to achieve it, high temperature and pressure of the catalyst equipment are required. In this situation, the catalyst must be operated in a stable condition. In this method, methane gas is modified by CO_2 thermal conversion processes. In the electrochemical method, CO_2 turns into products such as ethanol, methanol, oxalic acid, formic acid, formaldehyde, and several other products. The faradaic efficiency of this method depends on parameters such as pH, catalyst, electrode potential, etc. The advantage of this method is the need for low temperature and pressure, which has made this process popular. One of the disadvantages of this process is the production of hydrogen gas, along with high energy consumption during the chemical reaction of the cathode with water. Solvents other than water should be used to solve this problem [14]. In an industry, by performing catalytic processes, such as those of H_2, CO, and CO_2, a large amount of methanol can be produced. Other chemicals can be extracted from CO_2, such as formic acid, dimethyl carbonate, copolymers, ethylene carbonate, salicylic acid, formaldehyde, polymer building blocks, and cyclic carbonates [17]. The photocatalysts are one of the most prominent processes for converting CO_2 into fuel, because they directly convert CO_2 into usable fuel. The important components of these catalysts are light absorption panels, charge separation, and activation of CO_2 on the catalyst surface. CO_2 activation is one of the complex steps of this method and has not fully matured and needs more research and development. In biological conversion, CO_2 is processed by photosynthesis using living species, such as microalgae, plants, and cyanobacteria. In this process, CO_2 turns into sugars, proteins, lipids, and other things. The process of photosynthesis forms carbohydrates with the reaction of CO_2 and H_2O. CO_2 is absorbed by these organisms and turns into various products. The effective factors in this process are things such as the growth rate of organisms, sufficient light and temperature, required nutrients, and biomass collection [14]. Research has shown that the growth of microalgae leads to an increase in CO_2 concentration. By forming a biological process from microalgae, ethanol gas can be made and used in biodiesel fuel [17].

Recent research explores new technologies and processes using renewable energy for carbon sequestration and minimal carbon footprint. The electrochemical reduction of CO_2 is one of the processes in which compounds and products such as methane, methanol, ethylene, and formic acid can be obtained by using renewable energy sources. Carbon products, such as metal-free carbon and metal-based carbon catalysts, are suitable catalysts for use in these processes. These metallic carbon materials are widely used in the structure of catalysts due to their suitable properties, such as porous and adjustable structure, resistance to acids, abundance and availability in nature, temperature, and environmentally friendly characteristics. Various types of plasma, especially non-thermal plasma, are another technology used in microwave, radio frequency, glow, dielectric barrier, corona, and nanosecond pulse discharge options, as some examples. In the agricultural sector, algae photosynthesis can be used to convert these into biological products and biofuels, as well as the cultivation of agricultural products. Several studies have been carried out on the optimization of photobioreactors and algae cultivation conditions, as well as genetic engineering and cultivation of algae jointly with microorganisms. There are many limitations in the use of CO_2 that require more research and studies. Cost and economic issues are one of these things, so that each specific technology has its own costs. For electrochemical regeneration, choosing a suitable catalyst with high efficiency and low energy consumption is another factor that needs to be investigated. Photocatalysts have limited efficiency and scalability issues that require scientific research for the advancement of this technology and its optimization. In the agricultural sector, greenhouses and algae planting on a large scale require extensive investment in the infrastructure sector in this area [18].

4.3.3 CARBON STORAGE

One of the important matters that was considered in the 2015 Paris Agreement was CO_2 storage. To solve the existing solutions to and hurdles of CO_2 storage, examinations and research and development in this field are needed. CO_2 storage is done in two general ways, natural and artificial. *Terrestrial sequestration.* The absorption of CO_2 from the atmosphere by photosynthesis and its storage in soil and plant covers is called *terrestrial sequestration* [14]. *Ocean sequestration.* The best and most promising method for CO_2 storage is sequestration of the ocean. This type of storage has the ability to store CO_2 in high capacities to the extent that it can store about 90% of the available CO_2. It is injected and stored to a depth of about 1 km in the ocean or the sea through pipes and marine platforms. Due to the lower density of water, it is possible for CO_2 to dissolve in water, and there are environmental concerns in this method. Due to the high concentration of CO_2, it is possible to acidify the water and change the PH. Therefore, investigation and operational research of this method are very expensive and complicated. Storing CO_2 in this way has many risks, especially for marine life, so this method can only be used for the temporary storage of CO_2, which helps slow down global warming [17]. *CO_2 capture through gas hydrates.* Chlorate hydrates are ice crystals that are known as gas hydrates, and they are formed by different components, such as nitrogen, oxygen, hydrogen, and carbon, by water molecules. CO_2 capture through these gas hydrates has been considered an effective and efficient technique in the past decades. This type of CO_2 capture process is used for gas separation, transportation, refrigeration, and storage. In this method, CO_2 is used in the form of clathrate hydrates under the seabed for long-term storage in bulk. It should be noted that for CO_2 storage in this process, a special solution should be considered, and research and development around the issues should be fully investigated. Among these conditions are thermodynamic conditions, phase balance of CO_2 hydrates in the seabed, wave movement, sea depth, studies of seabed sand; all these issues are important in terms of the stability of hydrates, and sufficient studies should be done. *Geological sequestration.* In this method, CO_2 capture is done through convulsion, and it is one of the most used methods. By storing CO_2 in underground tanks, there is a possibility of leakage, and therefore, applications and security measures are necessary to prevent CO_2 leakage. Thousands of people have been killed due to the explosion of CO_2 tanks in Africa [2]. In this method, CO_2 is stored in underground beds, such as saltwater underground beds, abandoned oil and gas reservoirs, and abandoned coal underground beds, and this includes various techniques mentioned in the following [14].

Storage in saltwater aquifers. Formations and beds of salt water are considered as the best place to store CO_2 because these beds have a high potential capacity to store large amounts of CO_2 for long periods of time. Also, since these saltwater beds are not used for agriculture, industry, and consumption, they are a suitable option for CO_2 storage. For example, it has been estimated that Alberta's saltwater beds can store about 103 Gt of CO_2. *Abandoned and unusable gas and oil tanks.* With the passage of time and the extraction of oil and gas mines, and with the subsidence of these tanks and the reduction of pressure inside them, these can be used for storage. Discharged gas tanks are especially suitable for CO_2 storage due to their sealing and leakage properties, advanced hydrocarbon recovery, and suitable capacity. The estimated storage in this type of reservoirs is about 390 Gt [14]. Mineral storage or mineral carbonation storage is permanent, and the possibility of leakage is very low. Mineral carbonization is done both on-site and off-site and turns it into calcite ($CaCO_3$), magnesite ($MgCO_3$), and dolomite, and in this process, calcium and magnesium silicates are widely used around the world [17]. *Storage in coal beds.* During the recovery and extraction process of methane gas, seams and coal beds can be a good storage place for CO_2 due to the many fractures in them. By injecting CO_2 into these holes, the possibility of extracting CH_4 increases greatly. It is replaced by methane extraction and stored in coal seams. This storage is highly dependent on the porosity and permeability of coal. Numerous laboratory and field reports of this type of storage can be seen in the San Juan Basin, the Sydney Basin of Australia, and the deep coal seam in Alberta, Canada [14]. *CO_2 storage during enhanced oil recovery (EOR).* This method has been used in oil fields so that when CO_2 enters the tanks, it causes the oil in the tank to swell and reduces its surface tension and viscosity; this process causes the mixing of CO_2 with oil, and a single-phase fluid is formed. This mixing of CO_2 and oil inside the tank causes the recovery of hydrocarbons. At pressures higher than the mixing pressure, CO_2 mixes with oil. Many reservoirs in the world have been formed in this way, such as the Weyburn field in Canada, the Shengli oil field in China, and an oil field in West Texas [14]. Although this method is not a permanent method, CO_2 is very effective in oil recovery and plays an effective role in reducing atmospheric CO_2 immediately. Also, in gas treatment facilities, excess CO_2 is used to produce fertilizer and urea [17]. So far, a lot of research has been done in the field of CCS and CCU technologies in the field of laboratory and commercial science, and considering the scope of the subject and further commercialization of these processes, more research and development is needed in this field. Some of these key and important fields in the future can be summarized with these topics: examining economic issues and comparing costs, safety and health issues, designing advanced and innovative processes and evaluating their performance, investigating and evaluating environmental issues and advertising strategies, community incentives and their participation, regulations and guidelines, and calculated planning for CCU and CCS technologies [16].

There are several challenges for CO_2 storage, which include long-term storage and leakage risks, economic and government costs and related laws and regulations, suitable storage location, effective and continuous monitoring of storage holes, and suitable technologies for sealing and pressure management [18].

4.4 METHANE GENERATION USING GREEN H₂ AND CAPTURED CO₂

According to the environmental conditions, the primary energy source and the required technology for hydrogen production are divided into blue, gray, brown, black, and green colors. Based on this, *blue* hydrogen is obtained from natural gas vapor. During its chemical processes, natural gas is converted into CO_2 and hydrogen (H_2) derivatives. In this process, about 90% of CO_2 is absorbed, and the rest cannot be absorbed and is thus released into the atmosphere. Absorbed industrial carbon is stored in underground tanks, with necessary arrangements, and can be consumed for a long time. However, long-term storage has uncertain negative effects, and its leakage and release may have harmful effects on the environment. Fossil fuels, such as coal, brown coal, methane, and natural gas, produce *gray* hydrogen using thermal reformers. This process is similar to the type used for blue hydrogen, with a fundamental difference that the produced carbon is not absorbed and all of it is released into the atmosphere (for the production of 1 t of gray hydrogen, about 11 t of CO_2 enter the atmosphere). This type of process is the most widely used today, and hydrogen gas is produced this way. In contrast to these, *green* hydrogen is produced by using the electrolysis process of water and combining it with electricity. In this planet, the input is electricity and water, and the output is hydrogen (H_2) and oxygen (O_2). According to global decisions and the transition to green energy and CO_2 reduction, water electrolysis and green hydrogen production are an efficient and effective technology to reduce environmental pollution. And this technology has the ability to scale and is expanding day by day by using renewable energies, especially the sun and wind. By using water electrolysis technology and its combination with renewable energies, it can be done with chemical processes, and excess electrical energy can be produced from these primary sources of hydrogen and oxygen, thus establishing energy balance and security for electricity and energy demand. These produced oxygen and hydrogen can also be directly used in various sectors of the industry and transportation, such as fuel cell vehicles, and in various other industries as an energy carrier and primary source of energy. In addition to these, the produced hydrogen can be used as a raw material for the production of synthetic fuels and ammonia in the petrochemical and chemical industries. And research has shown that hydrogen energy storage systems have better efficiency and performance than batteries [19].

Although manufacturing industries and the chimneys of power plants have many environmental challenges on footprints, absorption, storage, and consumption of CO_2, emission reduction and energy efficiency in various industries have economic and environmental benefits. Therefore, it can be utilized together with catalyzers and methanation systems to produce useful chemicals and energy carriers. Greenhouse gases mainly include CO_2, methane (CH_4), nitrogen oxide (N_2O), chlorofluorocarbons (CFCs), and fluorocarbons (CFs). Among them, the concentration of CO_2 has increased greatly, causing pollutant substances. From chemistry points of view, CO_2 is a useful, available, abundant, and environmentally compatible material with the capability to be converted into other energy products, especially for storage. Among these products, salicylic acid (a medicinal substance), urea synthesis (for nitrogen fertilizers and plastic), and polycarbonates (for plastic) can be introduced. Also, CO_2 can be converted into fuels, such as methanol and synthetic gases, causing lower consumption of petroleum products and, consequently, reducing global warming [20]. Methane production from CO_2 was proposed by Paul Sabatier in 1902. Methane is a simple organic hydrocarbon substance which can be used as a raw material to produce acetylene, carbon monoxide, and formaldehyde. Its calorific value is high, and it can be used for small-scale heating and for fueling power plants, as well as in the transportation sector.

Dual-function materials (DFMs) are usually used for CO_2 capture and conversion into CH_4 inside a Sabatier reactor under high temperatures of about 320°C using catalysts [21]. During the CO_2 capturing process, CaO is commonly used as an adsorbent. Then, hydrogenation is carried out by either group VIII metals or nickel-based catalysts. After adsorbing CO_2 by CaO, CO_2 is reacted with H_2 inside Ni-based catalysts. To reach an optimum CH_4 adsorption temperature, a small amount of platinum group metals, such as Ru, Pt, and Pd, is added to Ni-based catalysts. In addition to Ni, Ru is a more active metal for CO_2 methanation. At low temperatures, Ni-based catalysts are deactivated due to the interactions of metal particles with carbon monoxide, while ruthenium has a more stable status [20]. In Ru-based catalysts, the energy efficiency of methane extraction is improved by adding MgO to Ru metals [21]. Platinum has also been used in catalysts for carbonization and methanation. Inside Pt-based catalysts, platinum nanoparticles are uniformly dispersed in titania nanotubes. Moreover, CO_2 is adsorbed by them for hydrogenation of CO_2 at low temperatures [20]. Ru, Ni, and Pt have higher energy efficiency in fast CH_4 extraction processes. Meanwhile, noble metals catalysts are more expensive than Ni-based ones. Catalysts are operated in the temperature range of 300–350°C, with average lifespan of less than five years. Poisoning and scorching of catalysts may reduce their life cycle.

Adiabatic fixed-bed methanation is one of the most common and newest thermochemical methanation methods. In this method, the particles in the catalyst are pushed into the reactor in millimeters. The methanation reaction is very exothermic, and temperature control is one of the upcoming challenges, which is maintained in the range of 300°C with appropriate control systems. To reach this suitable operating temperature, a series of reactors with interstage cooling is used. In addition to fixed-bed methanation, reactors with fluidized-bed technology have also been introduced, which are suitable for large-scale processes. In this method, fine catalyst particles are fluidized by gaseous

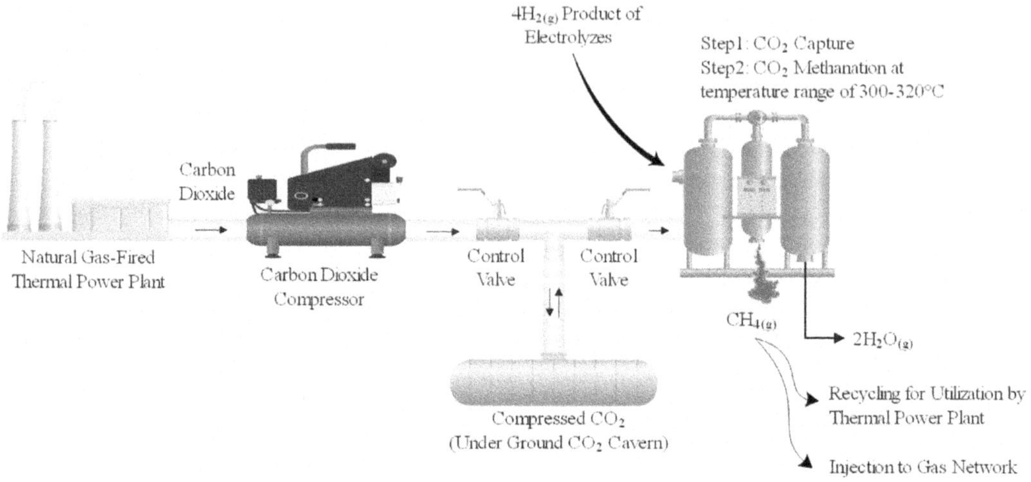

FIGURE 4.4 Schematic representation of carbon storage, capture, and utilization for hydrogenation and methane extraction.

reactants, and the methanation process takes place. There are isothermal conditions inside the reactor, and this is one of the advantages of this method that, due to the uniform mixing of the solid in the gas, the heat is also uniform and the design is relatively simple, and due to the uniform and low temperature, it is possible to use a reactor [22].

In the CCSU facility, the carbon dioxide gases emitted from the natural gas fired power plants are compressed and then stored inside an underground cavern. Methane and vapor are produced by integrating the hydrogen product of the electrolizing process with the carbon dioxide captured from the storage cavern at the temperature range of 300–320°C. The methane is either injected into the gas network or recycled to be utilized by the conventional thermal power plants.

4.5 CONCLUSIONS AND FUTURE TRENDS

For CCSU, about 70% of the total cost is associated with CO$_2$ absorption facilities, which should be reduced. Expensive, high-quality absorbents, and their lower capacity, which leads to higher energy consumption, have caused greater capital investment and operation and maintenance costs. The contamination of absorbent membranes is also one of the challenges ahead. Optimum operation and exploitation conditions require investigation in research and development of CCSUs. The development and optimization of the CCSU operation require the full support of CO$_2$ absorption infrastructures, transportation systems, transmission pipelines, and storage systems. Also, financial laws and regulations and social cultural incentives, such as subsidies, and other policy measures are necessary for the optimal use of CCSU systems so that society and the industry can fully understand their importance. Industries related to CCSU, such as chemical and petrochemical industries, power plants, iron and steel, cement, and others, can work on these programs and promote culturalization and commercialization.

Reduction, and control, of CO$_2$ is a global and urgent necessity because of its impact on climate change and global warming. Despite the many research and studies focusing on this field, suggestions and techniques have been made to stabilize this factor and control.

In the methanation process, the internal structure of the catalyst, and its components, has a significant effect on its performance and the methanation of energy efficiency. A catalyst with a high specific surface area and nanoparticles in its internal structure and active metal sites is more optimal. Therefore, for a more efficient and practical use of CO$_2$ and a better understanding of the molecular and atomic structure of catalysts, the use of new reactors and catalysts for low-temperature synthesis will be a new topic and research platform for development. Noble metal catalysts have received much attention and study due to their excellent performance, but their use imposes high costs. So far, ICCU methanation processes have been at the research and laboratory level, and the commercialization and operation of these processes on industrial and larger scales require more efforts. It is a promising technique for green and sustainable hydrogen production, steam methane reforming, and CO_2 absorption, but it also comes with challenges, such as:

- Development and improvement of catalyst structure
- Economic-technical and exergy calculations to optimize process and costs
- Application of heat recovery systems
- Minimization of energy requirement using renewable energies

In order to overcome these challenges, future research is expected to focus on the development of robust artificial intelligence and machine learning techniques. Such techniques are widely used in predicting the optimal properties of efficient catalysts.

However, many studies have been conducted about CCSU technologies. But its technical aspects, such as economic, energy optimization, and efficiency reduction, and its non-technical aspects, such as public culture, government laws and regulations, and assignment of economic subsidies, are barriers and challenges facing this industry and require more attention and investigation. Therefore, more suitable solvents and adsorbents should be used to increase efficiency and reduce energy consumption in the CO_2 absorption stage.

Subsurface CO_2 storage is a method that has been widely considered so far. Therefore, appropriate mechanisms, capacity, and storage location are important issues in planning studies. Safety issues and CO_2 leakage may cause ground-, sea-, and ocean water contamination. Although subsurface storage technology has matured, there is not much desire to use oceanic CO_2 storage due to leakage into the marine ecosystem. CO_2 transfer is done through pipelines in large volumes. To be cost-effective, CO_2 is usually supercritical and in the dense phase, so adjusting the pressure and temperature of the transfer path is necessary to achieve the desired density. Also, the main problem in pipe corrosion is the presence of impurities in CO_2. This issue has not been seriously investigated so far and needs more research.

The widespread application of machine learning for modeling, the improvement of various CCS techniques, exploration, reservoir location, and cost reduction is of interest. The subsurface storage of carbon dioxide is associated with uncertainties, and with the development of predictable models by machine learning algorithms, it is possible to make the necessary predictions and assessments to prevent future risks.

Hydrogen production through water electrolysis is a promising method for generating hydrogen, particularly for transitioning to low-carbon energy systems. In this process, water is separated into H_2 and O_2, making it an environmentally friendly alternative to traditional fossil fuel–based hydrogen production. Hydrogen production, consumption, and storage present several significant challenges that need to be addressed to facilitate its wider adoption as a clean energy source:

- Challenges in hydrogen production consist of high production cost, infrastructure limitations, and material and technological challenges.
- Challenges in hydrogen storage include storage density and volume, safety concerns, and technological development.
- Challenges in hydrogen consumption consist of fuel cell efficiency, market acceptance and infrastructure, and regulatory and standardization issues.

Addressing these challenges is crucial to commercialize H_2-fueled thermal processes, enabling its role in the decarbonization of the energy sector and enhancing water and energy security. Solutions to hydrogen challenges include improving green hydrogen production technologies, developing efficient transportation and storage methods, establishing necessary infrastructure, promoting hydrogen blending in natural gas pipelines, and increasing government support and collaboration. By addressing these challenges through technological innovations, efficient transportation and storage solutions, necessary infrastructure development, hydrogen blending, and increased government support, the hydrogen economy can overcome its current obstacles and pave the way for a cleaner, more sustainable energy future.

REFERENCES

[1] P. Saha, F. A. Akash, S. M. Shovon, M. U. Monir, M. T. Ahmed, M. F. H. Khan, S. M. Sarkar, M. K. Islam, M. M. Hasan, D. V. N. Vo, and A. A. Aziz, "Grey, blue, and green hydrogen: A comprehensive review of production methods and prospects for zero-emission energy," *International Journal of Green Energy*, vol. 21, no. 6, pp. 1383–1397, 2024.

[2] M. Talaie, F. Jabari, and A. A. Foroud, "Modeling and economic dispatch of a novel fresh water, methane and electricity generation system considering green hydrogenation and carbon capture," *Energy Conversion and Management*, vol. 302, p. 118087, 2024.

[3] A. Islam, T. Islam, H. Mahmud, O. Raihan, M. S. Islam, H. M. Marwani, M. M. Rahman, A. M. Asiri, M. M. Hasan, M. N. Hasan, and M. S. Salman, "Accelerating the green hydrogen revolution: A comprehensive analysis of technological advancements and policy interventions," *International Journal of Hydrogen Energy*, vol. 67, pp. 458–486, 2024.

[4] B. S. Zainal, P. J. Ker, H. Mohamed, H. C. Ong, I. M. R. Fattah, S. M. A. Rahman, L. D. Nghiem, and T. M. I. Mahlia, "Recent advancement and assessment of green hydrogen production technologies," *Renewable and Sustainable Energy Reviews*, vol. 189, p. 113941, 2024.

[5] T. Tiwari, G. A. Kaur, P. K. Singh, S. Balayan, A. Mishra, and A. Tiwari, "Emerging bio-capture strategies for greenhouse gas reduction: Navigating challenges towards carbon neutrality," *Science of The Total Environment*, p. 172433, 2024.

[6] A. Nemmour, A. Inayat, I. Janajreh, and C. Ghenai, "Green hydrogen-based E-fuels (E-methane, E-methanol, E-ammonia) to support clean energy transition: A literature review," *International Journal of Hydrogen Energy*, vol. 48, 2023.

[7] Z. Wang, B. Dong, J. Yin, M. Li, Y. Ji, and F. Han, "Towards a marine green power system architecture: Integrating hydrogen and ammonia as zero-carbon fuels for sustainable shipping," *International Journal of Hydrogen Energy*, vol. 50, 2024.

[8] M. Awad, A. Said, M. H. Saad, A. Farouk, M. M. Mahmoud, M. S. Alshammari, M. L. Alghaythi, S. H. E. Abdel Aleem, A. Y. Abdelaziz, and A. I. Omar, "A review of water electrolysis for green hydrogen generation considering PV/wind/hybrid/hydropower/geothermal/tidal and wave/biogas energy systems, economic analysis, and its application," *Alexandria Engineering Journal*, vol. 87, pp. 213–239, 2024.

[9] A. Risco-Bravo, C. Varela, J. Bartels, and E. Zondervan, "From green hydrogen to electricity: A review on recent advances, challenges, and opportunities on power-to-hydrogen-to-power systems," *Renewable and Sustainable Energy Reviews*, vol. 189, p. 113930, 2024.

[10] J. P. Viteri, S. Viteri, C. Alvarez-Vasco, and F. Henao, "A systematic review on green hydrogen for off-grid communities–technologies, advantages, and limitations," *International Journal of Hydrogen Energy*, vol. 48, 2023.

[11] C. Drawer, J. Lange, and M. Kaltschmitt, "Metal hydrides for hydrogen storage – Identification and evaluation of stationary and transportation applications," *Journal of Energy Storage*, vol. 77, p. 109988, 2024.

[12] R. Gao, L. Zhang, L. Wang, X. Zhang, C. Zhang, K. W. Jun, S. K. Kim, H. G. Park, Y. Gao, Y. Zhu, and T. Zhao, "A comparative study on hybrid power-to-liquids/power-to-gas processes coupled with different water electrolysis technologies," *Energy Conversion and Management*, vol. 263, p. 115671, 2022.

[13] S. Davoodi, M. Al-Shargabi, D. A. Wood, V. S. Rukavishnikov, and K. M. Minaev, "Review of technological progress in carbon dioxide capture, storage, and utilization," *Gas Science and Engineering*, p. 205070, 2023.

[14] M. Yusuf and H. Ibrahim, "A comprehensive review on recent trends in carbon capture, utilization, and storage techniques," *Journal of Environmental Chemical Engineering*, p. 111393, 2023.

[15] F. M. Kusin, S. N. M. S. Hasan, V. L. M. Molahid, and M. H. Soomro, "Dual adoption opportunities and prospects for mining and industrial waste recovery through an integrated carbon capture, utilization and storage," *Sustainable Production and Consumption*, vol. 48, 2024.

[16] S. C. Gowd, P. Ganeshan, V. S. Vigneswaran, Md. S. Hossain, D. Kumar, K. Rajendran, H. H. Ngo, and A. Pugazhendhi, "Economic perspectives and policy insights on carbon capture, storage, and utilization for sustainable development," *Science of the Total Environment*, p. 163656, 2023.

[17] M. I. Rashid, Z. Yaqoob, M. Mujtaba, M. Kalam, H. Fayaz, and A. Qazi, "Carbon capture, utilization and storage opportunities to mitigate greenhouse gases," *Heliyon*, vol. 10, 2024.

[18] G. Leonzio and N. Shah, "Recent advancements and challenges in carbon capture, utilization and storage," *Current Opinion in Green and Sustainable Chemistry*, vol. 46, p. 100895, 2024.

[19] S. S. Kumar and H. Lim, "An overview of water electrolysis technologies for green hydrogen production," *Energy Reports*, vol. 8, pp. 13793–13813, 2022.

[20] W. Wei and G. Jinlong, "Methanation of carbon dioxide: An overview," *Frontiers of Chemical Science and Engineering*, vol. 5, pp. 2–10, 2011.

[21] Z.-L. Guo, X.-L. Bian, Y.-B. Du, W.-C. Zhang, D.-D. Yao, and H.-P. Yang, "Recent advances in integrated carbon dioxide capture and methanation technology," *Journal of Fuel Chemistry and Technology*, vol. 51, no. 3, pp. 293–302, 2023.

[22] M. Götz, A. M. Koch, and F. Graf, "State of the art and perspectives of CO_2 methanation process concepts for power-to-gas applications," in *International Gas Union Research Conference*, vol. 13. Norway: International Gas Union Fornebu, 2014.

5 Synergetic Deployment of Power-to-X across Diverse Sectors

Samuel Raafat Fahim and Hany M. Hasanien

Abbreviations

AFID	Alternative Fuels Infrastructure Directive
ANSI	American National Standards Institute
CBAM	Carbon Border Adjustment Mechanism
CEN	European Committee for Standardization (Comité Européen de Normalisation)
CHP	combined heat and power
CO	carbon monoxide
CO_2	carbon dioxide
COP	coefficient of performance
EED	Energy Efficiency Directive
ESR	Effort Sharing Regulation
ETD	Energy Taxation Directive
EU	European Union
EU ETS	EU Emissions Trading System
EV	electric vehicle
FMEA	failure mode and effects analysis
FTA	fault tree analysis
H_2	hydrogen
ICC	International Code Council
IEC	International Electrotechnical Commission
ISO	International Organization for Standardization
LCA	life cycle assessment
LULUCF	land use, land-use change, and forestry
O_2	oxygen
PtG	power-to-gas
PtH	power-to-heat
PtL	power-to-liquid
PtX	power-to-X
PV	photovoltaic
QA	quality assurance
QC	quality control
R&D	research and development
RED	Renewable Energy Directive
RES	renewable energy system
SAF	sustainable aviation fuel
SCF	social climate fund

DOI: 10.1201/9781032719436-5

5.1 INTRODUCTION

The worldwide circumstances from the ongoing conflicts and the arising of new ones, in addition to the global existing burden of climate change, have boosted the transition into a decarbonized economy. The economy becomes greener by eliminating the need for carbon or at least lowering it to be independent of it. Renewable resources offer candidates with tremendous potential. Among these resources are the solar, wind, and ocean energies. Those resources can collaborate strategically to provide a comprehensive solution for societies. This can be done by interconnecting the different renewable energy conversion systems and applying proper storage, thereby creating sustainable and efficient solutions. This solution can be applied in diverse sectors ranging from residential to transportation and, also, industry [1].

Each PtX technology addresses a certain type of application. As for PtG, the main application is fueling operating fleets. This helps increase renewable energy in land, air, and marine sectors. The surplus of PtG can be directed toward the applications of heating/cooking in both the residential and industrial sectors. For PtL, methane and/or methanol can be used in industrial applications that need fossil fuel, which PtL or even PtG could replace. Finally, PtH technology targets heat recovery and electrical boilers [2].

To increase the feasibility of the PtX solutions, a sophisticated optimization process should be carried out. It shall consider both size and location, along with aligning policies, appealing incentives, and governmental regulations. Not forgetting to investigate the availability of synergy between various sectors, that is, cross-*sectoral synergy* [3, 4].

5.2 PtX TECHNOLOGIES OVERVIEW

The synergistic deployment of PtX technologies among various sectors offers a promising route toward a sustainable and decarbonized energy system. In principle, PtX converts electrical energy into other chemicals, such as hydrogen, methane, and methanol. Thus, energy can be stored and utilized later on by various sectors, including transportation, power generation, and industry. In this section, different PtX applications and technologies will be demonstrated, along with the benefits, opportunities, and challenges accompanying the adoption of PtX in various sectors [5].

The integration of PtX into the energy sector can provide many opportunities. Firstly, the penetration level of the renewable energy from solar and wind resources can increase significantly. This is by producing fuels that are both transportable and storable. Secondly, the grid's stability can be enhanced in conclusion to the produced fuels, which increases the flexibility of the grid service and makes it more balanced. Thirdly, the previous two points increase the economy's ability to decarbonize the hard-to-abate industries [6].

Meanwhile, the adoption of PtX into the energy system shall be handled with care. This is due to several reasons, including the elevated cost of PtX systems, the energy storage solutions being cost-effective and efficient, and the lack of both regulatory framework and supportive policies, in addition to the negative environmental impacts that result from PtX integration, which lies in the presence of some harmful gaseous emissions and the heavy use of water resources [7].

By balancing the challenges and opportunities of PtX technologies, the latter are much more significant. To maximize the potential opportunities, PtX adoption shall be synergistically coupled across diverse sectors. This section presents a brief overview of the state of the art of PtX technologies. Besides the potential applications, implementation is successful in addressing the challenges, thus creating a decarbonized, sustainable, secure, and resilient energy system [8]. This is depicted in Figure 5.1.

5.2.1 OVERVIEW OF POWER-TO-GAS, POWER-TO-LIQUID, AND POWER-TO-HEAT

The essence of PtX technologies is converting electrical power into various forms. Each of them is suitable for certain application. Thus, PtX technologies act as the linchpin that fosters both sustainability and efficiency. This could be achieved by optimizing the energy conversion and synergistic deployment of PtX properly among various sectors.

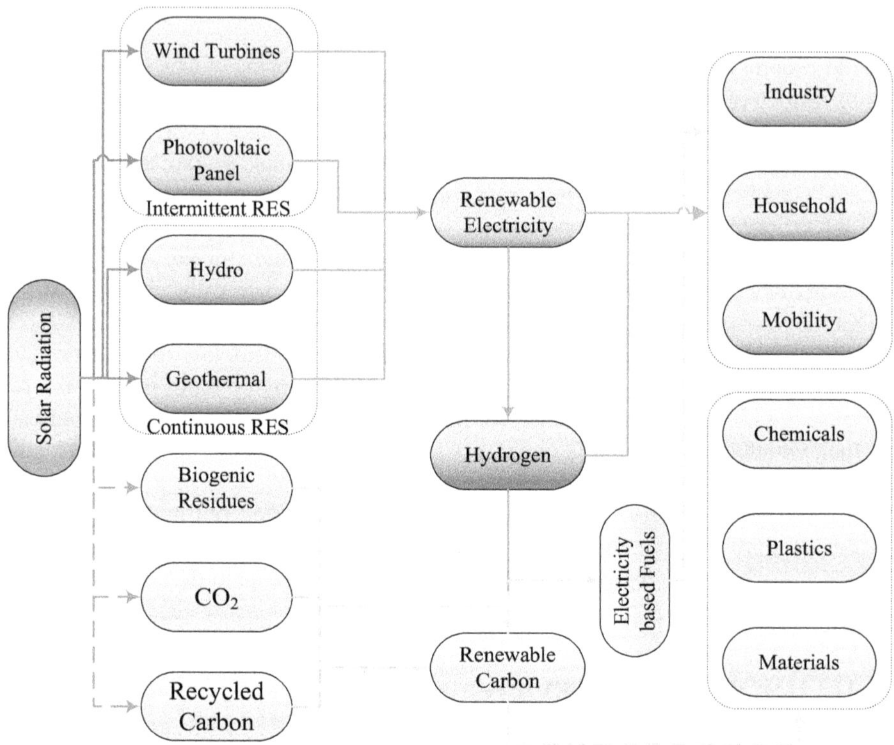

FIGURE 5.1 PtX schematic to decarbonize various sectors.

PtX technologies include power-to-gas, power-to-liquid, and power-to-heat. All three offer a layout of possibilities. Each technology has its opportunities and challenges. Integrating these technologies can revolutionize the residential, transportation, and industrial sectors.

In the residential sector, PtX integration will lead to increased sustainability. Also, it will make the decentralized energy solutions, that is, localized energy generation, more appealing. This will provide a higher degree of energy independence. PtX application can be extended beyond electric vehicles to aviation in the transportation sector. This can redefine the fuel sources available to aviation. In turn, this can reduce carbon emissions, thus impacting the aviation industry's environmental footprint. Finally, in the industrial sector, PtX impacts the manufacturing process, specifically in energy-intensive operations, where it can introduce a radical transformation. In addition, PtX can reduce the factors' reliance on conventional energy resources, as well as diminishing its ecological impacts [9, 10].

5.2.2 State-of-the-Art PtX Technologies and Their Relative Efficiency

Power-to-gas (PtG) technology. In this technology, electrical energy is converted into gases, specifically hydrogen or methane. To achieve this conversion, there are two main steps, *electrolysis* and *conversion*. In electrolysis, the electrical energy is used to split the water (H_2O) molecule into hydrogen (H_2) and oxygen (O). Meanwhile, conversion is the reverse operation, that is, converting gases into electrical energy again. In recent years, electrolysis efficiency has improved significantly, reaching 70–80%. However, the PtG system's overall efficiency depends on many factors, including electrolysis, compression, storage, and the conversion process. Therefore, PtG has an overall efficiency of 40–60%. This is due to the energy losses in various stages. Despite the losses, the PtG possesses several advantages, mainly in storing surplus renewable energy, along with the flexibility to distribute energy via existing gas infrastructure [11].

Power-to-liquid (PtL) technology converts electrical energy into methanol, ethanol, or hydrocarbon. All are considered synthetic liquid fuels. The PtL efficiency depends mainly on the conversion process. The common conversion process depends on electrolysis to produce hydrogen, and carbon dioxide (CO_2) is captured from the air directly or from industrial sites and converted into carbon monoxide (CO). Both H_2 and CO are synthesized into liquid hydrocarbon. The overall efficiency of the PtL ranges from 30% to 50%. The purity of the CO_2, the conversion technology, and the synthesized fuel influence these. PtL can produce liquid fuels that are rich in energy, with ease of storage, transportation, and utilization. This kind of fuel offers an appealing choice for aviation, heavy-duty, and energy-dense transportation [12].

Power-to-heat (PtH) technology. Here, the electrical energy is stored in the form of heat. Thus, it can be used later on in various applications. Among these are space heating, industrial processes incorporating heat, or district heating. PtH can be used to convert electricity into heat from heat bumps or electric boilers. Electric boilers have a high efficiency percentage, ranging from 90% up to 98%. This depends mainly on the operation conditions and the conversion technology. On the other hand, heat pumps can reach efficiencies close to 100%, as they use ambient heat sources, including air, water, or ground, to generate the heat. Heat pump efficiency is measured by coefficient of performance (COP). In this case, the heat pump has a COP of 3 to 5. This means that 3 to 5 units of heat are produced for every unit of the electrical energy used. Another major advantage of PtH is that the electricity/heat conversion process is direct, without further conversion steps. This incorporates in attaining remarkable efficiency percentages, making them suitable for several heating applications [13].

5.3 RENEWABLE ENERGY INTEGRATION

Renewable energy, such as solar, wind, tidal, etc., is intermittent by nature. This, in turn, implies that they are not dispatchable. PtX technologies offer a promising solution to mitigate intermittency and boost dispatchability. This is done by converting the surplus generated electrical energy from renewable resources into other forms that are both transportable and storable. This will be discussed in detail in this section [14].

5.3.1 PtX for Facilitating Renewable Energy Resources Integration

In some instances, there are sunny and windy days, meaning, there will be an excess of the electrical energy generated from renewable units. This means that there will be surplus electrical power compared to the consumer's demand. One of the main benefits of PtX technologies lies in their ability to store the surplus power generated. This stored energy can be used in other instances of renewable generation deficiency. PtG, for example, can store surplus electrical power in hydrogen or methane, as discussed in the previous section. And this can be stored in one of the existing gas infrastructures, which, in turn, plays a significant role in balancing the intermittent nature of the renewable generation against the fluctuating demand. In conclusion, this enhances dispatchability and system reliability [15, 16].

PtX technologies can also be incorporated into different sectors, which creates a new term, called *sector coupling*. This means green renewable energy is transferred to various sectors, decreasing and eliminating fossil fuel need. This is because PtX products are versatile, which enables them to be used in heating, electricity, transportation, industry, etc. [17].

In addition, PtX products are more easily transported and distributed than the conventional electrical form of generated renewable energy. Some renewable energy generation might occur in a remote photovoltaic substation in the desert or in an offshore wind farm. Therefore, transferring the generated hydrogen gas, produced from power-to-gas, can be easier through an existing natural gas infrastructure. This enables renewable generation to be geographically flexible and be able to integrate different renewable resources among multiple locations [18].

All the aforementioned factors make the PtX act as a buffer, thus ensuring a secure, steady, reliable, and continuous energy supply. Thus decreasing the need to restore conventional fossil fuel–based backup systems [19].

5.3.2　PtX for Mitigating Renewable Energy Curtailment

Renewable energy curtailment is an inherent nature of this type of generation. With renewable energy generation, the electrical grid is susceptible to a mismatch between the generation and the load required. Using PtX, in general, can help mitigate this energy curtailment, thus enhancing the usage of renewable energy resources. This usage boosts the performance of the electrical grid in terms of stability and reliability, especially in the instances of low renewable energy generation and high load demand [20].

The ease in transportation and distribution of PtX output can be transferred from areas with surplus generations and limited demand to other areas with limited generation but with high demand. These areas could be both foreign and domestic, which mitigates the curtailment locally and internationally, thus promoting international collaboration and making the electrical grids interconnected globally [21].

5.3.3　PtX Potentials in Providing Grid Stability and Flexibility

Another important issue is the promising role of PtX in maintaining both grid stability and flexibility.

As mentioned earlier, grid stability lies in storing and reusing surplus energy elsewhere and whenever needed, which can be called the decentralization of energy production and consumption. This, in turn, is reflected on decreasing the grid congestion and deferring grid infrastructure investments. On the other hand, PtX technologies work alongside energy storage, effectively time-shifting the demand peaks. Also, in case of low demand, the storage of surplus power soaks up one of the reasons that can destabilize the grid performance. This elevates grid stability to another level [22].

Grid flexibility lies mainly in demand-side management. PtX products can respond quickly to ramping up demand with low renewable energy generation. PtX's role is bidirectional, storing in case of surplus power and using PtX products to generate electricity in case of high demand. This makes the electrical grid more responsive and resilient to fluctuations. And it also decreases grid vulnerability to service disruptions [23].

5.4　PtX SECTORAL IMPACT ASSESSMENT: SYNERGIES AND BENEFITS

PtX, in general, can be coupled with various sectors of the economy, enabling the coupled sector to be decarbonized into a greener one. This impact can be maximized through sectoral synergy and renewable energy integration. This changes the landscape of energy management and its utilization [24].

This section focuses on three main sectors and selects an application from the sector. In the residential sector, the PtX impacts alternative energy resources. In the transportation sector, the emphasis is on the opportunities offered by PtX for both electric vehicles and aviation. Finally, the focus is on integrating PtX with the industrial sector. Then, PtX integration is addressed, along with the expected possible benefits [25].

5.4.1　Residential Sector: PtX for Alternative Energy Sources

In the residential sector, PtX enables the utilization of converted renewable energy in a decentralized way. The stored energy in PtX technologies acts as an energy storage solution. If a residential facility owns private renewable generation capabilities, solar PV, or small wind turbines, surplus generation can be stored in PtG (hydrogen) or PtL (synthetic fuel) forms. These forms enable homeowners to use energy-produced carriers later on and to provide a reliable backup and boost energy self-sufficiency. Also, it reduces the homeowner's dependency on the main grid. This decrease in reliance benefits them either in the case of low renewable energy generation or power outages in the main grid [26].

The excess stored energy, for example, hydrogen, can be used by homeowners in various applications, such as cooking, heating, electricity generation, and combined heat and power (CHP). Also, PtG can be used to fuel electric vehicles, especially types based on hydrogen (i.e., EV fuel cell). This reduces the EV owner's reliance on fossil fuel. In addition, synthetic fuel can be used in heating systems or as a replacement for fossil fuels. All this adds up to the efforts of decarbonization [27].

Homes in remote areas and far away from grids can suffer from energy poverty. Conventional energy facilities based on fossil fuels can present a serious challenge. Due to the scarcity of traditional fuel supply, PtX technologies can store surplus energy, harnessed locally, from installed renewables on the premises. The stored energy can be used to fulfil the energy need of an isolated community. This makes PtX able to provide a sort of energy independence, as well as sustainable development, which ultimately enhances the quality of life and eventually creates economic opportunities [28].

Finally, PtX can play a crucial role as an alternative energy source, by storing energy and powering various residential applications. Moreover, it enables coupling of the residential sector with the electrical grid via an effective, synergistic manner. The synergetic concept, in summary, is either in creating PtX locally or utilizing PtX products generated by the grid [29].

5.4.2 Transportation Sector: PtX for Electric Vehicles and Aviation

For the transportation sector, PtX offers fossil fuel alternatives that are based on renewable energy. With the same concept explained for personal EV, this could be expanded to transportation fleets. PtX, in general, addresses a major problem that faces EVs: the capacity of batteries. Not only are the batteries of limited capacity, but charging stations are also somehow scarce. From PtX technologies, PtL, in particular, produces methanol and liquid hydrocarbon, which can act as drop-in replacement for conventional fossil fuel. This makes the transition to hybrid EVs easier. Even if not, especially for large trucks, liquid hydrocarbon can help decarbonize a transportation fleet effectively. Plus, the usage of PtL liquids can benefit from the existing fueling infrastructure. Finally, PtL can provide an alternative to power hybrid EVs in case grid service is low or intermittent [30].

Regarding the aviation sector, the industry relies mainly on fuels that produce emissions with high concentrations levels of carbon dioxide. PtX technologies, in general, provide a significant cut-down in carbon emissions. These fuels pave the way to a new kind of fuel, called sustainable aviation fuel (SAF), that originates from renewable resources. The fuel produced via PtL technology meets aviation fuel requirements and specifications, thus making SAF an efficient replacement for traditional jet fuels. This, in turn, enables reducing greenhouse gases without significant modifications, either in the existing jet fueling infrastructure or in the aircraft itself. This ultimately cuts down a large carbon footprint from the aviation industry and adds up to the competing efforts against climate change [31].

Another added advantage to PtX usage in the transportation sector is energy diversification and energy security. This can be achieved when countries cut down their share from imported fossil fuel and enhance country self-sufficiency from energy, thereby achieving both energy resilience and independence for the transportation fleet. PtX not only uses surplus renewable energy to produce fuels but also does so by capturing CO_2 from either the air or the industrial process. This is called *carbon recycling*, which helps in closing the carbon loop. In the end, these, together, add to the previously mentioned efforts facing environmental impact [32].

In order to harness the benefits of PtX coupling with the transportation sector, there are some factors that should be cared about. They are the fuel production cost, infrastructure requirements, and scalability. Other challenging factors are PtX process efficiency and renewable energy surplus availability. Added to the previous challenging factors are supportive development and financial incentives. Not forgetting the importance of exerting efforts in research and development, which plays a crucial role in driving PtX to be deployed in a commercial level in the transportation sector [33].

Summing up, the introduction of PtX technologies into the transportation sector comes with benefits and challenges as well. Applying it to EVs and aviation can be of significant importance toward the decarbonization of the sector, this by shifting the dependence on low-carbon synthetic fuels, which aims to reduce carbon footprint, facing the challenges in aiming for benefits and energy diversification; hence, research and the development of supportive policies are essential to harness the full potential of PtX technologies, transforming the transportation sector toward sustainability [34].

5.4.3 Industrial Sector: PtX Integration in Manufacturing

Meanwhile, in the industrial sector, PtX makes the transition of the sector to renewable energy smoother. The industrial sector, by nature, is heavily dependent on carbon-rich fuels, which form the backbone of this sector. Similar to other sectors, PtX technologies offer an appealing alternative to decarbonize the sector and reduce greenhouse gas emission. This achieves a substantial carbon emission reduction, especially to the industries of steelmaking, cement facilities, or chemical manufacturing. This is because the electrification of these industries is not as easy to do as adopting PtX proper technology. This ultimately makes a significant reduction in carbon footprint [35].

PtX can play another important role, that is, enhancing energy efficiency and resource optimization. The storage of surplus power, and reusing it when needed, not only mitigates renewable curtailment but also enhances energy flexibility, which impacts, positively, energy management as well. In the long run, this improves the overall efficiency of the industrial sector. PtX also allows the integration of hydrogen and synthetic fuels and allows the adoption of fuel cells besides CHP systems. The integration of versatile energy vectors at the same time provides multiple energy outputs to various industrial processes [36].

Another added value to adopting PtX to the industrial sector is the circular economy, where carbon dioxide is captured from both industrial emissions and from the atmosphere directly. Then, it is turned into synthetic fuel via PtL technology. This helps in closing the carbon loop in industrial processes by using waste carbon efficiently. This promotes the sustainable use of the available resources [37].

Naturally, PtX technologies face challenges when deployed into the industrial sector. The challenges are related to the cost of PtX deployment and infrastructure requirements, in addition to the efficiency of the PtX process and the presence of supportive policies and governing regulations. Also, the existence of financial incentives that can give a push to the wide spreading of PtX among the industrial sector can be a barrier. Here comes the importance of the collaboration between industry and academics, along with policymakers. This collaboration is of vital importance to accelerating research and development (R&D), which eventually leads to enhancing PtX performance and unlocking their potential, making the adoption of PtX into industry an attractive alternative [38].

5.4.4 Cross-Sectoral Integration/Synergy by PtX

In general, integrating PtX into any sector brings multiple benefits, along with challenges, as aforementioned. But the concept of cooperation between various sectors, called *cross-sectoral synergy*, can maximize the overall benefits and mitigate the challenges on energy, environment, and economic levels. The positive impacts on both the energy and the environment have been mentioned previously. But economy is also important. The PtX's positive impact arises in attracting more investments to renewable energy systems and subsidiary technologies. In addition, it creates new markets and stimulates innovation [39].

The way to achieve any of the benefits should be done by many guarding concepts, among which are [40]:

Strategic planning and policy frameworks. This implies the importance of having roadmaps with long-term goals and objectives. Market mechanisms that encourage cross-sectoral collaboration are invented, and also to optimally allocate the resources available.

Technological advancements. The ongoing R&D efforts exerted in this regard aim to enhance many aspects, for example, PtX efficiency, scalability, and making PtX technologies cost-competitive. With regard, it also mitigates any negative impact on the environment from the technology.

Infrastructure. This aspect can either elevate or demote the adoption of, and give synergy as well to, PtX technologies. The infrastructure shall be both flexible and resilient, to adopt PtX easily and facilitate the spread of the benefits and make the challenges a minimum. This is by enabling the transportation of PtX products efficiently.

5.5 TECHNO-ECONOMIC FEASIBILITY AND CHALLENGES

One of the important aspects that face PtX feasibility is the availability, and scalability, of the energy produced from renewable energy resources. The process underlies the PtX technologies, demands a significant amount of renewable energy electricity in order to produce hydrogen and synthetic fuels in a usable quantity. The generated power from PVs and wind farms is geographically dependent and suffers from intermittency and imposes a challenge in providing the surplus, after satisfying the electrical demand, that is needed by the PtX processes. This challenge can be addressed technically by introducing proper energy management scenarios and adequate mechanisms for demand response. Also by ensuring a renewable electricity supply that is stable, reliable, and sufficient for PtX technologies. Another helping aspect lies in developing a robust renewable energy infrastructure, coupled with grid integration. In the end, this contributes to a widespread deployment of PtX technologies [41].

Another technical aspect is the energy efficiency of PtX processes. As the PtX process involves multiple stages of conversion, this makes the whole process more susceptible to energy losses. Logically, improving the efficiency of each conversion step will enhance the overall efficiency in turn. Additionally, this will reduce the associated costs, making PtX economically interesting. This can be done by investing in the R&D that will explore and introduce new advanced catalysts and materials, with the goal of minimizing the overall operational costs [42].

PtX technologies compete against well-established conventional fossil fuels, which is a major challenge, due to the economic viability of fossil fuels over PtX technologies. Also, the production cost of electricity from renewable energy resources forms the primary input in pricing the products of PtX technologies and plays a major role in determining PtX technologies' economic feasibility. However, despite the steady decline in the production cost of renewable electricity, it is still higher than the cost of their fossil fuel counterpart [43].

To boost the techno-economic feasibility of PtX technologies, it is crucial to cut down on the cost of the production of electricity from renewable resources—this by technological advancements and the introduction of supportive policies, along with the enhancement of the PtX process efficiency.

In accordance with the aforementioned, the techno-economic feasibility of PtX is affected by the presence of governing regulations and policies. The presences of regulatory frameworks encourages investments, and it will motivate innovation. Not to mention that the existence of carbon pricing and renewable energy generation standards can internalize carbon emission external costs. This can pull up the PtX technology market. Also, there will be a need for policies that address the R&D, and for technological demonstrations that can provide a leverage to overcome the barriers of the initial cost and thereby drive PtX technologies towards being more commercial [44].

Finally, PtX technology deployment requires the presence/development of a robust supply chain, market integration, and dedicated infrastructure. PtX carriers imply the establishment of infrastructure that contains pipelines, refueling stations, and storage facilities, thus making storage, transportation, and distribution more cost-effective. This requires the alliance of various parties, such as stakeholders, technology suppliers, energy producers, factory owners, and policymakers [45].

This section's emphasis is more into the techno-economic aspects facing the adoption of PtX technologies in terms of feasibility, and the relative challenges.

5.5.1 Technical Feasibility and Challenges Facing PtX Deployment

One of the major challenges that face PtX deployment is the production cost of the fuel. This is because the PtX process includes multiple steps, starting from the generation of electricity from renewables, then with the energy conversion stage following, and finally, fuel synthesis. All these steps add up to the PtX production cost as a whole. It is of crucial importance to continue lowering the price of renewably generated electricity via R&D to increase the efficiency of renewable units, and also to encourage the widespread adoption of the units, thus making them economically feasible. Following that, this challenge can be overcome [46].

The next challenge is the intermittency nature of renewable energy sources. The PtX process needs the renewable supply to be both steady and reliable. The fluctuating nature and the availability of renewable resources, such as solar irradiance and wind speed, are the primary reasons behind the generation intermittency. Energy storage could play an important role in stabilizing the renewable generation and mitigation of the supply intermittency [47]. This could provide a supply in which its characteristics are more continuous, steady, and reliable [48].

Another challenge is the scalability, which aims to achieve large-scale deployment. Mitigating this challenge needs large infrastructure, large renewable energy generation facilities, storage warehouses, and a reliable network for the transportation of PtX fuels. This needs extensive investment coordination between various stakeholders. On the other hand, the scaling-up of PtX technologies requires advanced manufacturing techniques and optimizing the production process, thus achieving a cost-effective scheme. The collaboration between academia, industry, and policymakers is vital to address this challenge [49].

Moreover, the presence of regulations and policies helps in making PtX deployment a techno-economic feasible alternative. The regulations shall have a clear framework, thus creating a long-term support to the PtX technologies in the market. Also, to encourage investment and bring up more innovation. On the other hand, the presence of policies that target carbon pricing, as well as renewable energy, can accelerate the deployment of PtX technologies, thereby creating favorable market conditions to accept the PtX. The harmonization and collaboration of the international regulations push the PtX technology deployment on a global scale, making PtX technologies more capable to cross the borders [50].

Furthermore, another challenge comes from the aim to integrate PtX technologies, in general, into existing energy systems and markets. The fuels produced from PtX technologies, like hydrogen and synthetic fuels, needs an infrastructure that secures its storage, transportation, and distribution. Typically, the founding of these infrastructures needs a remarkable amount of initial investments, along with coordination between owners of the energy production facilities, stakeholders, fuel suppliers, and end users. Plus, PtX fuels need to be produced in accordance with safety standards, thus ensuring both their compatibility and acceptance to be used by the market [51].

5.5.2 Potential Economic Benefits

The primary economic benefit of PtX is the ability to get usage from surplus renewable energy power by storing it and creating an additional revenue later on. This is by converting the energy into other products, like hydrogen, chemicals, or synthetic fuels, thus preventing wasting of the excess generated renewable energy and mitigating power curtailment. The created revenues are beneficial to energy producers and other grid operators. PtX technologies offer monetization of surplus generation, thus enhancing the economic viability of the renewable energy project, which contributes to increasing the profitability of the renewable energy generation facility [52].

Another economic benefit is the ability of PtX to act as energy storage that can balance the electrical grid performance. The variation of meteorological conditions affects the renewable energy production significantly. PtX stabilizes this by storing the excess generated power in low-demand instances, and it can be used later on in either low-energy availability or high load demand. This incorporates avoiding the unbalance between supply and demand, which, in turn, boosts grid reliability and capability in meeting consumer demands. Eventually, this defers to the need for installing peak power plants and/or investing in grid reinforcements [53].

By adopting PtX technology, energy freedom from the dependence on imported energy could also be achieved and level up the energy security. The reliance of countries on imported fossil fuels limits their economic freedom and makes them prone to geopolitical pressures. The adoption of PtX technologies domestically enhances the nation's self-sufficiency in the energy sector and mitigates economic volatility. In addition to strengthening the energy security, PtX technologies protect a country's government from fluctuations in the pricing of fossil fuel in the global market. Another economic benefit from PtX localization is the creation of new job opportunities, which foster the economic growth of the renewable energy market and PtX industries as well [54].

PtX technologies aid in the transition to a low-carbon (green) economy, which, in the long run, bares economic benefits. PtX technologies pave the way into decarbonizing the economy, this in terms of decreasing the emission of greenhouse gases among the sectors using PtX products, thus enabling the countries to achieve their global commitment toward protecting the environment. Not only does this mitigate climate change globally, but it also elevates the country's position in the forefront, in terms of many aspects, such as transitioning into clean energy, market leadership, export potentials, and attracting funds for investment [55].

5.5.3 Life Cycle Assessment of PtX

Life cycle assessment (LCA) is defined as the systemic analysis of the environmental impact over the entire life cycle of certain products or processes. The cycle starts from raw material extraction, up to the end of life of the final product. Performing LCA on PtX technologies is important to understand their environmental impact, which, in turn, helps in identifying the available opportunities to improve the PtX technology among all the life cycle steps. The steps will be discussed in detail in this section [56, 57].

The first stage in the LCA of PtX technologies is the production of electricity from renewable resources. This includes the assessment of associated environmental impacts on the whole process of renewable electricity production, starting from the manufacturing of the units, PV panels, and wind turbines; the installation and operation of the renewable substation; and the LCA assessment of the technologies and resources used in the production of the renewable units, which all influence the environmental footprint of the PtX technologies.

In the next stage, PtX technologies convert the renewable electricity into products, such as hydrogen or synthetic fuel. Therefore, LCA focuses on examining the energy efficiency of the conversion process, plus the amount of emissions generated during the conversion, and also the required resources for PtX production. Not forgetting to include the technologies and the materials used by the PtX process, like catalysts and/or electrolyzers, in regard to their environmental impacts.

Subsequently, LCA investigates the utilization of PtX end products. In case of hydrogen, as a transportation fuel, the LCA shall consider the environmental footprint of the distribution, storage, infrastructure, and the fuel cell/hybrid EV that will use this fuel. In the comparison between PtX products and fossil fuels, the LCA shall evaluate the efficiency and the associated emissions of PtX products, which is essential in building a complete and clear picture about the environmental benefits/drawbacks, which make the decision-making process easier.

Finally, the LCA explores the treatment of PtX products, and by-products, at their end of life. This assessment includes the expected emissions, the generated waste, and the strategies used in recycling/disposal. The management of by-products is of crucial importance, such as the captured carbon utilization or the utilization of waste heat. This also contributes to the overall environmental footprint of PtX technologies.

Conducting LCA studies to PtX technologies is a complex task to perform. It demands comprehensive yet meticulous data collection, modeling, and interpretation. Also, the associated uncertainties in the data arise in the availability of data, system boundaries, and assumption. The way of handling the uncertainties affects greatly both the accuracy and reliability of the LCA results. Therefore, R&D in the data collection field is important to improve the accuracy and robustness of the LCA studies of PtX technologies.

5.6 REGULATIONS PERSPECTIVE

PtX technologies, like any system, need regulations that can guide and encourage several aspects for implementation, like systems development, deployment, and safe operation. All this with regard to social benefits and environmental sustainability [58, 59].

Firstly, the developed regulations characteristics should be both clear and consistent. Considering the environmental footprint of products and processes involved in PtX technologies, the regulations should set clear limits concerning emission standards, pollution, and the responsible sourcing and utilization of products. With the existence of stringent regulations, governments can easily ensure that PtX technologies are an ally to the decarbonization goals, thus contributing in minimizing the negative environmental impacts and enhancing air and water resources quality [60].

Secondly, the potential risks accompanying PtX technologies should be addressed by these regulations to ensure safety, as PtX processes involve several risks sources, such as handling hazardous material, high pressure, and/or high temperature. The regulations shall be firm in enforcing workers toward protection, not only them, but also public personnel and the environment. Moreover, the regulations should regard PtX facilities in all phases, starting from design, through construction, and ending by operation. Not forgetting to mention the guidelines for both emergency response and risk management [61].

Thirdly, there are more aspects to be addressed by regulations, such as market integration, grid compatibility, and fair competition, as PtX technologies can disrupt the existing energy systems and markets. Thus, regulations should develop frameworks that ensure fair market access for the product of PtX technologies and enable fair competition, all while prohibiting unfair practices by setting safeguarding rules [62].

Fourthly, the regulations should give support to R&D efforts for their importance in developing PtX technologies. This is by providing mechanisms for funding and incentives and creating collaborative platforms. On the other side, governments have a great role in encouraging innovation, providing research grants, lowering taxes, and promoting public–private partnerships. Additionally, the regulations should facilitate knowledge sharing among academia, industry, and research institutes. This collaboration can accelerate the technological advancements, plus offering an ecosystem that supports PtX development [63].

Fifth, the regulation could be extended to cover international coordination. This harmonization between various regulations and standards will be beneficial on a global scale, especially in trades crossing borders. This harmonization diminishes the inter-trade barriers, encourages market growth, and makes PtX product exchange easier [64].

Sixth, and finally, the regulations should consider the social and ethical implications that result from PtX technological implementation. In terms of equity, accessibility, and public acceptance, regulations should ensure that PtX deployment will not aggravate any existing social inequalities or add burden on societies. In addition, regulations should enable equitable access to PtX benefits, such as the created jobs and/or energy affordability. One more crucial thing, the government should enable public engagement and participation in formulating the regulations, thus incorporating diverse perspectives and considering public concerns related to PtX technology deployment. This should promote a bridge of trust and transparency [65].

In summary, PtX technologies require comprehensive regulations and framework that address environmental footprint, safety, market integration, R&D, international coordination, development of support, and social consideration. These aspects pave the way for governments toward the more fruitful deployment of PtX technologies, which enables the unlocking of PtX potentials for the transition into a greener economy. The collaboration among various parties from academia, industry, and policymakers should consider the technological advancements in PtX technologies.

5.6.1 ANALYZING THE POLICIES AND REGULATIONS INFLUENCING PTX INTEGRATION

PtX technologies have gained a wide reputation aiming to transition trends toward sustainable energy and climate neutrality. The policies governing them are of multiple levels. In this section, the influence of national, regional, and international levels will be addressed.

For the national level, many policies and regulations encourage PtX integration. The encouragement comes in the terms of feed-in tariffs, quotas, and tax incentives. These can attract personal investments and initiate R&D for PtX technologies. The national strategy shall be of long-term nature, thus providing a stable yet predictable environmental policy for PtX development [66].

For the regional level meanwhile, there is the European Green Deal initiative as an example. It aims, by 2050, to achieve a low-carbon economy. These initiatives offer a framework for member countries in the EU to develop their own policies regarding PtX technologies [67].

Further, on the international level, there is the Paris Agreement, which addresses climate change, regarding which countries are mandated to cut down their greenhouse gas emission share, along with putting greater reliance on renewable energy resources. This adds to the global efforts toward mitigating carbon emissions. Here, PtX technologies' crucial role comes as a pathway to decarbonize various sectors that are coupled with PtX [68].

However, the poor formulation of regulations imposes challenges that impede the integration of PtX technology freely. This arises in regulations that are both unclear and inconsistent across different sectors. This applies also to the required permits, licenses, and uncertainty associated with the treatment of products derived from PtX technologies [69].

Finally, the formulation of PtX technologies' regulations shall consider both social acceptance and the engagement of the public opinion, besides the involvement of stakeholders. This participatory decision-making process, in which all the aforementioned are involved, leads to clearing the parties' concerns toward land use, ecological impacts, and any economic implications [70].

5.6.2 STANDARDS AND REGULATIONS FOR PTX INFRASTRUCTURE AND SAFETY

The existence of safety standards for the PtX industry is pivotal, as this industry includes handling flammable gases, high pressures, and hazardous chemical reactions. Therefore, risk assessment should be conducted thoroughly throughout the entire PtX value chain. This includes the production, storage, transportation, and utilization of PtX products [71].

A safe operation of PtX facilities shall be ensured. This is by developing a set of inclusive guidelines. In general, the guidelines shall consider several aspects, such as standards for design and construction. These are engineering directives for the PtX infrastructure, material selection, equipment specifications, and safety systems. The other aspects include procedures that govern the operation, maintenance protocols, and emergency response plans. This implies performing rigorous inspections, to be sure that, for example, a certain facility complies with the specifications. In case of passing, the facility could be granted certification(s) to start commissioning [72].

Another safety aspect includes the operational procedures for PtX facilities which aim to ensure minimum risk and maximum efficiency. The procedures shall include protocols for both safe start-up and shutdown, as well as requirements for personnel training and schedules for regular equipment maintenance. A stringent monitoring and control system shall be adopted to assess various parameters, among which are temperature, pressure, gas composition, and flow rates. With the spotting of any deviations or anomalies away from preset values/conditions, alarms shall be triggered, and in severe cases, shutdown mechanisms are needed, thereby preventing accidents and protecting the working personnel within the facility or infrastructure [73].

In addition to the safety issues are the maintenance protocols, which are crucial to PtX infrastructure concerning long-term integrity and reliability. To avoid any safety hazards, there are multiple measures that should be carried out, such as the regular inspection of the facility, the performance of preventive maintenance, and the checking of equipment calibration. Thus, any potential issues could be identified and addressed before escalation. Here, keeping comprehensive records is a vital action. It shall record the maintenance activities' dates and the equipment performance, thus enabling the facility to trace and investigate any incidents/accidents [74].

Here comes the importance of emergency response plans to the safety of PtX facilities, thus mitigating and managing any potential risks within the infrastructure. The plans put outlines to the necessary procedures to respond to various scenarios, such as leaks, fires, and/or natural disasters. The outlines include ordered measures for how to safely evacuate personnel, the isolation of affected areas, and coordination with emergency services. Regular trainings should be conducted to examine the effectiveness of these plans. Plus, personnel's level of preparation to deal with emergency situations [75].

The following are points concerning the same landscape regulations governing PtX infrastructure safety [76–83]:

- *International harmonization* is of crucial importance in order to facilitate the deployment of PtX infrastructure on a global scale. This includes the collaboration of countries, intercontinental firms, etc., thus exchanging the best practice.
- *Risk management* of the PtX infrastructure undergoes a comprehensive process of analysis and management of potential hazards, then the development of appropriate measures for hazard mitigation. Hazard identification studies include failure mode and effects analysis (FMEA) and fault tree analysis (FTA), which help in understanding the expected modes of failure and associated consequences. Risk mitigation strategies consider adopting redundant safety systems and procedures for shutdown in case of emergency, thus mitigating accidents' impact and likelihood.
- For *personnel training and competency* in PtX infrastructure safety, it should be ensured that each working person is following the safety culture. Therefore, training programs shall target education concerning certain hazards, the associated safety, and the best counter-practice.
- *Cybersecurity* shall be included as an important aspect in PtX infrastructure safety, as systems are increasingly interconnected and controlled digitally. This includes securing communication protocols, data encryption, and control of the access to the system. This will enhance protection of the PtX infrastructure from cyberthreats and maintain operational integrity.
- *Standardization organization.* PtX projects can use multiple standardization organizations, such as the International Organization for Standardization (ISO), the International Electrotechnical Commission (IEC), the American National Standards Institute (ANSI), and the European Committee for Standardization (CEN). Experts from both academia and industry collaborate in these organizations to establish regulations and standards to ensure the interoperability of PtX projects.
- *International codes and guidelines* contribute also, along with standards and regulations, to the safety standards of PtX infrastructure. As an example, the International Code Council (ICC) contributes on international building codes that aim to ensure safety and structural integrity. This, in turn, helps in formulating a framework for the construction of PtX facilities.

- *Quality assurance/quality control (QA/QC)* is one important measure that should be maintained throughout the PtX infrastructure to ensure safety, and also to ensure that the design, construction, and operation abide by the relevant standards. QA/QC programs include stringent processes for inspection, testing, and documentation, thus ensuring that the materials, equipment, and processes are as per specifications and guarantee the PtX infrastructure in the terms of reliability, durability, and safety.

5.6.3 Recommendations for Enabling and Reinforcement of Regulations

In this section, a list of proposed recommendations is presented in order to make regulations more effective. The recommendations shall cover all aspects that are expected to reinforce the application of the PtX technologies [84–89].

- *Technology-specific standards* should address all the PtX process stages, each specifically. This specification shall make the performance consistent, efficient, and safe. Also, the standards should cover various aspects, such as the design of equipment, the integration of systems, the operating parameters, and the quality of products.
- *Environmental regulation.* PtX processes impact the environment in many ways, especially in water usage and waste management. This makes PtX projects environmentally friendly. Therefore, regulations that address the environmental impacts are more acceptable than regulations that don't.
- *Permits and licensing processes* are another aspect that reinforces PtX regulations. The process of obtaining either a permit, license, and/or approvals shall be clear and efficient. Also, it shall consider coordination with other entities, such as energy utility, environmental agency, and the transportation department. In addition, the existence of a clear guidance is crucial, where the application processes should be standardized, and the timeline for decision-making must be reasonable.
- *Monitoring and reporting* is an additional aspect. It is essential for performance tracking, efficiency monitoring, and the environmental performance of PtX projects. The monitored parameters shall be defined, such as consumption of energy, quality of product(s), generated wastes, and emissions. Thus, any non-compliance can be identified, on time. All this adds up to reinforcing PtX-related regulations.
- *Research and development support.* Regulations that encourage R&D have a reinforced privilege over regulations that don't. This encouragement boosts the advancement of PtX technologies. Regulations should support research collaboration between institutions, industry, and funding agencies. This can include grant offers, the establishment of test bed facilities, and platforms for knowledge exchange.
- *Intellectual property protection.* Good regulations should consider the ways to protect innovations and intellectual property. Such protection provides a great incentive to R&D, which, in turn, will attract investments. The guidelines for patent rights should be clear, as well as licensing agreements and technology transfer between partners. Meanwhile, there should be a balance between intellectual property protection and knowledge sharing.

5.6.4 Regulatory Architecture and Policy Proposals

The proposal for a comprehensive regulation architecture is suggested. This is an example summarizing the required regulations that should be met in order to commercialize PtX technologies and their products into the European Union (EU). The regulations cover four main aspects: climate, energy, transportation, and social. Figure 5.2 depicts the architecture of the regulations and policy to achieve this vision by 2030 [90].

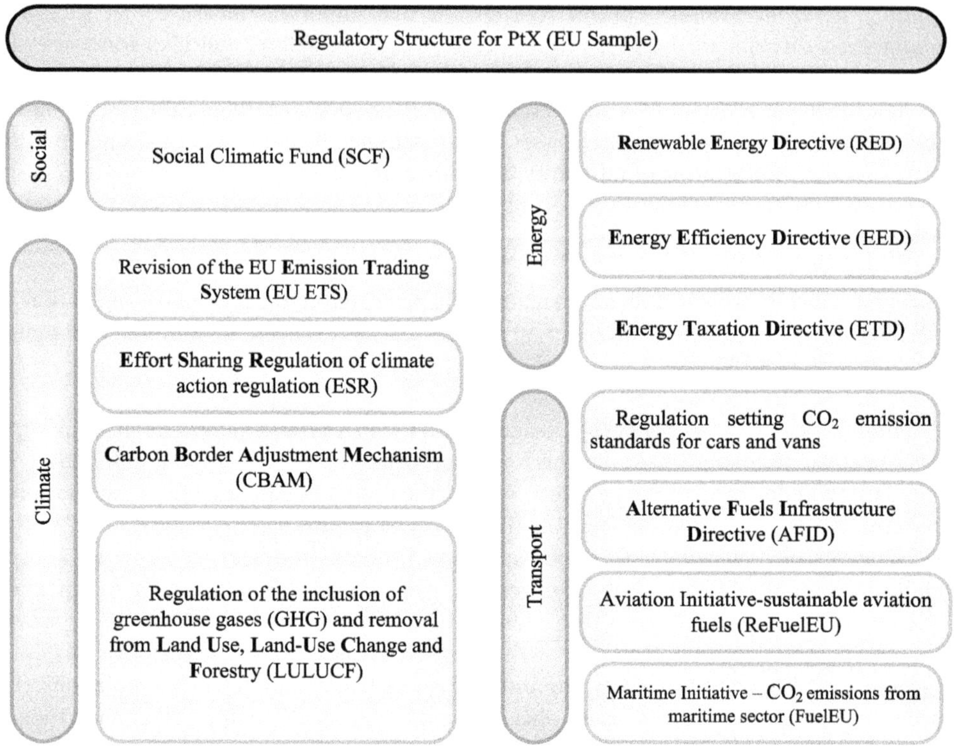

FIGURE 5.2   Proposed regulatory structure for PtX in EU.
Source: [90].

5.7   PROPOSED FRAMEWORK FOR COLLABORATIVE DEPLOYMENT

Through the previous sections in this chapter, it was demonstrated that each PtX technology targets a certain sector. Also, various sectors can be coupled, thus maximizing the benefits. Not only that, but PtX technology also facilitates energy transition and is capable of achieving sustainability goals. Therefore, this section gives highlights on the aspects that should be considered while developing a framework, thus achieving the aforementioned goals and, additionally, enabling various sectors in the same country, regionally, or internationally, and their relevant stakeholders, to foster exchange, coordinated actions, and sharing [91].

5.7.1   SYNERGIZED FRAMEWORK INCLUDING VARIOUS SECTORS

For a synergized framework, the following aspects are proposed. The demonstrated various points cover all the possible aspects, including technical, financial, and policy. By including collaboration in PtX technologies' deployment, the framework not only promotes synergy among sectors but is also comprehensive [92–101].

- A *comprehensive regulatory framework* is of crucial importance in order to deploy PtX technologies safely. The framework shall include technology standards, safety protocols, permits and licensing processes, and environmental impact assessment, along with coordination between industry stakeholders and research institutions, to ensure all associated challenges and offered opportunities from PtX technologies are addressed.
- *Multidisciplinary collaboration.* PtX technologies cover a wide range of sciences, such as engineering, chemistry, economics, environmental science, and policy. The framework, in

general, shall emphasize on a multidisciplinary collaboration, including academia, industry moguls, politicians, and society. This could be done via establishing research consortia, industrial groups, and public partnerships and allowing the exchange of experience, data, and best practices. When these diverse perspective come together, it will be easier to holistically solve problems, innovate, and make decisions based on diversified information.

- *Research and development coordination* should focus on achieving effective, coordinated efforts based on R&D. The framework shall find a way to establish research agenda(s), engage research facilities, and pave the way for collaborative funding, thus allowing for the alignment of research priorities, pooling of the scattered resources, and ultimately, avoiding duplication of efforts. The proposed coordination in the R&D sector enables the development of breakthroughs in PtX technologies and potentially reducing costs and generating the needed knowledge for market development. In addition, this enables the exchange of learned lessons and the sharing of skills and knowledge.

- *Policy and regulation alignment.* The developed framework shall emphasize on the unification and alignment of various policies across different jurisdictions and sectors. This alignment is crucial for harmonizing standards, setting clear permit processes, and eliminating legal barriers, which, in turn, mitigate market uncertainty and create a playing field for scaling up PtX technologies freely.

- *Technology transfer and capacity development* can be considered as one of the important goals behind developing a framework. The transfer is done from developed countries to developing ones in the aspects of knowledge, expertise, and technology. Practically, this could be via training programs, projects that demonstrate technology and technical assistance and support the establishment and development of the PtX infrastructure in developing nations. At the end, this boosts the global equity and ensures spreading the PtX benefits among both types of nations.

- *Demonstration projects and pilot initiatives.* These were discussed in brief in the previous point. In detail, pilot projects allow the testing of PtX technologies on various scales and enable performance assessment and the identification of potential challenges. Meanwhile, these projects enable the sharing of learned lessons and refining strategy deployment. This on all project phases, starting from planning and execution and ending with the evaluation of the projects.

- *Data sharing and digitalization.* Nowadays, aspects prove it is beneficial to all sectors; therefore, it should be included in the framework. PtX technologies can get use from the offered benefits of data sharing and digitalization. The framework should focus on establishing platforms for data sharing, standards for interoperable data, and management systems to secure data. The collaboration among various parties enables the exchange of available data, technology performance, and market dynamics. PtX deployment and operation may now be more effective, efficiently monitored, and optimized in real time thanks to digitalization.

- *Infrastructure planning and integration.* The developed framework should consider establishing a robust infrastructure that includes generated energy from renewable resources, storage facilities, and transmission networks. All these need a well-planned infrastructure for effective PtX technologies integration. The synergy among the energy, transportation, and industrial sectors should overcome the imposed infrastructure challenges as well as optimize PtX integration. Both coordination and planning of infrastructure offer an efficient utilization of resources and minimize environmental impacts and support PtX technologies' scalability.

- *Monitoring, evaluation, and learning.* These important aspects forge a successful framework which impacts the deployment of PtX technologies. This should be via establishing monitoring systems, performance indicators, and evaluation. Monitoring and evaluation use shared data, experiences, and best practices and thereby assess the impacts of PtX deployment on the environment, economics, and society. At the same time, the learning aspect enables the identification of factors of success, potential challenges, and underlying opportunities. Ultimately, it supports adaptive management, which helps in the continuous improvement of PtX deployment strategies.

- *Financing mechanisms and investment support.* Last but not the least, this aspect should be in the core of the framework to successfully address PtX technology deployment, as PtX projects require significant investments, which implies the development of financial supporting mechanisms. These mechanisms target the attraction of private funding as well as public ones. Collaboration efforts shall be coordinated among financial institutions, governments, and the business sector. Thus, to develop innovative financing models, these mechanisms encourage and support the deployment of PtX projects on various scales.

It is worth mentioning that the following aspects should be considered in the framework of the topics presented in the previous sections, such as life cycle assessment, risk management, international collaboration, public acceptance and stakeholder engagement, standardization, and certification.

5.7.2 IDENTIFYING THE COMMON POINTS AMONG SECTORS

In order to achieve synergy between various sectors that use different PtX technologies, the common points between the sectors should be identified. By leveraging the sectors' common points, the resources can be optimized, synergy may be fostered, and the adoption of PtX technologies is accelerated [102].

The first common point is resource availability, which means the renewable energy, which is the major input that is needed by PtX technologies to operate. Therefore, the distribution of renewable energy resources should be determined based on geographical mapping, and the areas with abundance should be identified. Then, the synergetic deployment of PtX technologies should be determined in terms of feasibility. For example, the areas with a high percentage of surplus energy could be assigned to energy-intensive sectors, such as industry and transportation. Ultimately, by aligning both resource availability and sectoral demands, an optimized utilization of renewable energy could be achieved, and the benefits of PtX technologies maximized [103].

The second common point is the utilization of specific PtX products among various sectors. For example, hydrogen possesses the potential to be used as an energy carrier. This means it can be used in multiple sectors—in the transportation sector, in electric vehicles with fuel cells, as well as in the industry sector, for both decarbonizing the sector and for heating proposes. Additionally, it is utilized in the energy sector for the storage of energy and balancing grid performance. The cross-sectoral collaboration may pave the way toward knowledge exchange and to explore the possibility for potential synergies. This collaborative integration promotes value chains of the PtX product, which means that the output from certain sectors becomes the input in another sector, thus creating an economy that is characterized by both sustainability and a circular economy [104].

The third common point is the formulation of a common policy and collaborative framework. By aligning the standards, regulations, and policies among various sectors, this facilitates the way to deploy PtX technologies easier. The creation of supportive policies diminishes barriers, and it provides incentives. In turn, a seamless integration can be found for the integration of PtX products into existing infrastructure and markets, which enhances its cross-sectoral adoption. The collaboration of different entities, from academia, industry, and politicians, helps in developing comprehensive frameworks and unlocking PtX technologies' full potential [105].

The fourth common point is R&D collaboration, which enables the identification of challenges and proposes innovative solutions. With the sharing of R&D findings, best practices, and advancements in technology, both researchers and experts from industry can tackle various sets of problems, which include technical, economic, and environmental aspects. Those hurdles are associated with the deployment of PtX technologies. The proposed collaboration in R&D enables the improvement of PtX system efficiency, cost reduction, product quality, and sustainability enhancement. Therefore, the resulting solutions are tailored to meet various sectors' needs [106].

Fifth, and the last, common point is public awareness and the engagement of stakeholders. The societal dialogue between personnel from academic, civil society, industry, and government spots both the associated risks and potential benefits. The transparency in this regard makes various voices express their concerns, encourages idea contribution, and unifies opinions, which all lead to spreading social acceptance by fostering trust and support among the public [107].

5.7.3 EVALUATING THE FRAMEWORK ADAPTABILITY AND SCALABILITY

In the last section of this chapter, the focus will be on emphasizing the adaptability and scalability of the proposed framework. Those two aspects will be highlighted in detail and from a perspective that was not discussed previously or was only mentioned transitorily [108].

The proposed framework promotes the adaptability concept, which is essential in general due to the rapid evolvement of PtX technologies in terms of innovation and discoveries that will emerge continuously. The adaptability imposes flexibility that will accommodate the changes and advancements. Achieving this implies a continuous review of regulations, policies, and technical standards, maintaining abreast to the developments in technology, in which the framework will harness the breakthroughs in PtX technologies and ensure that the integration to various sectors is smooth [109].

Furthermore, scalability is another critical aspect toward a PtX-synergized framework, thereby covering the energy demand of various sectors. The challenges that face scalability can be summarized into factors, such as requirements of infrastructure, availability of resources, and maturity of technology. The scalability concept is guarded by the necessary guidelines and strategies. Both enable the expansion of PtX projects and the continuation of previously successful initiatives. Collaboration, which is included in the proposed framework, ensures the development of scalable solutions that are both effective and efficient [110].

The proposed framework is focused on supporting the environmental guidelines for PtX technologies, which reflect on adaptability and scalability in general. This will also encourage the growth of PtX technologies in an environmentally friendly manner. But maintaining an ecosystem necessitates the presence of financial incentives, like grants and subsidies, and also, the participation of both the public sector and the government gives a forward push to the scalability of PtX projects [111].

Referring to the synergetic framework, the knowledge sharing and learning from previous experiences came in the forefront of priority. Achieving this requires meticulous documentation, dissemination of best practices, case studies, and learned lessons from existing PtX projects. Practically, conferences, workshops, and online repositories form platforms for sharing knowledge and expertise smoothly and effectively. In addition, the unsuccessful initiatives also bear lessons, as well as the successful ones. All these fuel the improvement of both adaptability and scalability of the framework over time [112].

Finally, the proposed framework implicitly promotes modularity and design flexibility, as well as enhancing adaptability. Unarguably, the infrastructure of PtX projects should consider a design that has the ability to be scaled up easily. If it is also adaptable to different contexts and sectors, this is an added advantage. In a practical scale, PtX technologies should be designed in a modular fashion. This standardization makes the replication of the units easy and enables the deployment and integration whatsoever of the application. In conclusion, the flexible design ensures that PtX technologies, in general, have an optimality that is based on specific requirements [113].

5.8 CONCLUSION

This chapter has discussed the synergetic deployment of PtX technologies among various sectors, starting from a quick overview of various PtX technologies. Then the benefits gained from PtX integration were discussed, followed by a discussion of the specific impacts of PtX technologies on the residential, transportation, and industrial sectors, and also its cross-sectoral integration, in addition to the techno-economic feasibility of PtX synergetic deployment. Subsequently, the chapter also discussed the regulations governing PtX integration and infrastructure. Finally, a comprehensive framework was proposed for the collaborative deployment of PtX technologies.

The achievements can be summarized in the following points:

- PtX technologies facilitate the transition into a decarbonized economy, where synergetic deployment among sectors leverages this potential.
- Another potential offered by PtX is its ability to better mitigate the renewable energy intermittency and thereby increase various sectors' dependence on renewable energy. It may also increase the penetration percentage of renewable energy due to the buffer capability of PtX technologies. Therefore, the electrical grid performance can be both balanced and resilient.
- However, the previous point implies creating a large infrastructure investment for synthesis plants, storage, and transportation. Here, policies can play a significant role in order to facilitate this issue.
- This could also be supported by a regulatory framework which should be clear and consistent regarding the technical and economic goals, like energy targets, carbon pricing, and emissions reduction.
- A collaborative effort from governments, academy, industry, and civil society is of crucial importance. The collaboration is not only a comprehensive framework but also promotes innovation and encourages strategic investment.
- International cooperation is beneficial for PtX technology integration. It makes resource sharing easier, creates new energy markets, and fosters relations between developed and developing countries.

REFERENCES

[1] R. Quitzow, *et al.*, "Positioning Germany in an international hydrogen economy: A policy review," *Energy Strat. Rev.*, vol. 53, pp. 1–24, May 2024.

[2] C. Breyer, *et al.*, "The role of electricity-based hydrogen in the emerging power-to-X economy," *Int. J. Hydrogen Energy*, vol. 49(D), pp. 351–359, Jan 2024.

[3] L. M. Pastore, *et al.*, "Synergies between power-to-heat and power-to-gas in renewable energy communities," *Renew. Energy*, vol. 198, pp. 1383–1397, Oct 2022.

[4] J. C. Osorio-Aravena, *et al.*, "Synergies of electrical and sectoral integration: Analysing geographical multi-node scenarios with sector coupling variations for a transition towards a fully renewables-based energy system," *Energy*, vol. 279, pp. 1–16, Sep 2023.

[5] B. S. Zainal, *et al.*, "Recent advancement and assessment of green hydrogen production technologies," *Renew. Sustain. Energy Rev.*, vol. 189(A), pp. 1–30, Jan 2024.

[6] R. Daiyan, *et al.*, "Opportunities and challenges for renewable power-to-X," *ACS Energy Lett.*, vol. 5, no. 12, pp. 3843–3847, Nov 2022.

[7] A. Risco-Bravo, *et al.*, "From green hydrogen to electricity: A review on recent advances, challenges, and opportunities on power-to-hydrogen-to-power systems," *Renew. Sustain. Energy Rev.*, vol. 189(A), pp. 1–53, Jan 2024.

[8] D. J. Arent, *et al.*, "Challenges and opportunities in decarbonizing the U.S. energy system," *Renew. Sustain. Energy Rev.*, vol. 169, pp. 1–30, Nov 2022.

[9] M. Sterner, *et al.*, "Power-to-gas and power-to-X—The history and results of developing a new storage concept," *Energies*, vol. 14, no. 20, pp. 1–18, Oct 2021.

[10] G. Buffo, *et al.*, "Power-to-X and power-to-power routes," in *Solar Hydrogen Production*. Academic Press, pp. 529–557, Jan 2019.

[11] M. Götz, *et al.*, "Renewable Power-to-Gas: A technological and economic review," *Renew. Energy*, vol. 85, pp. 1371–1390, Jan 2016.

[12] P. Schmidt, *et al.*, "Power-to-liquids as renewable fuel option for aviation: A review," *Chem. Ing. Tech.*, vol. 90, pp. 127–140, Jan 2018.

[13] A. Bloess, *et al.*, "Power-to-heat for renewable energy integration: A review of technologies, modeling approaches, and flexibility potentials," *Appl. Energy*, vol. 212, pp. 1611–1626, Feb 2018.

[14] C. Schnuelle, *et al.*, "Socio-technical-economic assessment of power-to-X: Potentials and limitations for an integration into the German energy system," *Energy Res. Soc. Sci.*, vol. 51, pp. 187–197, May 2019.

[15] G. Fambri, *et al.*, "Techno-economic analysis of Power-to-Gas plants in a gas and electricity distribution network system with high renewable energy penetration," *Appl. Energy*, vol. 312, pp. 1–20, Apr 2022.

[16] R. Habibifar, *et al.*, "Robust energy management of residential energy hubs integrated with Power-to-X technology," in *IEEE Texas Power and Energy Conference (TPEC)*, pp. 1–6, Feb 2021.

[17] J. Ramsebner, *et al.*, "The sector coupling concept: A critical review," in *Wiley Interdisciplinary Reviews: Energy and Environment*, vol. 10, no. 4, pp. 1–27, Jul 2021.

[18] C. Andrade, *et al.*, "The role of power-to-gas in the integration of variable renewables," *Appl. Energy*, vol. 313, pp. 1–25, May 2022.

[19] M. Qi, *et al.*, "Continuous and flexible renewable-power-to-methane via liquid CO_2 energy storage: Revisiting the techno-economic potential," *Renew. Sustain. Energy Rev.*, vol. 153, Jan 2022.

[20] M. J. Palys, *et al.*, "Power-to-X: A review and perspective," *Comput. Chem. Eng.*, vol. 165, Sep 2022.

[21] C. Breyer, *et al.*, "On the techno-economic benefits of a global energy interconnection," *Econ. Energy Environ. Policy*, vol. 9, no. 1, pp. 83–102, Mar 2020.

[22] Z. Chehade, *et al.*, "Review and analysis of demonstration projects on power-to-X pathways in the world," *Int. J. Hydrogen Energy*, vol. 44, no. 51, pp. 27637–27655, Oct 2019.

[23] F. Feijoo, *et al.*, "A long-term capacity investment and operational energy planning model with power-to-X and flexibility technologies," *Renew. Sustain. Energy Rev.*, vol. 167, pp. 1–15, Oct 2022.

[24] I. Kountouris, *et al.*, "Power-to-X in energy hubs: A Danish case study of renewable fuel production," *Energy Policy*, vol. 175, pp. 1–18, Apr 2023.

[25] D. Bogdanov, *et al.*, "Full energy sector transition towards 100% renewable energy supply: Integrating power, heat, transport and industry sectors including desalination," *Appl. Energy*, vol. 283, pp. 1–17, Feb 2021.

[26] L. M. Pastore, *et al.*, "Smart energy systems for renewable energy communities: A comparative analysis of power-to-X strategies for improving energy self-consumption," *Energy*, vol. 280, pp. 1–20, Oct 2023.

[27] L. Luo, *et al.*, "Cogeneration: Another way to increase energy efficiency of hybrid renewable energy hydrogen chain – A review of systems operating in cogeneration and of the energy efficiency assessment through exergy analysis," *J. Energy Storage*, vol. 66, pp. 1–38, Aug 2023.

[28] V. M. Maestre, *et al.*, "Challenges and prospects of renewable hydrogen-based strategies for full decarbonization of stationary power applications," *Renew. Sustain. Energy Rev.*, vol. 152, pp. 1–24, Dec 2021.

[29] M. Jayachandran, *et al.*, "Challenges and opportunities in green hydrogen adoption for decarbonizing hard-to-abate industries: A comprehensive review," *IEEE Access*, pp. 22363–23388, Feb 2024.

[30] A. Mansour-Saatloo, *et al.*, "Robust decentralized optimization of multi-microgrids integrated with Power-to-X technologies," *Appl. Energy*, vol. 304, pp. 1–51, Dec 2021.

[31] M. F. Rojas-Michaga, *et al.*, "Sustainable aviation fuel (SAF) production through power-to-liquid (PtL): A combined techno-economic and life cycle assessment," *Energy Convers. Manag.*, vol. 292, pp. 1–23, Sep 2023.

[32] S. Fuentes, *et al.*, "Composed index for the evaluation of the energy security of power systems: Application to the case of Argentina," *Energies*, vol. 13, no. 15, pp. 1–15, Aug 2020.

[33] H. Golmohamadi, "Flexible operation of Power-To-X energy systems in transportation networks," in *Interconnected Modern Multi-Energy Networks and Intelligent Transportation Systems: Towards a Green Economy and Sustainable Development*, pp. 117–164, 2024.

[34] D. Bogdanov, *et al.*, "Energy transition for Japan: Pathways towards a 100% renewable energy system in 2050," *IET Renew. Power Gen.*, vol. 17, no. 13, pp. 3298–3324, Oct 2023.

[35] F. Bauer F, *et al.*, "Assessing the feasibility of archetypal transition pathways towards carbon neutrality – A comparative analysis of European industries," *Resour. Conserv. Recycl.*, vol. 177, pp. 1–10, Feb 2022.

[36] N. Ahmad, *et al.*, "From smart grids to super smart grids: A roadmap for strategic demand management for next generation SAARC and European power infrastructure," *IEEE Access*, vol. 11, pp. 12303–12341, Feb 2023.

[37] D. Rusmanis, *et al.*, "Electrofuels in a circular economy: A systems approach towards net zero," *Energy Convers. Manag.*, vol. 292, pp. 1–13, Sep 2023.

[38] L. M. Pastore, *et al.*, "Heading towards 100% of renewable energy sources fraction: A critical overview on smart energy systems planning and flexibility measures," in *E3S Web of Conferences*, vol. 197, pp. 1–17, 2020.

[39] S. Arens, *et al.*, "Sustainable residential energy supply: A literature review-based morphological analysis," *Energies*, vol. 13, no. 2, pp. 1–28, Jan 2020.

[40] M. Salimi, *et al.*, "The role of clean hydrogen value chain in a successful energy transition of Japan," *Energies*, vol. 15, no. 16, pp. 1–19, Aug 2022.

[41] N. Souissi, *E-Diesel in the Shipping Sector: Prospects and Challenges*. Oxford: The Oxford Institute for Energy Studies, pp. 1–36, Mar 2024.

[42] M. Qi, *et al.*, "Strategies for flexible operation of power-to-X processes coupled with renewables," *Renew. Sustain. Energy Rev.*, vol. 179, Jun 2023.

[43] P. G. Panah, *et al.*, "Guerrero JM: Marketability analysis of green hydrogen production in Denmark: Scale-up effects on grid-connected electrolysis," *Int. J. Hydrogen Energy*, vol. 47, no. 25, pp. 12443–12455, Mar 2022.

[44] J. F. Wiegner, *et al.*, "Interdisciplinary perspectives on offshore energy system integration in the North Sea: A systematic literature review," *Renew. Sustain. Energy Rev.*, vol. 189, pp. 1–23, Jan 2024.

[45] J. M. Mendoza, *et al.*, "Technology-enabled circular business models for the hybridisation of wind farms: Integrated wind and solar energy, power-to-gas and power-to-liquid systems," *Sustain. Prod. Consump.*, vol. 36, pp. 308–327, Mar 2023.

[46] G. Herz, *et al.*, "Economic assessment of Power-to-Liquid processes – Influence of electrolysis technology and operating conditions," *Appl. Energy*, vol. 292, pp. 1–18, Jun 2021.

[47] S. R. Fahim, H. M. Hasanien, "Energy storage devices," in *Modernization of Electric Power Systems: Energy Efficiency and Power Quality*. Cham: Springer International Publishing, pp. 407–441, Jun 2023.

[48] O. Lugovoy, *et al.*, "Feasibility study of China's electric power sector transition to zero emissions by 2050," *Energy Economics*, vol. 96, pp. 1–38, Apr 2021.

[49] P. J. Ansell, "Review of sustainable energy carriers for aviation: Benefits, challenges, and future viability," *Progress in Aerospace Sciences*, vol. 141, Aug 2023.

[50] E. Panos, *et al.*, "Challenges and opportunities for the Swiss energy system in meeting stringent climate mitigation targets. Limiting Global Warming to Well Below 2°C," in *Energy System Modelling and Policy Development*, pp. 155–172, 2018.

[51] P. Colbertaldo, *et al.*, "Clean mobility infrastructure and sector integration in long-term energy scenarios: The case of Italy," *Renew. Sustain. Energy Rev.*, vol. 133, pp. 1–35, Nov 2020.

[52] S. Farah, *et al.*, "Cost and CO_2 emissions co-optimisation of green hydrogen production in a grid-connected renewable energy system," arXiv:2404.11995, pp. 1–16, Apr 2024.

[53] A. R. Dahiru, *et al.*, "Recent development in Power-to-X: Part I – A review on techno-economic analysis," *Journal of Energy Storage*, vol. 56, pp. 1–18, Dec 2022.

[54] M. Münster, *et al.*, "Perspectives on green hydrogen in Europe: During an energy crisis and towards future climate neutrality," *Oxford Open Energy*, vol. 3, pp. 1–10, oiae001, Jan 2024.

[55] H. Dai, *et al.*, "The impacts on climate mitigation costs of considering curtailment and storage of variable renewable energy in a general equilibrium model," *Energy Economics*, vol. 64, pp. 627–637, May 2017.

[56] J. C. Koj, *et al.*, "Environmental impacts of power-to-X systems: A review of technological and methodological choices in Life Cycle Assessments," *Renew. Sustain. Energy Rev.*, vol. 112, pp. 865–879, Sep 2019.

[57] M. H. Khan, *et al.*, "Strategies for life cycle impact reduction of green hydrogen production: Influence of electrolyser value chain design," *Int. J Hydrogen Energy*, vol. 62, pp. 769–782, Apr 2024.

[58] Q. Liu, *et al.*, "Making waves: Power-to-X for the water resource recovery facilities of the future," *Water Res.*, pp. 1–5, Apr 2024.

[59] S. A. Fitts, *et al.*, "Legal issues associated with renewable fuels," *J. Energy Nat. Resour. Law*, vol. 40, no. 4, pp. 495–500, Oct 2022.

[60] Ş. Kılkış, *et al.*, "Effective mitigation of climate change with sustainable development of energy, water and environment systems," *Energy Convers. Manag.*, vol. 269, Oct 2022.

[61] M. Munster, *et al.*, "Sector coupling: Concepts, potentials and barriers," in *AEIT International Annual Conference (AEIT)-IEEE*, pp. 1–6, Sep 2020.

[62] O. Cagdas Artantas, "Green electricity promotion in Germany," in *Promotion of Green Electricity in Germany and Turkey: A Comparison with Reference to the WTO and EU Law*. Cham: Springer Nature Switzerland, pp. 139–167, Nov 2023.

[63] F. Bovera, *et al.*, "From energy communities to sector coupling: A taxonomy for regulatory experimentation in the age of the European Green Deal," *Energy Policy*, vol. 171, Dec 2022.

[64] E. Rosales-Asensio, *et al.*, "Electricity balancing challenges for markets with high variable renewable generation," *Renew. Sustain. Energy Rev.*, vol. 189, pp. 1–11, Jan 2024.

[65] G. Chicco, *et al.*, "Challenges for a transition towards the smart grids," *Tech. Sci.*, vol. 3, no. 2, pp. 155–174, Jun 2018.

[66] M. Boulakhbar, *et al.*, "Towards a large-scale integration of renewable energies in Morocco," *J. Energy Storage*, vol. 32, pp. 1–17, Dec 2020.

[67] T. Schittekatte, *et al.*, "Making the TEN-E regulation compatible with the Green Deal: Eligibility, selection, and cost allocation for PCIs," *Energy Policy*, vol. 156, pp. 1–16, Sep 2021.

[68] J. Sillman, *et al.*, "Emission reduction targets and electrification of the Finnish energy system with low-carbon Power-to-X technologies: Potentials, barriers, and innovations – A Delphi survey," *Technol. Forecast. Soc. Change*, vol. 193, pp. 1–11, Aug 2023.

[69] W. Cheng, *et al.*, "How green are the national hydrogen strategies?," *Sustainability*, vol. 14, no. 3, pp. 1–33, Feb 2022.

[70] J. K. Kirkegaard, *et al.*, "Tackling grand challenges in wind energy through a socio-technical perspective," *Nat. Energy*, vol. 8, no. 7, pp. 655–664, Jul 2023.

[71] R. Alfasfos, *et al.*, "Lessons learned and recommendations from analysis of hydrogen incidents and accidents to support risk assessment for the hydrogen economy," *Int. J. Hydrogen Energy*, vol. 60, pp. 1203–1214, Mar 2024.

[72] J. F. Wiegner, *et al.*, "Interdisciplinary perspectives on offshore energy system integration in the North Sea: A systematic literature review," *Renew. Sustain. Energy Rev.*, vol. 189, pp. 1–23, Jan 2024.

[73] A. Team, *et al.*, "Business models for flexible production and storage," *Policy*, pp. 1–107, Dec 2015.

[74] A. N. Akpolat, *et al.*, "Modeling and operation of a fuel cell stack for distributed energy resources: A living lab platform," *Int. J. Hydrogen Energy*, Apr 2024.

[75] F. Alasali, *et al.*, "A review of hydrogen production and storage materials for efficient integrated hydrogen energy systems," *Energy Sci. Eng.*, pp. 1–35, Mar 2024.

[76] P. A. Ashari, *et al.*, "Pathways to the hydrogen economy: A multidimensional analysis of the technological innovation systems of Germany and South Korea," *Int. J. Hydrogen Energy*, vol. 49, pp. 405–421, Jan 2024.

[77] A. Moradi, *et al.*, "Risk-based optimal decision-making strategy of a Power-to-Gas integrated energy-hub for exploitation arbitrage in day-ahead electricity and Natural Gas markets," *Sustain. Energy Grids Netw.*, vol. 31, Sep 2022.

[78] H. Wojtaszek, *et al.*, "Hydrogen energy as a catalyst for sustainable development: A comparative analysis of policies, strategies, and implementation in Poland and Germany," in *Scientific Papers of Silesian University of Technology: Organization & Management/Zeszyty Naukowe Politechniki Slaskiej. Seria Organizacji i Zarzadzanie*, vol. 179, pp. 1–33, 2023.

[79] R. Hassink, *et al.*, "Exploring the scope of regions in challenge-oriented innovation policy: The case of Schleswig-Holstein, Germany," *Eur. Plan. Stud.*, vol. 30, no. 11, pp. 2293–2311, Nov 2022.

[80] O. Inderwildi, *et al.*, "The impact of intelligent cyber-physical systems on the decarbonization of energy," *Energy Environ. Sci.*, vol. 13, no. 3, pp. 744–771, 2020.

[81] S. Majer, *et al.*, "Gaps and research demand for sustainability certification and standardisation in a sustainable bio-based economy in the EU," *Sustainability*, vol. 10, no. 7, pp. 1–44, Jul 2018.

[82] G. Ulpiani, *et al.*, "Let's hear it from the cities: On the role of renewable energy in reaching climate neutrality in urban Europe," *Renew. Sustain. Energy Rev.*, vol. 183, pp. 1–15, Sep 2023.

[83] B. Nastasi, *et al.*, "Solar power-to-gas application to an island energy system," *Renew. Energy*, vol. 164, pp. 1005–1016, Feb 2021.

[84] M. Di Somma, *et al.*, "Integrated energy systems: The engine for energy transition," in *Technologies for Integrated Energy Systems and Networks*, pp. 15–40, May 2022.

[85] C. Carbone, *et al.*, "Potential deployment of reversible solid-oxide cell systems to valorise organic waste, balance the power grid and produce renewable methane: A case study in the Southern Italian Peninsula," *Frontiers in Energy Research*, vol. 9, pp. 1–17, Feb 2021.

[86] J. Hyvönen, *et al.*, "Possible bottlenecks in clean energy transitions: Overview and modelled effects – Case Finland," *J. Clean. Prod.*, pp. 1–15, Apr 2023.

[87] P. Schmidt, *et al.*, "Power-to-liquids as renewable fuel option for aviation: A review," *Chem. Ing. Tech.*, vol. 90, pp. 127–140, Jan 2018.

[88] S. S. Cordova, *et al.*, "What should we do with CO2 from biogas upgrading?," *J. CO$_2$ Util.*, vol. 77, pp. 1–12, Nov 2023.

[89] Q. Li, *et al.*, "Sustainable and market-oriented solutions for the electricity sales market reform," *Econ. Change Restruct.*, vol. 57, no. 2, pp. 1–31, Apr 2024.

[90] Commission, "Commission staff working document, impact assessment report part 1," COM (2024) 63 Final, Part 1 of 5.

[91] J. Y. Lim, *et al.*, "Nationwide sustainable renewable energy and Power-to-X deployment planning in South Korea assisted with forecasting model," *Appl. Energy*, vol. 283, Feb 2021.

[92] D. Schlund, *et al.*, "The who's who of a hydrogen market ramp-up: A stakeholder analysis for Germany," *Renew. Sustain. Energy Rev.*, vol. 154, pp. 1–36, Feb 2022.

[93] Y. Zhou, *et al.*, "Peer-to-peer energy sharing and trading of renewable energy in smart communities – Trading pricing models, decision-making and agent-based collaboration," *Renew. Energy*, vol. 207, pp. 177–193, May 2023.

[94] B. Delaval, *et al.*, *Hydrogen RD&D Collaboration Opportunities*. Australia: CSIRO, 2022.

[95] P. Lamers, *et al.*, "Linking life cycle and integrated assessment modeling to evaluate technologies in an evolving system context: A power-to-hydrogen case study for the United States," *Environ. Sci. Technol.*, vol. 57, no. 6, pp. 2464–2473, Feb 2023.

[96] C. Hackenesch, *et al.*, *Green Transitions in Africa–Europe Relations: What Role for the European Green Deal*. Brussel, Belgium: ETTG, pp. 1–20, Apr 2021.

[97] F. Schipfer, *et al.*, "Status of and expectations for flexible bioenergy to support resource efficiency and to accelerate the energy transition," *Renew. Sustain. Energy Rev.*, vol. 158, pp. 1–14, Apr 2022.

[98] J. Magyari, *et al.*, "Integration opportunities of power-to-gas and internet-of-things technical advancements: A systematic literature review," *Energies*, vol. 15, no. 19, pp. 1–19, Sep 2022.

[99] J. Vendrik, *et al.*, "Case study: Integrated infrastructure planning," *4i-TRACTION Deliv.*, vol. 4, no. 2, pp. 1–87, May 2023.

[100] S. Sillak, "All talk, and (no) action? Collaborative implementation of the renewable energy transition in two frontrunner municipalities in Denmark," *Energy Strat. Rev.*, vol. 45, pp. 1–14, Jan 2023.

[101] R. Huttunen, *et al.*, Carbon neutral Finland 2035 – National climate and energy strategy, pp. 1–223, 2022.

[102] M. A. Ancona, *et al.*, "Parametric thermo-economic analysis of a power-to-gas energy system with renewable input, high temperature co-electrolysis and methanation," *Energies*, vol. 15, no. 5, pp. 1–25, Feb 2022.

[103] J. Jurasz, *et al.*, "A review on the complementarity of renewable energy sources: Concept, metrics, application and future research directions," *Solar Energy*, vol. 195, pp. 703–724, Jan 2020.

[104] L. F. Vega, *et al.*, "Perspectives on advancing sustainable CO_2 conversion processes: Trinomial technology, environment, and economy," *ACS Sustain. Chem. Eng.*, vol. 12, no. 14, pp. 5357–5382, Mar 2024.

[105] R. Rodrigues, *et al.*, "Narrative-driven alternative roads to achieve mid-century CO_2 net neutrality in Europe," *Energy*, vol. 239, pp. 1–17, Jan 2022.

[106] Z. Csedő, *et al.*, "Hydrogen economy development opportunities by inter-organizational digital knowledge networks," *Sustainability*, vol. 13, no. 16, pp. 1–26, Aug 2021.

[107] J. J. Häußermann, *et al.*, "Social acceptance of green hydrogen in Germany: Building trust through responsible innovation," *Energy Sustain. Soc.*, vol. 13, no. 1, pp. 1–19, Jun 2023.

[108] A. A. Bouramdane, "Crafting an optimal portfolio for sustainable hydrogen production choices in Morocco," *Fuel*, vol. 358, Feb 2024.

[109] Z. M. Hou, *et al.*, "International experience of carbon neutrality and prospects of key technologies: Lessons for China," *Petrol. Sci.*, vol. 20, no. 2, pp. 893–909, Apr 2023.

[110] F. A. Plazas-Niño, *et al.*, "National energy system optimization modelling for decarbonization pathways analysis: A systematic literature review," *Renew. Sustain. Energy Rev.*, vol. 162, pp. 1–19, Jul 2022.

[111] B. J. Singh, *et al.*, "Green hydrogen production: Bridging the gap to a sustainable energy future," in *Challenges and Opportunities in Green Hydrogen Production*. Singapore: Springer Nature Singapore, pp. 83–124, May 2024.

[112] M. B. Rosendal, *et al.*, "Renewable fuel production and the impact of hydrogen infrastructure—A case study of the Nordics," *Energy*, vol. 297, pp. 1–14, Jun 2024.

[113] C. Varvoutis, *et al.*, "Recent advances on CO_2 mitigation technologies: On the role of hydrogenation route via green H_2," *Energies*, vol. 15, no. 13, pp. 1–39, Jun 2022.

Exploring the Role of Power-to-X in Energy Hubs

Amir Meydani, Hossein Shahinzadeh,
Gevork B. Gharehpetian, Mohammad Mohsen Hayati,
and Mehdi Abapour

6.1 INTRODUCTION

Energy security is a critical global concern, intensified by population expansion, rising energy consumption, and the depletion of fossil fuel reserves. The pursuit of energy that is secure, affordable, reliable, and environmentally friendly has become a fundamental aspect of both national security and sustainable development, highlighting the importance of resilient energy systems [1, 2]. Historically, energy production relied predominantly on fossil fuel–powered thermal power plants, which were not only inefficient but also detrimental to the environment. This reliance has led to significant energy losses and expensive infrastructure costs associated with transmitting electricity over long distances from centralized stations, rendering existing systems unsustainable for future energy needs. To address these challenges, the advancement and incorporation of innovative technologies—such as combined heat and power (CHP) systems, electric heat pumps (HP), and electric vehicles (EV)—have facilitated the establishment of more intricate and effective energy systems. These advancements underscore the significance of multi-carrier energy (MCE) systems, sometimes referred to as multi-energy systems (MES) or energy hubs (EH) [3]. MCE systems effectively manage and optimize various energy carriers, including electricity, natural gas (NG), heating, cooling, and renewable energy sources (RES), like solar and wind. The benefits of MCE systems are manifold, encompassing adaptability, durability, and efficiency, which enable dynamic control of energy supply and demand, thereby reducing costs and emissions.

An EH within an MCE system is a complex arrangement that integrates different energy converters, transmission lines, storage systems, and load requirements. This integration allows for the simultaneous management of multiple energy types, enhancing the overall reliability and effectiveness of the system. Integrated electricity–gas systems, for example, improve energy resilience by utilizing diverse resources to meet energy demands concurrently. EHs can address a wide range of energy needs—including electricity, heating, and cooling—through the use of RESs and advanced energy storage (ES) technologies, such as batteries, pumped hydro, and hydrogen (H2) storage. Essential components, like CHP units, gas boilers, fuel cells (FC), and power to-gas (PtG) technologies, facilitate the conversion of energy between different forms, thereby increasing system flexibility [4, 5]. However, MCE systems face challenges, particularly in managing parametric uncertainties that can impact energy costs, renewable outputs, and load demands. Employing management methodologies such as stochastic programming, resilient optimization, and fuzzy programming is crucial for maintaining system stability. Also, demand response management (DRM) plays a vital role by providing monetary incentives to consumers for modifying their energy consumption patterns [6], thereby aiding in load balancing and reducing operational expenses. MCE frameworks also address environmental concerns, including emissions from CHP and boiler systems, through strategies like carbon taxes and carbon capture technologies, enhancing the sustainability of these systems. Moreover, the multi-energy approach not only lowers operational costs but also improves system resilience, particularly during severe conditions, such as natural disasters. Resilience measures include reducing unsupplied loads, enhancing system preparedness, and ensuring rapid recovery times, all of which are essential for the sustainable and efficient functioning of modern energy systems. Consequently, MCE systems, utilizing the concept of EHs, offer a holistic approach to achieving a sustainable, efficient, and resilient energy future. In light of these challenges and opportunities within MCE systems, the evolving energy landscape necessitates innovative solutions to address the issues arising from the increasing integration of RESs and the consequent demand for a reliable and robust energy infrastructure. Power-to-X (PtX) technologies are at the forefront of this transformation, providing flexible methods to convert excess renewable electricity into various energy carriers, including H2, ammonia, methane, and other valuable products [7]. These technologies enhance the control of energy supply and demand, playing a vital role in stabilizing electricity markets and improving the resilience of EHs.

EHs, serving as centralized locations for the generation, transformation, storage, and distribution of energy, stand to gain substantial advantages from the incorporation of PtX systems. Through PtX technologies, these hubs can maximize the efficiency of RESs, address the intermittency issues commonly associated with RE, and ensure consistent and reliable energy output. This integration not only enhances the operational effectiveness of EHs but also contributes to the broader objective of reducing carbon emissions in the energy sector. PtX encompasses a range of applications beyond mere energy storage, with significant potential to revolutionize energy production and consumption. For example, power-to-hydrogen (PtH) generates H2 via electrolysis driven by RESs [8], which can be used as a fuel or as a feedstock for producing synthetic fuels and chemicals. Similarly, power-to-ammonia [9] and power-to-methane [10] facilitate the production of NH_3 and CH_4, respectively, both of which are crucial for various industrial and energy applications. The deployment of PtX systems also has profound implications for electricity markets. PtX technologies can aid in stabilizing prices, enhancing the overall efficiency of electricity markets, and regulating supply and demand by providing methods to store and utilize excess renewable electricity. This is particularly relevant in the context of peer-to-peer (P2P) energy trading, where PtX can offer novel opportunities for market participation and energy exchange [11–13]. Effective scheduling and resilience of PtX-based systems are paramount in energy centers. The optimal utilization of PtX technologies requires strategic scheduling that accounts for the demand patterns of the energy center and the variability of RESs. Furthermore, integrating PtX can bolster the resilience of these centers, ensuring a consistent energy supply under various scenarios and providing robust protection against disruptions.

This chapter delves into the multifaceted roles of PtX technologies within energy centers and electricity markets, exploring the potential of H2-based systems and the specific contributions of NH_3, CH_3OH, and CH_4. By understanding these dynamics, we can gain a deeper appreciation of the transformative impact PtX technologies have on the future of sustainable energy.

6.2   EH MODELING, CONCEPT, AND COMPONENTS

6.2.1   EH Concept

Under the "Vision of Future Energy Networks" initiative, researchers in Power Systems and the High Voltage Laboratory at ETH Zurich developed the concept of EHs. The objective of this initiative is to envision future energy systems over the next 20–30 years. The initiative specifies several critical attributes for forthcoming energy systems [3]:

- *Transition to multi-energy systems (MES).* MESs capitalize on the synergies among various energy carriers, thereby improving the overall reliability and efficiency of the system.
- *Implementation of non-hierarchical organizational structures.* Shifting from conventional hierarchical energy system models to more decentralized and adaptable configurations.

- ***Integration and interconnection.*** Creating interconnected energy systems that enable the joint transportation and management of multiple energy carriers over extended distances.
- ***Development of EHs.*** EHs are integrated devices designed to convert, store, and manage a variety of energy carriers. By integrating energy storage systems (ESSs), transmission infrastructure, generation sources, and consumer interfaces, they serve as central nodes in energy systems.

EHs interact with various energy carriers through conversion devices or direct connections. The matrix model of an EH is illustrated in equation 6.1, which displays the relationship between various input and output energy vectors across a coupling matrix. The intrinsic features of an EH, such as converters, and their efficiencies, are often represented using matrix formulations in modeling. Geidl and Andersson further refined the definition of an EH, describing it as a framework that offers inputs, outputs, ESSs, and conversion capabilities for multiple-energy carriers. Various energy infrastructures— like district heating (DH), NG, and electricity systems—are connected to network participants, including generators and consumers, through EHs. The term "hybrid" is occasionally used with EHs to emphasize the interaction among the numerous energy carriers within these systems. Essentially, an EH is a location where a variety of energy carriers are received, converted, stored, and consumed.

An EH is a versatile element that enables efficient interconnectivity of energy infrastructures by converting, storing, or transmitting various forms of energy. This model optimizes the utilization of the unique benefits provided by different energy carriers; for example, electricity can be transmitted over long distances with minimal losses, and cost-effective technologies can store chemical carriers like NG. Energy can be stored by increasing the pressure of chemical carriers transported through pipelines without the need for separate storage units. The capacity to optimize economic benefits by converting various forms of energy according to price disparities among carriers is a critical component of an integrated MES. Figure 6.1 demonstrates this adaptability, as inputs such as electricity, NG, heat, and biomass are ultimately converted into outputs such as cooling, heat, and electricity. Components such as transformers, combined heat and power (CHP) units, thermal furnaces, absorption chillers, electricity storage systems, and hot water storage are all part of the conversion process.

For hub modeling, the input energy carriers are defined by the vector P, and the output carriers by L, as shown in Figure 6.1. The mutual matrix C establishes a correlation between these two vectors according to:

$$\begin{bmatrix} L_\alpha \\ L_\beta \\ \vdots \\ L_\omega \end{bmatrix} = \begin{bmatrix} C_{\alpha\alpha} & C_{\beta\alpha} & \cdots & C_{\omega\alpha} \\ C_{\alpha\beta} & C_{\beta\beta} & \cdots & C_{\omega\beta} \\ \vdots & \vdots & \ddots & \vdots \\ C_{\alpha\omega} & C_{\beta\omega} & \cdots & C_{\omega\omega} \end{bmatrix} \times \begin{bmatrix} P_\alpha \\ P_\beta \\ \vdots \\ P_\omega \end{bmatrix} \tag{6.1}$$

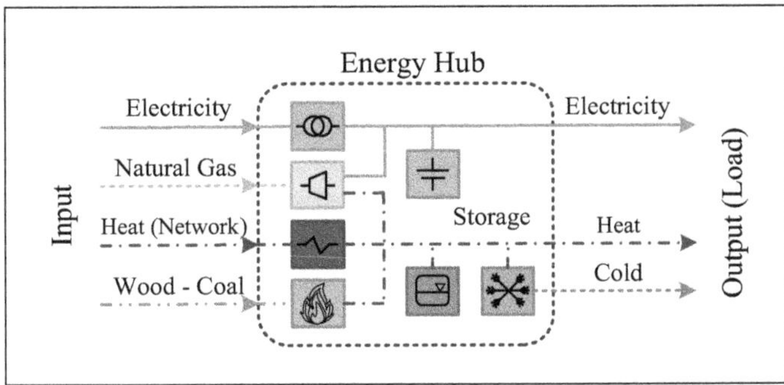

FIGURE 6.1 Energy hub components.

Each element of C reflects the correlation between the input and output carriers. Put simply, $C_{\alpha\beta}$ denotes the connection or association between two individuals who possess the characteristics α and β. EHs are highly adaptable systems engineered to manage various energy inputs and outputs, resulting in enhanced flexibility and efficiency in energy management (EM). These hubs can be supplied with a wide variety of energy sources, such as electricity, NG, district heat, wind, solar radiation, water, hydrogen (H_2), and biomass like wood chips. The common outputs encompass electricity, heating, and cooling. To transform these inputs into usable energy forms, EHs employ a range of technologies, including furnaces, solar thermal panels, heat pumps (HPs), CHP units, transformers, power-to-gas (PtG) units, electric chillers, photovoltaic (PV) panels, FCs, heat exchangers, electrical batteries, and storage systems for NG, cooling, and heat, as well as absorption chillers.

The main purpose of these EH structures is to efficiently and cost-effectively manage inputs to meet diverse output requirements. EHs provide an inherent flexibility that enables them to adjust to various load demands, thereby improving system reliability. This adaptability also allows maintenance operations to be carried out without interrupting the energy supply, enhancing the system's dependability. Moreover, the ability to balance energy supply and demand by flexibly integrating various energy carriers makes EHs a remarkably efficient solution for energy management. EHs can be utilized in systems of different scales, making them suitable for a wide array of applications. By combining different energy sources, EHs provide a holistic solution for modern energy requirements, promoting both efficiency and sustainability.

6.2.2 Hub Components

An *EH* is a complex and interconnected framework specifically designed to oversee the numerous processes involved in the production, conversion, transmission, storage, and consumption of different energy carriers. This framework functions as a central hub where various energy sources—including electricity, heat, NG, and hydrogen—come together. The EH facilitates the linking of different energy carriers, enabling their interaction through a variety of conversion and storage technologies. These technologies encompass options such as CHP systems, batteries, thermal energy storage, and electrolysis units.

The complexity of an EH arises from the interconnections between various energy carriers which are linked through conversion and storage processes. For instance, electricity generated from renewable sources can be transformed into thermal energy for district heating networks, stored as electrical energy in batteries, or converted into hydrogen through electrolysis. These interconnections enable changes or improvements in one component to have a ripple effect across the entire energy network. Effectively managing this intricate system necessitates precise simulation, control, and the utilization of advanced modeling tools and algorithms capable of handling the complex interactions between energy carriers and conversion technologies. The EH framework plays a pivotal role in the modeling and planning of integrated energy systems (IES) by representing these complex interactions. Through EH models, planners and policymakers can optimize resource allocation, assess infrastructure investments, and evaluate policy impacts. For example, EH models help identify the most efficient mix of energy sources and technologies to meet demand at the lowest cost and environmental impact. They also provide insights into the cost–benefit analysis of new infrastructure investments, such as RE installations, storage systems, or network expansions. EH models can simulate the effects of policies like carbon pricing, subsidies, or RE mandates on the energy system's performance. Due to the intricate nature of an EH, advanced optimization strategies are frequently utilized to achieve efficient energy management, ensuring that supply matches demand in a manner that is both cost-effective and environmentally friendly. As a platform for testing various scenarios, EHs enable researchers and engineers to explore more efficient integration of multiple energy carriers. This can result in strategies to minimize energy wastage, improve system reliability, and enhance the integration of RESs. The concept of EHs is particularly beneficial in the development of next-generation smart grids (SG) and energy systems, where flexibility, resilience, and sustainability are of paramount importance.

EHs are particularly valuable in managing energy systems under fluctuating market conditions, such as variable energy prices, demand uncertainty, and intermittent RE generation. Through real-time optimization, EHs enable dynamic operation strategies that adjust energy flows based on real-time data on prices and demand. Also, EHs facilitate the integration of DSM and DR programs, allowing the modulation of consumption patterns to match supply conditions and reduce costs during peak periods. Also, EHs support the integration of intermittent RESs by providing flexibility through energy storage and conversion technologies. Denmark has been a leader in integrating wind energy into its national grid. EH models have been employed to optimize the interaction between wind power, CHP units, and district heating systems. The primary objective was to manage the high variability of wind energy and minimize reliance on fossil fuels. By coordinating electricity and heat production and utilizing excess wind power for heating via heat pumps or thermal storage, Denmark's energy system has achieved improved flexibility and reduced curtailment of wind energy, resulting in higher RE utilization and lower GHG emissions. At the regional level, EHs support the planning and operation of interconnected energy systems across cities or districts, enabling coordinated energy management that transcends individual utilities or sectors. The EU's goal of creating an integrated energy market involves complex cross-border energy flows. EH models have been applied to analyze and optimize energy exchanges between countries, enhancing energy security, reducing costs, and meeting climate targets through efficient cross-border energy trade. This approach has helped identify beneficial trade patterns, infrastructure investment needs, and policy recommendations to facilitate a more integrated and efficient European energy market.

The versatility of EHs lies in their capacity to handle the increasing complexity of modern energy systems, which are characterized by the simultaneous management of multiple energy carriers, diverse technologies, and sector coupling. EHs integrate various generation sources (renewables, fossil fuels), storage options, and conversion devices while linking traditionally separate sectors, like transportation, heating, and power, to exploit synergies. This comprehensive integration allows for more effective and sustainable energy management. Rapid urbanization in China has led to increased energy demand and environmental concerns. EHs have been utilized to design sustainable urban energy systems that integrate distributed generation (e.g., rooftop solar PV), district heating, and EVs within urban microgrids (MG). These EH models have enhanced energy efficiency, reduced emissions, and provided a roadmap for scaling solutions to other urban centers, thereby contributing to sustainable urban development. Moreover, the EH framework supports decision-making processes by offering quantitative analysis tools for policy development, investment decisions, and operational strategies. For governments, EH models assist in crafting policies that promote sustainable energy practices. For utilities and investors, these models guide capital allocation toward the most effective technologies and infrastructure. For system operators, EH models enable the development of strategies that optimize system performance under varying conditions. In the United States, EH concepts have been applied to optimize MGs for campuses and military bases. The primary objectives have been to enhance energy security, reduce costs, and increase the utilization of renewables. EH models have been used to simulate the integration of solar PV, energy storage, CHP units, and DR programs, resulting in improved reliability and resilience of energy supply, cost savings, and increased RE penetration.

The EH framework is a powerful tool for modeling, planning, analyzing, and operating IESs across various scales and market conditions. Its ability to handle complex interactions between multiple energy carriers and technologies makes it indispensable for enhancing system efficiency, facilitating renewable integration, and supporting sustainable development by guiding policies and investments toward environmentally friendly energy solutions. As energy systems continue to evolve with the advent of SGs, increasing electrification, and the proliferation of RES, the EH framework will become even more critical. Future applications may include modeling the role of hydrogen as an energy carrier within EHs; expanding sector coupling to link transportation, heating, and industrial processes; and utilizing advanced control systems, such as artificial intelligence (AI) and machine learning (ML), to enhance EH operation.

TABLE 6.1

Energy Hub Components

Components	Description	Primary Functionality	Examples	Related Technologies	Operational Characteristics	Economic Considerations	Flexibility in Multi-Energy Integration	Integration Challenges and Requirements	Future Trends and Developments
Energy conversion systems	Devices that convert one form of energy into another to match supply with demand	Transform energy types for end use applications	Turbines, generators, inverters, heat exchangers	PtX, power electronics, thermoelectric converters, FCs	Efficiency, conversion rates, operational modes	Capital costs, maintenance expenses, ROI	High; can handle multiple energy inputs and outputs	Technical compatibility, control strategies, efficiency optimization	Advancements in conversion efficiency, integration with renewables, development of hybrid converters
Energy storage systems	Systems that store energy for later use	Balance supply and demand by storing excess energy and releasing it when needed	Batteries, pumped hydro storage, thermal storage units	Advanced battery technologies, flywheels, supercapacitors	Storage capacity, charge/discharge rates, efficiency	Cost per unit of energy stored, lifespan, maintenance costs	Medium to high; can store different forms of energy	Scalability, efficiency losses, integration with control systems	Development of higher-capacity batteries, reduced costs, improved storage technologies
Energy distribution units	Infrastructure for transmitting energy from sources to consumers	Distribute energy reliably and efficiently	Electrical grids, gas pipelines, heat networks	Smart grids, microgrids, distribution automation	Capacity, reliability, losses	Infrastructure investment, operation and maintenance costs	Medium; depends on infrastructure adaptability	Interoperability, regulatory compliance, upgrading legacy systems	Smart distribution networks, integration of distributed energy resources
Control and management systems	Systems that monitor and control energy flows within the hub	Optimize performance, ensure stability, manage resources	SCADA systems, energy management software	IoT, AI algorithms, predictive analytics	Responsiveness, scalability, reliability	Software and hardware costs, training, maintenance	High; can manage various energy types	Data integration, cybersecurity, standardization	AI-driven optimization, enhanced cybersecurity, advanced analytics

(Continued)

TABLE 6.1 (Continued)

Conversion interface	Interfaces that facilitate the exchange of energy between different systems	Enable interoperability between different energy systems	AC/DC converters, gas-to-power interfaces	Power electronics, interface protocols	Conversion efficiency, compatibility	Interface equipment costs, maintenance	High; crucial for multi-energy systems	Technical compatibility, standardization	Universal interfaces, improved conversion technologies
Information and communication systems	Systems that handle data communication and information exchange	Enable real-time monitoring, control, and data analytics	Communication networks, sensors, data analytics platforms	IoT, wireless communication, cloud computing	Bandwidth, latency, security	Infrastructure costs, data management expenses	High; essential for coordination	Interoperability, data standards, cybersecurity	5G integration, edge computing, enhanced IoT
RESs	Energy sources that are replenished naturally	Provide sustainable energy generation	Solar panels, WTs, hydroelectric generators	Photovoltaics, WTs technologies	Intermittency, capacity factors, site dependency	Installation costs, subsidies, feed-in tariffs	Medium; variable output requires integration solutions	Grid stability, storage needs, forecasting	Improved efficiency, cost reduction, advanced materials
Backup and auxiliary units	Systems that provide backup power or auxiliary services	Ensure reliability during peak demand or outages	Diesel generators, emergency batteries	Standby generators, UPS systems	Availability, response time, reliability	Capital and operational costs, fuel expenses	Low to medium; mainly for emergency use	Maintenance, emissions regulations	Cleaner backup solutions, integration with renewables
Demand-side management systems	Systems that manage and adjust consumer demand	Optimize energy use, reduce peak demand	Smart thermostats, load control devices	Smart appliances, dynamic pricing models	User engagement, automation capabilities	Cost savings, incentives, investment in smart devices	High; can adjust demand across energy types	Consumer acceptance, privacy concerns	AI-driven demand response, more intelligent devices
Energy market interaction units	Interfaces that connect the EH to the market	Enable trading of energy, price signals, and market participation	Market platforms, bidding systems	Blockchain, smart contracts	Market responsiveness, compliance	Revenue generation, transaction costs	Medium; depends on market structures	Regulatory compliance, data security	Peer-to-peer trading, decentralized markets

6.2.2.1 Resources (Inputs)

EHs utilize a range of resources—such as RESs, electricity, NG, and district heating networks (DHN)—to fulfill various consumer needs. Conventional energy supply systems, like thermal power plants, often have low efficiency, resulting in substantial energy wastage during production and transmission. In coal-fired thermal power plants, the amount of primary energy that actually reaches end customers is quite modest. EHs are incorporating distributed generation (DG) systems, which utilize technologies including waste heat recovery, micro-gas turbines, RE systems, and FCs. These systems are commonly located close to where energy is used, minimizing losses during transportation and enhancing energy efficiency. RESs are vital in DG applications as they facilitate the production of electricity and heat using geothermal, solar, and wind energy. Solar energy is a versatile resource that can be harnessed in both small-scale and large-scale applications. Integrating solar technologies—such as PV panels and concentrated solar power plants—with EHs can greatly improve their performance and efficiency. Wind energy is also being utilized as a reliable and sustainable power source in EHs. It is generally incorporated using stochastic methods that account for the unpredictable nature and variability of wind. The limitations and environmental impacts of fossil fuels are primary factors motivating the transition to RESs. Waste heat utilization in cogeneration and polygeneration systems, as well as technologies such as solar energy, biomass, and FCs, is vital for the transition to 100% RE systems. Studies have shown the feasibility of establishing a fully RE system in Europe by 2050 using current technologies.

Biomass can be converted into renewable fuels, such as hydrogen and ethanol. RESs can be employed to generate a variety of energy carriers, such as electricity and heat. Hydrogen is emphasized as a clean energy carrier that has the potential to improve EH performance, particularly when produced via electrolyzers and stored using appropriate storage technologies. However, the environmental challenges associated with producing hydrogen from fossil fuels have led to a focus on RES-based hydrogen production, despite higher costs. Utilizing municipal waste, biomass waste, and surplus RESs for hydrogen production has the potential to decrease expenses and enhance feasibility. Water-based energy inputs—including hydroelectric power and tidal power—are also essential to EHs. Hydroelectric power, in particular, is a RES that has been extensively used in both small- and large-scale applications. Wood and waste are employed as input sources for EHs in certain regions, offering renewable and locally available alternatives to conventional fuels. The efficacy and sustainability of EHs can be enhanced through the integration of various waste-to-energy technologies.

While electricity from the utility grid and NG is currently a primary input for EHs—primarily due to the efficiency and reduced pollution associated with NG in CHP applications—there are sustainability concerns arising from reliance on these conventional energy sources. NG, although cleaner than other fossil fuels, still contributes to GHG emissions. This underscores the necessity of incorporating alternative inputs, particularly RESs. District energy systems (DESs), which include DHNs, play a substantial role in EHs by enabling centralized production and distribution of energy. By integrating RESs with conventional energy sources, DESs can enhance efficiency and decrease the consumption of fossil fuels and their associated emissions. In particular, DHNs are essential for the development of sustainable energy systems in the future, as they can utilize excess heat and RESs, such as geothermal energy and biomass. Although electricity and NG presently dominate energy inputs, there is increasing awareness of the need to transition to more sustainable and clean energy sources. The EH model is vital for integrating diverse energy infrastructures and optimizing the synergy between different energy carriers. This integration should be expanded to include RESs and innovative technologies within distributed energy resources (DERs) and DES frameworks, in addition to traditional resources.

6.2.2.2 Converters and Transmitters

EHs serve as intermediaries between energy producers and consumers, capable of integrating a diverse array of energy sources—including electricity, NG, DHNs, fossil fuels, solar and wind power, hydrogen, biomass, biogas, geothermal heat, and municipal waste. These inputs can either be directly transmitted to outputs or converted to satisfy various requirements, such as electric power, heat, cooling, and compressed air. In EH models, the primary types of input–output connections are direct connections, which transfer energy without conversion, and converters, which transform energy carriers into other forms suitable for consumption. Technologies utilized in these conversions include electric machines, gas turbines, internal combustion engines (ICE), FCs, thermoelectric converters, pumps, transformers, inverters, and heat exchangers. The objective of EH models is to efficiently satisfy a variety of energy demands by managing and converting multiple energy carriers, thereby reducing primary energy consumption, lowering costs, and enhancing overall efficiency. This approach is exemplified by cogeneration systems, particularly CHP systems, which generate electricity and heat from a single energy source. CHP units have experienced substantial growth in the past two decades due to their exceptional efficiency and adaptability. As of recent data, the total installed capacity of CHP in the United States has significantly increased, with applications in commercial, industrial, and other sectors. The number of CHP units has grown from 640 in 1980 to over 4,600.

CHP operation is subject to feasible operating regions, and the electricity and thermal outputs are interdependent due to technical constraints. Multiple approaches are implemented to simulate CHP operation. One common method involves representing operating points within the feasible region as convex combinations of corner points (extreme points). This means that any achievable combination of power and heat output can be expressed as a weighted sum of these extreme points, providing a practical way to model the CHP's operational capabilities. The cost of operating a CHP system is often represented by a quadratic function, which is nonlinear and may be non-convex, depending on the system characteristics. To simplify optimization, the cost function is sometimes partitioned into components such as start-up and shutdown costs, maintenance expenses, fuel consumption, and pollution control. Efficient EM requires optimal scheduling of heat and power generation, which can be accomplished through economic dispatch (ED) methods. During off-peak periods, the main grid can supply electricity, whereas during peak demand or high electricity prices, CHP systems can generate electricity, with any surplus sold back to the grid. CHP systems are frequently implemented in large industrial and commercial structures, allowing surplus electricity to be sold in day-ahead energy markets. The management of CHP systems is further enhanced by implementing demand response (DR) programs and stochastic scheduling. There has been increasing interest in utilizing CHP systems in residential sectors due to their advantages in cost savings, energy efficiency, and emission reductions. Optimal planning models that incorporate CHP and other equipment, along with investment and operating costs, have been analyzed for residential demand. Micro-CHP technologies in residential buildings are designed to reduce emissions and operating costs, demonstrating promising results.

A *micro-turbine* is a type of small combustion turbine with power outputs ranging from 25 to 500 kW. Contrary to the text's claim, diesel generators often have higher efficiencies than micro-turbines. Micro-turbines are valued for their compact size, low emissions, and ability to quickly respond to load variations, making them suitable for maintaining frequency stability and tracking load changes. Accurate modeling and scheduling of CHP systems are vital for efficient EM. Traditionally, equipment like CHP units has been assumed to operate at constant efficiency, which can lead to inaccuracies. Recent developments have introduced models that incorporate variable efficiencies, providing more accurate cost estimates and operational results. Utilizing non-constant efficiency functions offers a more precise representation of actual systems, minimizing errors in energy center models. Advanced optimization methods, including meta-heuristic algorithms, contribute to achieving more accurate and acceptable results in CHP and EH systems. The utilization of surplus heat generated during electricity production is a key factor in the high efficiency of CHP systems. By using this waste heat to generate additional energy products, the system's overall efficacy is improved.

Expanding upon the CHP concept, trigeneration and polygeneration systems have been developed to generate power, heat, and other energy products, including domestic hot water and refrigeration [14].

These systems incorporate a variety of prime movers, such as ICEs, steam engines, gas turbines, and RE systems, like FCs and PV panels. Integrating polygeneration systems into DESs is vital for reducing energy imports and optimizing the use of local resources, a critical step toward sustainable energy systems. Trigeneration, or combined cooling, heating, and power (CCHP), involves producing refrigeration through absorption or electrical compression chillers, recovering waste heat, and generating electricity using a prime mover [15]. Absorption chillers are generally preferred due to their efficiency, primarily using heat with minimal electricity consumption. When recovered heat is insufficient, auxiliary boilers compensate for the deficit. The sustainable and efficient operation of a CCHP system depends on the selection and sizing of the prime mover and other components. Optimization models, often based on EH frameworks, identify the optimal configuration for specific applications by considering financial, environmental, and operational factors. Integrating CCHP systems with technologies like district heating and cooling (DHC) networks, PV systems, and various types of ESSs maximizes their benefits [16]. This integration can substantially reduce primary energy consumption and energy costs, fostering a more sustainable and efficient energy system. Additionally, incorporating solar collectors and ground source HPs into CCHP systems enhances fuel efficiency, reduces operating expenses, and increases potential revenue from the sale of surplus electricity. The selection and placement of storage systems in CCHP configurations significantly impact energy savings and overall system performance. Moreover, the operation and efficacy of a CCHP system, particularly when integrated as an EH, are influenced by climatic conditions. This necessitates the development of operational strategies specifically designed to optimize performance under various weather conditions. This adaptability ensures that CCHP systems can operate cost-effectively and efficiently across different climates, making them a sustainable and versatile energy production option. Another prevalent thermal source in MCE systems is the boiler, which generates heat by consuming fuel. Also, electric and absorption chillers produce cooling energy in MCE systems by converting electricity and heat into cooling energy.

FCs, particularly hydrogen-based ones, are emphasized as critical components in polygeneration systems due to their adaptability to various fuels, like hydrogen, biogas, methanol, and NG [17, 18]. They are efficient trigeneration converters, capable of generating electricity and heat, with water produced as a minor by-product. The primary benefits of FCs include low emissions, high efficiency, reliable operation, and ease of installation. They support the use of renewable fuels and decentralized energy generation, making FCs a promising option for enhancing energy efficiency and reducing environmental pollution. While FCs offer significant advantages, they are complementary to renewable technologies, like PV systems and wind turbines (WT), rather than direct alternatives [19]. Integrating FCs with PV systems or other RESs can further improve system efficiency and reduce operating costs. The excess heat generated by high-temperature FCs, such as SOFCs, can be utilized in bottoming thermodynamic cycles, like Brayton or Rankine cycles, potentially integrating with other technologies, such as gas turbines [20]. This integration demonstrates the versatility and potential of FCs in multigenerational technologies, facilitating the development of more intricate systems, like CCHP. The comparison of various FC technologies is illustrated in Table 6.2 [21–23].

In EH models, transformers are essential for adapting electricity from the grid, highlighting the systems' reliance on the electrical network. Absorption chillers are more prevalent than electric chillers in EH models because they generate refrigeration by utilizing waste heat, often in conjunction with CHP systems. The commercial availability and relatively lower installation costs of PV systems, small WTs, and solar collectors contribute to their widespread adoption, particularly for on-site power generation. Heat exchangers are frequently incorporated with DHNs in EH models. When combined with absorption systems, FCs are particularly advantageous, as they can generate electricity and heat, thereby contributing to trigeneration or polygeneration systems. These systems are promising for future MESs as they can efficiently meet various energy demands. HPs are another versatile technology within EH models, capable of providing both heating and cooling. An innovative application involves using HPs to convert surplus electricity into thermal energy, which

TABLE 6.2

Comparison of Various FC Technologies

FC Type	Electrolyte	Electrical Efficiency (%)	Anode/Cathode Catalysts	Cell Voltage	Typical Stack Size	Fuel	Operating Temperature (°C)	Charge Carrier	Interconnect Material	Major Contaminants
Solid oxide (SOFC)	Solid yttria-stabilized zirconia (YSZ)	55–65	Nickel–YSZ composite/ strontium-doped lanthanum manganite (LSM)	~0.7–1.1 V	1–250 kW+	Methane	800–1,000	O^{2-}	Ceramics	Sulfides
Phosphoric acid (PAFC)	Concentrated liquid phosphoric acid (H_3PO_4) in silicon carbide (SiC)	36–45	Platinum	1.1 V	5–400 kW	Hydrogen	160–220	H^+	Graphite	CO, siloxane, H_2S
High-temperature proton-exchange membrane (HT-PEM)	Solid composite Nafions and polybenzimidazole (PBI) doped in phosphoric acid	50–60	Platinum–ruthenium	1.1 V	<1–100 kW	Hydrogen	110–180	H^+	Graphite	CO and H_2S
Low-temperature proton-exchange membrane (LT-PEM)	Solid Nafions	40–60	Platinum	1.1 V	<1 kW–100 kW	Hydrogen	60–80	H^+	Graphite	CO
Alkaline (AFC)	KOH water solution and anion exchange membrane (AEM)	60–70	Nickel/silver	1 V	1–100 kW	Hydrogen	Below 0–230	OH^-	Metallic wires	CO_2

(Continued)

TABLE 6.2   (Continued)

Direct ethanol fuel cell (DEFC)	Solid Nafions, alkaline media, and alkaline–acid media	20–40	Platinum–ruthenium/ platinum	~0.5–0.6 V	1–50 kW	Liquid ethanol–water solution	Ambient–120	H$^+$orOH$^-$	Graphite	CO poisoning, intermediate oxidation products
Direct methanol fuel cell (DMFC)	Solid Nafions	35–60	Platinum–ruthenium/ platinum	~0.4–0.6 V	1–50 kW	Liquid methanol–water solution	Ambient–110	H$^+$	Graphite	CO
Molten carbonate fuel cell (MCFC)	Liquid alkali carbonate (Li_2CO_3, Na_2CO_3, K_2CO_3) in lithium aluminate ($LiAlO_2$)	55–65	Nickel chromium (NiCr)/lithiated nickel oxide (NiO)	0.7–1.0 V	300 kW–3 MW	Methane	600–700	CO_3^{2-}	Stainless steel	Sulfides and halides
Direct carbon fuel cell (DCFC)	YSZ, molten carbonate, and molten hydroxide	70–90	Graphite or carbon-based material/LSM	~0.7–1.0 V	1–100 kW	Solid carbon (e.g., coal, coke, biomass)	600–1,000	O^{2-} or CO_3^{2-}	N/A	Ash and sulfur

can then be stored. Although not yet widely implemented, this method improves flexibility and decreases operating costs in EH models. Biomass is employed in different capacities within EH models, including biomass furnaces for heat generation, biomass reactors for electricity production, and gasification reformers for biofuel production. While there has been significant emphasis on cogeneration systems that generate electricity and heat simultaneously, there is growing interest in trigeneration and polygeneration systems to meet a broader spectrum of energy needs. Real-world energy systems require a wider array of resources, including fuels, compressed air, and cooling, highlighting the importance of expanding EH models to incorporate these elements.

6.2.2.3 Energy Storage Systems (ESS)

RESs, multigeneration systems, and energy storage systems (ESSs) are essential components of the transition to sustainable energy systems. ESSs improve system efficiency, reduce emissions, and lessen dependence on fossil fuels by capturing energy during periods of low demand and storing it for future use. Among the various ESS options, thermal energy storage (TES) is widely employed, particularly in applications involving waste heat management and system reliability [3]. TES plays a crucial role in managing energy demand on the consumer side by storing excess thermal energy for later use, thereby preventing heat loss. Thermal energy is utilized in numerous applications, including industrial processes and building heating, ventilation, and air-conditioning (HVAC) systems. The investment cost, capacity, efficiency, and suitability of various ESS technologies can vary significantly. Selecting an appropriate ESS should align closely with the specific objectives of its application, which typically include:

1. *Enhancing system resilience and performance.* Improving the ability of the energy system to withstand and recover from disturbances.
2. *Facilitating smart energy systems and optimization objectives.* Enabling advanced energy management through SGs and optimization of resources.
3. *Increasing system reliability and integration of RESs.* Addressing the intermittency of RESs to ensure a stable and reliable energy supply [24].

One of the primary challenges faced by RESs is their intermittent and fluctuating output, which can result in imbalances between supply and demand. ESSs help mitigate these issues by providing storage capabilities that stabilize the energy supply, thereby increasing the dispatchability and reliability of RESs, particularly in off-grid and microgrid scenarios. ESSs also enable systems to respond more effectively to market price fluctuations, reducing operational costs through energy arbitrage—buying or storing energy when prices are low and selling or using it when prices are high. Studies have shown that it is possible to satisfy a significant portion of electrical demand by combining solar and wind resources without ESSs. However, the incorporation of efficient ESSs can substantially increase the percentage of demand met by RESs. For instance, in certain models, the penetration of renewables can increase from 50% to over 80% with the addition of ESSs, though actual figures depend on specific system characteristics and regional factors [25]. ESSs are also instrumental in mitigating power system fluctuations and uncertainties by providing ancillary services, such as frequency and voltage regulation, grid stability, and reserve capacity. They assist in load shifting by storing energy during off-peak periods for use during peak demand, optimizing infrastructure use, and reducing costs.

Another significant application of ESSs is in the development of smart energy systems. These systems utilize local ESS technologies in conjunction with smart management programs to maximize equipment performance and resource utilization. For example, ESSs can substantially improve a system's efficiency and its ability to respond to variable energy pricing within DR and DER programs. Plug-in electric vehicles (PEV) represent a novel application of ESS technology gaining traction. Through vehicle-to-grid (V2G) capabilities, PEVs can provide grid services by storing energy and supplying it back to the grid when needed. The potential of PEVs to reduce peak loads and overall system costs has been demonstrated in residential MGs, highlighting their value in peak load reduction [26]. Optimizing the scheduling, sizing, and design of ESSs within the system is crucial. The size of an ESS, encompassing both energy capacity (how much energy it can store) and power capacity (the rate at which it can charge or discharge), must be determined by multiple factors, including system demand profiles, RES generation patterns, and economic considerations. The location of the ESS within the system is also critical. ESSs can be positioned close to the site of energy consumption to reduce transmission losses or placed strategically within the grid to optimize efficiency and control. The comparison of various ESS technologies is shown in Table 6.3 [27–31].

Implementing an efficient control strategy for charging and discharging ESSs is imperative. This strategy should consider technical constraints, anticipated production capacities, energy market conditions, pricing plans, demand patterns, and weather forecasts. In off-grid power systems—which may consist of WTs, biomass generators, and diesel generators—ESSs are instrumental in stabilizing voltage and frequency. This is particularly important due to the variable nature of RESs and the dynamic response characteristics of diesel generators. For example, in systems that integrate solar and biomass resources, daily operational planning of ESSs can be more effective than weekly planning, although this may vary depending on specific system conditions. This underscores the potential for EH models to implement optimized ESS control strategies [32]. Energy storage can be situated on either the input side (e.g., gas storage) or the output side (e.g., thermal storage) of an energy center. Input-side storage retains energy for future use or supplies it to conversion units as needed. Output-side storage, such as TES, manages and stores heat, improving system efficiency and reducing costs. While increasing storage capacity can enhance system flexibility and potentially lower operational expenses, it also involves higher capital investment. Therefore, strategic planning is essential to balance the benefits and costs of additional storage capacity. Also, improving forecasting methods by extending the forecast horizon—predicting future energy demands and production more accurately—can reduce energy costs. Better forecasts enable more efficient operational planning, potentially reducing the reliance on increased storage capacity. This approach emphasizes the significance of strategic planning and the integration of ESSs within energy systems to improve efficiency and decrease overall costs.

6.2.2.4 Demand-Side Management (DSM)

In a multi-energy system (MES), various energy carriers are integrated to provide a range of services, meeting the diverse energy needs of consumers. Effective planning and management in these systems hinge on the coordination of production and consumption patterns to ensure that demand is consistently met. The unpredictable nature of consumer behavior and the uncertainty of demand present substantial challenges in demand forecasting. Accurate demand forecasts are essential, as they impact system modeling, capacity planning, and reliability. Techniques such as regression analysis, fuzzy logic, and time series analysis are implemented to anticipate consumption patterns and reduce forecast errors. Inadequate forecasting can lead to unrealistic system models, resulting in issues such as stability concerns and capacity shortages. Understanding the dynamics of various types of loads, including thermal and electrical, is imperative. Thermal loads require distinct modeling methodologies due to their slower rate of change compared to electrical loads. Traditional energy management systems (EMS) have primarily concentrated on production-side management, increasing their ability to manage system fluctuations and meet increasing demand. However, this method is not always effective and can result in increased costs and emissions, as well as reliability and stability issues, when production cannot keep up with demand fluctuations. The current trend in contemporary energy systems is a transition to flexible demand-side management (DSM), increasingly integrated with DERs [33, 34]. DSM is a critical component of optimal resource allocation, involving strategies to manage and control energy demand. It encompasses various elements, including energy efficiency measures, demand response (DR) programs, energy conservation efforts, and load shifting strategies. Energy efficiency initiatives encourage the implementation of advanced, more efficient technologies to deliver the same level of service with less energy. These DSM strategies are especially pertinent in the industrial and commercial sectors, as they have the potential to significantly influence system efficiency and overall energy consumption [35, 36]. Modern EM

TABLE 6.3

Comparison of Various ESS Technologies

ESS Technology	Cycle Efficiency (%)	Available Capacity (MW)	Lifetime (Cycles)	Response Time	Power Density (kW/m^3)	Energy Density (Wh/kg)	Energy Density (kWh/m^3)	Discharge Time	Capital Cost (Power) ($/kW)	Capital Cost (Energy) ($/kWh)
Li-ion batteries	70–100	0.1–100	250–10,000	20 ms –s	60–10,000	30–300	90–750	Min to hr	1,200–4,000	<500
Lead acid batteries	60–90	0.1–40	100–2,000	5–10 ms	10–700	10–50	25–90	Min to hr	300–600	<300
NiMH batteries	50–80	0.1–1	300–3,000	Ms to s	8–3,000	30–90	40–300	Min to hr	500–1,500	400–1,000
Hydrogen fuel cell	20–50	0.1–58.8	>1,000	<1 s	>500	800–10,000	500–3,000	Sec to 24 hr+	500–10,000	N/A
Thermal storage	30–60	—	—	Min to hr	—	80–250	80–500	Hr to days	200–300	3–60
Flywheel (FES)	70–96	C.1–20	20,000–100,000	<4 ms to s	40–5,000	5–200	0.3–400	Ms to 15 min	300–2,200	1,000–8,800
Magnetic (SMES)	75–99	C.1–10	10,000–100,000	<100 ms	300–4,000	0.3–75	0.2–14	Ms to 8 s	130–10,000	1,000–10,000
Compressed air (CAES)	41–90	≤–1,000	10,000–30,000	1–15 min	0.04–10	3–60	0.4–20	1–24 hr+	400–1,500	1–50
Pumped hydro	50–90	—	—	Min	0.01–1.5	0.2–2	0.2–2	Hr to days	500–4,600	<300
Capacitor	60–70	0.001–0.01	>50,000	μs to ms	>100,000	0.05–5	2–10	μs to s	200–400	500–1,000
Supercapacitor	65–100	0–0.3+	10,000–1,000,000	8 ms	40,000–120,000	1–15	1–35	Ms to 60 min	100–515	300–2,000

requires the implementation of DR programs, which are intended to modify electricity consumption patterns in accordance with supply conditions. By utilizing financial incentives or dynamic pricing mechanisms, these programs contribute to grid stabilization, improved reliability, and optimized energy consumption. DR programs are broadly classified into two categories: incentive-based (IB) and price-based (PB), each employing distinct strategies to influence consumer behavior.

Incentive-based programs include [37–40]:

- *Load management (LM).* Directly controlling electrical consumption to balance supply and demand. Techniques encompass direct load control (DLC), where utilities remotely manage high-consumption appliances during peak periods, and interruptible load control (ILC), which provides financial incentives to large consumers for reducing their load during critical periods.
- *Market-based initiatives.* Utilizing market mechanisms to promote load reductions. For example, demand-side bidding enables consumers or aggregators to bid on load reduction in response to market signals. Capacity market participation compensates consumers for their availability to reduce load during periods of system need, while providing ancillary services offers grid stability functions, such as voltage support or frequency regulation.

Price-based programs focus on modifying electricity prices to influence consumption:

- *Time-of-use (TOU) pricing.* Establishes varying rates for electricity based on the time of day, motivating users to shift their consumption to off-peak hours, when electricity is more cost-effective.
- *Real-time pricing (RTP).* Provides a dynamic pricing model that reflects the actual cost of electricity production, enabling consumers to adjust their utilization accordingly.
- *Critical peak pricing (CPP).* Significantly increases prices during periods of high demand or constrained supply, encouraging consumers to reduce their consumption.

The key objectives of DR programs are to mitigate peak demand, prevent outages, and alleviate grid strain. Also, they facilitate load shifting, promoting the use of electricity during off-peak hours, and valley filling, which involves increasing consumption during low-demand periods to more effectively balance supply and demand. DR programs can also offer ancillary services crucial for grid stability, including reserve power and frequency regulation. EHs enhance the capabilities of DSM by integrating various energy carriers and technologies. In this framework, consumers have the option to engage in DR not only by reducing consumption or shifting loads but also by utilizing technologies such as CHP systems to generate their own electricity or heat. This increased adaptability enables consumers to manage their energy usage more effectively, optimizing resources and reducing costs without sacrificing comfort. EHs provide utilities with valuable tools for reducing the need for costly infrastructure investments, improving grid stability, and smoothing out demand peaks. This approach to EM is a substantial stride toward the development of more sustainable and efficient energy systems. In conventional single-carrier systems, such as electricity infrastructure, DR involves reducing consumption during peak periods or shifting it to lower-cost periods. However, the benefits are amplified in EHs that employ various energy carriers. By optimizing production scheduling, utilizing a variety of conversion and storage technologies, and transitioning between energy carriers, EHs can participate in advanced DSM strategies. This not only increases consumer engagement in DSM programs but also provides greater EM flexibility, even for those with non-responsive loads. Consumers can reduce energy costs by sourcing energy from more cost-effective alternatives, such as CHP systems, during periods of high prices without sacrificing comfort. Ultimately, this benefits the entire energy system by reducing peak demand and establishing a more stable capacity curve for power companies.

6.3 POWER-TO-HYDROGEN (PtH) AND HYDROGEN-TO-X (HtX) TECHNOLOGIES

6.3.1 Power-to-Green Hydrogen (H_2)

In power-to-hydrogen (PtH) technology, the electrolysis process utilizes electrical energy to split water into hydrogen (H_2) and oxygen (O_2) molecules. The generated H_2 can be stored in various forms, such as a cryogenic liquid, high-pressure gas, or metal hydrides. This stored H_2 can then be converted back into energy using combustion technologies or FCs. Additionally, PtH plays a crucial role in producing diverse compounds by providing the necessary H_2, serving as a precursor for electrifying chemical production processes, thereby enabling the synthesis of NH_3, CH_4, and CH_3OH. This integration not only enhances the sustainability of chemical manufacturing but also supports the broader goal of decarbonizing energy systems.

6.3.1.1 Electrolyzers

Electrolysis is the process of separating water into H_2 and O_2 by applying an electrical current. This occurs in electrolysis cells composed of an anode and a cathode separated by an electrolyte. The specific reactions at the anode and cathode depend on the electrolysis technology employed. Alkaline water electrolyzers (AWE), proton-exchange membrane water electrolyzers (PEMWE), and solid oxide electrolyzer cells (SOEC) are the three primary forms of electrolysis employed in power-to-hydrogen (PtH) systems [41–43]. AWEs are currently the most prevalent technology. Based on the lower heating value (LHV) of H_2, AWEs achieve efficiencies of up to 85% when operating at temperatures ranging from 50 to 90°C. In AWEs, OH^- anions pass through an aqueous alkaline hydroxide electrolyte, typically NaOH or KOH. Both the cathode and anode are composed of nickel or nickel alloys (anode: Ni, Ni–Co alloys; cathode: Ni, Ni–Mo alloys). AWEs are preferred for their high technical maturity, low capital costs, extended operating lifetimes of up to 90,000 hr, and the absence of costly noble metals. However, AWEs have limited dynamic flexibility, with cold start-up times that can be lengthy and an operational lower bound of 10–40% of installed capacity. PEMWEs exhibit dynamic response characteristics that are substantially superior to those of AWEs, with rapid start-up and response times, often within seconds to a few minutes, and an operational lower bound of 0–10%. They can operate at pressures of up to 200 bar, making them suitable for high-pressure applications or gaseous hydrogen storage. PEMWEs employ a solid polymer membrane electrolyte, such as Nafion, to facilitate the passage of protons (H^+) from the anode to the cathode. The anodes are typically composed of iridium oxide (IrO_2), while the cathodes are made of platinum (Pt). PEMWEs are approximately twice as costly (around €1,390–2,320/kW$_{el}$) as AWEs (€620–1,200/kW$_{el}$) due to the use of these rare metals. The flexibility and ability to integrate with intermittent RESs have motivated efforts to reduce costs and extend their lifetimes, which currently range from 40,000 to 80,000 hr. At present, PEMWEs have a moderate market share; however, there are indications that they may become the dominant technology by 2030. SOECs are currently in the early stages of development and are primarily used in demonstration projects. They can achieve efficiencies approaching 100% of hydrogen's LHV due to their high operating temperatures of 600–1,000°C. SOECs employ yttria-stabilized zirconia (YSZ) as the electrolyte. O^{2-} ions are transported from a nickel-based cathode (Ni/YSZ) to a perovskite-type lanthanum strontium manganite (LSM) anode on YSZ. Although the capital costs of SOECs are presently high, they may be reduced through further development. However, the high operating temperatures present challenges, such as accelerated material degradation, leading to lifespans of less than 10,000 hr. Also, SOECs exhibit dynamic flexibility that is comparable to or inferior to that of AWEs, with start-up times that can be lengthy and operational lower bounds exceeding 30%.

6.3.1.2 Hydrogen Storage

Effective H_2 storage is essential for the practicality of PtH technology, as hydrogen has a low density of approximately 0.0899 kg H_2/m³ under normal conditions. Efficient storage solutions aim to enhance the energy density per unit volume while considering the costs and energy required to convert hydrogen into a storable state. Moreover, green hydrogen production facilities are usually situated in remote regions with ample RES. This necessitates the implementation of two storage methods, one on-site and another near the delivery location, to effectively handle the intermittent and fluctuating nature of RES. Storage is crucial for aligning supply with demand. Hydrogen can be stored using physical- or material-based approaches. Physical-based methods involve direct hydrogen storage, such as through compressed gas or liquefied hydrogen. Material-based techniques employ indirect storage of hydrogen in alternative compounds, such as liquid organic hydrogen carriers (LOHC) and metal hydrides (MH) [44, 45]. Table 6.4 presents a general overview of the different hydrogen storage methods, outlining their key characteristics and typical applications.

Compressed gas storage is the most advanced and well-established method of storing hydrogen. The process involves pressurizing hydrogen to a minimum of 350 bar and up to 700 bar, with many storage containers designed for energy purposes at approximately 700 bar [46]. Compressing hydrogen to 700 bar results in an energy loss of around 10% of the total energy it contains (as measured by the LHV). High-pressure containers feature a composite shell made of glass or carbon fiber infused with resin for mechanical strength, and a liner of metal or polymer to prevent hydrogen permeation. The investment costs for compressed hydrogen storage amount to $10–15/kWh, depending on the LHV. Hydrogen-powered vehicles often store 5–7 kg of hydrogen at a pressure of 700 bar, allowing them to travel a distance of 400–600 km [47]. For larger-scale stationary storage, salt caverns are utilized. Hydrogen can be stored at pressures of up to 200 bar, making it cost-effective for storage quantities over 20 t. The investment costs for storing over 100 t of hydrogen are approximately $1–1.5/kWh. However, geographical constraints and transportation difficulties make them less practical over moderate distances. Salt caverns, with capacities around 1 million cubic meters, provide a versatile and cost-effective solution for storing hydrogen [48]. They maintain storage pressures ranging from 45 bar (Teesside) to 202 bar (Spindletop) due to the impermeability of the surrounding rock layers. The technological viability and reliability of this storage method have been proven over decades. Notable salt cavern storage sites include Clemens Dome, Spindletop, and Moss Bluff in the United States, and Teesside in the UK. These sites have significant storage capacities and high storage efficiency, reaching approximately 95%, thus playing a crucial role in the hydrogen infrastructure. While Étzel in Germany and Ragusa in Italy are typically used for NG storage, they are currently being investigated for hydrogen storage potential as part of energy transition initiatives. Kingston in Ontario, Canada, is also being assessed for hydrogen storage capacity, reflecting the increasing global interest in using salt caverns for large-scale hydrogen storage to support the developing hydrogen economy.

Cryogenic liquid storage involves cooling hydrogen to a temperature of −253°C, typically at atmospheric pressure. This method achieves double the energy density compared to compressed gas and mitigates safety concerns associated with high-pressure storage. However, liquefying hydrogen consumes around 30–35% of the hydrogen's LHV. Recent developments in cryo-compressed storage use temperatures ranging from −240°C to −160°C and pressures between 50 and 700 bar. This approach reduces energy consumption for storage to approximately 25% of the LHV but adds the expense of compression [49]. Cryogenic storage is not suitable for stationary energy applications or hydrogen-fueled cars due to the need for expensive onboard cooling and insulation equipment. This technology is mainly used in situations where high volumetric energy densities are crucial, even at the expense of energy efficiency. Metal hydride storage is based on the exothermic absorption of hydrogen onto a metal surface and its subsequent absorption into the metal lattice. The absorbed hydrogen is released through an endothermic process when the temperature is raised. Magnesium hydride (MgH_2) is a promising option for hydrogen storage, capable of storing more than 7% hydrogen by weight at pressures lower than 20 bar. MgH_2 exhibits rapid charging and discharging

TABLE 6.4

Comparison of Hydrogen Storage Methods in PtX Systems

Storage Method	Principle	Energy Density	Operating Conditions	Energy Efficiency/ Losses	Advantages	Disadvantages	Typical Applications	Investment Cost per kWh
Compressed gas storage	Hydrogen stored as compressed gas in high-pressure tanks	• **Volumetric**: low • **Gravimetric**: high	• **Pressure**: 350–700 bar • **Temperature**: ambient	• ~10% energy loss due to compression	• Mature technology • Quick refueling • Well-understood safety protocols	• Low volumetric density • Heavy tanks • Safety concerns at high pressures	• FC vehicles • Small- to medium-scale stationary storage	$10–15 per kWh (LHV)
Salt cavern storage	Hydrogen stored underground in salt caverns	High volumetric capacity due to large cavern size	• **Pressure**: up to 200 bar • **Temperature**: varies (underground conditions)	• High storage efficiency (~95%)	• Large storage capacity • Low cost per unit • Proven technology	• Limited to specific locations • High infrastructure costs • Not suitable for small scale	• Grid balancing • Industrial hydrogen supply	$1–1.5 per kWh (for >100 t)
Cryogenic liquid storage	Hydrogen liquefied by cooling to −253°C and stored in insulated tanks	• **Volumetric**: higher than compressed gas (~8 MJ/L)	• **Temperature**: −253°C • **Pressure**: atmospheric	• Liquefaction consumes 30–35% of LHV • Boil-off losses due to evaporation	• Higher volumetric density • Reduced tank size compared to compressed gas	• High energy consumption • Requires advanced insulation • Boil-off losses	• Specialized transportation • Aerospace applications	Higher than compressed gas storage
Metal hydride storage	Hydrogen absorbed and released from metal hydrides	• **Volumetric**: high • **Gravimetric**: lower due to metal weight	• **Temperature**: 200–300°C (for hydrogen release) • **Pressure**: <20 bar	• Energy required for heating during release • Exothermic absorption	• Safe storage at low pressures • High volumetric density • Good for stationary use	• Heavy storage systems • Requires heating • Not suitable for mobile applications	• Stationary energy storage • Hydrogen buffering at fueling stations	Higher due to material costs
Liquid organic hydrogen carriers (LOHC)	Hydrogen chemically bound to liquid carrier molecules (e.g., toluene)	• **Hydrogen content**: ~6% by weight	• **Temperature**: >250°C for dehydrogenation • **Pressure**: ambient	• ~43% energy loss during hydrogenation/ dehydrogenation processes	• Compatible with existing liquid fuel infrastructure • Safe and stable at ambient conditions	• Low round-trip efficiency • Energy-intensive processes • Requires catalysts	• Long-distance hydrogen transport • Large-scale, long-term storage	Costs associated with conversion processes

capabilities, with proven stability over more than 2,000 cycles [50]. However, the high temperatures required for hydrogen release (200–300°C) and the bulky nature of the storage method make metal hydrides impractical for use in vehicles. Their ability to store hydrogen at low pressures is attractive for enhancing safety at fueling stations. LOHCs are substances that chemically store hydrogen within liquid molecules, such as toluene/methylcyclohexane and dibenzyltoluene. They have a hydrogen storage capacity of nearly 6% by weight [51]. This technique has garnered interest for long-distance transportation and extended storage periods because LOHCs remain stable at normal temperatures, unlike liquid hydrogen, which suffers from boil-off losses. An important benefit of LOHCs is their compatibility with existing crude oil infrastructure, allowing current facilities to be adapted for hydrogen storage and transport. LOHCs behave similarly to petroleum-based liquids, facilitating their handling using existing equipment and safety procedures. However, the conversion process of LOHCs is inefficient, resulting in an energy loss of around 43% during hydrogenation and dehydrogenation processes [52]. Despite this disadvantage, LOHCs are gaining significance in applications requiring reliable and long-term hydrogen storage, particularly in sectors such as stationary energy storage, where high energy density and stable storage conditions are vital.

6.3.2 Power-to-Methane (PtCH$_4$)

PtCH$_4$, a subset of the broader PtG concept, has garnered significant attention in Europe, with over 30 projects implemented in the past decade. PtCH$_4$ involves the synthesis of renewable methane (e-methane or synthetic methane) using electrical energy, typically sourced from RE systems. This synthetic methane can be seamlessly integrated into existing NG infrastructure for storage, distribution, and utilization in heat generation, in power production, or as a fuel for transportation. Methane synthesis through PtCH$_4$ can be achieved via catalytic or biological processes, both of which offer efficient and cost-competitive pathways. Catalytic methanation encompasses two primary methods: direct and indirect methanation [53]. Direct methanation, also known as the Sabatier reaction, involves the hydrogenation of CO2 to produce methane and water (H2O):

$$CO_2 + 4H_2 \rightarrow CH_4 + 2H_2O \quad \Delta H^0_{298} = -165\,kJ\big/mol \tag{6.2}$$

This exothermic reaction typically occurs at temperatures between 300 and 400°C and pressures ranging from 1 to 100 bar. Nickel (Ni) is the most commonly used catalyst due to its cost-effectiveness and satisfactory activity, although it is susceptible to deactivation by sulfur impurities present in the feedstock. Other catalysts, such as ruthenium (Ru), platinum (Pt), and cobalt (Co), offer higher activity but come with increased costs, limiting their widespread industrial adoption. Indirect methanation involves a two-step process. First, CO_2 is converted to CO via the reverse water–gas shift (RWGS) reaction:

$$CO_2 + H_2 \rightarrow CO + H_2O \tag{6.3}$$

Subsequently, the produced CO reacts with additional hydrogen in the methanation step:

$$CO + 3H_2 \rightarrow CH_4 + H_2O \tag{6.4}$$

This pathway offers greater flexibility in process conditions and catalyst selection but requires precise control to optimize overall efficiency.

Biological methanation operates at lower temperatures (below 100°C) and pressures (below 10 bar), utilizing methanogenic archaea to convert CO_2 and H_2 into CH_4 [54]. By injecting H2 into anaerobic digesters, methane yields can increase by over 20%, converting up to 90% of residual CO_2. However, this process can cause pH fluctuations that inhibit methanogenesis, which can be mitigated by conducting biological methanation in separate reactors. Although biological methods have lower space-time yields compared to catalytic processes, they offer a simpler and more cost-effective solution for small- to medium-sized anaerobic digestion facilities. Various reactor configurations are employed in catalytic methanation to optimize performance. Fixed-bed reactors are often used in multi-reactor setups with intercooling stages to manage the exothermic heat of reaction and prevent catalyst sintering, thereby improving conversion efficiency. Cooled reactors, such as multi-tubular reactors, utilize water vaporization or diathermic oil cooling to maintain near-isothermal conditions, facilitating high-equilibrium conversion to methane and minimizing side product formation. Fluidized-bed reactors are known for their excellent heat transfer due to high mixing rates, effectively preventing the formation of hotspots. However, they face challenges such as catalyst erosion and attrition. Advanced reactor designs, including catalytic membrane reactors and sorption-enhanced reactors, are at the forefront of process intensification efforts. Catalytic membrane reactors integrate a selective membrane within the reaction zone to continuously remove CH_4, thereby shifting the reaction equilibrium toward increased product formation [55]. This integration allows for more compact reactor designs and lower energy requirements, though challenges related to material compatibility, scalability, and initial costs must be addressed for industrial viability. Sorption-enhanced reactors incorporate sorbent materials to selectively absorb CO_2 or CH_4, enhancing reaction efficiency and enabling operation at lower pressures. These reactors offer benefits, such as enhanced reaction efficiency, lower pressure operation, and simplified processes, by reducing the number of required units. However, effective sorbent regeneration, material durability, and integrated reactor design remain significant challenges.

Process intensification strategies in methane synthesis aim to enhance sustainability, reduce costs, and improve efficiency in $PtCH_4$ systems. Key aspects include advanced heat management techniques to handle the exothermic nature of methanation and the development of selective, high-activity catalysts resistant to contaminants like sulfur. Research efforts are focused on hybrid nanomaterials and non-noble metal catalysts that perform efficiently over a wide range of temperatures and pressures, reducing production costs while maintaining high activity and selectivity. Additionally, integrating carbon capture within catalytic methanation using materials such as Ni/CaO adsorbents allows for the simultaneous capture and conversion of CO_2 to methane, reducing the number of required units and enabling operation under milder conditions. This integration contributes to overall cost savings and process efficiency. The environmental and economic benefits of $PtCH_4$ systems are substantial. By integrating RESs and utilizing captured CO_2 for methanation, $PtCH_4$ systems contribute to the reduction of greenhouse gas emissions and support the transition to a sustainable energy economy. Enhanced process efficiency and reduced energy consumption further improve the environmental impact and economic viability of methane synthesis processes. $PtCH_4$ not only leverages existing NG infrastructure but also provides a means to store excess RE, thereby addressing intermittency issues associated with renewable sources.

6.3.3 Power-to-Methanol ($PtCH_3OH$)

The $PtCH_3OH$ process combines hydrogen and CO2 to produce methanol, expanding upon the PtH concept. Methanol is easily stored as a liquid and can be utilized for energy generation through hydrogen decomposition, FCs, or combustion. This process sources CO_2 from various origins, including industrial waste, flue gas, or direct air capture. Since its development in the 1920s, catalytic methanol synthesis has remained the dominant industrial technology. Methanol production utilizes syngas—a mixture of H_2, CO, and CO_2—through a series of chemical reactions [56]:

$$+CO \rightarrow CH_3OH \quad \Delta H^0_{298} = -90.6\,{kJ}/{mol} \tag{6.5}$$

$$3H_2 + CO_2 \rightarrow CH_3OH + H_2O \quad \Delta H^0_{298} = -49.4\,{kJ}/{mol} \tag{6.6}$$

$$2H + CO_2 \rightarrow CO + H_2O \quad \Delta H^0_{298} = 41.6\,{kJ}/{mol} \tag{6.7}$$

TABLE 6.5

Overview of PtX Systems

PtX System	Process Description	End Product	Energy Efficiency	Key Technologies	Storage and Transport Considerations	Integration with Existing Infrastructure	Applications	Advantages	Challenges
Power-to-hydrogen (PtH$_2$)	Electrolysis of water using renewable electricity	Hydrogen gas (H$_2$)	• **Alkaline electrolyzers**: 65–82% • **PEMWEs**: 60–75% • **SOECs**: up to 100% (high-temp efficiency)	• AWEs • PEMWEs • SOECs	• Compressed gas storage • Liquefied hydrogen • Metal hydrides • LOHCs	• FC infrastructure • Industrial gas pipelines	• Energy storage • Transportation fuel • Industrial processes	• Zero-emission fuel • Versatile energy carrier • Supports grid balancing	• Storage challenges due to low density • Infrastructure investment needed
Power-to-methane (PtCH$_4$)	Methanation of CO$_2$ with H$_2$ via catalytic or biological processes	Synthetic methane (CH$_4$)	• **Catalytic methanation**: 60–80% • **Biological methanation**: 50–65%	• Sabatier reaction catalysts (Ni, Ru) • Biological reactors with methanogens	• Utilizes existing NG infrastructure • Underground storage	• Compatible with NG grids • Existing storage and transport systems	• Heating • Electricity generation • Transportation fuel	• Integrates with existing infrastructure • Reduces CO$_2$ emissions	• CO$_2$ source required • Lower overall efficiency compared to direct H$_2$ use
Power-to-liquids (PtL)	Fischer–Tropsch synthesis converting syngas to liquid hydrocarbons	Synthetic fuels (e.g., diesel, kerosene)	• Overall efficiency: 40–55%	• Electrolysis for H$_2$ production • CO$_2$ capture • Fischer–Tropsch reactors	• Liquid at ambient conditions • Existing fuel distribution networks	• Compatible with current fuel infrastructure • Use in existing engines	• Aviation fuel • Maritime fuel • Heavy transport fuel	• High energy density • Drop in replacement fuels • Reduces fossil fuel dependence	• Complex process • High capital costs • CO$_2$ sourcing

(Continued)

TABLE 6.5 (Continued)

Power-to-ammonia (PtNH₃)	Synthesis of ammonia from N_2 and H_2 via Haber–Bosch process	Ammonia (NH_3)	• Overall efficiency: 50–60%	• Electrolysis for H_2 production • Air separation for N_2 • Haber–Bosch reactors	• Liquid under mild pressure • Established transport methods	• Existing ammonia infrastructure • Fertilizer distribution networks	• Fertilizers • Energy carrier • Potential fuel for shipping	• High hydrogen density • Established technology • Easier to store than H_2	• Toxicity concerns • Energy-intensive production • Safety considerations
Power-to-methanol (PtMeOH)	Catalytic hydrogenation of CO_2 to produce methanol	Methanol (CH_3OH)	• Overall efficiency: 60–70%	• Electrolysis for H_2 production • CO_2 capture • Methanol synthesis reactors	• Liquid at ambient conditions • Easier to handle than H_2 or NH_3	• Can blend with fuels • Chemical industry feedstock	• Fuel additive • Chemical production • Energy storage	• Liquid fuel at room temperature • Versatile chemical feedstock	• CO_2 source required • Market competition with fossil methanol
Power-to-formic acid (PtFA)	Electrochemical reduction of CO_2 to formic acid	Formic acid (HCOOH)	• Variable efficiency: 30–50%	• Electrochemical cells with specialized catalysts	• Liquid at ambient conditions • Safe handling compared to other fuels	• Emerging infrastructure • Potential for FCs	• Hydrogen storage medium • FCs • Industrial uses	• Liquid hydrogen carrier • Lower toxicity • Easier storage	• Early-stage technology • Lower energy density • Limited infrastructure
Power-to-synthetic natural gas (PtSNG)	Methanation of syngas derived from biomass gasification	Synthetic natural gas (SNG)	• Overall efficiency: 50–65%	• Biomass gasification units • Methanation reactors	• Uses existing NG infrastructure • Underground storage	• Compatible with NG grids • Existing storage and transport systems	• Heating • Electricity generation • Industrial processes	• Renewable gas production • Utilizes biomass resources	• Biomass availability • Complex processing steps

The process typically operates under conditions of 50–100 bar and 200–300°C. Due to the exothermic nature of the reactions, efficient heat removal is essential to optimize methanol yield and minimize side products. Multi-tubular reactors are commonly employed for their ability to maintain near-isothermal conditions through water vaporization. Alternative reactor designs include coil-wound heat exchangers, radial flow reactors, and multi-stage adiabatic reactors with inter-bed cooling. Less common options, such as slurry, trickle-bed, and fluidized-bed reactors, face scalability and operational challenges. Single-pass conversion efficiencies range from 30% to 60%, necessitating the recycling of unreacted syngas after cooling and separation. Post-synthesis, methanol is purified through flash separation to remove light gases, followed by distillation to achieve energy or fuel-grade purity. To reduce costs and energy usage, advanced reactor technologies, such as membrane and sorption-enhanced reactors, are employed to remove methanol from the reaction zone, thereby shifting equilibrium toward higher conversions. Reactive distillation, which integrates reaction and separation steps, is also utilized to lower capital expenditures [57, 58].

Commercial methanol synthesis typically employs catalysts based on copper oxide (CuO) or zinc oxide (ZnO) supported on alumina (Al_2O_3), often with potassium promoters to enhance performance. In the context of co-electrolysis, catalysts like CuO-ZnO-Al_2O_3 are further enhanced with elements such as zirconium, gallium, or yttrium to improve CO_2 conversion and reduce water production [59–61]. Syngas-based synthesis generally achieves higher equilibrium conversions compared to direct CO_2 hydrogenation. Typically, CO_2 abatement involves a two-stage process: the water–gas shift (WGS) reaction partially converts CO_2 to CO, followed by methanol synthesis using the resultant syngas. Emerging technologies, such as co-electrolysis, offer promising but less-mature alternatives by simultaneously reducing water and CO_2 to produce H_2 and CO without the need for a WGS reactor, while generating O_2 at the anode. Direct conversion of CO_2 to methanol can potentially yield higher proportions of CH_3OH with fewer by-products compared to syngas-based methods. Additionally, lower heat release in direct conversion simplifies reactor and process designs, leading to reduced capital costs and enhanced adaptability.

6.3.4 POWER-TO-AMMONIA (PTA)

The PtA concept builds upon the PtH framework by combining hydrogen with nitrogen (N_2) sourced from the atmosphere to produce ammonia. *Ammonia* is a versatile liquid fuel that can be utilized for power generation and heat production through FCs or combustion technologies. The primary method for industrial NH_3 synthesis is the Haber–Bosch (HB) process, which has been in operation since the 1920s. This exothermic reaction involves the combination of H_2 and N_2 in a molar ratio of 3:1 to form NH_3 [62]:

$$3H_2 + N_2 \rightarrow NH_3 \quad \Delta H_{298}^0 = -91.8\,^{kJ}\!\!/_{mol} \tag{6.8}$$

The HB process typically operates under high pressures (100–250 bar) and elevated temperatures (400–500°C). Due to the reaction's equilibrium constraints, single-pass conversion rates are limited to approximately 15–20%, necessitating the recycling of unreacted H_2 and N_2. This recycling is facilitated by condensing the gases at temperatures below 0°C during synthesis. Modern HB plants employ multiple catalyst beds with intercooling stages to enhance conversion efficiency and reduce the volume of recycled gas. Intercooling is achieved through indirect cooling using heat exchangers or direct quench cooling with low-temperature supply gases, which helps shift the equilibrium toward NH_3 production by removing excess heat. Initially, iron-based catalysts were used for the HB process, but today enhanced iron catalysts, such as magnetite or wustite, are the industry standard. In recent decades, ruthenium (Ru) catalysts have been developed, offering faster reaction rates at lower temperatures and pressures. While Ru catalysts improve NH_3 production efficiency, their high cost limits widespread industrial adoption. Alternative separation methods, including the use of adsorbent beds like zeolites or supported metal halides, allow NH_3 production at lower pressures and ambient temperatures, potentially reducing costs. However, these methods remain largely experimental and are not yet commercially implemented.

Process intensification aims to optimize the HB process by integrating synthesis and separation stages, thereby increasing efficiency and reducing capital costs. Advanced technologies, such as catalytic membrane reactors and sorption-enhanced reactors, are at the forefront of this effort. Catalytic membrane reactors incorporate a selective membrane within the reaction zone to continuously remove NH_3, thereby driving the reaction forward and increasing conversion rates, while allowing for more compact reactor designs and lower energy requirements. Despite their potential, challenges related to material compatibility, scalability, and initial costs must be addressed for industrial viability. Sorption-enhanced reactors utilize sorbent materials to selectively absorb NH_3 as it is formed, shifting the equilibrium toward greater NH_3 production [63]. These reactors offer enhanced reaction efficiency, the ability to operate at lower pressures, and simplified processes by reducing the number of required units. Yet effective sorbent regeneration, material durability, and integrated reactor design remain significant challenges. Emerging technologies, such as electrochemical NH_3 synthesis, present alternative approaches to the HB process. In electrosynthesis cells, NH_3 is directly generated from nitrogen and water at ambient pressure. Various electrolytes, including hydrophobic ionic liquids, aqueous solutions, high-temperature molten salts, and solid-state proton-conducting membranes, are being explored. Each electrolyte type presents unique advantages and issues, such as selectivity, synthesis rates, energy efficiency, and membrane durability. While electrochemical synthesis offers benefits like low-temperature operation, modular scalability, and impurity tolerance, it is still in the research and development phase and is unlikely to be commercially viable for at least another decade. Table 6.5 offers a comprehensive review of the principal functions, applications, and advantages of various PtX technologies across numerous domains. It should be noted, however, that this overview is general, as there may be exceptions or new applications emerging with the continuous advancement of PtX technologies.

6.4 APPLICATIONS OF PTX SYSTEMS IN ENERGY HUBS

As the global community intensifies its efforts to transition toward sustainable and low-carbon energy systems, the concept of EHs has emerged as a pivotal framework for integrating diverse energy sources and technologies. Central to this transformation is the implementation of PtX systems, which play a critical role in enhancing the flexibility, efficiency, and resilience of energy networks. PtX technologies convert electrical power, often derived from renewable sources, into various forms of energy carriers, such as hydrogen, synthetic fuels, and chemicals, thereby facilitating the storage, transport, and utilization of energy across multiple sectors. The necessity of PtX systems within EHs stems from several interrelated factors. Firstly, the intermittent nature of RESs like wind and solar necessitates effective storage solutions to ensure a stable and reliable energy supply. PtX provides a versatile means of balancing supply and demand by transforming excess renewable electricity into storable and transportable energy carriers. Secondly, PtX enables sector coupling, bridging the gaps between the power, transport, industrial, and heating sectors. This interconnected approach not only optimizes energy use but also drives decarbonization across the entire energy ecosystem. Applications of PtX within EHs are diverse and multifaceted. In the transportation sector, PtX-derived hydrogen can be utilized in FCs for vehicles, offering a clean alternative to fossil fuels. In the industrial domain, synthetic fuels and chemicals produced through PtX processes can replace conventional, carbon-intensive inputs, thereby reducing GHG emissions. Additionally, PtX facilitates the production of renewable ammonia and methanol, which are essential for various industrial applications and energy storage solutions. Moreover, the integration of PtX systems enhances the overall resilience of EHs by providing multiple pathways for energy distribution and utilization. This flexibility is crucial in adapting to the fluctuating energy demands and mitigating the impacts of energy supply disruptions. By leveraging PtX technologies, EHs can achieve a higher degree of self-sufficiency and sustainability, aligning with global climate goals and fostering economic growth through innovation and technological advancement.

6.4.1 POWER SUPPLY AND GENERATION

PtX systems are becoming more critical in the energy sector as they provide a variety of applications that expand the value of RE across a variety of domains. PtX technologies facilitate the conversion of surplus electricity into chemical fuels, such as H_2, in both centralized and decentralized power facilities. These fuels can be utilized to generate electricity during periods of decreased renewable generation or peak demand. This capability is essential for the provision of ancillary services that are vital for the stability of the grid. PtX systems are capable of swiftly responding to fluctuations in grid frequency and voltage, thereby guaranteeing a consistent and dependable power supply. Also, PtX systems are instrumental in the re-electrification and storage of RE. This process involves the storage of excess RE in chemical form, which is subsequently converted back into electricity during periods of low generation or high demand. This is especially advantageous for remote and off-grid power supply in regions where utility connectivity is either nonexistent or limited, as it offers a sustainable and independent energy source. Also, PtX systems facilitate energy export by converting renewable electricity into transportable fuels, such as hydrogen or ammonia, which can be transported globally; hence, the market reach of RE is expanded. Also, PtX technologies are vital for emergency power solutions, as they provide dependable backup energy in the event of grid failures. They are being more frequently incorporated into data centers, where they provide primary and backup power, thereby reducing the carbon footprint of these energy-intensive facilities.

PtX in microgrids and distributed generation. [64] describes the development of a hydrogen-integrated microgrid (MG) by Xi'an Jiaotong University. The MG comprises a 640 kW PEM-FC system and a 1 MW PV plant. By integrating hydrogen production and supply, the MG aims to achieve near-zero carbon emissions, demonstrating hydrogen's potential as a clean and flexible energy carrier. Control strategies for distributed generation and bidirectional interlinking converters (BIC) ensure stable operation, allowing the system to effectively manage fluctuations in RE output and demand. In [65], a dynamic approach to planning integrated hydrogen–electrical (IHE) microgrids is investigated to facilitate the provision of carbon-neutral energy. A multi-stage stochastic mixed-integer program (MS-MIP) is formulated to develop a profitable and robust investment strategy addressing dynamic expansion, sizing, and siting of distributed energy resources. The model considers both fine-scale operational uncertainties and large-scale strategic uncertainties, enabling more accurate and flexible planning. A nested decomposition algorithm based on stochastic dual dynamic integer programming (SDDiP) is introduced to manage computational complexity. Case studies demonstrate that the SDDiP algorithm outperforms traditional solvers, effectively managing large-scale optimization problems. [66] employs the HOMER Pro software to simulate a hybrid renewable energy system (HRES) that includes hydrogen generation. The study compares two scenarios: one integrating wind generators and PV panels, and another relying solely on wind power. The integration of PV panels significantly reduces both the cost of energy (COE) and the net present cost (NPC), making the system more cost-effective than a wind-only setup. Both scenarios are viable for off-grid hydrogen production, highlighting the potential of combining multiple renewable sources with hydrogen storage to enhance energy reliability and economic efficiency.

PtX in IESs. [67] proposes a gas security management strategy for injecting hydrogen into NG systems, generated from surplus wind energy, within an integrated electricity and gas system (IEGS). To address gas security concerns such as combustion safety and appliance lifespan, a mixture of nitrogen and liquefied petroleum gas with hydrogen is employed to mitigate the risks associated with hydrogen injection. A distributionally robust optimization (DRO) model based on Kullback–Leibler divergence is implemented to manage uncertainties in wind power generation and ensure gas quality. The scheme effectively enhances the performance and security of the IEGS while maximizing RE utilization. In [68], a two-stage volt-VAR-pressure optimization (VVPO) model is constructed for PV-penetrated IESs to address challenges in voltage management and gas quality, particularly with hydrogen injection from power-to-gas (PtG) facilities. The model extends traditional volt-VAR optimization by incorporating PtG facilities to stabilize system voltages and manage fluctuating PV output. A two-stage DRO approach is employed to mitigate uncertainties in PV generation, and the model is formulated as a semidefinite programming problem. The VVPO effectively manages voltage levels, maintains gas quality, and achieves high economic efficiency, demonstrating adaptability for other green gas injections. [69] explores the potential of a shared hybrid hydrogen energy storage system (SHHESS) operating under a dynamic pricing paradigm within an IES alliance. A bi-level optimization model is developed, where the upper level optimizes SHHESS profits through dynamic pricing and capacity determination and the lower level minimizes the total operation costs of the IES alliance. The HESS significantly increases daily profits by up to 70.3% compared to battery-only systems and reduces RE curtailment by 80.93%, highlighting the benefits of shared energy storage in enhancing RE consumption and storage utilization.

PtX for energy storage and renewable energy integration. [70] examines the integration of a large-scale PEM water electrolysis (PEMWE) system with the APR1400 nuclear power plant to generate hydrogen. By simulating both subsystems using energy and exergy principles, the study assesses the system's performance under different operating conditions. The combined system achieves a thermal-to-hydrogen energy efficiency of 22.9% and an exergy efficiency of 53.52%. With a hydrogen generation rate of 7.544 kg/s (equivalent to 651 t per day), the integration demonstrates the feasibility of large-scale hydrogen production using nuclear power, contributing to energy storage and grid balancing. In [71], the integration of hydrogen energy storage into IESs is proposed to improve the utilization of wind power in China, addressing the challenge of wind curtailment. A bi-level optimal configuration model is developed, where the upper-level problem minimizes total configuration costs to determine the capacity of hydrogen storage devices and the lower-level problem minimizes operational costs considering variations in hydrogen production efficiency. A data-driven surrogate algorithm is used to solve the non-convex optimization problem, enhancing economic and environmental benefits and improving overall energy efficiency. [72] focuses on the energy management of a smart distribution network that integrates RESs with hydrogen storage. The scheme aims to optimize economic, operational, flexibility, and reliability objectives while adhering to system constraints and AC optimal power flow equations. A stochastic optimization approach mitigates uncertainties in renewable generation, energy prices, and demand. The inclusion of hydrogen storage significantly improves the economic viability of the distribution network by 46.8%, enhances operational efficiency, and achieves 100% flexibility in energy management.

PtX in emergency power and remote applications. [73] investigates the utilization of a PV–H_2 system in a mobile ad hoc mesh network to operate a portable emergency communication system (PECS). Using actual weather data and simulating emergency scenarios in Cancun, Mexico, the study designs a power system capable of ensuring consistent operation during hurricanes. The results indicate that the PV–H_2 system can operate with an efficiency of 79.14%, effectively preventing oversizing or undersizing and providing sufficient energy during emergencies. In [74], a multi-objective energy management strategy is introduced for hydrogen FC emergency power supply systems (FCEPSS), which are critical for disaster relief and outdoor power supply. The system addresses issues associated with prolonged operation, hydrogen supply challenges, and high maintenance costs. By formulating models to analyze hydrogen consumption and FC life degradation and employing an improved multi-objective optimization algorithm, the strategy reduces hydrogen consumption by 0.60% and life degradation by 5.65% over a 24 hr period compared to traditional methods.

PtX in data centers and communication systems. [75] conducts an economic analysis of hydrogen-powered data centers as a potential solution to the increasing electricity demand and CO_2 emissions from conventional data centers. The study compares hydrogen-powered configurations with traditional setups, highlighting the need of hydrogen costs and electricity price trends. The analysis suggests that hydrogen-powered data centers are economically viable if located near hydrogen sources and if hydrogen is produced from NG or industrial waste streams. Large-scale hydrogen-powered data centers are more likely to be cost-effective, contributing to reduced carbon footprints. [76] proposes a collaborative response framework for data centers that emphasizes the management of data traffic and hydrogen storage to increase flexibility and absorb excess RE. A global interval optimization approach is developed to solve the complex optimization problem under uncertain RE supply. By enhancing decision accuracy with a probability correction, the framework effectively optimizes RE utilization, as demonstrated in a case study on the IEEE 24-bus system.

PtX in carbon capture and decarbonization. [77] presents a decarbonization system that uses green ammonia to capture CO_2 from biomass flue gas, producing purified solid ammonium bicarbonate (NH_4HCO_3). Integrated with a biomass power plant, PV unit, and ammonia production system, the system captures 46,490.96 mt (metric tons) per year of CO_2 and generates 83,308.33 mt per year of NH_4HCO_3. The plant achieves a net CO_2 emission of $-1,105,965$ mt over 30 years, with a net present value expense of \$28.57 million. The decarbonization cost could be as low as \$20.70 per ton of CO2 if all products are sold. [78] examines the development of a multi-energy system (MES) in an EH to achieve carbon neutrality, exploring two pathways: power-to-gas-to-power and biomass-to-gas-to-power. Utilizing hydrogen and ammonia for energy storage, feedstock, and fuel, the operation of the EH is controlled using a combination of model-based mixed-integer linear programming (MILP) and model-free deep reinforcement learning (DRL) techniques. The BtXtP pathway is found to be more financially advantageous, with a net present value of \$550.76 million and a discounted payback period of five years.

Economic analyses and optimization strategies for PtX systems. [79] examines a techno-economic optimization tool designed to assess the costs of off-grid renewable hydrogen production. The tool optimizes the capacities of RE power plants, electrolyzers, storage systems, and hydrogen transport processes to minimize the levelized cost of hydrogen for a fixed demand. By considering various techno-economic factors and specific geographic RESs, the tool provides valuable insights for system development and implementation. A case study conducted in Argentina indicates that the minimum cost of hydrogen production ranges between \$3.2 and \$4.0 per kilogram, with total costs, including conversion, transportation, and storage, ranging from \$3.6 to \$5.6 per kilogram. In [80], a stochastic programming methodology is introduced to optimize the economic scheduling of a sustainable IES comprising hydrogen carriers, NG, and power. The primary objective is to reduce operational and emission costs while managing uncertainties in electricity prices, renewable generation, and demands through scenario-based stochastic programming. Downside risk constraints are implemented to mitigate risks, and the optimization problem is solved using MILP. The study highlights the trade-offs between expected costs and risk mitigation in system operation. [81] suggests an optimal scheduling strategy for mobile ESSs with variable-speed energy transmission, focusing on hydrogen carrier vessels in inter-island energy networks. By constructing a variable-speed transmission model and integrating it into an island energy MG operation framework, the strategy balances frequency response, power flow, and hydrogen supply. A stochastic dispatch strategy mitigates uncertainties in energy output and demand, and the model is solved using a two-layer algorithm, enhancing energy security and efficiency in island communities.

6.4.2 Heat and Power

In the current energy landscape, PtX technologies are vital for enhancing sustainability and efficiency in energy centers, especially within industrial process heat applications, district heating systems, and CHP systems. They convert surplus electricity from RESs into storable forms, like heat and power. In CHP systems, PtX allows the simultaneous generation of electricity and useful heat from the same source, increasing efficiency and reducing emissions. For district heating systems, PtX integrates RESs and enhances grid flexibility. Additionally, PtX decarbonizes high-temperature industrial processes traditionally reliant on fossil fuels. By adopting PtX technologies, energy centers can promote a more sustainable and resilient energy system, improve energy security, and optimize resource use.

PtX in CHP systems. [82] introduces a CCHP system integrating RESs, comprising hydrogen generation assisted by solar radiation, a SOFC–gas turbine subsystem, and a dual-effect lithium-bromide absorption chiller/heat pump (DAC/H) for cooling and heating. The system aims to maximize CH_4 conversion efficiency, hydrogen production, and CO_2 capture. Optimal conditions are a reforming temperature of 560°C, pressure of 1 atm, and specific ratios of S/C (3), CaO/CH_4 (1), and NiO/CaO (0.5). The system achieves a power generation efficiency of 53.23% and an exergy efficiency of 52.87%. [83] creates an optimization framework for enhancing a CH_3OH-steam-reforming PEM-FC system used in CCHP generation. By jointly optimizing system flowsheets and heat integration, it simplifies system design, involving the sizing of key components, like the reformer, FC stack, and absorption cooling apparatus, and simulating reactions and performance. Applied to a 1,000 kWe system, the optimized design achieved an LCOE of \$0.2374/kWh and an energy efficiency of 88.50%. Compared to conventional designs, this represents a 4.50% reduction in LCOE, a 5.45 percentage point increase in energy efficiency, and a 2.22 percentage point increase in exergy efficiency. [84] investigates optimizing a CHPS using multi-energy complementation by integrating hydrogen storage active load (HS-AL). By analyzing H_2 storage operations and the integrated demand response of electric and thermal loads, the study assesses HS-AL's demand elasticity. An optimal dispatch model incorporating HS-AL's power demand elasticity is developed. Simulations confirm that HS-AL elasticity enhances wind power integration and reduces regulatory pressure on coal-fired units at night.

PtX in district heating systems. [85] examines the potential of large-scale two-stage ammonia heat pumps (NH_3-HPs) used in Danish district heating to support grid balancing through primary frequency regulation. A dynamic Modelica model of the HP, validated with experimental data from a Copenhagen HP with variable-speed piston compressors, assessed the impact of various control strategies. The most effective control structure for rapid regulation included direct power uptake control, evaporation pressure control, source outlet temperature control, and vacuum line preheating. This configuration allowed the HP to adjust power input from 250 kW to 175 kW in 85 s, and to 100 kW in 140 s, meeting primary frequency regulation requirements. [86] introduces an ultra-low-temperature district heating (ULTDH) system to overcome the economic inefficiencies of booster heat pumps used for domestic hot water (DHW). The system uses the district heating supply line as a heat source, reducing capital costs and enhancing COP by integrating neighborhood-scale water source booster HPs with a triple-pipe network. A case study comparing ULTDH to low-temperature district heating (LTDH) sizes components based on DHW and space heating demands. Yearlong thermodynamic analysis shows ULTDH offers superior energy efficiency, with economic advantages increasing with higher electricity prices, outperforming LTDH, especially when electricity costs are elevated. [87] addresses surplus power in active distribution networks (ADN) from intermittent renewables and CHP plants by proposing a power-to-hydrogen-and-heat (PtHH) system with heat recovery to enhance efficiency. PtHH reduces CHP output by producing hydrogen and supplying heat to district heating networks, mitigating excess energy in ADNs. A T-H-H model integrating power-to-heat and power-to-hydrogen operations, along with a simplified dispatch model for coordinated management, is presented. Simulations on a modified IEEE 33-node system demonstrate that PtHH improves economic efficiency and system security.

PtX in industrial process heat applications. [88] aims to develop highly energy-efficient and economically viable methanol production processes using coke oven gas. Two process schemes are proposed, incorporating polygeneration of methanol, power, and heat. Rigorous process models were created to optimize methanol production or total energy output, determining optimal configurations and conditions. The proposed processes, evaluated based on methanol's minimum selling price, carbon balance, and energy efficiency, show significant energy efficiency ranging from 53% to 71% (compared to 38% for the base process). The processes are economically competitive with conventional methods, with methanol's minimum selling price between \$0.23 and \$0.29 per kg. [89] develops an environmentally friendly method to harness waste heat in NG power plants by converting flue gas into methanol. The system incorporates an organic Rankine cycle, an absorption chiller, heating units, an electrolyzer for H_2 production, and a methanol synthesis unit. Simulated with Aspen HYSYS, it achieves a methanol production rate of 2,712 kg/h at 99.97% purity and also generates hot water, chilled water, and power. The energy efficiency is 94.35%, with an exergy efficiency of 31.74%. Adjusting turbine pressure and fluid temperature can enhance exergy efficiency. The multigeneration scenario results in a carbon footprint of 0.1564 kg/kWh and a product cost of \$0.0485/GJ. [90] examines enviro-economic optimization for a biomass-gasification-driven combined heating, hydrogen, and electricity system, focusing on balancing uncertainties in infrastructure, resources, and technology adoption by incorporating hydrogen production. The optimization considers revenues from electricity and hydrogen, and expenses like capital, fuel, and emissions, using Weibull function modeling in a probabilistic process to determine optimal reliability, maintainability, and costs. Prioritizing hydrogen production is advantageous, achieving the lowest costs by favoring hydrogen modules over electricity generation.

Integration of hydrogen storage and energy management. [91] proposes a combined hydrogen, thermal, and power system incorporating PEM-FCs, metal hydride hydrogen storage, and AWE to mitigate the effects of RE volatility on the grid. Focusing on optimizing residual heat utilization from electrolysis, hydrogen adsorption, and PEM-FCs, four energy scheduling schemes are developed to enable near complete heat use. Heat priority, peak clipping, and load reduction strategies suit off-grid scenarios, while the peak–valley strategy is optimal for on-grid operation. Specifically, the heat priority strategy reduces residual electricity and heat, the peak clipping strategy maintains PEM-FC current density at 0.59 A/cm², and the load reduction strategy ensures power output meets demand 76.07% of the time during heat-led periods. [92] proposes a multisource energy storage system (MESS) integrating electricity, hydrogen, and thermal networks from the operator's perspective. After establishing the MESS framework and device model, a multi-objective optimal dispatch strategy considering time-of-use pricing is developed to balance stability, efficiency, and economy. Applied to MESS, the strategy effectively manages power fluctuations, reduces energy loss, and enhances economic benefits. Despite substantial energy loss, H2 and heat sales could be MESS's main revenue sources. The dispatch strategy optimizes benefits, promoting MESS's sustainable growth. [93] presents a rolling optimal dispatch method for an Antarctic IES incorporating an alert mechanism to address extreme weather and equipment failures. By modeling PEM-FC output using a linear P-H-T framework, operational flexibility is enhanced. A continuous dispatch strategy ensures system security through the alert mechanism. Normalized multiparametric disaggregation technology (NMDT) handles bilinear terms, converting MIQCP into MILP. A case study shows increased IES resilience, reduced load shedding during emergencies, and improved system economy.

Economic and environmental optimization of PtX systems. [94] proposes a carbon trading method for an integrated heat-power-hydrogen energy system using a Vickrey auction strategy to promote carbon neutrality and increase carbon trading participation. The model calculates carbon credits for hydrogen and renewables based on calorific value and emission factors, accounting for residual heat in hydrogen power generation. Traditional energy receives Chinese emission allowances (CEA) via a baseline approach, while renewables' CCERs are traded through a Vickrey auction. Economic benefits and reduced emissions are demonstrated across three scenarios. [95] develops an online probabilistic energy flow method to mitigate uncertainties from RE in an H_2-power-heat system. The system model addresses interdependencies among energy sources, incorporating hydrogen FCs converting hydrogen into power and heat. An efficient algorithm based on multi-parametric programming (MPP) enhances computational performance, suitable for online use. The method's accuracy and efficiency in managing uncertain wind power in energy flow calculations are validated on test systems.

6.4.3 Buildings

PtX technologies are transforming energy generation, storage, and utilization in the building sector, fostering more resilient and sustainable urban environments. By integrating renewable electricity into buildings' heating, ventilation, and power systems, PtX increases energy efficiency and reduces reliance on fossil fuels. Innovations like electrified heating systems (e.g., heat pumps) enable direct use of renewable electricity for space heating and hot water, cutting GHG emissions and energy costs. PtX also promotes building-level energy storage solutions, allowing surplus RE to be stored and used during low-generation periods, enhancing grid stability and energy independence. Smart buildings utilizing PtX can dynamically manage energy demand, supply, and storage in real time, optimizing consumption and turning buildings into active participants in the energy ecosystem, aiding broader climate and energy goals. Figure 6.2 demonstrates the integration of solar energy, NG, and ESSs within a residential EH, showcasing how multiple energy sources and storage technologies work together to balance and optimize the energy supply for residential demands.

Hybrid energy systems for residential power management. [96] proposes a hybrid energy system integrating a battery, DMFCs, and PV and solar thermal collectors for residential power management. DMFCs serve as a secondary power source to overcome batteries' limitations in meeting long-term power needs. Managed by a neural network (NN)–based adaptive control, the DMFC outperforms traditional PID control in handling set-point changes and disturbances. Incorporating DMFCs significantly reduces grid power requirements and extends battery life, especially when solar power is insufficient. An integrated solar energy system combined with a ground-sourced heat pump is proposed in [97] for standalone residential structures, providing power, heat, cooling, and domestic hot water. The system includes PEM-FCs, an anion exchange membrane electrolyzer for hydrogen storage, building-integrated photovoltaics (BIPV) with varying orientations, and a ground-sourced heat pump. The H_2 subsystem uses stored hydrogen during peak periods, while excess energy is used for hydrogen production. Thermodynamic analysis in five global cities shows that, in Ottawa, a 20-story building requires 550 kWp to 1,550 kWp of PV capacity, achieving energy and exergy efficiencies of 18.76% and 10.49%. In Istanbul, self-sufficiency is achieved with a 495 kWp BIPV façade and a 90 kWp rooftop PV.

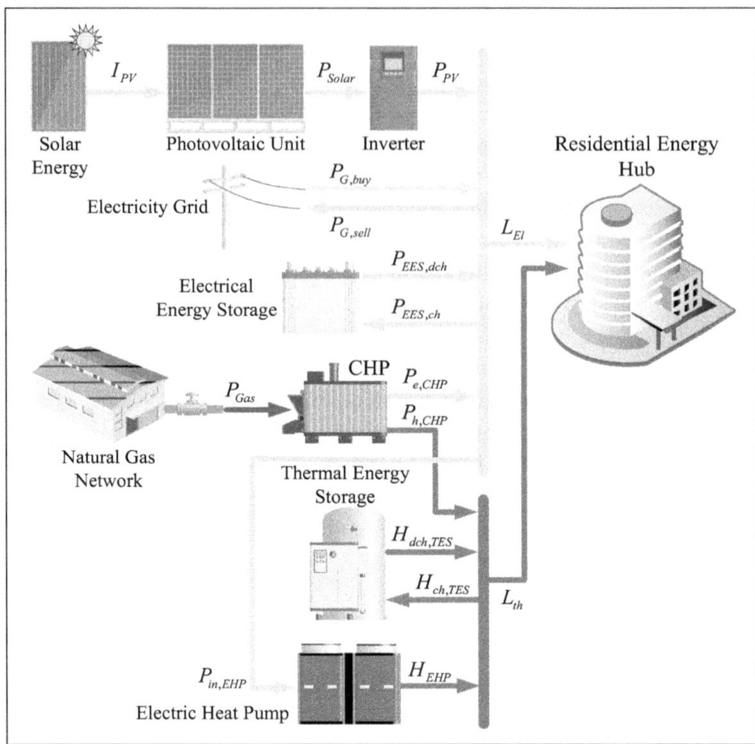

FIGURE 6.2 Concept of a residential energy hub integrating solar, natural gas, and storage systems.

Hydrogen storage and multi-energy systems in buildings. [98] develops a MES for district buildings that integrates hydrogen storage and electrochemical batteries to enhance energy flexibility, mitigate grid fluctuations, and facilitate RE integration. The system incorporates vehicle-to-building (VtB) and grid-to-building (GtB) interactions to boost self-consumption and reduce grid dependency. A stack voltage model assesses real-time FC degradation for cost analysis, while a low-grade heat recovery system improves energy efficiency. Implementing an off-peak power shifting strategy in Guangzhou reduces operating costs for hydrogen vehicles by 44.4% and total system costs by 40.6%, increasing RE utilization. However, the efficiency of power-to-hydrogen-to-power conversion is low, leading to an 88.22% increase in CO2 emissions. [99] focuses on the efficient operation of hydrogen-based multi-energy systems (HMES) by considering building thermal dynamics. Incorporating thermal dynamics optimizes thermal load flexibility, reducing operational costs. The study presents a complex cost-reduction problem addressed with a two-part algorithm: Lyapunov optimization decomposes the long-term problem into manageable subproblems, which are then solved using convex optimization and a multi-agent, attention-based deep deterministic policy gradient approach. Simulations with empirical data validate the method's effectiveness in lowering operational expenses.

Innovative approaches to net-zero energy buildings. [100] develops a tri-level multi-energy system planning approach for zero energy buildings (ZEB) to manage long- and medium-term uncertainties and achieve annual zero energy goals. The upper level focuses on optimal sizing of devices in an integrated electricity, thermal, and hydrogen system (EHT-MES). The middle level addresses long-term temperature fluctuations by creating representative scenarios, while the lower level handles short-term uncertainties, like solar radiation and load demands, using a hybrid stochastic and robust optimization (HSRO) method. A simplified version of the multi-level model is solved using an enhanced column-and-constraint generation (C&CG) method. Simulations confirm the strategy's effectiveness in optimizing multi-energy device deployment and ensuring resilience against uncertainties. [101] presents a multidimensional economy–durability (MDED) optimization method for the integrated energy and transportation system (IETS) of net-zero energy buildings (NZEB), addressing challenges from increased RE and zero-emission vehicles (ZEV). The approach integrates energy and transportation over day-ahead and intraday periods. A cost-optimal scheduling method utilizes an intelligent parking lot (IPL) for collaborative operation. In the intraday phase, synchronized scheduling of ZEVs and optimized energy strategies meet transportation and consumption demands. Results show the approach balances charging services and overall system costs, highlighting economic and coordination benefits. The utilization of a dynamic economy–health matrix (DEHM) in rolling optimization enhances cost-effectiveness and system longevity, outperforming traditional methods.

Energy management strategies for smart buildings. [102] presents a hybrid cloud-based framework for creating a self-sufficient, eco-friendly residential house stable against RE fluctuations. The system uses wind turbines and PV panels as primary power sources, supplemented by an FC, electrolyzer, and battery for storage. The framework employs deep learning to schedule components based on projected demand and renewable output, with a compensation system for real-time scheduling errors. Real-time simulation confirms its effectiveness for stable and efficient energy management in residential settings. [103] introduces an EMS for smart buildings utilizing an FC and battery hybrid energy storage (FBHES) system, developed using an improved game theory approach. The EMS aims to optimize power distribution, enhance efficiency, and extend the lifespan of both the FC and the battery. The revised max-min game theory effectively balances power, considering uncertainties in demand and storage performance. The method reduces FC degradation by 12.64% compared to standard approaches and increases battery life by up to 82.81% compared to other methods. The study highlights the EMS's ability to improve energy efficiency and reduce emissions in smart buildings.

Integration of PtX in urban energy systems. [104] develops energy management and optimization strategies for commercial buildings reliant on RE, incorporating pumped hydro storage and hydrogen taxis to achieve net-zero consumption. The approach manages energy use based on time-of-day targeting scenarios with high renewable output and shifted peak demand. Multi-objective optimizations determine optimal sizes for pumped hydro storage. Larger systems increase renewable self-consumption by 9.24%, improve demand satisfaction by 6.10%, and reduce annual electricity costs by 25.21%. The method achieves significant decarbonization, reducing carbon emissions by 95.14% compared to conventional grid power and 80.74% compared to petroleum taxis. Integrating decarbonized transportation into urban buildings offers valuable guidance for policymakers aiming for carbon neutrality by 2050.

PtX technologies in building environmental control. [105] investigates a method for eliminating atmospheric CH_4 by combining a small-scale solar chimney (SC) with photocatalytic reactors (PCR) in buildings. Two computational models were created to replicate the process of photocatalytic reaction and interior ventilation. These models were used to assess the efficiency of CH_4 removal and the performance of the ventilation system. The study conducted a comparison between honeycomb (HPCR) and plate (PPCR) reactors and concluded that HPCR is superior in terms of CH_4 removal efficiency. The CH_4 purification rate was optimized at 57.27 µg/s using particular parameters, including a 0.3 m air gap width, 1.5 m HPCR length, and 0.85 porosity. SCs equipped with PCRs provide a viable, energy-efficient alternative for removing low-concentration CH_4 in a decentralized manner, while also improving interior air quality.

6.4.4 TRANSPORTATION

PtX technologies are driving the adoption of hydrogen fuel cell electric vehicles (FCEV) and associated infrastructure, promoting cleaner and more sustainable mobility in the transportation sector. By generating green hydrogen from renewable electricity, PtX provides a zero-emission alternative to conventional internal combustion engine vehicles across various modes of transport, including buses, trucks, trains, and ships. PtX also facilitates the development of hydrogen refueling stations, ensuring consistent access to clean fuel and enabling a transition to a hydrogen-based transportation network. Incorporating PtX technologies allows the transportation sector to reduce reliance on fossil fuels, lower greenhouse gas emissions, and contribute to a sustainable and resilient energy future.

Integration of PtX in transportation systems. [106] presents an operational framework for hydrogen microgrids (H2-MG) integrated with traffic and power networks to improve decision-making amid increasing electric and hydrogen vehicle usage. Using a risk-averse approach with information gap decision theory, it addresses traffic flow prediction challenges impacting H2-MGs and power systems. A hybrid deep learning method combining 1-D CNN and bidirectional LSTM, enhanced by Bayesian hyperparameter tuning, accurately predicts traffic flow in Edmonton, Canada. This predictive capability optimizes EV and HV charging, enhancing energy efficiency and thermal demand support. The framework improves operational efficiency, routing, and charging, outperforming conventional methods in maintaining secure operations during disruptions and enhancing system robustness. [107] develops a techno-economic optimization model to determine the optimal size of an electrolyzer and hydrogen storage system in a virtual power plant (VPP) for producing green hydrogen for seaborne passenger transport. Focusing on the Danish island of Bornholm, the study uses three years of data from a RE-powered substation and ferry operations. An optimal setup with a 9.63 MW electrolyzer and 1.45 t hydrogen storage efficiently harnesses local surplus electricity, provides flexibility, supplies 26% of the ferries' hydrogen needs, and covers 21.4% of district heating requirements with electrolyzer waste heat. [108] introduces a real-time EMS for an off-grid smart home powered by wind turbines and photovoltaics, with an FC as backup and excess power used to produce hydrogen via electrolysis. Using data from a smart home at Yildiz Technical University, the EMS aims to minimize hydrogen consumption and optimize RE use by regulating controllable loads with a fuzzy logic controller, ensuring user comfort. Load shifting enhances wind and solar utilization by over 10%, reducing annual hydrogen consumption by 7.03% and increasing FC efficiency by 4.6%. [109] proposes a charging strategy for plug-in electric vehicles (PEVs) in a hydrogen energy–based microgrid to effectively manage power system loading and enhance customer satisfaction. The two-stage method first classifies PEV charging requests using an ensemble learning classifier for high accuracy, then schedules charging based on classifications and user demands. Testing shows significant improvements over unmanaged scenarios, with a 52.1% reduction in peak load and a 72.3% improvement in valley filling.

Feasibility and implementation of PtX in transportation modes. [110] assesses the feasibility of replacing long-range, high-capacity transport vehicles—including military combat vehicles, freight trains, and ships—with battery electric and hydrogen FC alternatives, evaluating if they can maintain range, mass, volume, and power ratios. The study finds that with specific technological advancements, vehicles like armored tanks, freight trains, oceangoing vessels, helicopters, prop planes, and jumbo jets can adopt clean energy technologies. Using the US Army as an example, such a shift could be equivalent to removing nearly 700,000 passenger cars from the road, significantly enhancing sustainability. [111] develops an electricity/hydrogen/gas/refueling station (EHGRS) integrating RESs to supply clean energy for electric, hydrogen FC, and NG vehicles in off-grid areas. Employing a hybrid stochastic/information gap decision theory (IGDT) optimization approach, it addresses scheduling under uncertainties in wind and solar power and demand. Two strategies—risk-averse and risk-seeker—are examined. The EHGRS efficiently provides multiple fuels simultaneously. Compared to pure stochastic optimization, the risk-averse strategy reduces earnings, while the risk-seeker strategy increases profits, highlighting the impact of risk attitudes on profitability. [112] investigates using ammonia and hydrogen in ICEs, specifically employing H_2 multiple-injection jet ignition (MIJI) for ammonia combustion. Tests on a heavy-duty engine show that while passive jet ignition requires at least 15% H_2 energy ratio for stable operation, active H_2 jet ignition significantly enhances combustion stability. A triple-injection technique adjusts the NH_3–H_2 mix for optimal combustion, achieving stable combustion with only 3% H_2 energy ratio and an indicated thermal efficiency of 42.5% by optimizing H_2 injection and spark timing. [113] aims to reduce nitrogen oxide (NO_X) emissions in hydrogen-fueled ICE vehicles by improving NO_X storage and reduction (NSR) capabilities of $LaMnO_3$ perovskite oxide catalysts. Using the Pechini method and pluronic 123 as a pore-forming agent, a porous $LaMnO_3$ catalyst (LMO-PC-P) with high surface area and abundant oxygen vacancies was developed. This catalyst achieved a NO_X storage capacity of 205 μmol/g, more than double that of catalysts made via the sol-gel method. In dynamic NSR tests, LMO-PC-P converted 84% of NO_X at 350°C, outperforming the sol-gel catalyst's 46% conversion. [114] designs and validates a thermoelectric generator (TEG) module for energy recovery from exhaust in a hydrogen internal combustion engine urban bus. The system includes 40 TEG modules with cooling fins and channels, arranged in a robust rectangular frame. A detailed model predicts energy recovery of 1.0–1.1% from engine exhaust and identifies an optimal load resistance of 3.2 Ω. The system recovers energy effectively under various operating conditions. [115] explores enhancing energy efficiency on liquefied natural gas (LNG)–powered ships by converting some boil-off gas into hydrogen via steam methane reforming (SMR), using it as an auxiliary power source through a high-temperature PEM-FC. The study finds methane conversion rates exceed 90% at temperatures above 973 K, S/C ratios over 3, and pressures below 2 bar. Optimal reformer design is proposed for temperatures over 1,023 K at 2 bar and 923 K at 4 bar, with a recommended reformer length under 2 m to fit onboard space constraints. [116] evaluates hydrogen storage technologies for aviation, focusing on impacts on aircraft design related to mass and volume. The study assesses storage methods suitable for aircraft like the Airbus A320 and Embraer E 190 for short- and medium-haul flights. Assuming the maximum takeoff mass remains constant, a tank concept compatible with conventional aircraft designs is developed, facilitating the replacement of fossil fuels with hydrogen to enhance aviation sustainability. [117] introduces advanced metal hydride (MH) hydrogen storage tanks for FC electric forklifts. The tanks feature a shell with six baffles and a tube containing 120 kg of AB5 alloy enhanced with copper fins for improved heat transfer. The AB5 alloy stores hydrogen at 1.6 wt% with low equilibrium pressure at room temperature. Finite-element models show tanks with corrugated fins reach 1.5 wt% saturation in 630 s, with diagonal baffles optimizing refueling time due to higher cooling water velocity. [118] examines the effect of hydrogen-blended NG on methane slip in two-stroke marine engines using simulation models. Increasing hydrogen molar fraction reduces methane slip significantly; a 40% hydrogen molar fraction cuts methane slip by 26.76% and reduces equivalent CO_2 emissions by 142.78 g/kWh. While increasing hydrogen content further decreases methane slip and CO_2 emissions, it slightly increases fuel consumption and reduces power output. [119] studies the safety and stability of a liquid hydrogen system for heavy-duty trucks under fire scenarios and pressure reduction valve failures. A new liquid hydrogen system model validated with test data shows that high temperatures can lead to overpressure and overheating, affecting system safety. Damage to pressure reduction valves can cause temperature and pressure imbalances, impacting FC performance. An optimal design strategy is proposed to enhance safety and stability.

[120] simulates and analyzes the performance of an FC bus using dynamic modeling and machine learning methods to address issues like HVAC demands, FC degradation, and battery aging. The model considers varying thermal loads and passenger numbers. Findings show that electric motors consume over 60% of energy, with energy use increasing by up to 28 MJ for every 40 passengers. FC degradation after 2,000 hr leads to a 10% increase in hydrogen consumption, and battery capacity decreases by 17.6% after a year at high ambient temperatures. [121] explores optimal sizing of a rural microgrid using a scenario-based approach and two-stage stochastic programming. The system includes multi-energy storage and various EV technologies with vehicle-to-grid operations. Hybrid storage systems are most cost-effective, with up to 20% cost reductions, by allowing a loss of power supply probability of 0.2. Incorporating vehicle-to-grid modes in both BEVs and FCEVs yields an additional 2% cost reduction, with BEVs contributing up to 90% of vehicle-to-grid power.

Optimization frameworks and alternative fuels. [122] introduces a day-ahead optimization framework for sustainable energy supply at a hydrogen refueling station (HRS) and EV charging facility within a local multi-energy system (LMES). An integrated demand response program with incentives reduces operational costs and enhances flexibility. A hybrid multi-objective information-gap decision theory/robust optimization (HMIRO) model addresses uncertainties without requiring probability distributions. By applying duality theory, the model simplifies to a single-level MILP, allowing simultaneous risk-averse strategies. This approach reduces risks associated with wind generation and HRS demand by 20.6% and 14.3%, respectively, with a 5% increase in daily operating costs. Optimal demand response implementation reduces daily energy costs across various loads. [123] proposes a cooperative operation model for a combined wind farm and HRS system to meet the growing HRS demand using wind energy for on-site hydrogen production and storage. Robust optimization mitigates uncertainties in wind power and market prices. The Nash–Harsanyi bargaining game theory ensures fair profit allocation among stakeholders. The problem is divided into energy trading and payment bargaining subproblems, solved using a column and constraint generation algorithm and a privacy-preserving distributed approach, respectively. [124] addresses onboard hydrogen storage limitations by introducing a vehicle powered by an ammonia solid oxide FC (NH_3-SOFC). A life cycle assessment evaluates environmental impacts and energy consumption. Photochemical ozone creation potential accounts for 62.65% of environmental impacts. The use phase significantly contributes to energy use and CO_2 emissions. The NH_3-SOFC vehicle's global warming potential is 0.124 kg CO_2-eq/km. Sensitivity analysis shows total environmental impacts could be reduced by 29% with increased life cycle mileage and component replacements. The NH_3-SOFC vehicle has lower GWP than other vehicle types when using renewable-generated ammonia.

Multi-energy microgrid optimization with EVs. [125] aims to enhance energy efficiency and reduce emissions by optimizing multi-energy microgrids (MEMGs) with high RE penetration. MEMGs use renewable sources, ESSs, and energy conversion systems to ensure sustainable supply. A two-stage stochastic programming method addresses uncertainties in energy demands and renewable intermittency. The model incorporates bilateral contracts and pool electricity markets, solved using MILP in GAMS. RESs reduce operation costs by 34.47%, and bilateral contracts cut risk metrics by 50.42%. Storage units and demand response programs further reduce purchased electricity costs and overall operation costs.

6.4.5 INDUSTRY

PtX technologies significantly contribute to decarbonizing energy-intensive industrial processes, such as chemical production, refineries, steel and metals production, glass manufacturing, synthetic materials, and waste management. By converting renewable electricity into H2, NH_3, and CH_3OH, PtX enables low-carbon production of chemicals and fuels. It supports refineries by producing H_2

for hydrotreating and hydrocracking, reducing reliance on fossil-based H_2. In steel production, PtX replaces carbon-intensive methods with hydrogen-based reduction, allowing for eco-friendly steel manufacturing. The glass industry benefits by using H_2 or synthetic fuels for high-temperature processes, lowering NG dependence and emissions. PtX also offers sustainable feedstocks for synthetic materials and innovative waste-to-energy solutions, promoting a circular economy. Overall, PtX reduces emissions, enhances resource efficiency, and drives a shift toward a sustainable industrial landscape.

PtX applications in industrial processes. [126] performs a techno-economic evaluation of a methanol production facility that uses electricity from a floating PV power plant to hydrogenate CO_2 captured from an oil refinery's waste gas in Libya. The facility captures 4,890 t of CO_2 per day, reducing emissions by 4,513 t. Over 30 years, it produces over 43 Mt (million tons) of methanol at a unit cost of $412.9 per ton, demonstrating the viability of integrating PtX technologies in chemical production. [127] presents a hybrid biomass conversion to ammonia system (HBCAS) employing a chemical looping process supported by solar and wind energy. The system achieves an ammonia selectivity of 79.36% and a production rate of 34.1 kmol/h, showcasing the potential of PtX in sustainable ammonia production. [128] introduces the solar oil refinery, integrating solar energy into the refining process. The system employs a solar-powered thermo-electrochemical method to crack residual oil, significantly enhancing solar utilization efficiency, cracking rates, and hydrogen yield. The solar cracking process operates at lower temperatures and achieves higher product yields compared to conventional methods. [129] investigates the use of surplus wind energy to produce hydrogen for oil refineries in Brazil. By replacing steam methane reforming (SMR) in hydrotreating units and supplying oxygen for oxy-combustion CO_2 capture, refineries can reduce GHG emissions by up to 22.11%. [130] examines the susceptibility of low-carbon pipeline steel to hydrogen embrittlement, crucial when using existing NG pipelines for hydrogen transport. The study highlights the importance of material considerations in safely implementing hydrogen-based processes in the steel industry. [131] explores the combined use of landfill gas (LFG) and residue-derived fuel (RDF) from excavated waste to produce methanol. The system achieves exergy efficiencies above 70% and significantly reduces the GWP, highlighting the environmental benefits of integrating PtX in waste management.

Advances in PtX technologies for industry. [132] proposes a hydrogen production system using SOECs powered by industrial waste heat. The system can produce up to 2.473 kmol/h of hydrogen at 800°C, demonstrating the potential of waste heat recovery in hydrogen production. Waste heat contributes 15.5% to 20.6% of the total energy consumption, enhancing efficiency. [133] assesses the technological and economic feasibility of producing green hydrogen directly at Mexican refineries using RESs. The study finds that solar photovoltaic energy could offer more affordable green hydrogen, improving the competitiveness of green hydrogen plants in the petrochemical industry. [134] investigates a carbon-negative method for producing methanol by combining bagasse pyrolysis, activation, chemical looping, and methanol synthesis. The process achieves negative GWP values, indicating net carbon sequestration, and presents economic viability with payback periods of 5.80 to 6.53 years. [135] presents a solar hydrogen production system that integrates chemical energy storage and spectral beam-splitting technology. The system achieves an energy efficiency of 32.08% and reduces CO2 emissions by 25.7% over a year, addressing efficiency and intermittency challenges in solar energy utilization. [136] aims to enhance the efficiency and stability of ammonia-fueled SOFCs by optimizing anode design to improve ammonia utilization. Empirical correlations for ammonia decomposition rates using low-cost metal catalysts contribute to better SOFC performance. [137] examines the use of stainless steel as an electrocatalyst in electrochemical processes for water oxidation, enhancing the selectivity of the oxygen evolution reaction (OER) while reducing CO_2 to methanol. Stainless steel shows high selectivity and could serve as an efficient anode for CO_2 electrolysis. [138] investigates hydrogen gas sensing capabilities of Pd-capped SnO_2 thin films deposited on various substrates. Sensors produced on quartz substrates exhibit superior sensitivity and stability, important for monitoring hydrogen gas in industrial settings. [139] develops plasticized solid biopolymer electrolytes (SBEs) utilizing carboxymethyl cellulose for solid-state hydrogen energy storage. The study demonstrates the potential of bio-electrolytes in solid-state hydrogen ionic cells with good rechargeability and open-circuit potential.

Integration of renewable energy in industrial processes. [140] examines the impact of hydrogen sulfide (H2S) on chemical looping hydrogen production processes. Controlling temperature and sulfur deposition is crucial to maintaining oxygen carrier stability and overall process efficiency. [141] presents a system utilizing solar and wind energy to produce liquid hydrogen and ammonia as energy carriers, meeting urban demands for electricity, cooling, heating, and fresh water. The optimized system achieves an overall energy efficiency of 56.78% and an exergy efficiency of 44.69%.

6.4.6 AGRICULTURE

PtX technologies are revolutionizing energy production, use, and integration in agriculture, enhancing sustainability and efficiency. A key application is the production of green NH_3 using renewable H_2, which reduces GHG emissions and supports eco-friendly fertilizer production. PtX also converts agricultural waste into valuable energy sources, like biogas or liquid biofuels, for machinery or electricity generation. Additionally, biogas can be upgraded to biomethane as a cleaner alternative to NG for power, heating, or vehicle fuel, increasing the value of agricultural waste. PtX further enables electrification and hydrogenation of farming equipment, offering cleaner options over diesel. By integrating PtX, the agriculture sector can cut emissions, generate energy, and support a more sustainable agricultural system.

Green ammonia production for sustainable fertilizers. [142] assesses the feasibility of a hydrogen-powered green ammonia plant in Nepal capable of producing 1,245 t per day of ammonia for urea production. Using simulations with DWSIM and Aspen Plus, the study estimates the capitalized costs and levelized costs of hydrogen and ammonia at $3,602/t and $826/t, respectively. Economic indicators show strong viability with an internal rate of return (IRR) of 26%, a return on investment (ROI) of 38%, and a payback period of three years. Sensitivity analysis highlights electricity costs as a significant factor influencing levelized costs. [143] investigates the economic and environmental benefits of the methane-assisted biomass-to-fuel (M-BtF) process, which converts organic solid waste into bio-oil using an artificial neural network (ANN) model to predict yields. The approach eliminates the need for costly pilot plant trials. The study finds that renewable diesel can be produced at $3.45/gal, 26% lower than industry average rates, with an IRR above 50%. Additionally, the process emits approximately 67% fewer GHGs compared to conventional methods.

Waste-to-energy conversion and biogas production. [144] evaluates Portugal's potential to generate biogas and biomethane from manure through anaerobic digestion (AD). Using spatial analysis with ArcGIS and thermodynamic modeling in MATLAB, the study identifies regions with the highest production potential. National production potentials are estimated at approximately 3.56 Mt per year for biogas and 1.79 Mt per year for biomethane, with energy efficiencies of 62.96% and 51.82%, respectively. Economic analysis suggests average selling prices of €0.8174/kg for biogas and €1.290/kg for biomethane.

[145] explores methods to increase methane production from palm leaflet waste (PLW) using thermal and ultrasound pre-treatment techniques in anaerobic digestion. The study finds that thermal pre-treatment yields the highest cumulative methane production (2,341 Nml), followed by ultrasound and untreated PLW. The enhanced biodegradation rates due to pre-treatment significantly improve methane yields. [146] examines the feasibility of producing bio-hythane—a mixture of methane and hydrogen—from agricultural residues through anaerobic digestion without pre-treatment. The study assesses the biochemical methane potential of various wastes and finds that sugarcane bagasse, mixed fruit waste, and mixed vegetable waste yield the highest bio-hythane proportions, indicating their suitability for bio-hythane production. [147] investigates methane production potential from biopolymers (cellulose, hemicellulose, lignin) and agricultural biomass using a two-stage pyrolysis-catalytic hydrogenation reactor. The presence of a 10 wt.% Ni/Al2O$_3$ catalyst increases methane yields, with hemicellulose and cellulose producing significantly more methane than lignin. The synergy between biopolymers enhances methane production, emphasizing the importance of biomass composition. [148] develops a light-assisted bioreactor using response surface methodology to address excessive ammonia inhibition in anaerobic digestion. Under optimal light conditions, methane yield doubles compared to dark reactors, demonstrating the practical feasibility of using sunlight to enhance methane production. Light stimulation improves enzyme activities and microbial communities, alleviating ammonia stress. [149] studies the impact of acyl homoserine lactone (AHL)–based regulation on anaerobic digestion of agricultural wastes. The timing of AHL addition significantly influences methane production, with certain conditions increasing methane yield by 47.71% compared to control. AHLs enhance hydrolysis, acidification, and methanogenic activity, promoting the hydrogenotrophic methanogenesis pathway. [150] evaluates the potential of anaerobic sludge from a rice parboiling industry to produce biomethane and biohydrogen, focusing on the effects of different substrate-to-inoculum (S/I) ratios. Optimal ratios result in substantial CH$_4$ and H$_2$ production, indicating that anaerobic sludge is an efficient inoculum for biofuel production in digestion and fermentation processes.

Renewable energy systems for agriculture. [151] investigates a novel, standalone hybrid energy system for an agricultural farm in Kairouan, Tunisia. The system integrates PV panels and FCs without battery storage, using solar-powered water electrolysis to produce and store hydrogen. Stored hydrogen is converted back to electricity via FCs when solar energy is unavailable. The system also desalinates saline well water for electrolysis and irrigation, offering a sustainable solution with low operational costs. [152] proposes integrating agrivoltaics (solar photovoltaics combined with crop production) with concentrated solar power (CSP) systems on agricultural land, termed "agri-CSP." Simulations for a California community suggest that agri-CSP can produce 1,716 MWh of electricity per MWp annually, along with significant amounts of fresh water and hydrogen. The system demonstrates high energy and exergy efficiencies, promoting sustainable, self-sufficient communities. [153] estimates hydrogen production potential from agricultural residues in Punjab, Pakistan, including maize, cotton, sugarcane, rice, and wheat. Using Aspen Plus simulations and GIS mapping, the study finds that Punjab could produce 2.62 million metric tons of hydrogen annually, with sugarcane trash being the largest contributor. This hydrogen could support industrial, transportation, and urea production sectors, offering a sustainable energy alternative. [154] assesses the feasibility of producing hydrogen and biogas from solid urban waste (SUW) and wine waste in Bento Gonçalves, Brazil. Three scenarios are explored, with the combined use of wine waste and SUW yielding the highest energy output. The biomethane produced can power the local bus fleet, and hydrogen generated via steam reforming can prevent significant CO_2 emissions while producing substantial electricity. [155] investigates the use of immobilized cell technology with carrageenan and gelatin in two-stage batch bioreactors to optimize biogas production from high-strength food waste hydrolysate. Immobilized cells improve microbial stability and activity, resulting in increased hydrogen and methane production rates and high removal efficiencies for total solids, volatile solids, and chemical oxygen demand.

6.5 PTX SYSTEMS IN THE ELECTRICITY MARKET

EHs serve as critical nodes in the energy network, enabling the conversion, storage, and distribution of multiple energy forms, such as electricity, heat, and gas. These hubs are strategically designed to optimize energy flows by considering real-time market signals, supply–demand dynamics, and operational constraints. With the aid of advanced algorithms and data analytics, EHs can effectively schedule energy transactions across various markets, thereby ensuring both cost-effectiveness and operational reliability. One of the primary strengths of EHs is their close alignment with market dynamics, allowing for optimal energy scheduling and sector coupling. EHs actively participate in day-ahead, intraday, and real-time electricity markets, adjusting their operations in response to price fluctuations. Through the integration of electricity with heat and gas markets, these hubs facilitate cross-sector energy transfers, thereby enhancing overall system efficiency. Additionally, EHs can offer flexibility services, such as frequency regulation and voltage support, which are crucial for maintaining grid stability.

The integration of PtX technologies within EHs further strengthens their role in energy markets. PtX technologies are essential for sector coupling, where electricity networks interconnect with other energy domains, such as transportation, heating, and industrial processes. By converting surplus electricity—often sourced from renewables—into alternative energy carriers, PtX systems can alleviate grid congestion and reduce RE curtailment. This integration supports market-based operations through arbitrage opportunities, enabling EHs to capitalize on price differentials by storing energy when prices are low and supplying it when prices are high. Additionally, PtX systems can engage in demand response activities by adjusting their operations in response to market signals, which allows for the reduction of energy consumption during peak price periods and an increase during low price periods. Consequently, these capabilities facilitate greater penetration of intermittent RESs within the market. From a planning and optimization perspective, EHs equipped with PtX technologies offer numerous advantages. They enable investment optimization by determining the ideal mix of PtX technologies and capacities to maximize economic returns. Moreover, these hubs enhance operational flexibility by providing the ability to respond to market uncertainties and demand variations. Finally, EHs support regulatory compliance by meeting emission targets and RE integration mandates through efficient energy conversion and utilization strategies.

Several real-world case studies illustrate the effective integration of PtX technologies within EHs. In Germany, for instance, EHs have been employed to incorporate PtX systems for hydrogen production. Excess RE is converted into hydrogen via electrolysis, which is then either stored or injected into the gas grid. The EH optimizes its operations based on electricity market prices and hydrogen demand, participating in both electricity and gas markets. This integration has led to reduced renewable curtailment, revenue generation from multiple energy markets, and enhanced grid stability through the provision of balancing services. Similarly, Denmark has implemented EHs that integrate PtX systems within district heating networks. Electric boilers and heat pumps convert electricity into heat, which is subsequently stored and distributed according to thermal energy demand. The EH schedules its operations based on electricity prices and heating needs, resulting in significant cost savings by utilizing low-price electricity and providing enhanced flexibility to shift between electricity and heat, depending on prevailing market conditions. These case studies underscore the potential of PtX-integrated EHs in advancing the efficiency and sustainability of energy systems across different regions.

PtG systems and energy storage. [156] explores the application of PtG energy storage systems that convert electricity into hydrogen or SNG. Instead of converting the gas back to electricity—which is inefficient due to energy losses—the study focuses on the direct use or sale of SNG. A model is proposed to optimize day-ahead scheduling of PtG storage and gas load management, aiming to minimize gas usage costs. The system includes a gas demand prediction algorithm and a reserve mechanism to handle gas consumption during power grid failures. Test cases demonstrate the model's effectiveness in optimizing gas load management and reducing expenses.

Local energy markets and peer-to-peer trading. [157] introduces a framework for a local energy market (LEM) that enables the exchange of electricity and hydrogen among participants, including renewable distributed generators (DGs), loads, hydrogen vehicles (HVs), and a hydrogen storage system (HSS). The LEM employs an iterative clearing mechanism based on the merit order concept, allowing participants to submit offers and bids aligned with their preferences and utility functions. This decentralized approach simplifies the process, maintains privacy, and avoids complexities associated with centralized decision-making. Case studies show that the LEM enhances local RE integration, reduces peak demand, and increases participant utility. [158] presents a peer-to-peer (P2P) framework for electricity–hydrogen trading in multi-microgrids (MMGs). It incorporates essential components like purification subsystems connected to electrolyzers. The framework allows non-purified hydrogen to participate in energy trading through a purification sharing mechanism using multi-material hydrogen pipes. Nash bargaining theory optimizes the trading process, ensuring profitability for all MGs. A novel self-adaptive mechanism accelerates convergence, reducing iterations by 50% and eliminating fluctuations. Numerical case studies demonstrate a 5.50% reduction in overall operating costs.

EHs and multi-energy systems. [159] proposes a probabilistic approach for coordinating energy management in EHs within a day-ahead market (DAM). The approach integrates electric, gas, and heat systems, considering inputs like gas and electricity and outputs like electricity and heat. The system includes CHP units, energy storage, boilers, RERs, and EVs. An objective function maximizes hub profits in the DAM while accounting for uncertainties in loads, power pricing, and RER output. The gray wolf optimization algorithm solves the problem, and simulations confirm the method's efficiency and reliability. [160] explores the integration of multiple-energy carriers in an EH managed by a hub operator to supply power, heating, and cooling to users. The energy storage system incorporates thermal, ice, and hydrogen storage systems to optimize economic benefits. The EH actively participates in electricity, gas, and heat markets, utilizing hydrogen power technology to enhance overall efficiency. A multi-energy DR model provides increased flexibility for energy management. The Latin hypercube sampling (LHS) method addresses uncertainties related to energy source availability and electricity price fluctuations. An enhanced whale optimization algorithm with a local search component achieves the optimization objective. Incorporating hydrogen storage with multi-energy DR reduces operational expenses by 6%.

Virtual power plants (VPP) and DR. [161] presents a method to evaluate the flexibility of VPPs by integrating distributed electricity–hydrogen interactions. The goal is to maximize economic benefits in the peak-regulation market (PRM) and increase RE integration. The study considers DERs like EVs, air-conditioning systems (AC), and electricity–hydrogen coupled hydrogen refueling stations (HRS) as virtual storage units. The combined flexibility is visually represented as a polytope, and the multidimensional flexibility region is determined using the Minkowski sum. A day-ahead peak regulation scheduling model incorporates conditional value at risk (CVaR). The method increases the VPP's power range and net revenue by 13.05% and 15.11%, respectively. [162] investigates competition in a retail energy market that includes an integrated DR (IDR) program aimed at reducing prices for prosumers and increasing retailer profits. Prosumers, equipped with EHs containing boilers and CHP systems, seek to minimize costs, while retailers aim to maximize profits. This interaction forms a multi-leader-follower game modeled as a bi-level program. The upper level focuses on retailer profit maximization, and the lower level minimizes prosumer costs. Mathematical programming with equilibrium constraints (MPEC) models retailer strategic behavior, and solving all MPECs simultaneously results in an equilibrium problem with equilibrium constraints (EPEC). Increased competition among retailers lowers prosumer costs, and the IDR program reduces prosumer expenses while boosting retailer earnings.

Market participation and strategic bidding. [163] proposes a method to enhance large-scale green hydrogen production by integrating wind farms with electrolysis devices, forming a wind–electrolysis joint system (WEJS). The system leverages the flexibility of the electrolysis system to participate in secondary frequency regulation (SFR), reducing hydrogen production costs. A cooperative power allocation strategy allows WTs and electrolyzers to respond to automatic generation control (AGC) signals, maximizing capacity utilization and increasing full-load operating hours. A robust optimization method for bidding in energy and regulation markets ensures adherence to contractual hydrogen production obligations. Case studies show that the method reduces the relative error of daily hydrogen production to 3.37% and increases electrolysis system efficiency by 50.21% compared to participating only in the energy market. [164] introduces a bi-level strategic bidding model for a profit-oriented power-to-hydrogen and methane (PtHM) plant that integrates electricity, NG, and hydrogen markets. The upper-level model maximizes the plant's profit by optimizing day-ahead bids for energy prices and quantities across markets while considering operational constraints. The lower-level model addresses market behaviors using optimal power and gas flows (OPGF) for electricity and NG markets and a Cournot competition model for the hydrogen market. The bi-level optimization problem is transformed into an MPEC and linearized using strong-duality theory. [165] proposes a strategy to enhance the market competitiveness of renewable energy generation companies (REGCs) by combining hydrogen energy storage with RESs, forming a hydrogen–wind–photovoltaic (HWP) system. The system aims to mitigate risks associated with RE output variability. It participates in both energy and flexible ramping markets, supplying flexible ramping products (FRPs) to maintain grid stability. A bi-level bidding strategy model allows the HWP system to optimize energy production and bidding strategies based on market conditions. Engaging in both markets increases total revenue by 12.22% compared to participating in a single market.

Renewable energy integration and hydrogen markets. [166] examines the capabilities of integrated power-to-hydrogen systems (IP2HS) as long-term energy storage solutions that enhance RE integration. A long-duration robust optimization (LDRO) framework optimizes energy arbitrage opportunities in power systems and H_2 supply networks. The framework includes a capacity bidding model for FCs and electrolyzers in ancillary service markets and uses a hybrid H_2 demand model based on the Cournot competition model. A direct approach combining column-and-constraint generation algorithms and strong duality theory addresses uncertainties in market prices. Case studies demonstrate the framework's effectiveness in managing price uncertainties and leveraging hybrid H_2 demand for energy arbitrage. [167] develops a model for a regional electricity–hydrogen market to manage uncertainties arising from increased RE penetration. Integrating hydrogen storage systems (HSS) with electrical systems stabilizes renewable generation and enhances energy

TABLE 6.6

Applications of PtX Systems in the Energy Market Sector

Application	PtX Technology	Role of PtX	Applications/Use Cases	Benefits	Challenges	Future Trends/ Opportunities
Energy trading and market integration	PtX with digital platforms	Enable trading of PtX products in energy markets	Trading renewable electricity, hydrogen, e-fuels	Market efficiency, price discovery, increased liquidity	Regulatory hurdles, standardization of products	Development of PtX exchanges, blockchain technology
Price arbitrage and peak shaving	PtG, PtHeat	Utilize PtX for energy storage to exploit price differences	Storing energy when prices are low, supplying when high	Revenue generation, grid balancing, optimized asset utilization	Energy conversion losses, market volatility	Advanced forecasting, dynamic pricing models
Renewable energy certificates (REC) and guarantees of origin	PtX certification	Provide traceability and certification for PtX products	Issuing RECs for green hydrogen, e-fuels	Transparency, consumer confidence, premium pricing	Certification costs, verification processes	Blockchain for certification, international standards
Carbon trading and emission reduction credits	PtX with carbon accounting	Generate carbon credits through PtX projects	Selling credits from CO_2 reduction via PtX applications	Additional revenue streams, compliance with emission targets	Verification complexity, market price fluctuations	Integration with carbon markets, supportive policies
Energy derivatives and financial instruments	PtX-linked financial products	Hedge against energy price risks, invest in PtX growth	Futures, options, swaps based on hydrogen or e-fuel prices	Risk management, investment opportunities	Market maturity, liquidity concerns	Development of PtX financial markets, institutional participation
DR programs	PtX-enabled flexibility	Participate in demand response to balance supply and demand	Adjusting PtX operations based on market signals	Revenue from grid services, reduced energy costs	Complexity in participation, regulatory requirements	Aggregation platforms, real-time pricing
VPP participation	PtX with digitalization	Aggregate PtX assets for market services	Selling aggregated PtX capacity into energy markets	Enhanced revenue, grid services provision	Coordination complexity, technology integration	AI-driven VPPs, regulatory support
P2P energy trading	PtX with blockchain	Facilitate direct trading between producers and consumers	Trading excess PtX-produced energy locally or regionally	Empowered consumers, efficient energy distribution	Regulatory barriers, need for secure platforms	Decentralized energy markets, technological advancements
Integration with ancillary services markets	PtX for grid services	Provide services like frequency regulation, voltage support	Rapid response using PtX storage solutions	Additional revenue, improved grid stability	Market access barriers, technology readiness	Expanded ancillary markets, supportive regulations

(Continued)

TABLE 6.6 (Continued)

Capacity markets participation	PtX assets as capacity providers	Commit to provide capacity during peak demand periods	Registering PtX plants in capacity markets	Revenue certainty, support grid reliability	Long-term commitments, performance penalties	Policy support, increasing capacity market roles
Green hydrogen and e-fuel certification schemes	PtX certification	Establish standards and certifications for green PtX products	Certification labels for green hydrogen, e-methane	Market differentiation, consumer trust, potential for higher prices	Certification costs, need for international alignment	Development of global certification schemes, regulatory mandates
Investment and financing of PtX projects	PtX project development	Attract investment through market mechanisms	Project financing via green bonds, crowdfunding	Access to capital, investor diversification	Financial risks, market uncertainties	Innovative financing models, government incentives
Integration with renewable energy auctions	PtX in procurement processes	Include PtX in RE procurement strategies	Bidding PtX projects in auctions for long-term contracts	Revenue stability, market visibility	Competitive pricing pressures, project execution risks	Inclusion of PtX in auction designs, policy support
Dynamic pricing models for PtX products	PtX with digital platforms	Adjust prices in real time based on market conditions	Time-of-use pricing for hydrogen, e-fuels	Optimal resource allocation, incentivize off-peak usage	Consumer acceptance, technology infrastructure	Smart contracts, advanced metering systems
Energy-as-a-service models	PtX service offerings	Offer PtX products as a service rather than commodity	Subscription models for hydrogen supply, heating services	Predictable revenue, customer retention	Service delivery complexities, up-front investment	Growth of as-a-service markets, bundling with other services
PtX integration in international energy trade	PtX for export/import	Facilitate cross-border trading of PtX products	Import/export of hydrogen, e-fuels between countries	Diversify energy supply, geopolitical benefits	Infrastructure needs, trade agreements	Development of global PtX trade networks, standardization
Development of PtX spot and futures markets	PtX-linked financial markets	Create liquid markets for immediate and future delivery	Spot trading of hydrogen, futures contracts for e-fuels	Price transparency, hedging opportunities	Market liquidity, regulatory oversight	Establishment of PtX exchanges, financial market participation
PtX aggregation platforms	PtX with digital platforms	Aggregate small-scale PtX producers for market access	Platforms connecting producers with buyers	Market access, better pricing for small producers	Platform development costs, competition	Technological innovation, supportive policies
Energy storage monetization	PtX storage solutions	Monetize storage capabilities in energy markets	Selling stored energy during peak prices, ancillary services	Revenue generation, grid services provision	Market rules, competition from other storage technologies	Expansion of storage markets, regulatory support

efficiency. The market structure includes trading mechanisms among participants, like RE generation companies, HSSs, and power distribution corporations. A two-stage market-clearing model optimizes trading quantities and profits, incorporating the Shapley value for profit allocation. Case studies reveal significant reductions in power deviations for RE companies and notable profits for HSSs and generation companies, indicating that coupling with the hydrogen market is a promising strategy for integrating large-scale RE into power systems. Table 6.6 provides a general overview of the role of PtX systems in the energy market, highlighting their key applications and contributions. However, this overview serves as a high-level summary (i.e., it is general), and specific details or additional roles may vary depending on evolving technologies and market conditions.

6.6   CONCLUSION

The integration of PtX technologies represents a transformative approach to advancing sustainable electricity markets and EHs. This chapter has examined the multifaceted applications and potential of PtX systems across various sectors, underscoring their critical role in the decarbonization of energy systems and the enhancement of energy efficiency. The conceptualization and modeling of EHs provide a comprehensive framework for understanding the dynamic interactions and optimization of multiple-energy carriers. By incorporating components such as energy storage, RESs, conversion technologies, and DSM, EHs can effectively manage energy flows, balance supply and demand, and ensure a reliable energy supply. Power-to-green hydrogen (PtH) and hydrogen-to-X (HtX) technologies are at the forefront of this transition, offering adaptable solutions for converting surplus RE into valuable chemical fuels. These technologies not only address the intermittency of RES but also provide sustainable alternatives for a variety of industrial processes and applications by producing green hydrogen, methane, methanol, and ammonia. Diverse applications of PtX systems within EHs include power supply and generation, CHP systems, residential buildings, transportation, industry, and agriculture. These systems enable the effective utilization of RE, mitigate GHG emissions, and enhance energy resilience across various sectors. For instance, the practical advantages of PtX technologies in terms of operational efficiency and carbon footprint reduction are illustrated by H2-powered vehicles and machinery in the transportation and agriculture sectors.

PtX technologies facilitate the transition to more resilient and flexible energy systems within electricity markets. By optimizing energy usage and stabilizing grid operations, PtX systems address the characteristics of RE-dominated markets, manage high variability, support strong policy-driven decarbonization efforts, and nurture integrated energy markets. The deployment of PtX in ancillary services markets, local energy markets, capacity markets, day-ahead markets, and P2P markets underscores its potential for establishing a sustainable and balanced energy landscape. The comprehensive benefits, issues, and future direction of integrating PtX systems in EHs are summarized in Table 6.7, which highlights how these technologies can enhance energy security, facilitate sector coupling, and contribute to economic development through cross-sectoral synergies and resource sharing. However, as outlined in the table, several challenges, such as high CAPEX, technological immaturity in some areas, and the need for clear regulatory frameworks, still need to be addressed for PtX to achieve its full potential in transforming EHs.

To achieve global decarbonization objectives and guarantee a sustainable energy future, it is vital to continue the development and integration of PtX technologies. Advancing these technologies, scaling up their deployment, and realizing their maximum potential in transforming sustainable electricity markets and EHs will require collaborative efforts among policymakers, industry stakeholders, and researchers. As outlined in Table 6.7, the future direction for PtX systems will involve improving technological efficiency, fostering multi-sectoral coordination, and creating robust market mechanisms to support their adoption. PtX technologies will unquestionably play a critical role in the development of future energy systems through strategic implementation and innovative solutions.

TABLE 6.7

Analysis of Benefits, Issues, and Future Directions for the Integration of PtX Systems in EHs

Category	Benefits	Issues	Future Direction
Energy storage and conversion	• Enables flexible and efficient energy storage solutions across multiple sectors (e.g., power-to-gas for electricity storage, power-to-heat for thermal storage)	• High energy losses during multi-stage conversion processes (e.g., electricity to gas and back to electricity or heat)	• Development of integrated PtX-EHs, focusing on improving the efficiency of multi-stage energy conversions
	• Supports energy balancing and load management within the EH, reducing the need for grid expansion	• Complexity in managing multiple energy flows and storage mediums within a single hub	• Advancement in smart grid and digital solutions for real-time optimization and control of PtX-based EHs
	• Enhances energy security and resilience of EHs by providing multiple-energy vectors (e.g., hydrogen, synthetic fuels)	• Limited availability of infrastructure for storage and transport of PtX-derived energy carriers	• Establishment of regional PtX hubs with integrated infrastructure for energy storage, transport, and distribution
Decarbonization in multi-sector EHs	• Provides decarbonization pathways for interconnected sectors (e.g., industry, transport, power) through RE integration	• High production costs of renewable hydrogen and synthetic fuels when compared to fossil-based counterparts	• Implementation of cost-reduction strategies for PtX technologies (e.g., scaling up production, improving efficiency)
	• Enables sector coupling, where RE from one sector (e.g., power) can be used to decarbonize other sectors (e.g., transport, industry)	• Need for policy and regulatory frameworks to support decarbonization efforts across interconnected sectors	• Introduction of harmonized policy frameworks that support cross-sector decarbonization and incentivize RE use in multiple sectors
	• Optimizes resource utilization by using waste or surplus energy from one sector as input for other sectors (e.g., using excess electricity for hydrogen production and subsequent industrial use)	• Complexity in measuring and reporting carbon savings across sectors	• Development of standardized carbon accounting and reporting frameworks for EHs
Sector coupling and integration	• Facilitates energy exchange and resource sharing between interconnected sectors, enhancing overall system efficiency (e.g., PtHeat, PtG, and gas-to-power)	• High complexity in designing and operating sector-coupled EHs, requiring advanced management systems	• Development of digital platforms and tools (e.g., blockchain, AI) for managing energy flows, transactions, and interactions in sector-coupled EHs
	• Reduces transmission and distribution losses by utilizing locally produced RE within the hub	• Coordination challenges between sectors with different energy demands and operational constraints	• Establishment of EH management entities or consortia to coordinate energy flows and sectoral interactions
	• Improves resilience and reliability of EHs by providing multiple options for energy supply (e.g., hydrogen, synthetic methane, electricity)	• Risk of sectoral interdependencies leading to cascading failures in the case of supply disruptions	• Integration of energy storage and redundancy mechanisms to minimize the impact of supply disruptions in multi-sector EHs

(Continued)

TABLE 6.7 (Continued)

Economic benefits and market potential of EHs	• Creates new market opportunities for PtX-based energy carriers (e.g., hydrogen, synthetic fuels, renewable chemicals) within EHs	• High initial investment costs for establishing PtX-based EH infrastructure	• Introduction of financial support mechanisms (e.g., subsidies, green certificates) and public–private partnerships for PtX-based EH development
	• Promotes local economic development through job creation and local energy production	• Need for long-term financial incentives and policies to support the growth of PtX markets	• Implementation of market-based mechanisms, such as carbon pricing, feed-in tariffs, and green power purchase agreements (PPA) for PtX-EHs
	• Supports participation in new energy markets, such as carbon trading and ancillary services, increasing revenue streams for EHs	• Lack of clear market signals and pricing mechanisms for PtX-based energy products and services	• Establishment of clear market signals and regulations that encourage PtX adoption and participation in EH operations
Technology and innovation for PtX	• Facilitates innovation by enabling the integration of advanced digital technologies (e.g., IoT, AI, digital twins) for energy flow management and optimization	• Low technological maturity of some PtX technologies (e.g., advanced electrolysis, CO_2 conversion processes)	• Increased R&D investment and pilot projects for demonstrating the potential of PtX technologies in EHs
	• Supports the development of integrated energy systems that utilize multiple PtX technologies for cross-sectoral energy optimization	• Need for knowledge transfer and upskilling of the workforce to operate and manage complex PtX-based EH systems	• Development of educational and training programs to build expertise in PtX technology and EH management
Environmental and sustainability benefits of PtX	• Reduces overall emissions and environmental footprint of EHs by utilizing RE and converting CO_2 into valuable products (e.g., synthetic fuels)	• High energy requirements for CO_2 capture and utilization technologies	• Development of low-energy CO_2 capture and utilization technologies, as well as exploring new materials and processes for sustainable PtX systems
	• Supports the circular economy by converting waste streams and emissions into energy carriers or chemicals	• Limited technological maturity of some waste-to-energy PtX processes	• Scaling up and commercialization of circular economy PtX projects (e.g., waste-to-hydrogen, CO_2-to-fuels)
Regulatory and policy considerations	• Encourages investment in EHs by creating stable policy and regulatory environments that support cross-sectoral integration	• Lack of harmonized regulatory frameworks and standards for PtX adoption in EHs	• Establishment of clear and consistent regulations and standards for PtX deployment and cross-sectoral EH development
	• Drives innovation and technology development by setting ambitious climate and energy targets	• Need for long-term policy support and coordination between different sectors and regulatory bodies	• Introduction of long-term policy roadmaps and incentives that encourage cross-sectoral integration and technology adoption for PtX in EHs
Societal and community benefits of PtX	• Supports local energy production, which enhances energy security and contributes to sustainable community development	• Potential social resistance to new infrastructure projects and safety concerns related to hydrogen storage and transport	• Community engagement and education initiatives to raise awareness and demonstrate the benefits of PtX technologies to local communities
	• Provides new opportunities for community participation in energy production and management (e.g., energy cooperatives, community energy projects)	• Need for social acceptance and building public trust in new PtX technologies and infrastructure	• Establishment of community-led energy projects that leverage PtX systems and contribute to local development plans

REFERENCES

[1] Axon CJ, Darton RC (2021) Sustainability and risk – A review of energy security. Sustainable Production and Consumption 27:1195–1204. doi: 10.1016/j.spc.2021.01.018.

[2] Siksnelyte-Butkiene I (2023b) Defining the perception of energy security: An overview. Economies 11:174. doi: 10.3390/economies11070174.

[3] Mohammadi M, Noorollahi Y, Mohammadi-Ivatloo B, Yousefi H (2017) Energy hub: From a model to a concept – A review. Renewable and Sustainable Energy Reviews 80:1512–1527. doi: 10.1016/j.rser.2017.07.030.

[4] Qi H, Yue H, Zhang J, Lo KL (2021) Optimisation of a smart energy hub with integration of combined heat and power, demand side response and energy storage. Energy 234:121268. doi: 10.1016/j.energy.2021.121268.

[5] Eveloy V, Gebreegziabher T (2018) A review of projected power-to-gas deployment scenarios. Energies 11:1824. doi: 10.3390/en11071824.

[6] Burre J, Bongartz D, Brée L, Roh K, Mitsos A (2019) Power-to-X: Between electricity storage, e-production, and demand side management. Chemie Ingenieur Technik 92:74–84. doi: 10.1002/cite.201900102.

[7] Hayati MM, Safari A, Nazari-Heris M, Oshnoei A (2024) Hydrogen-incorporated sector coupled smart grids: A systematic review & future concepts. Green Hydrogen in Power Systems. doi: 10.1007/978-3-031-52429-5_2.

[8] Acar C, Beskese A, Temur GT (2022) Comparative fuel cell sustainability assessment with a novel approach. International Journal of Hydrogen Energy 47:575–594. doi: 10.1016/j.ijhydene.2021.10.034.

[9] Mounaïm-Rousselle C, Valera-Medina A, Amer-Hatem F, Azad AK, Dedoussi IC, De Joannon M, Fernandes RX, Glarborg P, Hashemi H, He X, Mashruk S, McGowan J (2021) Review on ammonia as a potential fuel: From synthesis to economics. Energy & Fuels 35:6964–7029. doi: 10.1021/acs.energyfuels.0c03685.

[10] Sterner M, Specht M (2021) Power-to-Gas and Power-to-X—The history and results of developing a new storage concept. Energies 14:6594. doi: 10.3390/en14206594.

[11] Khani H, Sawas A, Farag HEZ (2021) An estimation-based optimal scheduling model for settable renewable penetration level in energy hubs. Electric Power Systems Research 196:107230. doi: 10.1016/j.epsr.2021.107230.

[12] Bjarghov S, Loschenbrand M, Saif AUNI, Pedrero RA, Pfeiffer C, Khadem SK, Rabelhofer M, Revheim F, Farahmand H (2021) Developments and challenges in local electricity markets: A comprehensive review. IEEE Access 9:58910–58943. doi: 10.1109/access.2021.3071830.

[13] Cui S, Wang Y-W, Xiao J-W (2019) Peer-to-peer energy sharing among smart energy buildings by distributed transaction. IEEE Transactions on Smart Grid 10:6491–6501. doi: 10.1109/tsg.2019.2906059.

[14] Rong A, Lahdelma R (2016) Role of polygeneration in sustainable energy system development challenges and opportunities from optimization viewpoints. Renewable and Sustainable Energy Reviews 53:363–372. doi: 10.1016/j.rser.2015.08.060.

[15] Wu DW, Wang RZ (2006) Combined cooling, heating and power: A review. Progress in Energy and Combustion Science 32:459–495. doi: 10.1016/j.pecs.2006.02.001.

[16] Ameri M, Besharati Z (2016) Optimal design and operation of district heating and cooling networks with CCHP systems in a residential complex. Energy and Buildings 110:135–148. doi: 10.1016/j.enbuild.2015.10.050.

[17] Fan L, Tu Z, Chan SH (2021) Recent development of hydrogen and fuel cell technologies: A review. Energy Reports 7:8421–8446. doi: 10.1016/j.egyr.2021.08.003.

[18] Sharaf OZ, Orhan MF (2014) An overview of fuel cell technology: Fundamentals and applications. Renewable and Sustainable Energy Reviews 32:810–853. doi: 10.1016/j.rser.2014.01.012.

[19] Mekhilef S, Saidur R, Safari A (2012) Comparative study of different fuel cell technologies. Renewable and Sustainable Energy Reviews 16:981–989. doi: 10.1016/j.rser.2011.09.020.

[20] Buonomano A, Calise F, D'Accadia MD, Palombo A, Vicidomini M (2015) Hybrid solid oxide fuel cells–gas turbine systems for combined heat and power: A review. Applied Energy 156:32–85. doi: 10.1016/j.apenergy.2015.06.027.

[21] Sharaf OZ, Orhan MF (2014) An overview of fuel cell technology: Fundamentals and applications. Renewable and Sustainable Energy Reviews 32:810–853. doi: 10.1016/j.rser.2014.01.012.

[22] Mekhilef S, Saidur R, Safari A (2012) Comparative study of different fuel cell technologies. Renewable and Sustainable Energy Reviews 16:981–989. doi: 10.1016/j.rser.2011.09.020.

[23] Manoharan Y, Hosseini SE, Butler B, Alzhahrani H, Senior BTF, Ashuri T, Krohn J (2019) Hydrogen fuel cell vehicles: Current status and future prospect. Applied Sciences 9:2296. doi: 10.3390/app9112296.

[24] Palizban O, Kauhaniemi K (2016) Energy storage systems in modern grids—Matrix of technologies and applications. Journal of Energy Storage 6:248–259. doi: 10.1016/j.est.2016.02.001.

[25] Weitemeyer S, Kleinhans D, Vogt T, Agert C (2015) Integration of renewable energy sources in future power systems: The role of storage. Renewable Energy 75:14–20. doi: 10.1016/j.renene.2014.09.028.

[26] Del Granado PC, Pang Z, Wallace SW (2016) Synergy of smart grids and hybrid distributed generation on the value of energy storage. Applied Energy 170:476–488. doi: 10.1016/j.apenergy.2016.01.095.

[27] Koohi-Fayegh S, Rosen MA (2020) A review of energy storage types, applications and recent developments. Journal of Energy Storage 27:101047. doi: 10.1016/j.est.2019.101047.

[28] Chen H, Cong TN, Yang W, Tan C, Li Y, Ding Y (2009) Progress in electrical energy storage system: A critical review. Progress in Natural Science Materials International 19:291–312. doi: 10.1016/j.pnsc.2008.07.014.

[29] Electrical Energy Storage. IEC. www.iec.ch/basecamp/electrical-energy-storage.

[30] Sabihuddin S, Kiprakis A, Mueller M (2014) A numerical and graphical review of energy storage technologies. Energies 8:172–216. doi: 10.3390/en8010172.

[31] Meishner F, Sauer DU (2019b) Wayside energy recovery systems in DC urban railway grids. eTransportation 1:100001. doi: 10.1016/j.etran.

[32] Ho WS, Macchietto S, Lim JS, Hashim H, Muis ZA, Liu WH (2016) Optimal scheduling of energy storage for renewable energy distributed energy generation system. Renewable and Sustainable Energy Reviews 58:1100–1107. doi: 10.1016/j.rser.2015.12.097.

[33] Suganthi L, Samuel AA (2012) Energy models for demand forecasting: A review. Renewable and Sustainable Energy Reviews 16:1223–1240. doi: 10.1016/j.rser.2011.08.014.

[34] Meyabadi AF, Deihimi MH (2017) A review of demand-side management: Reconsidering theoretical framework. Renewable and Sustainable Energy Reviews 80:367–379. doi: 10.1016/j.rser.2017.05.207.

[35] Panda S, Mohanty S, Rout PK, Sahu BK, Parida SM, Samanta IS, Bajaj M, Piecha M, Blazek V, Prokop L (2023) A comprehensive review on demand side management and market design for renewable energy support and integration. Energy Reports 10:2228–2250. doi: 10.1016/j.egyr.2023.09.049.

[36] Bakare MS, Abdulkarim A, Zeeshan M, Shuaibu AN (2023) A comprehensive overview on demand side energy management towards smart grids: Challenges, solutions, and future direction. Energy Informatics 6. doi: 10.1186/s42162-023-00262-7.

[37] Malehmirchegini L, Farzaneh H (2023) Incentive-based demand response modeling in a day-ahead wholesale electricity market in Japan, considering the impact of customer satisfaction on social welfare and profitability. Sustainable Energy Grids and Networks 34:101044. doi: 10.1016/j.segan.2023.101044.

[38] Alasseri R, Tripathi A, Rao TJ, Sreckanth KJ (2017) A review on implementation strategies for demand side management (DSM) in Kuwait through incentive-based demand response programs. Renewable and Sustainable Energy Reviews 77:617–635. doi: 10.1016/j.rser.2017.04.023.

[39] Yan X, Ozturk Y, Hu Z, Song Y (2018b) A review on price-driven residential demand response. Renewable and Sustainable Energy Reviews 96:411–419. doi: 10.1016/j.rser.2018.08.003.

[40] Sharma B, Gupta N, Niazi KR, Swarnkar A (2022) Estimating impact of price-based demand response in contemporary distribution systems. International Journal of Electrical Power & Energy Systems 135:107549. doi: 10.1016/j.ijepes.2021.107549.

[41] Li C, Baek J-B (2021) The promise of hydrogen production from alkaline anion exchange membrane electrolyzers. Nano Energy 87:106162. doi: 10.1016/j.nanoen.2021.106162.

[42] Chi J, Yu H (2018) Water electrolysis based on renewable energy for hydrogen production. Chinese Journal of Catalysis 39:390–394. doi: 10.1016/s1872-2067(17)62949-8.

[43] Dincer I, Acar C (2015) Review and evaluation of hydrogen production methods for better sustainability. International Journal of Hydrogen Energy 40:11094–11111. doi: 10.1016/j.ijhydene.2014.12.035.

[44] Simanullang M, Prost L (2022) Nanomaterials for on-board solid-state hydrogen storage applications. International Journal of Hydrogen Energy 47:29808–29846. doi: 10.1016/j.ijhydene.2022.06.301.

[45] Usman MR (2022) Hydrogen storage methods: Review and current status. Renewable and Sustainable Energy Reviews 167:112743. doi: 10.1016/j.rser.2022.112743.

[46] Barthelemy H, Weber M, Barbier F (2017) Hydrogen storage: Recent improvements and industrial perspectives. International Journal of Hydrogen Energy 42:7254–7262. doi: 10.1016/j.ijhydene.2016.03.178.

[47] Lipman TE, Elke M, Lidicker J (2018) Hydrogen fuel cell electric vehicle performance and user-response assessment: Results of an extended driver study. International Journal of Hydrogen Energy 43:12442–12454. doi: 10.1016/j.ijhydene.2018.04.172.

[48] Zivar D, Kumar S, Foroozesh J (2021) Underground hydrogen storage: A comprehensive review. International Journal of Hydrogen Energy 46:23436–23462. doi: 10.1016/j.ijhydene.2020.08.138.

[49] Yanxing Z, Maoqiong G, Yuan Z, Xueqiang D, Jun S (2019) Thermodynamics analysis of hydrogen storage based on compressed gaseous hydrogen, liquid hydrogen and cryo-compressed hydrogen. International Journal of Hydrogen Energy 44:16833–16840. doi: 10.1016/j.ijhydene.2019.04.207.

[50] Luo Q, Li J, Li B, Liu B, Shao H, Li Q (2019) Kinetics in Mg-based hydrogen storage materials: Enhancement and mechanism. Journal of Magnesium and Alloys 7:58–71. doi: 10.1016/j.jma.2018. 12.001.

[51] Chu C, Wu K, Luo B, Cao Q, Zhang H (2023b) Hydrogen storage by liquid organic hydrogen carriers: Catalyst, renewable carrier, and technology – A review. Carbon Resources Conversion 6:334–351. doi: 10.1016/j.crcon.2023.03.007.

[52] Zou YQ, Von Wolff N, Anaby A, Xie Y, Milstein D (2019) Ethylene glycol as an efficient and reversible liquid-organic hydrogen carrier. Nature Catalysis 2:415–422. doi: 10.1038/s41929-019-0265-z.

[53] Huynh HL, Yu Z (2020) CO_2 methanation on hydrotalcite-derived catalysts and structured reactors: A review. Energy Technology 8. doi: 10.1002/ente.201901475.

[54] Götz M, Lefebvre J, Mörs F, Koch AM, Graf F, Bajohr S, Reimert R, Kolb T (2016) Renewable power-to-gas: A technological and economic review. Renewable Energy 85:1371–1390. doi: 10.1016/j. renene.2015.07.066.

[55] Castro-Dominguez B, Mardilovich I, Ma L-C, Ma R, Dixon AG, Kazantzis NK, Ma YH (2016) Integration of methane steam reforming and water gas shift reaction in a Pd/Au/Pd-based catalytic membrane reactor for process intensification. Membranes 6:44. doi: 10.3390/membranes6030044.

[56] Nestler F, Schütze AR, Ouda M, Hadrich MJ, Schaadt A, Bajohr S, Kolb T (2020) Kinetic modelling of methanol synthesis over commercial catalysts: A critical assessment. Chemical Engineering Journal 394:124881. doi: 10.1016/j.cej.2020.124881.

[57] Demirhan CD, Tso WW, Powell JB, Heuberger CF, Pistikopoulos EN (2020) A multiscale energy systems engineering approach for renewable power generation and storage optimization. Industrial & Engineering Chemistry Research 59:7706–7721. doi: 10.1021/acs.iecr.0c00436.

[58] Bozzano G, Manenti F (2016) Efficient methanol synthesis: Perspectives, technologies and optimization strategies. Progress in Energy and Combustion Science 56:71–105. doi: 10.1016/j.pecs.2016.06.001.

[59] Sarvestani ME, Norouzi O, Di Maria F, Dutta A (2024) From catalyst development to reactor design: A comprehensive review of methanol synthesis techniques. Energy Conversion and Management 302:118070. doi: 10.1016/j.enconman.2024.118070.

[60] Dang S, Yang H, Gao P, Wang H, Li X, Wei W, Sun Y (2019) A review of research progress on heterogeneous catalysts for methanol synthesis from carbon dioxide hydrogenation. Catalysis Today 330:61–75. doi: 10.1016/j.cattod.2018.04.021.

[61] Guil-López R, Mota N, Llorente J, Millán E, Pawelec B, Fierro JL, Navarro RM (2019) Methanol synthesis from CO_2: A review of the latest developments in heterogeneous catalysis. Materials 12:3902. doi: 10.3390/ma12233902.

[62] Valera-Medina A, Xiao H, Owen-Jones M, David WI, Bowen PJ (2018) Ammonia for power. Progress in Energy and Combustion Science 69:63–102. doi: 10.1016/j.pecs.2018.07.001.

[63] Smith C, Torrente-Murciano L (2021) Exceeding single-pass equilibrium with integrated absorption separation for ammonia synthesis using renewable energy—Redefining the Haber-Bosch loop. Advanced Energy Materials 11. doi: 10.1002/aenm.202003845.

[64] Wang X, Huang J, Xu Z, Zhang C, Guan X (2024) Real-world scale deployment of hydrogen-integrated microgrid: Design and control. IEEE Transactions on Sustainable Energy 1–13. doi: 10.1109/ tste.2024.3418494.

[65] Sun X, Cao X, Zeng B, Zhai Q, Guan X (2023) Multistage dynamic planning of integrated hydrogen-electrical microgrids under multiscale uncertainties. IEEE Transactions on Smart Grid 14:3482–3498. doi: 10.1109/tsg.2022.3232545.

[66] Luta DN, Raji AK (2018) Decision-making between a grid extension and a rural renewable off-grid system with hydrogen generation. International Journal of Hydrogen Energy 43:9535–9548. doi: 10.1016/j. ijhydene.2018.04.032.

[67] Zhao P, Gu C, Hu Z, Xie D, Hernando-Gil I, Shen Y (2021) Distributionally robust hydrogen optimization with ensured security and multi-energy couplings. IEEE Transactions on Power Systems 36:504–513. doi: 10.1109/tpwrs.2020.3005991.

[68] Zhao P, Lu X, Cao Z, Gu C, Ai Q, Liu H, Bian Y, Li S (2021b) Volt-VAR-pressure optimization of integrated energy systems with hydrogen injection. IEEE Transactions on Power Systems 36:2403–2415. doi: 10.1109/tpwrs.2020.3028530.

[69] Xu F, Li X, Jin C (2024) Optimal capacity configuration and dynamic pricing strategy of a shared hybrid hydrogen energy storage system for integrated energy system alliance: A bi-level programming approach. International Journal of Hydrogen Energy 69:331–346. doi: 10.1016/j.ijhydene.2024.05.011.

[70] Alabbadi AA, AlZahrani AA (2024) Nuclear hydrogen production using PEM electrolysis integrated with APR1400 power plant. International Journal of Hydrogen Energy 60:241–260. doi: 10.1016/j.ijhydene.2024.02.133.

[71] Zeng G, Liu M, Lei Z, Zhang S, Chen Z (2024) Optimal configuration of hydrogen energy storage in an integrated energy system considering variable hydrogen production efficiency. Journal of Energy Storage 98:113044. doi: 10.1016/j.est.2024.113044.

[72] Liang H, Pirouzi S (2024) Energy management system based on economic flexi-reliable operation for the smart distribution network including integrated energy system of hydrogen storage and renewable sources. Energy 293:130745. doi: 10.1016/j.energy.2024.130745.

[73] Barbosa R, Sanchez VM, Escobar B, Cruz JC, Toral-Cruz H (2015) Sizing of a solar-hydrogen power source for a portable emergency communication system: Case study of hurricanes in Cancun, Mexico. International Journal of Hydrogen Energy 40:17361–17370. doi: 10.1016/j.ijhydene.2015.09.083.

[74] Zhou Z, Fu Z, Zhang L, Yu S, Zhao D, Fan J, Chen Q (2024) Multi-objective optimization for low hydrogen consumption and long useful life in fuel cell emergency power supply systems. International Journal of Hydrogen Energy 68:297–310. doi: 10.1016/j.ijhydene.2024.04.233.

[75] Xie Y, Cui Y, Wu D, Zeng Y, Sun L (2021) Economic analysis of hydrogen-powered data center. International Journal of Hydrogen Energy 46:27841–27850. doi: 10.1016/j.ijhydene.2021.06.048.

[76] Long X, Li Y, Li Y, Ge L, Gooi HB, Chung C, Zeng Z (2023) Collaborative response of data center coupled with hydrogen storage system for renewable energy absorption. IEEE Transactions on Sustainable Energy 1–15. doi: 10.1109/tste.2023.3321591.

[77] Wang S, Li T, Wang S, Pan P, Sun R, Zhang N, Ma X (2024) Performance assessments of an integrated system for post-combustion CO_2 capture and NH_4HCO_3 production in a biomass power plant based on green ammonia. Energy Conversion and Management 315:118752. doi: 10.1016/j.enconman.2024.118752.

[78] Wen D, Aziz M (2023) Data-driven energy management system for flexible operation of hydrogen/ammonia-based energy hub: A deep reinforcement learning approach. Energy Conversion and Management 291:117323. doi: 10.1016/j.enconman.2023.117323.

[79] Ibagon N, Muñoz P, Correa G (2023) Techno economic analysis tool for the sizing and optimization of an off-grid hydrogen hub. Journal of Energy Storage 73:108787. doi: 10.1016/j.est.2023.108787.

[80] Yi J, Wang J, Wei X (2024) Economic and risk based optimal energy management of a multi-carrier energy system with water electrolyzing and steam methane reform technologies for hydrogen production. International Journal of Hydrogen Energy 81:1034–1044. doi: 10.1016/j.ijhydene.2024.07.340.

[81] Sui Q, Zhang J, Sun L, Liang J, Wei F, Lin X (2024) Optimal scheduling of mobile energy storage capable of variable speed energy transmission. IEEE Transactions on Smart Grid 15:2710–2722. doi: 10.1109/tsg.2023.3329294.

[82] Huang S, Duan L, Wang Q, Huang L, Zhang H (2024) Thermodynamic analysis of a CCHP system with hydrogen production process by methane reforming enhanced by solar-assisted chemical looping adsorption. Applied Thermal Engineering 248:123137. doi: 10.1016/j.applthermaleng.2024.123137.

[83] Liang Z, Liang Y, Luo X, Wang H, Wu W, Chen J, Chen Y (2023) Integration and optimization of methanol-reforming proton exchange membrane fuel cell system for distributed generation with combined cooling, heating and power. Journal of Cleaner Production 411:137342. doi: 10.1016/j.jclepro.2023.137342.

[84] Lin L, Zheng X, Gu J (2022) Optimal dispatching of combined heat and power system considering the power demand elasticity of hydrogen storage active load. IEEE Transactions on Industry Applications 58:2760–2770. doi: 10.1109/tia.2021.3105618.

[85] Meesenburg W, Markussen WB, Ommen T, Elmegaard B (2020) Optimizing control of two-stage ammonia heat pump for fast regulation of power uptake. Applied Energy 271:115126. doi: 10.1016/j.apenergy.2020.115126.

[86] Topal HI, Arabkoohsar A (2024) Enhancing ultra low temperature district heating systems with neighborhood-scale heat pumps and triple-pipe distribution: A techno-economic analysis. Journal of Building Engineering 110316. doi: 10.1016/j.jobe.2024.110316.

[87] Li J, Lin J, Song Y, Xing X, Fu C (2019) Operation optimization of power to hydrogen and heat (P2HH) in ADN coordinated with the district heating network. IEEE Transactions on Sustainable Energy 10:1672–1683. doi: 10.1109/tste.2018.2868827.

[88] Kim S, Kim M, Kim YT, Kwak G, Kim J (2019) Techno-economic evaluation of the integrated polygeneration system of methanol, power and heat production from coke oven gas. Energy Conversion and Management 182:240–250. doi: 10.1016/j.enconman.2018.12.037.

[89] Faisal S, Abbas A, Eladeb A, Agrawal MK, Muhammad T, Ayadi M, Ghachem K, Kolsi L, Wang M, Mustafa A (2024) Innovative modification process of a natural gas power plant using self-sufficient waste heat recovery and flue gas utilization for a CCHP-methanol generation application: A comprehensive multi-variable feasibility study. Process Safety and Environmental Protection 183:801–820. doi: 10.1016/j.psep.2024.01.022.

[90] Rezaei M, Sameti M, Nasiri F (2024) Design optimization of an integrated tri-generation of heat, electricity, and hydrogen powered by biomass for cold climates. International Journal of Thermofluids 22:100618. doi: 10.1016/j.ijft.2024.100618.

[91] Zhao J, Tu Z, Chan SH (2024) Energy management strategy for accurate thermal scheduling of the combined hydrogen, heating, and power system based on PV power supply. International Journal of Hydrogen Energy 78:721–730. doi: 10.1016/j.ijhydene.2024.06.341.

[92] Hou H, Chen Y, Liu P, Xie C, Huang L, Zhang R, Zhang Q (2022) Multisource energy storage system optimal dispatch among electricity hydrogen and heat networks from the energy storage operator prospect. IEEE Transactions on Industry Applications 58:2825–2835. doi: 10.1109/tia.2021.3128499.

[93] Wang Y, Su J, Xue Y, Chang X, Li Z, Sun H (2024) Toward on rolling optimal dispatch strategy considering alert mechanism for antarctic electricity-hydrogen-heat integrated energy system. IEEE Transactions on Sustainable Energy 1–13. doi: 10.1109/tste.2024.3422236.

[94] Luo S, Li Q, Pu Y, Xiao X, Chen W, Liu S, Mao X (2023) A carbon trading approach for heat-power-hydrogen integrated energy systems based on a Vickrey auction strategy. Journal of Energy Storage 72:108613. doi: 10.1016/j.est.2023.108613.

[95] Zhou Y, Peng H, Yan M (2024) Online probabilistic energy flow for hydrogen-power-heat system based on multi-parametric programming. Applied Energy 372:123836. doi: 10.1016/j.apenergy.2024.123836.

[96] Jienkulsawad P, Eamsiri K, Chen Y-S, Arpornwichanop A (2022) Neural network-based adaptive control and energy management system of a direct methanol fuel cell in a hybrid renewable power system. Sustainable Cities and Society 87:104192. doi: 10.1016/j.scs.2022.104192.

[97] Temiz M, Dincer I (2023) Design and assessment of a solar energy based integrated system with hydrogen production and storage for sustainable buildings. International Journal of Hydrogen Energy 48:15817–15830. doi: 10.1016/j.ijhydene.2023.01.082.

[98] Zhou L, Zhou Y (2023) Study on thermo-electric-hydrogen conversion mechanisms and synergistic operation on hydrogen fuel cell and electrochemical battery in energy flexible buildings. Energy Conversion and Management 277:116610. doi: 10.1016/j.enconman.2022.116610.

[99] Yu L, Xu Z, Guan X, Zhao Q, Dou C, Yue D (2023) Joint optimization and learning approach for smart operation of hydrogen-based building energy systems. IEEE Transactions on Smart Grid 14:199–216. doi: 10.1109/tsg.2022.3197657.

[100] Sun Q, Wu Z, Gu W, Zhang XP, Liu P, Pan G, Qiu H (2023) Tri-level multi-energy system planning method for zero energy buildings considering long- and short-term uncertainties. IEEE Transactions on Sustainable Energy 14:339–355. doi: 10.1109/tste.2022.3212168.

[101] Li L, Han Y, Li Q, Chen W (2024) Multi-dimensional economy-durability optimization method for integrated energy and transportation system of net-zero energy buildings. IEEE Transactions on Sustainable Energy 15:146–159. doi: 10.1109/tste.2023.3275160.

[102] Tajalli SAM, Tajalli SZ, Homayounzadeh M, Khooban MH (2023) Zero-carbon power-to-hydrogen integrated residential system over a hybrid cloud framework. IEEE Transactions on Cloud Computing 1–11. doi: 10.1109/tcc.2023.3257995.

[103] Zou W, Li J, Yang Q, Duan Z (2023) An improved max-min game theory control of fuel cell and battery hybrid energy system against system uncertainty. IEEE Journal of Emerging and Selected Topics in Power Electronics 11:78–87. doi: 10.1109/jestpe.2022.3168374.

[104] Liu J, Zhou Y, Yang H, Wu H (2022) Net-zero energy management and optimization of commercial building sectors with hybrid renewable energy systems integrated with energy storage of pumped hydro and hydrogen taxis. Applied Energy 321:119312. doi: 10.1016/j.apenergy.2022.119312.

[105] Li A, Ming T, Xiong H, Wu Y, Shi T, Li W, de Richter R, Chen Y, Tang X, Yuan Y (2023) A high-performance solar chimney in building integrated with photocatalytic technology for atmospheric methane removal. Solar Energy 260:126–136. doi: 10.1016/j.solener.2023.05.035.

[106] Jadidbonab M, Abdeltawab H, Mohamed YA-RI (2024) A hybrid traffic flow forecasting and risk-averse decision strategy for hydrogen-based integrated traffic and power networks. IEEE Systems Journal 1–12. doi: 10.1109/jsyst.2024.3420237.

[107] Zepter JM, Engelhardt J, Marinelli M (2023) Optimal expansion of a multi-domain virtual power plant for green hydrogen production to decarbonise seaborne passenger transportation. Sustainable Energy Grids and Networks 36:101236. doi: 10.1016/j.segan.2023.101236.

[108] Boynuegri AR, Tekgun B (2023) Real-time energy management in an off-grid smart home: Flexible demand side control with electric vehicle and green hydrogen production. International Journal of Hydrogen Energy 48:23146–23155. doi: 10.1016/j.ijhydene.2023.01.239.

[109] Çakmak R, Meral H, Bayrak G (2024) A new intelligent charging strategy in a stationary hydrogen energy-based power plant for optimal demand side management of plug-in EVs. International Journal of Hydrogen Energy 75:400–414. doi: 10.1016/j.ijhydene.2024.02.132.

[110] Katalenich SM, Jacobson MZ (2022) Toward battery electric and hydrogen fuel cell military vehicles for land, air, and sea. Energy 254:124355. doi: 10.1016/j.energy.2022.124355.

[111] Xu X, Hu W, Liu W, Wang D, Huang Q, Huang R, Chen Z (2021) Risk-based scheduling of an off-grid hybrid electricity/hydrogen/gas/refueling station powered by renewable energy. Journal of Cleaner Production 315:128155. doi: 10.1016/j.jclepro.2021.128155.

[112] Wang Z, Qi Y, Sun Q, Lin Z, Xu X (2024) Ammonia combustion using hydrogen jet ignition (AHJI) in internal combustion engines. Energy 291:130407. doi: 10.1016/j.energy.2024.130407.

[113] Zou Y, Xu G, An Y, Zhang M, Sun Y, Liu Z, Yu Y, He H (2024) Modified $LaMnO_3$ perovskite oxide as NOx storage and reduction catalyst for emission control of hydrogen internal combustion engines. Fuel 375:132500. doi: 10.1016/j.fuel.2024.132500.

[114] Oh S, Ko K-H, Kim J (2024) Development of thermoelectric exhaust energy recovery system of a hydrogen internal combustion engine in a city bus using a three-dimensional multiphysics model. Energy Conversion and Management 300:118006. doi: 10.1016/j.enconman.2023.118006.

[115] Lim T-W, Hwang D-H, Choi Y-S (2024) Design and optimization of a steam methane reformer for ship-based hydrogen production on LNG-fueled ship. Applied Thermal Engineering 243:122588. doi: 10.1016/j.applthermaleng.2024.122588.

[116] Prewitz M, Schwärzer J, Bardenhagen A (2023) Potential analysis of hydrogen storage systems in aircraft design. International Journal of Hydrogen Energy 48:25538–25548. doi: 10.1016/j.ijhydene.2023.03.266.

[117] Wang H, Du M, Wang Q, Li Z, Wang S, Gao Z, Derksen J (2023) Enhancement of hydrogen storage performance in shell and tube metal hydride tank for fuel cell electric forklift. International Journal of Hydrogen Energy 48:23568–23580. doi: 10.1016/j.ijhydene.2023.03.067.

[118] Xin M, Gan H, Cong Y, Wang H (2024) Numerical simulation of methane slip from marine dual-fuel engine based on hydrogen-blended natural gas strategy. Fuel 358:130132. doi: 10.1016/j.fuel.2023.130132.

[119] Tian Y, Han J, Bu Y, Qin C (2023) Simulation and analysis of fire and pressure reducing valve damage in on-board liquid hydrogen system of heavy-duty fuel cell trucks. Energy 276:127572. doi: 10.1016/j.energy.2023.127572.

[120] Mousavi SB, Ahmadi P, Raeesi M (2024) Performance evaluation of a hybrid hydrogen fuel cell/battery bus with fuel cell degradation and battery aging. Renewable Energy 227:120456. doi: 10.1016/j.renene.2024.120456.

[121] Er G, Soykan G, Canakoglu E (2024) Stochastic optimal design of a rural microgrid with hybrid storage system including hydrogen and electric cars using vehicle-to-grid technology. Journal of Energy Storage 75:109747. doi: 10.1016/j.est.2023.109747.

[122] Shoja ZM, Mirzaei MA, Seyedi H, Zare K (2022) Sustainable energy supply of electric vehicle charging parks and hydrogen refueling stations integrated in local energy systems under a risk-averse optimization strategy. Journal of Energy Storage 55:105633. doi: 10.1016/j.est.2022.105633.

[123] Mi Y, Cai P, Fu Y, Wang P, Lin S (2022) Energy cooperation for wind farm and hydrogen refueling stations: A RO-based and nash-Harsanyi bargaining solution. IEEE Transactions on Industry Applications 58:6768–6779. doi: 10.1109/tia.2022.3188233.

[124] Liao C, Tang Y, Liu Y, Sun Z, Li W, Ma X (2023) Life cycle assessment of the solid oxide fuel cell vehicles using ammonia fuel. Journal of Environmental Chemical Engineering 11:110872. doi: 10.1016/j.jece.2023.110872.

[125] Nasir M, Jordehi AR, Tostado-Véliz M, Mansouri SA, Sanseverino ER, Marzband M (2023) Two-stage stochastic-based scheduling of multi-energy microgrids with electric and hydrogen vehicles charging stations, considering transactions through pool market and bilateral contracts. International Journal of Hydrogen Energy 48:23459–23497. doi: 10.1016/j.ijhydene.2023.03.003.

[126] Alsunousi M, Kayabasi E (2024) Techno-economic assessment of a floating photovoltaic power plant assisted methanol production by hydrogenation of CO_2 captured from Zawiya oil refinery. International Journal of Hydrogen Energy 57:589–600. doi: 10.1016/j.ijhydene.2024.01.055.

[127] Weng Q, Toan S, Ai R, Sun Z, Sun Z (2021) Ammonia production from biomass via a chemical looping–based hybrid system. Journal of Cleaner Production 289:125749. doi: 10.1016/j.jclepro.2020.125749.

[128] Li C, Wang M, Li N, Gu D, Yan C, Yuan D, Jiang H, Wang B, Wang X (2024) Solar oil refinery: Solar-driven hybrid chemical cracking of residual oil towards efficiently upgrading fuel and abundantly generating hydrogen. Energy Conversion and Management 300:117900. doi: 10.1016/j.enconman.2023.117900.

[129] Da Silva GN, Rochedo PRR, Szklo A (2022) Renewable hydrogen production to deal with wind power surpluses and mitigate carbon dioxide emissions from oil refineries. Applied Energy 311:118631. doi: 10.1016/j.apenergy.2022.118631.

[130] Zvirko O, Nykyforchyn H, Krechkovska H, Tsyrulnyk O, Hredil M, Venhryniuk O, Tsybailo I (2024) Evaluating hydrogen embrittlement susceptibility of operated natural gas pipeline steel intended for hydrogen service. Engineering Failure Analysis 163:108472. doi: 10.1016/j.engfailanal.2024.108472.

[131] Tang J, Tang Y, Liu H, Peng S, Sun Z, Liu Y, Deng J, Chen W, Ma X (2024) Renewable methanol production based on in situ synergistic utilization of excavated waste and landfill gas: Life cycle techno-environmental-economic analysis. Energy Conversion and Management 314:118727. doi: 10.1016/j.enconman.2024.118727.

[132] Wu C, Zhu Q, Dou B, Fu Z, Wang J, Mao S (2024) Thermodynamic analysis of a solid oxide electrolysis cell system in thermoneutral mode integrated with industrial waste heat for hydrogen production. Energy 131678. doi: 10.1016/j.energy.2024.131678.

[133] De La Cruz-Soto J, Azkona-Bedia I, Cornejo-Jimenez C, Romero-Castanon T (2024) Assessment of levelized costs for green hydrogen production for the national refineries system in Mexico. International Journal of Hydrogen Energy. doi: 10.1016/j.ijhydene.2024.03.316.

[134] Su G, Zulkifli NWM, Liu L, Ong HC, Ibrahim S, Yu KL, Wei Y, Bin F (2023) Carbon-negative co-production of methanol and activated carbon from bagasse pyrolysis, physical activation, chemical looping, and methanol synthesis. Energy Conversion and Management 293:117481. doi: 10.1016/j.enconman.2023.117481.

[135] Fang J, Yang M, Sui J, Luo T, Yu Y, Ao Y, Dou R, Zhou W, Li W, Liu X, Zhao K (2024) Enhancing solar-powered hydrogen production efficiency by spectral beam splitting and integrated chemical energy storage. Applied Energy 372:123833. doi: 10.1016/j.apenergy.2024.123833.

[136] Zheng K, Yan Y, Sun Y, Yang J, Zhu M, Ni M, Li L (2023) An experimental study of ammonia decomposition rates over cheap metal catalysts for solid oxide fuel cell anode. International Journal of Hydrogen Energy 48:19188–19195. doi: 10.1016/j.ijhydene.2023.01.312.

[137] Karthik E, Sangaletti L, Ferroni M, Alessandri I (2024) Investigation of electrocatalytic oxygen evolution reaction (OER) selectivity against methanol oxidation on stainless steel. Catalysis Science & Technology. doi: 10.1039/d4cy00030g.

[138] Kumar V, Adalati R, Gautam YK, Gautam D (2024) An investigation of glass, ITO, and quartz transparent substrates on Pd/SnO_2 hydrogen sensor structure and sensitivity. Materials Today Communications 109280. doi: 10.1016/j.mtcomm.2024.109280.

[139] Isa MIN, Sohaimy MIH, Ahmad NH (2021) Carboxymethyl cellulose plasticized polymer application as bio-material in solid-state hydrogen ionic cell. International Journal of Hydrogen Energy 46:8030–8039. doi: 10.1016/j.ijhydene.2020.11.274.

[140] Turap Y, Zhang Z, Wang Y, Wang Y, Wang Z, Wang W (2024) Deposition and release behavior of H_2S during chemical looping hydrogen production process. Chemical Engineering Journal 487:150621. doi: 10.1016/j.cej.2024.150621.

[141] Saray JA, Gharehghani A, Hosseinzadeh D (2024) Towards sustainable energy carriers: A solar and wind-based systems for green liquid hydrogen and ammonia production. Energy Conversion and Management 304:118215. doi: 10.1016/j.enconman.2024.118215.

[142] Devkota S, Ban S, Shrestha R, Uprety B (2023) Techno-economic analysis of hydropower based green ammonia plant for urea production in Nepal. International Journal of Hydrogen Energy 48:21933–21945. doi: 10.1016/j.ijhydene.2023.03.087.

[143] Zainul R, Basem A, Jasim DJ, Yadav A, Hasson AR, Logroño JP, Ajaj Y, Muzammil K, Islam S (2024) Predictive analysis of methane-enhanced conversion of organic waste into sustainable fuel: A machine learning approach. Process Safety and Environmental Protection. doi: 10.1016/j.psep.2024.06.129.

[144] Fernandes DJ, Ferreira AF, Fernandes EC (2023) Biogas and biomethane production potential via anaerobic digestion of manure: A case study of Portugal. Renewable and Sustainable Energy Reviews 188:113846. doi: 10.1016/j.rser.2023.113846.

[145] Kerrou O, Lahboubi N, Bakraoui M, Bari HE (2023) Improving methane production from palm leaflets waste with thermal and ultrasound pre-treatment. Physics and Chemistry of the Earth Parts A/B/C 132:103482. doi: 10.1016/j.pce.2023.103482.

[146] Rena N, Zacharia KMB, Yadav S, Machhirake NP, Kim SH, Lee BD, Jeong H, Singh L, Kumar S, Kumar R (2020) Bio-hydrogen and bio-methane potential analysis for production of bio-hythane using various agricultural residues. Bioresource Technology 309:123297. doi: 10.1016/j.biortech.2020.123297.

[147] Jaffar MM, Nahil MA, Williams PT (2020) Pyrolysis-catalytic hydrogenation of cellulose-hemicellulose-lignin and biomass agricultural wastes for synthetic natural gas production. Journal of Analytical and Applied Pyrolysis 145:104753. doi: 10.1016/j.jaap.2019.104753.

[148] Zhu Y, Liu Z, Zhang C, Ming J, Chen G, Yang Y (2022) Light triggers green recovery: Boosted biomethane production from ammonia-stressed anaerobic digestion through optimized illuminated bioreactor. Chemical Engineering Journal 450:138173. doi: 10.1016/j.cej.2022.138173.

[149] Liu Y, Zhao W, Xi Y, Wang S, Liang J, Zeng Y, Dong W, Chen K, Jia H, Wu X (2024) Acyl homoserine lactone-based regulation strategy for improved methane production in anaerobic digestion of agricultural wastes. Applied Energy 358:122621. doi: 10.1016/j.apenergy.2024.122621.

[150] Nadaleti WC, Gomes J, De Souza E, Santos M, Belli P, Borges A, Mohedano R, Libardi N, da Silva FM, Correa E, Vieira B (2024) Biomethane and biohydrogen production from an anaerobic sludge used in the treatment of rice parboiling effluent: Specific methanogenic and hydrogenic activity. International Journal of Hydrogen Energy 60:702–710. doi: 10.1016/j.ijhydene.2024.01.157.

[151] Farhani S, Barhoumi EM, Islam QU, Becha F (2024) Optimal design and economic analysis of a stand-alone integrated solar hydrogen water desalination system case study agriculture farm in Kairouan Tunisia. International Journal of Hydrogen Energy 63:759–766. doi: 10.1016/j.ijhydene.2024.03.043.

[152] Temiz M, Dincer I (2024) Development of concentrated solar and agrivoltaic based system to generate water, food and energy with hydrogen for sustainable agriculture. Applied Energy 358:122539. doi: 10.1016/j.apenergy.2023.122539.

[153] Irfan M, Razzaq A, Chupradit S, Javid M, Rauf A, Farooqi TJ (2022) Hydrogen production potential from agricultural biomass in Punjab province of Pakistan. International Journal of Hydrogen Energy 47:2846–2861. doi: 10.1016/j.ijhydene.2021.10.257.

[154] Nadaleti WC, Martins R, Lourenço V, Przybyla G, Bariccatti R, Souza S, Manzano-Agugliaro F, Sunny N (2021) A pioneering study of biomethane and hydrogen production from the wine industry in Brazil: Pollutant emissions, electricity generation and urban bus fleet supply. International Journal of Hydrogen Energy 46:19180–19201. doi: 10.1016/j.ijhydene.2021.03.044.

[155] Pomdaeng P, Kongthong O, Tseng C-H, Dokmaingam P, Chu CY (2024) An immobilized mixed micro-flora approach to enhancing hydrogen and methane productions from high-strength organic loading food waste hydrolysate in series batch reactors. International Journal of Hydrogen Energy 52:160–169. doi: 10.1016/j.ijhydene.2023.09.187.

[156] Khani H, Farag HEZ (2018) Optimal day-ahead scheduling of power-to-gas energy storage and gas load management in wholesale electricity and gas markets. IEEE Transactions on Sustainable Energy 9:940–951. doi: 10.1109/tste.2017.2767064.

[157] Xiao Y, Wang X, Pinson P, Wang X (2018) A local energy market for electricity and hydrogen. IEEE Transactions on Power Systems 33:3898–3908. doi: 10.1109/tpwrs.2017.2779540.

[158] Bo Y, Xia Y, Wei W, Li Z, Zhou Y (2023) Peer-to-peer electricity-hydrogen energy trading for multi-microgrids based on purification sharing mechanism. International Journal of Electrical Power & Energy Systems 150:109113. doi: 10.1016/j.ijepes.2023.109113.

[159] Gu S, Rao C, Yang S, Liu Z, Rehman AU, Mohamed MA (2023) Day-ahead market model based coordinated multiple energy management in energy hubs. Solar Energy 262:111877. doi: 10.1016/j.solener.2023.111877.

[160] Li Z, He T, Farjam H (2023) Application of an intelligent method for hydrogen-based energy hub in multiple energy markets. International Journal of Hydrogen Energy 48:36485–36499. doi: 10.1016/j.ijhydene.2023.03.124.

[161] Chen S, Zhang K, Liu N, Xie Y (2024) Unlock the aggregated flexibility of electricity-hydrogen integrated virtual power plant for peak-regulation. Applied Energy 360:122747. doi: 10.1016/j.apenergy.2024.122747.

[162] Aghamohammadloo H, Talaeizadeh V, Shahanaghi K, Aghaei J, Shayanfar H, Shafie-khah M, Catalão JP (2021) Integrated demand response programs and energy hubs retail energy market modelling. Energy 234:121239. doi: 10.1016/j.energy.2021.121239.

[163] Cheng X, Lin J, Liu F, Qiu Y, Song Y, Li J, Wu S (2023) A coordinated frequency regulation and bidding method for wind-electrolysis joint systems participating within ancillary services markets. IEEE Transactions on Sustainable Energy 14:1370–1384. doi: 10.1109/tste.2022.3233062.

[164] Pan G, Gu W, Lu Y, Qiu H, Lu S, Yao S (2021) Accurate modeling of a profit-driven power to hydrogen and methane plant toward strategic bidding within multi-type markets. IEEE Transactions on Smart Grid 12:338–349. doi: 10.1109/tsg.2020.3019043.

[165] Gong X, Li X, Zhong Z (2024) Strategic bidding of hydrogen-wind-photovoltaic energy system in integrated energy and flexible ramping markets with renewable energy uncertainty. International Journal of Hydrogen Energy 80:1406–1423. doi: 10.1016/j.ijhydene.2024.07.083.

[166] Gu Z, Pan G, Gu W, Qiu H, Lu S (2024) Robust optimization of scale and revenue for integrated power-to-hydrogen systems within energy, ancillary services, and hydrogen markets. IEEE Transactions on Power Systems 1–15. doi: 10.1109/tpwrs.2023.3323660.

[167] Zhu J, Meng D, Dong X, Fu Z, Yuan Y (2023) An integrated electricity: Hydrogen market design for renewable-rich energy system considering mobile hydrogen storage. Renewable Energy 202:961–972. doi: 10.1016/j.renene.2022.12.015.

7 Application and Advances in Decision Support Tools for Energy Grid Planning and Operation

Nawaf Nazir, Sarmad Hanif, and Adil Khurram

7.1 INTRODUCTION

7.1.1 EVOLUTION OF THE ENERGY GRID

The integration of renewables in the power grid has introduced additional variability and uncertainty. Also, the load is no longer static as a result of active participation of consumers reacting to prices and tariffs and adopting PV and storage technologies, including EVs. Furthermore, power-to-X (PtX) technologies that convert electricity from renewables into other forms of energy or products have gained increased attention in recent years. PtX has emerged as a viable means for storing excess renewable generation for subsequent dispatch for end use or to produce green fuels. Here, excess and underutilized solar and wind generation are used in the production of fuels, such as hydrogen, methane, hydrogen peroxide, and ammonia, among others [1]. As a result, PtX technologies have emerged as a significant driver in progressively pushing the transition toward a fossil-free energy scenario, by increasing the utilization of solar and wind power [2]. Another benefit of PtX technologies is the potential decoupling of power from the electricity sector for use in other sectors, utilizing power produced from renewables during excess generation periods [3]. Recent large-scale investment in PtX technologies shows the potential that it has in achieving strong decarbonization targets [4].

7.1.2 CHALLENGES IN GRID PLANNING AND OPERATION

The changes brought about by the introduction of new power generation and storage technologies require a paradigm shift in how the grid is operated. While renewable energy from solar and wind sources is key to a decarbonized future, not all industries can undergo electrification. Other low-emission energy sources and fuels are needed for sectors such as aviation, maritime shipping, railways, and heavy industries. PtX has the ability to bridge the gap in these sectors by shifting excess renewable generation into the production of low-emission fuels and other energy storage options [5]. This helps not only in the cost-effective production of low-emission fuels but also in the utilization and storage of excess renewable generation, making renewable generation even more economically viable. At the same time, the flexibility provided by PtX in terms of storing excess renewable generation and supplying energy in periods of low renewable generation can further improve the operational efficiency and resiliency of the energy grid. In lieu of all these recent developments, power system planning and operation has become a challenging exercise that requires new methodologies and tools to help utilities. Several reports have suggested that a scenario with large renewable penetration cannot be achieved without grid operators in control rooms equipped

DOI: 10.1201/9781032719436-7

with the necessary decision support tools [6]. For example, the North American Electric Reliability Corporation (NERC) has identified several new advanced decision support tools that are important to managing grids with large amounts of energy from renewables such as wind and solar. Such tools include voltage stability analysis, optimal power flow with wind forecasts, transient stability analysis, optimization-based transmission planning, probabilistic reserve allocation, dynamic security assessment with stochastic model, stochastic power flow, stochastic unit commitment, and price-responsive demand forecast.

With the challenges of aging infrastructure, increased capacity demands, and the impact of more extreme weather events, grid operators need intelligent decision support tools in order to operate at near zero downtime. However, most utilities have limited visibility of their grid. With smart monitoring tools, grid operators can identify locations prone to overload and signs of degradation before outages occur.

Modern power systems provide a plethora of data that can be harnessed for the design, monitoring, and control of the grid. Decision-making under uncertainty, and multi-criteria decision-making, is an important tool to capture the complexity and distributed nature of future electric power grids with a high level of renewable penetration.

The development of smart grid technologies, such as advanced metering infrastructure (AMI) and supervisory control and data acquisition systems (SCADA), needs decision support systems to help in the decision-making process [7].

Software solutions designed to manage challenges posed by the expanded use of intermittent power sources—like wind and solar—are a powerful way to modernize the energy grid. These solutions, called "distributed energy resource management systems," or DERMS, have proven effective in improving electric grid reliability while reducing operational costs.

When it comes to planning for new transmission lines, multi-criteria decision analysis (MCDA) offers a useful approach to determine the optimal path of future transmission lines with minimum impact on the environment, on the landscape, and on affected citizens [8].

The authors in [9] present a decision support system based on integer programming to determine cost-optimal grid reinforcements in order to deal with the stress on distribution grids due to distributed photovoltaic (PV) units. The support system provides guidelines for future grid planning.

Other decision support tools include NREL's Engage energy modeling tool that enables state, local, and tribal governments to visualize their communities' energy futures. It facilitates custom energy system planning, simulation, and data management based on user-supplied datasets.

The electric grid multi-objective decision plan (MOD-plan) helps grid planning frameworks better account for emerging objectives in energy justice and equity, resilience, and decarbonization, alongside traditional grid planning objectives. To do so, planning frameworks require new methods to be incorporated into current practices [10].

7.1.3 NECESSITY OF DECISION SUPPORT TOOLS

The energy grid is currently undergoing rapid transformation due to the increased adoption of renewable energy resources and distributed energy resources, including demand-side resources, such as electric vehicles. Furthermore, in recent years, various PtX technologies have started to be integrated into the energy grid, including power-to-heat and power-to-gas technologies. The variable and uncertain nature of renewable generation, coupled with the flexibility offered by PtX solutions, could lead to progressive loss in our ability to securely manage and operate the energy grid. In the face of these transformations, many conventional tools and models are woefully inadequate to provide rigorous decision support and policy analysis. Present tools lack the appropriate level of model resolution, operational detail, and system constraints to incorporate various PtX technologies within their studies. As a result, the need of the hour is to develop new decision support tools that can deal with the intermittency challenges brought forth by renewable generation, together with managing the flexibility offered by PtX technologies.

7.2   ADVANCES IN DECISION SUPPORT TOOLS

In general, dispatch tools solve either a planning or an operational problem:

- **Operation problems**, which describe the analysis of the system at operation timescale (5 min to 1 hr) with the purpose of providing operation decision support. Examples of this category are unit commitment problems, optimal control/model predictive control problems, and market-clearing problems. For the purpose of this book chapter, we focus on optimization problems, for which a market-clearing problem is an example of how operational decision-making is expressed as an optimization problem.
- **Planning problems**, which characterize design decisions for the energy system, that is, at planning timescale (days to months), with the goal of providing investment decision support. These studies can be addressed regarding simulation-based scenario analysis or optimal planning problems. For the context of this book chapter, the optimization-based planning problem can be thought of as optimal planning, which seeks to determine optimal values for the design decision variables, for example, component sizing and placement.

Theoretically, the optimization tools responsible for operation and planning decision-making determine state and control variables that optimize some objective subject to operational constraints. The complexity of the planning and operational tools relies on the type of mathematical model which governs the optimization algorithm and, consequently, the solution algorithm. Mathematical models for the electric grid are essentially obtained by aggregating the models of its subsystems, that is, generators, transmission systems, distribution systems, and DERs. To this end, complexity management is an essential aspect of power system dispatch tools which is done by managing the model's formulation efforts, model parameter data requirements, and computational limitations. The most important for the electric power grid is the "electrical" model, also known as the power flow model, which is expressed in the dispatch tools as:

- **Balanced AC model** highlights whether steady-state properties, that is, voltage, branch flow, and losses, can be represented for single-phase electric grids.
- **Multi-phase AC model** highlights whether steady-state properties, that is, voltage, branch flow, and losses, can be modeled for multi-phase unbalanced electric grids.
- **Dynamics model** describes the ability to model transient properties of the electric grid, in addition to steady-state properties.

Based on the aforementioned power flow model, the following numerical optimization models have been proposed to solve power system dispatch tools:

- **Local optimization models** are the optimization models used to solve distribution grid problems. Usually, multi-phase ACs are used in these models, along with available DER models. Some examples of these models are the Volt/Var optimization model, feeder reconfiguration, and feeder peak shaving and load shedding. Strictly speaking, these models are targeted toward the low-voltage portion of the grid, with a single node representing 10–100s of customers.
- **Global energy procurement models** are the global supply–demand optimization problem that is solved by an entity to maximize its profit. For example, a market-clearing engine solved by an ISO and a profit maximization load procurement problem solved by the LSE are a few examples of such optimization problems. Strictly speaking, these models are targeted toward the high-voltage portion of the aggregated, with a single node representing 10,000–100,000s of customers.

For the purpose of this chapter, we will focus on the local optimization problem.

7.2.1 Overview of Local Distribution Grid Optimization Problem— Distribution Grid Optimal Power Flow Problem (DOPF)

The theoretical research on the DOPF formulation and solution algorithms is an active research area. An electric model in the DOPF problems either uses AC branch power flow [11, 12], AC nodal power flow formulation [13, 14], or DC nodal power flow model [14, 15], where the non-convexity is simplified through convex relaxation methods [16] and convex restriction methods [17]. In addition, in distribution systems, the scheduling of discrete mechanical devices, such as transformers, switches, and capacitor banks, has been proposed to model as the continuous flexible loads to further simplify the problem [18–20]. As a tool, DOPF has been solved in a research-grade tool called *PowerModels* package [21] and the *PowerModelsDistribution* [22].

The next section of this chapter provides a generalized method for developing DOPF.

7.2.2 Process Grid Model

The grid to be optimized by models needs to be converted to an appropriate vector-and-matrices format. This allows the DOPF developer to utilize packages/libraries which can improve computational performances (e.g., see Python's SciPy library implementation [23]). Based on the input circuit model format and the information available in the files, an appropriate topology reading algorithm may be written to develop such functionality as follows. First, an injection vector determining how much load/generation is injected/withdrawn at an appropriate node is developed. Second, incidence matrices mapping controllable equipment to the appropriate node in the injection vector are determined. Third, a verification model to map the injections to the grid states is developed to verify that the model obtained matches the real-world grid states. This step basically translates the grid information into a form that is similar to power flow models, where admittance translates nodal power injections to voltage calculations. Consider a DOPF algorithm aiming to control load shedding p^{fl}, capacitor banks q^{cb}, tap settings t_{rg}, PV active power shed P_{pv} and PV reactive power injections q_{pv}, and power feed-in from the bulk power system P_0 / q_0. The injection vector to be controlled is then:

$$u = \left\{ p^{\text{fl}}, \mathbf{q}_{\text{cb}}, t_{\text{rg}}, \mathbf{P}_{\text{pv}}, q_{\text{pv}}, P_0, q_0 \right\} \tag{7.1}$$

In this case, x is the set of state variables (like voltages and resultant current magnitude flows), C_x includes the impact of the controllable vector on state variables, and C_u is the set of incidence matrices that map the controllable equipment list onto the appropriate location in the network node list. An example of the incidence matrix for the regulator follows. An incidence matrix C_r links regulator location to voltage changes such that $C_r(i, j) = 1$ if tap regulator j is upstream to node i; otherwise, $C_r(i, j) = 0$. The grid processing steps attempt to formulate the following equation to verify that the grid model is appropriate:

$$f\left(u, x, C_x, C_u\right) = 0 \tag{7.2}$$

Different from the power flow problems, the DOPF requires non-equality constraints to be formulated too, such as voltage limitations, current flow limitations, and controllable variables movement limitations. These constraints can be compacted in a similar manner to power flow model[1] as:

$$q\left(u, x, C_x, C_u\right) \leq 0 \tag{7.3}$$

7.2.3 Obtain Tractable Power Flow Form

The first step of the process is to develop a tractable form of power flow problem. This is done by simplifying a non-convex distribution power flow model, along with the discrete controllable equipment functional space, into a tractable model form [24–27]. The reason for performing this step is that distribution grids differ greatly in design and in the spirit of standardizing DOPF model development; a tractable power flow form must be made that can achieve a balance between the solution space and computational efforts. As a case of the simplest tractable form, a linear power flow model with continuous decision variables mapping can be picked. To ensure the tractable model form is able to be used in developing the DOPF algorithm, the model should be parameterized in the controllable equipment variables. This allows for validation of the errors on the approximation of the power flow, as the change in the controllable equipment is proposed by the DOPF solution. From the most general grid model described earlier, a tractable approximation (e.g., linear model) can be expressed as

$$\tilde{x} = Au + C_u u \tag{7.4}$$

where the preceding model is the linear mapping of control variables u to approximated grid states $\tilde{x}$ using a linear sensitivity matrix A and the incidence matrix C_u.

7.2.4 Implement Reliable Solution Algorithm

This step utilizes the processed grid model and its approximation to develop a reliable solution format. By "reliable" it is meant that the solution algorithm should provide acceptable results in a timely fashion. This can be achieved by implementing a solution algorithm using a standard, off-the-shelf solver, which can then be benchmarked against the individual algorithms developed by the user. This step is more than just converting power flow formulation to an optimization problem. This step is introduced in the DOPF core features to allow for testing various different algorithms and techniques. These are necessary, as a wide number of distribution grid operation techniques can be developed depending on the application and necessity of the operator, and such a solution algorithm development process can be used to obtain a reliable solution algorithm. Next, we present two examples where an approximate model (7.4) is shown to be utilized for setting up (1) a benchmark method that utilizes a standard DOPF formulation to be solved by an off-the-shelf solver and (2) an alternate method that solves the standard DOPF formulation using a different algorithm.

$$\min c\left(u, \tilde{x}\right) \tag{7.5a}$$
$$s.t.\, f\left(u, \tilde{x}\right) = 0 : \lambda \tag{7.5b}$$
$$g\left(u, \tilde{x}\right) \leq 0 : \mu \tag{7.5c}$$

where $c(\cdot)$ is the objective function term to be minimized and (7.5b) and (7.5c) are the approximated counterparts of (7.2) and (7.3), formulated in a similar manner as (7.4). Variables λ and μ are the Lagrange multiplier of the respective constraints. To complement the tractability of the optimization problem, similar to approximation of constraints, the objective function can also be made linear or quadratic.

The standard DOPF formulation of (7.5) could be solved using an off-the-shelf solver [28]. However, following is an example of an alternate solution algorithm, which can be developed and benchmarked against (7.5). Consider the Lagrangian of (7.5):

$$\mathcal{L}\left(u, \tilde{x}, \lambda, \mu\right) = c\left(u, \tilde{x}\right) - \lambda\left(f\left(u, \tilde{x}\right)\right) - \mu\left(g\left(u, \tilde{x}\right)\right) \tag{7.6}$$

which can be deployed to update primal and dual variables at each iteration k using the following rules:

$$\text{Primal Updates}: \boldsymbol{u}^{k+1} = \mathbf{L}_u(\cdot), \tag{7.7a}$$

$$x^{k+1} = \mathbf{L}_x(\cdot), \tag{7.7b}$$

$$\text{Dual Updates}: \boldsymbol{\lambda}^{k+1} = \boldsymbol{\lambda}^k - \alpha\Delta\mathbf{L}_\lambda(\cdot), \tag{7.7c}$$

$$\boldsymbol{\mu}^{k+1} = \boldsymbol{\mu}^k - \beta''\mathbf{L}_\mu(\cdot), \tag{7.7d}$$

until there are no variables updates. In (7.7), $''\mathbf{L}(\cdot)$ is the first-order derivative of $\mathbf{L}$ with respect to $(\cdot)$, and (α,β) are small positive scalars. The algorithm (7.7) is formally known as the gradient descent [29].

Traditionally, power system dispatch tools have not been designed specifically to consider PtX. However, they can be adapted to incorporate PtX technologies. Unfortunately, there seems to be limited literature on the dispatch tools and methods demonstrating the integration of PtX into the power system planning models, but some suggestions for them to be included in the aforementioned optimal power flow methodology are as follows:

- *For power-to-hydrogen.* Hydrogen production capacity can be included as a component in the supply-and-demand balance equation.
- *For power-to-ammonia.* Ammonia (NH_3) production facilities can be included as the extra demand, as well as its conversion and integration in the grid as the supply variable.
- *For power-to-methane (P_2M).* The optimal locations for P_2M facilities can be included, along with the consideration of associated factors, like grid infrastructure, renewable energy availability, and demand patterns in the optimal power flow problem.
- *For power-to-synthetic fuels.* The generation of synthetic fuel production can be included as a supply variable in the power system planning models.
- *For power-to-heat (P_2H).* The P_2H methods modeled in large buildings/facilities could be utilized also in power system planning models, after adjusting for the scale and efficiency.
- *For power-to-battery storage.* This one is widely studied, and the usual power system dispatch tools are starting to implement these on a regular basis and hence could be adopted from those models.
- *For power-to-electrofuels.* Such integration would require including electrofuel production variable as a component in power system dispatch models.

In summary, the goal of integrating PtX technologies into electric power system planning models would be to allow for a holistic evaluation of optimal PtX capacities, locations, and their impact on the future grid planned conditions.

7.2.5 Existing Challenges in Planning Energy Grid with PtX Technologies

However, there still exist many challenges to address these problems, including, but not limited to:

- *Efficiency losses.* PtX processes involve multiple energy conversions, leading to efficiency losses at each step. How to include these losses into the planning problems remains an open question.
- *Intermittency and grid integration.* Renewable energy sources (RES) are inherently intermittent. As PtX facilities depend on surplus electricity from RES, how to account for their coincidence with the demand remains an interesting challenge to solve.

- *Storage and transportation.* The traditional power system planning models don't consider constraints on complex storage and transportation processes such as observed in PtX, such as storing hydrogen, ammonia, or synthetic fuels requiring specialized infrastructure.
- *Lack of available models.* Planning software and tools used in power systems do not have PtX technology modules, at least in the aggregation of multiple technologies. Hence, improvements are needed in the field of analytical modeling of PtX for energy system software.

Hence, advancements are needed in the modeling community of PtX to consider them appropriately in the planning models.

7.3 DER AGGREGATION METHODOLOGIES

Distributed energy resources (DERs), such as thermostatically controlled loads (TCLs), battery energy storage systems (BESS), electric vehicles (EVs), smart inverters with solar/PV, etc., can be used for demand-side management for peak-load reduction, energy arbitrage, and demand charge reduction [30–47]. Furthermore, DERs can also be used to participate in energy markets for providing grid services, such as ancillary services, including frequency regulation [32–37] and spinning reserves [39–41]. The idea of flexibility comes from the fact that these DERs store energy in some form that the grid can harness [35]. For example, TCLs, such as electric water heaters; heating, ventilation, and air-conditioning (HVAC) systems; electric space heaters; etc., store thermal energy and are flexible since their power consumption can be shifted in time. Similarly, in case of electric vehicles, charging can be spread over time since, usually, the EV layover period is greater than the charging time needed to fully charge EVs. Residential EV charging stations and workplace EV charging stations are some of the examples of flexible EV charging stations, where aggregation can be used to harness flexibility [48, 49]. However, in order to harness DER flexibility, careful coordination is needed that enables DERs to adjust their behavior depending upon the needs of the grid.

7.3.1 DER AGGREGATION TO FORM VIRTUAL POWER PLANTS

Large fleets of DERs can be aggregated to form virtual power plants (VPPs) that can be controlled similar to a conventional generator. VPPs are responsible for minimizing electricity costs, providing grid services, and participating in energy markets while maintaining quality of service of the DERs. VPPs use aggregate models to obtain flexibility that can be achieved from a particular set of DERs. Multiple VPPs can further be coordinated together. However, coordinating large fleets of DERs requires communication between DERs and the DER coordinator.

To achieve scalable control of a DER fleet, aggregate models are needed that capture the behavior of the entire fleet. Consider a VPP consisting of TCLs, battery energy storage system (BESS), PV, EVs, as well as conventional generators; the VPP then aims to solve the following multi-period optimization problem [35]:

$$\min \ \mathcal{F}_t(\cdot) = \sum_{k=t}^{t+K} \Big(f_c\big(y_{grid}(k)\big) + f_{gen}\big(y_{gen}(k)\big) + f_{TCL}\big(x_{TCL}(k)\big) + f_{BESS}(x_{BESS}(k) + f_{EV}\big(x_{EV}(k)\big)\Big), \tag{7.8a}$$

$$y_{BESS}(k) + y_{PV}(k) + y_{grid}(k) + y_{gen}(k) = y_{TCL}(k) + y_{EV}(k) + y_{load}(k), \tag{7.8b}$$

$$x_{TCL}(k+1) = h_{TCL}\big(x_{TCL}(k), u_{TCL}(k)\big), \ y_{TCL}(k) = g_{TCL}\big(x_{TCL}(k)\big) \tag{7.8c}$$

$$x_{BESS}(k+1) = h_{BESS}\big(x_{BESS}(k), u_{BESS}(k)\big), \ y_{BESS}(k) = g_{BESS}\big(x_{BESS}(k)\big) \tag{7.8d}$$

$$x_{EV}(k+1) = h_{EV}\big(x_{EV}(k), u_{EV}(k)\big), \ y_{EV}(k) = g_{EV}\big(x_{EV}(k)\big) \tag{7.8e}$$

$$\underline{x} \le x(k) \le \bar{x}, \tag{7.8f}$$

$$\underline{u} \le u(k) \le \bar{u}, \tag{7.8g}$$

$$\underline{y} \le y(k) \le \bar{y}, \ \forall k = t, \ \ldots, \ t+K, \tag{7.8h}$$

$$x(t) = x_0, \tag{7.8i}$$

where t is the current time; K is the time horizon, which is typically set to 24 hr; and $F_t(\cdot)$ is the total cost function. In (7.8), the state variables, control inputs, and output variables are denoted as $x(k)$, $u(k)$, and $y(k)$, respectively, with appropriate subscripts, as explained next.

In the power balance constraint (7.8b), the total uncontrollable load to be served is denoted by $y_{load}(k)$, $y_{grid}(k)$ is the total grid import, $y_{gen}(k)$ is the conventional generation, and $y_{TCL}(k)$, $y_{BESS}(k)$ $y_{EV}(k)$, and $y_{PV}(k)$, respectively, are the total power consumption of TCLs, BESS, EVs, and PV. The aggregator uses the control input $u(k) = \left(u_{TCL}(k), \ u_{BESS}(k), u_{EV}(k), u_{PV}(k) \right)$ to control TCLs $\left(u_{TCL}(k) \right)$, BESS $\left(u_{BESS}(k) \right)$, EVs $\left(u_{EV}(k) \right)$, and PV $\left(u_{PV}(k) \right)$. The control input for PVs, $u_{PV}(k)$, has been added to allow the aggregator to curtail excess PV. Similarly, $x(k) = \left(x_{TCL}(k), x_{BESS}(k), x_{EV}(k) \right)$ is the state space corresponding to DERs and is obtained from aggregate DER models used to capture the dynamics of TCL (x_{TCL}), BESS (x_{BESS}), and EVs (x_{EV}). Further details about the aggregate dynamic models represented in (7.8c)–(7.8e) by $h_{TCL}(\cdot)$ and $g_{TCL}(\cdot)$ for TCLs, $h_{BESS}(\cdot)$ and $g_{BESS}(\cdot)$ for BESS, and $h_{EV}(\cdot)$ and $g_{EV}(\cdot)$ for EVs are presented in the next sections. The initial state is set to the known value x_0 at the start of the optimization horizon, and the desired final state at the end of the horizon is set equal to x_{end}. In (7.8f), $\underline{x}$ and $\bar{x}$ are the lower and upper limits on $x(k)$, respectively, and in (7.8g), $\underline{u}$ and $\bar{u}$, are the lower and upper limits on $u(k)$, respectively.

The cost of importing power from the grid is captured in the cost function $f_c\left(y_{grid}(k) \right)$. Difference rate structures can be implemented, for example, time of use pricing and demand charges, as shown in the following [49]:

$$f_c(y_{grid}) = \sum_{k=t}^{t+K} (\Delta t c_e(k) y_{grid}(k)) + c_{nc} \max\{y_{grid}(k)\}_{k=t}^{t+k} + c_{op} \max\{y_{grid}(k)\}_{k \in Iop} \tag{7.9}$$

where Δt is the discrete time step and c_e is the electricity cost, which could be variable at each k, as in the case of time-of-use pricing. Demand charges are calculated based on the single maximum electricity demand observed over a full month and consist of non-coincident demand charges and on-peak demand charges. The non-coincident demand charges correspond to single-peak demand observed during 24 hr in each day, and c_{nc} is the associated cost in \$/kWh. On-peak demand charges, on the other hand, only correspond to the time period between 4:00 p.m. and 9:00 p.m. in each day with cost c_{op}. In (7.9), I_{OP} captures the time indices corresponding to the on-peak period. Further details about demand charges are available in [48, 49].

The cost of internal generators is usually captured using quadratic cost functions in $f_{gen}(y_{gen})$. In case of DERs, f_{TCL}, f_{BESS}, and f_{EV} are added to the total cost function to fulfil DER-specific requirements, such as quality of service, so that the individual needs of the DERs are captured in the aggregate model, for example, all EVs are fully charged by the departure time.

DER aggregate models and the underlying coordination schemes are the two important aspects that define the nature of the VPPs and are discussed next, starting with the coordination schemes.

7.3.2 DER Coordination Schemes

DER coordination schemes can be divided into centralized, decentralized, and distributed schemes.

- **Centralized coordination** schemes consist of a single central decision-making entity and require that each DER transmit its current state (e.g., temperature or state of charge) to the coordinator at regular intervals. The coordinator then solves an optimization problem to determine and then send the control command for each DER. Communication only occurs between the DERs and the central entity; as a result, communication requirements are high and do not scale well with the number of DERs. Examples of central communication are the schemes proposed in [32, 42, 43, 48, 49].
- In **decentralized coordination** schemes, decision-making occurs locally at each DER but still involves communication with a central decision-making entity. The difference between centralized and decentralized control schemes is that each DER's local state is not required to be transmitted to the central decision-making entity, and only aggregate information, such as total power consumption of the fleet, is required. Furthermore, the central decision-making entity synthesizes a control signal which is used by DERs to make decisions locally, for example, to switch from *on* to *off* in case of TCLs, or kWh dispatched in case of batteries, etc. The global objectives, such as tracking a power reference signal, are known only to the central entity, which it achieves by solving an optimization problem that generates the control signal for DERs. The control signal can be broadcast to all DERs in a top-down approach, in which the same control signal is transmitted to all DERs. The DERs then make decisions based on their local state. Examples of top-down approach are [39, 40]. Alternatively, a decentralized scheme can also be implemented in a bottom-up approach, as in [33–38], in which a new scheme, called packetized energy management, is developed. In PEM, DERs request the aggregator to turn on for a specific fixed time interval, for example, 5 min. The aggregator collects requests from all DERs and then determines which requests to accept so that global objective is achieved.
- **Distributed coordination** schemes leverage the principles of distributed optimization, where multiple agents (DERs) cooperatively solve a global optimization problem. The global objective function is disaggregated into each agent's local objective function. Decision-making occurs locally at the agent, along with coordination with neighboring agents, to solve the optimization problem. Specifically, each agent iteratively shares local decision variables with neighboring agents to compute its next control input (e.g., on/off state of a TCL). Examples of this type of coordination include the Packetized Energy Management (PEM) of [33–38], decentralized scheme of [39, 40], and distributed scheme of [41].

Centralized coordination schemes are simpler to implement but require bidirectional communication between the VPP and DERs. Due to high communication requirements, this scheme does not scale well with the number of DERs in the VPPs [41, 47]. Furthermore, there are cybersecurity concerns with having information of all DERs contained in a single location. Distributed coordination is more scalable than centralized schemes and have lower computational overhead as well. Although information is exchanged between DERs, cybersecurity is improved since the information exchanged consists of derivatives of local objective function and does not contain critical information of the DER, such as its on/off state. After a coordination scheme has been chosen, the next step is to model the individual dynamics of each DER, as explained next.

7.3.3 Modeling Individual DER Dynamics

In order to obtain the aggregate flexibility from large fleets of DERs, the dynamics of an individual DER needs to be modeled first.

Thermostatically controlled loads (TCLs), such as electric water heaters, air conditioners, HVACs, etc., are designed to maintain temperature z_i within a narrow deadband $\left[\underline{z}_i, \overline{z}_i\right]$ around a user-specified set point z_{set}, where $\underline{z}_i$ and $\overline{z}_i$ are the lower and upper limits, respectively. The individual TCL dynamics are modeled using first-order difference equation, given by

$$z_i\left(k+1\right)=\eta_{sl}z_i\left(k\right)+\eta_c P_c^r \phi_i\left(k\right)+w_i\left(k\right) \tag{7.10}$$

where z_i is the temperature of the i th DER, η_{sl} is the standing loss, η_c is the heating or cooling efficiency, P_c^r is the rated power consumption, and $\phi_i\left(k\right)\in\left\{0,1\right\}$ is the discrete on/off state of the TCL. The discrete state $\phi_i\left(k\right)$ is determined by the thermodynamic switch, which is operated with the following logic:

$$\phi_i\left(k+1\right)=\begin{cases} 0, & z_i\left(k\right)\geq\overline{z}_i, \\ 1 & z_i\left(k\right)\leq\underline{z}_i, \\ \phi_i\left(k\right), & \text{otherwise.} \end{cases} \tag{7.11}$$

Finally, $w_i\left(k\right)$ is the end use consumption. The end use consumption is an uncontrollable process which captures the end use behavior, such as hot-water extraction in electric water heaters, random heat loss or gain from a room in air conditioners or electric space heaters, etc. Several approaches have been proposed to model the end use process and differ for each DER type. For example, the Gaussian process has been used for electric space heaters [30, 31] and air conditioners [33], and the Poisson random pulse process has been used for electric water heaters [33–38]. The modeling choice depends upon the nature of the end use consumption process and is further discussed later in this section.

For on/off loads, such as TCLs that can only consume power, (7.10) and (7.11) accurately capture their individual dynamics. However, in case of batteries that can both consume and discharge power into the grid, an additional term is added to (7.10), resulting in the following dynamic model:

$$z_i\left(k+1\right)=z_i\left(k\right)+\eta_c P_c^r \phi_i\left(k\right)+\eta_d P_d^r \phi_i\left(k\right)+w_i\left(k\right) \tag{7.12}$$

where η_d is the discharging efficiency, and P_d^r is the rated power when the battery is discharging. Standing losses are negligible in batteries and can be ignored, that is, $\eta_{sl}=1$ [33]. Furthermore, different from TCLs, $\phi_i\left(k\right)\in\left[-1,1\right]$ is allowed to vary continuously. Using the individual DER dynamic model, the three different types of aggregate models are discussed next.

7.3.4 Modeling Aggregate DER Dynamics

When modeling large fleets of DERs, using (7.9) or (7.10) requires keeping track of the state z_i of each DER, which does not scale well as the number of DERs increase. Therefore, aggregate models capture the behavior of DER fleets with a fewer number of states than the size of the fleets. These models consist of three different types, as described later.

7.3.4.1 State Bin Transition Models

Population models, or state bin transition models, are Markovian models that capture the dynamics of the entire fleet in discrete time [32, 33, 36, 37, 42]. The fleet can consist of thousands of DERs, but only a few states, typically fewer than 100, are required.

The state bin transition model is obtained by partitioning the continuous state space $z \in [\underline{z}, \overline{z}]$ into N discrete states, or bins, $b = (b_1, b_2, \ldots, b_N)$. The subscript i in z from the previous section has been dropped to highlight that the model derived is an aggregate model for the entire fleet. In each bin, let $x_{DER} = (x_{DER,1}, x_{DER,2}, \ldots, x_{DER,N})$, where $x_{DER,j}$ is the probability of being in state b_j ; then the population dynamics are given by

$$x_{DER}(k+1) = M^{\mathrm{nat}} M^{\mathrm{ctrl}}(u_{DER}(k)) x_{DER}(k) \tag{7.13a}$$

$$y_{DER}(k+1) = g_{DER}(q(k)) \tag{7.13b}$$

where $x_{DER}(k)$ is the probability distribution of the fleet; $u_{DER}(k)$ is the control input, which depends upon the DER coordination scheme; $M^{\mathrm{nat}} \in \mathbb{R}^{N \times N}$ is the transition probability matrix, in which $M^{\mathrm{nat}}(i,j)$ is the probability of a DER transitioning from state i to j ; $M^{\mathrm{ctrl}} \in \mathbb{R}^{N \times N}$ is the transition probability matrix that captures the effect of the DER coordination scheme under the action of the control input $u_{DER}(k)$. The right-hand side of the preceding equation forms the function $h_{DER}(x_{DER}(k), u_{DER}(k))$ in (7.8), where DER can be either TCL, BESS, or EVs. The function h_{DER} is a bilinear function due to the product of $x_{DER}(k)$ and $u_{DER}(k)$. The output $y_{DER}(k)$ depends upon the available information from the fleet, for example, if only the total power consumption of the fleet is measured, then $y_{DER}(k) = C_{dem} x_{DER}(k)$, where C_{dem} is a vector that maps the *on* TCLs in $x_{DER}(k)$ to the total power demand of the fleet. In case of batteries, C_{dem} maps the charging/discharging batteries to the total power demand of the fleet. Similarly, if individual temperature (TCLs) or SOC (batteries) is available, then it can be obtained as $y_{QOS}(k) = I_N x_{DER}(k)$, where I_N is the identity matrix of dimension N. Finally, the cost function $f_{DER}(x_{DER}(k))$ can be defined with respect to a desired nominal probability distribution x_{DER}^{nom}, for example, $f_{DER}(x_{DER}(k)) = x_{DER} - x_{DER2}^{nom}$, where $\cdot_2$ is the 2-norm.

Model accuracy depends upon the number of bins in the state bin transition model. Specifically, the model accuracy increases with the number of bins. There exists a trade-off between model accuracy and computational complexity, which depends upon the timestep, the width of each bin, and the DER type. For TCLs with slow dynamics (in time), such as HVACs, in which the temperature changes on the order of hours, the number of bins can be kept small, whereas for TCLs with fast temperature dynamics, such as residential electric water heaters, a larger number of states are required. Finally, the predictive capability of the state bin transition models improves as the fleet size grows, according to the law of large numbers. It is shown in [37] that 250 TCLs are sufficient to model a fleet of electric water heaters using state bin transition models.

State bin transition models are a scalable and powerful tool for capturing the dynamics of large fleets of DERs. This is because the model, once developed, is not required to be updated if the fleet size changes. Furthermore, $u_{DER}(k)$ is obtained in terms of aggregate quantities based on the normalized probability distribution of the fleet and is automatically translated into an appropriate control action for each DER by the underlying coordination scheme. The state bin transition model described previously is suitable for flexible loads, such as electric water heaters, space heaters, pool pumps, residential air conditioners, etc. These models can also be used for residential-scale BESS and residential EV chargers. For utility-scale batteries, state bin transition models are not suitable since these are few in number and the utilities may want to operate them directly.

7.3.4.2 Virtual Battery Models

Virtual battery models are low-order models compared to state bin transition models, in which the entire DER fleet is modeled using a single state, called the state of charge (SOC) [35, 39, 40]. This SOC allows the fleet to be represented as a virtual battery with energy capacity and power capacity similar to traditional batteries. With $\underline{u}_{DER}$ and $\bar{u}_{DER}$ as the minimum and maximum power capacity in kW, respectively, and $\underline{x}_{soc}$ and $\bar{x}_{soc}$ as the minimum and maximum energy capacity in kWh, respectively, of the virtual battery, the model is given by:

$$x_{soc}(k+1) = \alpha x_{soc}(k) + u_{DER}(k) \tag{7.14a}$$

$$\underline{u}_{DER}(k) \leq u_{DER}(k) \leq \bar{u}_{DER}(k) \tag{7.14b}$$

$$\underline{x}_{soc} \leq x_{soc}(k) \leq \bar{x}_{soc} \tag{7.14c}$$

The SOC, $x_{soc}(k)$, can be interpreted as the average temperature of a fleet of TCLs, or the average SOC in case of BESS. As in the previous section, u_{DER} can correspond to TCLs, BESS, EVs, etc. The power and energy capacity depends upon several factors, including fleet size, baseline power consumption, and DER dynamics. Power capacity is time-varying as well due to end use consumption. Energy capacity has been obtained by first considering an individual DER and then computing the energy required to fully charge the DER, which is then scaled by the size of the fleet. For TCLs, it is the energy required to change the temperature from one end of the deadband to the other. Obtaining $\bar{u}_{DER}(k)$ presents a significant challenge since SOC is an aggregate quantity and is not enough to obtain the power capacity of the fleet due to its dependence upon the individual SOC of DERs. Additional information is needed in the virtual battery model to get an accurate estimation of energy capacity. This can be in the form of including additional states in the model, as is done in [35], in which the underlying coordination scheme is incorporated into the model.

7.3.4.3 Set-Based Flexibility

An alternative to virtual battery models is the notion of set-based flexibility. DER flexibility is defined as a set of trackable power reference trajectories that is defined as:

$$\mathbb{U}_i = \{\phi_i \in \mathbb{R}^K \mid \phi_i(k) \in \left[\underline{\phi}, \bar{\phi}\right], \phi_i(0) = \phi_0, (10) \text{ or } (12), \forall k = t, \ldots, t+K\} \tag{7.15}$$

The aggregate flexibility is then obtained as the Minkowski sum [43]:

$$\mathbb{U} = \sum_{i=1}^{N} \mathbb{U}_i. \tag{7.16}$$

In order to implement the set-based flexibility model in the optimization formulation of (7.8), equations (7.8c)–(7.8e) are replaced by the constraint

$$u_{DER} \in \mathbb{U}, \tag{7.17}$$

where the set $\mathbb{U}$ captures the aggregate flexibility of the corresponding DER fleet. Set-based methods are computationally extensive since computing the aggregate flexibility set becomes intractable as the fleet size and time horizon increase. Efficient methods to compute the aggregate flexibility set have been developed in [43–46].

7.3.5 MODELING END USE CONSUMPTION OF TCLs FOR PREDICTING DEMAND

End use consumption in TCLs, $w(k)$, is an uncontrollable stochastic process and depends upon the end user behavior that is different for each DER. Different modeling techniques depending upon the DER type have been considered in the literature. In the case of heating or cooling loads such as air conditioners or electric space heaters, the end use process is the heat loss or gain due to the environment and can be modeled as a Gaussian process [30, 31].

The case is more complicated for residential electric water heaters, where the end use consumption consists of random water draws over a short time interval (e.g., 15 min), representing a hot-water shower event. The Gaussian process is not an accurate representation for the end use process in EWHs since the time duration between successive hot-water events and the transition between on/off states need not necessarily be Gaussian. Instead, the Markov renewal process is first developed in [31], which is then used in [38] to model and identify the Markov renewal process's parameters directly from EWH data. The Markov renewal process allows both the time duration between hot-water events and the transition between on/off states to belong to arbitrary probability distributions. Another method to model the EWH end use consumption is to use the Poisson random pulse process (PRP), which consists of random start/stop times and a random magnitude that models the intensity of hot-water drawn from the EWH tank. PRP has been used in [37] to obtain aggregate models for EWH.

7.3.6 Flexibility from Electric Vehicles

Electric vehicles are different from other DER types and are discussed separately in this section. This is because EV flexibility is governed by the driving patterns of the EV driver as well as the location and type of the EV charger. Two main types of EV charging stations that are suitable for DER aggregation are workplace EV charging stations and residential EV charging stations. For residential EV charging stations, charging mostly occurs overnight, when both the demand and the renewable generation are low, whereas in workplace EV charging stations, charging mainly occurs during the day, when both the demand and the renewable generation are high. Flexibility from both of these types of charging stations requires further modeling of the EV driving patterns. The following models have been used for capturing EV behavior:

- **Markov chain** models consider each EV to be either in standby state or in driving state. The resulting two-state Markov chain requires only the transition probabilities between the two states consisting. These probabilities can be obtained from historical data. The model then provides the steady-state probability distribution of the EV fleet, along with the mean driving time and the mean standby time that DER control algorithms can then use for scheduling charging. Markov chain models have been used in [37] to obtain the steady-state statistics of a fleet of residential EVs, and then used in a state bin transition model.
- **Direct forecasting** of EV at each charging station is another method for forecasting EV behavior. The driving pattern and charging needs of each EV are captured using three quantities, arrival time, departure time, and energy demand. A hybrid machine learning model has been developed in [49] for workplace EV charging in forecasting these three quantities. Historical data is used to train the machine learning model that can forecast EV demand for up to a week in advance.

7.3.7 Combined DER and PtX Dispatch

PtX technologies, in conjunction with DER flexibility, can be leveraged to handle the challenges that arise due to high penetration of intermittent renewable energy sources, such as solar and wind. In the presence of excess generation, when DER flexibility is not sufficient, energy storage, such as batteries, are needed. Batteries are charged with the excess energy and then discharged when the generation is low. With PtX technologies, instead, this excess energy can be used to generate and store green fuels, which can then be used for heating and providing power to loads. In a smart grid, where multiple VPPs and PtX technologies are available, co-dispatch of VPPs and PtX technologies can be achieved by solving the following multi-period optimization problem [50, 51]:

$$\min H_t\left(\cdot\right) = F_t\left(\cdot\right) + \sum_{k=t}^{t+K}\sum_{j=1}^{N_{PtX}}\left(G_{PtX}^j\left(e_j\left(k\right)\right)\right) \tag{7.18a}$$

$$y_{BESS}\left(k\right) + y_{PV}\left(k\right) + y_{grid}\left(k\right) + y_{gen}\left(k\right) + \sum_{j=}^{N_{PtX}} y_{PtX}^j\left(k\right) = y_{TCL}\left(k\right) + y_{EV}\left(k\right) + y_{load}\left(k\right), \tag{7.18b}$$

$$x_{PtX}^j\left(k+1\right) = h_{PtX}^j\left(x_{PtX}^j\left(k\right), e_j\left(k\right)\right), y_{PtX}^j\left(k\right) = h_{PtX}^j\left(x_{PtX}^j\left(k\right)\right) \tag{7.18c}$$

$$\underline{x}_{PtX}^j \le x_{PtX}^j\left(k\right) \le \overline{x}_{PtX}^j \tag{7.18d}$$

$$\underline{e}_j \le e_j\left(k\right) \le \overline{e}_j, \tag{7.18e}$$

$$\underline{y}_{PtX}^j \le y_{PtX}^j\left(k\right) \le \overline{y}_{PtX}^j \tag{7.18f}$$

$$x_{PtX}^j\left(t\right) = x_{PtX,0}^j, x_{PtX}^j\left(t+K\right) = x_{PtX,end}^j, \tag{7.18g}$$

$$f_{PtX}^j\left(e_j\right) \le 0, \tag{7.18h}$$

$$g_{PtX}^j\left(e_j\right) = 0, \quad \forall j = 1,\ldots,N_{PtX}, \tag{7.18i}$$

$$\left(8c\right) - \left(8i\right), \forall \; k = t, \ldots, t+K$$

For N_{PtX} PtX generators, where the energy balance equation from (7.8b) has been augmented to include the output y_{PtX}^j of PtX generators in (7.18b), e_j is the control input for the jth PtX generator; $h_{PtX}^j\left(.\right)$ and $h_{PtX}^j\left(.\right)$ capture the dynamics of the jth PtX generator with state vector $x_{PtX}^j\left(k\right)$; (7.18d)–(7.18f) capture the minimum and maximum limits on state, control input, and output, respectively; and (7.18g) captures the known initial state and the desired final states of PtX generators. The state vector $x_{PtX}^j\left(k\right)$ represents the energy stored in PtX generators, for example, in the form of green synthetic fuels. Additionally, $f_{PtX}^j\left(e_j\right)$ and $g_{PtX}^j\left(e_j\right)$ are the inequality and equality constraints that have been included to capture the constraints not handled in (7.18c)–(7.18g), such as time-varying electricity production potential of PtX generators. The optimization problem (7.16) can be solved both in a centralized manner or a distributed manner [47].

However, as described in Section 7.2, there are several challenges with PtX technologies, including efficiency losses and the time-varying electricity production potential of these technologies, which have to be modeled in (7.18). Furthermore, excess renewable energy is used to produce synthetic fuels; therefore, when optimizing over a horizon which is equal to a full day, renewable forecasts are needed, which make (7.18) prune to forecast errors. Stochastic versions of (7.18) need to be developed to minimize the impact of these errors.

7.4 CONCLUSION

Traditionally, power system dispatch tools have not been designed specifically to consider PtX. However, they can be adapted to incorporate PtX technologies. Unfortunately, there seems to be limited literature on the dispatch tools and methods demonstrating the integration of PtX into the power system planning models. Ideally, the goal of integrating PtX technologies into electric power system planning models would be to allow for a holistic evaluation of optimal PtX capacities, locations, and their impact on the future grid planned conditions. However, there still exists the need for analytical modeling of PtX technologies to be deployed in energy system planning models.

Flexibility from DERs can be harnessed for providing services to the grid by coordinating DERs in a centralized, decentralized, or distributed configuration. Aggregate models can be used to capture the aggregate behavior of large fleets of DERs. These models are agnostic to the coordination scheme. DER flexibility has been studied significantly in the literature. This chapter presented an overview of DER aggregation methods; however, the flexibility potential of DERs can be enhanced by integrating PtX technologies. A multi-period optimization formation is also presented in this chapter that can be used for the co-dispatch of DERs and PtX technologies but requires further exploration and validation.

NOTE

1 Where D_x is the inequality counterpart of C_x, just as line flow utilizes different admittance quantities than nodal admittance.

REFERENCES

[1] R. Daiyan, I. MacGill, and R. Amal, "Opportunities and challenges for renewable Power-to-X," *ACS Energy Letters*, vol. 5, no. 12, pp. 3843–3847, 2020. doi: 10.1021/acsenergylett.0c02249.

[2] B. R. de-Vasconcelos and J.-M. Lavoie, "Recent advances in power-to-X technology for the production of fuels and chemicals," *Frontiers in Chemistry*, vol. 7, 2019.

[3] S. Byfield and D. Vetter. "Flexibility concepts for the German power supply in 2050: Ensuring stability in the age of renewable energies," ACATECH – National Academy of Science and Engineering, Munich, Germany, 2016.

[4] W. Leitner, "Power-to-X: Entering the energy transition with kopernikus," RWTH Aachen University, 2016.

[5] K. S. Hedegaard, "See how Power-to-X could be a key component in the global energy transition," www.weforum.org/agenda/2023/11/power-to-x-a-key-component-in-the-global-energy-transition/.

[6] L. E. Jones, "Strategies and decision support systems for integrating variable energy resources in control centers for reliable grid operations," United States: N. p., 2011. Web. doi: 10.2172/1219362.

[7] C. H. Barriquello, V. J. Garcia, M. Schmitz, D. Bernardon, and J. S. Fonini, "A decision support system for planning and operation of maintenance and customer services in electric power distribution systems." *System Reliability*, pp. 355–370, 2017.

[8] S. Grassi, F. Veronesi, J. Schito, and M. Raubal. "An integrated GIS-based method for planning power lines targeting interactive public participation," in *1st SCCER-FURIES Annual Conference, SCCER-FURIES*, 2014.

[9] G. Gust, W. Biener, T. Brandt, K. Dallmer-Zerbe, D. Neumann, and B. Wille-Haussmann. "Decision support for distribution grid planning," in *2016 IEEE International Energy Conference (ENERGYCON)*, IEEE, 2016.

[10] B. Pierre, R. Broderick, M. DeMenno, J. Paladino, and J. Yoshimura, "MOD-plan: Multi-objective decision planning framework for electric grid resilience, equity, and decarbonization," United States: N. p., 2022. Web. doi: 10.2172/1885093.

[11] M. Baran and F. F. Wu, "Optimal sizing of capacitors placed on a radial distribution system," *IEEE Transactions on Power Delivery*, vol. 4, no. 1, pp. 735–743, January 1989. doi: 10.1109/61.19266.

[12] N. Nazir, P. Racherla, and M. Almassalkhi, "Optimal multi-period dispatch of distributed energy resources in unbalanced distribution feeders," *IEEE Transactions on Power Systems*, vol. 35, no. 4, pp. 2683–2692, July 2020. doi: 10.1109/TPWRS.2019.2963249.

[13] S. Hanif, K. Zhang, C. M. Hackl, M. Barati, H. B. Gooi, and T. Hamacher, "Decomposition and equilibrium achieving distribution locational marginal prices using trust-region method," *IEEE Transactions on Smart Grid*, vol. 10, no. 3, pp. 3269–3281, May 2019. doi: 10.1109/TSG.2018.2822766

[14] S. Hanif, M. Barati, A. Kargarian, H. B. Gooi and T. Hamacher, "Multiphase distribution locational marginal prices: Approximation and decomposition," in *2018 IEEE Power & Energy Society General Meeting (PESGM)*, Portland, OR, USA, 2018, pp. 1–5. doi: 10.1109/PESGM.2018.8585925.

[15] S. Hanif, H. B. Gooi, T. Massier, T. Hamacher, and T. Reindl, "Distributed congestion management of distribution grids under robust flexible buildings operations," *IEEE Transactions on Power Systems*, vol. 32, no. 6, pp. 4600–4613, November 2017. doi: 10.1109/TPWRS.2017.2660065.

[16] L. Gan and S. H. Low, "Convex relaxations and linear approximation for optimal power flow in multiphase radial networks," in *2014 Power Systems Computation Conference*, Wroclaw, Poland, 2014, pp. 1–9. doi: 10.1109/PSCC.2014.7038399.

[17] N. Nazir and M. Almassalkhi, "Grid-aware aggregation and realtime disaggregation of distributed energy resources in radial networks," *IEEE Transactions on Power Systems*, vol. 37, no. 3, pp. 1706–1717, May 2022. doi: 10.1109/TPWRS.2021.3121215.

[18] N. Nazir and M. Almassalkhi, "Receding-horizon optimization of unbalanced distribution systems with time-scale separation for discrete and continuous control devices," in *2018 Power Systems Computation Conference (PSCC)*, Dublin, Ireland, 2018, pp. 1–7. doi: 10.23919/PSCC.2018.8442555.

[19] N. Nazir and M. Almassalkhi, "Voltage positioning using co-optimization of controllable grid assets in radial networks," *IEEE Transactions on Power Systems*, vol. 36, no. 4, pp. 2761–2770, July 2021. doi: 10.1109/TPWRS.2020.3044206.

[20] N. Almassalkhi, S. Brahma, N. Nazir, H. Ossareh, P. Racherla, S. Kundu, S. P. Nandanoori, T. Ramachandran, A. Singhal, D. Gayme, and C. Ji, "Hierarchical, grid-aware, and economically optimal coordination of distributed energy resources in realistic distribution systems," *Energies*, vol. 13, p. 6399, 2020. doi: 10.3390/en13236399.

[21] C. Coffrin, R. Bent, K. Sundar, Y. Ng, and M. Lubin, "PowerModels. JL: An open-source framework for exploring power flow formulations," in *2018 Power Systems Computation Conference (PSCC)*, Dublin, Ireland, 2018, pp. 1–8. doi: 10.23919/PSCC.2018.8442948.

[22] D. M. Fobes, S. Claeys, F. Geth, and C. Coffrin, "PowerModelsDistribution.jl: An open-source framework for exploring distribution power flow formulations," *Electric Power Systems Research*, vol. 189, p. 106664, 2020. ISSN: 0378–7796. doi: 10.1016/j.epsr.2020.106664.

[23] E. Bressert, "SciPy and NumPy: An overview for developers," 2012.

[24] A. Bernstein, C. Wang, E. Dall'Anese, J.-Y. Le Boudec, and C. Zhao, "Load flow in multiphase distribution networks: Existence, uniqueness, non-singularity and linear models," *IEEE Transactions on Power Systems*, vol. 33, no. 6, pp. 5832–5843, November 2018. doi: 10.1109/TPWRS.2018.2823277.

[25] S. Bruno, S. Lamonaca, G. Rotondo, U. Stecchi, and M. La Scala, "Unbalanced three-phase optimal power flow for smart grids," *IEEE Transactions on Industrial Electronics*, vol. 58, no. 10, pp. 4504–4513, October 2011. doi: 10.1109/TIE.2011.2106099.

[26] M. Bazrafshan and N. Gatsis, "Comprehensive modeling of three-phase distribution systems via the bus admittance matrix," *IEEE Transactions on Power Systems*, vol. 33, no. 2, pp. 2015–2029, March 2018. doi: 10.1109/TPWRS.2017.2728618.

[27] L. E. Jones, "Strategies and decision support systems for integrating variable energy resources in control centers for reliable grid operations," United States: N. p., 2011. Web. doi:10.2172/1219362.

[28] Gurobi Optimization, LLC, "Gurobi optimizer reference manual," 2023. www.gurobi.com.

[29] S. P. Boyd and L. Vandenberghe, *Convex Optimization*, Cambridge University Press, 2004.

[30] R. Malhame and C.-Y. Chong, "Electric load model synthesis by diffusion approximation of a high-order hybrid-state stochastic system," *IEEE Transactions on Automatic Control*, vol. 30, no. 9, pp. 854–860, September 1985. doi: 10.1109/TAC.1985.1104071.

[31] S. El-Ferik and R. P. Malhame, "Identification of alternating renewal electric load models from energy measurements," *IEEE Transactions on Automatic Control*, vol. 39, no. 6, pp. 1184–1196, June 1994. doi: 10.1109/9.293178.

[32] J. L. Mathieu, S. Koch, and D. S. Callaway, "State estimation and control of electric loads to manage real-time energy imbalance," *IEEE Transactions on Power Systems*, vol. 28, no. 1, pp. 430–440, February 2013. doi: 10.1109/TPWRS.2012.2204074.

[33] A. Khurram, "Modeling and control for packetized energy management," PhD Thesis, The University of Vermont and State Agricultural College, United States—Vermont, 2021.

[34] M. Almassalkhi, L. A. Duffaut Espinosa, P. D. H. Hines, J. Frolik, S. Paudyal, and M. Amini, "Asynchronous coordination of distributed energy resources with packetized energy management." In *Energy Markets and Responsive Grids: The IMA Volumes in Mathematics and its Applications*, S. Meyn, T. Samad, I. Hiskens, and J. Stoustrup, Eds., vol. 162. New York, NY: Springer. doi: 10.1007/978-1-4939-7822-9_14.

[35] L. A. Duffaut Espinosa, A. Khurram, and M. R. Almassalkhi, "A virtual battery model for packetized energy management," in *2020 59th IEEE Conference on Decision and Control (CDC)*, Jeju, Korea (South), 2020, pp. 42–48. doi: 10.1109/CDC42340.2020.9304065.

[36] L. A. Duffaut Espinosa and M. Almassalkhi, "A packetized energy management macromodel with quality of service guarantees for demand-side resources," *IEEE Transactions on Power Systems*, vol. 35, no. 5, pp. 3660–3670, September 2020. doi: 10.1109/TPWRS.2020.2981436.

[37] L. A. D. Espinosa, A. Khurram, and M. Almassalkhi, "Reference-tracking control policies for packetized coordination of heterogeneous DER populations," *IEEE Transactions on Control Systems Technology*, vol. 29, no. 6, pp. 2427–2443, November 2021. doi: 10.1109/TCST.2020.3039492.

[38] A. Khurram, R. Malhamé, L. A. Duffaut Espinosa, and M. Almassalkhi, "Identification of hot water end-use process of electric water heaters from energy measurements," *Electric Power Systems Research*, vol. 189, p. 106625, 2020. ISSN 0378–7796. doi: 10.1016/j.epsr.2020.106625.

[39] N. Cammardella, J. Mathias, M. Kiener, A. Bušić, and S. Meyn, "Balancing California's grid without batteries," in *2018 IEEE Conference on Decision and Control (CDC)*, Miami, FL, USA, 2018, pp. 7314–7321. doi: 10.1109/CDC.2018.8618975.

[40] H. Hao, Y. Lin, A. S. Kowli, P. Barooah, and S. Meyn, "Ancillary service to the grid through control of fans in commercial building HVAC systems," *IEEE Transactions on Smart Grid*, vol. 5, no. 4, pp. 2066–2074, July 2014. doi: 10.1109/TSG.2014.2322604.

[41] T. Anderson, M. Muralidharan, P. Srivastava, H. V. Haghi, J. Cortés, J. Kleissl, S. Martinez, and B. Washom, "Frequency regulation with heterogeneous energy resources: A realization using distributed control," *IEEE Transactions on Smart Grid*, vol. 12, no. 5, pp. 4126–4136, September 2021. doi: 10.1109/TSG.2021.3071778.

[42] E. Benenati, M. Colombino, and E. Dall'Anese, "A tractable formulation for multi-period linearized optimal power flow in presence of thermostatically controlled loads," in *2019 IEEE 58th Conference on Decision and Control (CDC)*, Nice, France, 2019, pp. 4189–4194. doi: 10.1109/CDC40024.2019.9030029.

[43] H. Hao, B. M. Sanandaji, K. Poolla, and T. L. Vincent, "Aggregate flexibility of thermostatically controlled loads," *IEEE Transactions on Power Systems*, vol. 30, no. 1, pp. 189–198, January 2015. doi: 10.1109/TPWRS.2014.2328865.

[44] L. Zhao, H. Hao, and W. Zhang, "Extracting flexibility of heterogeneous deferrable loads via polytopic projection approximation," in *2016 IEEE 55th Conference on Decision and Control (CDC)*, Las Vegas, NV, USA, 2016, pp. 6651–6656. doi: 10.1109/CDC.2016.7799293.

[45] M. S. Nazir, I. A. Hiskens, A. Bernstein, and E. Dall'Anese, "Inner approximation of Minkowski sums: A union-based approach and applications to aggregated energy resources," in *2018 IEEE Conference on Decision and Control (CDC)*, Miami, FL, USA, 2018, pp. 5708–5715. doi: 10.1109/CDC.2018.8618731.

[46] F. A. Taha, T. Vincent, and E. Bitar, "An efficient method for quantifying the aggregate flexibility of plug-in electric vehicle populations," *IEEE Transactions on Smart Grid*. doi: 10.1109/TSG.2024.3384871.

[47] P. Srivastava, C.-Y. Chang, and J. Cortés, "Enabling DER participation in frequency regulation markets," *IEEE Transactions on Control Systems Technology*, vol. 30, no. 6, pp. 2391–2405, November 2022. doi: 10.1109/TCST.2022.3143711.

[48] A. Ghosh, M. Z. Zapata, S. Silwal, A. Khurram, and J. Kleissl, "Effects of number of electric vehicles charging/discharging on total electricity costs in commercial buildings with time-of-use energy and demand charges," *Journal of Renewable and Sustainable Energy*, vol. 14, no. 3, 2022.

[49] G. McClone, A. Ghosh, A. Khurram, B. Washom, and J. Kleissl, "Hybrid machine learning forecasting for online MPC of work place electric vehicle charging," *IEEE Transactions on Smart Grid*, vol. 15, no. 2, pp. 1891–1901, March 2024. doi: 10.1109/TSG.2023.3296014.

[50] S. Klyapovskiy, Y. Zheng, S. You, and H. W. Bindner, "Optimal operation of the hydrogen-based energy management system with P2X demand response and ammonia plant," *Applied Energy*, vol. 304, 2021, p. 117559. ISSN 0306–2619. doi: 10.1016/j.apenergy.2021.117559.

[51] L. B. Jaramillo and A. Weidlich, "Optimal microgrid scheduling with peak load reduction involving an electrolyzer and flexible loads," *Applied Energy*, vol. 169, 2016, pp. 857–865. ISSN 0306–2619. doi: 10.1016/j.apenergy.2016.02.096.

8 From Electrons to Solutions
Power-to-X Strategies for Energy Systems Integration

Reza Gharibi, Behrooz Vahidi,
Rahman Dashti, and Reza Khalili

Abbreviations

EU	European Union
HTECs	high-temperature electrolysis cells
IES	integrated energy systems
MSs	member states
NECPs	national energy and climate plans
PEM	proton-exchange membrane
PtM	power-to-methane
PtX	power-to-X
RCFs	renewable carbon fuels
RED II	Renewable Energy Directive II
RES	renewable energy sources
RFNBOs	renewable fuels of non-biological origin
SNG	synthetic natural gas
SOECs	solid oxide electrolysis cells

8.1 INTRODUCTION

The entire world is currently navigating an intricate path toward achieving a sustainable energy transition, driven by swift progress in production technology for different energy sources. With the growing quantity and variety of these transporters, there is a rising demand for complete integrated solutions to address energy concerns [1]. The core of these solutions is the concept of integrated energy systems (IES), which represents a change in behavior regarding how energy infrastructures are viewed, developed, and managed. The concept of IES provides the basis for rethinking conventional energy infrastructure, but its full potential is realized by adopting a comprehensive approach to energy management [2]. By adopting a comprehensive approach to energy management, IES goes beyond the limitations of conventional energy architecture. The goal is to maximize resource efficiency, improve system adaptability, and reduce environmental damage. The central focus of this conceptual framework is the establishment of power-to-X (PtX), which acts as a flexible platform for connecting different energy domains and promoting collaborative interactions among the power, heat, transport, and industrial sectors [3].

Exploring extensively into the field of IES, it is clear that PtX technologies are important for harnessing renewable energy sources (RES) that are intermittent and resolving the inherent difficulties they provide. In order to attain a sustainable energy future, it is important to efficiently utilize intermittent RES. PtX technologies play a vital role in this domain, offering state-of-the-art options for storing excess electricity in various forms of storable energy carriers [4]. PtX solutions encompass a

DOI: 10.1201/9781032719436-8

range of applications, including electrolysis-based hydrogen production and the synthesis of carbon-neutral fuels. These solutions effectively tackle intermittent issues and enable a smooth transition to renewable energy systems [5]. Furthermore, PtX technologies not only tackle the urgent requirement to decrease carbon emissions but also provide creative solutions to counteract the environmental consequences of traditional energy systems. The release of carbon and greenhouse gases, primarily caused by the use of fossil fuels and conventional energy systems, is one of the most significant environmental challenges [6]. The need to reduce carbon emissions becomes increasingly urgent as time passes. PtX technologies, by using their capacity and incorporating carbon emissions from many industries into the creation of synthetic fuels and other goods, make a substantial contribution to carbon reduction efforts and the achievement of a future without carbon emissions [7].

PtX technologies are emerging as customized solutions to fulfill the various energy needs and aspirations of different locations as the regional dynamics of energy systems are investigated. Energy systems possess a distinct regional nature, influenced by elements such as the presence of facilities and resources, energy consumption, types of energy, and the economic conditions specific to the region. PtX technologies provide customized solutions to meet specific energy requirements in different regions, optimize the balance between energy supply and demand, improve energy adaptability, and support energy initiatives that correspond with community goals [8]. Real-world case studies demonstrate the adaptability of PtX technology as a catalyst for regional energy change. However, the effective incorporation of PtX technologies into current energy systems relies not only on scientific progress but also on advantageous economic and regulatory conditions. The effective implementation of PtX technologies relies not only on scientific progress but also on advantageous economic and regulatory conditions. Evaluating the economic feasibility of PtX solutions, the effectiveness of regulatory frameworks, and incentives for technology advancement in multi-energy systems will determine the changing roles of stakeholders in influencing energy market dynamics. Recognizing the complex relationship between economics, politics, and technology is essential for creating an environment that supports the expansion of PtX deployment [9].

Looking ahead, the evolution of PtX technologies promises to usher in a new era of energy transmission, characterized by groundbreaking innovations and transformative advancements in electrochemical procedures and material sciences. When considering the future of energy transmission, it is evident that this process has evolved and is moving forward. Thus, by examining the future of PtX technologies and identifying rising trends, innovative developments, and disruptive technical advancements, we may gain valuable insights to guide us on an appropriate path. These technologies are ready to introduce the next phase of electrochemical procedures and sophisticated materials, profoundly transforming the energy landscape [10].

8.1.1 Contributions

This chapter focuses on the awareness and advancement of IES and the crucial significance of PtX technologies in the modern energy environment. It begins by presenting a thorough definition of IES, emphasizing its complex and diverse characteristics and significance in the modern energy landscape. It demonstrates how the interplay between various energy sources and infrastructure contributes to achieving energy flexibility, sustainability, and efficiency by utilizing IES as a fundamental framework. Additionally, the study investigates the crucial significance of PtX technologies in revolutionizing energy storage approaches. The importance of energy storage in addressing grid intermittency and managing excess electricity is emphasized, and many types of storable energy carriers made possible by PtX processes are identified. Furthermore, the crucial function of energy storage in enhancing the flexibility of the grid is determined. Subsequently, the convergence of PtX technologies with decarbonization initiatives is examined, highlighting how these inventive methods help the electrification of sectors that now lack access to electricity and enable the production and utilization of synthetic fuels. Its potential to lower greenhouse gas emissions and promote a sustainable energy ecosystem is highlighted by looking at the symbiotic relationship between PtX and the utilization of captured carbon. Furthermore, it examines the various complex consequences of PtX technologies on the energy requirements for various regions. This discussion highlights the capacity of technologies to effectively align supply with regional demand, therefore enhancing energy independence and flexibility. This study stresses the importance of PtX as a community energy solution by showcasing case studies that demonstrate its regional applicability. It enables energy regions to effectively utilize their renewable resources and enhance energy efficiency.

The elements influencing the acceptance and scalability of PtX technologies are clarified and investigated, along with economic and regulatory considerations. Through the evaluation of the economic feasibility of these technologies and the delineation of policy frameworks and incentives, valuable information is offered to policymakers, industry stakeholders, and investors. Furthermore, the discussion addresses the market dynamics and the influential role of stakeholders in determining the path of PtX implementation. It emphasizes the necessity of collaborative efforts to fully unleash its potential. Ultimately, this text offers a progressive viewpoint on the future of PtX technologies and the imminent technological advancements that will change the energy sector. By recognizing new and developing trends and advances in the conversion and storage of energy, it establishes the foundation for ongoing investigation and innovation in utilizing PtX capacity to tackle energy and environmental concerns. Overall, the chapter's contributions can be summarized as follows:

- A brief overview of IES is provided, and its significant function in contemporary energy systems is explained.
- PtX technologies are emerging as innovative methods for storing energy, enabling the management of surplus power and enhancing the stability of the grid.
- Decarbonization activities involve the conversion of non-electrified sectors to electricity and the production of synthetic fuels via PtX techniques.
- PtX technologies are used to address regional energy needs by matching supply with demand and boosting flexibility.
- The widespread implementation of PtX technologies depends on several crucial factors, including economic viability, policy frameworks, market dynamics, and stakeholder engagement.
- PtX technologies provide numerous opportunities to address local energy requirements, enhance energy independence, and improve flexibility.
- Technological advancements and upcoming trends in energy storage and conversion are discussed.

8.2 THE CONCEPT OF INTEGRATED ENERGY SYSTEMS (IES)

8.2.1 Defining IES

Integrated energy systems are defined as the combined and entangled framework for the production, distribution, and consumption of various energy carriers, as well as the simultaneous management of these sectors. The concept of energy system integration is based on future energy visions and clean energy regulations, which have grown in significance in recent decades as people recognize and comprehend the value of sustainable energy. One of the key cases addressing the future vision of energy was a project presented by a research team at the University of Zurich's Power Systems Laboratory, named "Vision of Future Energy Networks" (VOFEN) [11]. This project showed and discussed a long-term image (20–30 years) of clean energy and green field approaches, as well as techniques for connecting to future energy networks. The concept of an energy hub was introduced and detailed at the core of this project, with the goal of integrating various energy carriers to meet the need for varied energies. The presentation of the concept of an energy hub aimed to achieve a variety of objectives, including the transition to multi-energy and integrated systems, as well as non-hierarchical energy structures [12].

In addition to other notions that follow the vision of the future of clean, affordable, and non-hierarchical energy, utilizing diverse sources is a fundamental principle. In general, *energy system integration* refers to the coordination and combination of various technologies and energy carriers' infrastructures in order to improve stability and reliability, as well as the optimization and management of various production, distribution, and consumption sectors. The major objective of this approach is to develop a landscape in which energies are more connected and coordinated and it is possible to integrate energies such that when one of the energy carriers is diminished for whatever reason, the system's needs are satisfied without problem [13]. IESs include several aspects, depending on their purpose and description, some of which are mentioned in the following.

- **Different energy sources.** IES can employ either RES (solar, wind, etc.) and fossil energy. Furthermore, combining different energy carriers in response to alterations in energy demand increases adaptability and flexibility [14].
- **Modern infrastructure and smart grid.** The integration of various energies necessitates modern technologies. Modern energy networks and intelligent management provide coordination between producers, users, and the network itself, allowing for the most efficient access to energy [15].
- **Energy storage.** IESs use a storage device to store various energy carriers during periods of reduced energy usage. Furthermore, these storage devices could act as an interface for various energy conversion technologies. A further purpose of storage in these systems is to balance the intermittent nature of some RES, such as solar and wind [15].
- **Energy management systems.** In order to control communication between different carriers in IES, a complicated energy management system is necessary. These systems utilize modern algorithms and data analysis to optimize various parts, such as production, distribution, and consumption, as well as communication between them, while simultaneously pursuing economic, environmental, and other optimization goals [16].
- **Decentralized energy production.** One of the primary goals of integration is to transition to decentralized production and address the problems raised by the conventional energy monopoly system. These decentralizations improve reliability, lower casualties, and encourage the use of renewable resources [17].
- **Intersectoral integration.** The integration of energy systems goes beyond the power sector and includes other sectors, such as transportation. For example, the electrification of the transportation system and the steps toward green transportation systems are among the cases where transportation systems can be considered as consumers in energy systems. This provides opportunities to improve and optimize the use of various resources [18].
- **Environmental sustainability.** By integrating various forms of RES and decreasing dependence on fossil fuels, the integrated energy system reduces greenhouse gas emissions, achieving one of the system's primary purposes, especially environmental sustainability [16].

In brief, *energy system integration* is a comprehensive strategy aimed at increasing energy efficiency, reliability, and sustainability. This approach involves seamlessly integrating varied resources and novel technologies into an integrated and interconnected network. This strategy contributes significantly to establishing a global energy vision that prioritizes flexibility and sustainability by enhancing the synergy between various components. The integration of energy systems serves as a guiding principle in the transition to a future in which energy use is not only efficient but also consistent with environmental principles. Figure 8.2 is an example of a network integrated with various energies and communication of various carriers.

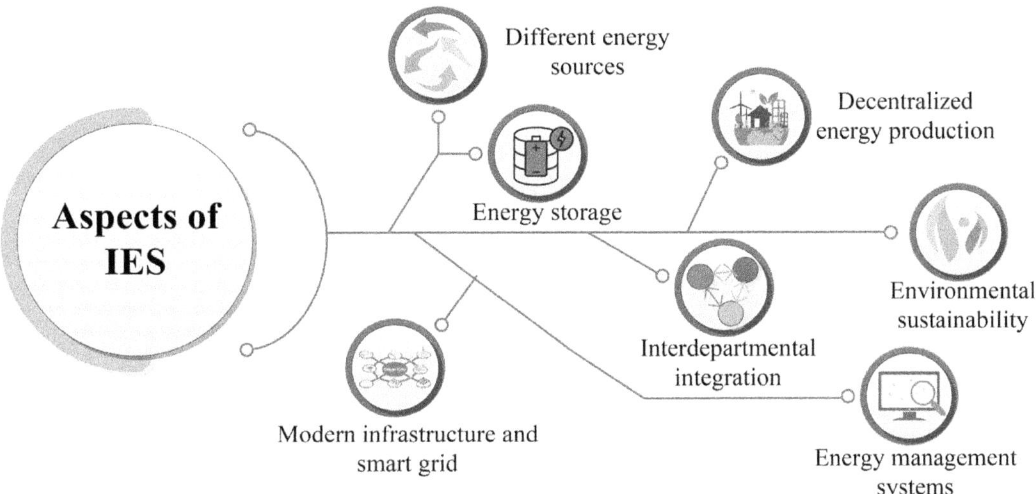

FIGURE 8.1 Different aspects of energy systems integration.

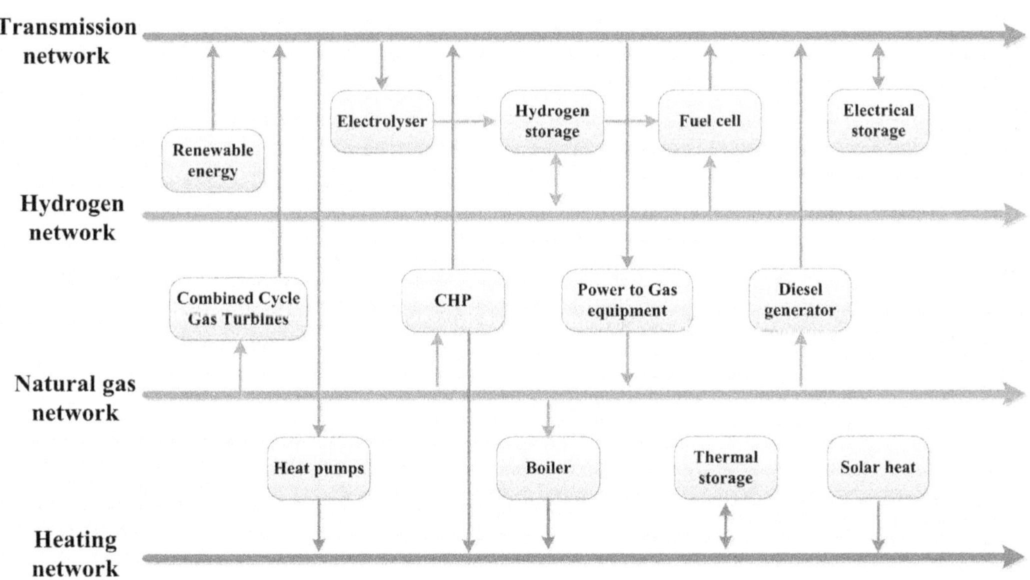

FIGURE 8.2 Example of how various energy carriers interact and communicate across an integrated energy system.

8.2.2 ROLE OF IES IN MODERN ENERGY SYSTEMS

Modern energy systems are a complex network of technologies and infrastructures that are controlled and operated in accordance with policies that promote the most optimal utilization of energy carriers. These systems, in addition to the previous monopolistic models designed to be centralized, incorporate a variety of energy sources that include RES, traditional fossil fuels, and advanced technology. The main challenge of these modern networks is achieving a balance between the expanding demand for various energy carriers while also reducing environmental consequences, increasing stability and flexibility [19].

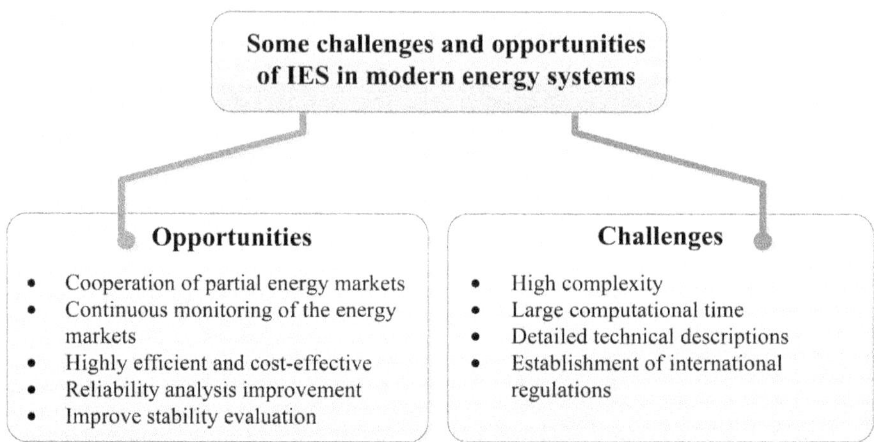

FIGURE 8.3   Some challenges and opportunities of IES in modern energy systems.

IESs play a critical role in addressing these challenges. IESs are known for their ability to combine and optimize various energy carriers, as well as their storage capacity. The comprehensive approach of IES that manages and controls a variety of energies at the same time improves the reliability and flexibility of modern systems while also achieving a more stable balance in the simultaneous usage of various energies. Furthermore, the use of IESs minimizes dependence on fossil fuels, reducing environmental consequences and increasing environmental compatibility. This viewpoint and its implications position IES as a catalyst for substantial changes in energy production and consumption. IES, as a strategic solution, plays an important role in responding to the environmental concerns and social acceptability of modern energy systems. By utilizing advanced control and monitoring systems and technologies for real-time optimization, IES not only improves the momentary response to energy demand but also allows for the efficient use of excess energy during periods of reduced consumption [2, 20].

Although the benefits of IES in current energy systems have been discussed, implementing these systems in modern networks presents challenges. To fully utilize IES's capabilities, relevant design and integration methodologies must be used, as well as supporting legal frameworks and policies. Figure 8.3 also illustrates, shortly and case by case, some of the challenges and opportunities that should be explored in IESs. These various challenges and opportunities must be carefully considered to guarantee that the IES is properly integrated and compatible with modern energy systems [2].

8.2.3   Power-to-X: Bridging Energy Systems

PtX is a sustainable energy system approach in which extra available energy, such as RES, and excess production are transformed into various other forms, such as hydrogen, fuels like ethanol, or another carrier. The primary purpose of PtX is to address the problem of renewable resource intermittency through storage [8]. The function of this system serves a useful purpose, including increasing response to production and demand balances as well as linking disparate energy systems. PtX's approach and technologies, which include various instruments for storing and transporting different energy carriers in various sectors, allow different renewable energies to be integrated into existing infrastructures. As a result, PtX serves as a fundamental principle in enhancing the future energy system, by increasing the interconnection of diverse energy carriers and improving their compatibility [21].

As previously stated, one of the primary issues in contemporary energy systems is an imbalance between energy production and demand, resulting in a scarcity of energy in the demand sector and energy waste in the production sector. In this regard, PtX, as an intermediary, can facilitate the movement of excess energy from surplus production sectors to the demand sector. For example, extra solar or wind energy can be converted into hydrogen to meet various industrial, transportation, and heating demands. Furthermore, PtX could assist in the integration of RES with traditional energy systems. By transforming renewable energies into carriers that are compatible with the traditional energy grid, PtX can facilitate the transition to a more sustainable and environmentally friendly energy system [22].

Because of these functionalities of PtX in energy systems, its bridging role extends beyond conventional energy system issues. In this approach, PtX promotes coordination among many sectors, such as transportation and industry, ultimately strengthening the establishment of an efficient energy environment. According to modern energy policies and a clean and sustainable energy perspective based on renewable energies, it is vital to harness the potential of PtX to connect various energy systems [23].

8.3 ENERGY STORAGE THROUGH POWER-TO-X

8.3.1 NECESSITY OF ENERGY STORAGE

Energy storage is a critical component in achieving the vision for the future of energy, particularly in the realm of PtX technology. The primary purpose of storage systems is to address the problem of intermittent RES, such as solar and wind. By collecting excess energy produced by RES during peak production times, storage systems provide energy without interruption during periods of increased use and reduced production from renewable sources. This characteristic and performance of storage devices improve network stability and the energy system's resistance to disruptions induced by production and supply imbalances, as well as changes in renewable resource production [24].

Another aspect of the significance of energy storage in multi-energy systems is its role in optimizing and managing the use of PtX technology. Storage devices not only store energy at peak energy production times but also have the ability to convert distinct energies into each other, increasing the possibility for optimization across different carriers. Increasing the potential for RES optimization and management using storage devices not only improves reliability but also makes it easier to implement a variety of PtX technologies and adds to sustainable energy infrastructures. With these interpretations, storage devices play an important role in energy networks. Some of the benefits of energy storage systems in energy networks are listed in what follows [25].

- **Sustainability and reliability.** Energy storage balances supply and demand, improving sustainability. Furthermore, the significance of storage systems as a buffer in decreasing swings in renewable production ensures a consistent and non-fluctuating supply [24].
- **Optimizing demand response.** Energy storage systems can store excess energy and compensate for energy reductions during peak use periods. This capability enables consumers to optimize their energy consumption while also reducing network load during peak periods [26].
- **Cost reduction and savings.** Because different energy markets have varying pricing at different times of day, customers can employ storage devices to provide the energy required during price reductions and use it at other times, lowering the total cost of energy [24].
- **Reduction of losses and transmission costs.** Storage is a strategy for lowering the cost of big expenditures in transmission infrastructure construction. Furthermore, when energy storage and consumption in local dimensions improve, there is less demand for long-distance energy transmission, resulting in lower energy losses [27].

8.3.2 POWER-TO-X AS STORAGE FOR EXCESS ELECTRICITY

Storage of excess electricity is a crucial element in energy management systems, particularly in systems incorporating renewable resources. Storage utilizes many methods and technologies, such as tiny batteries for quick energy supply, hydraulic pump storage, compressed air storage, and thermal storage to meet electricity demands over extended periods. PtX technologies provide alternative techniques for storing surplus electricity in the energy system, in addition to the usual storage methods. PtX is focused on converting electricity into more portable, practical, and controllable carriers for energy storage, rather than relying solely on traditional technologies. For example, using PtX technology, surplus electricity is utilized to turn water into hydrogen through an electrolyzer. The hydrogen generated through this method is regarded as an energy carrier with multiple applications and potential for future utilization. PtX technology could convert excess electricity into synthetic fuels, like methane, thus avoiding the wastage of excess power. Integrating this technology into traditional energy systems significantly affects parallel systems, like transportation, by requiring greater electricity storage and fuel production. Furthermore, these fuels can be transformed back into electrical power through relevant technology to support the power grid during periods of demand and power deficits [5].

8.3.3 Different Forms of Storable Energy Carriers

Energy carriers that can be stored play a critical function in modern energy systems. Storage is critical in modern systems that combine many energy carriers as well as intermittent renewable sources, such as wind and solar energy. The integration of energy systems with varying consumption demands and the presence of PtX technologies have resulted in various carriers in these systems. The existence of many carriers, as well as the issue of supply and demand balance in different carriers, emphasizes the need of storing storable carriers and consuming them when needed. With these characteristics, the storage of various carriers and their diversity play an important role in separating the energy production sector from immediate consumption, and it is seen as a strategic solution to the lack of synchronization between energy supply and demand [28]. The storage of various energy carriers is broadly categorized into four sectors: electrical, mechanical, thermal, and chemical. All of them are explained separately in what follows.

8.3.3.1 Electricity

Electricity is the most commonly used and stored energy carrier in multi-energy and integrated networks. Electricity storage is the most commonly employed approach in these networks. The majority of electric energy is stored in multi-energy systems using batteries, specifically lithium batteries. These batteries store surplus electricity created during periods of low usage, which is then released during peak demand, when production declines [24]. Electric energy storage also uses capacitors and supercapacitors for short-term energy storage [29]. Overall, storing the electrical energy carrier improves network flexibility and enables the smooth integration of renewable resources into modern energy networks.

8.3.3.2 Mechanical

Mechanical energy conversion and storage is a suitable method for energy storage in multi-energy networks. Excess electrical energy can be utilized during times of surplus generation and low consumption by storing water in raised reservoirs using hydroelectric pumps, effectively converting it into potential energy. The stored energy can be released and transformed into electrical energy during periods of high demand. Furthermore, flywheels are a novel method for storing mechanical energy, particularly suited for short-term storage needs. This approach involves storing kinetic energy in the rotational motion of the flywheel. Stored mechanical energy, using common methods, can be effectively converted to electrical energy when needed, increasing the flexibility and resilience of the multi-energy grid. Mechanical energy storage technologies are useful for stabilizing the variable output of RES and enhancing the dependability of energy provision in modern and changing energy networks [30].

8.3.3.3 Thermal

Thermal energy carriers are crucial in multi-energy networks because they respond to various users in different network sections. Thermal energy carriers in multi-energy networks serve as an interface for storing and transferring energy across diverse energy sources, like electricity and renewable energy. Thermal energy storage systems vary based on storage scale, duration, and desired effectiveness. Heat storage systems often use materials such as water, molten salts, or phase transition compounds. One type of heat storage is thermochemical storage, which involves utilizing chemical reactions to store and release heat. Heat storage devices, similar to other storage devices, enhance the flexibility of integrated networks by separating the production of energy from the consumption of energy [31].

8.3.3.4 Chemical

Chemical carriers are crucial elements in IES utilizing PtX technology and serve a significant function. Carriers consist of various chemical substances tailored to specific networks and purposes, serving a distinct function in storing and distributing energy. These carriers are created during periods of surplus electricity and renewable energy production through PtX methods. They are utilized across various sectors, like industry and transportation, and have the capacity to store and reuse energy for electricity generation. These carriers aid in achieving a stable and adaptable energy system by reducing production peaks, meeting consumer demands, and optimizing the energy system [21]. The chemical carriers used in IES vary depending on the technology and grid needs. Following are some examples:

- **Hydrogen (H_2).** Hydrogen is crucial in the transition toward sustainable energy systems. An established technique for its generation involves the electrolysis of water, which is a crucial step in PtX technology. In this process, an electric current is applied to water, resulting in the separation of its component elements: hydrogen and oxygen. This simple but essential method produces hydrogen of high purity, which may be used as a flexible and environmentally friendly fuel in several industries. Hydrogen has great potential in decreasing greenhouse gas emissions and promoting a more environmentally friendly future, whether it is used for electric vehicles, to fuel industrial operations, or to generate energy [21].
- **Synthetic natural gas.** Synthetic natural gas (SNG) is an essential element of the PtX system, namely in the power-to-methane (PtM) sector. PtM is dedicated to the sustainable production of methane (CH_4) by utilizing carbon dioxide (CO_2) and RES. The Sabatier process is a chemical reaction where hydrogen (H_2) and CO_2 combine to form CH_4 and water (H_2O). Methane is an important energy carrier because it has a larger energy content per unit volume compared to hydrogen. An important benefit of SNG is its capacity for storage, allowing for efficient storage and transportation and providing flexibility in the distribution and consumption of energy [32].
- **Synthetic fuels (e-fuels).** PtX technologies have the capacity to generate a range of synthetic fuels, including synthetic gasoline, diesel, and jet fuel. These fuels are generated by the conversion of hydrogen and carbon dioxide utilizing techniques like Fischer–Tropsch synthesis and hydrocracking. They serve as a replacement for traditional fossil fuels, facilitating a smooth transition to ecologically sustainable energy sources [10]. An important benefit of synthetic fuels is their high storage capacity. Compared to sporadic RES like solar or wind energy, synthetic fuels possess the capacity to efficiently store and transmit energy, offering flexibility in energy distribution and consumption. Due to their high energy storage capacity, they are crucial for applications that need a consistent or immediate energy supply. Additionally, it provides the capability to guarantee the reliability and

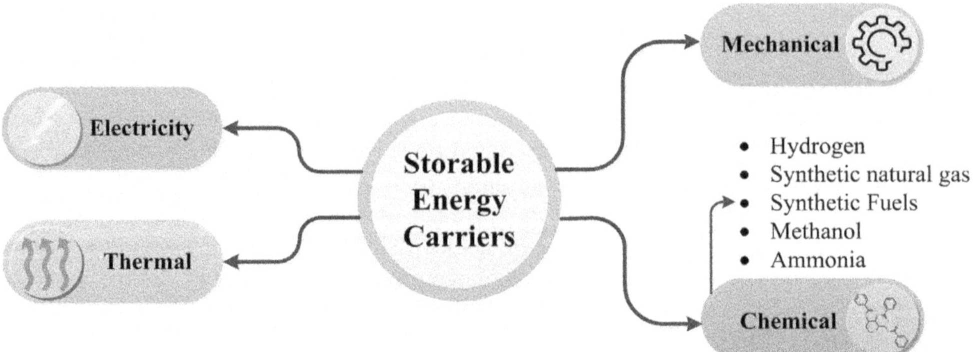

FIGURE 8.4   Variety of storable energy carrier forms.

dependability of energy systems while aiding in a transition toward reduced carbon emissions [33].

- **Methanol (CH$_3$OH).** The use of CO$_2$ in methanol synthesis is a valuable application of PtX technology due to the liquid fuel's high energy density. This characteristic enables convenient storage, transportation, and utilization in various chemical production processes. Methanol is an essential precursor in the production of olefins, dimethyl ether, and formaldehyde, which find use in many sectors, including textiles, packaging, and paint manufacturing. Furthermore, methanol serves as a highly adaptable alternative or supplement to traditional combustion engines and methanol fuel cells. Renewable methanol manufacturing systems, such as the ThyssenKrupp Uhde methanol technology, use renewable hydrogen generated by alkaline electrolysis and waste CO$_2$ in a hydrogenation unit [8].

- **Ammonia (NH$_3$).** PtX technologies play a role in developing the production of ammonia, which is a fundamental component of the $80 billion worldwide fertilizer industry. Historically, ammonia has been produced by the energy-intensive Haber–Bosch process, which depends on high pressures, temperatures, and hydrogen and nitrogen input obtained from fossil fuels. In order to eliminate carbon emissions from this business, intensive research is being conducted to investigate different methods of producing ammonia using renewable sources of electricity. These methods involve using renewable hydrogen in the Haber–Bosch process, converting pure nitrogen into ammonia through electrolysis (eNRR), converting air into ammonia through plasma-driven processes, oxidizing air to nitrate and nitrite and then reducing it to ammonia, and capturing NO$_x$ emissions from power plants and reducing them into ammonia (NORR). Currently, there is a growing worldwide trend toward using renewable hydrogen in the Haber–Bosch process which is leading to a substantial improvement in energy efficiency. In addition, NORR is becoming a strong competitor because of the accessibility of NO$_x$ from power plants, well-developed capture systems, and a high yield of ammonia synthesis [8].

8.3.4   ROLE OF ENERGY STORAGE IN GRID STABILITY

Energy storage devices are crucial for maintaining the stability and dependability of power grids, particularly in the intricate structure of small and integrated systems. The core of their performance is in their capacity to quickly absorb or release power while maintaining an optimal balance between immediate power production and consumption. These systems have a role in controlling voltage and frequency, acting as stabilizers and improving the overall performance of microgrids during changes in power. Energy storage systems are essential assets in situations when power output or demand levels are unexpected or changing. These devices have the ability to promptly

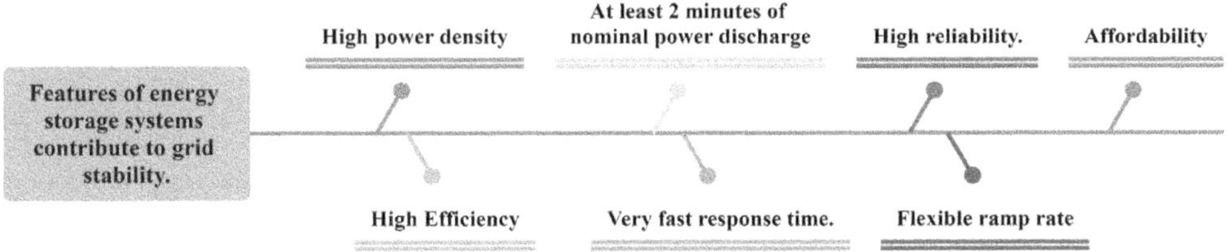

FIGURE 8.5 Critical characteristics of energy storage systems to improve grid stability.

respond to abrupt fluctuations in power use or generation, hence ensuring the stability and smooth operation of the grid [34].

To achieve maximum grid stability, energy storage systems require a set of necessary characteristics, each of which serves a separate but interconnected objective. These significant characteristics may be summarized in seven specific characteristics: (1) These systems are characterized by high power density, which is a crucial feature. This feature allows energy storage technologies to provide substantial power output within a small area. This characteristic is very significant in expediting prompt reactions to variations in demand or supply and guarantees timely interventions to maintain grid stability. (2) Efficiency is a crucial attribute that determines the capacity of storage devices to efficiently convert and retain energy throughout the process of charging and discharging. The enhanced efficiency not only reduces energy wastage but also optimizes the use of stored energy and improves the efficacy of grid stabilization efforts. (3) The reliability of storage systems is regarded as another crucial aspect. This characteristic ensures the consistent operation of energy storage systems under various operating situations. Systems that are strong and dependable instill trust in grid operators, decrease the potential of interruptions, and enhance the ability of energy networks to withstand challenges. (4) A stable discharge capacity at rated power levels for a minimum of 2 min is crucial as it enables the system to effectively manage prolonged changes in power demand or supply. This function reduces potential instabilities and guarantees continuous power supply for end users. (5) One crucial element of energy storage systems is the ability to quickly modify their output power in reaction to fluctuations in grid circumstances, which is known as flexibility in ramp rates. These technologies enhance network flexibility by offering quick reaction capabilities and effortlessly adjusting to changing power dynamics. (6) Ensuring minimal efficiency drop during discharge cycles is crucial for preserving the long-term performance and efficiency of energy storage technologies. (7) Eventually, ensuring affordability is a crucial economic element that ensures the long-term economic viability and sustainability of grid stabilization operations [35].

Taking use of these different properties, energy storage devices are emerging as critical enablers of grid sustainability, offering a versatile toolkit for navigating the complexity of today's energy landscape. Energy storage plays a critical role in increasing resilience and reliability in off-grid communities as well as urban microgrids. As technology progresses in this arena, the potential for energy storage to transform grid sustainability remains boundless [35].

8.4 DECARBONIZATION AND POWER-TO-X

8.4.1 Electrification of Non-Electrifiable Sectors

The integration of electrical solutions into sectors that have traditionally been resistant to such advancements poses numerous challenges, primarily due to the limitations of conventional methods and infrastructure. Industries such as heavy manufacturing, aviation, and long-haul shipping heavily depend on dense energy sources, like fossil fuels, because there are no viable electrical alternatives available. Nevertheless, attempts to electrify these sectors by conventional means encounter substantial barriers, such as infrastructure constraints, exorbitant costs, and inefficiency. Moreover, the ongoing dependence on fossil fuels exacerbates environmental problems, specifically by escalating carbon dioxide emissions and intensifying climate change [6].

Emerging PtX technologies offer a revolutionary way to address the enduring problems with electrifying sectors that were previously thought to be non-electrifiable. PtX technologies refer to a wide range of advanced processes designed to transform excess electricity into various forms of energy carriers, including hydrogen, synthetic fuels, and chemicals. Industries may effectively overcome the limits of direct electrification by utilizing PtX technology, enabling them to smoothly integrate RES into their processes. Hydrogen, produced via PtX techniques, is a crucial and sustainable energy carrier that has the potential to transform industrial operations, aviation, and maritime commerce. It can significantly reduce the dependence on fossil fuels and greatly decrease carbon dioxide emissions. Moreover, the synthesis of synthetic fuels through PtX methods not only presents a convincing substitute for traditional fossil fuels but also enables a gradual and fair shift toward sustainable energy systems. This highlights the transformative capacity of PtX technologies in forming the energy landscape in a positive manner [7].

Thus, there is a lot of promise for decreasing carbon dioxide emissions and speeding up the switch to sustainable energy systems by using PtX technology to electrify areas that are normally unfit for electricity. PtX overcomes the difficulties posed by conventional electrification technologies and offers a flexible solution that may significantly reduce the environmental impact of sectors dependent on fossil fuels. Adopting PtX technologies is an essential measure in achieving the goal of carbon neutrality and the worldwide struggle against climate change. However, utilizing these potential needs collaboration by governments, industry, and researchers to speed the advancement and adoption of PtX infrastructure and solutions [7].

8.4.2 Creation and Utilization of Synthetic Fuels

The prevalent utilization of conventional fuels, primarily sourced from fossil origins, has historically served as the principal energy reservoir for global domestic, industrial, and transportation needs. However, their pervasive application is accompanied by a pronounced environmental toll. The combustion processes of these fuels engender the release of a myriad of deleterious pollutants, including carbon dioxide (CO_2), sulfur oxides (SOx), particulate matter (PM10 and PM2.5), nitrogen oxides (NO_x), and volatile organic compounds (VOCs). These emissions constitute a grave threat to both ecological systems and human health, markedly exacerbating the phenomenon of environmental degradation. Notably, the transportation sector, heavily reliant on fossil fuel derivatives, assumes a prominent role in the emission landscape, contributing a substantial 16% share of global greenhouse gas emissions. This sector has witnessed a stark escalation in CO_2 emissions, surging by 45% from 1970 to 2007, with a further projected ascent of 40% by the year 2030. In 2020, transportation-related activities contributed to 27% of the 5,981 Mt of CO_2 emissions in the United States. Exacerbating this issue is the predominance of road traffic, responsible for an alarming 30–40% of CO_2 emissions attributable to transportation. Consequently, immediate and concerted efforts are imperative to mitigate the adverse ramifications wrought by the entrenched reliance on traditional fuels, both in terms of environmental preservation and safeguarding public health [18].

PtX technology has recently emerged as an achievable solution in the worldwide attempt to reduce carbon emissions by producing synthetic fuels. PtX, in its essence, entails the transformation of sustainable energy sources, such as solar or wind power, into synthetic fuels via a sequence of chemical reactions. Hydrogen, methane, and liquid hydrocarbons are sustainable alternatives to conventional fossil fuels. They effectively decrease carbon emissions in various sectors [21].

The process of producing synthetic fuels begins with electrolysis, which involves the separation of water into hydrogen and oxygen through the use of energy derived from renewable sources. Afterward, various techniques can process this hydrogen to produce synthetic fuels, like synthetic methane, or liquid hydrocarbons, like methanol or dimethyl ether. PtX enables the utilization of additional RES and eliminates the issue of irregular power generation commonly linked with solar

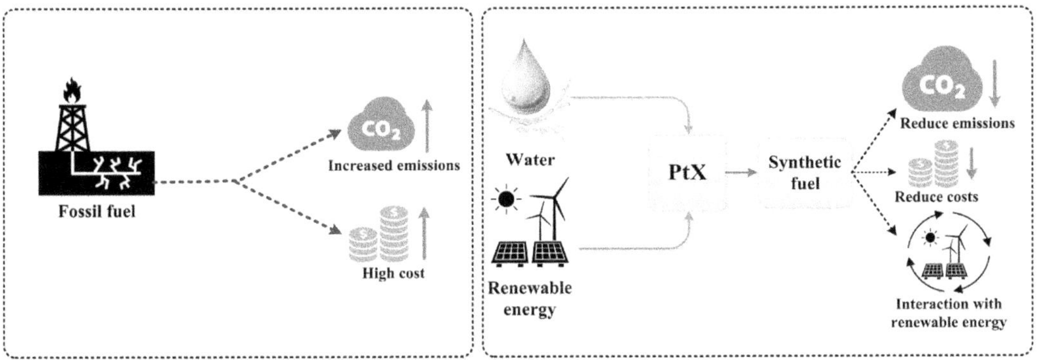

FIGURE 8.6 Synthetic fuels and decarbonization.

and wind energy. PtX reduces greenhouse gas emissions and promotes energy security and grid stability by storing renewable energy in the form of synthetic fuels [8, 36].

Moreover, the adaptability of synthetic fuels makes them well-suited for a diverse array of uses, extending from transportation to industrial activities. One instance is the utilization of synthetic methane as a renewable substitute for conventional natural gas through its injection into pre-existing pipelines. Furthermore, PtX has the capability of producing liquid hydrocarbons that can efficiently substitute gasoline, diesel, and aviation fuels. Synthetic fuels are produced by PtX technologies, which rely on RES, including wind, sun, and hydropower. The combination of renewable power generation with fuel manufacturing allows for the development of carbon-neutral or even carbon-negative fuels, which helps achieve emission reduction targets. This technique provides a seamless shift to RES without necessitating substantial modifications to existing infrastructure or vehicles [37].

Synthetic fuels not only give immediate decarbonization benefits but also promise long-term sustainability advantages through the facilitation of carbon recycling and the incorporation of circular economy principles. By employing PtX technology, carbon dioxide extracted from industrial activities or straight from the atmosphere can be turned into synthetic fuels. This novel process efficiently closes the carbon loop, mitigating greenhouse gas emissions at their initial sites. This closed-loop methodology harmonizes with worldwide plans to achieve carbon neutrality and face climate change head-on. It positions synthetic fuels as important factors in moving toward a low-carbon future, harmonizing with overall environmental goals on a global scale [38].

8.4.3 POWER-TO-X AND CARBON CAPTURE UTILIZATION

In the context of PtX technologies, the crucial procedure of carbon absorption involves the careful removal of carbon dioxide from many sources, including the emissions of industrial activities and the surrounding atmosphere. This precise extraction technique utilizes advanced technologies specifically engineered to efficiently and sustainably capture CO_2. After it is captured, this CO_2 turns into an adaptable raw material that may be used in a wide range of chemical processes. These pathways, which are carefully designed and coordinated, convert the captured CO_2 into a range of chemicals and fuels with added value through a series of precisely regulated reactions and transformations. One famous example of the many possibilities is the production of synthetic fuels, such as synthetic methane, or synthetic liquid hydrocarbons. These fuels, created with great care using techniques like Fischer–Tropsch synthesis or methanation, are powerful solutions at the forefront of sustainable energy [10]. These are not just replacements but, rather, significant advancements that can be easily incorporated into our current energy system. These synthetic alternatives play a leading role in the significant shift toward reducing carbon emissions by allowing for the smooth

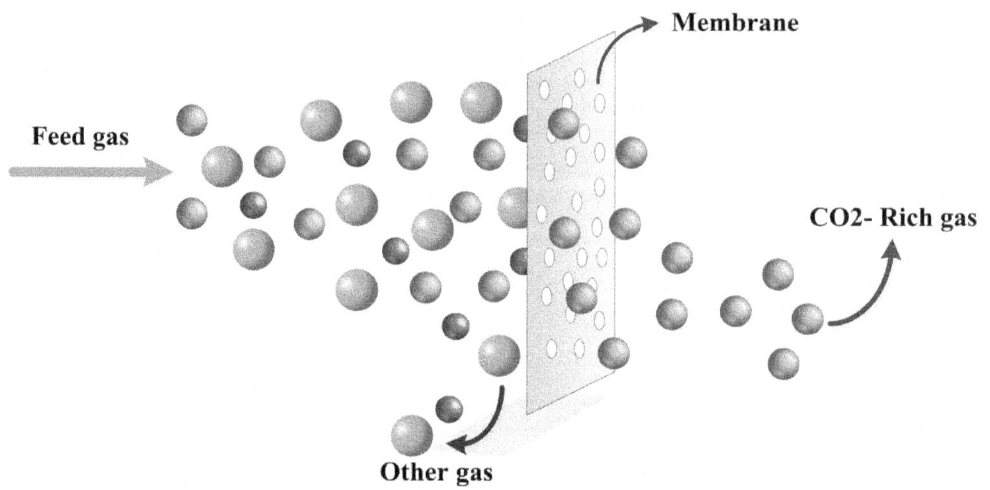

FIGURE 8.7   Carbon adsorption process using membrane separation technology.

replacement of traditional fossil fuels. They are especially influential in the fields of transportation and heating. Therefore, PtX technologies, with their cleverness and adaptability, play a significant role in the effort to address climate change and create a sustainable future [39].

The complex process of carbon consumption in PtX technologies involves a wide range of advanced processes, each carefully designed to collect carbon dioxide from different sources with great accuracy and effectiveness. Chemical absorption is a fundamental technique that uses carefully selected solvents or sorbents to capture CO_2 molecules from large gas streams. This creates a concentrated stream high in CO_2, which can then be processed and used [40]. Simultaneously, physical adsorption is a crucial method that utilizes the natural properties of porous materials, like activated carbon or zeolites, to effectively capture CO_2 molecules on their large surfaces through various physical interactions. This intricate process achieves a highly effective separation of CO_2. In the complex field of PtX methodologies, membrane-based separation plays a crucial role. It utilizes semi-permeable membranes to selectively transport CO_2 molecules while preventing other substances in the gas stream. This brings about a new era of accuracy and effectiveness in carbon capture and utilization efforts. PtX technologies combine many strategies to provide sustainable carbon management, leading to a future characterized by innovation, resilience, and environmental stewardship [41].

After the CO_2 is caught and concentrated, it might undergo additional processing steps based on the intended final products. In the creation of synthetic fuels, concentrated CO_2 is frequently coupled with hydrogen obtained from renewable sources using electrolysis. This combination leads to the formation of hydrocarbons through catalytic processes. Alternatively, the captured carbon dioxide can be employed in chemical synthesis processes to generate a diverse array of useful goods, such as polymers, chemicals, or construction materials. In PtX technologies, carbon absorption is essential as it allows for the use of CO_2 as a raw material to produce renewable fuels and chemicals. This contributes to the shift toward a carbon-neutral economy [39, 41].

8.5   POWER-TO-X FOR REGIONAL ENERGY NEEDS

8.5.1   MATCHING SUPPLY WITH REGIONAL DEMAND

In order to ensure a dependable and enduring energy provision in the current ever-changing environment, it is necessary to employ intricate technologies that can promptly adjust supply in response to demand. This holds particularly true in regions characterized by substantial variations in energy generation, which are influenced by factors such as weather conditions, daily cycles, and seasonal shifts [8]. The issues associated with unreliable supply and fluctuating energy production are multifaceted, encompassing not only technical aspects but also economic, environmental, and social factors. From an economic perspective, an unreliable energy supply can result in inefficiencies and higher expenses for both consumers and producers. Moreover, it has the potential to weaken investment in renewable energy infrastructure and impede the advancement toward sustainability objectives. From an environmental perspective, an imbalance between supply and demand could compel suppliers to utilize inefficient and highly polluting items, resulting in inadequate utilization of clean energy sources or in excessive dependence on polluting alternatives. In terms of social impact, inconsistent access to energy can have a disproportionate effect on marginalized populations, worsening the problem of energy poverty and inequality [42].

PtX technologies are being recognized as key solutions to tackle the issues arising from the imbalance between supply and demand. These technologies provide a flexible method for storing surplus energy generated during periods of excess production and utilizing it when demand surpasses supply. By incorporating PtX systems into energy infrastructure, regions may efficiently manage swings in renewable energy generation and assure a more consistent and dependable energy supply. Moreover, the utilization of PtX technology enables a separation of energy generation from consumption patterns, hence enhancing the adaptability of energy management and decreasing dependence on fossil fuel–dependent backup systems [43].

Furthermore, the incorporation of PtX technologies will enhance the region's independence and adaptability in terms of energy. Communities can diminish their dependence on centralized energy networks and susceptible supply chains by utilizing renewable resources that are readily accessible in their local areas. Decentralization enhances both energy security and the ability of regions to tailor energy production to their unique requirements and resources. Furthermore, the deployment of PtX solutions expedites economic growth by fostering the emergence of novel enterprises and employment prospects within the renewable energy sector. However, fully harnessing the potential of PtX demands the implementation of comprehensive policy frameworks and regulatory regimes that encourage investment and innovation. Effective collaboration across stakeholders, such as governments, business, and research institutions, is crucial to promote the mainstream acceptance of these revolutionary technologies and expedite the shift toward a sustainable and adaptable energy future [25].

8.5.2 POWER-TO-X AS A SOLUTION FOR ENHANCED FLEXIBILITY

PtX technology is recognized as a viable solution for enhancing flexibility of local energy needs. PtX includes a series of procedures that transform surplus electricity, typically derived from sustainable sources, like wind or solar power, into different types of energy carriers or commodities. The various types of energy encompass synthetic fuels, like hydrogen, methane, or ammonia, along with other valuable by-products, such as chemicals or heat. PtX's wide range of products and various energy forms make it flexible, making it an attractive choice for regions aiming to manage the variability of energy demand and renewable energy production [21].

An important benefit of PtX is its capacity to store renewable energy in a format that is simply transportable and applicable across several industries. PtX offers a solution for regions with ample renewable energy resources but limited grid capacity or cases of excess generation. It enables the conversion of excess power into fuels or materials that can be stored. These can be incorporated into current energy infrastructure, such as transportation, manufacturing, and heating systems, to enhance the overall flexibility and resilience of the system [44].

Furthermore, PtX technologies are crucial in the process of reducing carbon emissions in areas of the economy that are difficult to reduce carbon emissions. Industries such as aviation, shipping, and heavy manufacturing rely heavily on fossil fuels and currently lack viable low-carbon alternatives [45]. PtX technology offers a technique to create fuels and chemicals that are either carbon-neutral or even carbon-negative. This can help reduce greenhouse gas emissions from these industries, making a valuable contribution to global initiatives aimed at addressing climate change [40].

PtX technologies additionally provide flexibility and carbon reduction advantages, also enhancing energy system resilience and security. PtX contributes to the development of energy systems that are more resilient and less vulnerable to supply disruptions and price fluctuations by diversifying energy sources and decreasing reliance on imported fossil fuels. Moreover, PtX facilities can be dispersed throughout different areas, making use of renewable resources that are accessible in those locations. This improves energy self-sufficiency not only at the national level but also at the community level. In general, PtX has great potential to improve flexibility in energy systems and promote sustainability and resilience objectives [23].

8.5.3 Case Studies: Regional Applications of Power-to-X

PtX technologies have garnered significant interest in many global locations within the sustainable energy industry. This section presents two distinct case studies from Italy and Japan as examples of papers that investigate the regional applications and potential of PtX. The case studies, designated as Case 1 and Case 2, offer valuable insights into the integration and enhancement of PtX techniques within their respective regions' energy landscapes. The objective of presenting and summarizing this research is to provide a clear understanding of the many methods and outcomes associated with PtX, with a focus on highlighting its significance and potential influence on regional energy systems.

Case 1: The research paper [9] presented a case study that thoroughly analyzed the integration of green hydrogen production and PtX technologies in the Campania region of Italy, with the aim of achieving substantial sustainability objectives. The report provided and emphasized a prospective scenario for the year 2050, aiming to achieve an 80% reduction in CO_2 emissions compared to the levels recorded in 1990. This study posited that integrating alkaline electrolyzers into the energy system may steer the region toward a more sustainable development path, reduce greenhouse gas emissions, and offer a clean substitute for conventional internal combustion engines. The statistical analysis showed that by decreasing the capacity by 1,300 MW and storing 0.9 GWh, this system in this particular region has the capability to fulfill 10% of the transportation requirements. Additionally, it demonstrated the potential influence of hydrogen production on satisfying energy demands. Furthermore, the study argued that the integration of electrolyzers into the grid not only enhances energy security but also decreases dependence on a major energy source, particularly in regions that largely rely on imported fossil fuels. This study emphasizes the importance of broadening the range of internal energy sources in order to improve energy autonomy and robustness. Emphasizing energy efficiency measures and the installation of photovoltaic systems has been recognized as a crucial step to reduce emissions and transition toward a more sustainable energy combination. The study highlighted that the statistical insights offered provide a solid foundation for policymakers and stakeholders seeking to implement comparable methods in areas experiencing energy transition challenges. This paper argues for a comprehensive strategy for the energy transition and supports the incorporation of green hydrogen production as a crucial element in future energy policies. The study's findings demonstrate that regions can make a significant contribution to carbon emissions reduction and the transition to cleaner energy systems by using PtX strategies and electrolysis for hydrogen production.

Case 2: The study conducted in reference [46] explored the capabilities of flexible P2X technologies within the framework of renewable energy systems, with a specific focus on Japan as a case study. The objective of this study is to analyze the effects on the structure of a system and the associated energy expenses through the optimization of energy systems using P2X technologies, including water electrolysis, methanation, Fischer–Tropsch synthesis, and Haber–Bosch synthesis. The findings demonstrated that each P2X technology efficiently transmits electrical charges and reduces curtailment by over 80%. Water electrolysis is significant because of its cost-effectiveness and potential to be scaled up. The study demonstrated a noteworthy decrease in overall expenses by 35%, as well as a 41% reduction in energy supply costs and a 30% reduction in hydrogen supply costs. The investigation in this study revealed that the capacity factor of P2X plays a crucial role, with the most significant impact observed at a level of 30%. In order to be competitive with fossil fuels, synthetic hydrocarbons require a carbon price of €356/t of CO_2. Domestic electric fuels could encounter tough competition due to the high costs of global fuel. The study emphasized the financial benefit of P2X compared to other solutions for flexibility, demonstrating a 35% decrease in overall expenses. Furthermore, it was highlighted that P2X presents a competitive approach for enhancing the efficiency and cost-effectiveness of energy systems. This is achieved by diminishing the necessary capacities of power producers, stationary batteries, and transmission networks. Moreover, this study highlights the significance of incentive mechanisms in facilitating the integration of adaptable P2X technology into renewable energy systems. Potential methods to encourage P2X adoption include the utilization of real-time energy markets and aggregation systems. This study emphasized the need of using P2X operations to maintain a balanced grid system. It suggests that P2X operations should be seen as crucial regulators rather than solely devoted power supply buffers. In this study, the findings offered useful insights into the systemic impacts and competitiveness of P2X technologies with dynamic performance. They also gave a thorough understanding of how these technologies enhance the resilience and sustainability of regional energy systems.

8.5.4 Power-to-X as a Community Energy Solution

PtX technologies provide an opportunity to link decentralized renewable energy generation with local energy consumption. These technologies offer innovative methods to utilize excess renewable electricity and transform it into storable fuels or chemicals. These features enable residents and consumers to maximize the utilization of RES and enhance their capacity to manage energy disruptions and variations by incorporating technologies into the energy systems of the community. PtX is completely compatible with the community energy concept, which aims to encourage self-sufficiency and environmental sustainability at the local level [47]. PtX enables communities to decrease their dependence on centralized energy grids, thus enhancing local energy administration and mitigating the hazards linked to traditional energy infrastructure. The shift toward decentralized energy generation and storage not only promotes environmental sustainability but also empowers communities to exert significant control over their local energy supply and decrease their dependence on foreign sources [48].

The integration of PtX technology into community energy systems is advantageous since it efficiently addresses energy deficiencies in impoverished and marginalized communities. For instance, PtX projects, such as wind energy supply projects, have demonstrated notable efficacy in delivering dependable electricity to remote and underserved regions. Consequently, they can effectively address the issue of energy poverty and partially contribute to the concept of energy justice. By utilizing renewable resources available in their local area, these communities can liberate themselves from the constraints of depending on fossil fuels and provide a path toward attaining sustainable prosperity. Additionally, this approach can contribute to enhancing the environmental conditions of the region [49].

Beyond just producing energy, PtX technologies and community energy systems work well together. The approach includes a holistic strategy for sustainable development, incorporating

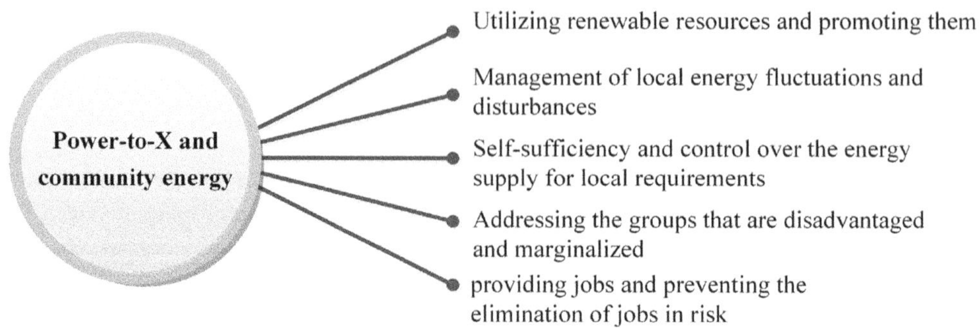

FIGURE 8.8   The advantages of power-to-X for local community energy.

elements such as the generation of employment opportunities, responsible management of the environment, and promotion of social unity. Furthermore, the efficacy of PtX in terms of carbon reduction and environmental enhancement ensures the preservation and revival of numerous endangered employment and economic prospects [50]. Furthermore, PtX-enabled community energy initiatives have the potential to expedite inclusive growth and allow communities to flourish in an unpredictable energy environment by promoting collaboration among many stakeholders, such as local governments, enterprises, and people [51].

8.6   ECONOMIC AND REGULATORY CONSIDERATIONS

8.6.1   Economic Viability of Power-to-X Technologies

PtX technologies provide favorable options for converting renewable energy into useful goods, such as synthetic fuels. These technologies offer the potential for both centralized and decentralized implementation. However, the economic viability of adopting PtX remains a significant worry, as it entails managing an intricate interplay of variables. The economic problem focuses on finding a careful balance between optimizing the operating efficiency of PtX power plants and effectively managing renewable energy resources. PtX plants function at peak efficiency year-round to maximize output and profitability. However, this objective contradicts the need for curtailment, which occurs when surplus renewable energy goes to waste and is not utilized [10].

A recent investigation has emphasized the significant relationship between operation times and costs in PtX systems. These studies highlight the crucial significance of operating hours in determining the economic viability of PtX power facilities. Ensuring that PtX facilities operate at their optimal levels for extended periods is crucial in order to enhance production and cost-effectiveness. Henning and Palzer's research demonstrates that the duration of operation has a substantial impact on the expenses associated with production in PtX systems. They demonstrate that operating for fewer than 2,000 hr per year can result in higher hydrogen production expenses. This phenomenon is caused by a reduction in economies of scale and an increase in fixed costs per unit of output that fall below specific thresholds. On the other hand, exceeding 5,000 hr per year might result in substantial cost savings. This is because greater utilization rates allow for more efficient distribution of capital expenses and operating costs, leading to better amortization [52].

Moreover, the economic prospects for PtX technologies are influenced by a range of market conditions and governmental considerations. Input electricity expenses for PtX processes might involve additional fees and taxes, especially if the fuel generated is not utilized for electricity generation. The presence of this regulatory framework increases the intricacy of PtX profitability calculations and underscores the significance of good market circumstances for maintaining sustainable operations. The introduction of decentralized PtX concepts increases the level of intricacy in the economic equation. Efficiently managing the input and output power capabilities is essential for maximizing the utilization rates in decentralized PtX systems, where PtX entities are directly linked to RES. Effective strategic planning and coordination are crucial to ensure the smooth integration of PtX technologies into the current energy infrastructure, while also maximizing economic advantages [23].

To effectively navigate the economic challenges associated with PtX technologies, a complete approach is necessary. This approach should take into account technology advancements, market dynamics, regulatory frameworks, and strategic deployment initiatives. By collectively addressing these challenges, stakeholders may fully exploit the potential of PtX technologies and expedite the transition to a sustainable energy future.

8.6.2 Policy Frameworks and Incentives

To accommodate the emergence of PtX technologies as feasible alternatives for decarbonizing numerous sectors, especially transportation, policy framework in Europe has experienced considerable changes in recent years. Before the release of the Renewable Energy Directive II (RED II) in 2018, PtX fuels had limited acknowledgment inside the European Union (EU) framework. The main emphasis was on biofuels, LNG, and direct hydrogen applications to reduce carbon emissions in transportation. RED II was a significant milestone since it introduced renewable fuels of non-biological origin (RFNBOs) and renewable carbon fuels (RCFs), which included PtX routes, such as hydrogen [53]. The revisions made to the directive in 2021 enhanced the regulatory framework, providing a more stable and predictable environment for PtX technology. In 2020, the European Commission's hydrogen strategy, which was issued as part of the Green Deal, emphasized the important role of hydrogen in the EU's integrated energy system. This approach prioritized the requirement for substantial financial commitments, with the goal of achieving a minimum of 40 GW (gigawatts) of renewable hydrogen electrolyzers and a maximum of 10 Mt of renewable hydrogen generation by 2030 [54].

In addition, the EU's Strategy for Energy System Integration, which was also adopted in 2020, recognized the significance of clean fuels, specifically electrofuels, in attaining an intelligent energy system. This strategy is in line with the fundamental principles of PtX since it facilitates the integration of different sectors and provides solutions for industries where electrification is difficult. The ReFuelEU Aviation and FuelEU Maritime initiatives, which are part of the Fit for 55 packages, highlight the EU's dedication to advancing sustainable aviation and marine fuels, such as electrofuels [55]. At the national level, the implementation of national energy and climate plans (NECPs) has allowed member states (MSs) to incorporate PtX technologies into their climate and energy objectives. The majority of MSs have included strategies for clean hydrogen in their national energy and climate plans (NECPs), specifically emphasizing the production of green or renewable hydrogen by electrolysis, which utilizes renewable electricity. In addition, multiple MSs have either created or are now in the process of creating national policies and roadmaps specifically focused on hydrogen or PtX technologies. The policies establish specific goals for the expansion of electrolysis capacity by 2030, indicating a definitive market trajectory [56].

The analysis of policy frameworks in different member states shows that there are different levels of recognition and adoption of PtX pathways. Several nations, such as Austria, Czechia, Denmark, Germany, Italy, Norway, Poland, Sweden, and the UK, have implemented comprehensive policy frameworks that include different PtX paths. However, there are other countries that are still in the early phases of developing such frameworks. At both the European and national levels, the changing policy landscape shows an increasing acknowledgment of PtX technologies as crucial elements in the shift toward a sustainable, low-carbon future [54].

Incentives are essential in enhancing the demand for PtX fuels and overcoming any possible challenges. The EU Commission at the European level is dedicated to proposing a fresh categorization and certification system for renewable and low-carbon fuels in order to encourage the use of clean fuels, such as green H2 and e-fuels [57]. Moreover, it is imperative to implement changes in pricing and grid connection models in order to ensure that PtX units become economically viable while encouraging the integration of RES. Member states have the ability to introduce measures like carbon-neutral footprint requirements for vehicles, technical standards, and public subsidies for vehicle conversions or purchases in order to stimulate demand for PtX fuels. As an illustration, Croatia intends to provide financial support for the transformation of current vehicle fleets to alternative fuels. Additionally, they give exemptions from registration taxes for vehicles that produce zero emissions in some countries such as Belgium, Spain, Norway, and the Netherlands [54].

However, there are still obstacles to overcome in order to achieve general acceptance of PtX fuels, particularly in transportation modes that go beyond personal vehicles. The implementation of explicit goals for e-fuels in the aviation and marine industries through RED II amendments and initiatives such as ReFuelEU Aviation and FuelEU Marine is an essential measure to encourage the adoption of these fuels in the market [54]. However, there are still worries over the prevalence of fossil fuel alternatives and the difference in cost between e-fuels and fossil fuels [58]. To address this gap, it may be necessary to implement policies such as levying taxes based on carbon intensity or providing subsidies to promote the competitiveness of green alternatives [59]. Thus, although there has been notable advancement in the support for PtX fuels, additional actions are required to address regulatory and economic obstacles and encourage the development of infrastructure and market acceptance. In order to fully harness the promise of PtX technologies in attaining a sustainable energy future, it is crucial to foster collaboration between EU institutions, member states, and industry partners.

8.6.3 Market Dynamics and the Role of Stakeholders

PtX technologies are leading the way in the renewable energy field and offer a creative solution to the challenges of intermittent energy supply and storage. The interplay among many components of the energy market and a diverse range of stakeholders is pivotal in defining the pattern of adoption and implementation of PtX technologies. The feasibility and expansion of PtX procedures are impacted by market dynamics, including factors such as technological advancements, policy frameworks, and economic viability. The participation of stakeholders, governments, industrial participants, research institutions, investors, and civil society is equally crucial in fostering market growth and enabling innovation [60].

Technological advancements are a significant factor in promoting innovation and cost reduction in the dynamic PtX market. As PtX technologies advance, they progressively challenge traditional energy sources and become more attractive to investors and industry participants. The deployment and market penetration of PtX depend on policy and regulatory frameworks influenced by government action and market incentives. Investment decisions and project development in the PtX sector are influenced by economic factors, such as capital costs, energy prices, and market demand. The market dynamics highlight the importance of players being flexible and adaptable in order to navigate the quickly evolving energy landscape, identify emerging investment opportunities, and manage possible risks [60, 61].

The stakeholders involved in the PtX strategy play various but interrelated roles in promoting performance, market expansion, and innovation. Government entities and regulatory bodies offer essential policy assistance and regulatory structures to encourage the implementation of PtX technology and ensure compliance with environmental and safety regulations. Stakeholders in the industrial sector, including makers of equipment, developers of energy, and producers of fuel, engage with shared projects and share experience to foster technical innovation. Research and academic institutions have a crucial role in advancing PtX by performing fundamental research, developing new technologies, and exploring innovative methodologies. Investors and financial institutions have a crucial role in providing money and financial expertise for PtX projects and impacting their ability to be commercialized. Civil society and community stakeholders have a crucial role in increasing awareness, tackling difficulties, and guaranteeing open decision-making

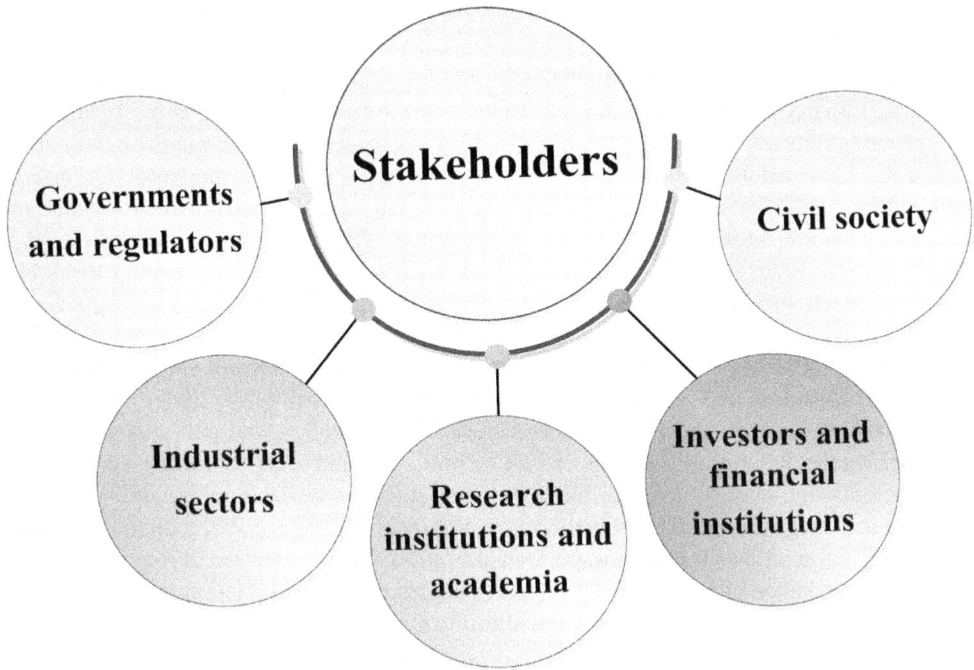

FIGURE 8.9 Several stakeholders involved in the power-to-X approach.

processes, hence facilitating the successful integration of PtX technologies into the wider energy landscape [62]. Therefore, it is crucial to have a well-coordinated link between market dynamics and stakeholders in order to influence the direction of PtX acceptance and implementation. Through comprehending the interdependence of these aspects, individuals or groups with an interest in the matter can efficiently tackle the intricacies of the PtX environment and foster the expansion and advancement of the market. Effective collaboration among governments, industry stakeholders, research institutions, investors, and civil society is essential to fully harness the capabilities of PtX technologies and expedite the shift toward a sustainable energy future.

8.7 FUTURE PERSPECTIVES AND TECHNOLOGICAL ADVANCEMENTS

8.7.1 Emerging Trends in Power-to-X Technologies

Examining the significance of technological advancements and the maturity of linked technologies is crucial in the constantly developing energy sector. The combination of tradition and innovation has emerged as a crucial catalyst for development in the attempt of a sustainable energy future. PtX technologies serve as a connection between conventional and sustainable energy systems, and the progress in equipment and techniques contributes to enhancing the prospects of energy in the future. A prominent trend in PtX technology is the swift progress and acceptance of novel electrolysis techniques. Electrolysis, which involves the use of electricity to separate water into hydrogen and oxygen, is the fundamental method employed in PtX routes. Efforts have been made in recent years to enhance the effectiveness and flexibility of electrolysis technologies, with the goal of facilitating the widespread implementation of PtX systems. Novel electrolysis technologies, including as solid oxide electrolysis cells (SOECs) and high-temperature electrolysis cells (HTECs), have the capability to achieve greater efficiencies and reduced prices in comparison to conventional alkaline and proton-exchange membrane (PEM) electrolyzers. These advancements encourage the development of a more economical and environmentally friendly method of producing hydrogen using renewable sources of electricity [10, 63].

Another emerging trend in PtX technologies is the exploration of alternative methods for producing hydrogen, moving beyond the traditional routes. While hydrogen remains the primary focus, there is a growing interest in utilizing PtX technologies to generate a diverse range of synthetic fuels and chemicals. These efforts involve the production of synthetic methane, methanol, ammonia, and other valuable products that can serve as environmentally friendly alternatives in various industries, such as transportation, heating, and chemical production. These alternative PtX pathways present opportunities to reduce carbon emissions in sectors that are difficult to electrify directly, thus expanding the reach and impact of PtX technologies in the transition toward a sustainable energy system [10, 64].

Furthermore, PtX technologies are more frequently included into current energy infrastructures to enhance the robustness and adaptability of the system. This involves the integration of PtX systems with renewable energy generation, energy storage technologies, and previous gas infrastructure. By utilizing the collaborative effects of PtX technologies and other renewable energy systems, such as solar and wind energy, it is feasible to enhance resource utilization, enhance grid stability, and facilitate the integration of RES. Furthermore, the incorporation of PtX technologies into current gas infrastructure allows for the storage, transportation, and distribution of renewable hydrogen and synthetic fuels, hence promoting their extensive utilization across different industries. The integration of technology with the gas network is a significant and developing aspect in the achievement of PtX objectives [10].

PtX technologies provide the potential to shift toward a low-carbon economy in the future. In order to fully harness the capabilities of PtX technologies and promote their widespread implementation, it is imperative to maintain ongoing research and development efforts, as well as provide favorable policies and investments. The dynamic field of PtX technologies, ranging from improved electrolysis systems to creative PtX pathways and IES, presents promising prospects for tackling climate change, enhancing energy security, and promoting sustainable development on a global scale. PtX technologies are expected to have a significant impact on future energy systems at a bigger scale as they continue to develop and expand [10, 64].

8.7.2 Innovations in Energy Conversion and Storage

Energy conversion and storage are crucial components in the pursuit of sustainable and environmentally friendly energy sources, particularly in the realm of PtX technology. By altering energy consumption methods, PtX has enabled the use of renewable electricity to generate versatile energy carriers, including hydrogen, synthetic fuels, and chemicals. PtX, at its essence, offers a strong and effective answer to the urgent issues of integrating renewable energy and ensuring the sustainability of the grid. However, the full realization of PtX's potential depends on constant improvements in energy conversion and storage technologies.

Electrolysis. Electrolysis is an innovative technique in PtX that utilizes electricity to separate water molecules into hydrogen and oxygen. Electrolyzers, which are crucial elements of this technology, are experiencing enormous developments due to enhanced efficiency, scalability, and cost-effectiveness. PEM and alkaline electrolysis technologies are being used more in electrolysis designs these days. Each has its own benefits when it comes to adaptability and operational parameters. Research endeavors have focused on enhancing the quality of catalytic materials used in electrolysis, aiming to maximize efficiency and sustainability while minimizing resource dependency [65].

Hydrogen storage. The success of PtX technologies largely depends on the efficient generation, storage, and transfer of hydrogen. The progress and achievement of numerous enduring goals of PtX technology depend on innovation and solving hydrogen storage challenges. Hydrogen storage innovations include a diverse range of processes, including conventional techniques like compression and liquefaction, as well as developing technologies like solid-state storage and chemical hydrogen carriers. Solid-state storage, especially with metal hydrides and porous materials, has the potential to develop secure and compact hydrogen storage solutions for various applications, like stationary energy storage and transportation [66].

Synthetic fuels. An alternative to carbon-neutral fuels from renewable sources is the production of synthetic fuels using PtX technology. This technology also presents a viable approach for reducing carbon emissions in challenging areas like aviation and heavy industry. Techniques such as Fischer–Tropsch synthesis and methanation can combine renewable hydrogen with absorbed CO_2 or carbon from biomass to produce synthetic hydrocarbons that closely resemble traditional fuels. These synthetic fuels have the potential to decrease greenhouse gas emissions and can be used with existing infrastructure, such as refineries and distribution networks, to produce and distribute sustainable energy. Developing advanced techniques for converting carbon into synthetic fuels, and effectively storing them, has the potential to be a highly beneficial approach for achieving sustainable energy in the future [10].

Network integration and energy conversions. The advancement of PtX technologies allows them to be integrated into existing energy infrastructures. The concept of multi-energy systems emerges as a fundamental principle from the mutually beneficial interaction among several energy sectors. The advancements in energy conversion and PtX-related inventions expedite and simplify the development of networked and adaptable energy systems. These mergers and conversions will achieve a balance between the supply and demand of various end customers while simultaneously optimizing renewable energy utilization. This highlights the crucial importance of developing advanced methods for conversion energy in PtX technology. These methods are essential for continuously enhancing processes to meet ever-changing energy requirements and laying the foundation for a sustainable energy future [67].

8.8 CONCLUSION

This chapter addressed the complex concept of IES and the crucial role of power-to-X (PtX) technologies within this framework. It clarifies and discusses the potential for change in combining various energy sources and storage systems. The sections of this chapter begin by presenting a thorough definition of IES as a comprehensive methodology for energy management, highlighting its ability to enhance efficiency and sustainability in modern energy systems. This study has shown that IES signifies a fundamental change in thinking that offers exceptional possibilities for reducing carbon emissions, ensuring the sustainability of the power system, and optimizing regional energy usage. The significance of energy storage was highlighted in the field of IES, with PtX technologies emerging as the fundamental component for enhancing electricity storage capacity. PtX addresses the inherent risk and problems of intermittency while facilitating the smooth integration of intermittent renewables into the grid by converting excess renewable energy into storage forms, like hydrogen or synthetic fuels. Furthermore, the importance of PtX in stimulating non-electrified industries was emphasized, offering novel avenues for decarbonization and diminishing dependence on fossil fuels.

Additionally, the study investigated the mutually beneficial connection between PtX and carbon capture utilization (CCU) as an effective means of decreasing carbon emissions while simultaneously generating important synthetic fuels and chemicals. PtX has been acknowledged as a means to synchronize supply with local demand, enhance the robustness of the power grid, and fortify energy resilience within communities. This chapter showcased case studies and practical examples from other research to illustrate how PtX empowers communities to actively engage in their energy transition process, therefore fostering local economic growth and energy self-sufficiency. Nevertheless, this chapter underscored the need to recognize that fully harnessing the capabilities of PtX depends not only on technological progress but also on economic and legal factors. An analysis was conducted on the economic feasibility of PtX technologies, highlighting the crucial influence of political frameworks, incentives, and market dynamics in directing investment and implementation. Effective collaboration among stakeholders, including governments, industry actors, and research institutions, is crucial for establishing a conducive climate that promotes the widespread implementation of PtX technology. Seeing the future, the prospects of PtX appear favorable, propelled by growing patterns and technical progress. Advancements in energy conversion and storage will enhance the effectiveness, flexibility, and cost-effectiveness of PtX, thereby creating new opportunities for sustainable energy generation and utilization. Overall, the integration of PtX into IES offers a revolutionary method to tackle the complex issues encountered by contemporary energy systems. By utilizing PtX technology, stakeholders may expedite the shift toward a more environmentally friendly, adaptable, and all-encompassing energy future.

REFERENCES

[1] Potrč, S., Čuček, L., Martin, M., and Kravanja, Z. (2021) Sustainable renewable energy supply networks optimization – The gradual transition to a renewable energy system within the European Union by 2050. *Renewable and Sustainable Energy Reviews*, **146**, 111186.

[2] Arvanitidis, A. I., Agarwal, V., and Alamaniotis, M. (2023) Nuclear-driven integrated energy systems: A state-of-the-art review. *Energies (Basel)*, **16**(11), 4293.

[3] Rahman, J., Jacob, R. A., and Zhang, J. (2022) Harnessing operational flexibility from power to hydrogen in a grid-tied integrated energy system. *International Design Engineering Technical Conferences and Computers and Information in Engineering Conference*, **86229**, V03AT03A022.

[4] Burre, J., Bongartz, D., Brée, L., Roh, K., and Mitsos, A. (2020) Power-to-X: Between electricity storage, e-production, and demand side management. *Chemie ingenieur technik*, **92**(1–2), 74–84.

[5] Sterner, M., and Specht, M. (2021) Power-to-gas and power-to-X: The history and results of developing a new storage concept. *Energies (Basel)*, **14**(20), 6594.

[6] Nadel, S. (2019) Electrification in the transportation, buildings, and industrial sectors: A review of opportunities, barriers, and policies. *Current Sustainable/Renewable Energy Reports*, **6**(4), 158–168.

[7] Araya, S. S., Cui, X., Li, N., Liso, V., and Sahlin, S. L. (2022) *Power-to-X: Technology Overview, Possibilities and Challenges*, AAU. p. 23. https://vbn.aau.dk/en/publications/power-to-x-technology-overview-possibilities-and-challenges

[8] Daiyan, R., MacGill, I., and Amal, R. (2020) Opportunities and challenges for renewable power-to-X. *ACS Energy Letters*, **5**(12), 3843–3847.

[9] Battaglia, V., and Vanoli, L. (2024) Power-to-X strategies for Smart Energy regions: A vision for green hydrogen valleys. *Energy Efficiency*, **17**(3), 1–17.

[10] Rego de Vasconcelos, B., and Lavoie, J.-M. (2019) Recent advances in power-to-X technology for the production of fuels and chemicals. *Frontiers in Chemistry*, **7**, 454241.

[11] Kienzle, F., and Andersson, G. (2009) A greenfield approach to the future supply of multiple energy carriers. In *2009 IEEE Power & Energy Society General Meeting*, 1–8.

[12] Geidl, M., Koeppel, G., Favre-Perrod, P., Klöckl, B., Andersson, G., and Fröhlich, K. (2007) The energy hub – A powerful concept for future energy systems. In *Third Annual Carnegie Mellon Conference on the Electricity Industry*, **13**, 14.

[13] Wu, J., Yan, J., Jia, H., Hatziargyriou, N., Djilali, N., and Sun, H. (2016) Integrated energy systems. *Applied Energy*, **167**, 155–157.

[14] Gharibi, R., and Vahidi, B. (2022) Coordinated planning of thermal and electrical networks. *Coordinated Operation and Planning of Modern Heat and Electricity Incorporated Networks*, 449–479.

[15] Abeysekera, M., Jenkins, N., and Wu, J. (2016) Integrated energy systems: An overview of benefits, analysis, research gaps and opportunities. *HubNet*, [Online]. https://orca.cardiff.ac.uk/id/eprint/133441/1/HubNET%20Position%20paper-Multi%20Energy-Revised%20version.pdf

[16] Yadollahi, Z., Gharibi, R., Dashti, R., and Jahromi, A. T. (2024) Optimal energy management of energy hub: A reinforcement learning approach. *Sustainable Cities and Society*, **102**, 105179.

[17] Cheng, Y., Zhang, N., Lu, Z., and Kang, C. (2018) Planning multiple energy systems toward low-carbon society: A decentralized approach. *The IEEE Transactions on Smart Grid*, **10**(5), 4859–4869.

[18] Gharibi, R., Vahidi, B., and Dashti, R. (2024) Green transportation systems. In *Interconnected Modern Multi-Energy Networks and Intelligent Transportation Systems: Towards a Green Economy and Sustainable Development*, 8–38.

[19] Daneshvar, M., Asadi, S., and Mohammadi-Ivatloo, B. (2021) *Grid Modernization: Future Energy Network Infrastructure,* Springer. https://doi.org/10.1007/978-3-030-64099-6

[20] Niu, H., Yu, F., Li, B., and Chen, J. (2019) Research on operation optimization of integrated energy system. *IOP Conference Series: Earth and Environmental Science*, **267**(3), 032094.

[21] Davoudi, S., Khalili-Garakani, A., and Kashefi, K. (2022) Power-to-X for renewable-based hybrid energy systems. In *Whole Energy Systems: Bridging the Gap Via Vector-Coupling Technologies*, Springer, pp. 23–40.

[22] Palys, M. J., and Daoutidis, P. (2022) Power-to-X: A review and perspective. *Computers & Chemical Engineering*, **165**, 107948.

[23] Schnuelle, C., Thoeming, J., Wassermann, T., Thier, P., von Gleich, A., and Goessling-Reisemann, S. (2019) Socio-technical-economic assessment of power-to-X: Potentials and limitations for an integration into the German energy system. *Energy Research & Social Science*, **51**, 187–197.

[24] Ibrahim, H., Ilinca, A., and Perron, J. (2008) Energy storage systems: Characteristics and comparisons. *Renewable and Sustainable Energy Reviews*, **12**(5), 1221–1250.

[25] Mansour-Saatloo, A., Pezhmani, Y., Mirzaei, M. A., Mohammadi-Ivatloo, B., Zare, K., Marzband, M., and Anvari-Moghaddam, A. (2021) Robust decentralized optimization of multi-microgrids integrated with Power-to-X technologies. *Applied Energy*, **304**, 117635.

[26] Huang, L., Walrand, J., and Ramchandran, K. (2012) Optimal demand response with energy storage management. In *2012 IEEE Third International Conference on Smart Grid Communications (SmartGridComm)*, 61–66.

[27] Mikulski, S., and Tomczewski, A. (2021) Use of energy storage to reduce transmission losses in meshed power distribution networks. *Energies (Basel)*, **14**(21), 7304.

[28] Bartolini, A., Carducci, F., Muñoz, C. B., and Comodi, G. (2020) Energy storage and multi energy systems in local energy communities with high renewable energy penetration. *Renewable Energy*, **159**, 595–609.

[29] Smaisim, G. F., Abed, A. M., Al-Madhhachi, H., Hadrawi, S. K., Al-Khateeb, H. M. M., and Kianfar, E. (2023) Retracted article: Graphene-based important carbon structures and nanomaterials for energy storage applications as chemical capacitors and supercapacitor electrodes: A review. *Bionanoscience*, **13**(1), 219–248.

[30] Biczel, P. (2008) Energy storage systems. In *Power Electronics in Smart Electrical Energy Networks*, Springer, pp. 269–302.

[31] Dincer, I., and Rosen, M. A. (2021) *Thermal Energy Storage: Systems and Applications*, John Wiley & Sons.

[32] De Saint Jean, M., Baurens, P., Bouallou, C., and Couturier, K. (2015) Economic assessment of a power-to-substitute-natural-gas process including high-temperature steam electrolysis. *International Journal of Hydrogen Energy*, **40**(20), 6487–6500.

[33] Rosa, R. N. (2017) The role of synthetic fuels for a carbon neutral economy. *C (Basel)*, **3**(2), 11.

[34] Eyer, J., and Corey, G. (2010) Energy storage for the electricity grid: Benefits and market potential assessment guide. *Sandia National Laboratories*, **20**(10), 5.

[35] Espinar, B., and Mayer, D. (2011) The role of energy storage for mini-grid stabilization. Report IEA-PVPS T11-0X:2011; MINES ParisTech/ARMINES. https://hal-mines-paristech.archives ouvertes.fr/hal-00802927

[36] Carmo, M., Fritz, D. L., Mergel, J., and Stolten, D. (2013) A comprehensive review on PEM water electrolysis. *International Journal of Hydrogen Energy*, **38**(12), 4901–4934.

[37] Schmidt, P., Batteiger, V., Roth, A., Weindorf, W., and Raksha, T. (2018) Power-to-liquids as renewable fuel option for aviation: A review. *Chemie Ingenieur Technik*, **90**(1–2), 127–140.

[38] Fonder, M., Counotte, P., Dachet, V., de Séjournet, J., and Ernst, D. (2024) Synthetic methane for closing the carbon loop: Comparative study of three carbon sources for remote carbon-neutral fuel synthetization. *Applied Energy*, **358**, 122606.

[39] Cormos, C.-C. (2023) Deployment of integrated power-to-X and CO_2 utilization systems: Techno-economic assessment of synthetic natural gas and methanol cases. *Applied Thermal Engineering*, **231**, 120943.

[40] Mostafa, M., Varela, C., Franke, M. B., and Zondervan, E. (2021) Dynamic modeling and control of a simulated carbon capture process for sustainable Power-to-X. *Applied Sciences*, **11**(20), 9574.

[41] Li, X., Zhang, L., Zhang, C., Wang, L., Tang, Z., and Gao, R. (2023) The efficient utilization of carbon dioxide in a power-to-liquid process: An overview. *Processes*, **11**(7), 2089.

[42] Spiecker, S., and Weber, C. (2014) The future of the European electricity system and the impact of fluctuating renewable energy – A scenario analysis. *Energy Policy*, **65**, 185–197.

[43] Ince, A. C., Colpan, C. O., Hagen, A., and Serincan, M. F. (2021) Modeling and simulation of Power-to-X systems: A review. *Fuel*, **304**, 121354.

[44] Qi, M., Vo, D. N., Yu, H., Shu, C.-M., Cui, C., Liu, Y., Park, J., and Moon, I. (2023) Strategies for flexible operation of power-to-X processes coupled with renewables. *Renewable and Sustainable Energy Reviews*, **179**, 113282.

[45] Sharmina, M., Edelenbosch, O. Y., Wilson, C., Freeman, R., Gernaat, D., Gilbert, P., Larkin, A., Littleton, E. W., Traut, M., and van Vuuren, D. P. (2021) Decarbonising the critical sectors of aviation, shipping, road freight and industry to limit warming to 1.5–2 C. *Climate Policy*, **21**(4), 455–474.

[46] Onodera, H., Delage, R., and Nakata, T. (2023) Systematic effects of flexible power-to-X operation in a renewable energy system-A case study from Japan. *Energy Conversion and Management: X*, **20**, 100416.

[47] Al Rafea, K., Elsholkami, M., Elkamel, A., and Fowler, M. (2017) Integration of decentralized energy systems with utility-scale energy storage through underground hydrogen–natural gas co-storage using the energy hub approach. *Industrial & Engineering Chemistry Research*, **56**(8), 2310–2330.

[48] Buffo, G., Marocco, P., Ferrero, D., Lanzini, A., and Santarelli, M. (2019) Power-to-X and power-to-power routes. In *Solar Hydrogen Production*, Elsevier, pp. 529–557.

[49] Breyer, C., Khalili, S., Bogdanov, D., Ram, M., Oyewo, A. S., Aghahosseini, A., Gulagi, A., Solomon, A. A., Keiner, D., and Lopez, G. (2022) On the history and future of 100% renewable energy systems research. *IEEE Access*, **10**, 78176–78218.

[50] Bobadilla, L. F., Azancot, L., Luque-Álvarez, L. A., Torres-Sempere, G., González-Castaño, M., Pastor-Pérez, L., Yu, J., Ramírez-Reina, T., Ivanova, S., and Centeno, M. A. (2022) Development of Power-to-X catalytic processes for CO_2 valorisation: From the molecular level to the reactor architecture. *Chemistry (Easton)*, **4**(4), 1250–1280.

[51] Kober, T., Bauer, C., Bach, C., Beuse, M., Georges, G., Held, M., Heselhaus, S., Korba, P., Küng, L., and Malhotra, A. (2019) *Perspectives of Power-to-X Technologies in Switzerland: A White Paper*, Technical Report, ETH Zurich. https://doi.org/10.3929/ethz-b-000352294

[52] Henning, H.-M. (2016) Was kostet die Energiewende? Wege zur Transformation des deutschen Energiesystems bis 2050.

[53] Commission, E. (2021) Proposal for a Directive of the European Parliament and of the Council Amending Directive (EU) 2018/2001 of the European Parliament and of the Council, Regulation (EU) 2018/1999 of the European Parliament and of the Council and Directive 98/70/EC of the European Parliament and of the Council as Regards the Promotion of Energy from Renewable Sources, and Repealing Council Directive (EU) 2015/652. *Off. J. Eur. Union.*

[54] Skov, I. R., and Schneider, N. (2022) Incentive structures for power-to-X and e-fuel pathways for transport in EU and member states. *Energy Policy*, **168**, 113121.

[55] Commission, E. (2021) Proposal for a Regulation of the European Parliament and of the Council on the Use of Renewable and Low-Carbon Fuels in Maritime Transport and Amending Directive 2009/16/EC, COM (2021) 562 Final. 2021. *COM (2021)*, **562**.

[56] Regulation, E.U. (2018) Regulation (EU) 2018/1999 of the European Parliament and of the Council of 11 December 2018 on the Governance of the Energy Union and Climate Action, amending Regulations (EC) No 663/2009 and (EC) No 715/2009 of the European Parliament and of the Council.

[57] Abad, A. V., and Dodds, P. E. (2020) Green hydrogen characterisation initiatives: Definitions, standards, guarantees of origin, and challenges. *Energy Policy*, **138**, 111300.

[58] Gozillon, D., Simon, V., Abbasov, F., Ambel, C. C., Earl, T., Buffet, L., Decock, G., and Marahrens, M. (2022) *FuelEU Maritime: T&E Analysis and Recommendations*. Ixelles, Belgium: Transport T&E Environment.

[59] Scheelhaase, J., Maertens, S., and Grimme, W. (2019) Synthetic fuels in aviation: Current barriers and potential political measures. *Transportation Research Procedia*, **43**, 21–30.

[60] Janke, L., McDonagh, S., Weinrich, S., Murphy, J., Nilsson, D., Hansson, P.-A., and Nordberg, Å. (2020) Optimizing power-to-H_2 participation in the Nord Pool electricity market: Effects of different bidding strategies on plant operation. *Renewable Energy*, **156**, 820–836.

[61] Feijoo, F., Pfeifer, A., Herc, L., Duić, N., and Groppi, D. A long-term energy planning model with endogenous capacity investment for energy transition pathways-interrelation between Power-to-X, demand response and market coupling.

[62] Emodi, N. V., Lovell, H., Levitt, C., and Franklin, E. (2021) A systematic literature review of societal acceptance and stakeholders' perception of hydrogen technologies. *International Journal of Hydrogen Energy*, **46**(60), 30669–30697.

[63] Dahiru, A. R., Vuokila, A., and Huuhtanen, M. (2022) Recent development in Power-to-X: Part I – A review on techno-economic analysis. *Journal of Energy Storage*, **56**, 105861.

[64] Chehade, Z., Mansilla, C., Lucchese, P., Hilliard, S., and Proost, J. (2019) Review and analysis of demonstration projects on power-to-X pathways in the world. *International Journal of Hydrogen Energy*, **44**(51), 27637–27655.

[65] Tomić, A. Z., Pivac, I., and Barbir, F. (2023) A review of testing procedures for proton exchange membrane electrolyzer degradation. *Journal of Power Sources*, **557**, 232569.

[66] Yan, X., Zheng, W., Wei, Y., and Yan, Z. (2024) Current status and economic analysis of green hydrogen energy industry chain. *Processes*, **12**(2), 315.

[67] Sorrenti, I., Rasmussen, T. B. H., You, S., and Wu, Q. (2022) The role of power-to-X in hybrid renewable energy systems: A comprehensive review. *Renewable and Sustainable Energy Reviews*, **165**, 112380.

9 Defossilization Dynamics
Exploring the Interplay of Power-to-X and Bioenergy

Danial Esmaeili Aliabadi, Matthias Jordan,
Andreas Meurer, and Niklas Wulff

9.1 INTRODUCTION

9.1.1 BACKGROUND AND RATIONALE

In order to stop climate change, the way we conduct day-to-day activities should be revised. All anthropogenic greenhouse gas (GHG) emissions caused by combusting fossil fuels should be replaced with clean alternatives by the mid-century [1]. Although direct electrification and improving efficiency can assist us in this mission, the historical evolution of technologies after the Industrial Revolution hinders decarbonizing the entire economy. Such a dramatic shift presents friction in all socio-political layers. The heterogeneity of players involved makes the coordination extremely difficult. Apart from social aspects in the acceptance of energy supply, distribution, and end use technologies, there are macro-level political concerns that can impact how each nation responds to coordinated actions [2]. The Franco-German grand bargain to label nuclear power as "green" demonstrates the strategic importance and long-term impact of these decisions [3]. As yet, carbon-based fuels are an indispensable part of hard-to-abate sectors; thus, the question is how humans can use clean or carbon-neutral sources instead of fossil fuels.

To make matters worse, shifting away from fossil fuels puts additional strain on the economy. Replacing fossil fuels with clean alternatives means opting for less cost-optimal solutions; however, the landscape is rapidly changing: the price of drilling and finding new reservoirs for crude oil and natural gas is on the rise [4], while the cost of renewable energies is dropping sharply with the increased expansion mainly of global solar and wind power capacities [5]. As reported by the International Energy Agency (IEA), the deployed capacity of solar and wind energy should be tripled by 2030 [6]. It should be noted that IEA constantly underestimated the real growth of these clean technologies [7] due to its tendency to emphasize the status quo.

Although the abundance of clean renewable electricity from wind and solar can assist us with transitioning to a net-zero economy, these technologies have a major drawback: the lack of control. Both solar and wind energies have an intermittent nature, which necessitates a drastic change in planning and managing procedures for the electricity and heat systems, implying increased flexibility demand to balance excess energy. To cope with the intermittency issue, experts propose expanding these technologies in a broader geographic scope, supporting the flow of electricity using a potent transmission network and increasing energy storage capacity [8, 9]; however, the current costs associated with battery energy storage systems (BESS) are high [10–12]. Another solution is to convert the renewable electricity to energy carriers that can store energy over longer periods of time. This solution can also benefit from already-available distribution infrastructure for natural gas and petroleum products. Transforming electricity to other energy carriers, the so-called power-to-X (or PtX in short), has another significant advantage. Many end use technologies, particularly in sectors like aviation or maritime, have undergone significant refinement through centuries of trial and error. By replacing

DOI: 10.1201/9781032719436-9

fossil fuels with synthetic chemically similar counterparts, society can bypass the necessity for extensive research and development of an entirely new portfolio of technologies. Blending synthetic fuel counterparts and the specific tuning of fuel properties ("designer fuels") are additional advantages.

While the conversion of renewable electricity into other storable energy forms helps alleviate stress on the grid, it is essential to acknowledge that this process is not flawless. Energy losses, often attributed to conversion into heat or the generation of unwanted by-products, can diminish the overall efficiency of these conversion pathways; therefore, more power is required for the production of energy in an appropriate form. Thus, this approach cannot be persuaded for all circumstances but rather for niche applications, when technical constraints are limiting factors for direct electrification.

Bioenergy is another clean concept, composed of technologies that can produce net-zero GHG power and fuels (and net-negative if it is combined with carbon capture and storage) in a suitable form, reducing the need for PtX. Unlike fossil fuels, bioenergy transforms biomass (carbon-based organic molecules) that is produced mostly by plants, (cyano)bacteria, and algae using photosynthesis to energy, thereby assumed to be a zero-emission technology. Bioenergy technology concepts are often categorized into two major groups: conventional and advanced bioenergy. Conventional bioenergy consumes energy crops to produce biogas and biofuels, while advanced bioenergy transforms non-food crops, lignocellulosic materials, as well as residues into energy. Since conventional bioenergy concepts compete with the food supply chain, policymakers prefer advanced bioenergy, the so-called second and third generations, over conventional bioenergy [13, 14].

9.1.2 OBJECTIVES OF THE STUDY

Building upon the prior discussion, one can suggest that there is no silver bullet to transform the entire economy, especially the heat and transport sectors, into a net-zero economy. Regarding the heterogeneity of stakeholders involved in energy supply, distribution, and consumption, several important questions remain:

- What is the priority of different options to reach planned climate targets?
- When should each strategy be followed, and what are the synergies, competitions, and cascading impacts?
- Which sectors are hard to defossilize even with all options combined?

In the following sections, we respond to these questions based on the outcomes of two projects focused on Germany: BEniVer [15] and SoBio [16–18]. In Section 9.2, we explain the complexity and challenges ahead in transitioning to a net-neutral energy system. Section 9.3 provides a generic technical understanding of various solutions that are currently available. Section 9.4 focuses on the role of bioenergy in reaching climate targets. In Section 9.5, we concentrate on two accomplished projects in Germany, as a case study, and discuss them in Section 9.6. Finally, Section 9.7 concludes.

9.2 ENERGY TRANSITION CHALLENGES

9.2.1 COMPLEXITY OF ENERGY SYSTEMS

The primary goal of energy systems is to reliably provide energy to end users with minimal cost, GHG emissions, and losses [19]. Figure 9.1 depicts a simplistic illustration of an energy supply chain where primary energy sources are converted to secondary/final energy carriers and used in different economic sectors. As depicted, energy systems are intertwined with earth and social systems. The GHG emissions from combusting fossil fuels can increase CO_2 levels in the atmosphere and the average global temperature, causing extreme events such as floods and drought. These extreme events subsequently degrade the soil and decrease the agricultural land potential to produce food and biomass, decreasing the bioenergy production level [20]. Consequently, the food crisis increases

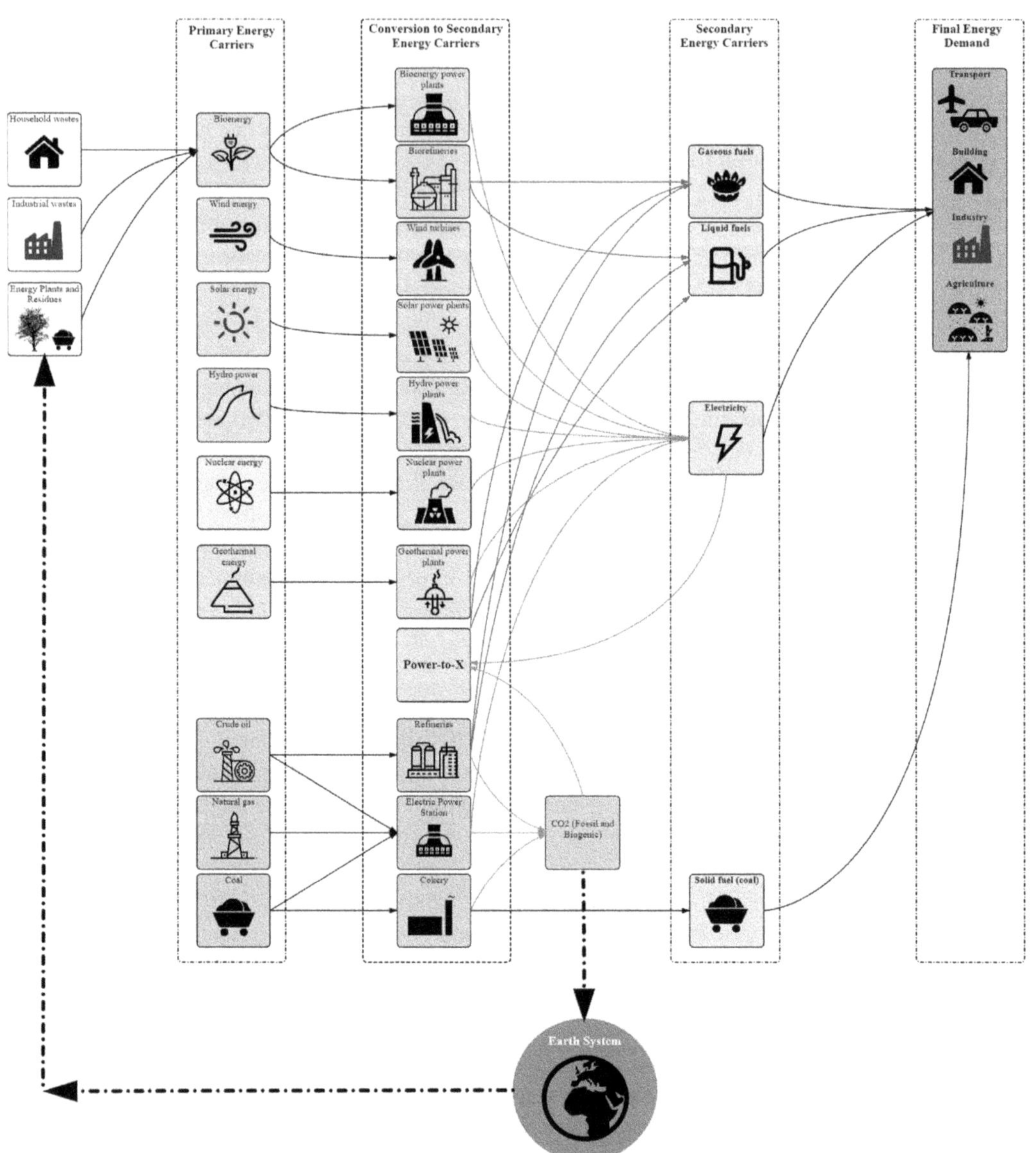

FIGURE 9.1   An illustration of energy supply chain impacting (and affected by) the environment.

Source: Graphic by authors.

the cost of living and dampens economic growth, affecting the interconnected socio-economic sub-systems (e.g., through mass migration) [21].

The heterogeneity of stakeholders in connected markets and systems, such as the capital market, labor market, and earth system, makes managing energy systems a complex task involving the interplay of systems-of-systems [22]. The investment decisions in the energy sector have short- and long-term impacts on these intertwined systems. Unfortunately, the overlap of decisions that can simultaneously benefit all systems is slim. For example, while burning inexpensive coal to produce electricity and heat generates short-term wealth and assists the economy, it concurrently damages the climate, health, and ecosystem. These trade-offs of multiple target dimensions are sometimes assessed in economics by *Pareto optimality*, meaning, no action or allocation can be found where all subsystems are at optimality.

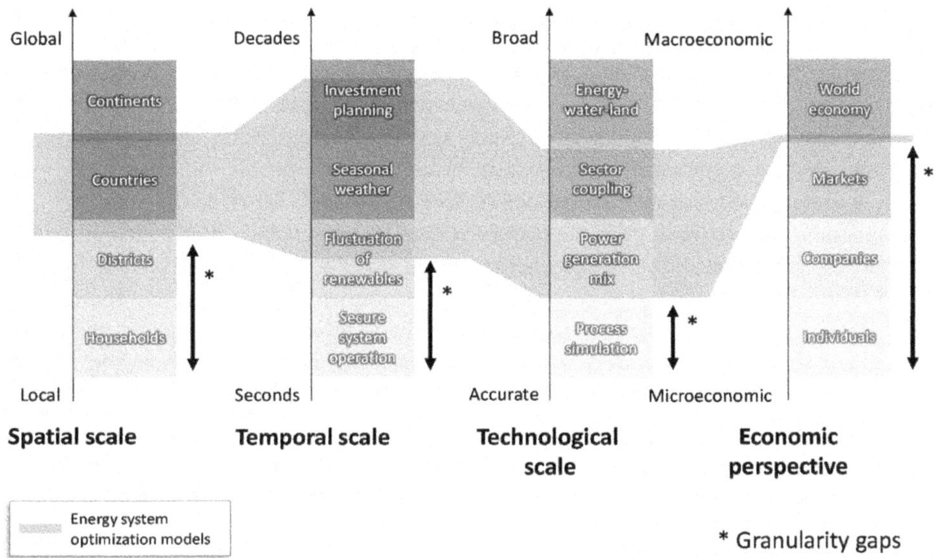

FIGURE 9.2 An illustration of the four dimensions space, time, technology, and economy, with energy system model resolutions.

Source: Cao et al. (2021) [23].

Energy system planning occurs on different scales, as exemplified in Figure 9.2. Modeling supply–demand balances can also comprise one or multiple scales in each dimension. Generally, scope and resolution are differentiated, referring to the model extent (scope) as well as how many elements are modeled within a dimension—for example, 16 federal states on NUTS1[1] level for Germany (resolution).

Two well-known methodologies to model energy systems on a national or supranational scale are integrated assessment models (IAMs) and energy system optimization models (ESOMs) [24]. IAMs often have lower temporal and spatial resolution, with a broader spatial scope, allowing them to consider the interplay between a broader range of economic sectors [25]. These models find an equilibrium solution in all markets and systems, enabling them to model a variety of intersectoral policy implications. On the other hand, ESOMs treat systems outside the energy system boundary as external, allowing for greater temporal, spatial, and technological details to account for wind and solar power intermittency and necessary supply- and demand-side flexibilities.

ESOMs gained popularity among policymakers in the last decades by providing actionable guidance. Unlike simulation models (and most IAMs), instead of merely illustrating "how the future system will look like," ESOMs offer actionable insights by addressing the critical question of "What should decision-makers do (and when) to guide the underlying system to reach certain goals?" This unique capability empowers policymakers with practical recommendations, enabling them to make informed decisions and strategically shape the desired outcomes. Additionally, ESOMs can present new technology concepts to find how they influence future energy market fundamentals, providing investors insights into the potential market size [24].

Modeling energy systems with increasing technological, spatial, and temporal details has revealed that carbon-neutral energy systems have to rely on a variety of technologies and require a fine-tuned planning process to account for the respective region's specificities (e.g., solar, wind, and water availabilities), economic and household demand distributions, as well as cultural factors. To address climate change, a portfolio of technologies combined with behavioral change is essential [26]; therefore, in the next section, we review GHG reduction strategies, focusing on the German case as an example.

9.2.2 Emission Mitigation Strategies in Energy Systems

Energy economy companies that only produce heat and power are still responsible for 30% of Germany's CO_2 emissions, similar to the average worldwide value of 29%.[2] This share increases to 90% when also accounting for energy-related end use sectors, such as industry, households, and transport, which are sometimes involved in broader definitions of the term *energy systems*.[3] Therefore, the first necessary step toward a net-zero energy system is to transform the energy supply by switching from fossil fuels as primary energy sources to renewable ones: hydro, biomass, wind, and solar energy as the backbone of the energy transition. The share of these four major providers of low-CO_2 energy provision is an instance of the aforementioned respective regional resource specificity. While countries such as Ethiopia, Venezuela, Afghanistan, Paraguay, and Switzerland can supply major shares of their energy from hydropower,[4] this is not always possible. Especially in countries with high industrial demands (e.g., the USA, China, and Germany), intermittent solar and wind power will have to supply the majority of electricity supply.

Switching from temporally reliable but CO_2-emitting power plant technologies to clean but partly intermittent ones dictates a necessary temporal flexibilization of all other parts of modern energy systems—from energy supply over transmission to demand. Intra- and international power grid expansions play a crucial role in circumventing spatial and temporal reliance on local variable resources that may be unavailable for weeks. As recent energy system model studies show, the planned grid reinforcements for Germany may still be too low for a GHG-neutral energy system. This refers to both increasing necessary interconnector capacities (to 80–110 GW compared to the currently planned 40 GW) and increasing Germany-internal transmission grid expansion by 60–100% compared to current reinforcement plans [27, 28].

Storage-coupled direct electrification of energy demands, especially in private vehicle mobility and household heating, is among the cheapest options for replacing fossil fuel–based energy service provision due to their higher efficiency, and thus lower cost, compared to hydrogen- or synthetic fuel–based energy service provision. However, the aforementioned techniques can merely supply short-term flexibility up until a week at most but cannot provide energy systems with adequate backup capacity to balance seasonal fluctuations of supply and demands. Thus, energy planning for 100% renewable energy systems must go beyond electric vehicles and heat pumps.

While it has been consensual that Germany will remain a net energy-importing country even in a future state of carbon neutrality (see, for example, Sensfuß et al. [27]), two aspects motivate a reduction, or at least diversification, of imported energy carriers: first, the suspension of Russian gas imports following the Ukraine war has strong economic impacts, increasing energy prices and requiring coal power plants to get re-activated from their reserve status. Secondly, imported energy carriers may reproduce and solidify unequal international trade and power structures.

A central challenge to rapid transformation are expansion gradients. Due to energy systems being an infrastructure element for modern economic activity, planning cycles are historically long. Thus, rapid changes in the supply structure may be infeasible. This is specifically true for the two sectors of transport and electricity that may be increasingly intertwined through direct or indirect electrification [29].

Increasing biofuel supply may act as an alleviating measure to defossilize passenger and goods transportation until sufficient variable renewable energy (VRE) capacities are available; however, care has to be taken about the fuel compatibility with end use technology stock. It is unlikely that car companies will restart developing flex-fuel vehicles, as biofuels are a controversial feedstock for energy provision. Thus, biofuels must be technically harmless for engines—as is the case for biodiesel, but not for bioethanol. Especially for clean kerosene, biofuel may be a replacement for electricity-based kerosene, which will also be required in the long term.

Regarding the impact of user behavior and efficiency, Rodrigues et al. [30] carried out a scenario analysis of energy systems in the EU-28 countries with three different models. They exhibit that the lack of willingness to change end use behavior implies heavy cost increases on the energy supply side, which could be alleviated by "deep renovation, lifestyle changes, energy savings, circular economy, and shared mobility." For this to happen, carbon prices would have to drastically rise to between €250 and €400/t CO_2 in 2030, €250 and €625/t CO_2 in 2040, and €360 and €950/t CO_2 in 2050. Arnz et al. [31] exhibit that energy supply infrastructure has to still undergo drastic transition even for a scenario of mobility lifestyle change incorporating avoidance and modal shifts; however, final energy reductions may reduce power plant capacity expansion needs by around 200 GW compared to the reference scenario.

Despite direct electrification's prominent role in energy system transformation, there might be social, physical, and economic obstacles resisting full electrification, the extent of which is unknown [32]. Schreyer et al. [33] estimate that the EU electricity demand can grow up to 180%, which can introduce many challenges in different tiers of the renewable energy supply chain.

All in all, it is important to consider all possible options to reduce the risk of failure when dealing with an extremely complex, intricate, and interdisciplinary subject like climate change. To take into account the stochasticity of energy systems when proposing a solution, policymakers prioritize concepts from "no-regret " options to "extremely risky " options. No-regret options are considered to be suitable responses under a wide range of circumstances.

9.2.3 Diverse Demands of Different Sectors

Although reliable forecasts for energy demand do not exist for carbon-neutral energy systems due to their chaotic nature, researchers can still explore relevant possibilities via scenarios [34]. To generate scenarios, the most comprehensive approaches include top-down computable general equilibrium models [35]; accounting frameworks [36]; bottom-up simulations, including agents' choices (e.g., for vehicle purchases); as well as expert estimates on technology development (e.g., aviation technologies). A portfolio of mentioned methods has been applied in the German long-term scenarios[5] and in Wulff et al. [28].

Analyses of a scenario focusing on improved strategies with limited end user behavior changes demonstrate that the transport sector's final energy demands, including international aviation and shipping, may be split into at least one-third e-kerosene, one-third e-diesel, and one-third electricity [28]. The remaining demands for clean fuel in GHG neutrality are mainly driven by aviation and shipping; however, current policies on the EU level in both sectors fail to target full defossilization by 2050.

Road transportation, as the most relevant subsector in terms of distance traveled and carbon dioxide emissions, has to be analyzed in the medium and long term. Synthetic diesel and gasoline may be valuable transitioning options for the decade 2030 to 2040, allowing economies of scale to make battery electric vehicles (BEVs) available at cheaper prices to a broader end user group.

In order to analyze the future role of PtX and bioenergy in the energy system, the sectors that are difficult to electrify are of particular interest in energy system models. In transport, these sectors are aviation, shipping, and heavy goods transport. The future transformation in these sectors is driven on the one hand by the development of the vehicle fleet, and on the other by the competition between fuels. Both are mutually dependent.

In the future electricity sector, PtX and bioenergy will not be able to compete with inexpensive VREs; however, they can play an important role in providing dispatchable power to fulfill the positive residual load demand. Millinger et al. [37] calculate future residual load demand scenarios in Germany based on the development of the future overall electricity demand and the expected expansion of fluctuating renewables (wind and PV) calculated in other studies, for example, in Purr et al. [38]. This result can serve as the demand for dispatchable power provision in energy system models focusing on PtX and bioenergy analyses [39, 40].

In the heating sector, the areas that are difficult to electrify are primarily in various sectors of the high-temperature industry. It is therefore necessary to consider not only the different demands in the temperature levels but also the respective energy carriers. In some industries, it is not possible to substitute natural gas applications with direct electrification, solid renewable fuels, or the direct use of hydrogen. Some applications require homogeneous heat radiation at a very high level, which is only possible by burning a renewable gas, such as biogas or power-to-gas (PtG) [41]. Traditionally, biomass is also used in the building sector. In order to identify the optimal areas of application in the future, it makes sense to cluster different building types according to renovation status and demand. An ideal model, therefore, should simulate the future refurbishment of the building stock in various scenarios and thus provides the change in demand for these clusters over time as a result, as done, for example, in Koch et al. [42].

9.3 SYNTHETIC FUELS AS AN ALTERNATIVE APPROACH

This section provides an overview of various alternatives for fossil fuels, outlining their pros and cons. We commence with the definitions to establish a shared foundation for the forthcoming discussions. Following this, the production methods of various energy carriers will be briefly reviewed. Finally, we will examine and compare the energy requirements of the mentioned fuel types.

9.3.1 Definition and Characteristics

In order to provide a common ground regarding different energy carriers, the following terminologies, which are used in subsequent sections, are offered.

The term "fuels" comprises all gaseous and liquid energy carriers primarily intended for end use oxygenation in various applications; thus, liquid organic hydrogen carriers (LOHC) are not considered in this group, since they are intended to transport energy. Figure 9.3 categorizes various

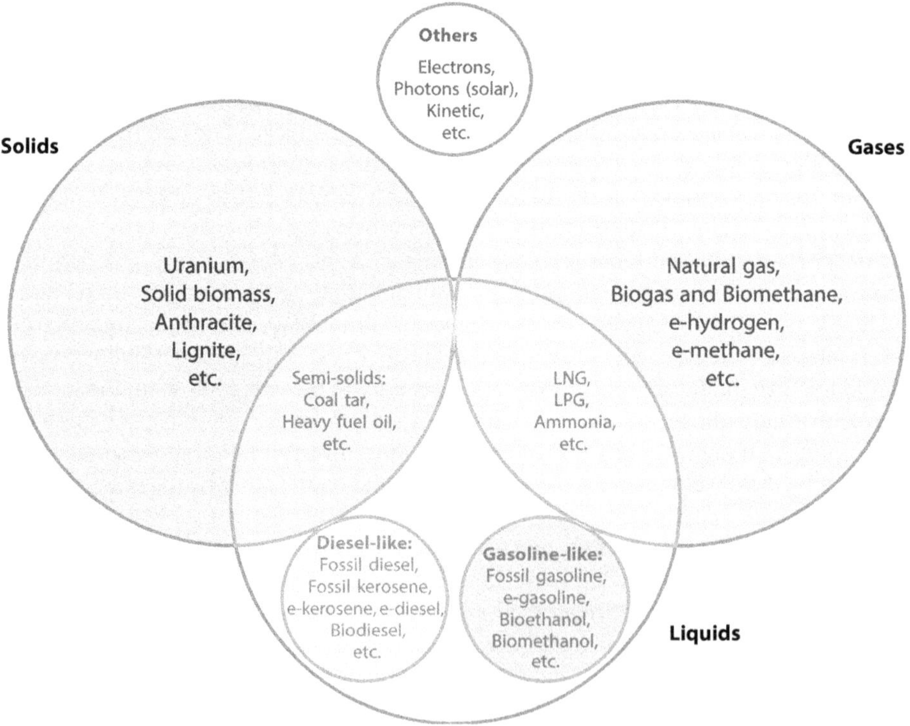

FIGURE 9.3 Venn diagram of different energy carriers. (Readers should note that this graph does not represent the entire spectrum of fuels.)

Source: Graphic by authors.

energy carriers based on chemical and physical properties. Clean fuels comprise non-fossil fuels that are based on non-fossil primary energy consisting of renewable fuels of non-biological origin (RFNBOs) and biofuels (and mixtures of the two), but not of blends between fossil fuels and RFNBOs or biofuels. Those would be considered gray fuels. RFNBOs consist of gases (mainly hydrogen and methane) and liquid fuels (e-crude, e-kerosene, e-diesel, e-gasoline, e-methanol, etc.). So far, solid RFNBOs are not being discussed but are technically possible.

Biofuels describe gases, liquids, and solids based on biogenic primary energy. They can be based on a multitude of feedstocks. Biofuels based on waste feedstocks, such as urban grass clippings, wood scrap, and manure from livestock, are referred to as advanced biofuels. First-generation biogenic feedstocks are usually based on agricultural energy crops, controversially discussed (tank vs. plate debate), and limited by EU policies, also because of competing interests from non-energy sectors, such as base chemical synthesis.

Operating synthesis of biogenic feedstocks including clean hydrogen as co-educt is referred to as power-biomass-to-liquid (PBtL) fuel.

Hydrocarbons are in another end use application-oriented category, comprising fossil hydrocarbons, e-hydrocarbons, and biogenic hydrocarbons. They can be gaseous, liquid, or solid.

Sector-specific terms comprise sustainable aviation fuels (SAF). *Jet fuel* or *Jet A-1 fuel* is a synonym for kerosene and describes norm-conform fuel independent of the primary energy source. The term *e-fuel* is usually used for liquid RFNBOs for combustion in internal combustion engine (ICE) vehicles.

9.3.2 PRODUCTION METHODS

The production chains of relevant sustainable synthetic fuels mostly follow the same general sequence, which can be separated into four major steps: the provision or production of required feedstocks, the pre-processing, the synthesis, and lastly, the post-processing and refining. The necessity and complexity of any of the process steps are determined by the type of fuel and also the desired final fuel quality, which might be affected by regulatory or product certification criteria.

Hydrogen is a component that is shared by all types of sustainable synthetic fuels. Following the definition of synthetic fuels provided earlier, which excludes the usage of biogenic raw materials, water electrolysis operated with sustainably generated electricity can be used for hydrogen production. Currently, there are mainly two types of electrolyzers in industrial operation, which both are highly proven technologies, alkaline electrolysis (AEL) and proton-exchange membrane electrolysis (PEMEL) [43]. A more recent technology, solid oxide electrolysis (SOEC), benefits from elevated operation temperatures, which increases the electrical efficiency in comparison with the AEL and PEMEL. However, as a trade-off, the elevated temperatures induce additional challenges on the material, affecting both construction and operation [43]. Active research and development on the SOEC is ongoing, improving the technology readiness level (TRL) and making it an expected relevant technology also for industrial applications in the upcoming years. Further information on the efficiencies of the different electrolysis technologies is displayed in Table 9.1. The produced hydrogen can either already serve as gaseous synthetic fuel or be used as a component for other process routes requiring a synthesis gas.

With the exception of ammonia, all further synthetic fuels relevant in the current context are hydrocarbon chains of different lengths and structures, mainly consisting of hydrogen and carbon atoms. Whereas fossil-based synthetic production processes are usually based on coal or natural gas, sustainable synthetic processes require CO_2 for the provision of the carbon atom. There are different carbon sources for CO_2 available, using either CO_2 from ambient air or CO_2 extracted from combustion processes exhaust gas streams, benefiting from an increased CO_2 concentration. For the extraction of CO_2 from ambient air, direct air capture (DAC) is currently the prevalent technology, although there are only a few such plants in operation. As DAC is a very recent technology, there is still a lack of reliable efficiency and cost expectations for the long term. For CO_2 sources with high concentration, typically industrial facilities or power plants, post-combustion capture is currently the most used technology, using amine-based adsorption processes in the CO_2-rich flue gas stream of the combustion [44].

With hydrogen and carbon dioxide as a basis, a pre-processing of those gases is necessary to provide a usable gas for the downstream synthesis process. The synthesis itself requires a gas mixture mainly consisting of hydrogen and carbon dioxide (CO_2) or carbon monoxide (CO) in different ratios, depending on the process. There are currently two processes available to provide the carbon monoxide for such synthesis gas:

- A reversed water–gas shift (rWGS) reaction can be carried out, where the feed stream of H_2 and CO_2 is partially converted into CO and H_2O. The outlet stream can subsequently additionally be mixed with pure hydrogen to obtain the desired ratio between H_2 and CO. The equilibrium of the water–gas shift reaction is highly temperature-dependent. To shift the equilibrium toward the rWGS, temperatures of above 900°C are required if no catalytic materials are used.
- The second option for the production of syngas combines the generation of H_2 and CO in a single process step, the high-temperature electrolysis presented earlier operated as a co-electrolysis. If, additionally to H_2O, CO_2 is added to the feed stream of the electrolyzer, the outlet stream contains a mixture of the relevant syngas components H_2 and CO. Although the mechanism taking place is not clearly resolved yet, current studies explain the formation of CO in the SOEC as a combination of CO_2 electrolysis and the rWGS reaction. Although there is currently no industrial co-SOEC unit in operation yet, future application of this technology could lead to a reduction of plant complexity, especially benefiting highly automated or decentralized plants.

Subsequent to the generation of syngas components, a gas cleanup system can be used to reduce the amount of trace components, impurities, or unreacted feed stream components which could potentially be harmful for the catalyst material in the synthesis step or reduce its reactor activity.

The conversion from synthesis gas to the synthetic crude or already final product represents the core of the entire process chain. Depending on the desired product, there are different synthesis options available. One is the Fischer–Tropsch (FT) synthesis, which originates from coal liquefaction but also comes with the opportunity of synthesizing sustainable syngas to a broad variety of different hydrocarbon chains. The FT synthesis is a catalytic conversion process which can be separated into different categories, depending on the main catalyst material, either cobalt or iron, and the general temperature level. During the conversion from synthesis gas to the hydrocarbon chains, water is produced as a side-product, which is the main reason for the energetic losses during synthesis, as hydrogen atoms of the syngas are bound to H_2O, which has no further usage in the context of fuels. The operation of FT synthesis offers high flexibility regarding the product output composition, which can mainly be governed by the catalyst material and the operation temperature, affecting the type and average length of the hydrocarbon chains. The outgoing stream usually represents a synthetic crude, as its composition is not yet complying with the composition and properties required for specific fuel types. A post-processing, and refining, of the synthetic crude is therefore required. FT crude can be used as a basis for most currently relevant liquid fuels for road, air, and maritime transport.

Another synthesis process, the Sabatier reaction, produces the shortest existing hydrocarbon chain, methane, which can be used as a gaseous fuel to replace, for example, fossil-based natural gas. This synthesis process is based on H_2 and CO_2 as syngas educts. Whereas the FT synthesis provides a crude product with a broad and complex spectrum of components, the Sabatier reaction is highly specialized exclusively for methane production.

Methanol can be directly used as a fuel or fuel additive and is produced via methanol synthesis, a catalytic process where syngas, composed mainly of H_2 and CO_2, is converted. Different technology types mainly vary in the range of operation pressure and the selection of catalyst material. If methanol is directly used for combustion, a post-processing is usually not necessary. Especially in the maritime sector, applications for direct methanol combustion are currently gaining interest. Aside its direct use as fuel, methanol can also serve as a base component for more complex fuel types, which are produced via a series of different refining steps.

The post-processing and refining of such base components like methanol or FT crude is usually carried out in complex refineries and comprises a multitude of possible processing steps. Those aim at the conversion to different hydrocarbon species to improve specific fuel properties and quality and thus enable the application as certified fuels. A current refining setup using methanol as basis is the methanol-to-gasoline process, producing liquid fuels typically in the range of gasoline fractions. The post-processing and refining of FT crude offers a broader variety and is not restricted to a certain fuel type. Based on the FT crude composition, different refining setups are applied in current FT syncrude refining. They all aim at the production of different fuel types to increase the usage of crude to a maximum and are especially tailored to the FT crude which is processed [45].

The efficiencies and achievable conversion rates of the process steps mainly responsible for the efficiency losses throughout the process chains are shown in Table 9.1. The conversion rates for the synthesis processes already include hydrogen losses for the conversion from CO_2 to CO and represent the ratio between the energy content of the outgoing product or crude stream divided by the energy content of the ingoing H_2 based on the lower heating values. Additional process-related losses due to flue gas streams or electricity consumption of auxiliary components are not

TABLE 9.1
Efficiency Overview of Various Processes

Process	Efficiency$_{LHV}$ Energy Demand	Notes
Electrolysis		
PEMEL$_{2020}$	56–60% [43]	
PEMEL$_{2050}$	67–74% [43]	
AEL2020	63–70% [43]	
AEL2050	70–80% [43]	
SOEC$_{2020}$	74–81% [43]	
SOEC$_{2050}$	77–90% [43]	
CO_2 capture		Data for low-temperature technology.
DAC$_{2020}$	250 kWh$_{el}$/t$_{CO2}$ [46] 1,750 kWh$_{th}$/t$_{CO2}$ [46]	
DAC$_{2050}$	180 kWh$_{el}$/t$_{CO2}$ [46] 1,100 kWh$_{th}$/t$_{CO2}$ [46]	
Synthesis		
Fischer–Tropsch	75–82%	H_2 losses due to chemical reaction. With increasing hydrocarbon chain length, the conversion efficiency is reduced. The efficiencies define theoretical upper limits.
Methane	82%	As the shortest hydrocarbon chain, the theoretical maximum can be reached.
Methanol	87%	Higher efficiencies are achievable for the synthesis of alcohol due to comparatively less H_2 losses.

considered. The electrical efficiencies for the electrolysis and DAC technologies consider values for 2020 and expected efficiency ranges for the year 2050.

From a systemic perspective, the need for refining, especially for FT crude, not only results in additional product costs and efficiency losses due to the additional processing steps but also increases infrastructure requirements. Current FT-based fuel applications depend on processing in large-scale refineries, which might be scarce in certain countries or world regions and where existing refinery sites are usually highly specialized to a certain set of output products.

Looking at the whole process chain of synthetic fuel production, from renewable energy generation, feedstock supply and synthesis, to refining, it is mainly refining that needs to be highlighted as the limiting process step, as it currently relies on a highly centralized infrastructure. While in the long term it is generally possible to adapt the refining infrastructure toward new refinery locations or decentralization toward highly specialized smaller refineries, the current highly centralized infrastructure is an important limiting factor for the market introduction of complex synthetic fuels in the short term.

9.4 BIOENERGY AS A CLEAN ENERGY PATHWAY

9.4.1 Overview and Concept

Bioenergy is an established method for the provision of energy. Nowadays, it is mostly used as a clean renewable energy source to substitute fossil fuels. However, in the future, the expectations from bioenergy are higher [47, 48]. It needs to be used in an efficient and system-integrated way to provide flexibility, serve as solutions for hard-to-electrify areas of the energy system, or even provide negative emissions [6].

Major resources for bioenergy are solid biomass (originally stemming from forests or short-rotation coppice), agricultural biomass, and biomass from waste and residues. Since biomass is a limited resource [49], its production and consumption should follow sustainability standards to prevent loss of biodiversity, soil, and nutrition content, especially when it comes to agricultural and forest biomass [50]. Besides the direct competition with nature conservation targets, an increasing demand for biomass products in the bioeconomy is expected in the future. Consequently, there are (and will be) competing demands for the use of biomass. The German government is thus developing a biomass strategy stating that under the condition of complying with nature conservation targets, the use of biomass should be prioritized in the following order: first, for food and feed; second, for chemical and material use; and third, for bioenergy. Consequently, within the idea of a circular economy, waste and residues will be the main resources for bioenergy in the future, at least in the German and European context [51].

Today, solid biomass and biogas are mostly used to provide electricity to cover parts of the base load demand in the power sector [47]. In Germany, however, a gradual shift toward the flexible utilization of bioenergy can be observed. In a future energy system which is affected by fluctuating sources, such as wind and PV, this flexible provision of energy will become the field of application for bioenergy. The other notable competitors to bioenergy are H_2 or PtX energy carriers, which can provide similar properties as bioenergy required for balancing fluctuating renewable sources. Of course, flexible bioenergy provision is more expensive. The full load hours of the plants decrease, and higher plant and storage capacities are required to provide energy in the time of peak demand. For example, possible concepts are biogas plants, biomethane-fired combined heat and power (CHP) units, biomethane-fired gas and steam turbines, wood-fired cogeneration plants, wood gasifier CHP plants, or special applications, such as paper pulp CHP plants.

For heating, mostly monovalent bioenergy systems are in place today, especially in households. Common technical concepts are woodstoves and wood pellet systems. For larger buildings, in industry, and for district heating, wood chip heating systems are also commonly used. However, in an integrated efficient future energy system, monovalent systems appear to be inefficient. The limited biomass potential should solely be used for peak demands in the depth of the winter season. Consequently, hybrid heating systems are developed where heat pumps or solar thermal technologies are combined with biomass boilers, providing energy at peak demands [52].

In transport, biofuels (mainly conventional) are blended into diesel or gasoline fuels all over the world. The alternative would be advanced biofuels stemming from waste and residues. In Europe and Germany, political ambitions are high to redirect production and imports from conventional to advanced biofuels due to nature conservation targets [53]. Nearly all biobased feedstocks can be used in transport; however, fatty biomasses are particularly advantageous in the transport sector, partly because they are not applicable in the electricity and heating sector. Common biofuels that are produced today are biomethane, bioethanol, FAME (fatty acid methyl ester), HEFA (hydro-processed esters and fatty acids), BtL (lignocellulosic biomass to liquid, gasification, and Fischer–Tropsch synthesis), and biomethanol. Besides the classical fields of application (diesel, gasoline, CNG, and LNG), biofuels are also optimized for use in sustainable aviation (SAF) and the use in shipping and are beginning to establish themselves on the markets [54, 55].

The case of Germany can deliver interesting insights as it has a long history of bioenergy use. For decades, attempts have been made to integrate bioenergy in a system-friendly way using suitable policy instruments. The promotion of biogas plants by the government in the 2000s led to the installation of nearly 10,000 biogas plants [56]. A biomass strategy is currently being developed in Germany to utilize biomass as system-efficiently as possible in the future [51], as set out in the *smart bioenergy* concept [47], for example. Efforts are also being made to develop strategies for a system-friendly integration of negative emission technologies (NETs) into the system, in which bioenergy with carbon capture and storage (BECCS) can play an essential role.

9.5 CASE STUDIES

This section is based on the results of two recently conducted projects, *BEniVer* and *SoBio*, which focused on integrating PtX and bioenergy within energy systems. In the BEniVer project, multiple accounting, simulation, and optimization models were soft-linked [57] to accurately formulate the dynamics between PtX and biofuels for the fulfillment of energy and GHG reduction targets in the transport sectors. Table 9.2 summarizes the list of involved models in BEniVer with their type, scope, and exchanged information. Although several models were involved in this project, in this study, we focus on the two optimization models (REMix and BENOPTex) as they consumed other models' outputs.

On the other hand, in SoBio, emphasis was on preparing high-quality data [62] on bioenergy technologies and biomass feedstocks for the downstream modeling of the German energy system with the BENOPTex model to find the cost-optimal distribution of biomass in Germany's future energy system until 2050.

In the remainder of this section, we first introduce the building blocks (i.e., utilized optimization models) in the mentioned projects. Then, the devised methodology and focus of each project are explained. Finally, the scenario results are explained in detail.

9.5.1 ENERGY SYSTEM MODELS

9.5.1.1 REMix Overview

The case study presented here relies on the application of a model derived from the renewable energy mix (REMix) energy system modeling framework. REMix provides the basis for analyzing energy system transformation scenarios in spatial and temporal resolution. Originally limited to the power sector [63], the framework has been continuously enhanced to include flexible coupling with the heating and transport sectors. The models built from REMix were initially used primarily to analyze the European electricity system [63, 64]. REMix has undergone validation through structured

TABLE 9.2

Involved Models in the BEniVER Project

Model Name	Method	Temporal Resolution	Scope and Spatial Resolu-tion	Input and Output (Unit)
REMix [58]	Optimization	4 years (2020, 2030, 2040, 2045), 4-hourly	Europe and Germany, Germany in **nine** nodes	Technology park (GW for power plants, grid technologies, and TWh for storages), technology dispatch (power production, charge, discharge)
BENOPTex [40]	Optimization	Hourly for power from 2020 until 2050	Germany in one node, annual transport sector fuel allocation	Sectoral biofuel supply (PJ/a)
Vector21 [59]	Simulation	Annual be-tween 2020- and 2050	Germany, street sector (passen-ger vehicles and commercial vehicles in respective submodels), four scenarios; see Table 9.3	Main inputs: transport useful energy demands, and user preferences for vehicle characteristics Outputs: new vehicle sales
4DRACE [60]	Accounting	Annual between 2019 and 2050	European airports, fuel demand, two scenarios	Annual energy demand at airport: jet fuel (kg), hydrogen (kg), electricity (kWh), etc.
SZEM	Accounting	Annual between 2019 and 2050	Europe, NUTS0 (34 countries), four scenarios, various subsector differentiations	Inputs: energy balance statistics Outputs: detailed technology and fuel allocations
TEPET [61]	Simulation	2018, 2030, and 2050	Chemical plant project in Germany; three processes (Fischer–Tropsch synthesis, methanol plant, methanation plant)	

comparisons with analogous frameworks, affirming its capacity to accurately capture the dynamics of technology deployment within energy systems [65, 66]. The framework employs a multi-node approach wherein nodes can be interconnected via various forms of transport infrastructure. Within each node, all units of a single technology are consolidated and managed as a unified entity.

REMix utilizes a linear cost minimization strategy aiming to optimize its objective function, which considers the annuities proportionate investment costs and fixed operation and maintenance (O&M) costs of endogenously added capacities, variable O&M costs of all assets, fuel costs, optional emission costs, and penalty costs for unsatisfied energy demand. The REMix framework is implemented in the general algebraic modeling system (GAMS) and, in this study, solved using CPLEX. The construction of energy system infrastructures and their hourly operation over the course of a year are optimized integrally in one model run with perfect foresight and from a macroeconomic planner's perspective.

Gils et al. [58] described the REMix model in detail for the power sector. The interested readers are invited to check Gils et al. [67] for the heating sector, and Luca et al. [68] for BEVs, and Gils et al. [69] for gas infrastructure modeling.

For the preparation of this case study, the framework has been further extended to include a simplified representation of synthetic fuel refineries and cost-quantity-steps of import cost potentials based on data of national bids from the project MENAFuels. [70]

9.5.1.2 BENOPTex Overview

The extended bioenergy optimization model (BENOPTex) [71] is an umbrella of standalone models [39, 40, 72, 73] that are developed in different projects to address different research questions. These island models are all linear optimization models with two objective functions: maximizing GHG abatement and minimizing total system costs. Figure 9.4 depicts various characteristics of standalone models which have been developed in different projects. By hard-linking different island models, we reviewed, revised, and combined assumptions, constraints, and experts' views from various disciplines to provide an integral outlook of bioenergy/bioeconomy.

BENOPTex benefits from a user interface that has been developed in MATLAB, and it is expected to be combined with a new front end [71], which visualizes the regional characteristics of the energy systems using GIS datasets. The optimization model (i.e., back end) is developed in GAMS, which is the standard in the field of operations research.

9.5.2 Methodology and Approach

Modeling energy systems with heterogeneous players often requires linking a multitude of models in order to capture the diverse set of objectives across the board [24]. Modelers are either soft-link or hard-link standalone models.

In the *BEniVer* project [15], we soft-linked several simulation, optimization, and accounting models to capture the technical, behavioral, and social aspects of transformation in different transportation modes [57]. For aviation, we used the results of an accounting model (i.e., 4D-Race [60]), while the results of a simulation model (i.e., Vector21 [59]) are employed for road transport. The complementary role of bioenergy and synthetic fuels was investigated in a sequential run between two optimization models: REMix [58] and BENOPTex. Doing so enabled us to take into account

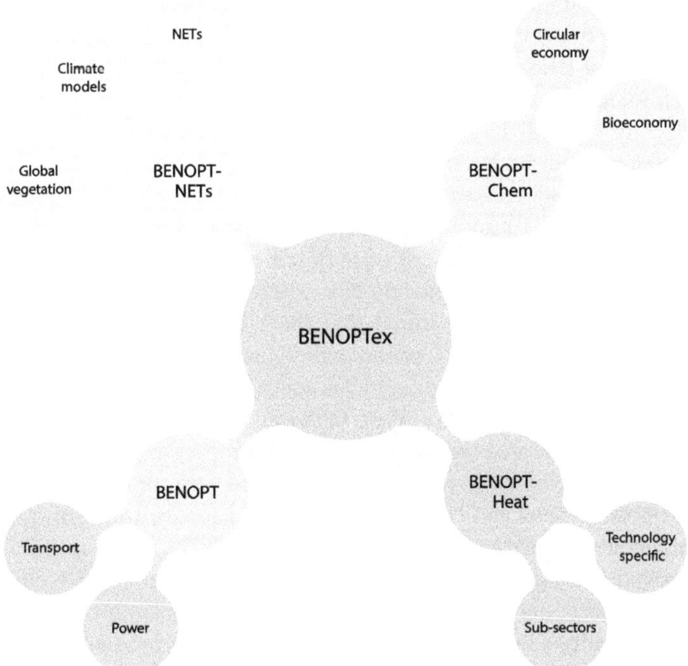

FIGURE 9.4 Hard-linking BENOPT submodels in order to paint a panorama view of bioenergy and the bioeconomy.

Source: Graphic by authors.

different policies regarding the renewable energy directives [74] as well as the expansion plan for transmission lines (the so-called Netzentwicklungsplan, or NEP).

On the other hand, the *SoBio* project took another approach by formulating and hard-linking bioenergy expert views in the BENOPTex model through integrating high-quality data [62] on bioenergy technologies and biomass feedstocks. As mentioned earlier, the SoBio project was investigating the future cost-optimal allocation of biomass within the energy system. Consequently, the focus of the investigation was mainly on bioenergy-relevant sectors, on a detailed representation of biomass feedstock prices and availabilities, conversion technologies, possible applications, and the competing technologies in the demand sectors. The utilized data values, as well as assumptions, were provided by interdisciplinary experts from various fields of bioenergy. In regard to non-biomass data, the investigation relied on existing studies.

In both BEniVer and SoBio, modelers used scenario analysis to capture future uncertainties when modeling cost-optimal (or GHG-optimal) biomass transition pathways.

9.5.3 SCENARIO RESULTS OF SOBIO AND BENIVER

Using scenario analysis via the BENOPT model, Millinger et al. [37] found that the use of excess electricity, representing 33% of the assumed electricity demand, for the production of e-fuels can avoid GHG emissions of 74 $MtCO_2eq$ yearly until 2050 for €250–320/tCO_2eq. As things stand, in situations where sufficient green hydrogen exists, there is an unsatisfied demand for clean fuel, and renewable energy would otherwise be curtailed, it is preferable to use electricity to produce clean fuels replacing fossil fuels than to use carbon capture and storage [75]. We want to reiterate that these conditions occur only in specific and potentially locally limited situations, such as when wind power plants and electrolyzers are expanded much faster than grid transmission capacities. From a German investor's perspective, it is questionable whether these situations alone justify high investments in synthetic fuel production.

Esmaeili et al. [11] also investigate the complementary role of utility-scale BESS and bioenergy in future German transportation. Although BESS can improve the GHG abatement level, their high investment costs make them economically infeasible when the intermittency of RES is ignored [11]. The high investment cost of BESS has also been underlined as a major impeding factor in other studies [76, 77]. While anticipated advancements in wind and solar power forecasting [78] decrease the significance of BESS for intraday balancing, vRES capacity expansion is expected to increase arbitrage potential for planned diurnal balancing. Moreover, BESS is found to have a high potential for providing frequency containment reserve [76], which can also technically be provided by BEVs [79].

Esmaeili et al. [74] analyzed multiple possible futures for renewable energy directives where the declared targets and mechanisms were formulated endogenously. The authors show that the German transport sector cannot be completely defossilized without imported non-fossil liquid and gaseous hydrocarbons. Moreover, they found that there exist two technological and managerial obstacles that should be addressed. First, policymakers need to step up their efforts to ensure that the widespread electrification of passenger vehicles succeeds as early as possible. The second challenge appears when policymakers impose more stringent GHG quota requirements. This will still require a major effort in many areas to research and develop new environmentally friendly commercial transportation technologies (e.g., electric aircraft, electric or hybrid vessels).

Considering the impact of seasonality while employing the just-in-time philosophy, Sadr et al. [80] show that the German energy system can consume more woody biomass until 2045 due to its longer shelf life and improved storability. They also exhibit that processing all biomass when available can overwhelm biorefineries; hence, they suggest leveling the consumption pattern by distributing leisure travel across the entire year. It is anticipated that the cost of biodiesel and bioethanol will rise in the coming years, primarily due to the displacement of conventional biofuels and the adoption of advanced biofuel technologies, which are more expensive than conventional ones. Although this change alleviates the conflict between food and energy supply chains, it means consumers should expect to see higher prices for these biofuels in the future (see Figure 9.5) [81].

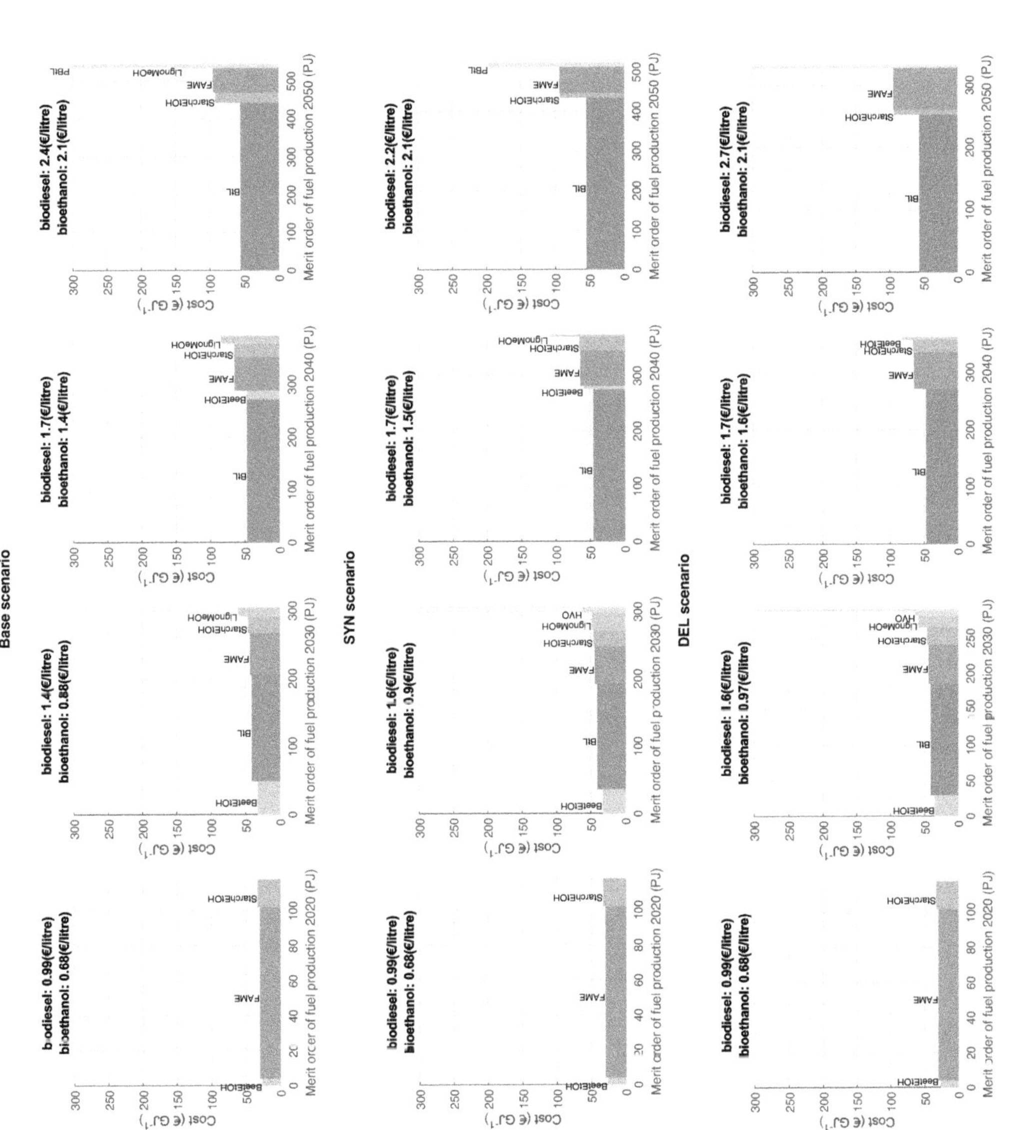

FIGURE 9.5 The production cost of each technology to produce bioethanol and biodiesel in 2020, 2030, 2040, and 2050 under various scenarios.

Source: Esmaeili Aliabadi and Jordan (2023) [81].

Wulff et al. [28] exhibit that increasing conversion depth and higher shares of final energy demand supplied by hydrogen or its derivatives mainly affect grid expansion, hydrogen production, and hydrocarbon refinery capacities. Furthermore, the authors find that biofuel substitution of demands for renewable hydrocarbons may allow the distribution of renewable energy expansion more evenly across years, reduce the necessary pace of German electric grid expansion, and significantly shift needed hydrogen production capacity from the medium term to the long term. This effect occurs independent from import fuel availability that was included in the assessment.

Consequentially, biomass as a feedstock may have two implications. In the medium term, biofuel may supply diesel, gasoline, and kerosene, alleviating a strongly increasing demand for electricity-based fuels in the decade 2030–2040. Hydrogen production ramp-up can be shifted further to the future by this substitution. However, in the long term, direct electrification prevails in passenger transport, shifting fuel demand to diesel for shipping and kerosene for aviation. Biofuels substitute parts of synthetic fuel imports in the long term but cannot replace them due to resource restrictions.

The analysis of the SoBio scenario results shows that from a cost-optimal system perspective, the limited biomass in Germany should be allocated to sectors where direct electrification is not possible (or only possible at high costs) and to provide flexibility toward climate neutrality in 2045 [18]. The relevant target sectors are high-temperature heat applications, maritime, and aviation. Also, in case of high biomass availability in the future, biogas plants are viable options for providing flexibility in the power sector. The amount of biomass used in these sectors varies partially according to certain factors, such as biomass availability, future energy demands, and the degree of technological development. Complementary, the results exhibit that solid biomass has its cost-optimal use in heating and partially in transport, while fatty biomass should be processed into biofuels utilized in transport. On the other hand, digestible biomass has broader target markets in power, heating, and transport. In the future, biomass, H_2, and its PtX derivates will compete to provide flexibility over the same target markets within the energy sector and the areas that are hard to electrify. Based on outcomes of scenarios, hydrogen plays a robust role in flexibility provision in the electricity sector and in transforming the steel industry. The results also emphasize that power-to-liquid is an unavoidable option in air transport to achieve net-zero emissions. Biomass can only cover a small proportion of the demand in air transport. Comparatively small proportions of PtG will be needed in the long term in niche applications in the high-temperature industry.

The direct influence on the demand for H_2 and PtX imports becomes more visible when varying the future availability of biomass between scenarios. For instance, phasing out the use of energy crops leads to an increased demand for H_2 and PtX imports by 1,400 PJ per year (see Figure 9.6), which is associated with costs of €14–25 billion annually in the long term [18]. This phaseout is directly influencing the future transformation strategy for high-temperature industrial heat applications, where solid biomass is identified as the future cost-optimal solution to fully transform it;

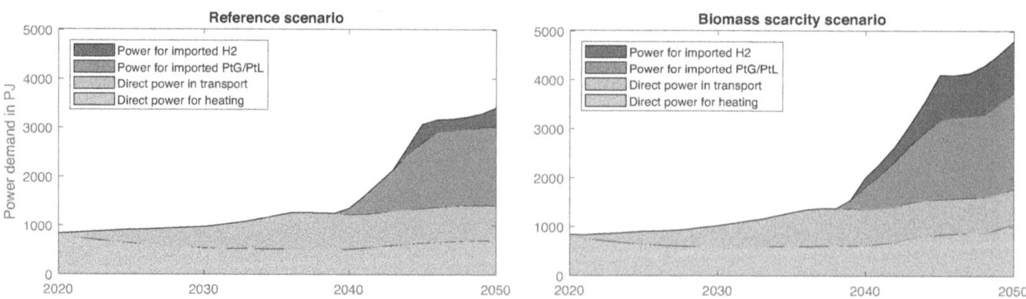

FIGURE 9.6 Power demand for H_2, PtX, and direct electrification in a reference and a biomass scarcity scenario.

Source: Jordan et al. [18].

however, we should note that a decrease in biomass potentials alters the cost-optimal allocation priorities completely in this sector.

Additionally, the current bioenergy-related policy instruments were investigated in short-term energy scenarios [82]. The analysis reveals that, for example, the GHG-quota instrument in transport does not promote the use of biofuels in the hard-to-electrify areas of the transport sector, where they should be cost-optimally allocated according to long-term energy scenarios. This might lead to counterproductive developments in the passenger road sector but simultaneously helps ramp up biofuel capacities required in shipping and aviation in the long term.

9.6 INSIGHTS INTO THE INTERPLAY

Based on the outcomes of conducted projects and the literature review, we can derive the following responses to the research questions:

- ***What is the priority of different options to reach planned climate targets, and when should they be followed?***
  According to the results of both projects, it can be concluded that direct electrification has the highest priority in terms of efficiency; however, due to techno-economic barriers, it is not always viable to opt for direct electrification, considering utility-scale BESS is expensive as yet. In navigation and aviation, e-fuels and biofuels remain the most viable options as of now. When excess renewable electricity (ERE) is available in ample quantity, converting it into e-fuels is an option, given that utility-scale electricity storage is not economically possible. This can serve as a medium-term defossilization option until end use demand services are adequately electrified (e.g., using BEVs instead of ICE vehicles). Meanwhile, biofuels can assist in the medium term to substitute domestic hydrogen production and downstream hydrocarbon syntheses, thereby alleviating power sector and hydrogen infrastructure expansion needs.

  As the costs of liquid and gaseous e-fuel products are projected to decrease with electrolysis production upscaling [70, 83], stricter policies are anticipated to raise associated costs and decrease resource potential of biofuels [81]. Biofuels may act as a medium-term option for cleaner primary energy until sufficient power sector and hydrogen infrastructure is available. Hence, biofuels can be seen as a bridging option, reducing emissions in the transport sector while alleviating infrastructure expansion requirements. The trilateral relationship between imported e-fuels, domestically produced e-fuels, and biofuels is, however, complex. While e-fuel production abroad will reach the lowest production costs and can potentially draw from high renewable potential, it is necessary to apply lessons learned from past economic import dependencies to newly established trade relationships.

- ***Which sectors are hard to defossilize even with all options combined?***
  In transportation, end use technologies are available for direct electrification of passenger and railway sectors. However, battery-electric aviation is expected to gain market shares only in short-distance travel for aviation beyond 2040. In the heat sector, low-temperature applications (such as district heating, low-temperature industry, and households) can, in most cases, be covered using heat pumps, geothermal energy, or solar thermal energy; however, the renewable technological options for high-temperature heat supply are scarce, and the quantities required in Germany are large. According to the results in SoBio, solid and digestible biomasses are the cost-optimal solution if the biomass potential is sufficient. Otherwise, PtX energy carriers need to be substituted. For synthetic kerosene and diesel, there is a contradiction between market readiness with industry scale yet to be reached and resource availability. To what extent actors will take risks to scale up productions in lesser-known but lower-cost resource availability regions is yet to be seen.

- ***What are the synergies, competitions, and interactions between direct electrification, the use of synthetic fuels, and biofuels?***
Defossilizing transport and high-temperature heat provision is and will be costly both in financial terms and in terms of scarce energy resources. This is especially true in countries that are densely populated and suffer from limited hydro potential and agricultural land (i.e., clean primary energy resources). The first pre-condition for transport defossilization is the availability of renewable energy. VRE expansions are robust across all scenarios (i.e., no-regret strategy), and their importance for the energy transition cannot be overestimated.

The second pre-condition is the availability of end use technologies compatible with fossil-free energy sources. Here, internal combustion engines, ship motors, and kerosene turbines already exist and can, to some extent, be used by synthetic fuels based on electricity or biogenic feedstocks. However, while end use sector technologies need lower investments, costs for the provision of synthetic fuels will be higher up to fourfold compared to electricity. This is the main trade-off between direct electrification and synthetic fuels without yet specifying the primary source.

Choosing from electricity or biogenic sources has to be analyzed sector-specifically. While gasoline vehicles can only refuel with ethanol or chemically similar substances to a limited degree (up to 10%), diesel engines are technically capable of burning mixtures of fossil, synthetic, and biodiesel fuels fairly flexibly. Thus, the potential of cheaper (both in terms of energy and costs) fuel blends will be decreasing faster for gasoline than for diesel vehicles as substitution through BEVs occurs at the same time.

Nevertheless, using biomass as a feedstock for bio-crude via Fischer–Tropsch synthesis, which is then further processed in a refinery with a multitude of post-processing steps, could potentially be used flexibly independent of the primary feedstock. Further analysis is required here, such as an analysis of central vs. distributed plant capacities. Crude refinement plants usually have large output quantities to keep entropy creation minimal, by integrating energy sources and sinks and minimizing energy transfer across temperature-level pinch points. To this end, integrating exergoeconomic analyses with ESOMs seems an interesting pursuit capable of capturing scale effects as well as VRE balancing.

Despite being costly, biogas-fired power plants can provide heat and power flexibly (as peaker plants) and may even have carbon capture and storage capabilities; hence, there might be positive synergies between bioenergy and the electrified energy sector.

One main synergy between RFNBOs and biofuels is the shared utilization of the already-existing transmission and distribution infrastructure. However, due to the relatively low losses for transporting electricity and significantly lower costs, electric grids will most likely be reinforced to distribute increasing electricity to households. The already-existing fuel distribution infrastructure may help, especially in regions where end users do not have the capacity for electrifying end use technologies before 2040. Still, to avoid social conflict and energy or mobility poverty, synthetic fuels may have to be subsidized to keep end use prices viable for poorer households.

Bioenergy and PtX are competing on the hard-to-electrify energy demand and in providing flexibility. The competitiveness is strongly dependent on future price developments, especially for the production of PtX. Another important factor might be the availability of both biomass and PtX. In Germany, there will not be enough renewable electricity available in the future to produce sufficient amounts of PtX, so a large proportion of the required quantities will have to be imported [84]. However, the future amount of available biomass is just as uncertain and depends on future forest management, the mobilization rate of biogenic residues, and the political decision as to whether Germany wants to continue using energy crops. The study by Jordan et al. [18] shows in a scenario comparison what additional costs and import demands will arise if significantly less biomass is available in the system than today.

9.7　CONCLUSION

This chapter outlined the different technology pathways and options that can assist in defossilizing the economy. Given that there is no silver bullet (i.e., single technology) by which societies can achieve climate targets, combining these options, and prioritizing them for different applications, becomes significant. To this end, we reported the results of two projects focusing on Germany. Although different angles of the German energy system were the foci of this chapter, the general trade-offs between domestic RFNBOs, imported RFNBOs, and biofuels may apply to other countries that face a similar challenge of high energy demand and relatively low renewable energy potential. Based on the findings of both projects, Germany should prioritize direct electrification in sectors where direct electrification is possible. For this, the power sector should be transformed by replacing fossil fuel–burning power producers with renewable ones as the main backbone of energy transition. Next, end users need to be assisted in switching to still more expensive heat pumps and BEVs.

For sectors where direct electrification is currently impractical, the exploration and adoption of bioenergy and synthetic fuels are recommended. The choice between these alternatives should be guided by the specific characteristics of each sector and the availability of renewable electricity, as one may offer benefits over the other depending on these factors. Timing is key: biofuels may act as bridging technologies in the short to medium term, profiting from existing distribution infrastructure and end use technology fleets until sufficient renewable energy and hydrogen production infrastructure is available. In the long term, domestic VRE resources will not be sufficient to supply the demand, making biofuels a substitute for imported fuels competing with educts for the circular economy.

Concluding, we want to point toward aspects on the edge of our system boundaries. Beyond the techno-economic considerations, addressing socio-political dimensions is challenging but necessary. First, for biofuel extraction in Germany, the perspectives of and implications for surrounding communities have to be taken into account. Secondly, in deepening trade relationships for the import of alternative fuels, the first point is also true for foreign communities abroad surrounding fuel production sites. Thirdly, scenarios of more radical changes in user behavior were not assessed here. Shifting from vehicles to public transportation may, in the end, be a cheaper option, however, with strong industrial policy implications.

9.8　ACKNOWLEDGMENTS

The BEniVer project has been supported by the German Federal Ministry for Economic Affairs and Energy (BMWi) under grant numbers 03EIV116A, 03EIV116F. Also, the SoBio project is funded via an internal strategic project of the German Biomass Research Center.

9.9　CONFLICT OF INTEREST

The authors declare that they have no competing interests.

1　NUTS stands for the Classification of Territorial Units for Statistics.
2　For more information, please visit https://rhg.com/research/global-greenhouse-gas-emissions-2022/.
3　See www.umweltbundesamt.de/themen/klima-energie/treibhausgas-emissionen.
4　For more information, please check https://ourworldindata.org/grapher/share-electricity-hydro.
5　Policy document: https://langfristszenarien.de/enertile-explorer-de/index.php.

REFERENCES

[1] M.T. Huang and P.M. Zhai, Achieving Paris Agreement temperature goals requires carbon neutrality by middle century with far-reaching transitions in the whole society, Advances in Climate Change Research 12 (2021), pp. 281–286.

[2] R. Antes, B. Hansjürgens, P. Letmathe, and S. Pickl, *Emissions Trading*. Springer Berlin Heidelberg, 2011. https://doi.org/10.1007/978-3-642-20592-7

[3] R.K. Wurzel, M. Di Lullo, and D. Liefferink, The European council, council and member states: Jostling for influence, in *Handbook on European Union Climate Change Policy and Politics*, Edward Elgar Publishing, 2023, pp. 38–52.

[4] M.Z. Lukawski, B.J. Anderson, C. Augustine, L.E. Capuano Jr, K.F. Beckers, B. Livesay, and J.W. Tester, Cost analysis of oil, gas, and geothermal well drilling, Journal of Petroleum Science and Engineering 118 (2014), pp. 1–14.

[5] G. Luderer, S. Madeddu, L. Merfort, F. Ueckerdt, M. Pehl, R. Pietzcker, M. Rottoli, F. Schreyer, N. Bauer, and L. Baumstark, Impact of declining renewable energy costs on electrification in low-emission scenarios, Nature Energy 7 (2022), pp. 32–42.

[6] IEA, Net zero roadmap: A global pathway to keep the 1.5°C goal in reach (2023). Available at www.iea. org/reports/net-zero-roadmap-a-global-pathway-to-keep-the-15–0c-goal-in-reach.

[7] K. Mohn, The gravity of status quo: A review of IEA's World Energy Outlook, Economics of Energy & Environmental Policy 9 (2020), pp. 63–82.

[8] P. Denholm, D.J. Arent, S.F. Baldwin, D.E. Bilello, G.L. Brinkman, J.M. Cochran, W.J. Cole, B. Frew, V. Gevorgian, J. Heeter, and B.M.S. Hodge, The challenges of achieving a 100% renewable electricity system in the United States, Joule 5 (2021), pp. 1331–1352.

[9] M. Aldeman, J. Jo, D. Loomis, and B. Krull, Reduction of solar photovoltaic system output variability with geographical aggregation, Renewable and Sustainable Energy Transition 3 (2023), p. 100052.

[10] W.J. Cole, D. Greer, P. Denholm, A.W. Frazier, S. Machen, T. Mai, N. Vincent, and S.F. Baldwin, Quantifying the challenge of reaching a 100% renewable energy power system for the United States, Joule 5 (2021), pp. 1732–1748.

[11] D.E. Aliabadi, M. Jordan, and D. Thrän, The complementary role of utility-scale battery energy storage systems and bioenergy in future German transportation, in *2023 19th International Conference on the European Energy Market (EEM)*, IEEE, 2023, pp. 1–6.

[12] A. Louwen and M. Junginger, Deriving experience curves and implementing technological learning in energy system models, in *The Future European Energy System: Renewable Energy, Flexibility Options and Technological Progress*, Springer International Publishing, 2021, pp. 55–73.

[13] European Parliament, Directive (EU) 2018/2001 of the European Parliament and of the Council of 11 December 2018 on the promotion of the use of energy from renewable source, Tech. Rep., Publications office of the European Union, Luxembourg, 2018.

[14] E. Commission, D.G. for Energy, S. Haye, Y. Panchaksharam, E. Raphael, L. Liu, J. Howes, A. Bauen, S. Searle, Y. Zhou, K. Casey, J. O'Malley, C. Malins, S. Alberici, M. Hardy, W. Elbersen, I. Vural Gursel, B. Elbersen, M. Rudolf, N. Hall, and B. Armentrout, Assessment of the potential for new feedstocks for the production of advanced biofuels: Final report, Publications Office of the European Union, 2022.

[15] M. Aigner, D.E. Aliabadi, A. Amri-Henkel, J. Anderson, L. Becker, M. Bergfeld, U. Brand-Daniels, M. Braun-Unkhoff, S. Brinkop, A. Brosowski, O. Deniz, R.U. Dietrich, C. Eisenmann, D. Ennen, J. Eschmann, B. Frieske, W. Grimme, S. Haas, S. Hassel- wander, E. Hauser, S. Heib, N. Heimann, J. Hendricks, J. Hildebrand, C. Hoyer-Klick, M. Jordan, S. Kronshage, C. Lutz, H. Mantke, P. Matschoss, M. Mertens, A. Neitz- Regett, I. Österle, M. Oswald, S. Pichlmaier, J. Prause, T. Pregger, M. Raab, S. Richter, E.S.A. Ruiz, N. Wulff, and B. Zeck, Roadmap für strombasierte Kraftstoffe. Gesamtbericht zur Begleitforschung Energiewende im Verkehr (BEniVer), Zenodo, 2023. https://doi.org/10.5281/ zenodo.10208039.

[16] Deutsches Biomasseforschungszentrum gemeinnützige GmbH, SoBio: Scenarios of optimal energetic biomass usage until 2030 and 2050. Available at www.dbfz.de/sobio (accessed on 9 August 2024).

[17] K. Meisel, M. Jordan, M. Dotzauer, J. Schröder, V. Lenz, K. Naumann, K.F. Cyffka, N. Dögnitz, H. Schindler, J. Daniel-Gromke, and G.C. de Paiva, Quo vadis, biomass? Long-term scenarios of an optimal energetic use of biomass for the German energy transition, International Journal of Energy Research 2024 (2024), p. 6687376.

[18] M. Jordan, K. Meisel, M. Dotzauer, J. Schröder, K.F. Cyffka, N. Dögnitz, C. Schmid, V. Lenz, K. Naumann, J. Daniel-Gromke, G.C. de Paiva, H. Schindler, D. Esmaeili Aliabadi, N. Szarka, and D. Thrän, The controversial role of energy crops in the future German energy system: The trade offs of a phase-out and allocation priorities of the remaining biomass residues, Energy Reports 10 (2023), pp. 3848–3858.

[19] T. Bruckner, I.A. Bashmakov, Y. Mulugetta, H. Chum, A. De la Vega Navarro, J. Edmonds, A. Faaij, B. Fungtammasan, A. Garg, E. Hertwich, and D. Honnery, *Energy systems*, Cambridge University Press, 2014.

[20] D.K. Ray, P.C. West, M. Clark, J.S. Gerber, A.V. Prishchepov, and S. Chatterjee, Climate change has likely already affected global food production, PLoS One 14 (2019), p. e0217148.

[21] S.L. Perch-Nielsen, M.B. Bättig, and D. Imboden, Exploring the link between climate change and migration, Climatic Change 91 (2008), pp. 375–393.

[22] S. Hadian and K. Madani, A system of systems approach to energy sustainability assessment: Are all renewables really green? Ecological Indicators 52 (2015), pp. 194–206.

[23] K.K. Cao, J. Haas, E. Sperber, S. Sasanpour, S. Sarfarazi, T. Pregger, O. Alaya, H. Lens, S.R. Drauz, and T.M. Kneiske, Bridging granularity gaps to decarbonize large-scale energy systems – The case of power system planning, Energy Science & Engineering 9 (2021), pp. 1052–1060.

[24] D.E. Aliabadi, E. Çelebi, M. Elhüseyni, and G. Şahin, Modeling, simulation, and decision support, in *Local Electricity Markets*, Elsevier, 2021, pp. 177–197.

[25] J. Sachs, D. Moya, S. Giarola, and A. Hawkes, Clustered spatially and temporally resolved global heat and cooling energy demand in the residential sector, Applied Energy 250 (2019), pp. 48–62.

[26] IEA, *World Energy Outlook 2022*, Paris, France: International Energy Agency (IEA), 2022.

[27] F. Sensfuß, B. Lux, C. Bernath, C. Kiefer, B. Pfluger, C. Kleinschmitt, K. Franke, J. Fragoso Garcia, G. Deac, W. Männer, H. Brugger, T. Fleiter, M. Rehfeldt, A. Herbst, P. Manz, M. Neuwirth, M. Wietschel, T. Gnann, D. Speth, M. Krail, P. Mellwig, S. Blömer, S. Köppen, B. Tersteegen, C. Maurer, A. Ladermann, T. Dröscher, S. Willemsen, J. Müller-Kirchenbauer, M. Evers, O. Akça, D. Jiang, J. Hollnagel, J. Giehl, and T. Mielich, Long term scenarios for the transformation of the energy system in Germany green house gas neutral scenarios t45. (original, German title: Langfristszenarien für die Transformation des Energiesystems in Deutschland. Treibhausgasneutrale Szenarien T45) (2022). Available at https://langfristszenarien.de/enertile-explorer-wAssets/docs/LFS3_T45_Szenarien_15_11_2022_final.pdf.

[28] N. Wulff, D. Esmaeili Aliabadi, S. Hasselwander, T. Pregger, O. Deniz, H. Gils, S. Kronshage, E. Ruiz, W. Grimme, J. Horst, and P. Jochem, Energy system implications of demand scenarios and supply strategies for renewable transportation fuels (2023). Available at http://doi.org/10.2139/ssrn.4820179.

[29] O. Ruhnau, S. Bannik, S. Otten, A. Praktiknjo, and M. Robinius, Direct or indirect electrification? A review of heat generation and road transport decarbonisation scenarios for Germany 2050, Energy 166 (2019), pp. 989–999.

[30] R. Rodrigues, R. Pietzcker, P. Fragkos, J. Price, W. McDowall, P. Siskos, T. Fotiou, G. Luderer, and P. Capros, Narrative-driven alternative roads to achieve mid-century CO_2 net neutrality in Europe, Energy 239 (2022), p. 121908.

[31] M. Arnz, L. Göke, J. Thema, F. Wiese, N. Wulff, M. Kendziorski, K. Hainsch, P. Blechinger, and C. von Hirschhausen, Avoid, shift or improve passenger transport? Impacts on the energy system, Energy Strategy Reviews 52 (2024), p. 101302.

[32] D. Rapson and J. Bushnell, The limits and costs of full electrification, Review of Environmental Economics and Policy 18 (2024).

[33] F. Schreyer, F. Ueckerdt, R. Pietzcker, R. Rodrigues, M. Rottoli, S. Madeddu, M. Pehl, R. Hasse, and G. Luderer, Distinct roles of direct and indirect electrification in pathways to a renewables-dominated European energy system, One Earth 7 (2024), pp. 226–241.

[34] C. Dieckhoff and A. *Leuschner, Die Energiewende und ihre Modelle: Was uns En- ergieszenarien sagen können-und was nicht*, Transcript Verlag, 2016.

[35] S. Fujimori, T. Hasegawa, T. Masui, K. Takahashi, D.S. Herran, H. Dai, Y. Hijioka, and M. Kainuma, SSP3: AIM implementation of shared socioeconomic pathways, Global Environmental Change 42 (2017), pp. 268–283.

[36] T. Naegler, C. Sutardhio, A. Weidlich, and T. Pregger, Exploring long-term strategies for the German energy transition: A review of multi-sector energy scenarios, Renewable and Sustainable Energy Transition 1 (2021), p. 100010.

[37] M. Millinger, P. Tafarte, M. Jordan, A. Hahn, K. Meisel, and D. Thrän, Electrofuels from excess renewable electricity at high variable renewable shares: Cost, greenhouse gas abatement, carbon use and competition, Sustainable Energy & Fuels 5 (2021), pp. 828–843.

[38] K. Purr, J. Günther, H. Lehmann, and P. Nuss, Resource-efficient pathways towards greenhouse-gas-neutrality: Rescue: Summary report (2019). Available at www.umweltbundesamt.de/sites/default/files/medien/376/publikationen/rescue_kurzfassung_eng.pdf.

[39] M. Jordan, M. Millinger, and D. Thrän, Benopt-heat: An economic optimization model to identify robust bioenergy technologies for the German heat transition, SoftwareX 18 (2022), p. 101032.

[40] M. Millinger, P. Tafarte, M. Jordan, F. Musonda, K. Chan, K. Meisel, and D.E. Aliabadi, A model for cost-and greenhouse gas optimal material and energy allocation of biomass and hydrogen, SoftwareX 20 (2022), p. 101264.

[41] Ö. Mutlu, M. Jordan, T. Zeng, and V. Lenz, Competitive options for bio-syngas in high-temperature heat demand sectors: Projections until 2050, Chemical Engineering & Technology 46 (2023), pp. 559–566.

[42] M. Koch, K. Hennenberg, K. Hünecke, M. Haller, and T. Hesse, Role of bioenergy in the electricity and heating market until 2050, taking into account the future building stock (rolle der bioenergie im strom-und warmemarkt bis 2050 unter einbeziehung des zukünftigen gebäudebestandes) (2018). Available at www.energetische-biomassenutzung.de/fileadmin/Steckbriefe/dokumente/03KB114_Bericht_Bio-Strom-W%C3%A4rme.pdf.

[43] International Energy Agency, *The Future of Hydrogen – Seizing Today's Opportunities*.

[44] S. Nagireddi, J.R. Agarwal, and D. Vedapuri, Carbon dioxide capture, utilization, and sequestration: Current status, challenges, and future prospects for global decarbonization, ACS Engineering Au (2023).

[45] A. de Klerk, Fischer – Tropsch fuels refinery design, Energy & Environmental Science 4 (2011).

[46] M. Fasihi, O. Efimova, and C. Breyer, Techno-economic assessment of CO_2 direct air capture plants, Journal of Cleaner Production 224 (2019), pp. 957–980.

[47] D. Thrän (ed.), *Smart Bioenergy*, Springer International Publishing, Cham, 2015.

[48] G. Klepper and D. Thrän, Biomass: Striking a balance between energy and climate policies: Strategies for sustainable bioenergy use (2019). Available at www.acatech.de/publikation/biomasse-im-spannungs-feld-zwischen-energie-und-klimapolitik-strategien-fuer-eine-nachhaltig-download-pdf/?lang=en.

[49] K. Hansen, B.V. Mathiesen, and I.R. Skov, Full energy system transition towards 100% renewable energy in Germany in 2050, Renewable and Sustainable Energy Reviews 102 (2019), pp. 1–13.

[50] P.K. Ramanujam, B. Parameswaran, B. Bharathiraja, and A. Magesh, Bioenergy: Impacts on environment and economy, in *Energy, Environment, and Sustainability*, Springer, Singapore, 2023.

[51] Eckpunkte für eine Nationale Biomassestrategie: NABIS (2022). Available at www.bmwk.de/Redaktion/DE/Publikationen/Wirtschaft/nabis-eckpunktepapier-nationale-biomassestrategie.html.

[52] M. Jordan, V. Lenz, M. Millinger, K. Oehmichen, and D. Thrän, Future competitive bioenergy technologies in the German heat sector: Findings from an economic optimization approach, Energy 189 (2019), p. 116194.

[53] Revision of the Renewable Energy Directive (2021). Available at www.europarl.europa.eu/legislative-train/carriage/revision-of-the-renewable-energy-directive/report?sid=7201.

[54] F. Müller-Langer, K. Oehmichen, S. Dietrich, K.M. Zech, M. Reichmuth, and W. Weindorf, PTG-HEFA hybrid refinery as example of a SynBioPTx concept – Results of a feasibility analysis, Applied Sciences 9 (2019), p. 4047.

[55] Neste Rotterdam, Netherlands: Press Release (2022). Available at www.safinvestor.com/project/141929/neste-rotterdam-netherlands/.

[56] Installierte elektrische Leistung der Biogasanlagen in Deutsch-land in den Jahren 2001 bis 2022 (2023). Available at https://de.statista.com/statistik/daten/studie/167673/umfrage/installierte-elektrische-leistung-von-biogasanlagen-seit-1999/.

[57] D. Esmaeili Aliabadi, N. Wulff, M. Jordan, K.F. Cyffka, and M. Millinger, Soft-coupling energy and power system models to analyze pathways toward a de-fossilized German trans- port sector, in *International Conference on Operations Research*, Springer, 2022, pp. 313–320.

[58] H.C. Gils, Y. Scholz, T. Pregger, D.L. de Tena, and D. Heide, Integrated modelling of variable renewable energy-based power supply in Europe, Energy 123 (2017), pp. 173–188.

[59] P. Mock, Development of a scenario model for the simulation of future market shares and CO_2 emissions from vehicles (VECTOR21) (2010).

[60] J.D. Scheelhaase, K. Dahlmann, M. Jung, H. Keimel, H. Nieße, R. Sausen, M. Schaefer, and F. Wolters, How to best address aviation's full climate impact from an economic policy point of view? Main results from AviClim research project, Transportation Research Part D: Transport and Environment 45 (2016), pp. 112–125.

[61] F.G. Albrecht, D.H. König, N. Baucks, and R.U. Dietrich, A standardized methodology for the techno-economic evaluation of alternative fuels – A case study, Fuel 194 (2017), pp. 511–526.

[62] M. Dotzauer, K.S. Radtke, M. Jordan, and D. Thrän, Advanced SQL-database for bioenergy technologies - A catalogue for bio-resources, conversion technologies, energy carriers, and supply applications, Heliyon 10 (2024).

[63] Y. Scholz, Renewable energy based electricity supply at low costs: Development of the REMix model and application for Europe (2012).

[64] M. Robinius, A. Otto, P. Heuser, L. Welder, K. Syranidis, D.S. Ryberg, T. Grube, P. Markewitz, R. Peters, and D. Stolten, Linking the power and transport sectors – Part 1: The principle of sector coupling, Energies 10 (2017), p. 956.

[65] H.C. Gils, T. Pregger, F. Flachsbarth, M. Jentsch, and C. Dierstein, Comparison of spatially and temporally resolved energy system models with a focus on Germany's future power supply, Applied Energy 255 (2019), p. 113889.

[66] H.C. Gils, H. Gardian, M. Kittel, W.P. Schill, A. Zerrahn, A. Murmann, J. Launer, A. Fehler, F. Gaumnitz, J. van Ouwerkerk, and C. Bußar, Modeling flexibility in energy systems – Comparison of power sector models based on simplified test cases, Renewable and Sustainable Energy Reviews 158 (2022), p. 111995.

[67] H.C. Gils, Balancing of intermittent renewable power generation by demand response and thermal energy storage (2015).

[68] D. Luca de Tena and T. Pregger, Impact of electric vehicles on a future renewable energy-based power system in Europe with a focus on Germany, International Journal of Energy Research 42 (2018), pp. 2670–2685.

[69] H.C. Gils, H. Gardian, and J. Schmugge, Interaction of hydrogen infrastructures with other sector coupling options towards a zero-emission energy system in Germany, Renewable Energy 180 (2021), pp. 140–156.

[70] J. Horst and U. Klann, Mena-fuels: Analyse eines globalen marktes für wasserstoff und synthetische energieträger hinsichtlich künftiger handelsbeziehungen. bericht aus dem teil- projekt b.ii: Mena-O¨konomie (2022). Available at https://wupperinst.org/fileadmin/redaktion/downloads/projects/MENA-Fuels_Teilbericht_12_Handelsmodell.pdf.

[71] D. Esmaeili Aliabadi, D. Manske, L. Seeger, R. Lehneis, and D. Thrän, Integrating knowledge acquisition, visualization, and dissemination in energy system models: BENOPTex study, Energies 16 (2023), p. 5113.

[72] F. Musonda, M. Millinger, and D. Thrän, Greenhouse gas abatement potentials and economics of selected biochemicals in Germany, Sustainability 12 (2020), p. 2230.

[73] M. Sadr, D. Esmaeili Aliabadi, M. Jordan, and D. Thrän, A bottom-up regional assessment of bioenergy with carbon capture and storage in Germany (2024) (submitted).

[74] D.E. Aliabadi, K. Chan, N. Wulff, K. Meisel, M. Jordan, I. Österle, T. Pregger, and D. Thrän, Future renewable energy targets in the EU: Impacts on the German transport, Transportation Research Part D: Transport and Environment 124 (2023), p. 103963.

[75] A. Sternberg and A. Bardow, Power-to-what? Environmental assessment of energy storage systems, Energy & Environmental Science 8 (2015), pp. 389–400.

[76] N. Şensoy, Determination of optimal battery and renewable capacity investment for hybrid energy storage systems, in *2023 19th International Conference on the European Energy Market (EEM)*, IEEE, 2023, pp. 1–6.

[77] Y. Hu, M. Armada, and M.J. Sánchez, Potential utilization of battery energy storage systems (BESS) in the major european electricity markets, Applied Energy 322 (2022), p. 119512.

[78] M.L. Sørensen, P. Nystrup, M.B. Bjerregård, J.K. Møller, P. Bacher, and H. Madsen, Recent developments in multivariate wind and solar power forecasting, Wiley Interdisciplinary Reviews: Energy and Environment 12 (2023), p. e465.

[79] J. Schlund, R. Steinert, and M. Pruckner, Coordinating e-mobility charging for frequency containment reserve power provision, in *Proceedings of the Ninth International Conference on Future Energy Systems*, 2018, pp. 556–563.

[80] M. Sadr, D. Esmaeili Aliabadi, B. Avşar, and D. Thrän, Assessing the impact of seasonality on bioenergy production from energy crops in Germany, considering just-in-time philosophy, Biofuels, Bioproducts and Biorefining 18 (2024), pp. 883–898.

[81] D. Esmaeili Aliabadi and M. Jordan, Final report of the UFZ subproject for BEniVer– accompanying research on energy transition in transport (2023), p. 31. Available at https://doi.org/10.57699/f04y-7207.

[82] M. Jordan, K. Meisel, M. Dotzauer, H. Schindler, J. Schröder, K.F. Cyffka, N. Dögnitz, K. Naumann, C. Schmid, V. Lenz, J. Daniel-Gromke, G.C. de Paiva, D.E. Aliabadi, N. Szarka, and D. Thrän, Do current energy policies in Germany promote the use of biomass in areas where it is particularly beneficial to the system? Analysing short-and long-term energy scenarios, Energy, Sustainability and Society 14 (2024), p. 32.

[83] Y. Zhou, S. Searle, and N. Pavlenko, Current and future cost of e-kerosene in the United States and Europe, I: International Council on Clean Transportation Working Paper 14 (2022).

[84] The National Hydrogen Strategy (2020). Available at www.bmwk.de/Redaktion/EN/Publikationen/Energie/the-national-hydrogen-strategy.pdf?20blob=%publicationFile&v=6.

10 Artificial Intelligence and Renewable Energy Integration in Optimizing Green Hydrogen Production

Ashkan Safari and Arman Oshnoei

10.1 INTRODUCTION

Artificial intelligence (AI) is leading the way in current technological progress, providing significant potential for transformation in several areas. Its significance is varied, including the improvement of efficiency and production, as well as transforming decision-making processes and problem-solving skills. AI-driven solutions in sectors including healthcare, finance, transportation, and manufacturing simplify processes, optimize resource allocation, and allow predictive analytics, resulting in substantial cost savings and better results. AI enables the creation of autonomous systems, such as self-driving automobiles, unmanned aerial aircraft, and robotic helpers, which improve convenience and support safety and sustainability goals. AI is crucial in enhancing the efficiency and dependability of renewable energy sources, including solar, wind, and hydroelectric power, for electricity production. AI uses sophisticated forecasting algorithms to improve the prediction of renewable energy supply, leading to increased grid stability and easier integration into current energy systems. AI-powered smart grids enhance energy distribution and consumption patterns, promoting increased resilience and sustainability. Hydrogen generation is becoming an essential element in the shift toward a low-carbon economy, alongside renewable energy sources. Utilizing AI algorithms enhance hydrogen production processes by improving efficiency, cost-effectiveness, and environmental sustainability via process optimization, resource management, and predictive maintenance. The combination of AI, renewables, and hydrogen production shows great potential in addressing climate change, decreasing reliance on fossil fuels, and promoting a more sustainable energy system, leading to a more environmentally friendly future energy sector.

By the side of literature review in this field, electricity generation is a crucial indicator of a country's progress index, and renewable energy resources are gaining recognition as vital alternatives for clean electricity production. For instance, [1] assesses wind and solar data to produce hydrogen as a clean fuel, highlighting the potential for sustainable energy production. As well, the research conducted in [2] demonstrates the feasibility of wind and solar power generation for cost-effective hydrogen production, emphasizing the importance of multi-criteria decision-making methods in site selection. The authors in [3] evaluate different configurations of hybrid solar/wind systems for hydrogen production. Regarding renewable hydrogen production technologies, [4] compares various methods and electrolyzer types, emphasizing the economic viability of PV/H_2 and $wind/H_2$ systems for sustainable energy applications. Innovative approaches such as integrating wind, ocean thermal, and solar energy for hydrogen production are explored in [5], which underscores the potential of clean energy systems for practical applications. Assessing country-level hydrogen potential, [6] utilizes GIS-based analyses to evaluate technical feasibility, providing insights into Canada's renewable energy potential and infrastructure constraints. [7] investigates suitable locations for renewable hydrogen production using

DOI: 10.1201/9781032719436-10

multi-criteria decision-making methods, aiding regional energy planning efforts. A techno-economic analysis in Queensland, Australia, conducted in [8], demonstrates the feasibility of large-scale renewable hydrogen projects, offering insights for policymakers and industry stakeholders.

Consequently, [9] assesses wind and solar energy variability in Australia and computes hydrogen energy storage requirements for grid stability, highlighting hydrogen's role in achieving net-zero commitments. Exploring Canada's green hydrogen production potential, [10] evaluates solar energy resources for electrolysis, providing valuable information for strategic planning and renewable energy development. The research performed in [11] focuses on offshore electrolysis facilities, examining challenges and technologies for hydrogen production and suggesting future research directions for large-scale hydrogen production.

Assessing the feasibility of a standalone hybrid renewable energy system in Egypt, [12] optimizes system components for electricity and hydrogen generation, offering insights for national-scale hydrogen production projects. This review categorizes hydrogen production methods by color codes, emphasizing green hydrogen from renewable sources, and highlights recent advancements in [13]. Investigating renewable energy integration for food waste drying systems, [14] assesses the performance and economics of hybrid systems, providing valuable insights for sustainable waste management. Proposing a capacity optimization method for utility-scale wind–photovoltaic–electrolysis–battery systems, [15] aims to reduce curtailment and produce green hydrogen, offering a roadmap for large-scale renewable hydrogen production. Evaluating a standalone hybrid renewable energy system for a rural community in Nigeria, [16] optimizes system components using local meteorological data, demonstrating the feasibility of cost-effective energy solutions for off-grid communities. Finally, assessing green hydrogen production in Oman's combined hydroelectric-photovoltaic power station, [17] optimizes system components for economic viability, providing visions for sustainable energy development. As well, [18, 19] present the concepts of hydrogen-incorporated smart grids and their importance in future power systems.

By the outline of this chapter, the importance of hydrogen production, its role in resilient smart grids, and AI's role in green hydrogen production is drawn to investigation using theories and practical simulations. By the side of chapter structure, Section 10.2 manifests the green hydrogen concept, its production, and its economy. For the next, artificial intelligence, the BiLSTM model, and KPIs are described in Section 10.3. As well, Sections 10.4, 10.5, and 10.6 anticipate the results, future perspectives, and drawn conclusion of the work, respectively.

10.2   GREEN HYDROGEN (GH$_2$)

GH$_2$ is produced using electrolysis, a method that uses electricity from renewable sources, like solar, wind, or hydropower, to separate water molecules into hydrogen and oxygen, as conceptualized in Figure 10.1.

As illustrated in Figure 10.1, hydrogen electrolysis is a process that involves the splitting of water molecules (H$_2$O) into hydrogen (H$_2$) and oxygen (O$_2$) using electricity (equations 10.1–10.2). It begins with the immersion of two electrodes, typically made of metals like platinum or nickel, into a water-based electrolyte solution. When an electric current is passed through the electrolyte, it triggers a chemical reaction. At the anode (positive electrode), water molecules lose electrons and decompose into oxygen gas and positively charged hydrogen ions (protons). Meanwhile, at the cathode (negative electrode), these protons gain electrons, forming hydrogen gas. The produced hydrogen can be captured and stored for various applications, making electrolysis a crucial process in the production of green hydrogen, particularly when powered by renewable energy sources, like solar or wind power, as it eliminates carbon emissions associated with traditional hydrogen production methods.

$$4H^+ + 4e^- \rightarrow 2H_2 \tag{10.1}$$

$$2H_2O \rightarrow O_2 + 4H^+ + 4e^- \tag{10.2}$$

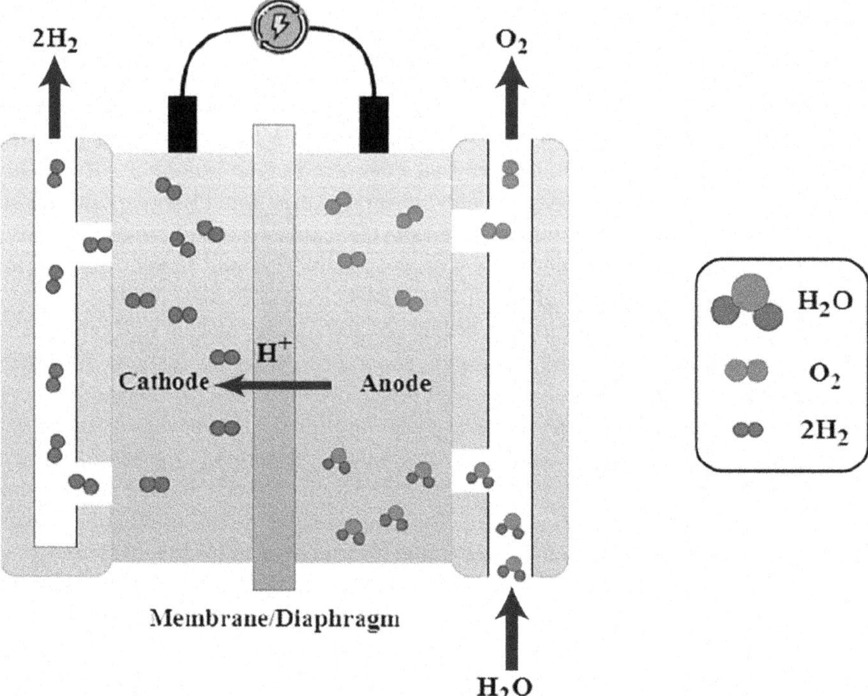

FIGURE 10.1 The electrolysis process of H_2.

This approach is environmentally friendly because it does not generate any greenhouse gases during production, unlike traditional methods that depend on fossil fuels. The process begins by producing renewable electricity to operate an electrolyzer that separates water into hydrogen and oxygen. The produced hydrogen can be stored long-term and used for purposes like fuel for automobiles, in industrial activities, or as a feedstock for chemical manufacture. The benefits of this technology include environmental advantages through carbon emission reduction, energy storage capabilities, adaptability in various sectors like transportation and industry, and the potential to decarbonize challenging sectors, such as heavy industrial and long-haul transportation. Challenges lie ahead due to the high costs associated with current production methods, which are influenced by the expensive nature of renewable electricity and the electrolyzer technology. However, as renewable energy costs decline and technology progresses, the overall cost is anticipated to decrease. Furthermore, the advancement of infrastructure requires substantial investment in electrolyzers, hydrogen storage facilities, and transportation infrastructure. Expanding output significantly to have a substantial effect on carbon emissions and energy transition necessitates the widespread use of renewable energy sources and favorable legislation. Governments, industries, and researchers are increasingly acknowledging green hydrogen as crucial for transitioning to a low-carbon economy. This requires investments in research, development, and infrastructure to fully utilize its potential and tackle climate change and energy security issues.

10.2.1 Hydrogen Electricity Production

A hydrogen production system utilizing wind and solar power presents a sustainable energy solution with minimal environmental impact. By harnessing renewable energy sources, such as wind and solar power, the system can produce hydrogen without relying on fossil fuels. This process helps reduce greenhouse gas emissions and dependence on non-renewable resources. Additionally, hydrogen can be stored and used as a versatile energy carrier, providing a reliable source of power

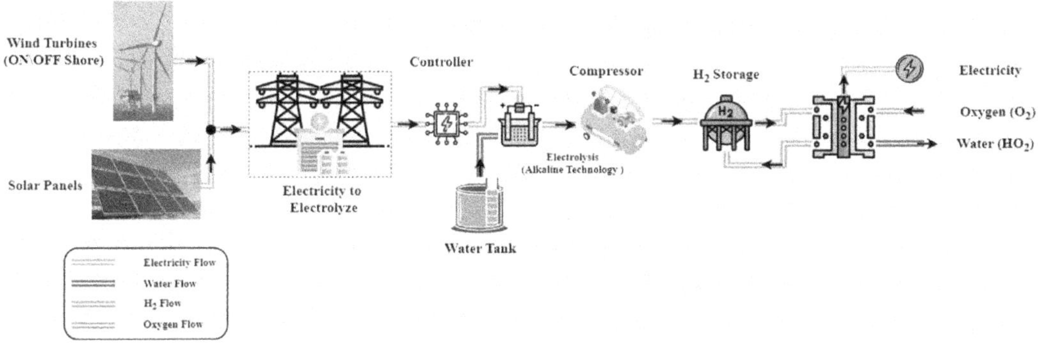

FIGURE 10.2   The concept of the hydrogen production bench.

for various applications, as a multi-carrier source. Consequently, the concept of this system is illustrated in Figure 10.2.

Depending on Figure 10.2, the process begins with the conversion of renewable energy sources into electrical power. Wind turbines capture kinetic energy from the wind, rotating blades that drive a generator to produce electricity. Solar panels, on the other hand, utilize photovoltaic cells to convert sunlight directly into electricity through the photovoltaic effect. Once electricity is generated, it is directed to an electrolyzer to convert electrical current to split H_2O into H_2 and O_2 molecules through electrolysis. This electrolysis process occurs within the electrolyzer, which typically consists of two electrodes submerged in an electrolyte solution. When electricity passes through the electrolyte, it induces a chemical reaction. At the anode, water molecules lose electrons, forming oxygen gas and positively charged hydrogen ions (protons). Simultaneously, at the cathode, these protons gain electrons to form hydrogen gas. The produced hydrogen is then captured and stored for later use. When electricity demand exceeds supply or when renewable sources are not actively generating power, the stored hydrogen can be converted back into electricity using a fuel cell. A fuel cell operates by combining hydrogen with oxygen from the air to produce electricity, heat, and water vapor. This process, known as electrochemical conversion, is highly efficient and emits no greenhouse gases or pollutants, making it a clean and sustainable energy solution.

As the PVs and wind turbines are the renewables of this bench, they are mathematically modulated accordingly:

$$P_{PV} = GHI \cdot n_m \cdot A_m \cdot \eta_{PV} \cdot \eta_{inv} \cdot f_{PV} \tag{10.3}$$

$$\eta_{PV} = \eta_{PV,\,NOM} \left[1 + \gamma \left(T_c - T_{c,\,ref} \right) \right] \tag{10.4}$$

$$T_c = T_a + \left(T_{noct} - T_{a,noct} \right) \frac{GHI}{GHI_{noct}} \frac{U_{L,noct}}{U_L} \left[1 - \frac{\mu_{PV}}{\tau \cdot \alpha} \right] \tag{10.5}$$

where the effective area (A_m) and the quantity of PV panels (n_m) are defined, respectively. The inverter's efficiency (η_{inv}) and the derating factor (f_{PV}) are specified. As the environment has an impact on global irradiance, η_{PV} is considered as (10.4) to integrate the impact on simulation. Additionally, T_C, T_a, and T_{noct} are the PV's actual operating temperature, the surrounding air temperature, and the nominal cell operating temperature, respectively. As well, the global irritation and nominal/actual heat transfer are symbolized by GHI_{noct}, U_L, and $U_{L,noct}$.. Finally, γ, $T_{c,ref}$, and $\tau \cdot \alpha$ are the temperature coefficient, reference condition operating cell temperature, and transmittance-absorptance coefficient, respectively.

For the side of wind turbine, its output depends on the wind speed at the hub. Output power can be determined based on the wind speed at the hub height and the fan power curve. The wind speed data included in this chapter were gathered at a height of 10 m above the ground. Wind speed at the hub height can be calculated using (10.6), and the fan's power output can be determined by integrating the power curve.

$$\frac{V_2}{V_1} = \left(\frac{h_2}{h_1}\right)^k \tag{10.6}$$

$$P_{WT} = \begin{cases} 0 & V_t < V_c \\ P_R^e\left(\dfrac{V_t - V_c}{V_r - V_c}\right) & V_c \leq V_t \leq V_r \\ P_R^e & V_r \leq V_t \leq V_f \\ 0 & V_t > V_f \end{cases} \tag{10.7}$$

in which V_1 represents the surface wind speed in meters per second, V_2 represents the wind speed at a height of 70 m above the ground in meters per second, and k is the wind shear coefficient with a value of 0.14. V_c represents the initial wind speed at which the fan starts, measured in meters per second. V_r is the fan's rated wind speed, also measured in meters per second. V_f is the wind speed at which the fan cuts off, measured in meters per second. P_R^e represents the rated output power of the fan, in megawatts, while P_{WT} stands for the actual output power of the fan, in megawatts.

Getting to the electrolyzer (10.1–10.2), its operating voltage is determined by:

$$V = V_{ocv} + V_{act} + " V + V_{ohm} \tag{10.8}$$

That is, V_{act} denotes the open circuit voltage, while V_{ocv} signifies the theoretical minimal electrolytic voltage at which water electrolysis arises. $\Delta V, and\ V_{ohm}$ represent the equivalent and ohmic overpotentials, caused by diffusion and the proton-exchange membrane, respectively.

As the next step, the hydrogen storage should be determined. Hydrogen storage is essential for using hydrogen as a clean energy source, especially in fuel cell cars and grid storage. Hydrogen storage is often achieved by high-pressure tanks constructed from composite materials or metal alloys capable of withstanding the necessary high pressures. The tanks hold hydrogen gas at pressures between 350 and 700 bar, depending on the specific use. Liquid hydrogen storage is a process that involves maintaining hydrogen in its liquid form by using cryogenic temperatures. Liquid hydrogen has a better energy density than gaseous storage but necessitates energy-intensive liquefaction procedures and well-insulated tanks to avoid boil-off. Furthermore, researchers are working on developing solid-state hydrogen storage materials, such as metal hydrides and carbon-based materials, to provide safe and efficient storage of hydrogen under normal circumstances. These materials can collect and release hydrogen via chemical processes, offering a viable substitute for storing hydrogen in compressed or liquid form. A hydrogen tank comprises various essential components, such as an outer tank for structural support and protection, an inner tank for high-pressure resistance, a filling and emptying coupling, a heater for temperature maintenance (especially in cryogenic systems), a heat exchanger for temperature control, a cryogenic filling valve for liquid hydrogen, a cryogenic return valve for excess gas or liquid release, a pressure regulation valve, an isolation shut-off valve, a boil-off valve, and a safety relief valve. Every part has a crucial function in securely storing and treating hydrogen gas, whether in liquid or gaseous state, to ensure effective operation and reduce dangers related to hydrogen management.

$$E_{HT}(t) = E_{HT}(t-1) + \left(P_{PEME-HT}(t) - \frac{P_{HT-FC}(t)}{\dot{}_{storage}}\right)" t \tag{10.9}$$

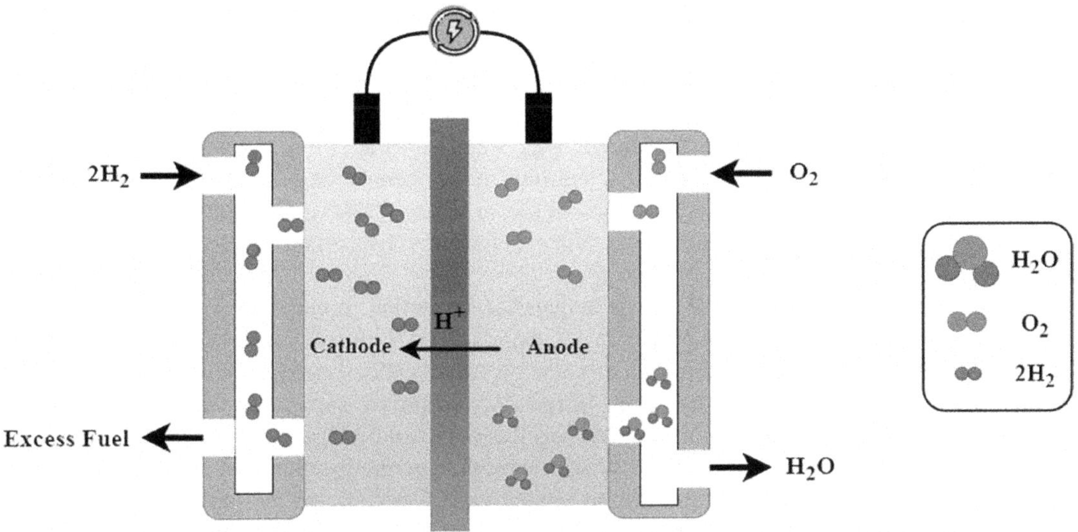

FIGURE 10.3 The structure of a fuel cell.

$$M_{HT}(t) = \frac{E_{HT}(t)}{HHV_{H2}} \tag{10.10}$$

$$E_{\text{HT,min}} \leq E_{\text{HT}}(t) \leq E_{\text{HT,max}} \tag{10.11}$$

The variable $P_{HT\text{-}FC}$ represents the power transmitted from the hydrogen tank to the fuel cell. The storage efficiency, denoted as *ηstorage*, is assumed to be 95%, accounting for losses during transit or storage. $M_{HT}(t)$ is an abbreviation. At each given time step t, the mass of stored hydrogen is equivalent to 39.7 *kW h/kg*, which represents the higher heating value (HHV) of hydrogen. The stored hydrogen mass cannot exceed the tank's rated capacity. Nevertheless, a small fraction of the hydrogen (specifically 5%) may not be retrieved for problems as of hydrogen pressure loss.

As the final phase of the process, the fuel cell should be manifested. A *fuel cell* is an electrochemical device that converts the chemical energy of hydrogen into electricity, with water and heat as by-products, as conceptualized in Figure 10.3.

The process involves the electrochemical reaction of hydrogen and oxygen, anticipated by Figure 10.3, typically from the air, to produce electricity, which can be used to power electric vehicles, buildings, and other applications. The key components of a fuel cell include an electrolyte, an anode, and a cathode. When hydrogen is supplied to the anode of the fuel cell, it is split into protons and electrons. The protons pass through the electrolyte to the cathode, while the electrons flow through an external circuit, generating an electrical current. At the cathode, the protons, electrons, and oxygen from the air combine to form water and heat. The overall chemical reactions are manifested as the following:

$$H_2 \rightarrow 2H^+ + 2e^- \tag{10.12}$$

$$2H^+ + \frac{1}{2}O_2 + 2e^- \rightarrow H_2O \tag{10.13}$$

$$H_2 + \frac{1}{2}O_2 \rightarrow H_2O + \text{Electricity} \tag{10.14}$$

As the output voltage of the fuel cell is provided in (10.16),

$$U = U_{nernst} - U_{ohm} - U_{act} - U_{con} \tag{10.15}$$

in which U_{nernst} represents the Nernst voltage, U_{ohm} stands for the ohmic overvoltage, U_{act} denotes the activation overvoltage, and as well, U_{con} signifies the concentration overvoltage, respectively.

10.2.2 Hydrogen Economy

A linear economy is characterized by the sequential extraction, processing, utilization, and dispersal of resources, which in turn results in the depletion of resources and detrimental ecological impacts. On the contrary, the notion of a circular economy endeavors to achieve sustainability by reducing emissions and pollution by means of reusing and repurposing materials at various stages of their life cycle; a particular roadmap to hydrogen is depicted by Figure 10.4.

At present, nevertheless, a mere 8.6% of the worldwide economy functions in accordance with this circular framework. For instance, approximately 10 $Gm^3/year$ of hydrogen is lost annually through industrial effluent streams in Europe [20]. In order to bolster the circular economy's dependence on renewable energy sources, the utilization and production of biohydrogen from refuse via microbial processes present itself as a potentially fruitful approach owing to its flexibility, energy conservation, and feasible operational demands. Such sustainable practices are consistent with the emission reduction targets and UN Sustainable Development Goals (UN-SDGs). However, obstacles, including costs, waste management, storage, transportation, and safety, must be considered. Electrolytic hydrogen production is anticipated to gain prominence in the coming years due to the growing availability and decreasing expenses of renewable energy sources. Green hydrogen, being a highly adaptable energy carrier, plays a pivotal role in a hydrogen-based economy across multiple sectors, including reserve energy systems, electricity and thermal generation, power-to-gas technologies, and diverse modes of transportation.

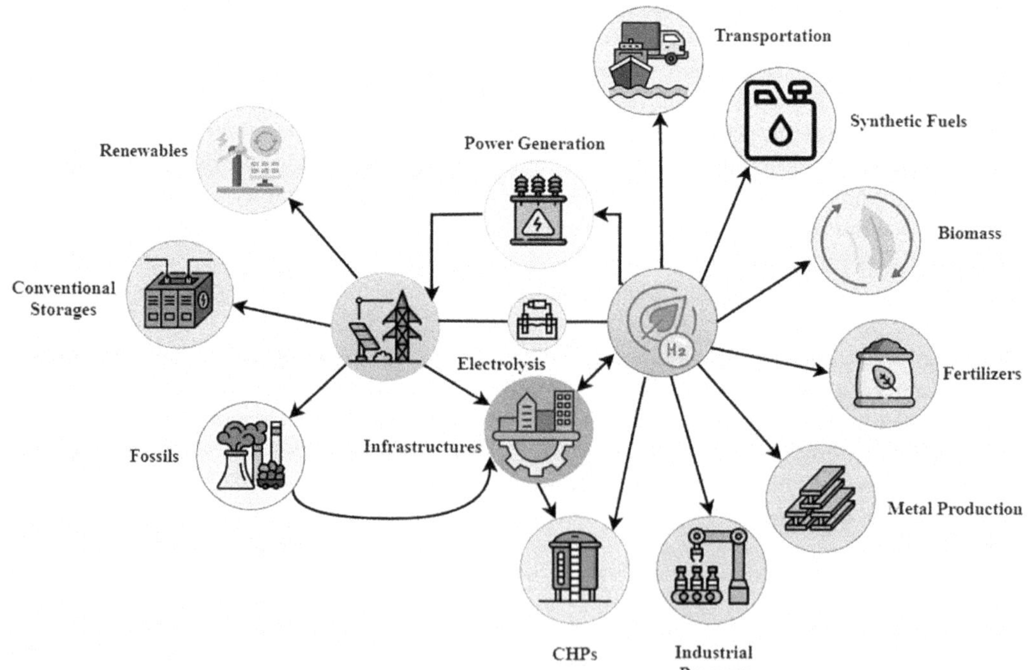

FIGURE 10.4   The representation of the hydrogen economy.

10.3 INTELLIGENT MODELS IN GH$_2$-INCORPORATED MICROGRIDS

The use of hydrogen into smart grids manifests considerable effect in the constantly changing energy generation and distribution sector. Hydrogen emerges as a versatile and clean energy carrier with significant promise as the globe moves toward sustainable energy solutions. Maximizing the benefits and maintaining optimal utilization inside smart grids require the use of advanced technologies, with AI and intelligent control approaches being key among them. The core of hydrogen-integrated smart grids requires smooth coordination and optimization of associated processes. AI is crucial in this sector as it provides advanced analytics, predictive skills, and decision-making algorithms that can adjust to changing circumstances. AI utilizes machine learning algorithms to assess extensive data from smart grid sensors, energy demand patterns, and production capacity in order to enhance the efficiency of hydrogen production, storage, and distribution. Intelligent control techniques enhance AI by providing the structure for monitoring and managing smart grid operations in real time. These methods use AI-generated predictions to implement accurate control strategies that enhance the stability, reliability, and effectiveness of hydrogen-based energy systems. Intelligent control approaches provide the flexibility and responsiveness needed for effective grid management, such as modifying electrolyzer output according to changing renewable energy availability and optimizing hydrogen storage and distribution to match different demand profiles. AI and intelligent control methods allow for autonomous decision-making in smart grids, decreasing the need for manual intervention and improving system resilience. By continuously learning and adapting, these technologies can predict possible problems, find the best solutions, and prevent dangers in advance. The combination of AI and intelligent control methods promotes innovation and scalability in hydrogen-based smart grids. Utilizing AI-driven perspectives allows stakeholders to discover potential for optimization, reveal hidden efficiencies, and enable new possibilities for sustainable energy consumption. Utilizing AI-driven innovation and intelligent control tactics can expedite the shift toward a sustainable and linked energy ecology. In this regard, a BiLSTM intelligent model is applied to the hydrogen-incorporated system presented in this chapter to analyze the results.

10.3.1 BIDIRECTIONAL LSTM

BiLSTM is a recurrent neural network (RNN) architecture commonly employed in natural language processing (NLP) and jobs using sequential input. The model improves typical LSTM networks by integrating information from past and future time steps, enabling it to detect more intricate patterns in sequential data, as conceptualized in Figure 10.5.

Based on Figure 10.5, the BiLSTM network comprises two LSTM layers, one operating on the input sequence in a forward manner, and the other in a backward manner. Bidirectional processing allows the network to gather information from both past and future contexts, leading to a more thorough comprehension of the input sequence. As of the overall process, in BiLSTM networks, input encoding involves converting input sequences, such as sentences or time series data, into numerical vectors using techniques like word embeddings or one-hot encoding. These encoded sequences are then fed into two LSTM layers, one processing the sequence in a forward direction, and the other in a backward direction, each consisting of memory cells and gates that regulate information flow. As the input sequence is processed, hidden states are computed at each time step, capturing learned representations and contextual information from both past and future contexts. Concatenating the hidden states from both directions provides a richer encoding. The concatenated states are passed through an output layer, typically comprising fully connected layers, followed by an activation function, to transform them into the desired output format, such as class probabilities or continuous values. During training, parameters are optimized using techniques like backpropagation through time or gradient descent to minimize a loss function. Once trained, the network can make predictions on new input sequences by passing them through the trained network, with the output interpreted according to the specific task, such as text classification or

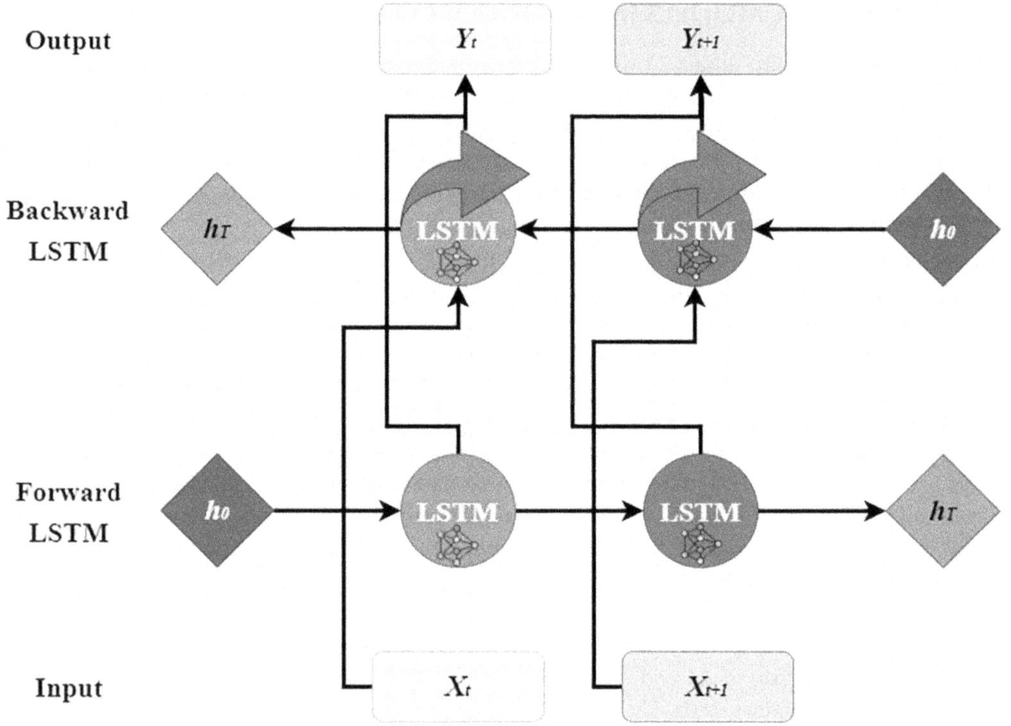

FIGURE 10.5   The structure of the utilized BiLSTM.

time series forecasting. As the next step, LSTM models should be formulated to set up BiLSTM. The *LSTM* model is calculated as:

$$i_{1t} = \sigma\left(W_{1i}h_{1t-1} + U_{1i}D_{In1t} + b_{1i}\right) \tag{10.16}$$

$$f_{1t} = \sigma\left(W_{1f}h_{1t-1} + U_{1f}D_{In1t} + b_{1f}\right) \tag{10.17}$$

$$O_{1t} = \sigma\left(W_{1o}h_{1t-1} + U_{1o}D_{In1t} + b_{1o}\right) \tag{10.18}$$

$$\tilde{C}_{1t} = \sigma\left(W_1 h_{1t-1} + U_1 D_{In1t} + b_1\right) \tag{10.19}$$

$$C_{1t} = \left(f_{1t} \odot C_{1t-1}\right) + \left(i_{1t} \odot \tilde{C}_{1t}\right) \tag{10.20}$$

where i_{1t} and f_{1t} compensate the input gate and the forget gate, respectively. O_{1t} and $\tilde{C}_{1t}$ are utilized as the output gate and memory cell candidate of the LSTM network. As the output vector of the LSTM, h_{1t} is defined from the output of the network as:

$$h_{1t} = O_t \odot \tan h(C_{1t}) \tag{10.21}$$

For the bidirectional operation, the forward, backward, and concatenated LSTMs are constructed as:

$$H_f, C_f = LSTM\left(X_b\right) \tag{10.22}$$

$$H_b, C_b = LSTM\left(X_b^{rev}\right) \tag{10.23}$$

$$\vartheta_{Con} = Concatenate\left(H_f, H_b\right) \tag{10.24}$$

in which H_f, C_f and H_b, C_b are the hidden and cell states of forward and backward LSTMs, respectively. As well, ϑ_{Con} is the concatenated version of states. To have the fully connected layers, output is determined as follows:

$$\vartheta_d = \sigma_d \left(\vartheta_f \cdot W_d + b_d \right) \tag{10.25}$$

$$Y = \vartheta_d \cdot W_{out} + b_{out} \tag{10.26}$$

In sequence, ϑ_d, σ_d and ϑ_f are the dense layer representation, activation function, and previous layer tensor input, respectively. As well, dense weight matrices and bias vectors are defined by W_d and b_d. Y, W_{out}, and b_{out} are the final output of the system, the output layer weight matrices, and the bias vectors.

10.3.2 Key Performance Indicators

KPIs are quantitative metrics used to assess the performance, efficacy, and efficiency of these models in meeting their goals. These indicators are essential for evaluating many elements of intelligent models in numerous fields, including machine learning algorithms and artificial intelligence applications. KPIs present comparative information on the model's performance, pinpointing areas that need enhancement and directing decision-making; also, they are fitted references for comparing the performance of different models. Consequently, the KPIs of MSE, MAE, RMSE, and R^2 are utilized in this chapter to assess the performance of the BiLSTM model, as follows [21–25].

Mean squared error (MSE) computes the mean of the squared differences between predicted and actual data. The calculation involves averaging the squared discrepancies between the anticipated values and the true values. Mathematically, it is expressed as

$$MSE = \frac{1}{n} \sum (\psi_{Observed} - \psi_{Predicted})^2 \tag{10.27}$$

where n is the number of data, and $\psi_{Observed}$ and $\psi_{Predicted}$ are the observed and predicted values of the features and target variables, respectively.

Mean absolute error (MAE) computes the average of the absolute discrepancies between anticipated and observed values. The calculation involves averaging the absolute differences between the anticipated values and the true values. Mathematically, it is expressed as:

$$MAE = \frac{1}{n} \sum \left| (\psi_{Observed} - \psi_{Predicted}) \right| \tag{10.28}$$

Root mean square error (RMSE) is the square root of the MSE and quantifies the average magnitude of the mistakes. The calculation involves finding the square root of MSE as:

$$RMSE = \sqrt{\frac{1}{n} \sum (\psi_{Observed} - \psi_{Predicted})^2} \tag{10.29}$$

R^2 indicates the amount of variance in the dependent variable that can be explained by the independent variables. It quantifies the accuracy of the regression model. R^2 values vary from 0 to 1, with 1 representing a perfect match and 0 indicating no enhancement above the mean forecast. R^2 is mathematically performed as:

$$R^2 = 1 - \frac{\sum_{i=1}^{n} \left(\psi_{i,predicted} - \psi_{i,actual} \right)^2}{\sum_{i=1}^{n} \left(\psi_{i,actual} - \overline{\psi_i} \right)^2} \tag{10.30}$$

It is ideal for MSE, MAE, and RMSE to be minimized as they quantify the discrepancy between expected and actual values. A lower value signifies superior performance, with 0 denoting flawless predictions. A higher coefficient of determination is preferred since it indicates a greater proportion of the variance in the dependent variable is accounted for by the independent variables. An R^2 value of 1 signifies a perfect fit of the regression model to the data, whereas lower values imply a less-accurate match. Thus, for R^2, greater values indicate superior model performance.

10.4 SIMULATIONS AND DERIVED RESULTS

In order to assess the performance of the BiLSTM model, the model presented in Figure 10.3 is simulated in MATLAB/Simulink, then the generated data is analyzed by BiLSTM. The model applied to a dataset containing various parameters related to a system's operation. The dataset includes metrics such as SoC ranging from 51.02% to 100%, solar current fluctuating between −1.23 A and 662.55 A, energy consumption varying from 46.09 kWh/kg to 85.85 kWh/kg, and other parameters like mass rate, generator current, electrolyzer current, storage current, and hydrogen mass. The dataset consists of 36,793 data points in total, the details of which are provided in Table 10.1, and the MATLAB/Simulink results are manifested in Figures 10.6 and 10.7.

Regarding Figure 10.6, the electricity currents are derived from wind and solar power electrolyzers, which produce hydrogen through electrolysis, with the current determining the rate of hydrogen generation. Energy storage systems like batteries play a pivotal role, storing excess renewable energy for later use and regulating grid stability.

The generator voltage fluctuates within the range of 0 to 250 volts, as in Figure 10.7a, reflecting the variability in power output from renewable resources, such as wind turbines or solar panels. This voltage variation is indicative of the intermittent nature of renewable energy generation, influenced by factors like wind speed and solar irradiance. Concurrently, the electrolyzer voltage remains relatively stable at around 250 V, reflecting its operational requirements for efficient electrolysis to produce hydrogen from water. As for the SoC, Figure 10.7b of the microgrid, it initially stands at 100% and gradually decreases to 50% as the grid is charged, indicating the utilization of stored energy from batteries or hydrogen fuel cells to meet demand. Consequently, the amount of produced hydrogen and its factors are manifested in Figure 10.8.

As manifested in Figure 10.8, over a 24 hr simulation period, the hydrogen-incorporated system produced a total of 50 kg of hydrogen, indicative of the electrolyzer's continuous operation and hydrogen generation. The energy input into the system from wind and electrolyzers varied dynamically, spanning from 0 to 2,000 kWh, reflecting the fluctuating nature of renewable energy generation and electrolysis processes. This variability represents the need for efficient energy management and utilization strategies within the microgrid. Additionally, the mass rate, representing the rate

---

TABLE 10.1

Details of the Generated Data

	Max	Min	Mean	Variance	Total Count
SoC (%)	100	51.02029	75.8702	175.192	36,793
Solar current (A)	662.5508	−1.22935	165.0126	51,912.57	
Energy consumption (kWh/kg)	85.85487	46.08816	47.89052	13.69754	
Mass rate (kg/hr)	6.083731	0.523402	1.906145	1.190985	
Generator current (A)	1667.071	−1.5E–06	320.91	131,044.1	
Electrolyzer current (A)	1085.348	181.0695	348.3774	37,462.91	
Storage current (A)	294.5976	−860.46	95.1637	31,316.28	
Hydrogen mass (kg)	48.58048	0	25.33139	220.6565	

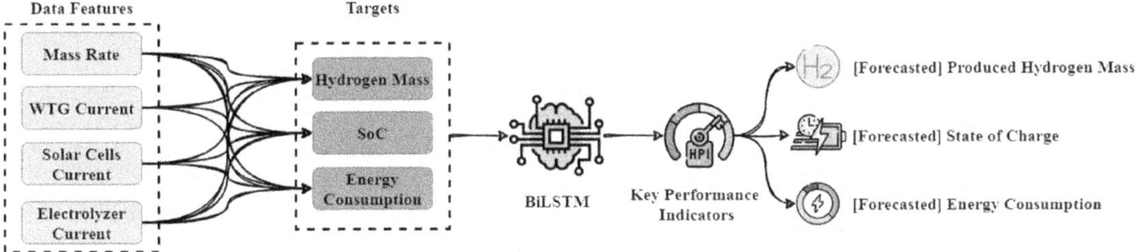

FIGURE 10.9 The overall process of prediction by BiLSTM over the H_2-incorporated microgrid.

TABLE 10.2

KPI Results Derived by BiLSTM

Metrics	Hydrogen Mass	SoC	Energy Consumption
MSE	0.000005614	0.002559	0.000523
MAE	0.001939	0.043361	0.010934
RMSE	0.002369	0.050585	0.022864
R^2	0.859	0.855	0.69

According to Figures 10.9 and 10.10, the model was assessed using various metrics to forecast three main components of the system: H_2 mass, SoC, and energy consumption. The model showed remarkable accuracy in predicting H_2 mass, with an MSE of 0.000005614, providing little difference between anticipated and real values. The model's precision in calculating hydrogen mass is highlighted by the MAE of 0.001939. The RMSE of 0.002369 indicates a minimal margin of error in the BiLSTM's predictions. The model's performance is indicated by the high R^2 value of 0.859, showing a significant connection between predicted and observed hydrogen mass. The BiLSTM model showed strong performance in predicting SoC, with MSE, MAE, and RMSE values of 0.002559, 0.043361, and 0.050585, respectively. The R^2 value of 0.855 signifies the model's capacity to precisely capture fluctuations in battery SoC. The BiLSTM model showed competitive performance in forecasting energy consumption with MSE, MAE, and RMSE values of 0.000523, 0.010934, and 0.022864, respectively. The R^2 score of 0.69 indicates a rather excellent fit between projected and actual energy consumption values, albeit being slightly lower than other indicators.

10.5 FUTURE PERSPECTIVES

The future prospect of AI- and H_2-integrated smart grids shows considerable potential for developing a sustainable and efficient energy environment. AI's progress allows for its incorporation into smart grid technologies, enhancing the monitoring, control, and optimization of energy systems. AI algorithms can process large volumes of data instantly, allowing for predictive maintenance, demand prediction, and adaptive energy management tactics. Furthermore, AI enables the creation of smart energy trading platforms and decentralized energy markets, allowing consumers to engage in energy exchange, as peers, and enhance their consumption habits. Simultaneously, smart grids that integrate hydrogen present an affordable renewable energy alternative that encompasses substantial potential for energy sector decarbonization and climate change mitigation. Hydrogen functions as a multi-functional energy carrier, possessing the ability to both store and convey renewable energy derived from intermittent sources, such as wind and solar power. Hydrogen technologies

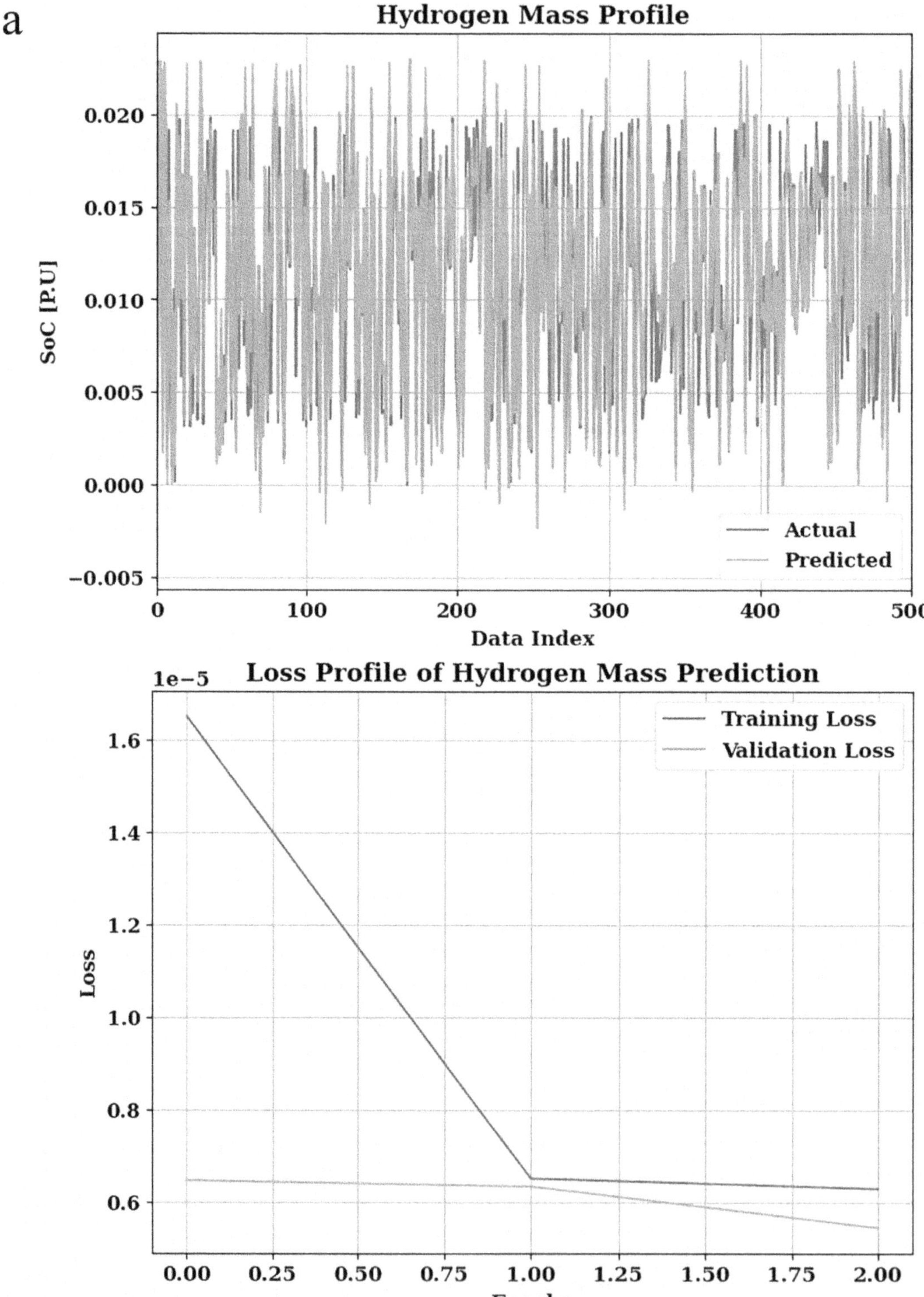

FIGURE 10.10   The prediction profile and the errors of (a) hydrogen mass, (b) SoC, and (c) consumed energy.

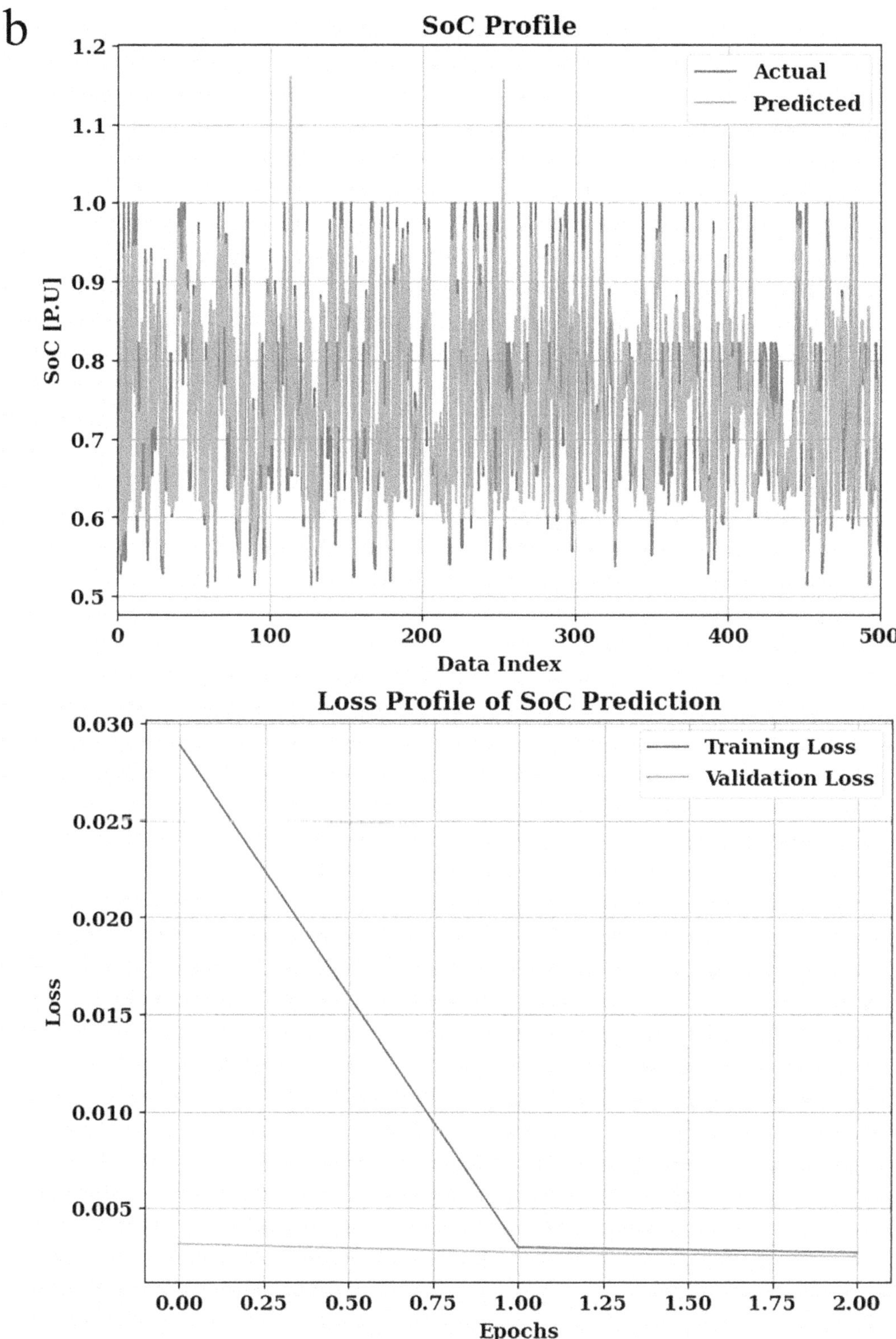

FIGURE 10.10 (Continued)

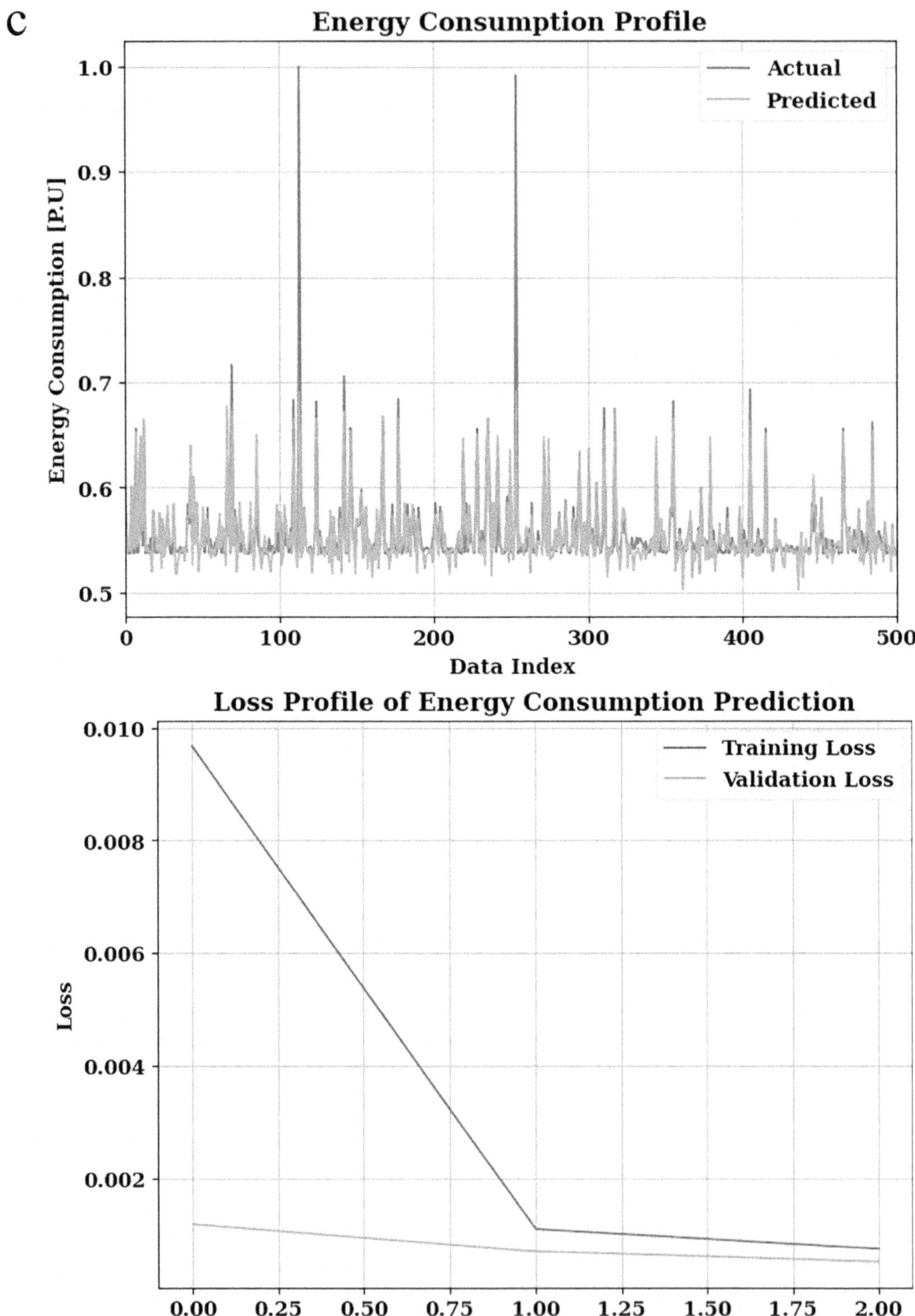

FIGURE 10.10   (Continued)

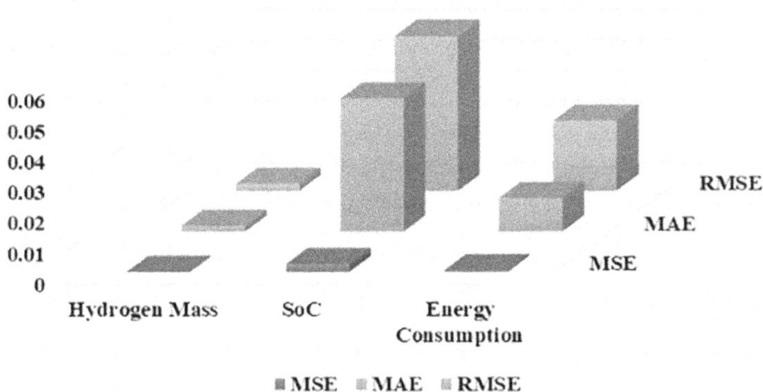

FIGURE 10.11 The KPIs of the model applied to the system.

support the electrification of diverse sectors, including industry and transportation, by facilitating the integration of renewable energy resources, enhancing the resilience of smart grid infrastructure, and incorporating hydrogen production, storage, and distribution.

Finally, AI- and hydrogen-powered smart grids provide prospects for energy sustainability and advancement. Machine learning algorithms have the capability to enhance the efficiency and reliability of smart grids by optimizing hydrogen production processes, forecasting energy demand patterns, and dynamically managing energy flows. In addition, predictive analytics enabled by AI can maximize the efficiency of hydrogen infrastructure deployment, detect potential cost-saving opportunities, and expedite the transition toward an economy reliant on hydrogen.

10.6 CONCLUSION

The capability to optimize hydrogen and electricity production processes, thereby improving efficiency and sustainability during the transition to renewable energy systems, is the significance of AI in hydrogen-incorporated smart grids. Consequently, the use of MATLAB/Simulink simulations for hydrogen synthesis from wind and solar power, combined with alkaline technology, and analyzed using the BiLSTM model, represents considerable progress in the renewable energy sector. Accurate forecasts were generated for hydrogen production, electricity consumption, and SoC levels across a 24 hr timeframe using this method. The evaluation of the BiLSTM model encompassed the assessment of its performance using KPIs, such as MAE, MSE, RMSE, and R^2. The model's results validated its effectiveness in capturing and forecasting system dynamics. This technique improves knowledge of hydrogen-integrated smart grid systems and provides perspectives for improving energy production, consumption, and storage.

REFERENCES

[1] Nematollahi O, Alamdari P, Jahangiri M, Sedaghat A, Alemrajabi AA. A techno-economical assessment of solar/wind resources and hydrogen production: a case study with GIS maps. Energy. 2019;175:914–930.

[2] Ahmadi MH, Hosseini Dehshiri SS, Hosseini Dehshiri SJ, Mostafaeipour A, Almutairi K, Ao HX, Rezaei M, Techato K. A thorough economic evaluation by implementing solar/wind energies for hydrogen production: A case study. Sustainability. 2022 Jan 20;14(3):1177.

[3] Hasan MM, Genç G. Techno-economic analysis of solar/wind power based hydrogen production. Fuel. 2022 Sep 15;324:124564.

[4] Benghanem M, Mellit A, Almohamadi H, Haddad S, Chettibi N, Alanazi AM, Dasalla D, Alzahrani A. Hydrogen production methods based on solar and wind energy: A review. Energies. 2023 Jan 9;16(2):757.

[5] Ishaq H, Dincer I. A comparative evaluation of OTEC, solar and wind energy based systems for clean hydrogen production. Journal of Cleaner Production. 2020 Feb 10;246:118736.

[6] Okunlola A, Davis M, Kumar A. The development of an assessment framework to determine the technical hydrogen production potential from wind and solar energy. Renewable and Sustainable Energy Reviews. 2022 Sep 1;166:112610.

[7] Almutairi K, Mostafaeipour A, Jahanshahi E, Jooyandeh E, Himri Y, Jahangiri M, Issakhov A, Chowdhury S, Hosseini Dehshiri SJ, Hosseini Dehshiri SS, Techato K. Ranking locations for hydrogen production using hybrid wind-solar: A case study. Sustainability. 2021 Apr 16;13(8):4524.

[8] Rezaei M, Akimov A, Gray EM. Economics of renewable hydrogen production using wind and solar energy: A case study for Queensland, Australia. Journal of Cleaner Production. 2024 Jan 5;435:140476.

[9] Boretti A, Castelletto S. Hydrogen energy storage requirements for solar and wind energy production to account for long-term variability. Renewable Energy. 2024 Feb 1;221:119797.

[10] Karayel GK, Dincer I. Green hydrogen production potential of Canada with solar energy. Renewable Energy. 2024 Feb 1;221:119766.

[11] Niblett D, Delpisheh M, Ramakrishnan S, Mamlouk M. Review of next generation hydrogen production from offshore wind using water electrolysis. Journal of Power Sources. 2024 Feb 1;592:233904.

[12] Elminshawy NA, Diab S, Yassen El SY, Elbaksawi O. An energy-economic analysis of a hybrid PV/wind/battery energy-driven hydrogen generation system in rural regions of Egypt. Journal of Energy Storage. 2024 Mar 1;80:110256.

[13] Manoo MU, Shaikh F, Kumar L, Arıcı M. Comparative techno-economic analysis of various stand-alone and grid connected (solar/wind/fuel cell) renewable energy systems. International Journal of Hydrogen Energy. 2024 Jan 2;52:397–414.

[14] Zainal BS, Ker PJ, Mohamed H, Ong HC, Fattah IM, Rahman SA, Nghiem LD, Mahlia TI. Recent advancement and assessment of green hydrogen production technologies. Renewable and Sustainable Energy Reviews. 2024 Jan 1;189:113941.

[15] Deymi-Dashtebayaz M, Abadi MK, Asadi M, Khutornaya J, Sergienko O. Investigation of a new solar-wind energy-based heat pump dryer for food waste drying based on different weather conditions. Energy. 2024 Mar 1;290:130328.

[16] Li R, Jin X, Yang P, Zheng M, Zheng Y, Cai C, Sun X, Luo Z, Zhao L, Huang Z, Yang W. Capacity optimization of a wind-photovoltaic-electrolysis-battery (WPEB) hybrid energy system for power and hydrogen generation. International Journal of Hydrogen Energy. 2024 Jan 2;52:311–33.

[17] Modu B, Abdullah MP, Bukar AL, Hamza MF, Adewolu MS. Operational strategy and capacity optimization of standalone solar-wind-biomass-fuel cell energy system using hybrid LF-SSA algorithms. International Journal of Hydrogen Energy. 2024 Jan 2;50:92–106.

[18] Hayati MM, Safari A, Nazari-Heris M, Oshnoei A. Hydrogen-incorporated sector coupled smart grids: A systematic review & future concepts. Green Hydrogen in Power Systems, 2024.

[19] Safari A, Hayati MM, Nazari-Heris M. *Hydrogen-Combined Smart Electrical Power Systems: An Overview of United States Projects*. Green Hydrogen in Power Systems, 2024 Mar 12:321–340.

[20] Kannaiyan K, Lekshmi GS, Ramakrishna S, Kang M, Kumaravel V. Perspectives for the green hydrogen energy-based economy. Energy. 2023 Dec 1;284:129358.

[21] Hodson TO. Root mean square error (RMSE) or mean absolute error (MAE): When to use them or not. Geoscientific Model Development Discussions. 2022 Mar 11;2022:1–10.

[22] Hodson TO, Over TM, Foks SS. Mean squared error, deconstructed. Journal of Advances in Modeling Earth Systems. 2021 Dec;13(12):e2021MS002681.

[23] Asuero AG, Sayago A, González AG. The correlation coefficient: An overview. Critical reviews in analytical chemistry. 2006 Jan 1;36(1):41–59.

[24] Karunasingha DS. Root mean square error or mean absolute error? Use their ratio as well. Information Sciences. 2022 Mar 1;585:609–29.

[25] Safari A, Daneshvar M, Anvari-Moghaddam A. Energy Intelligence: A Systematic Review of Artificial Intelligence for Energy Management. Applied Sciences. 2024 Nov 28;14(23):11112.

11 Intelligent Techniques of Demand-Side Management for Economic Flexibility

Younes Nourolahi, Shiva Ansaripour, and Mahshid Noorollahi

Abbreviations

AI   artificial intelligence
ANN   artificial neural network
DAP   day-ahead pricing
DNOs   distribution network operators
DR   demand response
DSM   demand-side management
PSO   particle swarm optimization
RES   renewable energy system
RTP   real-time pricing
TOU   time of use
VPPs   virtual power plants

11.1 INTRODUCTION

The large penetration of renewable energy sources (RES) and electro-mobility has made power systems' flexibility a challenging topic. On the other side, flexibility can be achieved by customers participating in demand response programs. To tackle these challenges, a variety of sophisticated computational techniques are considered a key solution to shift the flexibility as a duty of the power system to the customer side. Some optimization algorithms, such as customer behavior learning, optimal pricing, power flow, and device scheduling, can handle the reliability of the system [1].

DSM (demand-side management) aims to optimize energy consumption in alignment with energy supply through different strategies and technologies. By influencing consumer behavior and energy usage patterns, DSM enhances efficiency and stability within energy systems. It includes a broad spectrum, from high-energy-consuming devices in residential settings to complex dynamic load management systems. Importantly, DSM does not solely focus on reducing the demand; it also tries to shift the load consumption patterns to better match the energy supply.

In this context, power-to-X (PtX) technologies play a vital role in easing the conversion of surplus renewable energy power into other usable energy forms, such as hydrogen or other synthetic fuels. These advantageous technologies are crucial for renewable energy systems' integration into the energy system and contributing to energy transition goals. Furthermore, demand response initiatives, underpinned by intelligent management systems, are critical to optimize PtX applications. By adjusting energy demand dynamically in response to energy generation conditions, these systems improve the overall energy use efficiency, ensuring that PtX technologies can efficiently respond to variable energy sources and demand patterns. Together, these elements underscore the significance of DSM in encouraging a flexible and resilient energy system.

DOI: 10.1201/9781032719436-11

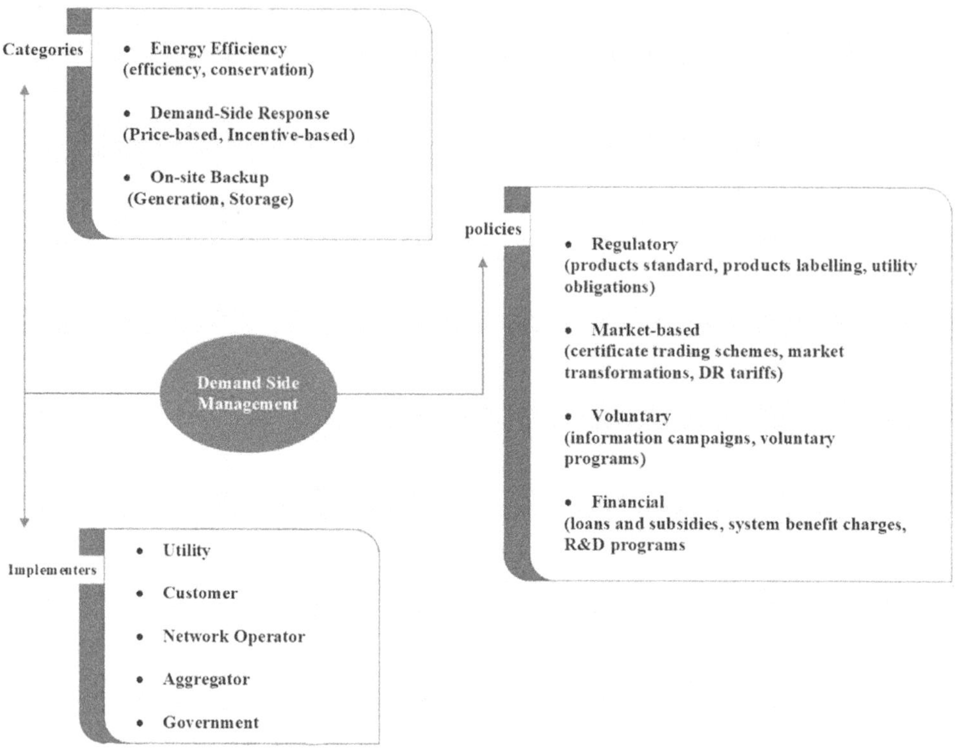

FIGURE 11.1 Demand-side management (DSM).

Source: [2].

With regards to an overview in [2], DSM in the UK can be defined in three main topics, as shown in Figure 11.1. Energy efficiency mostly concentrates on overall consumption reduction, while demand response mostly aims to make a balanced system by peak shaving programs. On-site backup can also be called a physical demand response and acts in peak hours by activating a diesel generator and using storage. In terms of DSM implementers, it can be arranged by the government and applied by utilities—to be more exact, distribution network operators (DNOs). Briefly, the DNO is responsible for safeguarding the system against imbalances by using different pricing systems and issuing regulatory signals [3]. Also, the policies branch is sorted in Figure 11.1. The aggregator's role is to link the customers and manage the programs by the following actions: registration and communication with users, making real-time measurements, implementing standards in the measurements for consumption verification, and calculating user engagement based on the input data [4].

The main customers in the DR programs may include residential, industrial, and also commercial users. Additionally, DR services can manifest in the form of power companies and act as virtual generators in the next-day market [5].

The intersection of PtX and DSM has significantly increased as energy systems evolve toward greater sustainability. Several studies have demonstrated how effectively DSM can enhance the performance of PtX systems, particularly by influencing demand as a result of flexibility increase to accommodate variable renewable energy supply. In other words, DSM in a power-to-X system involves optimizing energy consumption to align with production capabilities from renewable sources and enhancing overall system efficiency. This arrangement is crucial for maximizing the utilization of green energy, and waste minimization, to ensure that the energy demand curve closely mirrors the flexible supply generated by renewable energy sources. To achieve this, innovative technologies such as smart meters and advanced energy management systems are hired, allowing for real-time monitoring and control of energy use across various sectors. By merging these technologies, consumers can actively participate in the energy transition, shifting their consumption patterns in response to available renewables. Furthermore, sophisticated algorithms can predict energy supply and consumption trends, enabling proactive adjustments and better coordination between users and suppliers.

Collaboration between utilities, technology providers, and end users is also essential, strengthening a community-oriented approach to energy management. Through demand response programs, participants can receive incentives for modifying their consumption during peak times or when energy from renewables is abundant, thereby promoting a more adaptable network.

Eventually, the success of DSM in PtX systems is subject to a comprehensive understanding of both consumer behavior and technological improvements. By prioritizing flexibility and responsiveness within the energy landscape, DSM not only supports the integration of renewable resources but also paves the way for a sustainable, efficient, and resilient energy future.

Along with electrical grid development, several DSM approaches have been discussed; in the 1920s, direct load management was started in Germany. In 1956, in France, demand responses were applied for peak shaving. An initial term of DSM was presented in the USA around 1984 by Clark Gellings [6]. The wide use of demand response programs was developed around the globe from 1990 to 2010. Some methods, like real-time trading, storage systems, and flattening of the load curve, have been intensely expanded since 2010 [4].

The US Department of Energy has defined *demand response* as

> a tariff or program established to motivate changes in electric use by end-use customers in response to changes in the price of electricity over time, or to give incentive payments designed to induce lower electricity use at times of high market prices or when grid reliability is jeopardized.
>
> [7]

The main objective of demand response (DR) is to make consumers active in participating in the electric grid stability in coordination with systems conditions and economic signals. Based on DR programs, two main categories can be found:

1. **Price-based.** When different prices are defined for energy, consumers can reduce or shift their loads to lower energy bills.
2. **Incentive-based.** Mechanisms that rely on consumption modifications induced by non-price signals, for which the involved users receive compensations.

There are also other sights to classify the DR [8], including:

- **Market demand response.** Includes price signals, incentives, and real-time pricing mechanisms. This DR relies on certain markets which are not expected to be quick in changes, but in real-time pricing, the figures of the network will be forwarded to the end users without delay. As a matter of fact, due to the limited elasticity of consumers and other physical conditions, the price alone cannot be efficient for load shedding in a grid. That is the reason for implementing the physical DR.
- **Physical demand response (PDR).** It refers to the reduction or adjustment of electricity consumption by consumers in response to external signals, such as grid reliability or price change concerns. Unlike other forms of demand response, it might rely mostly on financial incentives or consumer behavior changes. PDR involves a tangible decrease in energy usage or a shift in energy consumption patterns during peak demand periods.

Research has shown that DSM plays a key role as an integrator of renewable energy into PtX systems. For instance, Van et al. [9] examined case studies where reinforced DSM strategies improved the efficiency of electronic-based processes for hydrogen production. By aligning energy demand with supply, the economic viability of hydrogen production from renewables was significantly enhanced. They indicate that personalized feedback and dynamic pricing can motivate consumers to participate actively, resulting in better load management and overall system efficiency in the PtX system.

Advanced technologies such as smart metering, IoT devices, and blockchain have been recognized as catalysts for effective DSM. A study by Palys et al. [10] discussed how these technologies can facilitate real-time data exchanges between energy providers and consumers, enhancing response times and decision-making in PtX applications. The economic implications of combining PtX with DSM are underscored by various analyses. They pointed out that policies promoting demand-side flexibility can lead to cost reductions in PtX projects by regulatory frameworks defendants that support investment in DSM technologies.

Barriers to effective DSM include technological limitations, consumer reluctance, and regulatory constraints. Research by Good et al. [11] discusses these challenges and suggests potential pathways for overcoming them, such as creating robust incentive structures for consumer participation.

It can be remarked that the literature on PtX underscores the critical role of demand-side management in optimizing power-to-X systems. Integrating these strategies enhances grid stability, promotes the use of renewable energy, and encourages sustainable energy practices. Future research should focus on developing comprehensive frameworks that incorporate evolving technologies and consumer behaviors to further refine and enhance the synergy between PtX and DSM.

11.2 DSM INSTRUMENTS

Energy savings is a way to reduce overall consumption and promote long-term stability. Higher energy efficiency is the way that helped most states in the EU to achieve objectives like reduction in greenhouse gas emissions and energy security [14]. Consequently, DR is a tool for energy efficiency, and these two areas should be assessed in parallel [15]. On the other hand, in the energy market, marginal and factual electricity consumption would be variable during the day, and end user tariffs may encourage the consumers to control their loads.

As mentioned in Figure 11.2, a wide range of optimization deployment approaches can be particular to demand-side management systems. These methods can be recognized by the type of user interaction, the timescale, and the method of optimization.

Demand response programs can be grouped into three sorts, based on the main goals.

11.2.1 DR Based on the Control Mechanism

DR based on control mechanisms refers to strategies that adjust the demand of consumers for electricity in response to supply characteristics, energy prices, or grid reliability. Specifically, control mechanisms can be categorized broadly into two main types, including manual control mechanisms and automated control mechanisms [16].

11.2.2 DR Based on Offered Motivations

This category comprises systems with price-based and incentive motivations or reward programs in subdivisions shown in Figure 11.2 [17]. A price-based approach can rely on different principles, like time-of-use (TOU) pricing, day-ahead pricing (DAP), and real-time pricing (RTP). In TOU, a static price table is applied to the electricity, and the price in peak hours is expected to be the

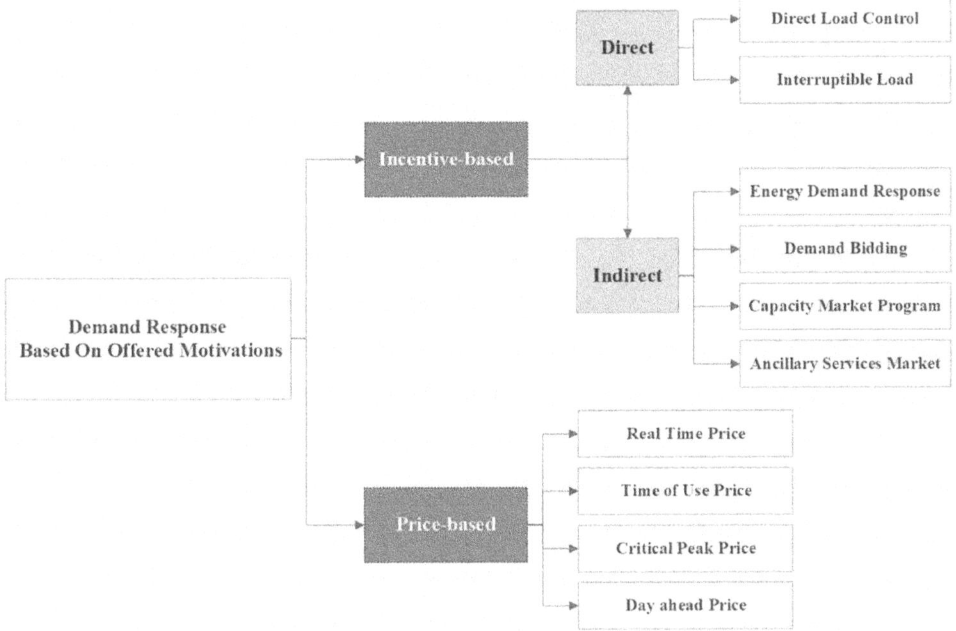

FIGURE 11.2 DR classification based on offered motivation.

highest rate during one day [18]. DAP is a short consumption prediction of 24 hr to highlight price based on that demand profile. However, the RTP pricing method is a dynamic scheme with up to 1 hr prediction [19].

11.2.3 DR Based on Decision Variable

This category distinguishes the methods by variables to determine task scheduling and power control DR systems [4]. Intense development focused on end user behavior projects is the most challenging part of DSM, as it is relatively complex and considered a stochastic variable that directly impacts the DR programs [20].

11.2.4 Machine Learning–Based DSM

Along with this, artificial intelligence (AI) response strategies offer smart and autonomous controllers and advanced data management software. The adequate output of consumption patterns, energy demand forecasting, and energy market participation can result in accurate real-time simulation [21].

Machine learning includes a set of algorithms following the input data to draw information for different purposes. The main usage of machine learning in DSM can be summarized in optimal power flow, optimal device scheduling, optimal pricing, and customer behavior learning [1, 19, 22]. Artificial neural network (ANN)–based algorithms are applied in [23] to make short-term forecasting in a price-sensitive environment by the implementation of fuzzy logic. Furthermore, deep learning algorithms were used for DSM and customer behavior learning in [24, 25] in the residential sector. Particle swarm optimization (PSO) is used in [26, 27] for optimal device scheduling, and in [28] for OPF. Reinforcement learning [29, 30], robust optimization, and Bayesian models also are used in DSM programs. Figure 11.4 disputes the forecasting role in an energy system and machine learning implementation in this field [31].

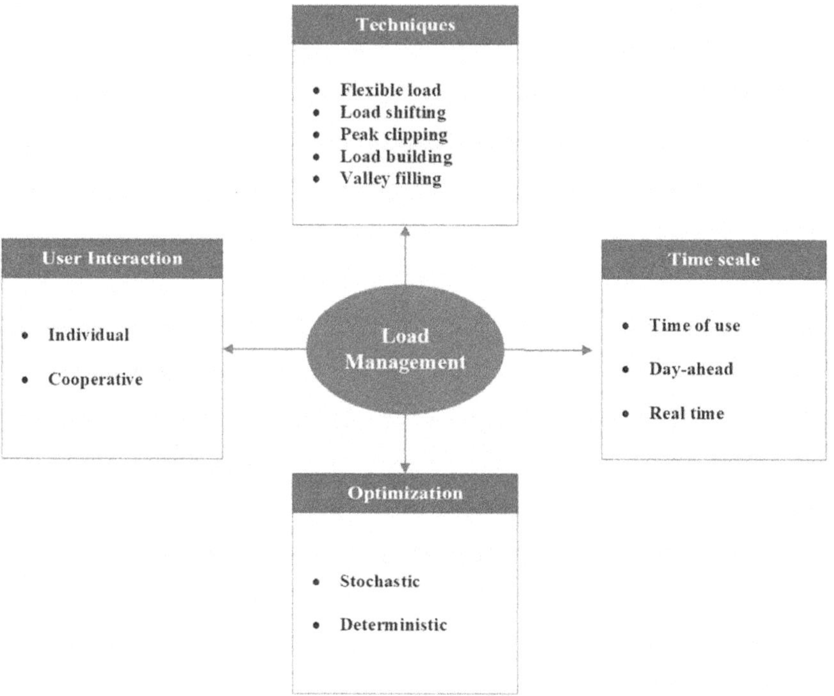

FIGURE 11.3 Load management techniques and aspects.

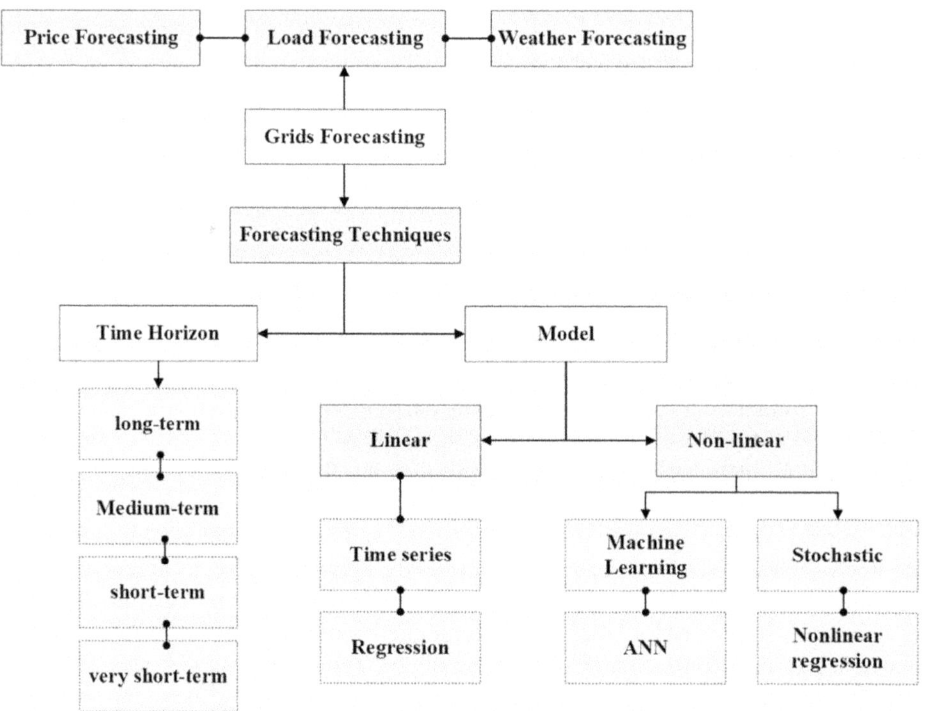

FIGURE 11.4 Categories of forecasting and methods in an energy system.

11.3 ENERGY DEMAND FORECAST USING MACHINE LEARNING

Customer behavior learning in different time horizons is a significant branch of DSM, and in the following, a review of forecasting methods is done to investigate the machine learning algorithms and also the criteria to be considered in multi-variate energy demand forecasting.

A comprehensive study encompassing 50 countries worldwide, including Iran, has thoroughly studied the interrelationships among population growth, gross domestic product (GDP), and the installation of renewable energy power plants concerning air pollution levels. The findings revealed a notable correlation wherein both GDP and population growth contribute positively to CO_2 emissions. This proposes that as a nation's economy develops and its population increases, the subsequent consumption of energy and resources may lead to heightened levels of air pollution, despite potential advancements in renewable energy installations. Subsequently, this study underscores the need for integrated policy approaches that balance environmental sustainability and economic development to moderate the negative effects of rising air emissions [32].

According to Ding's study [33], a novel self-adapting intelligent gray model is a better approach than competing natural gas forecasting methods. Using this model, Spain's primary energy consumption, combined with input–output models, resulted in the best combination for the gray model, as done by Liu et al. [34]. The logistic model has been applied in [35] to long-term natural gas consumption forecasting in China. For getting the parameters of the logistic model, the Levenberg–Marquardt algorithm is adopted. In [36], the autoregressive moving-average model with exogenous inputs (ARMAX) model has been developed for residential and commercial energy demand forecasting in Iran. Alcaraz and Villalvazo [37] presented the natural gas estimation shortage with an econometric analysis based on panel data and analyzed the natural gas interconnection shortage and GDP. In 2017, Scarpa and Bianco investigated long-term natural gas consumption, considering heating degree days, natural gas prices, and GDP per capita. To obtain this goal, they used the regression algorithm and the Kalman filter method [38]. The relationship between price and income with natural gas consumption has been shown in [34] by Liu et al. The generalized least squares method was used in this study. Wang et al. [39] studied the natural gas consumption model with high accuracy by a MAPE value of 2.32% with a hybrid model based on the particle swarm optimization–wavelet neural network (PSO-WNN). In [40], artificial neural networks (ANN), multiple linear regression (MLR), and support vector regression (SVR) were presented for forecasting natural gas consumption in Istanbul with these criteria: seasonal index, temperature, price of natural gas, population, and 12-year history of natural gas consumption. In [11, 12 and 41], electricity consumption forecasting in a building is done by using the GPR method. Mahdi Sharifzadeh et al. [43] did a comparative study of ANN, SVR, and GPR for forecasting residential electricity demand. Further, the Gaussian process quantile regression is used in [42, 44] to predict power load probability density. In some cases, the GPR method is preferred in wind forecasting [13, 15] because it is flexible to provide uncertainty representations [45, 46].

The assessment of various predicting methods for NG and electricity consumption highlights the effectiveness of advanced energy models, such as the self-adapting gray model and GPR, in producing precise forecasts. The utilization of hybrid methods and different algorithms shows a trend toward integrating various methods to boost precision. These studies mutually emphasize the importance of considering various socio-economic and environmental factors in developing more accurate forecasting models, ultimately contributing to better energy planning and management approaches.

Researchers have proposed some methods to decrease CO_2 emissions [47]. The share of renewable energy sources by 2030 in six European countries (France, Germany, Italy, United Kingdom, and Turkey) was forecasted by Utkucan Şahin, calculating the highest amount for Germany with 53.8 Mtoe renewable energy production. Dominkovic et al. studied southeast Europe to make a 100% renewable energy system for 2050 to achieve a zero–carbon energy community. For getting this object, biomass and other renewable energies have been used. Also, improving energy efficiency is considered a reformation strategy for decreasing CO_2 [48]. Daví-Arderius et al. expounded on the impact of electricity losses on CO_2 decrement. The construction of DGs near consumers and covering different generations with renewable energy are the two most essential policies suggested in [49]. Technical progress, energy structure, and economic level are the variables that are considered in [50] to analyze their impact on CO_2 emission in China. Also, it is concluded that technological advances have about a 1% effect on China's CO_2 emission. Supporting, 14 years' (2000–2014) research in China indicates that technological developments had a remarkable impact on CO_2 emission [51]. The South Asian Association for Regional Cooperation (SAARC) countries have been weighted and ranked based on the CO_2 emission issue by gray relational analysis (GRA) [52]. The results have shown substantial pollution problems in India with the first rank; hence, renewable energy installation and the adoption of ISO14001 certification were introduced to solve this problem. Toward the reduction of carbon emission, [53] offered three assortments that contain adding a clean energy supply, developing energy conservation, and negative emission strategies, like using CCS technology.

Figures 11.5 and 11.6 respectively show the most important criteria for demand forecasting and the different stages that should be considered in a machine learning algorithm.

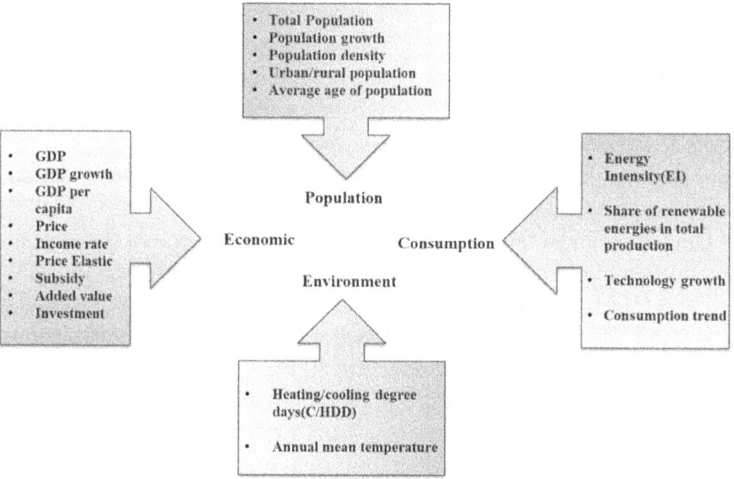

FIGURE 11.5 Criteria and variables in demand forecasting.

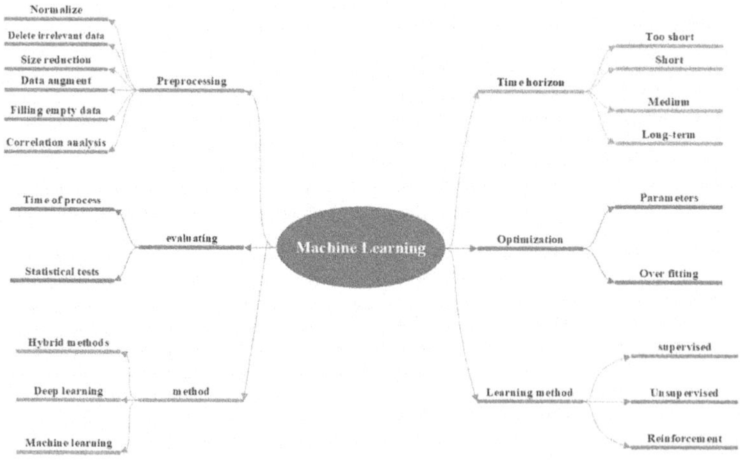

FIGURE 11.6 Machine learning principles.

11.4 ARCHITECTURE AND COMPONENTS OF A DSM FOR PTX

A well-structured architecture for DSM in PtX applications generally consists of several interrelated components, which can be categorized into three primary layers: the data acquisition layer, the decision-making layer, and the implementation layer. The *data acquisition layer* is responsible for collecting real-time data from different sources, such as energy demand patterns, weather condition forecasting, and market change indications. Advanced IoT devices and metering infrastructure can play a vital role in accumulating information on energy supply and demand, providing the necessary insights for knowledgeable decision-making. Furthermore, time series data analysis allows for the identification of demand trends and the estimate of peak demand scenarios. The *decision-making layer* is the heart of any DSM framework, which leverages data analytics and machine learning techniques to assess the collected data and develop strategies that endorse optimized energy demand. For PtX applications, this may involve development processes to overlap with periods of high renewable energy generation, thus minimizing dependence on fossil fuels. The *implementation layer* includes the technologies and systems employed to execute the decisions made in the aforementioned phase. This includes automated controls for demand response mechanisms, direct load control systems, and equipment used in PtX processes. Moreover, the integration of energy storage solutions, including thermal storage or batteries, can increase the flexibility and resilience of operations, enabling PtX facilities to shift energy demand based on grid situations.

11.5 CHALLENGES AND BARRIERS OF DSM TO PTX

The integration of DSM with PtX technologies presents several outstanding challenges and barriers that must be addressed for successful implementation, such as:

- The primary difficulties lie in the coordination between various stakeholders, including energy producers, consumers, and regulatory bodies, which can lead to inefficiencies and misalignment of objectives.
- The technological complexity involved in seamless integration often necessitates substantial investment in infrastructure and innovative solutions, posing financial challenges for both public and private entities.
- Regulatory frameworks may not yet be fully adapted to accommodate the dynamic nature of PtX systems, resulting in potential legal and operational hurdles.

Addressing these issues requires a concentrated effort toward association, investment in research and development, and the establishment of strong policies that ease the harmonious coexistence of DSM and PtX initiatives.

11.6 RESULT AND DISCUSSION

This chapter discussed the intersection of power-to-X technologies and demand-side management, emphasizing the importance of optimizing energy consumption to better unify renewable energy sources and enhance PtX systems. It highlights various flexibility strategies in DSM, such as load shifting and demand response, which improve the efficiency and economic viability of processes like hydrogen production from renewables. Both consumer engagement and technological advancements, including smart metering and IoT, are identified to be crucial for effective DSM. The text also addresses economic implications and challenges, advocating for robust regulatory frameworks to promote flexibility and investment in these technologies. Overall, the synergy between PtX and DSM plays a leading role in achieving sustainable energy practices.

Despite several intelligence DSM methods, there are some fundamental challenges to selecting an appropriate system:

1. Continuous input data for real-time decisions
2. The limited time of process
3. Encountering various periods rather than just a daily slot

The successful integration of DSM with PtX applications is essential for optimizing energy usage and encouraging sustainability. Nevertheless, substantial challenges remain, including the need for effective coordination among diverse stakeholders and the technological complexities that require substantial investment. Additionally, existing regulatory frameworks often lag behind the innovative nature of PtX systems, creating potential barriers to implementation. To overcome these issues, it is vital to adopt collaboration among all parties involved, invest in research and development, and establish supportive policies that facilitate the integration of DSM and PtX technologies.

By addressing these challenges, we can unlock the full potential of DSM in proceeding in the transition to a more sustainable energy future.

11.7 CONCLUSION

An energy system with a high penetration of renewable energy sources (RES) needs a well-integrated and bidirectional demand model to compensate for the fluctuations of RES and to keep the grid flexible.

In this context, some control strategies for traditional power systems are going to be changed. As evident, it is less expensive to intelligently control the load than to install new storage systems or power plants.

Demand-side management was developed for demand and supply balance by adding a certain portion of "smart" to the loads and participating consumers as active users. In this system, customers would rearrange their habits and schedules to minimize their bills.

Additionally, demand response (DR) is widely known as a critical element not only as a control strategy that regulates the system but also as a significant plan for the next generation of virtual power plants (VPPs).

In addition, the economic implications of effective DSM in PtX systems are significant. By peak demand and flattening the load curve, overall operational costs can be minimized, leading to lower energy prices for consumers. Moreover, a well-implemented DSM strategy contributes to a more stable energy market, reducing the risks associated with supply shortages and unpredictable prices.

REFERENCES

[1] T. A. Nakabi, K. Haataja, and P. Toivanen, "Computational intelligence for demand side management and demand response programs in smart grids," in *8th International Conference, BIOMA 2018*, Paris, France, 2018. [Online]. www.researchgate.net/publication/331382569_Computational_Intelligence_for_Demand_Side_Management_and_Demand_Response_Programs_in_Smart_Grids.

[2] P. Warren, "A review of demand-side management policy in the UK," *Renew. Sustain. Energy Rev.*, vol. 29. pp. 941–951, 2014. doi: 10.1016/j.rser.2013.09.009.

[3] L. Cioara, T. Anghel, I. Pop, C. Bertoncini, M. Croce, V. Ioannidis, and D. D'Oriano, "Enabling new technologies for demand response decentralized validation using blockchain," in *IEEE International Conference on Environment and Electrical Engineering and 2018 IEEE Industrial and Commercial Power Systems Europe (EEEIC/I&CPS Europe)*, Palermo, Italy, 2018, pp. 1–4.

[4] V. Stanelyte, D. Radziukyniene, and N. Radziukynas, "Overview of demand-response services: A review," *Energies*, vol. 15, no. 5, 2022.

[5] S. Radenkovi´c, M. Bogdanovi´c, Z. Despotovi´c-Zraki´c, M. Labus, and A. Lazarevi´c, "Assessing consumer readiness for participation in IoT-based demand response business models," *Technol. Forecast. Soc. Change*, vol. 150, p. 119715, 2020.

[6] C. W. Gellings, "The concept of demand-side management for electric utilities," *Proc. IEEE*, vol. 73, no. 10, pp. 1468–1470, 1985. doi: 10.1109/PROC.1985.13318.

[7] J. S. Vardakas, N. Zorba, and C. V. Verikoukis, "A survey on demand response programs in smart grids: Pricing methods and optimization algorithms," *IEEE Commun. Surv. Tutor.*, vol. 17, no. 1, pp. 152–178, 2015. doi: 10.1109/COMST.2014.2341586.

[8] P. Palensky and D. Dietrich, "Demand side management: Demand response, intelligent energy systems, and smart loads," *IEEE Trans. Ind. Inform.*, vol. 7, no. 3, pp. 381–388, 2011. doi: 10.1109/TII.2011.2158841.

[9] L. P. Van, K. Do Chi, and T. N. Duc, "Review of hydrogen technologies based microgrid: Energy management systems, challenges and future recommendations," *Int. J. Hydrogen Energy*, vol. 48, no. 38, pp. 14127–14148, 2023. doi: 10.1016/j.ijhydene.2022.12.345.

[10] M. J. Palys and P. Daoutidis, "Power-to-X: A review and perspective," *Comput. Chem. Eng.*, vol. 165, p. 107948, 2022. doi: 10.1016/j.compchemeng.2022.107948.

[11] N. Good, K. A. Ellis, and P. Mancarella, "Review and classification of barriers and enablers of demand response in the smart grid," *Renew. Sustain. Energy Rev.*, vol. 72, pp. 57–72, 2017. doi: 10.1016/j.rser.2017.01.043.

[12] A. R. Kumar, L. P. Raghav, D. K. Raju, and A. R. Singh, "Intelligent demand side management for optimal energy scheduling of grid connected microgrids," *Appl. Energy*, vol. 285, no. 116435, 2021.

[13] B. Pavithra and N. Esther, "Residential demand response using genetic algorithm," in *2017 Innovations in Power and Advanced Computing Technologies (i-PACT2017)*, pp. 1–4, 2017.

[14] S. Olkkonen, V. Rinne, S. Hast, and A. Syri, "Benefits of DSM measures in the future Finnish energy system," *Energy*, vol. 137, pp. 729–738, 2017.

[15] W. Wohlfarth, K. Worrell, and E. Eichhammer, "Energy efficiency and demand response—Two sides of the same coin?," *Energy Policy*, vol. 137, no. 11070, 2020.

[16] J. Hui, H. Ding, Y. Shi, Q. Li, F. Song, and Y. Yan, "5G network-based Internet of Things for demand response in smart grid: A survey on application potential," *Appl. Energy*, vol. 257, no. 113972, 2020.

[17] K. T. Wu and Y. K. Tang, "Frequency support by demand response—Review and analysis," *Energy Proc.*, vol. 156, pp. 327–331, 2019.

[18] F. M. T. Voulis, N. Van Etten, M. J. J. Chappin, É. J. L. Warnier, and M. Brazier, "Rethinking European energy taxation to incentivise consumer demand response participation," *Energy Policy*, vol. 124, pp. 156–168, 2019.

[19] H. Bergaentzlé, C. Clastres, and C. Khalfallah, "Demand-side management and European environmental and energy goals: An optimal complementary approach," *Energy Policy*, vol. 67, pp. 858–869, 2014.

[20] G. T. Chatzigeorgiou, I. M. Manolas, D. Gkaragkouni, and T. Andreou, "Demand response in Greece: An introductory mobile application," in *Proceedings of the 2018 IEEE International Conference on Environment and Electrical Engineering and 2018 IEEE Industrial and Commercial Power Systems Europe (EEEIC/I&CPS Europe)*, Palermo, Italy, 2018, pp. 1–5.

[21] H. R. K. Kiran Chaurasia, "New approach using artificial intelligence-machine learning in demand side management of renewable energy integrated smart grid for smart city," in *4th International Conference on Innovative Computing and Communication*, 2021.

[22] O. Hirotaka, T. Naoto, T. Shou, and K. Atsumi, "A design method for incentive-based demand response programs based on a framework of social welfare maximization," *IFAC-PapersOnLine*, vol. 51, no. 28, pp. 374–379, 2018.

[23] A. Khotanzad, E. Zhou, and H. Elragal, "A neuro-fuzzy approach to short-term load forecasting in a price-sensitive environment," *IEEE Trans. Power Syst.*, vol. 17, no. 4, pp. 1273–1282, 2002. doi: 10.1109/TPWRS.2002.804999.

[24] R. L. Ilcng Shi and M. Xu, "Deep learning for household load forecasting – A novel pooling deep RNN," *IEEE Trans. Smart Grid*, vol. 1, no. 99, 2017.

[25] D. Zhang, S. Li, M. Sun, and Z. O'Neill, "An optimal and learning-based demand response and home energy management system," *IEEE Trans. Smart Grid*, vol. 7, no. 4, pp. 1790–1801, 2016. doi: 10.1109/TSG.2016.2552169.

[26] H. A. Ehsan Dehnavi, "Optimal pricing in time of use demand response by integrating with dynamic economic dispatch problem," *Energy*, vol. 109, pp. 1086–1094, 2016.

[27] S. Sofana Reka and V. Ramesh, "Demand side response modeling with controller design using aggregate air conditioning loads and particle swarm in advanced computing and communication technologies optimization," in *Advanced Computing and Communication Technologies: Proceedings of the 9th ICACCT*. Singapore: Springer, pp. 573–581, 2016.

[28] M. F. Abdi and E. Dehnavi, "Dynamic economic dispatch problem integrated with demand response (DEDDR) considering non-linear responsive load models," *IEEE Trans. Smart Grid*, vol. 7, no. 6, pp. 2586–2590, 2015.

[29] J. W. S. K. Dehghanpour, M. Hashem Nehrir, and N. C. Kelly, "Agent-based modeling of retail electrical energy markets with demand response," *IEEE Trans.on smart grid*, 2016.

[30] Y. Liang, L. He, X. Cao, and Z. J. Shen, "Stochastic control for smart grid users with flexible demand," *IEEE Trans. Smart Grid*, vol. 4, no. 4, pp. 2296–2308, 2013. doi: 10.1109/TSG.2013.2263201.

[31] S. R. Battula, A. R. Vuddanti, and S. Salkuti, "Review of energy management system approaches in microgrids," *Energies*, vol. 14, no. 17, pp. 1–32, 2021.

[32] L. F. Anny Key de Souza Mendonça, G. de Andrade Conradi Barni, M. F. Moro, A. C. Bornia, and E. Kupek, "Hierarchical modeling of the 50 largest economies to verify the impact of GDP, population and renewable energy generation in CO_2 emissions," *Sustain. Prod. Consum.*, vol. 22, pp. 58–67, 2020. doi: 10.1016/j.spc.2020.02.001.

[33] S. Ding, "A novel self-adapting intelligent grey model for forecasting China's natural-gas demand," *Energy*, vol. 162, pp. 393–407, November 2018. doi: 10.1016/j.energy.2018.08.040.

[34] J. Liu, G. Dong, X. Jiang, Q. Dong, and C. Li, "Natural gas consumption of urban households in China and corresponding influencing factors," *Energy Policy*, vol. 122, pp. 18–26, 2018.

[35] Q. Shaikh and F. Ji, "Forecasting natural gas demand in China: Logistic modelling analysis," *Int. J. Electr. Power Energy Syst.*, vol. 77, pp. 25–32, 2016.

[36] A. Shakouri and G. H. Kazemi, "Selection of the best ARMAX model for forecasting energy demand: Case study of the residential and commercial sectors in Iran," *Energ. Effic*, vol. 9, pp. 339–352, 2016.

[37] S. Alcaraz, and C. Villalvazo, "The effect of natural gas shortages on the Mexican economy," *Energy Econ.*, vol. 66, pp. 147–153, 2017.

[38] V. Scarpa and F. Bianco, "Assessing the quality of natural gas consumption forecasting: An application to the Italian residential sector," *Energies*, p. 10, 1879, 2017.

[39] H. Wang, D. Liu, Y. Wu, Z. Fu, H. Shi, and Y. Guo, "Scenario analysis of natural gas consumption in China based on wavelet neural network optimized by particle swarm optimization algorithm," *Energies*, vol. 11, p. 825, 2018.

[40] O. F. Beyca, B. C. Ervural, E. Tatoglu, P. G. Ozuyar, and S. Zaim, "Using machine learning tools for forecasting natural gas consumption in the province of Istanbul," *Energy Econ.*, vol. 80, pp. 937–949, May 2019. doi: 10.1016/j.eneco.2019.03.006.

[41] D. W. Van Der Meer, M. Shepero, A. Svensson, J. Widén, and J. Munkhammar, "Probabilistic forecasting of electricity consumption, photovoltaic power generation and net demand of an individual building using Gaussian processes," *Appl. Energy*, vol. 213, pp. 195–207, January 2018. doi: 10.1016/j.apenergy.2017.12.104.

[42] M. Shepero, D. Van Der Meer, J. Munkhammar, and J. Widén, "Residential probabilistic load forecasting : A method using Gaussian process designed for electric load data," *Appl. Energy*, vol. 218, pp. 159–172, March 2018. doi: 10.1016/j.apenergy.2018.02.165.

[43] M. Sharifzadeh, A. Sikinioti-lock, and N. Shah, "Machine-learning methods for integrated renewable power generation: A comparative study of artificial neural networks, support vector regression, and Gaussian process regression," *Renew. Sustain. Energy Rev.*, vol. 108, pp. 513–538, February 2019. doi: 10.1016/j.rser.2019.03.040.

[44] Y. Yang, S. Li, W. Li, and M. Qu, "Power load probability density forecasting using Gaussian process quantile regression," *Appl. Energy*, vol. 213, 2017, pp. 499–509, August 2018. doi: 10.1016/j.apenergy.2017.11.035.

[45] X. Ma, F. Xu, and B. Chen, "Interpolation of wind pressures using Gaussian process regression," *J. Wind Eng. Ind. Aerodyn.*, vol. 188, pp. 30–42, February 2019. doi: 10.1016/j.jweia.2019.02.002.

[46] H. Cai, X. Jia, J. Feng, W. Li, Y. Hsu, and J. Lee, "Gaussian process regression for numerical wind speed prediction enhancement," *Renew. Energy*, vol. 146, pp. 2112–2123, 2020. doi: 10.1016/j.renene.2019.08.018.

[47] E. Ahmadi, Y. Noorollahi, B. Mohammadi-Ivatloo, and A. Anvari-Moghaddam, "Stochastic operation of a solar-powered smart home: Capturing thermal load uncertainties," *Sustainability*, vol. 12, no. 5089, pp. 1–18, 2020.

[48] D. F. Dominković, I. Bačeković, B. Ćosić, G. Krajačić, T. Pukšec, N. Duić, and N. Markovska, "Zero carbon energy system of South East Europe in 2050," *Appl. Energy*, vol. 184, pp. 1517–1528, 2016. doi: 10.1016/j.apenergy.2016.03.046.

[49] E. T.-B. Daniel Daví-Arderius, M.-E. Sanin, and E. Trujillo-Baute, "CO_2 content of electricity losses," *Energy Policy*, vol. 104, pp. 439–445, 2017. doi: 10.1016/j.enpol.2017.01.011.

[50] L. X. Zheng, Y. F. Li, H. T. Wu, and R. J. Wang, "Impact of technology advances on China's CO_2 emission reduction," *Chin. Sci. Bull.*, vol. 55, no. 19, pp. 1983–1992, 2010. doi: 10.1007/s11434-010-3241-1.

[51] B. Wang, Y. Sun, and Z. Wang, "Agglomeration effect of CO_2 emissions and emissions reduction effect of technology: A spatial econometric perspective based on China's province-level data," *J. Clean. Prod.*, vol. 204, pp. 96–106, 2018. doi: 10.1016/j.jclepro.2018.08.243.

[52] M. Ikram, Q. Zhang, R. Sroufe, and S. Z. Ali Shah, "Towards a sustainable environment: The nexus between ISO 14001, renewable energy consumption, access to electricity, agriculture and CO_2 emissions in SAARC countries," *Sustain. Prod. Consum.*, vol. 22, pp. 218–230, 2020. doi: 10.1016/j.spc.2020.03.011.

[53] J. Jiang, B. Ye, and J. Liu, "Research on the peak of CO_2 emissions in the developing world: Current progress and future prospect," *Appl. Energy*, vol. 235, October 2018, pp. 186–203, 2019. doi: 10.1016/j.apenergy.2018.10.089.

12 Cybersecurity Challenges in Power-to-X Integration

Ehsan Naderi and Arash Asrari

12.1 INTRODUCTION

12.1.1 DEFINITIONS AND MOTIVATION

The integration of renewable energy resources (RERs), along with energy storage systems, into the body of traditional energy networks has demanded optimal configurations for renewable energy communities as a step toward carbon-neutral power systems [1]. In this regard, the concept of power-to-X is the missing chain to link flexibility, reliability, and resilience to the traditional power systems via promoting reconversion pathways in order to safely utilize the extra electric energy generated by RERs [2]. To optimally meet the requirements of implementing power-to-X technologies, such as power-to-gas, power-to-heat, power-to-vehicle, and power-to-power, into modern power systems, the integral assets of the resultant modern power system (i.e., energy converters, storage devises, and also renewable energy units) need to be intertemporally monitored and controlled via ICT infrastructure [3]. Nevertheless, such integrated energy systems are vulnerable to data breach, paving the way for catastrophic cyberattacks targeting power-to-X components and causing a range of operational issues including but not limited to cascading failures. The result of such cyberattacks can be from a simple overloading issue and/or power shortage, reducing the social welfare of end use customers, to a nationwide blackout. Therefore, it is essential to scrutinize the cybersecurity challenges in power-to-X integration. There have been a number of standards and frameworks developed to address malicious cyberattacks (e.g., denial of service [DoS], distributed DoS [DDoS], and false data injection [FDI] attacks, as well as malwares and ransomwares), including the NIST Cybersecurity Framework and IEC 62443 series, which aim to help manage cybersecurity risks and ensure energy infrastructure resilience. However, the scope of this chapter is associated with FDI cyberattacks, where power system operators put themselves in the attackers' shoes to model different scenarios of cyberattacks that can potentially target power-to-X components in the concept of smart power systems.

12.1.2 LITERATURE REVIEW

It is not uncommon for energy systems to face significant cybersecurity challenges, such as sophisticated cyberattacks, vulnerabilities in legacy systems, and the integration of multiple technologies, such as power-to-X. In order to ensure the safe, reliable, and efficient operation of power-to-X components in modern power systems, cybersecurity analysis is crucial to their success and sustainability in the long run. Herein, brief discussions of the selected frameworks applied to power-to-X integration, along with the cybersecurity challenges of modern power systems, are included.

Related works on power-to-X integration. In [4], a set of system-wide constraints associated with power-to-X solutions for bulk penetration of RERs were discussed in Denmark's power system to push the country toward fossil-free energy by 2050. In [5], power-to-gas solutions were proposed to highlight the role of electrochemical energy storage technologies in providing daily and annual flexible services, maximizing the integration of RERs. In [6],

DOI: 10.1201/9781032719436-12

power-to-hydrogen and hydrogen-to-X technologies were investigated to directly transform the power generated via RERs into green hydrogen to enhance the promotion of carbon-free energy and transportation systems in Europe. An overview of recent research insights on the power-to-X economy, and also the role of hydrogen as an integral element for future power systems, was provided in [7] from different perspectives, such as efficiency, cost, data-driven, etc. In [8], different power-to-X technologies, along with their technical constraints, were taken into account to assess the merits and demerits of such technologies to enhance the efficiency of renewable energy communities to maximize the local energy self-consumption. In [9], the systematic effects of dynamically operated power-to-X components were discussed to introduce flexibility into large-scale and bulk renewable energy systems, where load curtailment was decreased significantly focusing on the Japanese power system. Moreover, a comprehensive review on power-to-X systems was accomplished in [10], highlighting the role of flexible energy storage systems in power-to-X integration.

Related works regarding cybersecurity of modern power systems. In [11], the deep learning concept was utilized to assess and prioritize the cyber events on the communication platform associated with a modern power grid from different stances, including, but not limited to, categorical data types used in anomaly detection systems. In [12], an adversarial deep learning algorithm reinforced by black box optimization was introduced to evade the existing intrusion detection mechanisms in the body of modern energy systems integrated with renewable energy while maximizing damage to the grid. A fast-triggered detection algorithm against time synchronization cyberattack was developed in [13], where the satellite signal–based time synchronization of a hybrid renewable power system was impacted by the cyberattack. In [14], the adversarial capabilities of RERs in designing cyberattacks compromising distributed energy resources were investigated from different standpoints, materializing protocols and device-level vulnerabilities into devastating cyberattacks targeting modern power systems. In [15], a cyberdefense mechanism based on cost–benefit analysis was modeled as a voltage-sourced converter commitment problem for a hybrid AC/DC power system, where a spatial-temporal dual framework was developed to evaluate cyberattacks targeting smart power systems. The main objective function of the proposed mixed-integer second-order cone programming in [15] was to achieve the trade-off between maximizing the cyberdefense benefits and minimizing the cyberdefense costs. In [16], a real-time test bed capable of simulating FDIAs on generic object-oriented substation events and sampled values protocols was introduced to study the negative impact of false data injection cyberattacks on the modern power grid. In [17], the feasibility of developing a physics-constrained adversarial attack to undermine data-driven power system stability assessment was introduced, focusing on decision tree–based stability assessment. In addition, three comprehensive surveys from different aspects of attack models, vulnerabilities of modern power system, detection mechanisms, and countermeasures against cyberattacks targeting energy systems, were presented in [18–20].

Earlier steps of this research. In the first attempt, a lab-scale hybrid renewable-based smart microgrid was developed to experimentally analyze the cybersecurity of a physical case study [21, 22]. Then, different models of FDIAs oriented toward hardware-in-the-loop systems were applied to the lab-scale smart microgrid, causing a range of operational issues including but not limited to power shortage and frequency imbalance [23]. Next, different types of artificial intelligence–based false data detection mechanisms were developed to track the measurements and alert the microgrid operator in case of malicious data systematically injected into the smart meters distributed in the physical layer of the microgrid [24–27]. Finally, to fully protect the developed lab-scale microgrid against severe FDIAs bypassing the detection stage, different remedial action frameworks were proposed to recover the targeted microgrid to the normal operation in a timely manner and enhance the social welfare of the end use customers [28, 29].

12.1.3 Research Question of This Chapter

Despite the extensive research efforts in the field of cyberattacks targeting modern power networks integrated with renewable energy resources (e.g., [11–29]), the following research question is yet to be addressed in the existing literature:

- How to model false data injection attacks (FDIAs) targeting power-to-X components in a hybrid renewable energy system to result in a range of operational issues (e.g., shortage of power, frequency imbalance, and system congestion, to name but a few)?

12.2 DEVELOPED FRAMEWORK AND PROBLEM FORMULATION

12.2.1 Cybersecurity Framework for Power-to-X Components in Hybrid Renewable Energy Systems

The FDIA framework is illustrated in Figure 12.1. From this figure, it can be perceived that a modern power system modified to contain renewable-based distributed generation units (e.g., PV modules and wind turbines) and power-to-X components is targeted by different FDIAs to result in a range of operational issues including but not limited to system congestion, power shortage, etc. In this regard, a team of attackers (i.e., the system operator being in the attackers' shoe) determine the most vulnerable regions of the power system to power outage, etc. Then, the attackers minimize the amount of false data to be injected into the load canters, to result in different rates of intended operational issues (e.g., system congestions and probable power outages).

12.2.2 Introducing Power-to-X Technologies

Power-to-gas. In small-scale applications, the surplus power of RERs can be converted into a gaseous fuel such as hydrogen or alternative fuels (e.g., natural gas); nevertheless, the process of producing electrofuels is not efficient in small scales [8]. In this chapter, the focus will be on small-scale units distributed in a smart distribution system.

Power-to-heat. In this process, the extra energy from the distributed RERs is directly converted into thermal energy for different applications ranging from heating to cooling through heat pumps [9].

Power-to-vehicle. Taking advantage of the dual role of electric vehicles (EVs) in smart distribution systems as loads and portable energy storage units, the vehicle-to-grid mechanism necessitates bidirectional power flow between power grid and EV [8, 9].

Power-to-power. In this process, the extra electric energy for RERs can be stored in either forms of mechanical or chemical energy to be used in the future, when the distribution system lacks enough power to serve the demand. A proper energy storage system (e.g., battery banks, flywheel, and pumped hydroelectric stations) completes the chain in power-to-power technology [7].

12.2.3 Targeting Power-to-X Components in a Modern Distribution System via FDIAs, Leading to a Range of Operational Issues (System Operator Being in the Attacker's Shoe)

According to Figure 12.1, one can infer that Attacker #1 targets a subset of smart meters associated with the electric load centers within the modern distribution system via injecting malicious load data optimally obtained through a bi-objective optimization problem. The same scenario will happen to two more subsets of smart meters associated with heat/thermal load centers (targeted by Attacker #2) and power aggregator for distributed renewable energy units (targeted by Attacker #3) throughout the

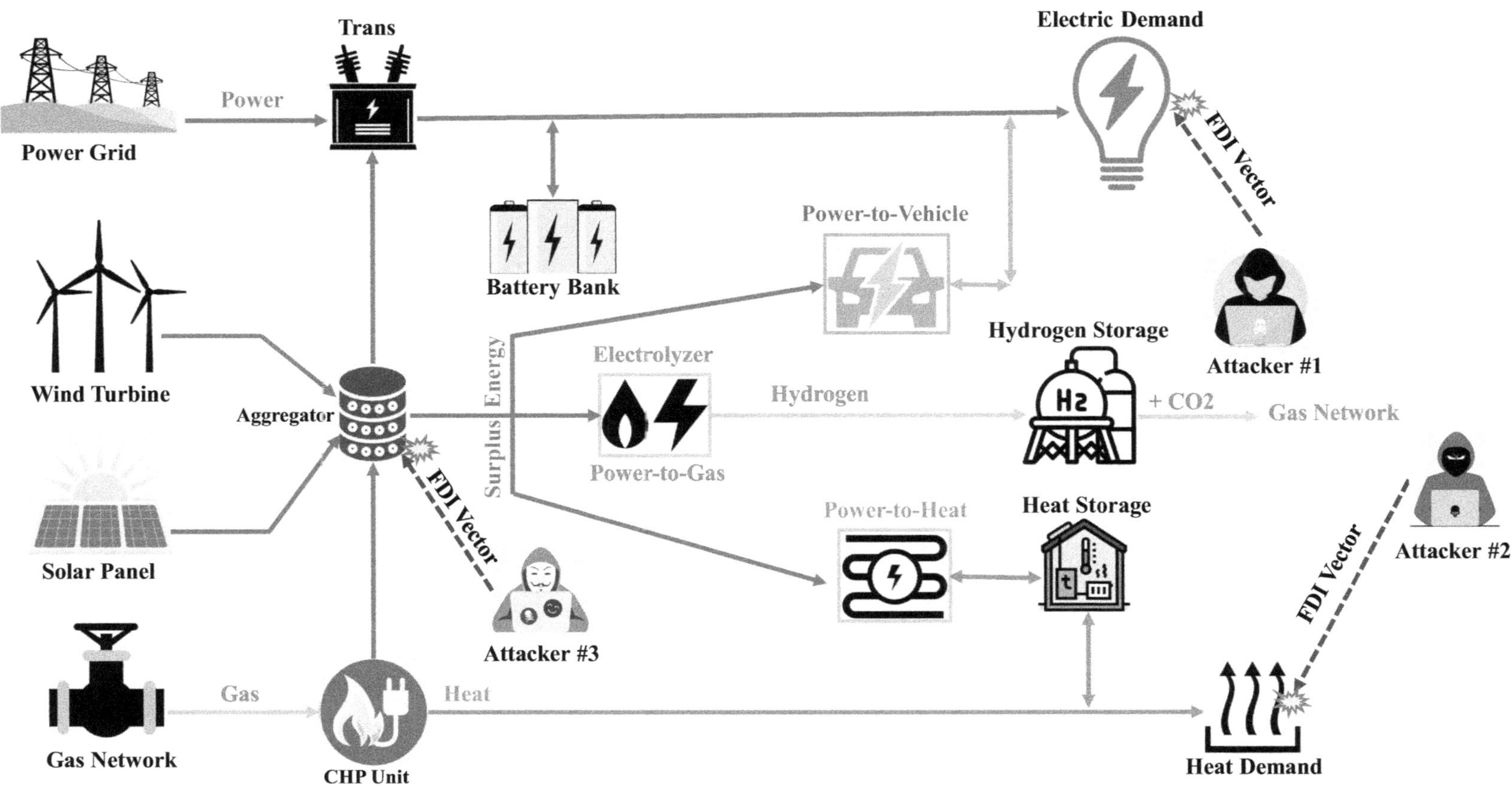

FIGURE 12.1 The FDIA framework targeting power-to-X technologies in a modern power system.

distribution system. More information about the process of optimally obtaining the malicious data in this coordinated FDIA will be provided in Section 2.4, and the result of such a coordinated FDIA is to disrupt the power balance in the distribution system and impose severe congestions in distribution branches within the communities, and also the ones connecting the adjacent communities.

12.2.4 PROBLEM FORMULATION

It is noted that, for each FDIA scenario targeting modern power systems integrated with power-to-X components, the attackers solve a bi-objective optimization problem to identify the severity of the operational issues (e.g., higher rate of system congestions experienced) to push the power system toward critical situations. Detailed information about each attack scenario, along with its technical constraints, will be provided in the following subsections.

The main objective functions of the FDIA are presented in (12.1) and (12.2), where indices e and g are, respectively, associated with the electrical system and the gas system.

$$\max\left\{P_e^{Grid}\times\left(\rho_e^{Grid}+\rho_e^{Emission}\right)+\sum_{i=1}^{N_{CHP}}P_{g,i}^{CHP}\times\left(\rho_g^{Network}+\rho_g^{Emission}\right)+\sum_{j=1}^{N_{Storage}}P_{e,j}^{Storage}\times\left(\rho_e^{BB}+\rho_e^{EV}\right)\right\} \quad (12.1)$$

$$\min\left\{\sum_{b=1}^{N_{Bus}}\Delta P_e^{Demand}+\Delta P_e^{RG}+\rho_{eg}\times\Delta P_g^{Demand}\right\} \quad (12.2)$$

where P_e^{Grid}, $P_{g,i}^{CHP}$, and $P_{e,j}^{Storage}$ are, respectively, the power related to the main grid, the ith CHP unit, and the jth energy storage system (i.e., battery banks and EV batteries); ρ_e^{Grid}, $\rho_g^{Network}$, ρ_e^{BB}, and ρ_e^{EV} are the power price associated with the main power grid, the gas price associated with the main gas network, the aggregated price of power associated with all static battery energy storage units in the system, and the price of power related to the mobile energy storage units (i.e., the battery of EVs), respectively; $\rho_e^{Emission}$ and $\rho_g^{Emission}$ are, respectively, the equivalent emission price associated with the main power grid and the main gas network; ΔP_e^{Demand} and ΔP_g^{Demand} are, respectively, the data manipulation associated with electric and thermal demands; ΔP_e^{RG} is the data manipulation associated with the renewable aggregator to falsify the aggregated power from the PV modules and wind turbines; N_{CHP}, $N_{Storage}$, and N_{Bus} are the number of CHP units, energy storage units, and system buses; and ρ_{eg} is the penalty factor to aggregate the electric and thermal demands' manipulations.

From (12.1)–(12.2), one can infer that the team of attackers (see Figure 12.1) targets the modern distribution system to maximize the overall operation cost of the system via minimizing the amount of malicious data to be injected into the smart meters. The result of such an FDIA will show up as shortage of power, system congestion, or power outage in the distribution system. More information about such operational issues will be provided in Section 12.4. It is noted that objective functions (12.1) and (12.2) need to be simultaneously optimized subject to meeting a set of techno-economic constraints, which are presented in (12.3)–(12.17), associated with the physics of the power grid and gas network, generation–demand equilibrium, and operational limitations of all the assets of the system, as depicted in Figure 12.1.

$$P_e^{Grid}+\sum_{i=1}^{N_{CHP}}P_{g,i}^{CHP}+\sum_{j=1}^{N_{Storage}}P_{e,j}^{Storage}+\sum_{k=1}^{N_{PV}}P_{e,k}^{PV}+\sum_{h=1}^{N_{WT}}P_{e,h}^{WT}=P_e^{Demand}+P_e^{Loss} \quad (12.3)$$

$$P_{elec}=P_e^{Grid}\times\eta_{Trans}+P_e^{CHP}\times\eta_{CHP}+P_e^{WT}\times\eta_{WT}+P_e^{PV}\times\eta_{PV}+P_e^{DBB}-P_e^{CBB}+P_e^{DEV}-P_e^{CEV} \quad (12.4)$$

$$P_{Ther}=P_g^{CHP}\times\eta_{CHP}+P_e^{DHS}-P_e^{CHS}+P_e^{DH2S}-P_e^{CH2S} \quad (12.5)$$

$$P_e^{Grid,min}\le P_e^{Grid}\le P_e^{Grid,max} \quad (12.6)$$

$$P_e^{CHP,min}\le P_e^{CHP}\le P_e^{CHP,max} \quad (12.7)$$

$$P_g^{CHP,min} \leq P_g^{CHP} \leq P_g^{CHP,max} \tag{12.8}$$

$$P_{Trans} \leq P_{Trans}^{Max} \tag{12.9}$$

$$P_e^{Storage} \leq P_e^{Storage,max} \tag{12.10}$$

$$P_e^{EV} \leq P_e^{EV,max} \tag{12.11}$$

$$P_{Ther}^{Storage} \leq P_{Ther}^{Storage,max} \tag{12.12}$$

$$P_{H2}^{Storage} \leq P_{H2}^{Storage,max} \tag{12.13}$$

$$P_{e,t}^{ESS} = P_{e,t-1}^{ESS} + \omega_C \times P_{e,t}^{ESS,C} - \frac{1}{\omega_D} \times P_{e,t}^{ESS,D} \tag{12.14}$$

$$\Delta P_e^{Demand} \leq \psi_1 \times P_e^{Demand} \tag{12.15}$$

$$\Delta P_g^{Demand} \leq \psi_2 \times P_g^{Demand} \tag{12.16}$$

$$\sum_{b=1}^{N_B} \Delta P_{e,b}^{Demand} = 0 = \sum_{b=1}^{N_B} \Delta P_{g,b}^{Demand} \tag{12.17}$$

where $P_{e,k}^{PV}$ and $P_{e,h}^{WT}$ are, respectively, the electric power produced by kth PV panel and hth wind turbine; P_e^{Demand} and P_e^{Loss} are, respectively, the total demand and power loss related to the distribution system; P_{elec} and P_{Ther} are the electric and thermal powers, respectively; η_{Trans}, η_{CHP}, η_{WT}, and η_{PV} are, respectively, the efficiency of the transformer, CHP unit, PV panel, and wind turbine; P_e^{DBB} and P_e^{CBB} are, respectively, the equivalent powers associated with discharging and charging processes of the static energy storage system (i.e., battery banks); P_e^{DEV} and P_e^{CEV} are the equivalent powers associated with discharging and charging processes of the mobile energy storage system (i.e., EV batteries); P_e^{DHS} and P_e^{CHS} are, respectively, the equivalent electric powers associated with discharging and charging processes of the thermal storage units; P_e^{DH2S} and P_e^{CH2S} are, respectively, the equivalent electric powers associated with discharging and charging processes of the hydrogen storage units; $P_e^{Grid,min}$ and $P_e^{Grid,max}$ are, respectively, the minimum and maximum boundaries for the power purchased from the main grid; $P_e^{CHP,min}$ and $P_e^{CHP,max}$ are, respectively, the minimum and maximum boundaries for the power purchased from the CHP units; $P_g^{CHP,min}$ and $P_g^{CHP,max}$ are, respectively, the minimum and maximum boundaries for the gas purchased from the CHP units; P_{Trans} and P_{Trans}^{Max} are the power and its maximum limit for the transformer; $P_e^{Storage}$ and $P_e^{Storage,max}$ are, respectively, the power and its maximum limit associated with energy storage systems (i.e., battery banks and EV batteries); P_e^{EV} and $P_e^{EV,max}$ are the available aggregated power for the EV fleet and its maximum boundary, respectively; $P_{Ther}^{Storage}$ and $P_{Ther}^{Storage,max}$ are, respectively, the equivalent power and its maximum limit associated with thermal storage units; $P_{H2}^{Storage}$ and $P_{H2}^{Storage,max}$ are, respectively, the equivalent power and its maximum limit associated with hydrogen storage units; $P_{e,avi}^{ESS}$ and $P_{e,ini}^{ESS}$ are, respectively, the available and initial electric storage capacities of all energy storage systems (i.e., battery banks and EV batteries); ω_C and ω_D are, respectively, the charging and discharging efficiencies for energy storage systems; ψ_1 and ψ_2 are the attack coefficients, controlling the amount of malicious demand information to be injected into the load centers, to limit the data manipulation through optimizing objective functions (12.1) and (12.2); and N_{PV} and N_{WT} are, respectively, the number of PV modules and wind turbine throughout the distribution system.

In order to obtain the Pareto optimal fronts from the developed bi-objective optimization problem, objective functions (12.1) and (12.2) need to be optimized concurrently, which provides ranges for the negative impacts of the FDIAs. To this end, a trapezoidal membership function (i.e., (12.18)) is utilized to convert the numerical values of the objective functions, which are not necessarily in accord with each other, into range 0–1. According to (12.18), it can be concluded that if the numerical value of the membership function (i.e., Θ_n) is equal to 1.00, the decision-maker will be fully satisfied, whereas 0 indicates no satisfaction in the process.

$$\Gamma_n = \begin{cases} 0 \rightarrow OF_n^{max} \leq OF_n \\[2mm] \dfrac{OF_n^{max} - OF_n}{OF_n^{max} - OF_n^{min}} \rightarrow OF_n^{min} \leq OF_n \leq OF_n^{max} \\[2mm] 1 \rightarrow OF_n^{min} \geq OF_n \end{cases} \qquad (12.18)$$

where Γ_n indicates the fuzzy set for the nth objective function, and the number of objective functions in this study is 2 (i.e., (1) and (2)); OF_n, OF_n^{min}, and OF_n^{max} are, respectively, the nth objective function and its minimum and maximum boundaries, which can be obtained via optimizing each objective function individually.

To store the non-dominated solutions obtained from simultaneously optimizing (12.1) and (12.2), along with satisfying the relevant technical constraints (i.e., (12.3)–(12.17)), and determine the best compromise solution (BCS) between both objective functions, a Pareto archive evolutionary strategy (PAES) that is based on the dominance concept is utilized [30]. To this end, (12.19) presents a dominance factor for any non-dominated solution stored in the repository.

$$\Theta_{n,s} = \frac{\sum_{n=1}^{N_{OF}} w_n \times \Gamma_{n,s}}{\sum_{s=1}^{N_{NDS}} \sum_{n=1}^{N_{OF}} w_n \times \Gamma_{n,s}} \qquad (12.19)$$

where Θ denotes the value of membership function (12.18); w_n is the weigh index for the nth objective function, representing the importance of the objective function in the optimization problem; and N_{OF} and N_{NDS} are, respectively, the number of objective functions (i.e., 2 in this study) and the number of non-dominated solutions stored in the repository. It is noted that w_n for $n = 1, 2$ needs to satisfy (12.20); hence, any combinations can be taken into account by the attackers to control the severity of the FDIAs. More information about the importance of such a weight factor in the developed optimization problem will be provided in Section 12.4.

$$\sum_{n=1}^{2} w_n = 1 \qquad (12.20)$$

12.3 DESCRIPTION OF THE CASE STUDY AND MODELED ATTACK SCENARIOS

Figure 12.2 illustrates the IEEE 13-bus test system, modified as a microgrid, that has been selected as the case study of this chapter in order to highlight the impacts of the organized cyberattack on the operation of the power system (refer to Figure 12.1). In this regard, a PV plant with the capacity of 220 kW and also a wind turbine, manufactured by Wind World, with a rated power of 200 kW and blade diameter of 29.2 m are, respectively, added to bus #6 and bus #7. The solar irradiance at standard test conditions and the certain irradiance point are equal to 1,000 W/m^2 and 120 W/m^2, respectively. To handle the uncertainty of the RERs (i.e., PV modules and wind turbines), (1) real-world solar and wind data have been extracted from [31, 32] and (2) a prediction technique is adopted from [33] to forecast the output of PV modules and wind turbines for the sake of optimization problems. It is noted that the concept of renewable power predicting is not in the scope of this chapter; therefore, interested readers are directed to [34–36] for detailed information. The hourly active and reactive power profiles in a 24 hr horizon are adopted from [37], which are depicted in Figure 12.3.

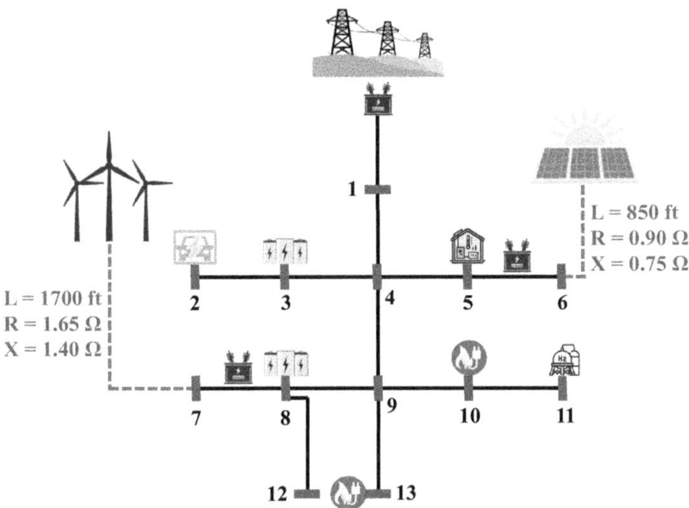

FIGURE 12.2 Modified IEEE 13-bus distribution system integrated with power-to-X technologies.

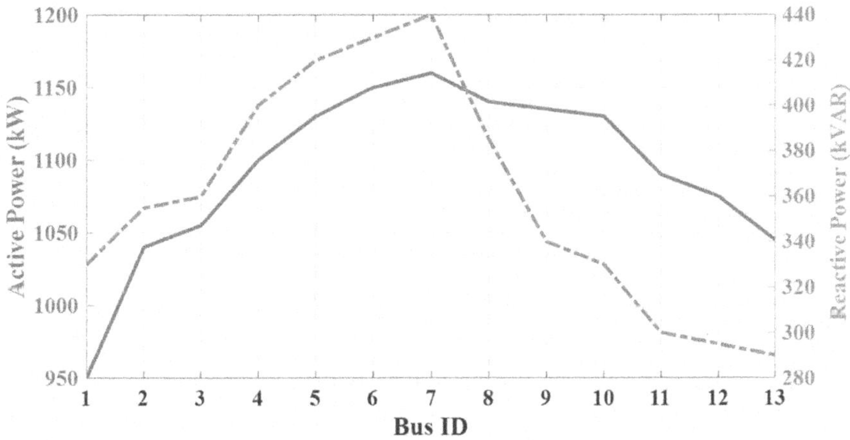

FIGURE 12.3 Active and reactive power profiles associated with modified IEEE 13-bus test system, depicted in Figure 12.2.

12.4 SIMULATION RESULTS AND ANALYSIS

To illustrate the negative impacts of the modeled FDIAs, the developed attack framework (see Sections 12.2.1–12.2.3) and the bi-objective optimization problem, introduced in Section 12.2.4, were simulated via the optimization toolboxes of MATLAB R2020b as well as the available functions and system descriptions of the MATPOWER package [38, 39] on an Intel® Core™ CPU i7–13700 at 2.10 GHz and 32 GB of RAM.

To show the impacts of the modeled cyberattacks on the normal operation of modern energy systems integrated with power-to-X components (see Figure 12.2), the load centers (i.e., both electric and thermal) within the distribution system are compromised via injecting false data (see objective function (12.2)) in order to push the power system toward a higher operational cost, as represented in objectibve function (12.1)), and operational issues as a consequence, such as overloading of the system's assets, system congestion, and probabale power outages. Thus, three scenarios of FDIA are considerred in this study, which are elaborated upon hereafter.

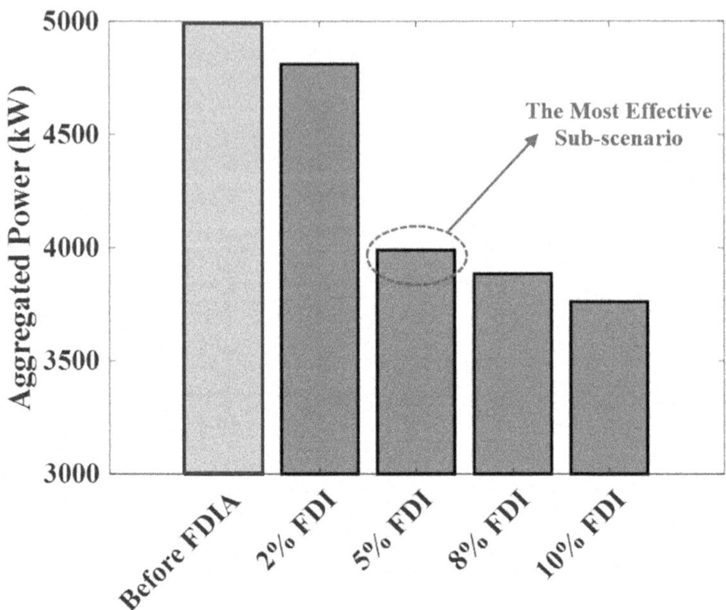

FIGURE 12.4 Aggregated power of the assets before and after the FDIA targeting smart meters of the electric demand in Scenario 1.

- **Scenario 1:** Compromising only electric load centers via injecting false data limited up to 10% of the rated power.
- **Scenario 2:** Compromising only thermal load centers via injecting malicious information up to 10% of their ratings.
- **Scenario 3:** Compromising the smart meters associated with both electric and thermal demand throughout the system.

Figure 12.4 depicts the aggregated power of the renewable energy sources (i.e., PV modules and wind turbines), static energy storage units (i.e., battery banks), and mobile energy storage units (i.e., EV batteries) after compromising the electric demand of the buses having load centers by up to 10% of the rated power. According to this figure, which also compares the level of power before and after the FDIA, one can infer that among four different sub-scenarios (see the red bars displayed in Figure 12.4), only the second sub-scenario (i.e., compromising the load centers by injecting 5% of the rated power of each bus, either addition or subtraction) is the most effective one, since the aggregated power associated with this sub-scenario is reduced by 20.25%, which is due to the fact that the attack objective functions are optimized cuncurrently. It is noted that the aggregated power of the PV modules and wind turbines, battery banks, and EV batteries was reduced after the 2%, 5%, 8%, and 10% FDIAs by, respectively, 3.41%, 20.25%, 21.98%, and 24.48%. Therefore, if attackers restrict the level of manipulation to 5% of the rated power of each bus, the launched cyberattack has a higher chance to remain undetectable from the sytem operator's standpoint. The amounts of false data (i.e., ΔP_e^{Demand}), obtained after solving the introduced optimization problem, injected into the system buses are presented in Table 12.1, where a positive sign indicates addition to the original load and a negative sign denotes subtraction from the original load of the bus.

To obtain a better perspective about the negative impact of the FDIA compromising electric load data, Figure 12.5 presents the power shortage throughout the IEEE 13-bus test system after launching the cyberattack. From this figure, it can be concluded that the system overall suffers from 17.55% power shortage in serving the electric load. As a case in point, although bus #9 does not have any renewable units, it experienced the maximum energy paucity compared to other buses,

TABLE 12.1

False Data Injected into the System Buses for Scenario 1 and the Most Effective Sub-Scenario*

Bus #	Original Active Power (kW)	FDI Vector (kW)
1	950	+4.1129
2	1,040	+6.0550
3	1,055	+4.3872
4	1,100	−5.4430
5	1,130	+1.0217
6	1,150	−4.0022
7	1,160	−3.1561
8	1,140	−1.8000
9	1,135	+8.7779
10	1,130	+0.6819
11	1,090	−10.9348
12	1,075	−0.9604
13	1,045	+1.2599

Note: That is, compromising the load centers by injecting 5% of the rated power of each bus.

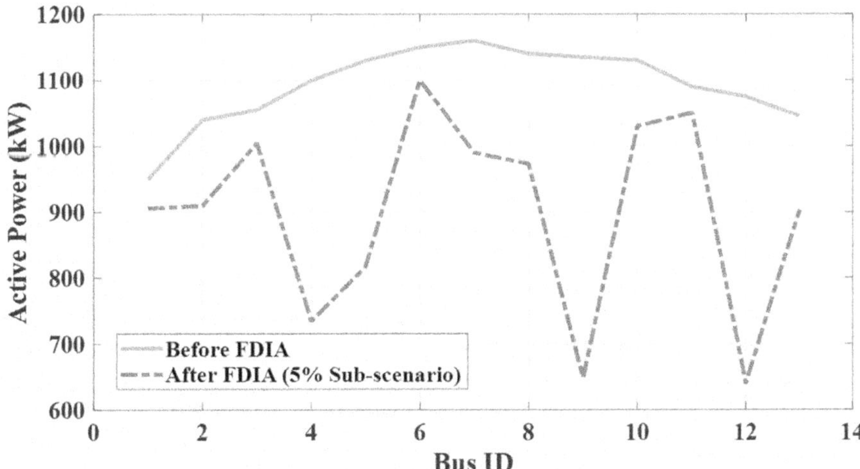

FIGURE 12.5   Active demand profile of the IEEE 13-bus test system before and after the FDIA targeting smart meters of the electric demand in Scenario 1.

equivalent to 42.82%. This is where the significance of appropriate remedial actions comes under the spotlight to recover the targeted system to the normal operation in a timely manner and enhance the social welfare of the end use customers, which will be the next step of this research.

Figure 12.6 displays the power flow through distribution branches after launchnig the FDIA targeting thermal demand throughout the system by injecting malicious data equivalent to 5% of the ratings. It can be seen from this figure that the cyberattcak caused overloading in majority of the distribution branches; however, the highest rate of overloading, leading to power outage, was associated with branch #5. This is due to the fact that line #5 connects the top region of the

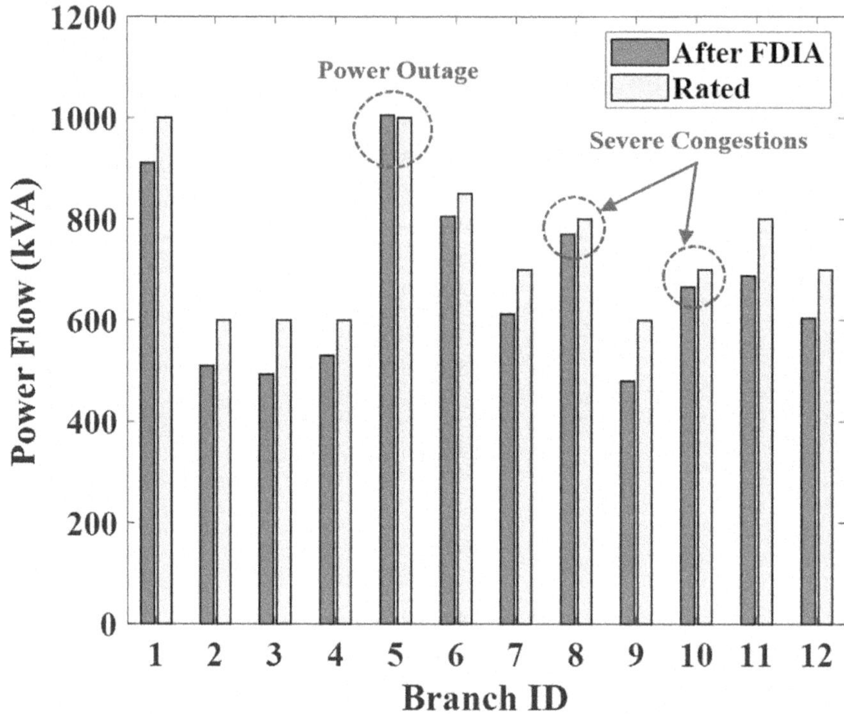

FIGURE 12.6 Active power flow in the distribution branches after launching FDIA asscoaited with Scenario 2.

distribution system (i.e., bus #1 to bus #6) to the bottom region of the system (i.e., buses #7 through bus #13). Compared to Figures 12.4 and 12.5, Figure 12.6 shows lower rates of negative impacts posed by the cyberattack, which correspond to the BCS (i.e., green hexagram shown in Figure 12.7) between objective functions (12.1) and (12.2), obtained after applying (12.19) with weight factors $w_1 = w_2 = 0.5$. Hence, it can be concluded that if attackers modify the weight factors to $w_1 = 0.9$ and $w_2 = 0.1$, the false data vector will be extremely minimized, whereas the operational cost of the system will be extremely maximized. This FDIA, which is assocaited with the yellow diamond in Figure 12.7, represents a weak cyberattack since the total operational cost of the system will also be minimized (see the yellow diamond shown in Figure 12.7). Therefore, attackers can have a control mechanism to adjust the severity of the attack consequences.

Figure 12.8 compares the active power flowing into all distribution branches after launching FDIAs associated with Scenarios 1–3. According to this figure, one can perceive that when both electric and thermal load centers were falsified via injecting false data vectors, the IEEE 13-bus test system experienced more severe system congestions such that five branches, #3, #5, #6, #8, and #10, experienced power outages, compared to Figure 12.6 presenting the active power flow in the distribution branches after launching FDIA targeting only thermal load points, leading to power outage for branch #5. It is noted that branch #5 was loaded up to 145% after the FDIA associated with Scenario 3.

Finally, to show the economic loss imposed by the cyberattacks targeting electric and thermal demands, Table 12.2 provides the total operation cost of the IEEE 13-bus test system before attack and after launching FDIAs asscoaited with Scenarios 1–3. From this table, it can be inferred that targeting both electric and thermal load centers throughout the IEEE 13-bus test system can push the system toward higher costs, leading to more severe operational issues, such as system congestions and power outages.

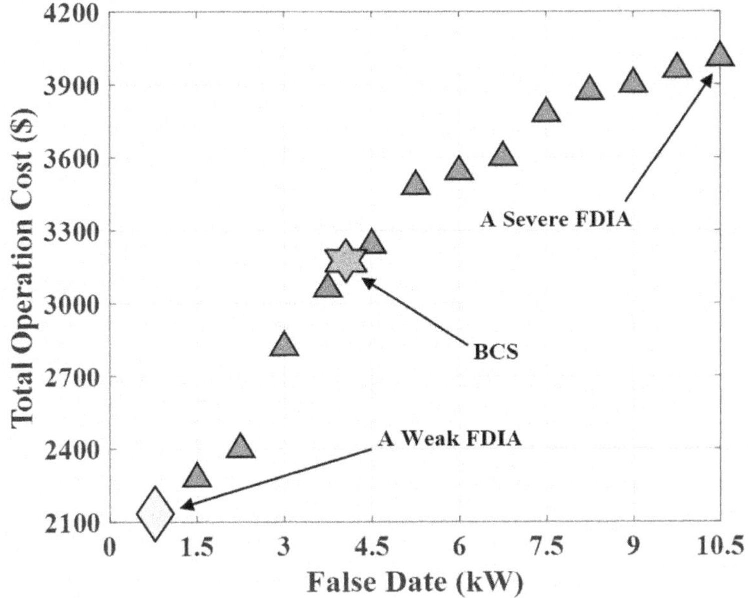

FIGURE 12.7 2D Pareto optimal front obtained after optimizing objective functions (12.1) and (12.2) for Scenario 2.

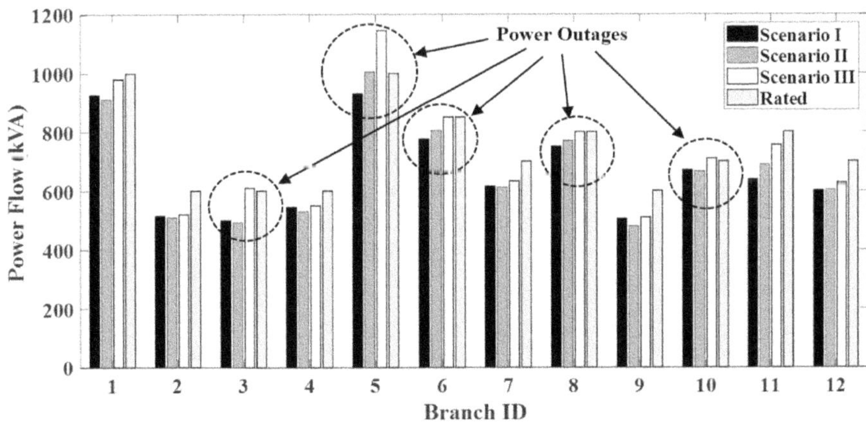

FIGURE 12.8 Active power flow in the distribution branches after launching FDIAs associated with Scenarios 1–3.

TABLE 12.2
Total Operation Cost of the System before and after the FDIAs Associated with All Three Attack Scenarios

	Total Operation Cost	Emission Cost
Scenario 1	285,412	122,674
Scenario 2	288,635	123,406
Scenario 3	313,019	148,987

12.5 CONCLUSIONS AND FUTURE WORK

This chapter scrutinized power-to-X integration targeted by false data injection attacks (FDIAs) resulting in operational issues (e.g., overloading, system congestions, power outages, to name but a few) in the modified version of the IEEE 13-bus test system. The main objective of the cyberattack targeting load centers (i.e., both electrical and thermal demand) was to push the distribution system toward higher operational cost and instability. To keep the attack undetectable from the system operator's standpoint, the FDIA was modeled as a bi-objective optimization problem aiming at (1) maximizing the total operational cost of the distribution system integrated with power-to-X technologies and (2) minimizing the amount of malicious data injected into the smart meters associated with load centers. The following main conclusions are drawn from the simulation results:

- In Scenario 1, where only the electric load centers were the target of FDIA, the consequence of the cyberattack (i.e., the power shortage) was noticeable when the load data were manipulated by 5% of the rated power. In other words, increasing the false data vectors from 2% to 5% increased the power shortage 170 kW to 990 kW, equivalent to more than a 480% increase. This amount of power shortage is quite unacceptable from the power system operation perspective. Increasing the level of false data after 5% did not significantly increase the power shortage beyond 480%.
- In Scenario 2, where only the thermal load centers were the target of FDIA, the IEEE 13-bus distribution system experienced two severe congestions in branches #8 and #10 and one power outage in branch #5.
- Scenario 3 represented a more severe FDIA compared to Scenarios 1 and 2 since the IEEE 13-bus network experienced a system-wide congestion in almost all buses, and also five power outages in branches #3, #5, #6, #8, and #10. This is where the importance of remedial action schemes recovering the targeted system to the normal operation in real time comes under the spotlight.

In the future step of this research, a remediation framework will be proposed to assist the system operator to react to the successful FDIAs in a timely manner and recover the system to normal operation through performing online versions of network reconfigurations with different objective functions to significantly reduce the total operation cost of the system and enhance the social welfare of end use customers.

REFERENCES

[1] H. Holttinen, J. Kiviluoma, D. Flynn, J. C. Smith, A. Orths, P. B. Eriksen, N. Cutululis, L. Söder, M. Korpås, A. Estanqueiro, and J. MacDowell, "System impact studies for near 100% renewable energy systems dominated by inverter based variable generation," *IEEE Transactions on Power Systems*, vol. 37, no. 4, pp. 3249–3258, July 2022. doi: 10.1109/TPWRS.2020.3034924.

[2] D. V. Pombo, J. Martinez-Rico, M. Carrion, and M. Cañas-Carretón, "A computationally efficient formulation for a flexibility enabling generation expansion planning," *IEEE Transactions on Smart Grid*, vol. 14, no. 4, pp. 2723–2733, July 2023. doi: 10.1109/TSG.2022.3233124.

[3] X. Teng, Z. Gao, Y. Zhang, H. Huang, L. Li, and T. Liang, "Key technologies and the implementation of wind, PV and storage co-generation monitoring system," *Journal of Modern Power Systems and Clean Energy*, vol. 2, no. 2, pp. 104–113, June 2014. doi: 10.1007/s40565-014-0055-1.

[4] Y. Zong, J. Wang, S. You, W. Su, V. F. Øhlenschlæger, S. B. Petersen, L. Wang, and J. Xiao, "Identifying the system-related conditions and consequences of power-to-X solutions for a high renewables penetration in Denmark," in *IEEE 4th Conference on Energy Internet and Energy System Integration (EI2)*, Wuhan, China, 2020, pp. 955–960. doi: 10.1109/EI250167.2020.9346867.

[5] L. Marchisio, C. Alvaro, S. Cerchiara, M. Sisinni, and A. Siviero, "The role of power-to-gas and electrochemical storage systems in a climate-neutral energy system," *AEIT International Annual Conference (AEIT)*, Catania, Italy, 2020, pp. 1–6. doi: 10.23919/AEIT50178.2020.9241118.

[6] M. Genovese, A. Schluter, E. Scionti, F. Piraino, O. Corigliano, and P. Fragiacomo, "Power-to-hydrogen and hydrogen-to-X energy systems for the industry of the future in Europe," *International Journal of Hydrogen Energy*, vol. 48, no. 44, pp. 16545–16568, May 2023. doi: 10.1016/j.ijhydene.2023.01.194.

[7] C. Breyer, G. Lopez, D. Bogdanov, and P. Laaksonen, "The role of electricity-based hydrogen in the emerging power-to-X economy," *International Journal of Hydrogen Energy*, vol. 49, no. 2, pp. 351–359, January 2024. doi: 10.1016/j.ijhydene.2023.08.170.

[8] L. M. Pastore, G. L. Basso, G. Ricciardi, and L.D. Santoli, "Smart energy systems for renewable energy communities: A comparative analysis of power-to-X strategies for improving energy self-consumption," *Energy*, vol. 280, p. 128205, October 2023. doi: 10.1016/j.energy.2023.128205.

[9] H. Onodera, R. Delage, and T. Nakata, "Systematic effects of flexible power-to-X operation in a renewable energy system—A case study from Japan," *Energy Conversion and Management: X*, vol. 20, p. 100416, October 2023. doi: 10.1016/j.ecmx.2023.100416.

[10] S. Nady, H. E. Fadil, M. Koundi, A. Hamed, and F. Giri, "Power-to-X systems: State-of-the-art (PTX)," *International Federation of Automatic Control*, vol. 55, no. 12, pp. 300–305, August 2022. doi: 10.1016/j.ifacol.2022.07.328.

[11] J. R. Feijoo-Martínez, A. Guerrero-Curieses, F. Gimeno-Blanes, M. Castro-Fernández, and J. L. Rojo-Álvarez, "Cybersecurity alert prioritization in a critical high-power grid with latent spaces," *IEEE Access*, vol. 11, pp. 23754–23770, 2023. doi: 10.1109/ACCESS.2023.3255101.

[12] E.-N. S. Youssef, F. Labeau, and M. Kassouf, "Adversarial dynamic load-altering cyberattacks against peak shaving using residential electric water heaters," *IEEE Transactions on Smart Grid*, vol. 15, no. 2, pp. 2073–2088, March 2024. doi: 10.1109/TSG.2023.3300239.

[13] W. Gao, H. Li, M. Zhong, and M. Lu, "An underestimated cybersecurity problem: Quick-impact time synchronization attacks and a fast-triggered detection method," *IEEE Transactions on Smart Grid*, vol. 14, no. 6, pp. 4784–4798, November 2023. doi: 10.1109/TSG.2023.3258963.

[14] I. Zografopoulos, N. D. Hatziargyriou, and C. Konstantinou, "Distributed energy resources cybersecurity outlook: Vulnerabilities, attacks, impacts, and mitigations," *IEEE Systems Journal*, vol. 17, no. 4, pp. 6695–6709, December 2023. doi: 10.1109/JSYST.2023.3305757.

[15] J. Hou, S. Lei, Y. Song, L. Zhu, W. Sun, and Y. Hou, "The cost and benefit of enhancing cybersecurity for hybrid AC/DC grids," *IEEE Transactions on Smart Grid*, vol. 14, no. 6, pp. 4758–4771, November 2023. doi: 10.1109/TSG.2023.3255250.

[16] S. Hussain, S. M. Suhail Hussain, M. Hemmati, A. Iqbal, R. Alammari, S. Zanero, E. Ragaini, and G. Gruosso, "A novel hybrid cybersecurity scheme against false data injection attacks in automated power systems," *Protection and Control of Modern Power Systems*, vol. 8, no. 3, pp. 1–15, July 2023, doi: 10.1186/s41601-023-00312-y.

[17] Z. Zhang, K. Zuo, R. Deng, F. Teng, and M. Sun, "Cybersecurity analysis of data-driven power system stability assessment," *IEEE Internet of Things Journal*, vol. 10, no. 17, pp. 15723–15735, September 2023. doi: 10.1109/JIOT.2023.3264492.

[18] R. Fu, M. E. Lichtenwalner, and T. J. Johnson, "A review of cybersecurity in grid-connected power electronics converters: Vulnerabilities, countermeasures, and testbeds," *IEEE Access*, vol. 11, pp. 113543–113559, 2023. doi: 10.1109/ACCESS.2023.3324177.

[19] Y. Li and J. Yan, "Cybersecurity of smart inverters in the smart grid: A survey," *IEEE Transactions on Power Electronics*, vol. 38, no. 2, pp. 2364–2383, February 2023. doi: 10.1109/TPEL.2022.3206239.

[20] D. Said, "A survey on information communication technologies in modern demand-side management for smart grids: Challenges, solutions, and opportunities," *IEEE Engineering Management Review*, vol. 51, no. 1, pp. 76–107, March 2023. doi: 10.1109/EMR.2022.3186154.

[21] E. Naderi, K. C. Bibek, M. Ansari, and A. Asrari, "Experimental validation of a hybrid storage framework to cope with fluctuating power of hybrid renewable energy-based systems," *IEEE Transactions on Energy Conversion*, vol. 36, no. 3, pp. 1991–2001, September 2021. doi: 10.1109/TEC.2021.3058550.

[22] E. Naderi and A. Asrari, "Experimental validation of grid-tied and standalone inverters on a lab-scale wind-PV microgrid," *IEEE International Power and Renewable Energy Conference (IPRECON)*, Kollam, India, 2021, pp. 1–6. doi: 10.1109/IPRECON52453.2021.9640998.

[23] E. Naderi and A. Asrari, "Hardware-in-the-loop experimental validation for a lab-scale microgrid targeted by cyberattacks," *International Conference on Smart Grid (icSmartGrid)*, Setubal, Portugal, 2021, pp. 57–62. doi: 10.1109/icSmartGrid52357.2021.9551023.

[24] E. Naderi and A. Asrari, "Detection of false data injection cyberattacks: Experimental validation on a lab-scale microgrid," *IEEE Green Energy and Smart System Systems (IGESSC)*, Long Beach, CA, USA, 2022, pp. 1–6. doi: 10.1109/IGESSC55810.2022.9955337.

[25] E. Naderi, A. Asrari, and B. Ramos, "Moving target defense strategy to protect a PV/wind lab-scale microgrid against false data injection cyberattacks: Experimental validation," *IEEE Power & Energy Society General Meeting (PESGM)*, Orlando, FL, USA, 2023, pp. 1–5. doi: 10.1109/PESGM52003.2023.10252369.

[26] E. Naderi and A. Asrari, "Toward detecting cyberattacks targeting modern power grids: A deep learning framework," *IEEE World AI IoT Congress (AIIoT)*, Seattle, WA, USA, 2022, pp. 357–363. doi: 10.1109/AIIoT54504.2022.9817309.

[27] E. Naderi, A. Aydeger, and A. Asrari, "Detection of false data injection cyberattacks targeting smart transmission/distribution networks," *IEEE Conference on Technologies for Sustainability (SusTech)*, Corona, CA, USA, 2022, pp. 224–229. doi: 10.1109/SusTech53338.2022.9794237.

[28] E. Naderi and A. Asrari, "Experimental validation of a remedial action via hardware-in-the-loop system against cyberattacks targeting a lab-scale PV/wind microgrid," *IEEE Transactions on Smart Grid*, vol. 14, no. 5, pp. 4060–4072, September 2023. doi: 10.1109/TSG.2023.3253431.

[29] E. Naderi, M. Ansari, J. P. Trimble, and A. Asrari, "Experimental validation of a market-based remedial action coping with cyberattackers targeting renewable-based microgrids," *IEEE PES/IAS PowerAfrica*, Nairobi, Kenya, 2021, pp. 1–5. doi: 10.1109/PowerAfrica52236.2021.9543289.

[30] E. Naderi, L. Mirzaei, J. P. Trimble, and D. A. Cantrell, "Multi-objective optimal power flow incorporating flexible alternating current transmission systems: Application of a wavelet-oriented evolutionary algorithm," *Electric Power Components and Systems*, vol. 52, no. 5, pp. 766–795, August 2023. doi: 10.1080/15325008.2023.2234378.

[31] "Solar power data for integration studies." Accessed: January 5, 2024. [Online]. www.nrel.gov/grid/solar-power-data.html.

[32] PJM Website. Accessed: January 5, 2024. [Online]. www.pjm.com/.

[33] E. Naderi and A. Asrari, "A remedial action scheme to mitigate market power caused by cyberattacks targeting a smart distribution system," *IEEE Transactions on Industrial Informatics*, vol. 20, no. 3, pp. 3197–3208, March 2024. doi: 10.1109/TII.2023.3304049.

[34] V. V. S. Lanka, M. Roy, S. Suman, and S. Prajapati, "Renewable energy and demand forecasting in an integrated smart grid," *Innovations in Energy Management and Renewable Resources*, Kolkata, India, 2021, pp. 1–6. doi: 10.1109/IEMRE52042.2021.9386524.

[35] O. Rubanenko, S. L. Gundebommu, M. Cosovic, and V. Lesko, "Predicting the power generation from renewable energy sources by using ANN," *20th International Symposium INFOTEH-JAHORINA (INFOTEH)*, East Sarajevo, Bosnia and Herzegovina, 2021, pp. 1–6. doi: 10.1109/INFOTEH51037.2021.9400527.

[36] N. N. Mwanza, P. M. Moses, and A. M. Nyete, "Short-term forecasting for integrated load and renewable energy in micro-grid power supply," *IEEE PES/IAS PowerAfrica*, Nairobi, Kenya, 2020, pp. 1–5. doi: 10.1109/PowerAfrica49420.2020.9219859.

[37] "Commercial and residential hourly load profile." Accessed: January 5, 2024. [Online]. https://openei.org/doe-opendata/dataset/commercial-and-residential-hourly-load-profiles-for-all-tmy3-locations-in-the-united-states.

[38] R. D. Zimmerman, C. E. Murillo-Sanchez, and R. J. Thomas, "MATPOWER: Steady-state operations, planning and analysis tools for power systems research and education," *IEEE Transactions on Power Systems*, vol. 26, no. 1, pp. 12–19, February 2011. doi: 10.1109/TPWRS.2010.2051168.

[39] R. D. Zimmerman and C. E. Murillo-Sanchez. MATPOWER (Version 7.1). Accessed: January 5, 2024. [Online]. https://matpower.org/download/.

13 Autonomous Intelligent Power-to-Mobility of Electric Vehicles Using Distributed Multi-Agent ADMM Framework for Grid Ancillary Services

Towfiq Rahman and Zhihua Qu

13.1 INTRODUCTION TO ELECTRIC VEHICLES (EVS) TECHNOLOGY IN THE GRID

Over the last decade, there has been a dramatic rise in the number of electric vehicle (EV) users in our community. Regular working people are buying EV as their primary vehicle due to the push from the government for a "greener" Earth and a reduction in greenhouse gas emissions. It became more lucrative when leading automotive industry giants paid attention to developing charging infrastructure, upgrading battery capacity, and having vehicle models at a reasonable price [1]. Due to these reasons, globally, nearly 14 million new EVs were registered in 2023, bringing the total number of EVs on the road to 40 million. The rapid growth of the electric vehicle market is evident, with sales in 2023 surpassing those of 2018 by six times and accounting for nearly one-fifth of all cars sold [2]. As a consequence, there is a large number of EVs connected to the distribution grid, which is causing a substantial degradation in the efficiency and reliability of the distribution power grid [3]. The authors in [4] and [5] showed that the charging of a large number of EVs puts additional strain on the grid and contributes to many negative effects, such as increased distribution energy losses and voltage deviation with unpredictable and variable load profile. But not all are bad with EVs in the distribution grid, as EVs can be used as a valuable resource that can be controlled to benefit the grid [6]. EVs can be used as a flexible grid-connected energy resource, and to effectively use their full potential, EVs can be controlled to not only charge but also discharge and return energy to the grid [7]. To this end, a multi-layer distributed alternating direction method of multiplier (ADMM) algorithm is developed which can handle dynamic EV constraints like the state of charge (SOC) to optimize and control EVs in the grid to properly utilize them for voltage regulation and, at the same time, ensure that the owners are compensated by maximizing their utility function and having the desired state of charge at the end of their EVs' charging period. To analyze the multi-layer distributed ADMM, one must understand the basics of the original ADMM, which is presented in Section 13.2. In Section 13.3, the developed distributed ADMM algorithm is introduced, and its convergence properties are discussed. In Section 13.4, a problem is formulated of voltage deviation minimization and EV utility maximization for a distribution system network with EV penetration and to implement the proposed algorithm and present simulation scenarios to illustrate the results. And finally, the chapter is concluded in Section 13.5.

DOI: 10.1201/9781032719436-13

13.2 ALTERNATING DIRECTION METHODS OF MULTIPLIERS: AN OVERVIEW

Let us consider the following problem:

$$
\begin{aligned}
\min_{x,z} \quad & f(x) + g(z) \\
s.t. \quad & Ax + Bz = c
\end{aligned}
\tag{13.1}
$$

where variables $x \in \mathcal{R}^n$ and $z \in \mathcal{R}^m$, $A \in \mathcal{R}^{p \times n}$, $B \in \mathcal{R}^{p \times m}$, and $c \in \mathcal{R}^p$ are the matrices and vectors in the linear constraint. We will assume f and g are convex and differentiable, and that their gradients are locally Lipschitz. The problem can be thought of as a general convex linear equality-constrained problem, except for the fact that the main optimization variable has been split into two parts, namely, x and z in this case, with objective function separable across this splitting, and thus, the algorithm is distributed in nature [8]. As with any primal-dual convex optimization, we form the Lagrangian as follows [9]:

$$
L_\rho(x,z,y) = f(x) + g(z) + y^T(Ax + Bz - c) + \frac{\rho}{2} \| Ax + Bz - c \|_2^2 .
\tag{13.2}
$$

The Lagrangian in this case is called the "augmented Lagrangian" due to the fact that an additional penalty term with multiplier $\rho > 0$ is added to the objective function. The problem is solved using ADMM with the following iterations:

$$
x^{k+1} := \arg\min_x L_\rho\left(x, z^k, y^k\right)
\tag{13.3a}
$$

$$
z^{k+1} := \arg\min_z L_\rho\left(x^{k+1}, z, y^k\right)
\tag{13.3b}
$$

$$
y^{k+1} := y^k + \rho\left(Ax^{k+1} + Bz^{k+1} - c\right)
\tag{13.3c}
$$

The ADMM algorithm consists of an x-minimization step (13.3a), followed by a z-minimization step (13.3b) and a dual variable update (13.3c). The primal variables x and z are updated in an alternating fashion, thus the name "alternating direction." The ADMM algorithm of this type can only deal with static linear constraints but falls short when the constraint becomes dynamic in nature. This is very true when we deal with a distribution network with EV penetration, since EV itself has a state-of-charge (SOC) parameter which evolves with time. To tackle the problem of this nature, in the next section, we will develop our proposed ADMM algorithm with dynamic constraints in the continuous-time domain.

13.3 DEVELOPMENT OF DISTRIBUTED MULTI-AGENT NETWORKED ADMM

In this section, we develop a continuous-domain real-time ADMM algorithm with a dynamic constraint that can be implemented to a broad class of networked multi-agent distributed optimization and control problems.

13.3.1 NETWORKED MULTI-AGENT SYSTEM

Let us consider a networked multi-agent system which is characterized by a bidirectional graph $G = (N, E)$, where $N = \{1, 2, \cdots, N\}$, N is the number of agents, and E represents the set of edges

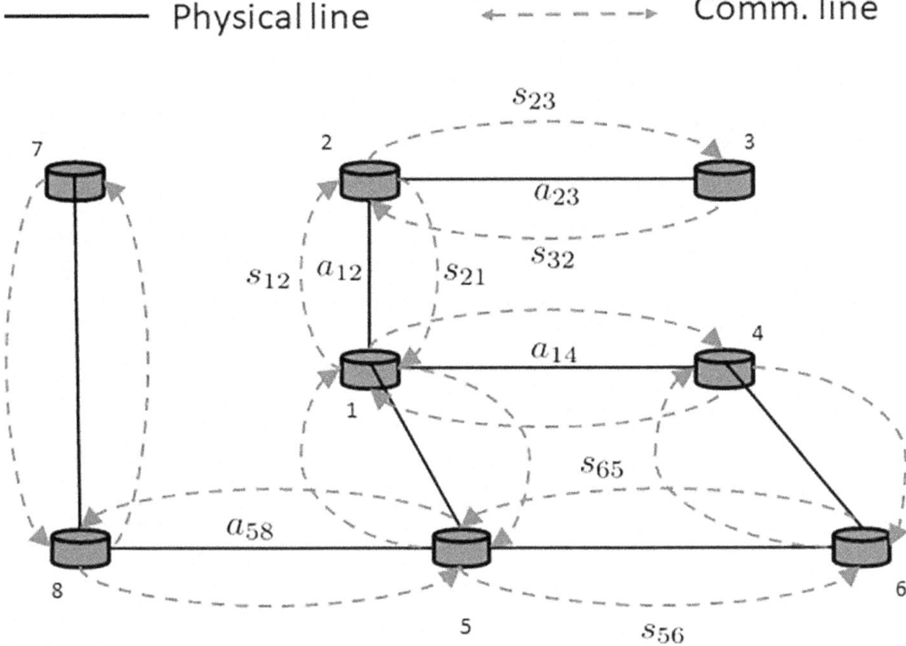

FIGURE 13.1 A simplified distribution network showing power flow between two buses.

between them. Also, let node 1 be the virtual leader of the network, which means, through the communication network, node 1 can obtain the information of all the nodes in the network, if necessary [10]. We can define the network interconnection with the following binary matrix:

$$S = \begin{bmatrix} 1 & 1 & \cdots & 1 \\ 1 & 1 & \dots & S_{2N} \\ \vdots & \vdots & \vdots & \vdots \\ 1 & S_{N2} & \cdots & 1 \end{bmatrix} \tag{13.4}$$

where $s_{ij} = 1$ if and only if $(i \leftrightarrow j) \in E$, and $s_{ij} = 0$ if otherwise. That means that if agent i can communicate with agent j, and vice versa, we have an entry of 1 in the position (i, j) in the communication matrix. The matrix S has 1 in the diagonal as every agent knows its information. We assume that the communication matrix S is irreducible, that is, the communication graph is strongly connected.

Figure 13.1 shows the network of multi-agent system. The agents can be connected physically, as we will see in the next subsection, and it will also be required when we implement the algorithm in a power distribution system.

13.3.2 DISTRIBUTED DYNAMIC ADMM

In this subsection, we will formulate a distributed networked multi-agent problem with a dynamic constraint. The work in [11] formulated a similar kind of problem without the dynamic constraint, and the solution with convergence proof was provided in the discrete-time domain. We will, in this chapter, add a dynamic constraint and tackle the problem in the continuous-time domain.

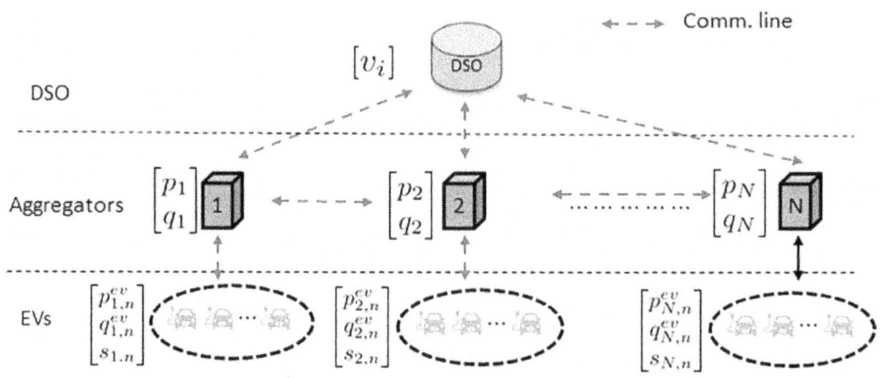

FIGURE 13.2 A networked multi-agent system having nodes with both physical and communication layers.

Consider the following distributed optimization problem:

$$\min_{y_i} \sum_{i \in N} f_i(y_i)$$

s.t.
$$\sum_{j \in N_i} A_{ij} z_{ji} = 0, \quad \forall i \in N; \quad y_i = z_{ij}, \quad \forall i \in N, j \in N_i,$$

$$\dot{x}_i = F_i(x_i) + G_i(x_i) u_i, \quad y_i = H_i(x_i),$$

where

- $x_i \in R^{n_i}$ is the state of the ith agent;
- $y_i \in R^{l_i}$ is the output of the ith agent;
- $f_i(y_i) : R^{l_i} \to \in R$ is the objective function of agent i;
- $u_i \in R^{m_i}$ is the control (or decision variable) vector; and
- A_{ij} are constant matrices of appropriate dimension.

We assume that the functions $f_i(y_i)$ are convex and differentiable, and their gradients are locally Lipschitz. In the preceding problem, y and z are the two primal variables used in the original ADMM. The matrix A_{ij} represents the physical interconnection between the agents, as shown in Figure 13.1. To tackle the problem, let us define the following design principles:

1. Let

$$u_i = U_i(x_i) + \omega_i$$

where local feedback control U_i is designed so that the subsystem of x_i is input-to-state stable, and that if $\omega_i \to c_i$ for any constant $c_i, y_i \to c_i$.

2. Input ω_i is chosen as the ADMM law.

Using the communication matrix S, let us also define a gain matrix D, whose values are calculated according to the following equation:

$$D = [d_{ij}] \in \mathbb{R}^{N \times N}, \quad d_{ij} = \frac{s_{ij} \beta_{ij}}{\sum_{l=1}^{N} s_{il} \beta_{il}}, \tag{13.5}$$

where $\beta_{ij} > 0$ are piecewise-constant scalar gains. The matrix D is a non-negative, row-stochastic (having row sum equal to 1) and diagonally positive matrix.

With the preceding design principle, we can redefine the optimization problem as follows:

$$\min_{\omega_i} \sum_{i \in N} f_i(y_i) \tag{13.6a}$$

$$\text{s.t.} \ \sum_{j \in N_i} A_{ij} z_{ji} = 0 \ \ j \in N_i, i \in N \tag{13.6b}$$

$$y_i = z_{ij} \ \ \ j \in N_i, i \in N \tag{13.6c}$$

$$\dot{x}_i = F_i(x_i) + G_i(x_i) u_i \tag{13.6d}$$

$$u_i = U_i(x_i) + \omega_i \tag{13.6e}$$

Now we can form the so-called augmented Lagrangian as follows:

$$L(u, z, \lambda) = \sum_{i \in \mathcal{N}} L_i(u_i, z_{ij}, \lambda_{ij})$$

$$L_i = f_i(y_i) + \sum_{j \in \mathcal{N}_i} \left[d_{ij} \lambda_{ij}^T (y_i - z_{ij}) + \frac{d_{ij}}{2} \| y_i - z_{ij} \|^2 \right] \tag{13.7}$$

It should be noted that only the consensus constraint (13.6c) is used in the augmented Lagrangian since it is the only constraint that contains both the primal variables. The rest of the constraints are taken into account while solving the individual subproblems. Also, the penalty parameter term in the augmented Lagrangian is replaced with entries d_{ij} from the D matrix defined in the design principle [11]. This enables the penalty term to conform with the actual interconnection of the agents. For this reason, the dual variable λ is also scaled by the d_{ij} term. The augmented Lagrangian (13.7) is solved using ADMM by solving the following subproblems in an alternating sequential manner:

1. For any $i \in N$, ω_i is updated according to

$$\dot{\omega}_i = \arg \min_{x_i \in \mathbb{R}^n} L(y, z^-, \lambda) \tag{13.8}$$

2. For any $i \in N$ and for $j \in N_i$, z_{ji} is solved as

$$\dot{z}_{ji} = \arg \min_{z_{ji} \in \mathbb{R}^n} L(y, z, \lambda) \tag{13.9}$$

$$\text{s.t.} \ \sum_{j \in \mathcal{N}_i} A_{ij} z_{ji} = 0 \tag{13.10}$$

3. For any $i \in N$ and for $j \in N_i$, λ_{ji} evolves as

$$\dot{\lambda}_{ji} = \arg \max_{\lambda_{ji} \in \mathbb{R}^n} L(y, z, \lambda) \tag{13.11}$$

where $z^- \triangleq z(t^-)$ is the immediate past solution to the problems of (13.10). We use the delayed version of z to mimic the alternating behavior of the ADMM algorithm. Using the convex properties and techniques from [11], we obtain the continuous-time dynamics including dynamic constraints for the solution of subproblems (13.8)–(13.11) as follows:

$$\dot{\omega}_i = -\alpha_i \left[\nabla_{y_i} f_i(y_i) + \sum_{j \in \mathcal{N}_i} d_{ji} \lambda_{ij} + \sum_{j \in \mathcal{N}_i} d_{ij} \left(\xi_i - z_{ij}^- \right) \right] \tag{13.12a}$$

$$\dot{z}_{ji} = -\alpha_i \left[-d_{ij} \lambda_{ji} - d_{ji} \left(\xi_j - z_{ji} \right) + A_{ij}^T \mu_i \right] \tag{13.12b}$$

$$\dot{\mu}_i = w_i \sum_{j \in N_i} A_{ij} z_{ji} \tag{13.12c}$$

$$\dot{\lambda}_{ji} = d_{ji} \left(\xi_j - z_{ji} \right). \tag{13.12d}$$

where μ_i is the dual variable associated with the constraint (13.10) of the z-minimization subproblem. The variable ξ_i replaces ω_i in all the dynamics equations, which is to be designed using the passivity-short framework [12]. Let us also define the error states as $\tilde{e}_i = e_i - e_i^*$, where $e = \left\{ u_i, z_{ij}, \lambda_{ij}, \mu_i, x_i, y_i \right\}$. With this, the error dynamics are given as

$$\dot{\tilde{\omega}}_i = -\alpha_i \left[\left(\nabla_{y_i} f_i(y_i) - \nabla_{y_i^*} f_i(y_i^*) \right) + \sum_{j \in N_i} d_{ij} \tilde{\lambda}_{ij} + \sum_{j \in N_i} d_{ij} \left(\tilde{\xi}_i - \tilde{z}_{ij}^- \right) \right] \tag{13.13a}$$

$$\dot{\tilde{z}}_{ji} = -\alpha_i \left[-d_{ji} \tilde{\lambda}_{ji} - d_{ji} \left(\tilde{\xi}_j - \tilde{z}_{ji} \right) + A_{ij} \tilde{\mu}_i \right] \tag{13.13b}$$

$$\dot{\tilde{\mu}}_i = w_i \sum_{j \in N_i} A_{ij} \tilde{z}_{ji} \tag{13.13c}$$

$$\dot{\tilde{\lambda}}_{ji} = d_{ji} \left(\tilde{\xi}_j - \tilde{z}_{ji} \right). \tag{13.13d}$$

We present theorem 1 in what follows, which shows that the error state dynamic equation (13.12) obtained converges to the optimal solutions.

Theorem 1: Consider the statically and dynamically constrained optimization problem (13.6). Suppose that U_i is designed such that for constant matrices C_{il}, there is an individual positive definite Lyapunov function $V_i'(x_i, \omega_i)$ satisfying the following inequality:

$$\left[\nabla_{x_i} V_i'(x_i, \omega_i) \right]^T \left[F_i(x_i) + G_i(x_i) U_i + G_i(x_i) \omega_i \right] + \left[\nabla_{\omega_i} V_i(x_i, \omega_i) \right]^T \dot{\omega}_i$$
$$\leq -\sum_l k_i \left\| \omega_i - C_{il} x_i \right\|^2 + \left[\sum_l (\omega_i - C_{il} x_i) \right]^T \dot{\omega}_i + (y_i - v_i) \dot{\omega}_i \tag{13.14}$$

Then, continuous-time ADMM algorithm (13.13) with

$$\xi_i = y_i + \sum_l (\omega_i - C_{il} x_i)$$

is globally convergent to the optimal solution.

The proof of the theorem is provided in the appendix at the end of the chapter. Using the set of dynamic equations (13.12), we will be solving the distribution grid problem with EV penetration in the following section.

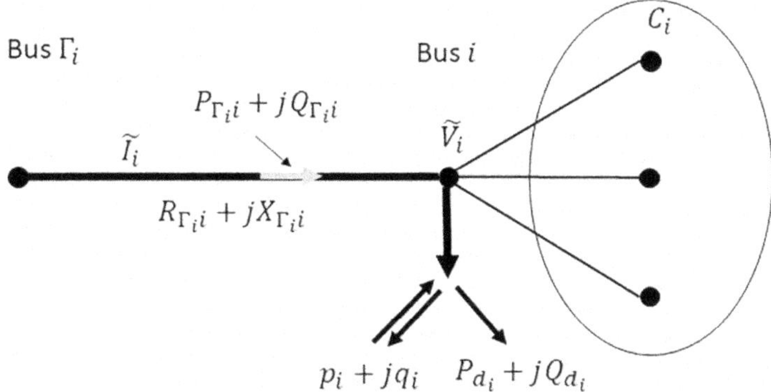

FIGURE 13.3 The multi-layer representation of distribution grid with associated variables. In the topmost layer, we have the DSO, who is in charge of overall operation of the distribution grid. The next layer consists of aggregators which collect EV information and communicate with them. Finally, in the bottom layer, we have the EVs connected to the grid through their charging ports.

13.4 OPTIMIZING EV PARTICIPATION IN GRID ANCILLARY SERVICES

In our proposed approach, we have divided the distribution system into three separate layers. On the top layer is the distribution system operator (DSO), who is in charge of the overall distribution grid. The DSO is responsible for maintaining optimal power flow in the grid, as well as maintaining system parameters like voltage within the desired tolerable range. In the next layer, we assume the existence of a third-party entity, generally termed as the *aggregator* in the literature, which collects EV information at each node and can also send/receive input/output signals [13, 14]. And in the final layer, we have all the consumers with the EVs who agree and sign a contract with the aggregators, allowing them to use the EVs for grid services in exchange for financial compensation. We assume, at any instance of time, that if the EV owners permit through the contract, aggregators can gather information on EVs' active and reactive power injection at each bus through the existing sensor network between them in real time [15, 16]. Figure 13.2 shows the structure of the multi-layer distribution grid and the associated variables which are defined in the next subsection.

13.4.1 BRANCH FLOW MODEL

For the branch flow model, consider a radial distribution network by a directed graph $G = (N, E)$, where $N := \{1, \cdots, N\}$ represents the set of buses and E represents the set of distribution lines connecting the buses in N. Without any loss of generality, the substation of the radial network is indexed by 1. Each node $i \in N$, 1 has a unique parent node Γ_i and a set of children nodes, denoted by C_i, as shown in Figure 13.3. We assume each directed line points toward its children, that is, power flows from parent Γ_i to node i.

For each bus $i \in N$, let N_i be the set of neighboring nodes, including node i and its parent, that is, $N_i = \Gamma_i \cup \{i\} \cup C_i$. Also, let V_i be its voltage, with $P_{d_i} + j Q_{d_i}$ being the load demand and $p_i + j q_i$ the aggregated active and reactive power injection by the EVs. For the branch $\Gamma_i \to i$, let I_i be the current flowing through it, with $R_{\Gamma_i i} + j X_{\Gamma_i i}$ being the impedance of the line and $P_{\Gamma_i i} + j Q_{\Gamma_i i}$ being the complex power flowing from the parent Γ_i to node i. For bus $i \in N$, the power balance equations are given by [17]:

$$P_{\Gamma_i i} = p_i + P_{d_i} + \sum_{j \in C_i} \left(P_{ij} + R_{ij} l_{ij} \right) \quad i \in N \tag{13.15a}$$

$$Q_{\Gamma_i i} = q_i + Q_{d_i} + \sum_{j \in C_i} \left(Q_{ij} + X_{ij} l_{ij} \right) \quad i \in N \tag{13.15b}$$

where $l_{ij} = I_i^2$ is the square of current magnitude, which is defined as

$$l_{ij} = \frac{P_{ij}^2 + Q_{ij}^2}{v_i} \tag{13.16}$$

where $v_i = V_i^2$ is the square of the voltage magnitude. Equation (13.16) is a nonlinear equation which can be linearized around current operating points P_{ij}^o, Q_{ij}^o, v_i^o, and l_{ij}^o, as follows [18]:

$$l_{ij} - l_{ij}^o = \frac{2P_{ij}^o}{v_i^o}\left(P_{ij} - P_{ij}^o\right) + \frac{2Q_{ij}^o}{v_i^o}\left(Q_{ij} - Q_{ij}^o\right) - \frac{\left(P_{ij}^o\right)^2 + \left(Q_{ij}^o\right)^2}{\left(v_i^o\right)^2}\left(v_i - v_i^o\right) \tag{13.17}$$

The power flow on all lines $(i,j) \in E$ is expressed as:

$$v_i - v_j = 2\left(R_{ij}P_{ij} + X_{ij}Q_{ij}\right) + \left(R_{ij}^2 + X_{ij}^2\right)l_{ij} \tag{13.18}$$

13.4.2 EV PARAMETERS

In our structure, EV owners who are interested to sign up for grid services are connected to aggregators at each node. Thus, the aggregated power of all EVs connected to node i at time t is

$$p_i(t) = \sum_{n=1}^{E_i} p_{i,n}^{ev}, \quad \forall i \in N \tag{13.19}$$

where E_i denotes the number of EVs controlled by the aggregator $i \in N$ in charge of bus $i \in N$, and $p_{i,n}^{ev} \in \left\{0, \overline{p}_{i,n}^{ev}, \underline{p}_{i,n}^{ev}\right\}$ is the charging/discharging power of EV n. Let us denote the time of arrival of nth EV as $t_{a,n}$, and departure time as $t_{d,n}$. The EV owner registers this time with the aggregator so that they can be used in the continuous-domain real-time optimization framework. The nth EV only charges/discharges itself between this timeframe. Thus, we can write:

$$p_{i,n}^{ev} = \begin{cases} 0 & t \notin \left[t_{a,n}, t_{d,n}\right] \\ \left[\overline{p}_{i,n}^{ev}, \underline{p}_{i,n}^{ev}\right] & t \in \left[t_{a,n}, t_{d,n}\right] \end{cases} \tag{13.20}$$

The state-of-charge dynamics $s_{i,n}$ of the EV can be represented by the following first-order differential equation:

$$\dot{s}_{i,n} = \frac{\eta_n p_{i,n}^{ev}}{B_{i,n}} \tag{13.21}$$

At the end of the charging period, the EV consumers want their EVs to be charged to a specific state of charge, specifically:

$$s_{i,n}\left(t_{d,n}\right) = s_{i,n}^d, \quad n \in E_i, i \in N \tag{13.22}$$

where $s_{i,n}^d$ is the desired SOC.

13.4.3 Objective Functions for DSO and EV Owners

For the distribution system operator (DSO), the objective would be to minimize the generation cost and maintain the voltage close to unity at every bus over the time horizon, mathematically:

$$f_i(p_i, v_i) = C_i(p_i) + H_i(v_i) \quad i \in N \tag{13.23}$$

where $C_i(p_i(t))$ is the cost function for generation production and $H_i(v_i(t))$ is the voltage regulation penalty function.

As for the EV owners, their aim would be to minimize their EV charging cost and also to maximize their gain by letting aggregators use the EV battery for grid services according to the agreed contract between them. Thus, we can express their objective function as follows:

$$W_{i,n}\left(p_{i,n}^{ev}\right) = \psi_{i,n} p_{i,n}^{ev} - U_i\left(p_{i,n}^{ev}\right) \quad n \in E_i, i \in N \tag{13.24}$$

where $\psi_{i,n}$ is a constant for the nth EV connected to bus i and $U_i\left(p_{i,n}^{ev}\right)$ is the utility function which represents the gain of each EV owner. We also have a terminal condition, where at the end of a charging period, the SOC of the EV should be at the desired SOC, that is:

$$S_{i,n}\left(s_{i,n}^{d}\right) = k_{i,n}\left(s_{i,n} - s_{i,n}^{d}\right)^2 \tag{13.25}$$

where $k_{i,n} > 0$ is the weight on terminal condition.

13.4.4 Problem Formulation and Solution

Let's stack all the state variables into a vector ω_i, that is:

$$\omega_i = \left[v_i, p_i, q_i, P_{\Gamma_i}, Q_{\Gamma_i}, l_{\Gamma_i}, p_{i,n}^{ev}, q_{i,n}^{ev}, s_{i,n}\right]^T \quad \forall n = [1, \cdots, E_i], i \in N.$$

Also, let's introduce an observation vector z_{ji}, which represents the variables of node j at node i, that is, $z_{ji} = \left[v_j^i, p_j^i, q_j^i, P_{\cdot\cdot j}^i, Q_{\cdot\cdot j}^i, l_{\cdot\cdot j}^i, p_{j,n}^{ev,i}, q_{j,n}^{ev,i}, s_{j,n}^{ev,i}\right]^T$. With these definitions, the optimization problem can be formulated according to the developed dynamic ADMM as follows:

$$\min_{\omega_i} \sum_{i=1}^{N} \sum_{n=1}^{E_i} \left[\phi_i(\omega_i)\right] \tag{13.26a–26b}$$

$$\text{s.t.} \quad \sum_{j \in N_i} A_{ij} z_{ji} + m_{ij} = 0 \quad \forall i \in N \tag{13.26c}$$

$$\dot{x}_i = A_i(x_i) + B_i(x_i)u_i \tag{13.26d}$$

$$y_i = z_{ij} \tag{13.26e}$$

$$y_i = \omega_i \tag{13.26f}$$

$$u_i = U_i(x_i) + \omega_i \tag{13.26g}$$

where $\phi_i(\omega_i) = \left[S_{i,n}(s_{i,n}) + f_i(p_i, v_i) + W_{i,n}(p_{i,n}^{ev})\right]$, $x_i = \begin{bmatrix} 0 & 0 & 0 & 0 & 0 & 0 & 0 & 0 & s_i \end{bmatrix}^T$, and $U_i(x_i) = 1$. It should be noted that the objective function is not summed over time, since the problem is solved in real time with continuous-domain dynamics, where the arrival and departure times can be tackled by the individual EV dynamics. Based on what agent j represents, the matrix A_{ij} and vector m_{ij} take the following form:

$$A_{ii} = \begin{bmatrix} 0 & 1 & 0 & -1 & 0 & 0 & 0 & \cdot & 0 & 0 & \cdot & 0 & 0 & \cdot & 0 \\ 0 & 0 & 1 & 0 & -1 & 0 & 0 & \cdot & 0 & 0 & \cdot & 0 & 0 & \cdot & 0 \\ 1 & 0 & 0 & 2R_{\Gamma_i} & 2X_{\Gamma_i} & \left(R_{\Gamma_i}^2 + X_{\Gamma_i}^2\right) & 0 & \cdot & 0 & 0 & \cdot & 0 & 0 & \cdot & 0 \\ 0 & 0 & 0 & \dfrac{2P_{\Gamma_i}^o}{v_{\Gamma_i}^o} & \dfrac{2Q_{\Gamma_i}^o}{v_{\Gamma_i}^o} & -1 & 0 & \cdot & 0 & 0 & \cdot & 0 & 0 & \cdot & 0 \end{bmatrix},$$

$$A_{ij} = \begin{bmatrix} 0 & 0 & 0 & 1 & 0 & R_{ij} & 0 & . & 0 & 0 & . & 0 & 0 & . & 0 \\ 0 & 0 & 0 & 0 & 1 & X_{ij} & 0 & . & 0 & 0 & . & 0 & 0 & . & 0 \\ 0 & 0 & 0 & 0 & 0 & 0 & 0 & . & 0 & 0 & . & 0 & 0 & . & 0 \\ 0 & 0 & 0 & 0 & 0 & 0 & 0 & . & 0 & 0 & . & 0 & 0 & . & 0 \end{bmatrix}, \ j \in C_i,$$

$$A_{ij} = \begin{bmatrix} 0 & 0 & 0 & 0 & 0 & 0 & 0 & \cdot & 0 & 0 & \cdot & 0 & 0 & \cdot & 0 \\ 0 & 0 & 0 & 0 & 0 & 0 & 0 & \cdot & 0 & 0 & \cdot & 0 & 0 & \cdot & 0 \\ -1 & 0 & 0 & 0 & 0 & 0 & 0 & \cdot & 0 & 0 & \cdot & 0 & 0 & \cdot & 0 \\ -\dfrac{\left(P_{ij}^o\right)^2 + \left(Q_{ij}^o\right)^2}{\left(v_{\Gamma_i}^o\right)^2} & 0 & 0 & 0 & 0 & 0 & 0 & \cdot & 0 & 0 & \cdot & 0 & 0 & \cdot & 0 \end{bmatrix}, \ j = \Gamma_i.$$

and

$$m_{ii} = \begin{bmatrix} P_{d_i} & Q_{d_i} & 0 & 0 \end{bmatrix}^T, \ m_{ij} = \{0 \ \ 0 \ \ 0 \ \ 0]^T, j \in C_i\} \text{ and}$$

$$m_{ij} = \left\{ \left[\begin{bmatrix} 0 & 0 & 0 & \left(-\dfrac{2\left(P_{ji}^o\right)^2 v_j^o + 2\left(Q_{ji}^o\right)^2 v_j^o + \left(P_{ji}^o\right)^2 - \left(Q_{ji}^o\right)^2}{\left(v_j^o\right)^2} - l_{ji}^o\right) \end{bmatrix}\right]^T, j \in \Gamma_i \right\}.$$

Following the procedures developed in Section 13.2 and the dynamic equation set (13.12), the following update dynamics for the system were obtained:

$$\dot{v}_i = -\alpha_i \left[\nabla_{v_i} \phi_i(\omega_i) + \sum_{j \in N_i} d_{ij} \left[\lambda_{ij}(1) + \left(v_i - v_i^{j-}\right)\right]\right] \tag{13.27a}$$

$$\dot{p}_i = -\alpha_i \left[\nabla_{p_i} \phi_i(\omega_i) + \sum_{j \in N_i} d_{ij} \left[\lambda_{ij}(2) + \left(p_i - p_i^{j-}\right)\right]\right] \tag{13.27b}$$

$$\dot{q}_i = -\alpha_i \left[\nabla_{q_i} \phi_i(\omega_i) + \sum_{j \in N_i} d_{ij} \left[\lambda_{ij}(3) + \left(q_i - q_i^{j-}\right)\right]\right] \tag{13.27c}$$

$$\dot{P}_{\Gamma_i} = -\alpha_i \left[\nabla_{P_{\Gamma_i}} \phi_i(\omega_i) + \sum_{j\in N_i} d_{ij} \left[\lambda_{ij}(4) + \left(P_{\Gamma_i} - P_{\Gamma_i}^{j-} \right) \right] \right] \tag{13.27d}$$

$$\dot{Q}_{\Gamma_i} = -\alpha_i \left[\nabla_{Q_{\Gamma_i}} \phi_i(\omega_i) + \sum_{j\in N_i} d_{ij} \left[\lambda_{ij}(5) + \left(Q_{\Gamma_i} - Q_{\Gamma_i}^{j-} \right) \right] \right] \tag{13.27e}$$

$$\dot{p}_{i,n}^{ev} = -\alpha_i \left[\nabla_{p_{i,n}^{ev}} \phi_i(\omega_i) + \sum_{j\in N_i} d_{ij} \left[\lambda_{in}(7) + \left(p_{i,n}^{ev} - p_{i,n}^{ev,j-} \right) \right] \right] \tag{13.27g}$$

$$\dot{q}_{i,n}^{ev} = -\alpha_i \left[\nabla_{q_{i,n}^{ev}} \phi_i(\omega_i) + \sum_{j\in N_i} d_{ij} \left[\lambda_{in}(8) + \left(q_{i,n}^{ev} - q_{i,n}^{ev,j-} \right) \right] \right] \tag{13.27h}$$

$$\dot{s}_{i,n} = -\alpha_i \left[\nabla_{s_{i,n}} \phi_i(\omega_i) + \sum_{j\in N_i} d_{ij} \left[\lambda_{in}(9) + \left(s_{i,n} - s_{i,n}^{j-} \right) \right] \right] \tag{13.27i}$$

where $n \in [1,\cdots,E_i]$, $j \in N_i$, and $i \in N$. The z_{ji} is given as:

$$
\begin{bmatrix} \dot{v}_j^i \\ \dot{p}_j^i \\ \dot{q}_j^i \\ \dot{P}_{\Gamma_j}^i \\ \dot{Q}_{\Gamma_j}^i \\ \dot{l}_{\Gamma_j}^i \\ \dot{p}_{j,n}^{ve,n} \\ \dot{q}_{j,n}^{ev,i} \\ \dot{s}_{j,n}^i \end{bmatrix}
= \alpha_i d_{ji}
\begin{bmatrix} \lambda_{ji}(1) \\ \lambda_{ji}(2) \\ \lambda_{ji}(3) \\ \lambda_{ji}(4) \\ \lambda_{ji}(5) \\ \lambda_{ji}(6) \\ \lambda_{jn}(7) \\ \lambda_{jn}(8) \\ \lambda_{jn}(9) \end{bmatrix}
+ d_{ji}
\begin{bmatrix} v_j - v_j^i \\ p_j - p_j^i \\ q_j - q_j^i \\ P_{\Gamma_j} - P_{\Gamma_j}^i \\ Q_{\Gamma_j} - Q_{\Gamma_j}^i \\ l_{\Gamma_j} - l_{\Gamma_j}^i \\ p_{j,n}^{ev} - p_{j,i}^{ev,i} \\ q_{j,n}^{ev} - q_{j,i}^{ev,i} \\ s_{j,n}^{ev} - s_{j,n}^i \end{bmatrix}
- A_{ij}^T
\begin{bmatrix} \mu_i(1) \\ \mu_i(2) \\ \mu_i(3) \\ \mu_i(4) \\ \mu_i(5) \\ \mu_i(6) \\ \mu_i(7) \\ \mu_i(8) \\ \mu_i(9) \end{bmatrix}
$$

The dual updates are given as:

$$
\begin{bmatrix} \dot{\mu}_i(1) \\ \dot{\mu}_i(2) \\ \dot{\mu}_i(3) \\ \dot{\mu}_i(4) \\ \dot{\mu}_i(5) \\ \dot{\mu}_i(6) \\ \dot{\mu}_i(7) \\ \dot{\mu}_i(8) \\ \dot{\mu}_i(9) \end{bmatrix}
= \sum_{j\in N_i} A_{ij}
\begin{bmatrix} v_j^i \\ p_j^i \\ q_j^i \\ P_{\Gamma_j}^i \\ Q_{\Gamma_j}^i \\ l_{\Gamma_j}^i \\ p_{i,n}^{ev,n} \\ q_{i,n}^{ev,j} \\ s_{i,n}^j \end{bmatrix}
$$

$$
\begin{bmatrix}
\dot{\lambda}_{ji}(1) \\
\dot{\lambda}_{ji}(2) \\
\dot{\lambda}_{ji}(3) \\
\dot{\lambda}_{ji}(4) \\
\dot{\lambda}_{ji}(5) \\
\dot{\lambda}_{ji}(6) \\
\dot{\lambda}_{jn}(7) \\
\dot{\lambda}_{jn}(8) \\
\dot{\lambda}_{jn}(9)
\end{bmatrix}
= d_{ji}
\begin{bmatrix}
v_j - v_j^i \\
p_j - p_j^i \\
q_j - q_j^i \\
P_{\Gamma_j j} - P_{\Gamma_j j}^i \\
Q_{\Gamma_j j} - Q_{\Gamma_j}^i \\
l_{\Gamma_j j} - l_{\Gamma_{jj}}^i \\
p_{j,n}^{ev} - p_{j,n}^{ev,i} \\
q_{j,n}^{ev} - q_{j,i}^{ev,i} \\
s_{j,n} - s_{j,n}^i
\end{bmatrix}
$$

13.5 CASE STUDY: VOLTAGE REGULATION AND SOC MANAGEMENT

The dynamic equations developed in the previous chapter were implemented on the IEEE 123 bus distribution system in a co-simulation with MATLAB and OpenDSS. The following tables show the EV type and EV charging parameter used in the simulation.

In the simulation of the proposed algorithm on the distribution system, we defined $C_i\left(p_i\right) = a_i^p p_i^2$, $H_i\left(v_i\right) = \left(1 - v_i\right)^2$, and $U_i\left(p_{i,n}^{ev}\right) = a_{p_n} log\left(p_{i,n}^{ev} + 1\right)$, and the parameters of the coefficients were taken from [20]. The price of electricity was fixed at an average of 15¢/kWh; 150 EVs were designed by utilizing the data from Table 13.1 and Table 13.2. A random number generator was designed which randomly selected 40 buses in the grid to place the EVs. Using the same random generator, among those 40 buses, these 150 EVs were randomly connected to the distribution system. The simulation was run for a 5 hr study horizon. In this horizon, EVs were given random arrival and departure times. It was fixed that by the end of their charging period, the EV owners want their vehicles to be charged to at least 90% of its capacity. The value of step-size ai was fixed at 0.01 for all the nodes.

TABLE 13.1

EV Charging Categories

Charging Method	Voltage	Max Current	Input Power
AC level 1	120 V	12 A	1.4 kW
AC level 2	208–240 V	32 A	7.2–19.2 kW
DC charging	400–1,000 V	300 A	50–150 kW

Source: [19].

TABLE 13.2

EV Types

Brand Name	Battery Capacity
Nissan Leaf	40–62 kWh
Toyota RAV4-EV	41.8 kWh
BMW i3	42.2 kWh
Tesla Model 3	75 kWh
Tesla Model S	100 kWh

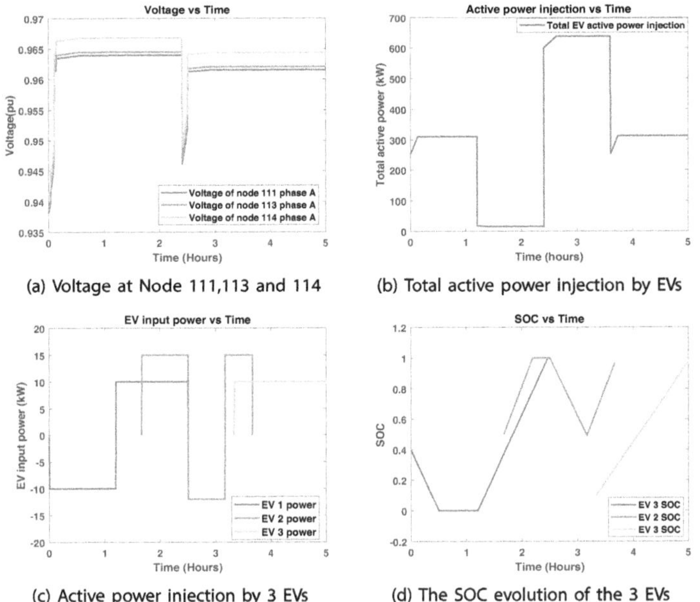

(a) Voltage at Node 111,113 and 114   (b) Total active power injection by EVs

(c) Active power injection by 3 EVs   (d) The SOC evolution of the 3 EVs

FIGURE 13.4   Simulation results demonstrating the impact of the algorithm on IEEE 123-bus system.

Initially, a base case was analyzed to identify buses with the lowest voltage when EVs were connected to the grid. The results highlighted buses 111, 113, and 114 as having the lowest voltages. Following the implementation of the algorithm, the voltage levels at these three buses were monitored throughout the study horizon, as depicted in Figure 13.4a. To evaluate the algorithm's effectiveness, a controlled voltage drop was applied to several nodes in the grid approximately halfway through the study period.

The total active power injection by the EVs in response to these voltage fluctuations is illustrated in Figure 13.4b. The data indicates a notable increase in the total active power injection at the moment the voltage drop occurred, demonstrating the EVs' responsiveness to grid voltage changes. Further analysis focused on three specific EVs—one from each of the selected buses (111, 113, and 114). Figure 13.4c and Figure 13.4d provide insights into these EVs' active power injections and their state-of-charge (SOC) evolution, respectively. The figures show the EVs' arrival and departure times and their responses to the voltage fluctuations. Notably, despite the voltage disturbances, the SOC of these EVs exceeded 90% by the end of their charging periods.

Overall, the results confirm that the implemented algorithm effectively utilizes EVs to mitigate voltage drops in the grid, enhancing grid stability while ensuring that the EVs maintain a high state of charge.

13.6   CONCLUSION

This chapter introduced a novel continuous-domain distributed multi-agent ADMM algorithm designed to handle dynamic constraints in real time. Unlike conventional discrete-time iterative methods, where solution accuracy and optimality are influenced by sampling rates and convergence times, the proposed algorithm operates efficiently in real-time environments. This capability is particularly beneficial for distribution power systems with intermittent energy resources, such as electric vehicles (EVs). The algorithm's convergence is rigorously established using the Lyapunov direct method. The efficacy of the proposed approach is demonstrated through analytical results and practical testing on the IEEE 123 bus test system. The chapter includes illustrative results that showcase the performance of the algorithm.

13.7 APPENDIX

Consider the following properties of the convex objective function $f_i(y_i)$:

The function $f_i(y_i)$ is convex and differentiable. In particular, the gradient of a convex function is a global underestimator, as:

$$f_i(y_i) - f_i(y_i^*) \geq \nabla_{y_i^*} f_i(y_i^*)(\tilde{y}_i). \tag{13.28}$$

and

$$f_i(y_i^*) - f_i(y_i) \geq \nabla_{y_i}^T f_i(y_i)(-\tilde{y}_i) \tag{13.29}$$

Adding the preceding two inequalities yields

$$\left[-\nabla_{y_i}^T f_i(y_i) + \nabla_{y_i^*}^T f_i(y_i^*) \right]^T \tilde{y}_i \leq 0. \tag{13.30}$$

The gradient of $f_i(y_i)$ (denoted by $\nabla_{y_i} f(y_i)$) is Lipschitz, that is, $\left\| \nabla_{y_i} f(y_i) - \nabla_{y_i^*} f(y_i^*) \right\| \leq l_i \left\| y_i - y_i^* \right\|$, where l_i is the Lipschitz constant.

From (13.30), we get $\left[\nabla_{y_i}^T f_i(y_i) - \nabla_{y_i^*}^T f_i(y_i^*) \right]^T \tilde{y}_i$, which is positive definite with respect to $\tilde{y}$.

Lemma 1: If the gradients denoted by $\nabla_{y_i} f_i(y_i)$ are locally Lipschitz, the sum

$$-\sum_l k_i \epsilon_{il}^2 - \alpha_i \left[\sum_l \epsilon_{il} \right]^T \nabla_{y_i} f_i(y_i) - \alpha \tilde{y}_i^T \nabla_{y_i} f_i(y_i) \tag{13.31}$$

is negative definite with respect to ∂_{il} and $\tilde{y}_i$ for small step-size α_i and for all gain $k_i > 0$ above a certain threshold.

Consider the Lyapunov function:

$$V = \frac{1}{2} \sum_{i=1}^N \left\{ 2V_i'(x_i, \omega_i) + \|\tilde{\omega}_i\|^2 + \frac{1}{w_i} \|\tilde{\mu}_i\|^2 + \sum_{j \in \mathcal{N}_i} \left[\frac{1}{\alpha_i} \|\tilde{z}_{ij}\|^2 + \|\tilde{\lambda}_{ij}\|^2 \right] \right.$$
$$\left. + d_{ij} \int_{t-\delta}^t \|\tilde{z}_{ij}(\tau)\|^2 \, d\tau \right\}$$

It follows from (13.14) that

$$\dot{V} \leq \sum_{i=1}^N \left\{ -\sum_l k_i \|\tilde{\omega}_i - C_{il}\tilde{x}_i\|^2 + \left[\sum_l (\tilde{\omega}_i - C_{il}\tilde{x}_i) \right]^T \dot{\tilde{\omega}} + \tilde{y}_i^T \dot{\tilde{\omega}}_i + \tilde{\mu}_i^T \dot{\tilde{\mu}}_i \right.$$
$$\left. + \sum_{j \in \mathcal{N}_i} \left[\tilde{z}_{ij}^T \dot{\tilde{z}}_{ij} + \tilde{\lambda}_{ij}^T \dot{\tilde{\lambda}}_{ij} + \frac{1}{2} d_{ij} \left[\|\tilde{z}_{ij}\|^2 - \|\tilde{z}_{ij}^-\|^2 \right] \right] \right\}$$
$$= \sum_{i=1}^N \left\{ -\sum_l k_i \|\omega_i - C_{il}\tilde{x}_i\|^2 - \alpha_i \left[\sum_l \omega_i - C_{il}\tilde{x}_i \right]^T \nabla_{y_i} f_i(y_i) \right.$$
$$- \alpha \tilde{y}_i^T \nabla_{y_i} f_i(y_i) + \sum_{j \in \mathcal{N}_i} \left[\tilde{\mu}_i A_{ji} \tilde{z}_{ij} + \tilde{\xi}_i^T \left[-d_{ij}\tilde{\lambda}_{ij} - d_{ij} \left(\tilde{\xi}_i - \tilde{z}_{ij}^- \right) \right] \right.$$
$$+ \tilde{z}_{ij}^T \left[d_{ij}\tilde{\lambda}_{ij} + d_{ij} \left(\tilde{\xi}_i - \tilde{z}_{ij} \right) - A_{ji}^T \tilde{\mu}_i \right] + d_{ij} \tilde{\lambda}_{ij}^T \left[\tilde{\xi}_i - \tilde{z}_{ij} \right]$$
$$\left. \left. + \frac{1}{2} d_{ij} \|\tilde{z}_{ij}\|^2 - \frac{1}{2} d_{ij} \|\tilde{z}_{ij}^-\|^2 \right] \right\}$$

$$= \sum_{i=1}^{N} \left[\underbrace{-\sum_l k_i \left\| \omega_i - C_{il}\tilde{x}_i \right\|^2 - \alpha_i \left[\sum_l \omega_i - C_{il}\tilde{x}_i \right]^T \nabla_{y_i} f_i(y_i) - \alpha \tilde{y}_i^T \nabla_{y_i} f_i(y_i)}_{\text{n.d. according to (31)}} \right.$$

$$- \frac{1}{2} d_{ij} \left\| \tilde{\xi}_i - \tilde{z}_{ij} \right\|^2 - \frac{1}{2} d_{ij} \left\| \tilde{\xi}_i - \tilde{z}_{ij}^- \right\|^2$$

which is negative definite with respect to $\left(\tilde{\xi}_i - \tilde{z}_{ij} \right)$ as well as $\left(\tilde{\xi}_i - \tilde{z}_{ij}^- \right)$. This concludes the proof.

REFERENCES

[1] L. Cai, J. Pan, L. Zhao, and X. Shen, "Networked electric vehicles for green intelligent transportation," IEEE Communications Standards Magazine, vol. 1, no. 2, pp. 77–83, 2017.

[2] IEA, *Global EV outlook 2024*, 2024. Available: www.iea.org/reports/global-ev-outlook-2024.

[3] Y. Yu, D. Reihs, S. Wagh, A. Shekhar, D. Stahleder, G. R. C. Mouli, F. Lehfuss, and P. Bauer, "Data-driven study of low voltage distribution grid behaviour with increasing electric vehicle penetration," IEEE Access, vol. 10, pp. 6053–6070, 2022.

[4] M. Curcic, S. Bajic, S. Aubert, and A. Besner, "Optimal EV charging schedule for managing distribution network over-load during CLPU," in *2024 IEEE/PES Transmission and Distribution Conference and Exposition (T&D)*, IEEE, 2024, pp. 1–5.

[5] H. Jahangir, S. S. Gougheri, B. Vatandoust, M. A. Golkar, A. Ahmadian, and A. Hajizadeh, "Plug-in electric vehicle behavior modeling in energy market: A novel deep learning-based approach with clustering technique," IEEE Transactions on Smart Grid, vol. 11, no. 6, pp. 4738–4748, 2020.

[6] A. Asrari, M. Ansari, J. Khazaei, and P. Fajri, "A market framework for decentralized congestion management in smart distribution grids considering collaboration among electric vehicle aggregators," IEEE Transactions on Smart Grid, vol. 11, no. 2, pp. 1147–1158, 2020.

[7] N. Nimalsiri, E. Ratnam, D. Smith, C. Mediwaththe, and S. Halgamuge, "A distributed coordination approach for the charge and discharge of electric vehicles in unbalanced distribution grids," IEEE Transactions on Industrial Informatics, vol. 20, no. 3, pp. 3551–3562, 2024.

[8] S. Boyd, N. Parikh, and E. Chu, *Distributed optimization and statistical learning via the alternating direction method of multipliers*, Now Publishers Inc, 2011.

[9] S. Boyd and L. Vandenberghe, *Convex Optimization*, Cambridge University Press, 2004.

[10] Z. Qu, *Cooperative Control of Dynamical Systems*, Springer, 2009.

[11] T. Rahman, Z. Qu, and T. Namerikawa, "Improving rate of convergence via gain adaptation in multi-agent distributed ADMM framework," IEEE Access, vol. 8, pp. 80480–80489, 2020.

[12] R. Harvey and Z. Qu, "Cooperative control and networked operation of passivity-short systems," in *Control of Complex Systems*, Elsevier, 2016, pp. 499–518.

[13] X. Shi, Y. Xu, G. Chen, and Y. Guo, "An augmented Lagrangian-based safe reinforcement learning algorithm for carbon-oriented optimal scheduling of EV aggregators," IEEE Transactions on Smart Grid, vol. 15, no. 1, pp. 795–809, 2024.

[14] W. Wu, J. Zhu, Y. Liu, T. Luo, Z. Chen, and H. Dong, "A coordinated model for multiple electric vehicle aggregators to grid considering imbalanced liability trading," IEEE Transactions on Smart Grid, vol. 15, no. 2, pp. 1876–1890, 2024.

[15] L. Wang, J. Kwon, N. Schulz, and Z. Zhou, "Evaluation of aggregated EV flexibility with TSO-DSO coordination," IEEE Transactions on Sustainable Energy, vol. 13, no. 4, pp. 2304–2315, 2022.

[16] C. Qi, C.-C. Liu, X. Lu, L. Yu, and M. W. Degner, "Transactive energy for EV owners and aggregators: Mechanism and algorithms," IEEE Transactions on Sustainable Energy, vol. 14, no. 3, pp. 1849–1865, 2023.

[17] T. Rahman, Y. Xu, and Z. Qu, "Continuous-domain real-time distributed ADMM algorithm for aggregator scheduling and voltage stability in distribution network," IEEE Transactions on Automation Science and Engineering, vol. 19, no. 1, pp. 60–69, 2022.

[18] C. Ai, G. Zhou, W. Gao, J. Guo, G. Jie, Z. Han, and X. Kong, "Research on quasi-synchronous grid-connected control of hydraulic wind turbine," IEEE Access, vol. 8, pp. 126092–126108, 2020.

[19] B. Powell and C. Johnson, "Impact of electric vehicle charging station reliability, resilience, and location on electric vehicle adoption," 2024.

[20] Y. Okawa, T. Namerikawa, and Z. Qu, "Passivity-based stability analysis of dynamic electricity pricing with power flow," in *2017 IEEE 56th Annual Conference on Decision and Control (CDC)*, IEEE, 2017, pp. 813–818.

14 Power-to-Gas Integration
in Energy Systems

Hamed Barfan and Mehrdad Setayesh Nazar

14.1 INTRODUCTION

In recent years, there has been a rapid increase in the integrated planning and operation of gas and electricity systems, treating them as multi-carrier energy networks with gas and electricity sources [1, 2]. The optimal operational planning of electrical and gas systems can highly increase the efficiency of these systems based on the fact that the gas and electrical systems are highly coupled [3]. Statistical reports have shown that approximately 38% of natural gas consumption in the USA in 2020 was attributed to power plants [4]. This surge can be attributed to the high penetration of gas-fired power plants, multi-energy storage systems, and flexible technologies, such as P2G systems [5, 6]. Additionally, there has been a growing penetration of renewable energy resources in energy markets. However, integrating intermittent energy sources into power systems has presented numerous challenges [7]. P2G facilities play a crucial role in addressing the intermittency challenge inherent in renewable energy sources. By converting surplus renewable electricity into hydrogen or methane through electrolysis or methanation processes, respectively, P2G facilitates the transformation of excess energy into storable forms. These storable gases can then be readily injected into existing gas grids, providing a flexible and scalable approach to energy storage. The P2G technology has undergone significant development in recent years. While the technical challenges have largely been addressed, the primary hurdles facing P2G are now economic [8].

In recent years, significant studies have focused on P2G technology and its impact on energy networks. Each of these studies examines various aspects of P2G units. Ref. [9] proposes a model that utilizes P2G technology, natural gas storage, photovoltaic (PV) systems, and peer-to-peer (P2P) market strategies to reduce operational costs. In ref. [10], CO_2-based electrothermal storage devices and wind turbines are employed to enhance the efficiency of P2G units in multi-carrier energy systems, thereby reducing operation costs. Ref. [11] introduces a novel method for the optimal operation of multi-carrier energy systems, incorporating energy storage systems, wind turbines, PV panels, and P2G facilities to supply gas, thermal, and electrical loads, with the aim of reducing pollution and increasing social welfare. The study demonstrates that energy storage devices enhance the efficiency of P2G technology. Ref. [12] proposes a stochastic MILP model for the optimal operation of an energy hub, incorporating energy storage systems and P2G facilities, with a demand response (DR) program applied for peak shaving and cost reduction. Results indicate that integrating P2G technology with DR programs further reduces operational costs and improves P2G efficiency. Ref. [13] investigates the integrated operation of power and gas networks in the presence of P2G facilities and gas storage, using a simulation model for day-ahead unit commitment. Ref. [14] presents an optimal model for an integrated power and gas system with P2G and hydrogen energy storage, highlighting the benefits of hydrogen energy storage in cost reduction, peak load modification, and system flexibility. Ref. [15] offers a planning and operation model for a multi-energy system featuring hydrogen and thermal storage, scheduling for both day-ahead and real-time stages. Ref. [16] proposes a stochastic day-ahead unit commitment model incorporating natural gas networks, power systems, various power generators, P2G, gas storage, plug-in electric vehicles (PEVs), and renewable energies while addressing power and gas network system limits. Ref. [17] introduces

DOI: 10.1201/9781032719436-14

a multi-objective dynamic framework for designing an energy hub incorporating various energy storages and applications, with P2G facilities and demand response programs enhancing system flexibility and reducing green gas emissions by 10%.

The review of previous studies underscores the pivotal role of P2G technology in enabling sector coupling and facilitating the seamless integration of renewable energy across diverse sectors, such as power generation, industrial processes, and heating. Hydrogen generated through P2G processes can serve as a clean fuel for transportation or as a feedstock for industrial applications, thereby mitigating greenhouse gas emissions and bolstering energy efficiency. The P2G technology is based on the principle of converting surplus renewable electricity into storable gases, such as hydrogen or methane, via electrolysis or methanation processes, respectively. These gases can be injected into existing gas grids, stored for future use, or distributed across multiple energy networks, including gas, power, and heating systems. The inherent flexibility and scalability of P2G technologies render them invaluable assets in addressing the intermittency challenges associated with renewable energy sources. As a result, P2G plays a crucial role in enhancing grid stability and fortifying the resilience of energy systems.

Recent research on P2G technology highlights its numerous advantages for energy networks, including enhanced system security, flexibility, and adaptability. P2G also offers lower operational costs and facilitates load balancing for system operators. Moreover, P2G facilities are environmentally friendly and well-suited for long-term storage, surpassing other energy storage options, like batteries. These combined advantages position P2G technology as a versatile and promising solution for improving energy storage, grid flexibility, and sustainability in the transition toward a low-carbon future. According to current papers, the advantages of P2G facilities for energy storage compared to other techniques include the following:

1. Firstly, the sole by-product of hydrogen combustion in the P2G process is water. The technology for producing methane and other carbonaceous fuels via hydrogen conversion is highly advanced. This process is carbon-neutral, as the captured carbon dioxide is completely utilized during conversion. Consequently, hydrogen-based fuels are expected to gain increased prominence in the near future. On the other hand, batteries create polluted waste once their lifespan ends. Therefore, overall, it can be said that P2G systems are less polluting compared to other storage solutions.
2. P2G systems offer higher storage capacity, greater energy conversion efficiency, and operational flexibility. Consequently, P2G systems can be effectively integrated into the power grid to complement wind, photovoltaic, and hydropower generation.
3. P2G facilities offer distinct advantages for long-term energy storage solutions. Their ability to convert surplus renewable energy into storable gases, such as hydrogen or methane, positions them as key players in addressing the intermittency of renewable sources. Moreover, the scalability and adaptability of P2G systems make them well suited for meeting future energy demands, particularly during periods of high demand or when renewable energy generation is low. As a result, P2G facilities emerge as promising assets for ensuring grid stability and reliability over the long term.

P2G technology presents a promising solution for efficiently storing and utilizing renewable energy. However, its widespread implementation encounters challenges across technical, regulatory, and economic domains. Technically, optimizing the efficiency and scalability of P2G systems remains a key hurdle. While these systems can convert surplus renewable electricity into hydrogen or methane, enhancing conversion efficiency and developing cost-effective, durable electrolyzers are ongoing priorities. Regulatory barriers also impede P2G deployment. Existing frameworks often lack clarity on integrating P2G into energy infrastructure and grid operations. Resolving issues related to grid connection, system certification, and market participation is crucial for P2G's regulatory acceptance. Moreover, economic viability poses a significant challenge. Despite offering long-term storage and grid-balancing benefits, high up-front costs and revenue uncertainties deter investments. Clarifying revenue models and providing policy support are essential to attract investment in P2G projects. Despite challenges, continued R&D efforts and collaborative policymaking can surmount barriers to P2G adoption. By addressing technical, regulatory, and economic concerns, P2G holds immense potential in advancing a sustainable energy transition.

The main contributions of this chapter can be summarized as follows:

- Presenting a comprehensive mathematical model for electricity and gas networks and incorporating the technical constraints of both systems. The AC power flow method is employed to enhance the accuracy of the results.
- Introducing an optimal day-ahead scheduling methodology for an integrated energy network aimed at minimizing total operating costs. This includes an examination of the impacts of P2G technology and hydrogen energy storage on overall operational expenses.
- Considering various power generators and facilities to illustrate the effects of coordinated operation on generator outputs and wind power utilization rates.
- Proposing a stochastic programming approach to model the problem uncertainties. Further, the local heat networks are considered to justify the utilization of CHP units.

The organization of this chapter is as follows: the mathematical modeling and formulation of the proposed model are given in Section 14.2. The results and analysis with different case studies are presented in Section 14.3. Finally, Section 14.4 discusses the book chapter's conclusions.

14.2 PROBLEM MODELING AND FORMULATION

This section outlines the formulation of the proposed model, encompassing the primary objective function, constraints of the gas network and power system, as well as operational limitations of units.

14.2.1 OBJECTIVE FUNCTION

The core aim of the model is cost minimization, succinctly expressed in equation (14.1). This equation encapsulates various cost components essential for system operation. In the first term, the formulation addresses the operational costs of non-gas-fueled units, encompassing generation costs as well as start-up and shutdown expenditures. The second term presents the cost of gas production. Notably, this includes the generation costs of gas-fueled units, as they rely on gas for power generation. Finally, the third term reflects the charge and discharge expenses associated with hydrogen energy storage units, while the fourth term accounts for the operational costs linked to natural gas storage.

$$MIN \sum_s PROB_s \left(\sum_t \left[\sum_i \left(C_{i,s,t}^{NGU} + SU_{i,s,t} + SD_{i,s,t} + \sum_{gsp} C_{gsp}^{gas} G_{x,s,t} \right) + \sum_{hs} \left(C_{h,i,t}^{P2H} - C_{h,i,t}^{h2p} \right) + \sum_{gs} C_{gs} Q_{gs,t,s}^{out} \right] \right) \tag{14.1}$$

The specified objective function highlights the economic facets of the proposed model, recognizing economic justification as the foremost consideration in every operational and planning scenario. This aspect has also been taken into account in the following discussed case studies.

14.2.2 Unit Commitment

In this subsection, the formulation of power generation is delineated. Utilizing an AC power flow model, both reactive and active power constraints are integrated. Equations (14.2) and (14.3) impose limitations on the output power of the generators, while equation (14.4) governs the power output of the wind turbine.

$$Pg_i^{min} onoff_{i,s,t} \leq Pg_{i,s,t} \leq Pg_i^{max} onoff_{i,s,t} \quad \forall i, \forall s, \forall t \tag{14.2}$$

$$Qg_i^{min} onoff_{i,s,t} \leq Qg_{i,s,t} \leq Qg_i^{max} onoff_{i,s,t} \quad \forall i, \forall s, \forall t \tag{14.3}$$

$$0 \leq Pw_{i,s,t} \leq Pw_i^{max} \quad \forall i, \forall s, \forall t \tag{14.4}$$

The operation of CHP plants entails additional complexity due to the interdependence of power and heat production. This model employs a convex model for CHP, as outlined in Ref. [18]. As depicted in Figure 14.1, CHP plants operate within a constrained area known as the feasible operation region (FOR). Equations (14.5)–(14.8) establish the relationship between the electrical and thermal outputs of the CHP units.

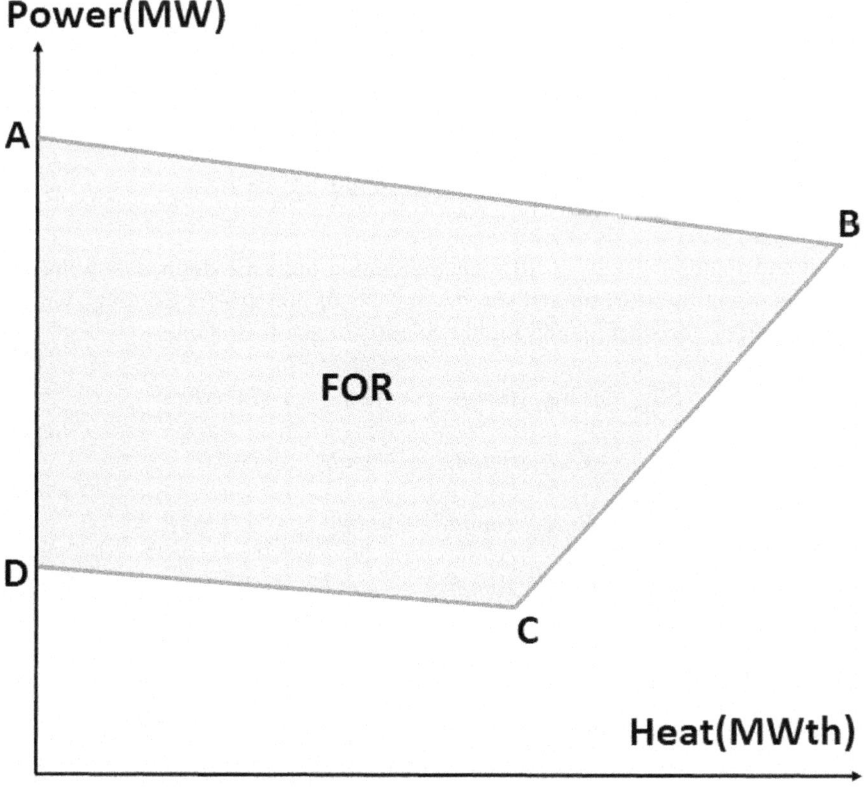

FIGURE 14.1 Feasible operation region of the CHP unit.

$$Pg_{i,s,t} - Pg_i^A - \frac{Pg_i^A - Pg_i^B}{H_i^A - H_i^B} \times (H_{i,S,t} - H_i^A) \leq 0 \quad i \in CH \tag{14.5}$$

$$Pg_{i,s,t} - Pg_i^B - \frac{Pg_i^B - Pg_i^C}{H_i^B - H_i^C} \times (H_{i,S,t} - H_i^B) \geq -(1 - onoff_{i,s,t}) \times M \quad i \in CH \tag{14.6}$$

$$Pg_{i,s,t} - Pg_i^C - \frac{Pg_i^C - Pg_i^D}{H_i^C - H_i^D} \times (H_{i,S,t} - H_i^C) \geq -(1 - onoff_{i,s,t}) \times M \quad i \in CH \tag{14.7}$$

$$0 \leq H_{i,s,t} \leq H_i^A \times onoff_{i,s,t} \quad i \in CH \tag{14.8}$$

Further constraints of generation units are modeled as follows: equations (14.9)–(14.10) specify the maximum ramp-up/ramp-down rates of generation units, respectively, indicating that unit output changes cannot exceed these rates. Additionally, equation (14.11) and equation (14.12) outline the minimum up/downtimes of the power plants, delineating the minimum operating times of the unit after activation and the minimum duration the unit must remain offline before it can be reactivated.

$$Pg_{i,s,t} - Pg_{i,s,t-1} \leq [1 - onoff_{i,s,t}(1 - onoff_{i,s,t-1})]RU_i + onoff_{i,s,t}(1 - onoff_{i,s,t-1})Pg_i^{\min} \tag{14.9}$$

$$Pg_{i,s,t-1} - Pg_{i,s,t} \leq [1 - onoff_{i,s,t-1}(1 - onoff_{i,s,t})]RD_i + onoff_{i,s,t-1}(1 - onoff_{i,s,t})Pg_i^{\min} \tag{14.10}$$

$$(Y_{i,s,t-1}^{on} - UT_i)(onoff_{i,s,t-1} - onoff_{i,s,t}) \geq 0 \tag{14.11}$$

$$(Y_{i,s,t-1}^{off} - DT_i)(onoff_{i,s,t} - onoff_{i,s,t-1}) \geq 0 \tag{14.12}$$

The start-up and shutdown costs of different generation units are defined as follows: equations (14.13)–(14.14) represent the start-up and shutdown costs of non-gas-fueled units, while equations (14.15)–(14.16) illustrate the start-up and shutdown costs of gas-fueled units.

$$SU_{i,s,t} \geq Csu_{i,s,t}(onoff_{i,s,t} - onoff_{i,s,t-1}) \quad i \in GF \tag{14.13}$$

$$SD_{i,s,t} \geq Csd_{i,s,t}(onoff_{i,s,t-1} - onoff_{i,s,t}) \quad i \in GF \tag{14.14}$$

$$SUG_{i,s,t} \geq Csug_{i,s,t}(onoff_{i,s,t} - onoff_{i,s,t-1}) \quad i \in NGF \tag{14.15}$$

$$SDG_{i,s,t} \geq Csdg_{i,s,t}(onoff_{i,s,t-1} - onoff_{i,s,t}) \quad i \in NGF \tag{14.16}$$

14.2.3 THERMAL ENERGY STORAGE MODELING

To accommodate heat storage, a heat buffer tank is incorporated into the model. The quantity of stored heat in the buffer tank is depicted in equation (14.17), which accounts for heat charge/discharge across various periods, heat losses during these processes, and the charge/discharge efficiency of the heat storage. The minimum and maximum capacity of thermal storage is constrained by equation (14.18). Equations (14.19)–(14.20) define the charge/discharge constraints for the heat buffer.

$$H_{hs,s,t} = (1 - \beta_{hs})H_{hs,s,t-1} + \beta_{hs}^{ch}H_{hs,s,t}^{ch} - \frac{H_{hs,s,t}^{dc}}{\beta_{hs}^{dc}} - H_{loss}SU_{hs,s,t} + H_{gn}SD_{hs,s,t} \tag{14.17}$$

$$H_{hs}^{\min} \leq H_{hs,s,t} \leq H_{hs}^{\max} \tag{14.18}$$

$$H_{hs,s,t} - H_{hs,s,t-1} \leq H_{hs}^{\max ch} \tag{14.19}$$

$$H_{hs,s,t-1} - H_{hs,s,t} \leq H_{hs}^{\max dc} \tag{14.20}$$

14.2.4 P2G and Hydrogen Energy Storage Modeling

As mentioned previously, the hydrogen produced by the P2G units can be utilized in gas-fired power plants and industrial applications [19]. It is assumed that the excess hydrogen can be stored in hydrogen energy storage devices and supplied to the network as needed. The operational formulation of hydrogen energy storage units is detailed in equations (14.21)–(14.26). Equation (14.21) represents the level of hydrogen energy storage, while equation (14.22) constrains the minimum and maximum levels of hydrogen in these units. Equation (14.23) models the initial and final levels of stored hydrogen. Furthermore, equations (14.24)–(14.25) restrict the amount of charge/discharge for hydrogen energy storage, while equation (14.26) prohibits simultaneous charging and discharging.

$$HES_{h,s,t} = HES_{h,s,t-1} + \lambda_h^{ch}PHES_{h,s,t}^{ch} - \frac{PHES_{h,s,t}^{dc}}{\lambda_h^{dc}} \tag{14.21}$$

$$HES_h^{\min} \leq HES_{h,s,t} \leq HES_h^{\max} \tag{14.22}$$

$$HES_{h,s,t-0} = HES_{h,s,t} = 24 = HES_{h,s,in} \tag{14.23}$$

$$PHES_h^{\min ch}onoff_{h,s,t}^{ch} \leq PHES_{h,s,t}^{ch} \leq PHES_h^{\max ch}onoff_{h,s,t}^{ch} \tag{14.24}$$

$$PHES_h^{\min dc}onoff_{h,s,t}^{dc} \leq PHES_{h,s,t}^{dc} \leq PHES_h^{\max dc}onoff_{h,s,t}^{dc} \tag{14.25}$$

$$onoff_{h,s,t}^{dc} + onoff_{h,s,t}^{dc} \leq 1 \tag{14.26}$$

14.2.5 Demand Response Program Modeling

The demand response program serves as a flexible tool for the system operator, aiding in cost reduction and improving network operation. To define the demand response program in this study, equations (14.27)–(14.30) are presented [20]. Equation (14.27) imposes a limit on the maximum power that can be shifted to electrical demand within a single hour, while equation (14.28) constrains the maximum decremental amount of load shift. Equation (14.29) ensures that the amount of load decreases and increases becomes zero within the operation interval, meaning, that shifted load should be supplied at other times. Finally, equation (14.30) illustrates the electrical demand after the demand response program is implemented.

$$0 \leq DRP_{l,s,t}^{up} \leq ZDRP.Pl_{l,s,t} \tag{14.27}$$

$$0 \le DRP^{dn}_{l,s,t} \le ZDRP.Pl_{l,s,t} \tag{14.28}$$

$$\sum_t DRP^{up}_{l,s,t} = \sum_t DRP^{dn}_{l,s,t} \tag{14.29}$$

$$Pl^{DR}_{l,s,t} = Pl_{l,s,t} - DRP^{dn}_{l,s,t} + DRP^{up}_{l,s,t} \tag{14.30}$$

14.2.6 POWER SYSTEM MODELING

In this section, the following equations pertain to electrical system constraints, with the application of AC power flow in the model. Equations (14.31)–(14.32) describe the AC power flow of active and reactive power in the electrical network. The capacity limitations of lines are addressed in equation (14.33), while equation (14.34) specifies allowable voltage fluctuations, and equation (14.35) sets constraints on the maximum and minimum levels of voltage angle. Finally, equations (14.36)–(14.37) consider the balance between the production and consumption of active and reactive power [21]. Notably, the P2G unit is treated as a load in the power network, as it utilizes electrical energy to produce hydrogen.

$$P\ln_{i,j,s,t} = \frac{V^2_{i,s,t}}{Z_{i,j}}\cos(\theta_{i,j}) - \frac{V_{i,s,t}V_{j,s,t}}{Z_{i,j}}\cos(\sigma_{i,t} - \sigma_{j,t} + \theta_{i,t}) \tag{14.31}$$

$$Q\ln_{i,j,s,t} = \frac{V^2_{i,s,t}}{Z_{i,j}}\sin(\theta_{i,j}) - \frac{V_{i,s,t}V_{j,s,t}}{Z_{i,j}}\sin(\sigma_{i,t} - \sigma_{j,t} + \theta_{i,t}) - \frac{bV^2_{i,j}}{2} \tag{14.32}$$

$$P\ln^2_{i,j,s,t} + Q\ln^2_{i,j,s,t} \le S^2_{i,j} \tag{14.33}$$

$$V^{min}_i \le V_{i,s,t} \le V^{max}_i \tag{14.34}$$

$$\sigma^{min}_i \le \sigma_{i,s,t} \le \sigma^{max}_i \tag{14.35}$$

$$Pg_{i,s,t} + Pw_{i,s,t} - Pp2g_{i,s,t} - Pl^{DR}_{i,s,t} = \sum_i^{j \in cx} P\ln_{i,j,s,t} \tag{14.36}$$

$$Qg_{i,s,t} + Qw_{i,s,t} - Ql_{i,s,t} = \sum_i^{j \in cx} Q\ln_{i,j,s,t} \tag{14.37}$$

14.2.7 GAS NETWORK MODELING

This section outlines the gas network constraints utilized in the study. The Panhandle A equation is employed to describe the behavior of the gas network [22]. Equation (14.38) restricts the amount of gas injection, while equation (14.39) ensures the balance of gas flow at each node in the gas network. Equation (14.40) provides the equation for gas flow through the pipelines. Natural gas compressors are employed in the gas network to increase pressure between two gas nodes. The compressor utilized in this study is of the gas-fueled type, and its gas consumption is incorporated into the gas balance equation. The power consumption of the compressor is detailed in equation (14.41). Equations (14.42)–(14.44) outline the operational constraints of the compressor, including pressure, flow capacity, and the maximum available power of the gas compressor. The limitations of pipelines and gas node pressures are defined by equations (14.45)–(14.46).

$$G_x^{\text{sup min}} \leq G_{x,s,t}^{\text{sup}} \leq G_x^{\text{sup max}} \tag{14.38}$$

$$G_{x,s,t}^{\text{sup}} - G_{p,s,t}^{line} - G_{c,s,t}^{cp} + PHES_{h,s,t}^{dc} + GP2G_{x,s,t}^{inj} + Gloss_{n,s,t}$$
$$Q_{gs,t,s}^{dc} - Q_{gs,t,s}^{ch} = Gl_{n,s,t}^{gas} \tag{14.39}$$

$$(\rho_{p,s,t}^{out})^2 - (\rho_{p,s,t}^{in})^2 = \frac{18.43 \times L_p}{(\lambda_{lp})^2 . DI_P^{4.854}} . (G_{p,s,t}^{line})^{1.854} \tag{14.40}$$

$$P_{c,s,t}^{cp} = \frac{\beta_{cp} . G_{c,t,s}^{cp}}{\lambda_{cp}} \left[\left(\frac{\rho_{c,s,t}^{out}}{\rho_{c,s,t}^{in}} \right)^{\frac{1}{\beta_{cp}}} - 1 \right] \tag{14.41}$$

$$\frac{\rho_{c,s,t}^{out}}{\rho_{c,s,t}^{in}} \leq \rho r^{\max} \tag{14.42}$$

$$G_{c,s,t}^{cp} \leq G_c^{cp\,\max} \tag{14.43}$$

$$P_{c,s,t}^{cp} \leq P_c^{cp\,\max} \tag{14.44}$$

$$\rho_x^{\min} \leq \rho_{x,s,t} \leq \rho_x^{\max} \tag{14.45}$$

$$G_p^{line\,\min} \leq G_{p,s,t}^{line} \leq G_p^{line\,\max} \tag{14.46}$$

Furthermore, *line pack* refers to natural gas stored within the pipes of a gas transmission or distribution system. Gas system operators utilize line packs to balance the system or meet customer demand, especially when the supply delivered to the system on a given day does not match consumption. System operators continuously manage the amount of gas in their pipes to ensure customer demands can be met without exceeding safe pressure levels [23]. Equation (14.47) represents the equation for line pack through the pipes under normal conditions. However, to account for the dynamic conditions arising from various gas flow inlet and outlet points in the pipelines, equation (14.48) should be incorporated into the model.

$$LP_P = \frac{\rho_p^{av} . Gv_P}{rho^{normal} . Z.R.T^{normal}} \tag{14.47}$$

$$LP_{P,s,t} = LP_{P,s,t}^0 + \sum_t \left(G_{p,s,t}^{line\,in} - G_{p,s,t}^{line\,out} \right) \tag{14.48}$$

14.2.8 COUPLING CONSTRAINTS

Given that gas-fueled power plants consume gas for operation, they are appropriately treated as loads in the gas balance equation. Additionally, P2G facilities serve as both a load in the power grid and a gas supplier in the gas network. Equations (14.49)–(14.52) outline the consumption of natural gas for CHP units and other gas-fueled units, illustrating their connection to the gas network.

$$C_{i,s,t}^{CHP} = c_i + b_i Pg_{i,s,t} + a_i (Pg_{i,s,t})^2 + d_i H_{i,s,t} + e_i (H_{i,s,t})^2$$
$$+ f_i H_{i,s,t} Pg_{i,s,t} + SU_{s,i,t} + SD_{s,i,t} \quad i \in CHP \tag{14.49}$$

$$C_{i,s,t}^{GF} = c_i + b_i Pg_{i,s,t} + a_i (Pg_{i,s,t})^2 + SU_{s,i,t} + SD_{s,i,t} \quad i \in GF \tag{14.50}$$

$$Gl_{n,s,t} = C_{i,s,t}^{CHP} \quad \forall n = i, ..., CHP \tag{14.51}$$

$$Gl_{n,s,t} = C_{i,s,t}^{GF} \quad \forall n = i, ..., GF \tag{14.52}$$

The P2G facility produces hydrogen by consuming electrical power. The generated hydrogen can either be stored in hydrogen energy storage or injected into the gas network. Equation (14.53) defines the connectivity between the gas network and power system for modeling P2G units. Finally, equation (14.54) demonstrates the balance between the produced hydrogen, stored amount, and injected hydrogen.

$$Pp2g_{i,s,t} = \lambda_{i,x}^{P2G} \phi^{P2G} Gp2g_{x,s,t} \quad \forall x = i, ..., P2G \tag{14.53}$$

$$Gp2g_{x,s,t} = PHES_{h,s,t}^{ch} + Gp2g_{x,s,t}^{inj} \tag{14.54}$$

14.3   NUMERICAL SIMULATIONS

14.3.1   INPUT DATA

To evaluate the proposed method, an integrated energy network comprising power, gas, and thermal energy carriers has been incorporated into the model. The model, depicted in Figure 14.2, encompasses a power network with six buses, a six-node gas system, and a local heat network. Residential gas loads are situated at nodes 1, 2, and 3, while electrical loads are connected to buses 3, 4, and 5. The model includes a CHP unit and a gas-fueled generator, both connected to the gas and electrical networks as they consume gas for operation. Additionally, non-gas-fueled units, a wind turbine, and a P2G facility are integrated into the network. Storage facilities such as gas storage, thermal storage, and hydrogen energy storage are utilized. The different unit parameters are outlined in Table 14.1, while the hourly data of thermal, gas, and electrical loads are summarized in Table 14.2. Furthermore, a PEV charging station is added to bus 5. Various case studies have been analyzed and compared to evaluate the proposed framework.

- Case 1: In this study, the day-ahead unit commitment of the integrated network is explored without accounting for flexible facilities, such as P2G units, demand response programs, and storage units.
- Case 2: A P2G facility is added to case 1, although energy storage units are not investigated in this case.
- Case 3: Energy storage units are added to case 2, and a demand response program has been considered to maximize the network efficiency.
- Case 4: The utilization of wind energy in different wind energy penetration levels is investigated in this case study.
- Case 5: Stochastic optimization is conducted to address uncertain parameters, such as wind power, electrical demand, and PEVs' load.

TABLE 14.1
Characteristics of the Units

Parameter	CHP	Gas-Fueled
a (m³/MW²h)	0.48705	0.0708
b (m³/MWh)	203.88096	250.60368
c (m³/h)	1563.228944	1945.5057
d (m³/MWthh)	0.42475	–
e (m³/MWth²h)	59.46528	–
f (m³/MWMWth)	0.87782	–
$Pg_i^{\min}$ (MW)	80	10
$Pg_i^{\max}$ (MW)	205	60
$Qg_i^{\min}$ (MW)	−50	−10
$Qg_i^{\max}$ (MW)	192	50
$H_i^{\min}$ (MWth)	0	–
$H_i^{\max}$ (MWth)	160	–
DT_i (h)	1	1
UT_i (h)	1	1
RU_i (MW/h)	55	20
RD_i (MW/h)	60	20

Parameter	Non-gas-fueled	Parameter	Hydrogen storage
a ($/MW²h)	0.001	$HES_h^{\min}$ (MW)	10
b ($/MWh)	32.65	$HES_h^{\max}$ (MW)	210
c ($/h)	129.97	λ_h^{ch}, λ_h^{dc}	0.7, 0.8
$Pg_i^{\min}$ (MW)	10	$PHES_h^{\max dc}$ (MW)	30
$Pg_i^{\max}$ (MW)	100	$PHES_h^{\max ch}$ (MW)	10
$Qg_i^{\min}$ (MW)	−20		**Gas storage**
$Qg_i^{\max}$ (MW)	100	$Q_{gs}^{dc\max}, Q_{gs}^{ch\max}$ (kcm)	22.65, 22.65
DT_i (h)	3	$\beta_{gs}^{ch}, \beta_{gs}^{dc}$	0.8, 0.8
UT_i (h)	2	$EQ_{gs}^{\min}, EQ_{gs}^{\max}$ (kcm)	0, 85
RU_i (MW/h)	50		**P2G**
RD_i (MW/h)	50	ϕ^{P2G} (MW/KCM)	3.66

(Continued)

TABLE 14.1 (Continued)

Parameter	CHP		Gas-Fueled	
	Heat storage		λ_i^{P2G}	0.65
$H_{hs}^{\min}$ (MWth)	0		$Pp2g_i^{\max}$ (MW)	50
$H_{hs}^{\max}$ (MWth)	75			Gas supplier
$H_{hs}^{\max ch}$ (MWth)	20		$G_1^{\sup\min}, G_1^{\sup\max}$ (kcm)	4.2, 155
$H_{hs}^{\max dc}$ (MWth)	20		$G_2^{\sup\min}, G_2^{\sup\max}$ (kcm)	2.8, 185
$\beta_{hs}^{ch}, \beta_{hs}^{dc}$	0.9			
β_{hs}	0.95			

TABLE 14.2

Expected Value of Input Data

Time	Heat Load	Electrical Load	Gas Load	Wind Power	PEV
1	150.478	177.293	54,769.62	128.363	13
2	150.472	167.131	59,268.72	127.136	2.95
3	150.370	160.574	55,873.14	130	0.374
4	155.259	156.586	56,765.76	124.272	0.261
5	157.649	156.920	49,076.4	126.787	1.157
6	156.454	162.405	67,410	117.318	1.889
7	154.064	175.470	66,741.21	107.090	1.278
8	149.283	192.684	72,752.79	93.181	3.750
9	144.283	208.026	73,680	72.727	4.501
10	124.94	219.806	76,291	55.136	5.121
11	119.402	231.353	78,894.63	38.29	8.589
12	111.036	238.933	83,361.1	19.154	11.235
13	100.279	245.086	86,709.69	32.217	15.350
14	100.152	246.523	95,422.73	38.217	18.112
15	95.06	251.846	110,311.9	30.127	24.257
16	107.012	258.859	115,499.3	45.509	32.832
17	121.793	259.072	135,681.7	46.545	42.268
18	127.796	249.700	108,681.7	38.394	39.812
19	132.55	248.921	102,759.6	36.318	30.500
20	139.721	240.198	81,799.2	46.454	41.842
21	142.112	240.157	80,117.1	29.363	40.910
22	144.502	229.865	78,436.5	26.909	42.005
23	145.697	203.462	72,894	22.821	39.034
24	148.088	199.111	66,907.23	29.272	33.768

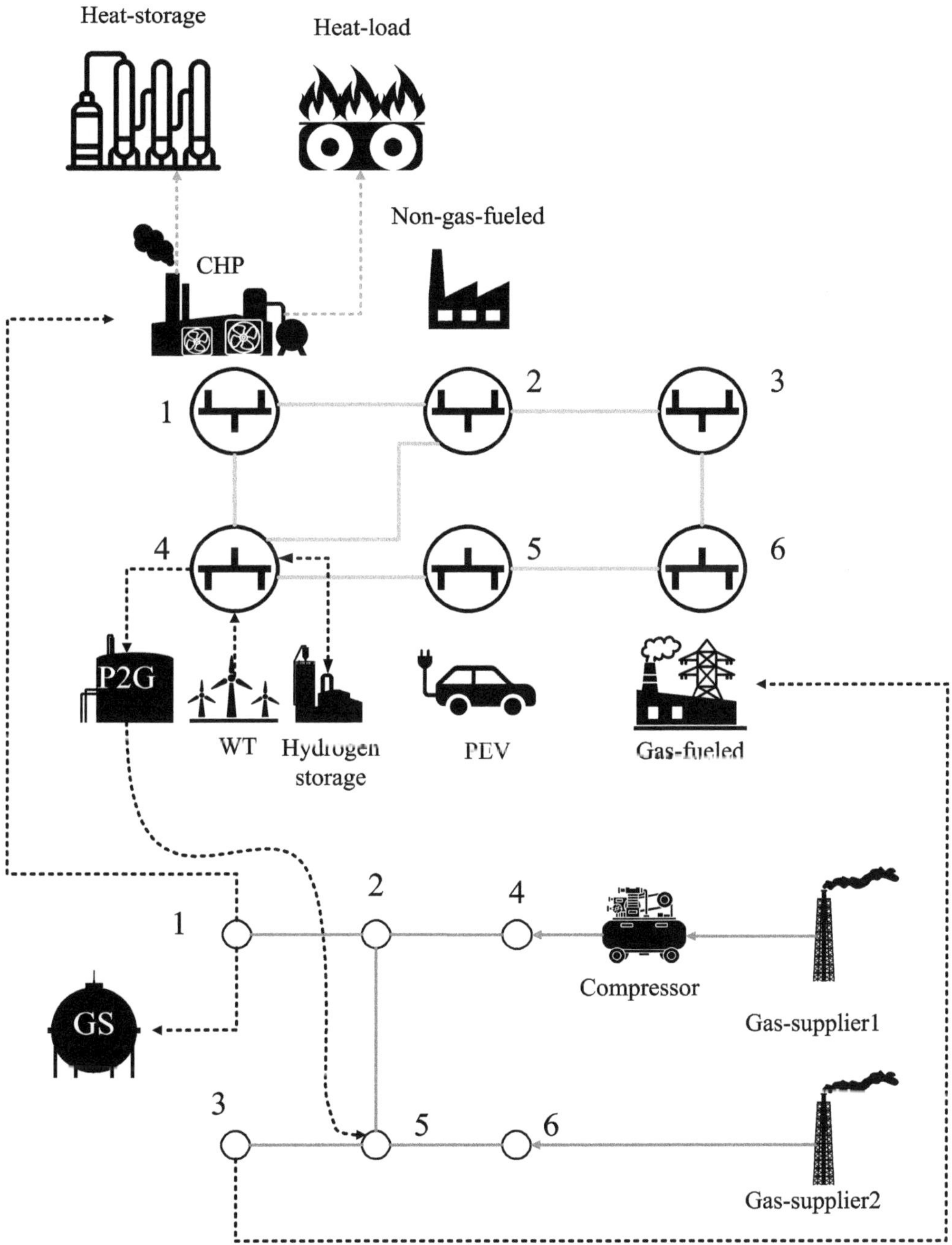

FIGURE 14.2 The overall structure of the integrated electricity, gas, and heating network.

The focus of the mentioned case studies is on investigating three major areas:

1. Contribution to reducing operation costs
2. Supporting gas and electricity network
3. Utilization of renewable energy resources

14.3.2 Obtained Results and Discussions

Case Study 1. In this initial case study, the optimal day-ahead scheduling of the integrated network without considering the influence of energy storage and P2G facilities is explored. The generator output over a 24 hr period is depicted in Figure 14.3, highlighting the pivotal role of the CHP unit. Renowned for its cost-effectiveness and high efficiency, the CHP unit serves as a primary source for both electrical and thermal demands. As electricity demand escalates, gas-fueled units come online to fulfill the increasing requirements. However, during hours 16–24, the activation of expensive non-gas-fired power plants is observed. This deviation is necessitated by the surge in gas demand during these hours, diverting gas supply away from gas-fired power plants. The question may arise as to why the gas allocation does not prioritize the CHP power plant. This allocation challenge stems from gas network limitations during peak hours, rendering it infeasible to meet the CHP plant's gas requirements. Consequently, the gas is redirected to alternative gas-fueled generators located on different nodes. As a result, the system operator resorts to more expensive units to satisfy power, gas, and thermal demands, elevating the overall system operation cost. In Figure 14.4, the disparity between available wind energy and actual wind power utilization is observed based on the fact that mismatches of wind power plant with electrical load lead to suboptimal efficiency for this facility. The total operation cost for this period amounts to \$3,647,936, incorporating the operation costs of non-gas-fueled units and gas production for household and gas-fueled unit consumption. Since no storage units are utilized in this scenario, their operation costs are presumed to be zero. Despite the comparatively lower operation costs of gas-fueled units, gas network constraints necessitate the activation of non-gas-fueled generators.

As can be concluded from the preceding case study, it is crucial to have an integrated view of the exploitation of energy networks, as these networks are interdependent. For instance, the output of gas-fueled units is contingent upon the limitations of gas networks. During periods of peak gas demand, the gas network may be unable to supply the necessary fuel for these units, thereby requiring the activation of non-gas-fueled units to maintain operational stability. On the other hand, heat demand is directly related to the output of CHP units, as these units operate within a specific range

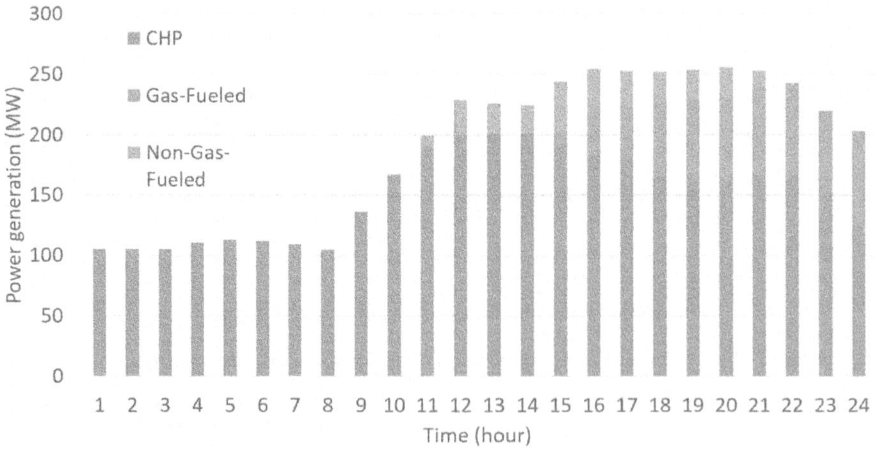

FIGURE 14.3 Power generation of power plants in case 1.

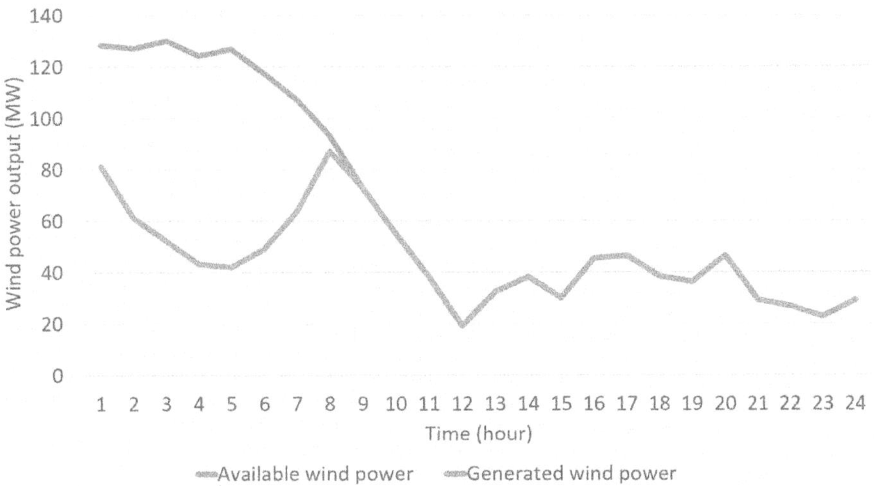

FIGURE 14.4 Wind turbine output in case 1.

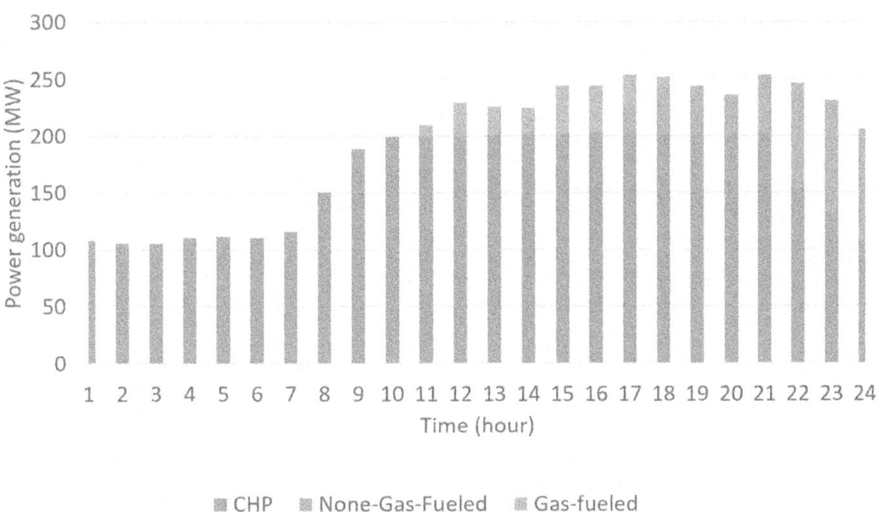

FIGURE 14.5 Power generation of power plants in case 2.

of output. These examples indicate that having an integrated perspective on energy systems is essential for ensuring both system security and efficiency.

Case Study 2. In the following case study, the impact of P2G technology on integrated network operation is investigated. P2G technology converts excess power produced by renewable energy sources into gas and hydrogen fuel for industrial use, particularly during periods of low electricity demand. The gas produced by P2G can be supplied to the gas network for domestic consumption and gas-fired power plants. Figure 14.5 illustrates the output of power plants in the second case study. It can be observed that the output power of the CHP unit in this case has increased compared to the first case study, attributed to the delivery of gas produced by P2G to the gas network. The increased output power of cheaper units such as P2G leads to a reduction in operating costs. Figure 14.6 displays wind power generation in case 1 and case 2. The utilization of renewable energies in case 2 is notably improved compared to case 1. This difference is particularly evident during the early hours, when electrical demand is low and available wind power is high. In other words, the presence of the P2G unit minimizes energy loss in the second case. As shown in Figure 14.7, the P2G unit operates primarily during hours 1–10, when electric demand is low. Although there is no

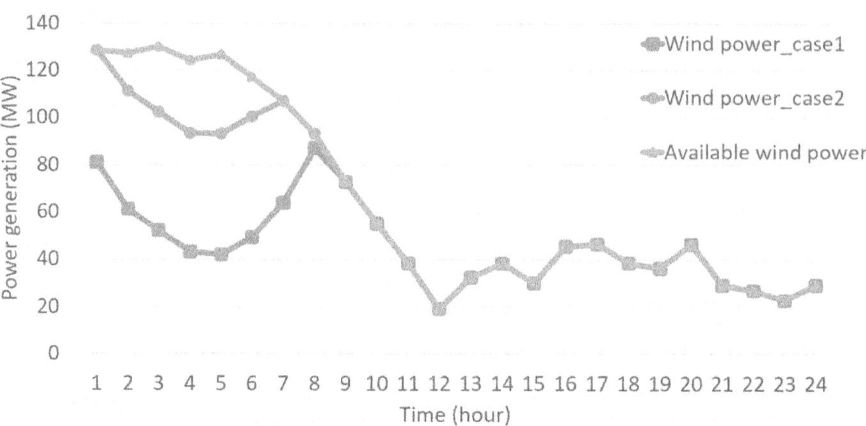

FIGURE 14.6   Wind turbine output in case 2.

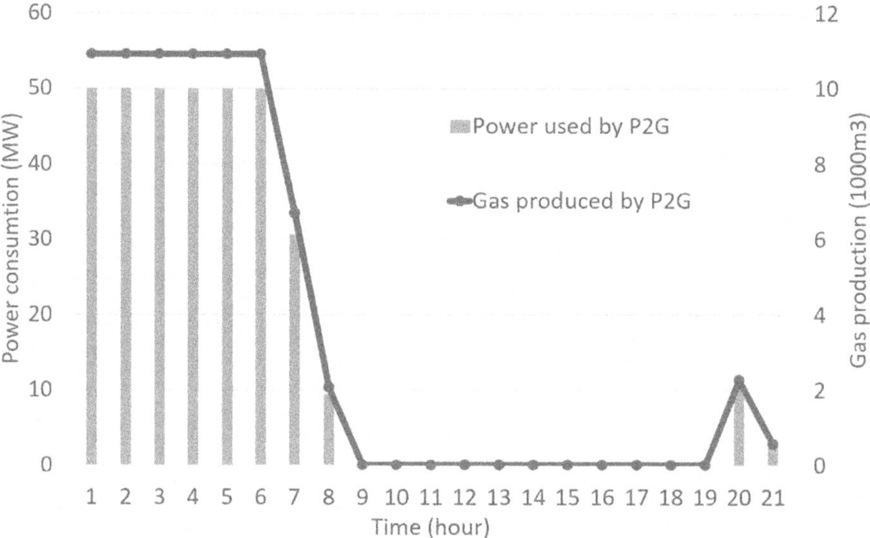

FIGURE 14.7   P2G hourly operation in case 2.

gas storage, the gas produced by the P2G unit can be consumed instantly. However, due to the line pack characteristic of the pipelines, a portion of the generated gas can be stored within the pipelines.

The operation cost of the network in this case study is $346,515, which is approximately $19,000 lower than in case 1. This reduction in costs can be attributed to two major reasons:

1. Enhanced efficiency of harnessing wind energy
2. Enhanced utilization potential of cost-efficient power plants, particularly during peak load periods

Thanks to the presence of P2G units, both factors mentioned are enhanced, further substantiating the economic justification behind these facilities. Figure 14.8 provides information about the gas consumed by the units as well as the gas produced by the P2G facility. As deduced from Figure 14.8, during peak gas load, CHP gas consumption decreases. However, the other gas-fueled generator begins power generation. This is primarily due to constraints in gas pipelines, which cannot provide

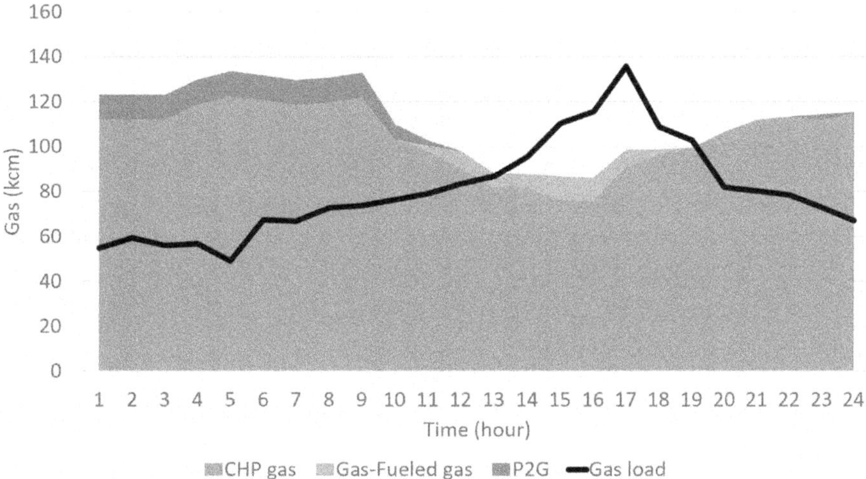

FIGURE 14.8 Units' gas consumption in case 2.

the required gas to the CHP power plant. It is noteworthy that the P2G unit operates mostly between hours 1 and 13, when electrical demand is low and wind power generation is high. In summary, the most significant impacts of P2G on the power and gas networks can be outlined as follows:

1. Reduction in operational costs
2. Enhanced harnessing of wind energy, thereby improving the efficiency of wind turbines
3. Strengthening the support for both gas and electrical networks

It is noteworthy that, according to the proposed model, load shedding is not expected to occur. However, in the event that load shedding does happen due to unforeseen circumstances, the presence of the P2G facility can help mitigate it.

Case Study 3. This subsection discusses the impacts of energy storages as well as demand response programs on system operation. Energy storage units such as heat storage and gas storage bring many advantages to the energy carriers. In order to achieve an in-depth review of these facilities, heat storage, natural gas storage, and additionally, hydrogen storage are considered in the model. The output of power generators in this case study is determined in Figure 14.9. A similar analysis of generators' output in case 2 can be performed for case 3. The first significant change that stands out in this figure compared to previous studies is that non-gas-fueled power plants are not producing and all electricity demand throughout the operational period is provided by CHP and the gas-fired power plants. This means that by supplying the required gas at all hours, the network has been relieved of the need for more expensive units, resulting in cost reduction. Similar to previous studies, the CHP unit is present at all hours as the primary provider of electricity and heat, although in this study, this power plant has been utilized more extensively. Contrary to the second case study, where the gas produced by the P2G facility could not be stored, in this case study the gas produced by P2G can be stored as either hydrogen or methane and injected into the network when needed. It is worth noting that natural gas storage facilities can also store gas produced by gas suppliers for times of low demand and supply it to the gas-fueled units or consumers at other times. This leads to greater overall performance for natural gas producers, P2G units, and high-efficiency power plants, like CHP units. Figure 14.10 illustrates the charging and discharging times of gas storage facilities. Since all produced hydrogen is entirely available to the gas network and not sold to the industries, the charging/discharging of both natural gas and hydrogen storage units is depicted together. The hydrogen storage facilities have a lower capacity compared to the natural gas storage units. Therefore, the hydrogen produced by P2G is converted to methane after being stored in hydrogen

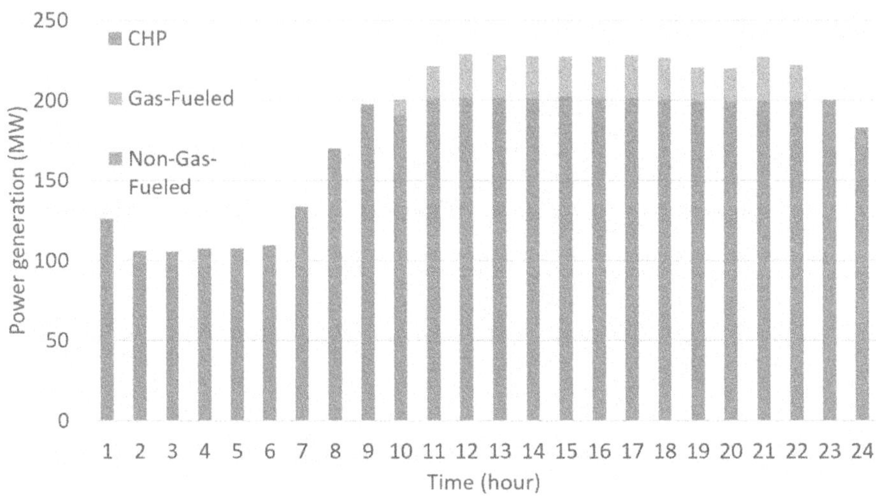

FIGURE 14.9 Power generation of power plants in case 3.

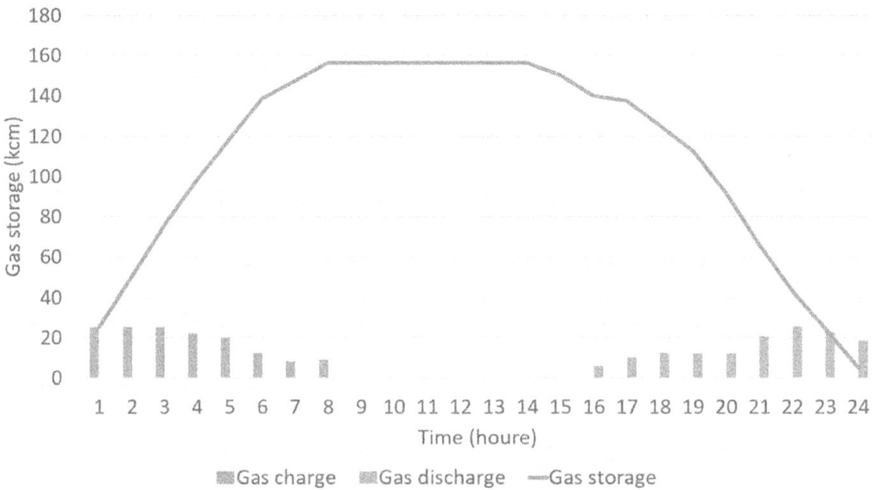

FIGURE 14.10 Gas storage operation in case 3.

storage and then delivered to natural gas storage units. As evident from the figure, during the initial 12 hr, when both the output of gas-fueled power plants and the demand for household gas load are low, these storage facilities begin to charge. In the subsequent hours, they inject the stored gas into the gas network.

On the other hand, the presence of thermal storage facilities is beneficial for two reasons:

1. When gas resources are limited, reducing gas consumption by CHP units and utilizing stored heat in storage facilities help alleviate the need for gas consumption by CHP units. As stated in equation (14.49), CHP units consume more gas for heat production, and the presence of heat storage facilities can mitigate this demand.
2. Additionally, CHP units operate within a defined operation area (FOR), meaning, their electrical and thermal outputs are interconnected. Therefore, the presence of heat storage facilities enables more efficient and effective operation of CHP units. For example, during certain hours when heat demand is low, it may be necessary to reduce the electrical power output of the power plant or the produced heat may be wasted, which is not cost-effective. However, the existence of heat storage facilities prevents this problem.

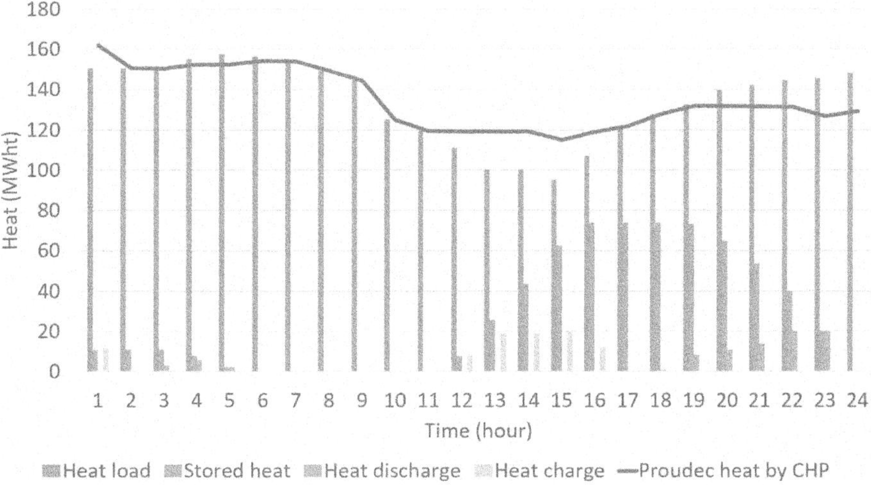

FIGURE 14.11 Heat storage and heat generation of CHP in case 3.

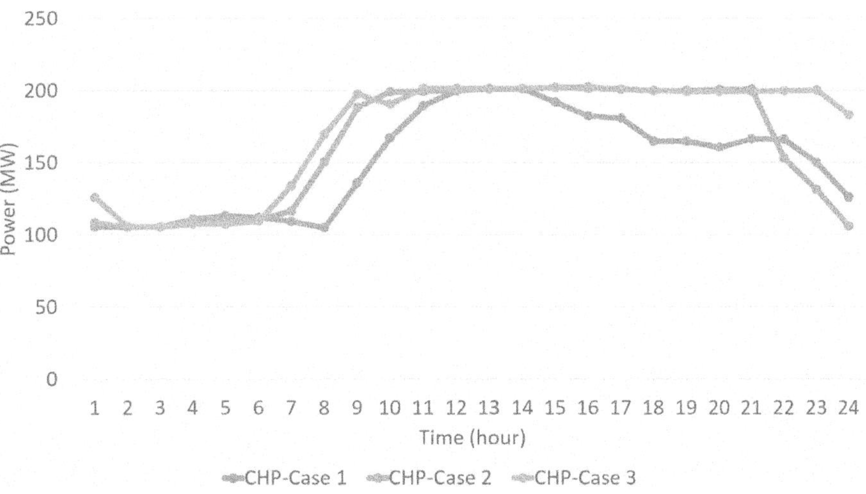

FIGURE 14.12 Power generation of CHP in all cases.

Figures 14.11 and 14.12 depict the heat storage charging/discharging state, produced heat by CHP, and local heat demand. Figure 14.12 also shows the power generation of the CHP unit in different case studies. A similar analysis akin to the second case can also be provided for this case. As previously mentioned, one of the reasons for cost reduction with the presence of energy storage systems is the more efficient operation of cost-effective power plants, such as CHP units. Figure 14.13 illustrates this point. As observable, in this case study, the gas network can provide more required gas to CHP units, especially during peak hours.

Another goal of this case study is to emphasize the impacts of demand response programs on system operation and efficiency. The demand response program, as a valuable tool, provides great flexibility for system operators by shaving peak load while enhancing renewable power usage. Responsive electrical loads are one of the practical alternatives to flatten the peak electricity demand [23]. The simultaneous use of demand response programs along with energy storage devices and P2G technology increases the effectiveness of this kind of program.

In this case study, the implementation of a demand response program has led to improved system performance and reduced costs by shifting electrical loads to designated times. The total energy

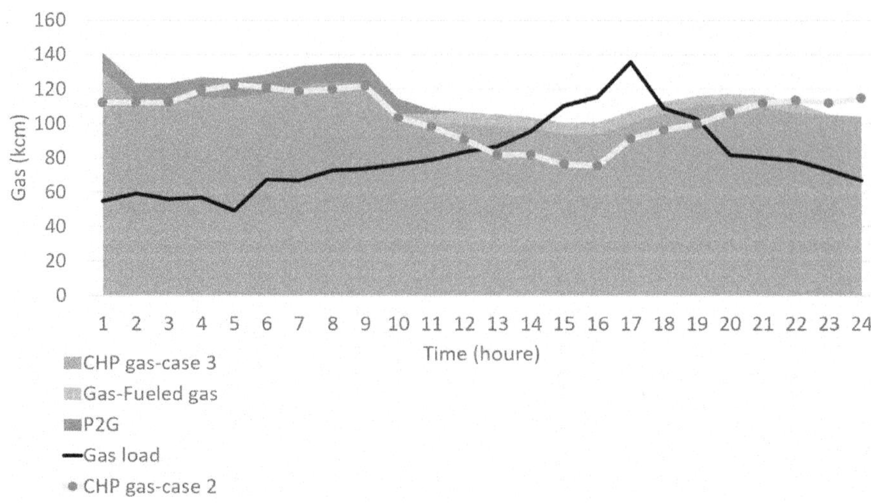

FIGURE 14.13   Units' gas consumption in case 3.

TABLE 14.3
Demand Response Implementation

Demand Response Level (%)		Total Cost ($)
Case Study		
1	0%	$330,285
2	5%	$327,972
3	10%	$325,771
4	15%	$323,825
5	20%	$322,143

supply cost within 24 hr in this case study has amounted to $325,771, representing a significant decrease compared to previous case studies. Table 14.1 briefly compares the different levels of the demand response program. As depicted, a higher level of demand response leads to higher cost reduction.

Case Study 4. As mentioned before, one of the key aspects for the growth and development of P2G technology in the future is the high penetration of renewable energy resources. For this purpose, two sensitivity analyses have been conducted in this study.

The proposed model considers a maximum power output of 130 MW for wind turbines per hour, which varies over a 24 hr period. However, this case study examines the different capacities of wind turbines and calculates the energy loss observed in previous case studies. Essentially, it compares the unused energy from wind turbines across each case. Table 14.4 presents the total energy loss for various wind turbine capacities in each case study. As shown, the integration of P2G technology (case 2) nearly halves the energy loss over a 24 hr period. Furthermore, the implementation of energy storage systems (case 3) shows potential for enhancing wind energy utilization and reducing energy loss.

The second sensitivity analysis focuses on different levels of P2G unit capacity in conjunction with various capacities of wind turbines. As illustrated in Table 14.5, increasing the P2G power by 10 MW in each increment results in a notable reduction in energy loss across different wind turbine capacities. For example, with a 10% increase in wind power capacity, the amount of unused energy decreases by approximately 67 MWh in the proposed model. However, when the P2G installed

TABLE 14.4
Wind Energy Harnessing

Wind energy integration	Energy Loss (MWH)			
	0%	10%	20%	30%
Case study 1	417.27	499.35	603.8	670
Case study 2	127.48	203.66	290.29	376.420

Note: Wind capacity varies.

TABLE 14.5
Wind Energy Harnessing

P2G Capacity (MW)	Energy Loss (MWh)			
Base case	38.93	105.64	180.206	264.05
60	3.46	51.05	120.63	195.54
70	0	11.79	63.51	135.67
80	0	0	20.37	76.33

Note: P2G capacity varies.

TABLE 14.6
The Operation Cost of Each Scenario

Scenarios	Total Cost
1	$329,846
2	$328,549
3	$326,768
4	$325,313
5	$324,172
Mean	$326,432

capacity reaches 80 MW, the energy loss is reduced to zero. In essence, the utilization of P2G technology significantly enhances the harnessing of wind energy.

Case Study 5. The primary objective of this case study is to tackle stochastic parameters such as wind turbine output, electrical and gas demand, and PEVs' charging stations, which introduce uncertainty into the model. Table 14.6 illustrates the total operation cost across various scenarios, arranged from the worst-case scenario to the best-case scenario. The variance in operation costs among scenarios reflects the uncertainty associated with these parameters. Given that each scenario has a different likelihood of occurrence, for a more precise estimation of the final cost, the cost of each scenario should be multiplied by its corresponding probability.

14.4 CONCLUSIONS

In recent years, the landscape of energy technology has witnessed a significant transformation catalyzed by the convergence of advancements in power-to-gas (P2G) technologies and the declining costs of renewable energy sources. This synergy has propelled the widespread adoption of P2G facilities on a global scale.

The results obtained from numerical studies highlight the importance of integrated scheduling of electricity and gas networks. Having an integrated perspective for operating energy systems not only enhances system performance and security but also is essential and inevitable. The total operating cost has amounted to \$364,793. The wasted wind energy, or in other words, the unused wind energy in this case, has amounted to 417.27 MWh. The reason for this energy loss is the discrepancy between the output power of the wind turbines and the electrical consumption between hours 1 and 8. Additionally, between hours 12 and 20, when the gas load is high, the inability of the gas network to supply the necessary gas for the gas-fueled units has forced the operator to resort to using expensive power plants. Furthermore, as evident from the findings in the second case study, the presence of P2G technology in the network has led to a reduction in operational costs. The total operating cost in this case is \$346,515, which represents a reduction of approximately 6% compared to the first case study. The presence of P2G technology enhances the performance of gas-fueled units, such as CHPs, and increases the overall efficiency of the gas network. Moreover, these results demonstrate the impact of P2G devices in maximizing the utilization of electricity generated by wind turbines. In this case study, the wasted wind energy has decreased significantly to 127.48 MWh compared to the first case study. This emphasizes the pivotal role of P2G technology in optimizing the utilization of renewable energy sources. P2G is primarily active during hours 1 to 8, which coincides with the peak wind energy hours. This underscores the importance of deploying such devices in regions with surplus electrical energy. Additionally, the output capacity of CHP has increased in this study compared to the first case, with differences in output capacity reaching up to 40 MW during certain hours.

Regarding the third case study, the impacts of hydrogen, gas, and thermal storages have also been investigated. The presence of these storage facilities alongside P2G technology enhances the effectiveness of P2G technology, resulting in a further reduction in system operating costs compared to the second case study. The total operating cost in this study has amounted to \$325,771, representing a reduction of approximately 11% compared to the first case study. Additionally, the presence of storages significantly contributes to the improvement of the performance of generating units. As a result, there is no need to activate expensive power plants, and the gas network will not face any constraints in supplying gas to the generators. Furthermore, the load response program has also enhanced the overall flexibility of the network. In this case, wind energy has been utilized almost entirely, or in other words, wind energy losses have been minimized to a significant extent. The gas storage facility also begins storing gas between hours 1 and 8, reaching its maximum capacity of 159 kcm. Subsequently, the stored gas is injected into the network between hours 16 and 21. This significantly aids in increasing the efficiency of units and enhancing the network's flexibility. The increase in CHP output has also been made possible thanks to the reduction in gas pipeline congestion, compared to the first and second studies. This increase is most noticeable during hours 20 to 24. All these factors, coupled with the increased utilization of wind energy, demonstrate the positive impact of these devices on network operation. The fourth case study provides a comprehensive examination of the most critical factor in the integration of P2G units into energy systems: the incorporation of renewable energy resources. The results indicate that with a 10%, 20%, and 30% increase in wind power plant capacity, the energy losses in the first case study (without P2G) amount to 499.35 MWh, 603.8 MWh, and 670 MWh, respectively. Meanwhile, these losses in the second study amount to 203.66 MWh, 290.29 MWh, and 376.420 MWh, demonstrating a halving of energy losses. The results derived from this case study underscore that the coexistence of P2G devices with renewable energy generators, such as wind turbines, enhances the efficiency of these resources and minimizes energy wastage. Consequently, this addresses a fundamental concern linked to renewable energy sources and their intermittent output and a common disparity with electrical demand. Essentially, P2G technology substantially alleviates this challenge and facilitates the expansion and integration of renewable energy sources alongside its advancement. Nevertheless, formidable challenges persist, necessitating robust solutions across multiple fronts. Paramount among these challenges is the imperative for supportive policy frameworks that incentivize investment in infrastructure and surmount technological barriers impeding large-scale implementation. Additionally, while the environmental benefits of P2G facilities are substantial, particularly in their capacity to offset greenhouse gas emissions, nuanced considerations surrounding the sustainability of P2G operations, notably concerning the source of electricity utilized for electrolysis, demand meticulous attention. An advantage of P2G facilities lies in their capacity to mitigate the seasonal misalignment between energy demand and renewable energy generation. By capitalizing on surplus renewable energy during periods of abundance and subsequently releasing it during periods of scarcity, P2G technologies bolster grid stability and fortify the resilience of energy systems. Furthermore, P2G plays a pivotal role in fostering sector coupling, facilitating the seamless integration of renewable energy across diverse domains encompassing transportation, industrial processes, and heating. For instance, consider a scenario where excess solar energy generated during daylight hours is harnessed by P2G facilities to produce hydrogen, which can then be utilized as a clean fuel for fuel cell electric vehicles or as a feedstock for industrial applications, such as ammonia production. This exemplifies the versatility and applicability of P2G in diversifying energy usage across sectors while reducing reliance on fossil fuels and mitigating environmental impacts.

In the author's opinion, the research gaps and recommendations for future studies in the field of P2G can be broadly categorized into three main areas:

1. *Cost.* One of the most significant challenges is the high investment and operational costs of P2G devices. These costs include the provision of infrastructure such as electrolyzers, methane production units, and sometimes hydrogen storage systems, as well as modifications to the gas network. Additionally, the processes of electrolysis and methane production require substantial amounts of energy, making it feasible only in regions with sufficient energy resources.
2. *Efficiency.* Improving the efficiency of these devices is crucial for reducing energy wastage and fostering their future development. Despite ongoing research in this field, the energy conversion rates of these devices remain low. Energy losses can occur in various stages, including electrolysis of water, methanation, and storage.
3. Policy and regulation are essential in motivating the development of these devices, including the following elements:

   - Offering incentives and financial support
   - Setting and enforcing required standards
   - Addressing safety and environmental issues
   - Solving market-related challenges for P2G services and products

In summation, the trajectory of P2G technology underscores a promising trajectory toward achieving a more sustainable, flexible, and interconnected energy landscape. With continued innovation and strategic deployment, P2G stands poised to be a cornerstone of the transition to a low-carbon future, underpinning energy security and resilience while fostering economic prosperity.

Nomenclature

i,j	index of electrical system buses
t	index of time
s	index of scenarios
x	index of gas network nodes
gsp	gas suppliers
h	P2G unit

hs	heat storage
p	index of pipelines
n	gas loads
$PROB_s$	probability of each scenario
C_{gsp}^{gas}	gas price
C_{gs}	operation cost of gas storage
Pg_i^{min}, Pg_i^{max}	minimum and maximum active power of generators
Qg_i^{min}, Qg_i^{max}	minimum and maximum reactive power of generators
RU_i, RD_i	maximum ramp-up/down rate of units
UT_i, DT_i	minimum up/downtime of units
Csu_i, Csd_i	shutdown and start-up costs of gas-fueled units
$Csug_i$, $Csdg_i$	shutdown and start-up costs of non-gas-fueled units
β_{hs}	heat storage efficiency
β_{hs}^{ch}, β_{hs}^{dc}	heat storage charge/discharge efficiency
H_{hs}^{min}, H_{hs}^{max}	maximum and minimum of stored heat in thermal storage
$H_{hs}^{max\,ch}$, $H_{hs}^{max\,dc}$	the upper bound of charge/discharge of the thermal storage
HES_h^{min}	minimum level of stored hydrogen in hydrogen energy storage
HES_h^{max}	maximum level of stored hydrogen in hydrogen energy storage
$PHES_h^{min\,ch}$, $PHES_h^{max\,ch}$	maximum and minimum rate of hydrogen energy storage charging
$PHES_h^{min\,dc}$, $PHES_h^{max\,dc}$	maximum and minimum rate of hydrogen energy storage discharging
$Pp2g_i^{max}$	maximum power of P2G
$Q_{gs}^{dc\,max}$, $Q_{gs}^{ch\,max}$	maximum rate of gas storage charging/discharging
β_{gs}^{ch}, β_{gs}^{dc}	gas storage charge/discharge efficiency
$G_x^{sup\,min}$, $G_x^{sup\,max}$	maximum and minimum production of gas suppliers
$Pl_{l,s,t}$	hourly electrical demand
$ZDRP$	demand response program rate coefficient
$Z_{i,j}$	lines impedance matrix
$\theta_{i,j}$	lines impedance angle
V_i^{min}, V_i^{max}	maximum and minimum of the allowable voltage level in each bus
σ_i^{min}, σ_i^{max}	maximum and minimum of the allowable voltage angle in each bus
$G_x^{sup\,max}$, $G_x^{sup\,min}$	maximum and minimum of the gas suppliers output
$Gl_{n,s,t}^{gas}$	gas load at each node
DI_p	pipeline diameter
L_p	pipeline length
λ_{cp}	compressor efficiency
$\lambda_{i,n}^{P2G}$	P2G efficiency
ϕ^{P2G}	P2G conversion factor (MW/MCM)
β_{cp}	gas turbine fuel rate coefficient of a compressor
ρr^{max}	compressor maximum ratio
$G_c^{cp\,max}$	gas compressor maximum gas consumption
$P_{c,s,t}^{cp\,max}$	maximum power of gas compressor
ρ_n^{min}, ρ_n^{max}	minimum and maximum pressure at each node
$G_p^{line\,min}$, $G_p^{line\,max}$	minimum and maximum of gas level through each pipeline
LP_P	line pack through the pipeline
Gv_P	volume of gas
rho^{normal}	gas density in standard condition
Z	compressibility factor

R	gas constant for natural gas
T^{normal}	gas temperature in standard condition
$C^{NGU}_{s,i,t}$	operation cost of non-gas-fueled units
$SU_{s,i,t}$	total start-up cost of each unit
$SD_{s,i,t}$	total shutdown cost of each unit
$C^{P2H}_{h,i,t}$	charge cost of hydrogen energy storage
$C^{h2p}_{h,i,t}$	discharge cost of hydrogen energy storage
$G^{sup}_{x,s,t}$	gas produced by gas suppliers
$Pg_{i,t}$	active power of generators
$Qg_{i,s,t}$	reactive power of generators
$Q^{out}_{gs,t,s}$	gas output of gas storage unit
$onoff_{i,s,t}$	binary variable indicates unit condition
$H_{i,S,t}$	heat produced by CHP unit
$Y^{on}_{i,s,t}, Y^{off}_{i,s,t}$	time that unit is on/off
$H_{hs,s,t}$	stored heat in thermal storage
$H^{ch}_{hs,s,t}, H^{dc}_{hs,s,t}$	heat storage charge/discharge power
$HES_{h,s,t}$	stored hydrogen in hydrogen energy storage
$PHES^{ch}_{h,s,t}$	charge rate of hydrogen energy storage
$PHES^{dc}_{h,s,t}$	discharge rate of hydrogen energy storage
$DRP^{up}_{l,s,t}$	power of shifted-up load
$DRP^{dn}_{l,s,t}$	power of shifted-down load
$Pl^{DR}_{l,s.t}$	electrical load after applying demand response program
$\sigma_{i,t}$	bus voltage angle
$Q\ln_{i,j,s,t}$	lines reactive power transmission
$P\ln_{i,j,s,t}$	lines active power transmission
$V_{i,s,t}$	bus voltage
$Pw_{i,s,t}$	wind power output
$Pp2g_{i,s,t}$	P2G power consumption
$G^{line}_{p,s,t}$	gas flow through the pipelines
$G^{cp}_{c,s,t}$	compressor gas consumption
$Gloss_{n,s,t}$	not supplied gas
$GP2G^{inj}_{n,s,t}$	P2G injected gas
$\rho_{n,s,t}$	pressure at node n
$P^{cp}_{c,s,t}$	power of compressor
$C^{CHP}_{i,s,t}$	CHP operation cost
$C^{GF}_{i,s,t}$	gas-fueled unit operation cost

REFERENCES

[1] Qiu, R., Zhang, H., Wang, G., Liang, Y., & Yan, J. (2023). Green hydrogen-based energy storage service via power-to-gas technologies integrated with multi-energy microgrid. *Applied Energy* (Vol. 350, p. 121716).

[2] Ghahramani, M., Nazari-Heris, M., Zare, K., & Mohammadi-Ivatloo, B. (2022). A two-point estimate approach for energy management of multi-carrier energy systems incorporating demand response programs. *Energy* (Vol. 249, p. 123671).

[3] Yarmohammadi, H., & Abdi, H. (2023). A comprehensive optimal power and gas flow in multi-carrier energy networks in the presence of energy storage systems considering demand response programs. *Electric Power Systems Research* (Vol. 214, p. 108810).

[4] Nazari-Heris, M., Mohammadi-Ivatloo, B., & Asadi, S. (2020). Optimal operation of multi-carrier energy networks with gas, power, heating, and water energy sources considering different energy storage technologies. *Journal of Energy Storage* (Vol. 31, p. 101574).

[5] Nazari-Heris, M., Mirzaei, M. A., Mohammadi-Ivatloo, B., Marzband, M., & Asadi, S. (2020). Economic-environmental effect of power to gas technology in coupled electricity and gas systems with price-responsive shiftable loads. *Journal of Cleaner Production* (Vol. 244, p. 118769).

[6] Ma, Y., Wang, H., Hong, F., Yang, J., Chen, Z., Cui, H., & Feng, J. (2021). Modeling and optimization of combined heat and power with power-to-gas and carbon capture system in integrated energy system. *Energy* (Vol. 236, p. 121392).

[7] Cai, T., Dong, M., Liu, H., & Nojavan, S. (2022). Integration of hydrogen storage system and wind generation in power systems under demand response program: A novel p-robust stochastic programming. *International Journal of Hydrogen Energy* (Vol. 47, Issue 1, pp. 443–458).

[8] Zeng, Q., Zhang, B., Fang, J., & Chen, Z. (2017). A bi-level programming for multistage co-expansion planning of the integrated gas and electricity system. *Applied Energy* (Vol. 200, pp. 192–203).

[9] Basnet, A., & Zhong, J. (2020). Integrating gas energy storage system in a peer-to-peer community energy market for enhanced operation. *International Journal of Electric Power Energy System* (Vol. 118, p. 105789).

[10] Cheng, Y., Liu, M., Chen, H., & Yang, Z. (2021). Optimization of multi-carrier energy system based on new operation mechanism modeling of power-to-gas integrated with CO_2-based electrothermal energy storage. *Energy* (Vol. 216, p. 119269).

[11] Eladl, A. A., El-Afifi, M. I., Saeed, M. A., & El-Saadawi, M. M. (2020). Optimal operation of energy hubs integrated with renewable energy sources and storage devices considering CO_2 emissions. *International Journal of Electric Power Energy System* (Vol. 117, p. 105719).

[12] Yuan, Z., He, S., Alizadeh, A., Nojavan, S., & Jermsittiparsert, K. (2020). Probabilistic scheduling of power-to-gas storage system in renewable energy hub integrated with demand response program. *Journal of Energy Storage* (Vol. 29, p. 101393).

[13] Mirzaei, A., Nazari-Heris, M., Mohammadi-Ivatloo, B., Zare, K., Marzband M., & Anvari-Moghaddam, A. (2020). A novel hybrid framework for co-optimization of power and natural gas networks integrated with emerging technologies. *IEEE Systems Journal* (Vol. 14, No. 3, pp. 3598–3608).

[14] Yang, Z., Gao, C., & Zhao, M. (2019). Coordination of integrated natural gas and electrical systems in day-ahead scheduling considering a novel flexible energy-use mechanism. *Energy Conversion and Management* (Vol. 196, pp. 117–126). Elsevier BV. doi: 10.1016/j.enconman.2019.

[15] Shi, M., Wang, H., Lyu, C., Xie, P., Xu, Z., & Jia, Y. (2021). A hybrid model of energy scheduling for integrated multi-energy microgrid with hydrogen and heat storage system. *Energy Reports* (Vol. 7, pp. 357–368). Elsevier BV. doi: 10.1016/j.egyr.2021.

[16] AlHajri, I., Ahmadian, A., & Elkamel, A. (2021). Stochastic day-ahead unit commitment scheduling of integrated electricity and gas networks with hydrogen energy storage (HES), plug-in electric vehicles (PEVs) and renewable energies. *Sustainable Cities and Society* (Vol. 67, p. 102736). Elsevier BV.

[17] Mansouri, S. A., Nematbakhsh, E., Ahmarinejad, A., Jordehi, A. R., Javadi, M. S., & Matin, S. A. A. (2022). A multi-objective dynamic framework for design of energy hub by considering energy storage system, power-to-gas technology and integrated demand response program. *Journal of Energy Storage* (Vol. 50, p. 104206). Elsevier BV.

[18] Nazari-Heris, M., Mohammadi-Ivatloo, B., & Asadi, S. (2020). Optimal operation of multi-carrier energy networks with gas, power, heating, and water energy sources considering different energy storage technologies. *Journal of Energy Storage* (Vol. 31, p. 101574). Elsevier BV.

[19] Shahbazbegian, V., Shafie-khah, M., Laaksonen, H., Strbac, G., & Ameli, H. (2023). Resilience-oriented operation of microgrids in the presence of power-to-hydrogen systems. *Applied Energy* (Vol. 348, p. 121429). Elsevier BV. doi: 10.1016/j.apenergy.2023.

[20] M. Ghahramani, M. Nazari-Heris, K. Zare, and B. Mohammadi-Ivatloo, "A two-point estimate approach for energy management of multi-carrier energy systems incorporating demand response programs," Energy, vol. 249. Elsevier BV, p. 123671, Jun. 2022

[21] Alavi, S. A., Ahmadian, A., & Aliakbar-Golkar, M. (2015). Optimal probabilistic energy management in a typical micro-grid based-on robust optimization and point estimate method. *Energy Conversion and Management* (Vol. 95, pp. 314–325). Elsevier BV.

[22] Shabazbegian, V., Ameli, H., Ameli, M. T., & Strbac, G. (2020). Stochastic optimization model for coordinated operation of natural gas and electricity networks. *Computers & Chemical Engineering* (Vol. 142, p. 107060). Elsevier BV. doi: 10.1016/j.compchemeng.2020.

[23] Romanchenko, D., Nyholm, E., Odenberger, M., & Johnsson, F. (2021). Impacts of demand response from buildings and centralized thermal energy storage on district heating systems. *Sustainable Cities and Society* (Vol. 64, p. 102510). Elsevier BV. doi: 10.1016/j.scs.2020.

15 Optimization Models for Operation of Power-to-X Technologies

Marialaura Di Somma, Christina Papadimitriou,
Nicola Bianco, and Giorgio Graditi

Abbreviations

AC	absorption chiller
BESS	battery energy storage system
CCHP	combined cooling heat and power
CHP	combined heat and power
CI	carbon intensity
DHC	district heating and cooling
EV	electric vehicle
EZ	electrolyzer
FC	fuel cell
G2V	grid-to-vehicle
GB	gas-fired boiler
GT	gas turbine
HP	heat pump
ICE	internal combustion engine
MCES	multi-carrier energy system
mGT	micro-gas turbine
PtX	power-to-X
PV	photovoltaic
RES	renewable energy sources
SOC	state-of-charge
ST	solar thermal
TES	thermal energy storage
V2G	vehicle-to-grid

15.1 INTRODUCTION

To contrast global warming, it is necessary to set ambitious targets for the reduction of greenhouse gas emissions. Without ambitious emissions reduction policies, it is foreseen that the average global temperature will increase between 1.5°C and up to 5°C over the course of this century according to different projected scenarios [1]. Limiting warming to 1.5°C implies reaching net-zero CO_2 emissions globally by 2050. It is thus clear that the model of economic development pursued till now needs to be restructured by considering the compelling need to decouple the economic growth from the use of fossil fuels and by pursuing a transition that is climate-neutral according to the Sustainable Development Goals of the United Nations [2] and the European Green Deal [3].

DOI: 10.1201/9781032719436-15

The energy system represents the key factor toward climate neutrality and net-zero emissions by 2050, according to the relevant plans adopted by the majority of G20 members and by over 120 countries worldwide. Reaching the goal of a full decarbonization of the energy system is possible, but the entire system architecture needs to be changed, by deleting dependence on fossil fuels and substituting them with renewable energy sources (RES). However, RES alone cannot contribute to achieving this ambitious goal. Electrification can help this process, being another central pillar of the clean energy transition: according to the IEA Sustainable Development Scenario (SDS) to 2070 [4], the share of RES in the global power generation mix needs to reach 86%, whereas the share of primary energy dedicated to electricity generation needs to jump from 38% today to over 60%. The electrification across all energy sectors represents an economically viable option to reduce by almost 30% the annual CO_2 emissions in 2070, as defined in the IEA SDS. *Electrification* refers to the process of transitioning from traditional, fossil fuel–based energy sources to the electricity energy carrier for various applications. It involves replacing the direct use of fossil fuels with electricity as the primary source of energy. An electrified future scenario will allow increasing the RES penetration for all final energy use, while also increasing the energy efficiency in the related energy conversion processes. In fact, another tangible benefit associated with the electricity energy carrier is that it can be converted into any other form of energy, thereby fostering the integration with other energy carriers (e.g. gas, heating, cooling, hydrogen, mobility, etc.), helping achieve decarbonization of all energy sectors. Exploiting synergies coming from the interplay of different energy carriers not only ensures maximization of RES penetration but also guarantees higher flexibility for the whole system, while reducing the need to reinforce existing network infrastructures [5].

Sector coupling is today considered as one of the key enablers for achieving climate neutrality before the end of the century, through the possibility offered for increased RES penetration levels. Sector coupling with **power-to-X** (PtX) provides options to "absorb" excess electricity supply from RES, to store energy, and to convert it into any other form of energy in times of high demand and prices. Some examples are provided in what follows (Figure 15.1).

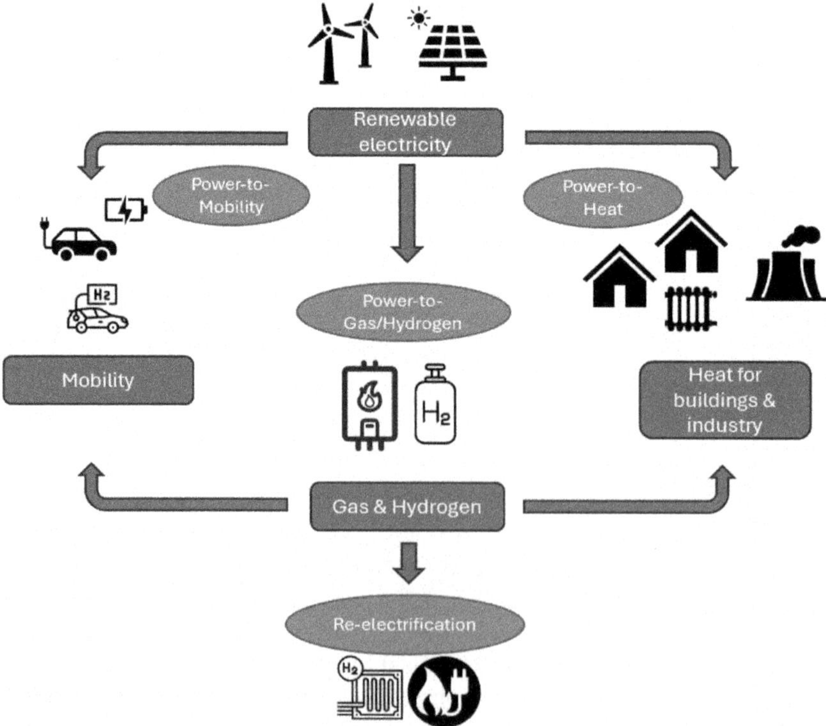

FIGURE 15.1 Scheme of energy conversion processes allowed by sector coupling.

Power-to-gas technologies can act as an important sink for surplus renewable electricity and provide low-carbon energy for industrial and transport applications; gas storage could be used to cope with seasonal variations in demand and renewable energy supply; and renewable gas can also be used in gas-fired power plants or fuel cells, providing low-carbon backup capacity to generate electricity when other RESs are unavailable. Power-to-gas also includes power-to-hydrogen, relying on green hydrogen, which can, in turn, be used for a wide number of applications, including industry, mobility, heating/cooling, and re-electrification. The benefits of power-to-gas for decarbonization across different energy sectors are multiple. Among others, it allows for the consistent reduction of renewable electricity curtailment, thanks to the ability to store for long time horizons, going from weeks to seasons, and renewable electricity during periods of excess production, and to use it also much later, especially during times when end user demand exceeds renewable generation. Moreover, energy conversion technologies such as electrolyzers and fuel cells could gradually replace conventional power generation units in the provision of ancillary services and capacity reserve, thereby contributing to the reliability of the power network [6]. This replacement is also foreseen at the local level, with hydrogen becoming the most sustainable energy carrier for satisfying multiple energy demands, including heating, cooling, and mobility.

Another practical example that is most valid at the local level is represented by **power-to-heat** technologies, such as heat pumps, which represent a cost-effective and more efficient alternative to conventional gas–fired boilers for heating purposes in buildings, thanks to their high conversion efficiency. Moreover, when combined with thermal storage, they offer the possibility to shift production of thermal energy when renewable electricity is in excess, thereby representing another option for reducing RES curtailment.

PtX may also refer to **power-to-mobility** through smart charging of battery electric vehicles [7], which allows minimizing the RES curtailment and using electric vehicles (EVs) as distributed storage while also meeting the mobility demand.

Despite all the benefits discussed related to the deployment of PtX solutions, they define new challenges for the energy system operation, among others the interdependency among energy carriers and related technologies and processes that needs to be properly modeled and optimized for an efficient operation, by serving different sustainability objectives.

This chapter aims to present a comprehensive analytical framework for the operation optimization of sector coupling and PtX technologies in the context of multi-carrier energy systems (MCESs) by considering multiple objectives. After a discussion of the main challenges posed by interdependencies among energy carriers and technologies and the objectives of optimizing sector coupling and PtX operation, the mathematical framework is presented by considering all layers of the dependency, including the system components and intersectoral aspects. The effectiveness of the proposed optimization models is demonstrated through two case studies with different degrees of complexity, both in terms of energy carriers considered and end users involved.

15.2 CHALLENGES AND OBJECTIVES

15.2.1 Challenges Related to the Operation of PtX Technologies in MCESs and Related Application Concepts

As mentioned in Section 15.1, operating PtX technologies within MCESs comes with several challenges. Within this subsection, some key challenges will be discussed:

Intermittency, variability, and availability. All PtX technologies mentioned earlier, that is, power-to-gas, power-to-heat, and power-to-mobility, intensively rely on intermittent and variable RES, like wind and solar, especially when it comes to regional local systems. This way, PtX processes within MCESs contribute to achieving environmental goals while keeping the system cost-effective. Nevertheless, managing the intermittent nature of these energy sources within MCESs requires effective energy storage and demand-side management strategies [8]. As far as the availability is concerned, PtX technologies require large amounts of resources, such as, for example, water for electrolysis and biomass for biofuel production. Ensuring sustainable sourcing of these resources without causing environmental harm is also crucial [9, 10].

Energy conversion efficiency. PtX processes involve energy conversion steps (e.g., electrolysis, synthesis) which, as expected, suffer from energy losses. Maximizing the overall efficiency of PtX systems while integrating them into MCESs is crucial to ensure cost-effectiveness and environmental sustainability. It is important to mention that efficiency can be treated as a global index for MCESs, making the optimal operation as a whole of crucial importance [11].

Grid integration. Integrating PtX technologies into existing energy grids and MCESs requires sophisticated control, communication systems, and collaboration among stakeholders coming from different sectors. Ensuring compatibility with grid infrastructure and standardized communication, maintaining grid stability, and managing bidirectional energy flows through the different systems are significant challenges. On top of these aspects, PtX integrated into the grid can put further stress due to capacity limitations, so careful planning and efficient control are needed [12, 13].

Policy and regulatory frameworks. PtX technologies are still emerging, and the regulatory framework may not be well-developed or supportive. Developing appropriate policies, incentives, and regulations to promote the adoption of PtX within MCES is crucial. The report [14] examines the local MCESs within the European political and regulatory landscape, focusing on PtX technologies as a topic of interest that could leverage these systems. Through power-to-gas and power-to-heat technologies, electricity facilitates the seamless interchange of energy carriers and enables the transportation of substantial energy volumes across Europe. Research evaluating the feasibility of power-to-gas technologies indicates that their competitiveness, at the time being, is limited to selected European countries with access to low-cost renewable electricity generation [15].

Scale-up and cost reduction. Central to the previous vision is the reliance on renewable hydrogen, aligning with the EU long-term objectives of climate neutrality and zero pollution. However, this vision faces challenges due to the current high production costs associated with renewable hydrogen. PtX technologies need to be scaled up to meet the demand for green fuels and chemicals effectively. However, scaling up while reducing costs remains a significant challenge. Economies of scale, technological advancements, and favorable policies, as mentioned before, are essential for cost reduction [16].

Considering these challenges, a bottom-up approach focusing on local MCESs offers several advantages. Local MCESs can be tailored to the specific needs, resources, and constraints of their communities, allowing for greater flexibility and adaptability in implementing sector coupling and PtX technologies. By involving local stakeholders in the planning and decision-making process, MCESs foster community engagement and ownership. This can lead to greater acceptance of renewable energy projects, PtX facilities, and sector-coupled systems.

PtX technologies leverage applications that are important for the energy vision of the future, as already analyzed in Section 15.1. In specific:

Sector coupling and integrated systems. PtX technologies enable the coupling of different energy sectors, such as electricity, heat, transportation, etc. [17]. By producing renewable gases or fuels, PtX facilitates the integration of these sectors, enabling greater flexibility and efficiency within MCESs.

Energy storage. PtX processes can serve as a form of energy storage by converting surplus renewable electricity into storable heat, fuels, or chemicals. These can be stored and later converted back to electricity, contributing to grid stability and energy security, but also better efficiency, or used as transportation fuels exploited in different sectors. It is important to underline that PtX processes can even offer seasonal energy storage that helps the aforementioned qualities even more. Storage of energy can contribute to more flexibility and allows participation in demand response schemes [8, 18].

Decentralized regional energy systems. PtX technologies enable the development of regional energy systems, where energy production and consumption are distributed, for example, across an energy community. This fostered bottom-up approach can enhance resilience, reduce transmission losses, and empower local communities to participate in energy production and management [19, 20].

Electrification of transport. This concept goes beyond EVs. PtX producing fuels such as hydrogen or synthetic fuels can be used to further decarbonize transportation, particularly in sectors where electrification is challenging, such as aviation and shipping. This application concept contributes to reducing greenhouse gas emissions and dependence on fossil fuels [21].

15.2.2 CHALLENGES POSED BY INTERDEPENDENCIES AMONG ENERGY CARRIERS AND TECHNOLOGIES, AND THE OBJECTIVES OF OPTIMIZING SECTOR COUPLING AND PTX

As could be realized from Section 15.1, PtX technologies allow carriers to interact and couple with each other. But there are hurdles to overcome. While laws and rules are starting to make way for integrated energy systems, like the European Green Deal and the Clean Energy for All Europeans package [22], there is still a lot to do from a technical standpoint. One substantial challenge is understanding how different energy sources (PtX technologies) and systems depend on each other. This is called *carrier interdependency.*

Imagine energy carriers as messengers carrying energy from one place to another, like electricity or heat. These carriers are all connected, so operation pattern in one carrier can affect another. For example, whether we use a certain type of energy in one place might depend on what is happening with a different type of energy somewhere else in the energy system. This interconnection spreads throughout the whole energy system, from where the energy is produced to where it is used.

Now, there are different types of dependencies to consider, as described in [23]. *Internal dependencies* happen within a local energy network and are put into motion by devices such as energy converters that facilitate change of one type of energy into another. *External dependencies*, on the other hand, are influenced by how people choose to use energy and technologies. They happen outside the control of energy system operators and are affected by consumer preferences and technology availability. Understanding these dependencies is crucial for overcoming challenges of MCES, but also formulating the appropriate management strategy.

Having said that, energy management and optimization of use of PtX technologies under these interdependencies are crucial. This issue is intricate due to the many technology options available, fluctuations in energy prices, and significant variations in energy consumption on both a daily and annual basis. Moreover, as environmental concerns grow in importance within these analyses, they must be factored in through relevant variables and constraints. Indeed, managing such a complex problem can lead to conflicting objectives, as, for instance, economic vs. environmental objectives.

Research literature explores various optimization approaches, typically focusing on single- or multi-objective functions, by combining two or three objectives together. In Figure 15.2, the main combination of the multi-objective functions based in the literature is shown. Most studies aim to minimize two objective functions, typically including a cost-related objective and an environmental objective or an energy efficiency objective. Rarely, all three types of objectives are combined, as then the problem becomes complex and the operational solution can be challenging. The primary emphasis is given on minimizing system costs and environmental emissions, particularly CO_2

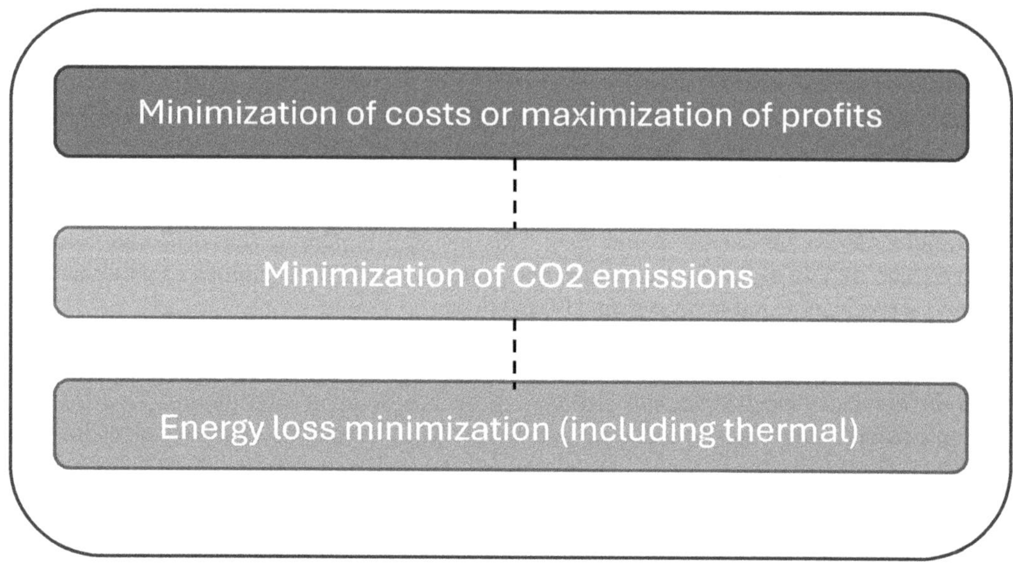

FIGURE 15.2 The multi-objective functions combination found in literature for operation optimization of MCESs.

reduction, while the energy loss (electrical or thermal) is usually combined with the cost-related objective in a two-objective problem formulation.

In some cases, the primary objective of costs can be seen related with an objective that is driven by the specificities of the energy system itself, such as minimization of water extracted from reservoirs or maximization of grid integration. The related bibliography is presented in the review [24].

15.3 SECTOR COUPLING AND PTX TECHNOLOGIES WITH OPERATIONAL PRINCIPLES AND PATTERNS IN MCES

In Figure 15.3, a general layout of an MCES with the related PtX is presented. Based on this, the potential sector coupling solutions are analyzed.

15.3.1 ELECTRICITY–HEATING/COOLING

One of the primary operational principles involves converting surplus electricity into heating or cooling energy through technologies like **heat pumps, electric chillers**, or **electric boilers**. Heat pumps (HPs) utilize electricity to extract heat/cooling from the environment (air, water, or ground) and transfer it indoors for space heating/cooling or water heating. Electric chillers utilize electricity to extract cooling and process cooling applications. Conversely, electric boilers directly convert electricity into heat for space heating or water heating purposes. These technologies offer flexibility in adjusting the heating or cooling output based on demand and electricity availability [25, 26].

Based on the operational pattern, **thermal energy storage systems** can be involved to store surplus heat or cold generated from electricity through renewables during periods of low demand. This stored energy can then be utilized later, when demand is high or when renewable electricity generation is low, as a smart arbitrage between the two carriers.

District heating and cooling (DHC) networks can play a crucial role in sector coupling by distributing centralized heating and cooling energy to multiple buildings or facilities within a local area, as an energy community, for instance. Electricity-driven heat pumps or electric boilers can feed into DHC networks to supply heating or cooling energy efficiently as well. Conversely, waste heat from

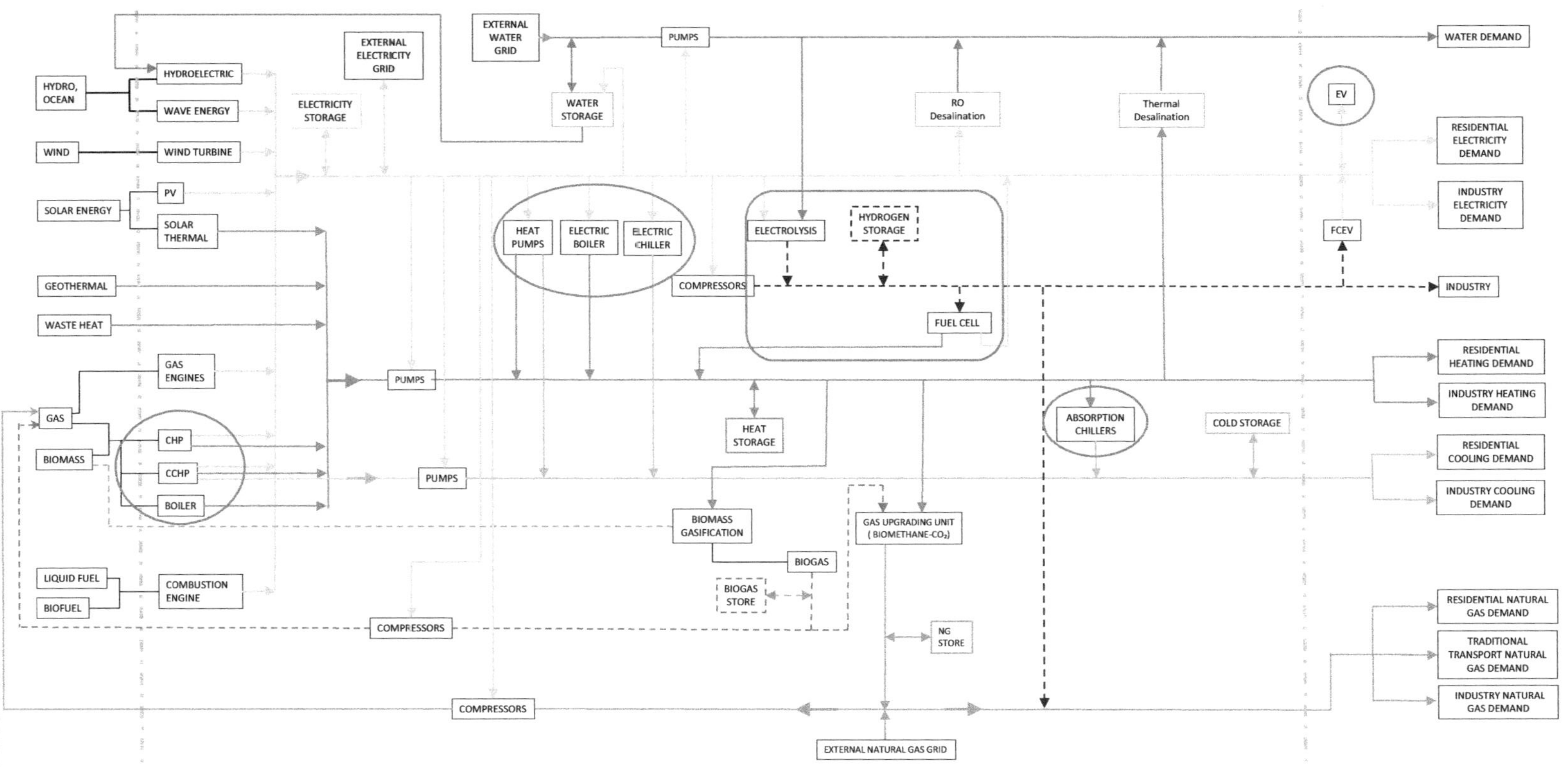

FIGURE 15.3 The overview of an MCES with representative PtX discussed in the chapter.

electricity generation or industrial processes can be utilized for district heating, enhancing overall energy efficiency and reducing environmental impact [27].

Demand-side management strategies play a significant role in optimizing the operation of electricity to heating/cooling technologies within MCESs. Demand response programs allow end users to adjust their heating or cooling consumption based on real-time electricity prices or grid conditions, enabling load shifting and peak shaving. Smart thermostats, building automation systems, and energy management systems facilitate demand-side flexibility and coordination, ensuring efficient use of energy resources.

15.3.2 Electricity–Hydrogen

While the interdependence of electricity and natural gas infrastructures has been widely studied, incorporating hydrogen into these systems has recently become a trend in the field of MCESs and sector coupling, especially after the RePowerEU plan [28]. This trend has sparked interest in power-to-gas technologies, such as electrolyzers, which involve converting surplus renewable energy into gaseous fuels like hydrogen through water electrolysis. This conversion process allows for long-term (seasonal) electricity storage or hydrogen storage and enhances the capacity to store excess renewable energy, thus supporting a transition to a fully renewable energy production system with less volatility [29]. Electrolyzers introduce interdependencies among electricity, gas, and water carriers, as seen in the previous figure. So electrolyzers are electrochemical devices that convert electrical energy into chemical energy by splitting water into hydrogen and oxygen. Due to their electrochemical nature, they are modular, are scalable, and can operate flexibly with minimal performance loss at partial loads. Large-scale electrolyzers facilitate the integration of significant amounts of renewable electricity into the energy system via hydrogen, especially in sectors difficult to electrify directly, such as the industries. From a demand response perspective, electrolyzers can be considered dispatchable loads that can be managed based on overall energy management strategies.

15.3.3 Electricity–Hydrogen–Heating/Cooling

Hydrogen can be further exploited for heating and cooling purposes through fuel cells (FCs), which are electrochemical devices that convert the chemical energy of a fuel gas directly into electrical energy (typically hydrogen, but also other fuel gas mixtures, depending on the FC technology). This direct conversion process allows FCs to achieve high electrical efficiency (up to 40–70%, surpassing the Carnot limit) and moderate thermal efficiency (20–30%), resulting in exceptionally high total efficiency with a high power-to-heat ratio. Because they rely on electrochemical conversion, FCs are modular and suitable for flexible operation, with the added benefit of being quiet and vibration-free.

FCs are primarily categorized into low-temperature and high-temperature types based on the materials used. Low-temperature FCs, such as alkaline and proton-exchange membrane types, operate within a temperature range of 60–100°C. They have an electrical efficiency of 40–50% but offer lower efficiency than high-temperature FCs, like solid oxide and molten carbonate types, which operate between 500 and 900°C, with an electrical efficiency of 60–70%. High-temperature FCs also provide high-grade heat output—when in cogenerative mode—due to their higher process temperatures, making them suitable for a broader range of thermal applications (e.g., space heating, district heating, process heating) compared to low-temperature FCs, which are more suited for domestic hot water production, due to the lower temperature of exhaust heat. However, high-temperature FCs have larger thermal inertia than low-temperature FCs, which limits their flexible operation and their ancillary services potential. For cooling needs coverage, the absorption chiller is needed to further operate [30, 31].

15.3.4 Electricity–Mobility

EVs are progressively replacing internal combustion engine (ICE) powertrains to achieve zero emissions in the mobility sector [32, 33]. The rate of electrification varies across different mobility segments. Small and medium-sized vehicles for short distances (such as those in the automotive sector, lightweight urban mobility vehicles, and last-mile delivery) can be powered directly by batteries with capacities up to 180–300 kWh. In contrast, heavy-duty vehicles for long-distance transportation require higher battery and charging capacities (over 1,000 kWh and ultrafast high-power charging stations), for which alternatives like hydrogen, sustainable fuels, or other carriers might be more suitable [34]. For railway transportation, particularly non-electrified lines, using hydrogen FCs may be more cost-effective than the high investment required for railway electrification. In the maritime sector, direct electrification of ships is challenging (except for short-range shuttle ferries), so decarbonization is primarily achieved through energy efficiency measures, alternative propulsion systems, or low-carbon shore connections. The aviation sector faces similar challenges as maritime transport [35, 36].

Technical constraints, electricity prices, and CO_2 taxation policies significantly influence the implementation of EVs, potentially affecting the competitiveness of different solutions. Charging stations consist of elements and equipment (hardware and software) required to connect EVs for charging. The interface between the EV charging point and the grid is typically hardwired to a control device and protection box for single users (wall box or charging post) but can also be designed as charging islands for multiple EVs (charging hubs for up to 10–12 EVs simultaneously). The power output is limited by the battery's charge acceptance, the charger's nominal ratings, the connector, and the cable between the vehicle and the charger. High charging currents require larger cable diameters to prevent overheating. Power output is mainly defined by local regulations and utility procedures (IEC61851), as well as the local distribution network, power supply characteristics, and constraints of electrical equipment.

15.3.5 Gas–Electricity–Heating/Cooling

Combined heat and power (CHP) systems simultaneously produce electricity and useful heat from a single fuel source, typically natural gas. These systems can be integrated into MCESs to efficiently utilize primary energy sources and reduce overall energy consumption and emissions. By supplying both electricity and heat, CHP can meet the combined needs of residential, commercial, and industrial sectors served by an MCES. Heat is obtained either from the hot exhaust streams or from the process heat, and it is conveyed to a heat transfer fluid in the form of sensible heat (e.g., hot water) or latent heat (e.g., steam). Heat is subsequently used locally or distributed to nearby users via district heating networks. The main CHP technologies powered by gas applied on both large and small scale are the following:

- ICE is based on an internal combustion cycle of a fuel. This technology is typically used in small-scale applications (between 1 and 50 kW), but it can also be scaled up and used in medium- and large-scale ones (up to the MW scale) [37, 38].
- Gas turbines and micro-gas turbines (GTs and mGTs) are based on the Brayton cycle typically fueled by natural gas. GTs present a higher electrical efficiency as compared to ICEs (around 30–40%), with a similar thermal efficiency at the expense of a higher complexity concerning ICEs [39].
- Steam turbines are based on the Hirn or Rankine cycle. The steam is produced in a steam generator fueled by solid or liquid fuels. They are typically used for utility-scale stationary power plants (multi-MW) with large and centralized configurations [40].

- Combined cycles are based on coupled top-bottoming cycles, recycling the high temperature of exhausts coming from a top cycle and used as a heat source for the bottoming one, improving the energy efficiency of the system (an electric efficiency of 40–55%, and a thermal efficiency of 35–40%) [41].

Combined cooling, heat, and power (CCHP) technology is like CHP, with the addition of an energy conversion step to convert all/part of the thermal energy to cooling energy. The main CCHP technology is based on a CHP device coupled with an absorption chiller, which exploits waste heat to provide chilled water for cooling purposes [42]. Similarly, cold energy can either be used locally or distributed to nearby users.

15.4 ANALYTICAL FRAMEWORK AND OPTIMIZATION MODELS FOR PTX TECHNOLOGIES IN MCESS

This section presents the analytical framework addressing the layers of dependency in MCESs, including system components and intersectoral aspects, by also presenting the mathematical models for optimizing the operation of sector coupling and PtX technologies, considering multiple objectives and the related adopted solution methodology.

15.4.1 ANALYTICAL FRAMEWORK ADDRESSING THE LAYERS OF DEPENDENCY IN MCES

Figure 15.4 shows the MCES used for the formulation of the analytical framework, with the distributed resources that can take the role of *energy carrier convertor* or *interconnector between energy carriers*. Convertors are the components that take one energy carrier by converting it into another form of energy carrier, whereas interconnectors convert an energy carrier into the same form for different usage.

The details of this dependency for each of the components are discussed in what follows, along with the optimization models and operational constraints that need to be considered. Additional constrains needed for the operation optimization of the MCES under study are presented in the appendix.

15.4.1.1 Combined Heat and Power Fueled by Natural Gas

As evident from the figure, CHP fueled by natural gas allows linking gas with electricity and a thermal energy carrier, thereby playing the role of a convertor in the process *gas–electricity–heating*. The prime mover of the CHP could be an ICE or a mGT. One of the most important constraints to consider for all distributed technologies and for the CHP is the capacity constraint that ensures that the power provided by the CHP in the MCES is limited by its minimum part load and the size, if the technology is on, that is, the binary decision variable x_{CHP} is equal to 1.

$$\underline{P_{CHP}} x_{CHP,t} \leq P_{CHP,t} \leq \overline{P_{CHP}} x_{CHP,t}$$

$$(15.1)$$

Another constraint to consider is the ramp rate constraint that limits the power generation change between two successive time steps to be within the ramp-down and ramp-up rates, that is:

$$DR_{CHP} \leq P_{CHP,t} - P_{CHP,t-1} \leq UR_{CHP}$$

$$(15.2)$$

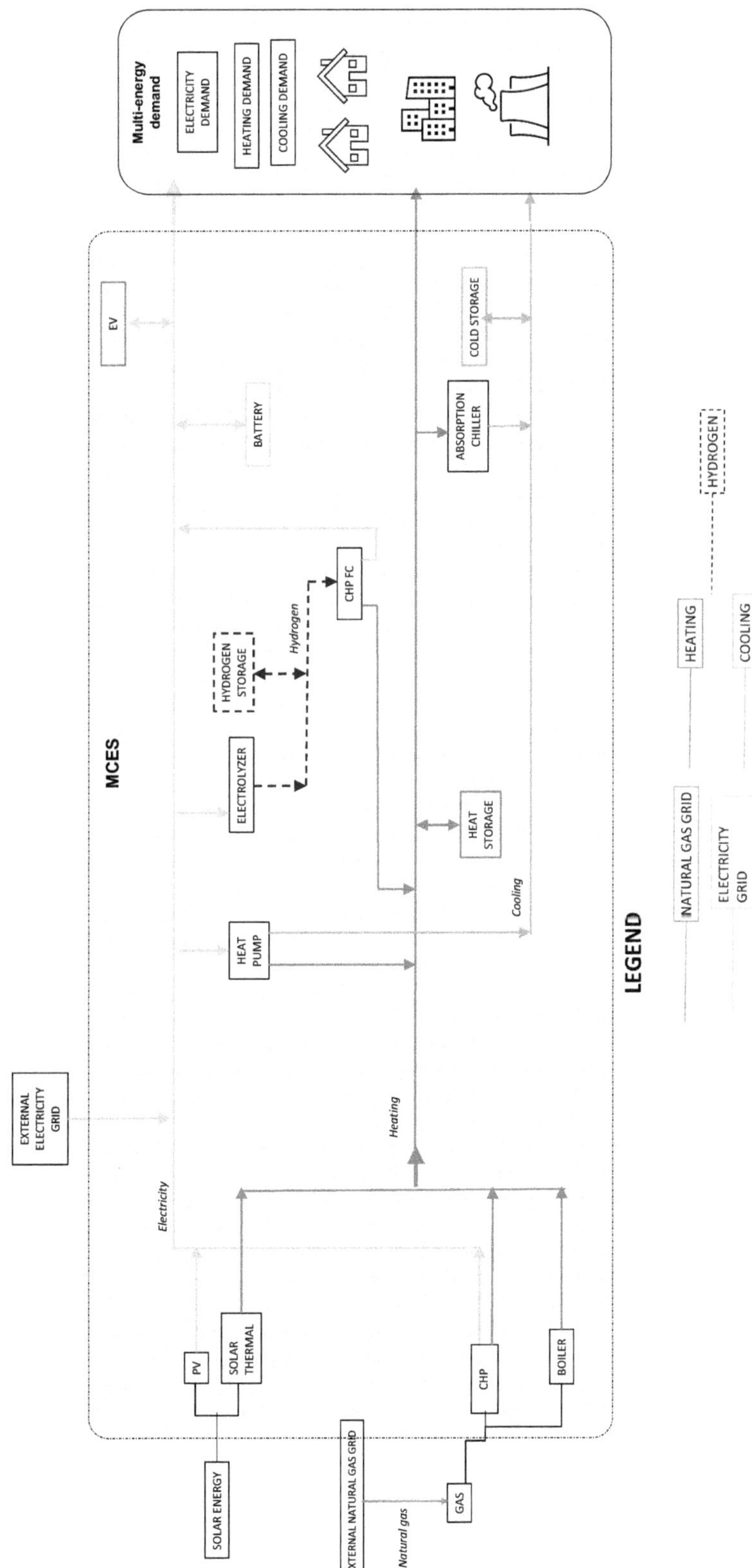

FIGURE 15.4 MCES for the optimization problem formulation.

The amount of natural gas required by the CHP to produce the power is formulated as:

$$G_{CHP,t} = \frac{P_{CHP,t}}{\eta_{e,CHP}\, LHV_{NG}} \tag{15.3}$$

where $\eta_{e,CHP}$ is the electrical efficiency of the CHP and LHV_{NG} is the lower heat value of natural gas. The thermal power recovered from the prime mover is formulated as:

$$H_{CHP,t} = \frac{P_{CHP,t}\, \eta_{th,CHP}}{\eta_{e,CHP}} \tag{15.4}$$

where $\eta_{th,CHP}$ is the thermal efficiency of the CHP. This thermal energy can be used to satisfy the thermal demand and to activate absorption chillers to produce cooling energy. In this latter case, CHP coupled with the absorption chiller is the convertor in the process *gas–electricity–heating/cooling*.

15.4.1.2   Hydrogen-Based Combined Heat and Power

Hydrogen-based CHP systems—as FC—allow linking hydrogen with electricity and thermal energy carrier, thereby playing the role of a convertor in the process *electricity–hydrogen–heating*. The capacity constraint formulated in equation (15.1) and the ramp-rate constraint in equation (15.2) are valid, and the following equations can be considered to model the CHP FC and electrolyzer (EZ):

$$H_{2CHP\,FC,t} = \frac{P_{CHP\,FC,t}}{\left(\eta_{e,CHP\,FC}\, LHV_{H_2}\right)} \tag{15.5}$$

$$P_{EZ,t}^{req} = \frac{H_{2CHP\,FC,t}\, LHV_{H2}}{\eta_{e,EZ}} \tag{15.6}$$

Equation (15.5) allows calculating the amount of hydrogen needed by the CHP FC to produce power and depends on the electrical efficiency of the FC and the LHV of hydrogen. Equation (15.6) allows calculating the power required by the EZ to produce hydrogen and depends on the EZ electrical efficiency. Often, to ensure that only green hydrogen is used to power the FC, in the presence of renewable power generation, the power required by the EZ to produce hydrogen is imposed to be lower or equal to the renewable generated power (for instance, by PV in this case, as shown in Figure 15.4).

The thermal power recovered from the CHP FC can be formulated as in equation (15.4). This thermal energy can be used to satisfy the thermal demand and to activate absorption chillers to produce cooling energy. In this latter case, CHP FC coupled with the absorption chiller is the convertor in the process *electricity–hydrogen–heating/cooling*.

15.4.1.3   Boiler

In the MCES, gas-fired boilers (GBs) are mainly used as auxiliary technologies to support thermal energy production from CHPs or RES plant to satisfy the thermal user demand. In the context of the MCES, they are *convertors*, by converting gas into thermal energy. The capacity constraint formulated in equation (15.1) is still valid. The amount of natural gas required by a gas-fired boiler to produce the thermal power is formulated as:

$$G_{GB,t} = \frac{H_{GB,t}}{\eta_{th,GB}\, LHV_{NG}} \tag{15.7}$$

where $\eta_{th,GB}$ is the GB thermal efficiency.

15.4.1.4 Renewable Energy Sources

Photovoltaic systems (PV) and solar thermal (ST) for domestic hot water production are the RES here considered, as shown in Figure 15.4. In the context of the MCES, they are *convertors*, by converting solar energy into electricity and thermal energy, respectively.

15.4.1.4.1 Solar PV

Solar PV can be used to satisfy the user electricity demand within MCES or to power other energy conversion technologies, such as EZ. The model representing the PV system is expressed thus:

$$P_{PV,t} = A_{PV}\eta_{e,PV}I_t \qquad (15.8)$$

where $P_{PV,t}$ is the generated power at time t, depending on the installed PV area, A_{PV}, the electric efficiency of the PV panels, $\eta_{e,PV}$, and the solar irradiance at time t, I_t.

15.4.1.4.2 Solar Thermal

ST can be used to satisfy the user domestic hot water demand within MCES. The model representing the ST system is expressed thus:

$$H_{ST,t} = A_{ST}\eta_{th,ST}I_t \qquad (15.9)$$

where $H_{ST,t}$ is the generated thermal power at time t, depending on the installed ST area A_{ST}, the thermal efficiency of the solar collectors $\eta_{th,ST}$, and the solar irradiance at time t.

15.4.1.5 Heat Pump

HPs allow linking electricity with the thermal energy carrier, thereby playing the role of a convertor in the process *electricity–heating*. In case the technology is reversible, it is the convertor in the process *electricity–heating/cooling*. The capacity constraint formulated in equation (15.1) is still valid. Considering the heating mode, the HP model can be expressed as:

$$P_{HP,t}^{HM,req} = \frac{H_{HP,t}^{HM}}{COP_{HP}^{HM}} \qquad (15.10)$$

which links the power required by the HP, $P_{HP,t}^{HM,req}$, to the thermal power provided, $H_{HP,t}$, through its coefficient of performance, COP_{HP}^{HM}. The model is similar for the HP operating in cooling mode, by considering the COP of the HP in the cooling mode. As the optimization model is formulated for optimizing the operation strategies of the MCES on a daily basis, there is no need to switch the operation of the heat pump due to the change of the season.

15.4.1.6 Absorption Chiller

Absorption chillers (ACs) can be used to meet the user cooling demand within MCES and can be powered by the CHPs and boilers in the MCES. They can thus be considered as *interconnectors*, converting a thermal energy carrier (heating) into the same form (cooling) for different usage. The AC is expressed thus:

$$C_{AC,t} = \left(\sum_{CHP}H_{CHP,t}^{SC} + \sum_{GB}H_{GB,t}^{SC}\right)COP_{AC} \qquad (15.11)$$

where $C_{AC,t}$ is the cooling power provided by the AC, depending on the thermal power for space cooling purposes provided by the CHP fueled by natural gas and by the CHP FC and the thermal power for space cooling purposes provided by the boiler.

15.4.1.7　Electrical Energy Storage

Electrical energy storage systems consist of a battery energy storage system (BESS), such as lithium-ion. In the MCES, it can be considered as an *interconnector*, converting electricity into the same form for different usage.

The BESS model is formulated thus:

$$SOC_{BESS,t} = SOC_{BESS,t-1} + \frac{P_{BESS,t}^{Ch}\eta_{BESS}^{Ch}\Delta t}{Cap_{BESS}} - \frac{P_{BESS,t}^{Disch}\Delta t}{\eta_{BESS}^{Dis}Cap_{BESS}} \tag{15.12}$$

which defines the battery state-of-charge (SOC) dynamics, where $P_{BESS,t}^{Ch}$ and $P_{BESS,t}^{Disch}$ are the charging and discharging power, respectively; η_{BESS}^{Ch} and η_{BESS}^{Dis} are the charging and discharging efficiencies, respectively; and Cap_{BESS} is the battery capacity.

This model is completed with the following additional equations:

$$\underline{SOC_{BESS}} \leq SOC_{BESS,t} \leq \overline{SOC_{BESS}} \tag{15.13}$$

$$0 \leq P_{BESS,t}^{Ch} \leq x_{BESS,t}^{Ch}\,\overline{P_{BESS}^{Ch}} \tag{15.14}$$

$$0 \leq P_{BESS,t}^{Disch} \leq x_{BESS,t}^{Disch}\,\overline{P_{BESS}^{Disch}} \tag{15.15}$$

$$x_{BESS,t}^{Ch} + x_{BESS,t}^{Disch} \leq 1 \tag{15.16}$$

The upper and lower limits of SOC are enforced by equation (15.13), whereas the battery charging/discharging power limits are enforced in equations (15.14), (15.15). Equation (15.16) ensures that the charging and discharging processes do not occur simultaneously.

15.4.1.8　Thermal Energy Storage

In the MCES, thermal energy storage (TES) for heating and cooling purposes can be considered as an *interconnector*, converting thermal energy into the same form for different usage.

As for the TES for heating, the state dynamic is modeled as:

$$H_{TES-Th,t} = H_{TES-Th,t-1}\left(1-\varphi_{TES-Th}\left(\Delta t\right)\right)+(H_{TES-Th,t}^{Ch} - H_{TES-Th,t}^{Disch})\Delta t \tag{15.17}$$

Meaning, that the energy stored at time t, $H_{TES-Th,t}$, depends on the non-dissipated energy stored at time $(t-1)$ $H_{TES-Th,t-1}$ (based on the loss fraction of the storage, $\varphi_{TES-Th}\left(\Delta t\right)$), and on the net energy flow $(H_{TES-Th,t}^{Ch} - H_{TES-Th,t}^{Disch})$. The model for TES for cooling is similar.

15.4.1.9　Hydrogen Storage

In the MCES, hydrogen storage (Hsto) can be considered as an *interconnector*, converting hydrogen energy carrier into the same form for different usage.

The state dynamic of the hydrogen storage can be modeled as:

$$H_{2Hsto,t}^{sto} = H_{2Hsto,t-1}^{sto}\eta_{Hsto} +\left(H_{2Hsto,t}^{Ch} - H_{2Hsto,t}^{Disch}\right)\Delta t \tag{15.18}$$

which relates the amount of hydrogen stored at time t with the one stored at previous time $(t-1)$ that depends on the efficiency of the hydrogen storage, η_{Hsto}. The operation constraints formulated in equations (15.14)–(15.16) can be formulated also for the hydrogen storage.

15.4.1.10 Electric Vehicles

In the MCESs, EVs can act as distributed storage with bidirectional charging services, namely, grid-to-vehicle (G2V) and vehicle-to-grid (V2G). They act as convertors in the process *electricity–mobility*.

The EV modeling is discussed in what follows [43]. The EV is characterized by specific value for (1) battery capacity, (2) arrival and departure times at/from charging stations, (3) SOC at the arrival time, and (4) SOC desired at the departure time.

Before the arrival and after the departure at/from the charging stations, the EV can be modeled as:

$$\begin{cases} P_{EV,t}^{Ch} = 0 \\ P_{EV,t}^{Disch} = 0 \end{cases} \text{, If } t \left\langle t_{EV}^{arrival}, and\, t \right\rangle t_{EV}^{departure} \tag{15.19}$$

where $P_{EV,t}^{Ch}$ is the charging power for the EV at time t, $P_{EV,t}^{Disch}$ is the discharging power of EV at time t, and $t_{EV}^{arrival}$ and $t_{EV}^{departure}$ are the arrival and departure times, respectively, at/from the charging stations for the EV.

During the parking period, the EV can be modeled as:

$$\begin{cases} SOC_{EV,t_{EV}^{arrival}} = SOC_{EV}^{initial} \\ SOC_{EV,t} = SOC_{EV,t-1} + \dfrac{P_{EV,t}^{Ch} \eta_{EV}^{Ch} \Delta t}{Cap_{EV}} - \dfrac{P_{EV,t}^{Disch} \Delta t}{\eta_{EV}^{Disch} Cap_{EV}}, \\ SOC_{EV,t_{EV}^{departure}} \geq SOC_{EV}^{Desired} \end{cases} \quad \text{if } t_{EV}^{arrival} < t < t_{EV}^{departure} \tag{15.20}$$

where $SOC_{EV,t}$ is the SOC of the battery of EV at time t, $SOC_{EV}^{initial}$ is the SOC of the battery at arrival time at charging station, η_{EV}^{Ch} and η_{EV}^{Disch} are the charging and discharging efficiencies, respectively; Cap_{EV} is the battery capacity of EV; $SOC_{EV}^{Desired}$ is the desired SOC at departure time from the charging station.

In addition to the EV operating mode expressed in equations (15.19) and (15.20), additional operation constraints for EV are defined thus:

$$\begin{cases} 0 \leq P_{EV,t}^{Ch} \leq x_{EV,t}^{Ch}\, \overline{P_{EV}^{Ch}} \\ 0 \leq P_{EV,t}^{Disch} \leq x_{EV,t}^{Disch}\, \overline{P_{EV}^{Disch}} \\ x_{EV,t}^{Ch} + x_{EV,t}^{Disch} \leq 1 \\ \underline{SOC_{EV}} \leq SOC_{EV,t} \leq \overline{SOC_{EV}} \end{cases} \text{, if } t_{EV}^{arrival} < t < t_{EV}^{departure} \tag{15.21}$$

which allow limiting the charging and discharging power for the EV's battery and the battery SOC.

15.4.2 Objective Functions

Traditionally, optimal operation problems related to MCESs considered an economic objective only. As the desired outcome is more complex, the objective functions formulation has later been extended to also account for other factors, such as environmental outcomes.

Regarding the *economic objective function*, the usual metric is the total MCES costs, considering operational costs of the system.

In the analytical framework established here, the economic objective to minimize is formulated thus:

$$F_{obj,eco} = \Delta t \left[\sum_{i \in \{CHP, GB\}} \sum_t \left(Pr_{NG} G_{i,t} \right) + Pr_{PG,t} P_{PG,t} - \sum_{EV} \sum_t \left(Pr_{PG,t} P_{EV,t}^{Disch,sell} \right) - \sum_t \left(Pr_{PG,t} P_t^{Tot,sell} \right) \right] \quad (15.22)$$

This objective function represents the total daily net energy cost to minimize and takes into account (1) the total cost of gas, calculated by multiplying the gas price by the total amount of gas consumed by the CHP and boiler, and (2) the total cost of grid power, calculated by multiplying the time-varying grid power price and the total amount of electricity taken from the grid, minus (3) the profit for selling electricity discharged from EVs and (4) the profit for selling electricity from CHPs and PV into the grid.

As for the *environmental objective*, the usual metric is the total carbon emissions of the MCES, related to its operation. The environmental objective to minimize is then formulated as:

$$F_{obj,env} = \Delta t \left[\sum_{i \in \{CHP, GB\}} \sum_t \left(CI_{NG} LHV_{NG} G_{i,t} \right) + \sum_t \left(CI_{PG} P_{PG,t} \right) \right] \quad (15.23)$$

The total daily carbon emission is thus calculated by considering (1) the carbon emissions associated with the gas consumed by the CHP and boiler, depending on the carbon intensity (CI) of gas, and (2) the total carbon emission associated with the grid power, depending on the carbon intensity of the power grid which the ILEC is connected to.

15.4.3 OVERVIEW OF SOLUTION METHODS

With the economic and environmental objectives formulated earlier, the operation optimization problem has two types of objective function to be minimized. To solve this multi-objective optimization problem, several methods can be used, such as the ε-constrained method or the weighted-sum method, with the latter chosen in this case, as formulated in what follows.

$$Fobj = c\omega F_{obj,eco} + (1 - \omega) F_{obj,env}, \quad (15.24)$$

where c is a constant scaling factor to keep the two objectives at the same order of magnitude, and ω is the weight for the economic objective function, varying in the range of 0–1. When $\omega = 1$, it is to find the solution that minimizes the total daily energy cost of the MCES, and when $\omega = 0$, it is to find the solution that minimizes the total daily carbon emission. When varying the weight ω in the range of [0, 1], the Pareto frontier between economic and environmental objectives can be found.

The problem formulated is linear and involves both discrete and continuous variables. To solve the problem efficiently, branch-and-cut, which is powerful for mixed-integer linear programming problems, can be used.

15.5 CASE STUDIES

The effectiveness of the proposed optimization models is demonstrated through two case studies with different degrees of complexity both in terms of energy carriers considered and end users involved, discussed in the following.

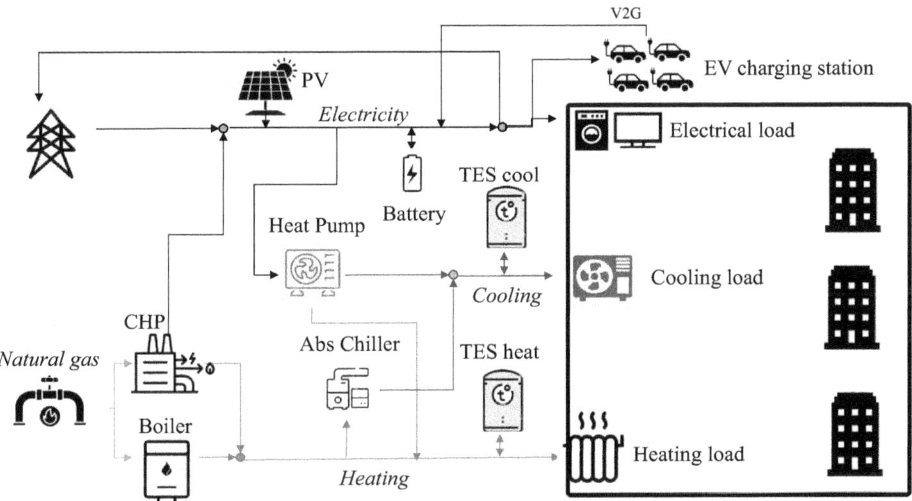

FIGURE 15.5 Scheme of the multi-carrier energy system for case study 1.

15.5.1 Case Study 1: MCES Associated with an Office Building Cluster with Three Interdependent Layers, Including Power-to-Mobility

In this case study, reference is made to an Italian context, with an MCES constituted of office buildings (100 small offices) in the city of Turin, shown in Figure 15.5. Electrical energy (black line) is provided by power grid, PV, CHP, battery, and EVs in V2G mode. This electrical energy can be used to power the electric HP and satisfy the cluster's electrical load, including charging needs for the EVs. The flexibility collected from CHP, PV, and EVs operating in V2G mode can be sold into the wholesale market. Thermal energy (red line) is provided by the CHP, GB, HP, and TES and is used to satisfy the cluster's heating load. Cooling energy (blue line) is provided by the AC, powered by the thermal energy from CHP and GB, the HP and TES, and is used to satisfy the cluster's cooling load.

This case study involves the following interdependent layers:

- Electricity–mobility
- Electricity–heating
- Gas–electricity–heating/cooling

The optimization model is implemented by using IBM ILOG CPLEX Optimization Studio V 12.6. A typical winter day of January is considered with 1 hr as time-step.

The technical data of technologies in the MCES and EVs is shown in Table 15.1 and Table 15.2, respectively.

Simulation results show that under economic optimization, the daily net costs are minimum and equal to €1,574, against maximum carbon emission equal to 20,514 kg CO_2. Under environmental optimization, the contrary occurs, with a maximum net daily cost of €1,995 against minimum carbon emission equal to 14,342 kg CO_2.

The optimized strategies for electricity energy carrier obtained under the economic and environmental optimization are shown in Figure 15.6. Under the economic optimization, grid power is mostly used to satisfy the total electrical load, whereas PV power is used during the central hours of the day, and whereas the battery is discharged to satisfy the load at hour 19, when the grid price is high. Under environmental optimization (Figure 15.6b), the power discharged from EVs is used only for self-use in the MCES, and this allows to minimize environmental impacts by using a carbon-free source to

TABLE 15.1

Technical Data of Energy Technologies in the MCES in Case Study 1

Energy Technology	Size	Efficiency	
		Electrical	*Thermal*
CHP	2.3 MW	0.38	0.41
GB	1.0 MW		0.9
PV	2,000 m²	0.14	
HP	1.2 MW		3.5
TES	2.5 MWh		0.9
BESS	1.0 MWh	$\eta_{BESS}^{Disch} = 0.75$ $\eta_{BESS}^{Ch} = 0.75$	

Source: [43].

TABLE 15.2

Technical Data of EVs in the MCES

EVs Group No.	Number of EVs	Battery Capacity (kWh)	Parking Duration		SOC	
			Arrival	*Departure*	*Initial*	*Desired*
1	30	50	9:00	18:00	0.2	0.8
2	20	50	8:00	17:00	0.2	0.8
3	10	80	10:00	19:00	0.2	0.8

Source: [43].

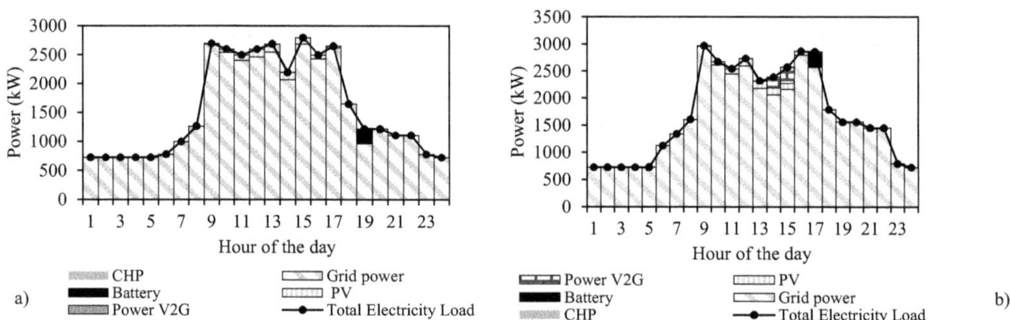

FIGURE 15.6 Optimized results for case study 1 obtained under economic optimization for electricity: (a) economic optimization; (b) environmental optimization.

satisfy the user's load. The CHP is never used, and this result highlights that in order to minimize CO_2 emissions, grid power is very convenient due to the low-carbon intensity of the Italian power grid.

Figure 15.7 shows the flexibility sold into the wholesale market under economic optimization. The power provided by the CHP is never used for self-use in the MCES, and it is all sold into the wholesale market when the electricity market price is high. The same occurs for the power discharged by EVs in V2G mode, which is all sold into the market in correspondence with high electricity market prices. These strategies allow maximizing the revenue for selling flexibility into

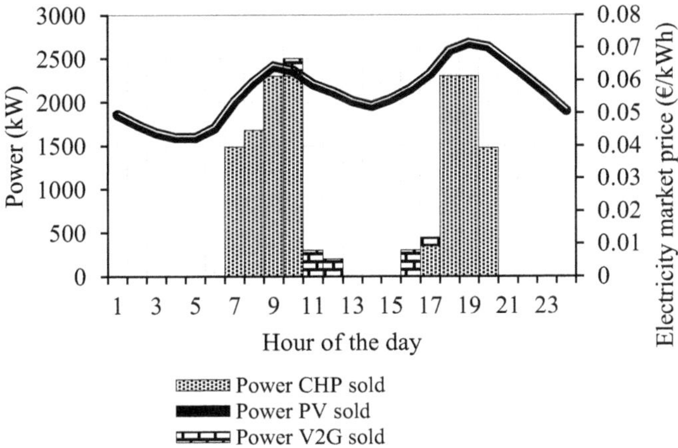

FIGURE 15.7 Flexibility sold into the wholesale market under economic optimization.

the market, thereby allowing minimizing net daily costs for the operator. Under environmental optimization, no flexibility is sold into the wholesale market, as this allows to minimize carbon emissions from the MCES.

By analyzing the case in the absence of EVs (without power-to-mobility), the net daily costs under the economic optimization result to be higher than in the previous case and equal to €1,884, increasing at about 20%, highlighting the contribution of power-to-mobility to unlock flexibility for valorizing it into the market. Under environmental optimization, CO_2 emissions reduce by about 6% as compared to the emissions in the previous case. This reduction is due to the fact that although the usage of power discharged from EVs represents a carbon-free source—contributing to the reduction of carbon emissions—the presence of EVs represents an electrical load to satisfy during the charging phase, with consequent larger usage of energy resources and related emissions.

15.5.2 Case Study 2: MCES Associated with a Residential Building Cluster with Three Interdependent Layers, Including Power-to-Hydrogen–Heating/Cooling

In this case study, the proposed methodology is employed for an MCES for a residential buildings cluster located in Torino, Italy, composed of 100 housing units.

This case study involves the following interdependent layers:

- Electricity–hydrogen–heating/cooling
- Electricity–heating/cooling
- Gas–electricity–heating/cooling

Figure 15.4 well represents the case study, by noting that EVs are not present here. Input data to the model include the aggregated users' energy demands, solar irradiance profiles, energy prices, technical characteristics of the technologies in the MCES (Table 15.3), and the carbon intensity of the energy carriers input to the MCES.

Reference is made to a hot season day, with 1 hr as time-step. Also, in this case, the optimization model is implemented by using IBM ILOG CPLEX Optimization Studio Version 12.6.

The most representative results are presented in what follows with reference to environmental optimization. The total daily cost for the hot season day is equal to €552, whereas carbon emission is equal to 1,279 kg CO_2. Figure 15.8 shows the optimized results for case study 2 obtained under environmental optimization for electricity. It can be noted that most of the total electrical load

TABLE 15.3

Technical Data of Energy Technologies in the MCES

Energy Technology	Size	Efficiency	
		Electrical	*Thermal*
CHP ICE	26 kW	0.32	0.64
CHP FC	16 kW	0.55	0.35
EZ	22 kW	0.70	
GB	84 kW		0.90
PV	1,260 m²	0.14	
ST	240 m²		0.60
AC	100 kW		0.8
HP	1.3 MW		3.5
TES	36 kWh		0.9
H₂sto	87 kWh		1.0

Source: [44].

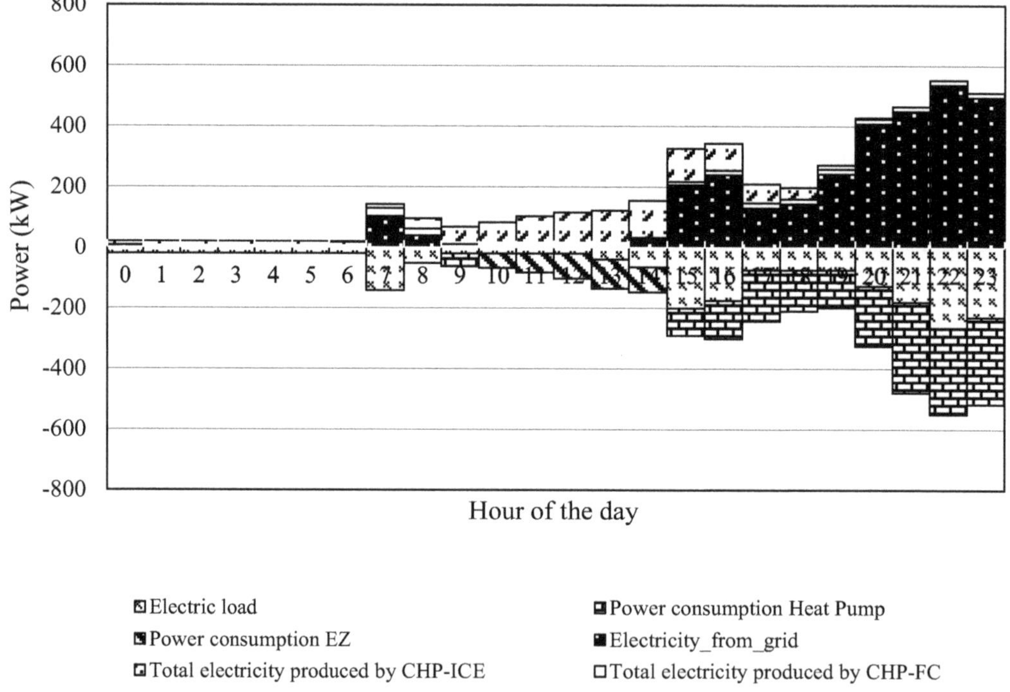

FIGURE 15.8 Optimized results for case study 2 obtained under environmental optimization for electricity.

(electrical demand summed to electricity required by the HP) in the MCES is covered by grid power due to the low carbon intensity of the Italian electricity grid. PV power is used in central hours of the day in the presence of solar irradiance. From hour 10 to hour 14, it is also used to power the electrolyzer and produce H_2 that is stored in the associated H_2 storage and then used to power the CHP FC to cover part of the electrical load, as occurs from hour 16 till the end of the day.

15.6   CONCLUSIONS

This chapter presented an analytical framework for the operational optimization of sector coupling in the context of MCES via PtX technologies.

The different PtX technologies have been discussed under the interdependencies of MCES, highlighting challenges such as the ones that are related to the operation of the complex system, the optimization of objectives that can be contradicting and regulatory gaps.

In the next, their mathematical modeling and a multi-objective optimization framework have been provided by considering all layers of the dependency, including the system components and intersectoral aspects. The effectiveness of the proposed optimization model has been demonstrated through two case studies.

It can be seen that PtX technologies and their optimal management can contribute to a more economic and environmentally friendly integrated energy system.

APPENDIX

ADDITIONAL CONSTRAINTS FOR THE DISTRIBUTED TECHNOLOGIES

$$P_{CHP,t} = P_{CHP,t}^{sell} + P_{CHP,t}^{self} \tag{A1}$$

$$H_{CHP,t} = H_{CHP,t}^{Th} + H_{CHP,t}^{SC} \tag{A2}$$

$$P_{CHP\ FC,t} = P_{CHP\ FC,t}^{sell} + P_{CHP\ FC,t}^{self} \tag{A3}$$

$$H_{CHP\ FC,t} = H_{CHP\ FC,t}^{Th} + H_{CHP\ FC,t}^{SC} \tag{A4}$$

$$P_{PV,t} = P_{PV,t}^{sell} + P_{PV,t}^{self} \tag{A5}$$

$$P_{EV,t}^{Disch} - P_{EV,t}^{Disch,sell} + P_{EV,t}^{Disch,self} \tag{A6}$$

$$P_t^{Tot,sell} = P_{CHP,t}^{sell} + P_{CHP\ FC,t}^{sell} + P_{PV,t}^{sell} + P_{EV,t}^{Disch,sell} \tag{A7}$$

SYSTEM BALANCE CONSTRAINTS (POWER AND THERMAL ENERGY BALANCES)

$$P_{dem,t} + P_{HP,t}^{req} + P_{EZ,t}^{req} = P_{CHP,t}^{self} + P_{CHP\ FC,t}^{self} + P_{PV,t}^{self} + P_{PG,t} + P_{BESS,t}^{Disch} + \sum_{EV} P_{EV,t}^{Disch,self} - P_{BESS,t}^{Ch} - \sum_{EV} P_{EV,t}^{Ch} \tag{A8}$$

$$H_{dem,t} - H_{CHP,t} + H_{CHP\ FC,t} + H_{GB,t} + H_{HP,t}^{HM} + H_{ST,t} + H_{TES-Th,t}^{Disch} - H_{TES-Th,t}^{Ch} \tag{A9}1$$

Nomenclature
Decision Variables

$H_{2CHP\ FC,t}$	H_2 produced by the CHP FC (m^3)
$H_{2Hsto,t}^{Ch}$	H_2 charging for H_2 storage (m^3/h)
$H_{2Hsto,t}^{Disch}$	H_2 discharging for H_2 storage (m^3/h)
$C_{AC,t}$	cooling rate provided by absorption chiller (dependent variable) (kW)

$F_{obj,eco}$	economic objective function (€)
$F_{obj,env}$	environmental objective function ($kgCO_2$)
$G_{CHP,t}$	gas volumetric flow rate consumed by CHP (dependent variable) (Nm^3/h)
$G_{GB,t}$	gas volumetric flow rate consumed by gas boiler (dependent variable) (Nm^3/h)
$H2_{Hsto,t}^{sto}$	H_2 stored in H_2 storage (m^3)
$H_{CHP\,FC,t}$	heat rate provided by CHP FC (dependent variable) (kW)
$H_{CHP,t}$	heat rate provided by CHP (dependent variable) (kW)
$H_{GB,t}$	heat rate provided by gas boiler (kW)
$H_{HP,t}^{HM}$	heat rate provided by the heat pump in heating mode (kW)
$H_{ST,t}$	heat rate generated by ST (kW)
$H_{TES-Th,t}$	thermal energy stored in TES (kWh)
$H_{TES-Th,t}^{Ch}$	charging heat rate to TES (kW)
$H_{TES-Th.t}^{Disch}$	discharging heat rate from TES (kW)
$P_{BESS,t}^{Ch}$	charging power for battery (kW)
$P_{BESS,t}^{Disch}$	discharging power for battery (kW)
$P_{CHP\,FC,t}$	power provided by CHP FC (kW)
$P_{CHP,t}$	power provided by CHP (kW)
$P_{EV,t}^{Ch}$	charging power for the EV (kW)
$P_{EV,t}^{Disch,sell}$	discharging power rate for the EV for selling back to the grid (kW)
$P_{EV,t}^{Disch}$	discharging power for the EV (kW)
$P_{EZ,t}^{req}$	power required by the electrolyzer (dependent variable) (kW)
$P_{HP,t}^{HM,req}$	power required by the heat pump in heating mode (dependent variable) (kW)
$P_{PG,t}$	grid power (kW)
$P_{PV,t}$	power provided by PV (kW)
$P_t^{Tot,sell}$	total power from CHPs and PV sold back to the grid (kW)
$SOC_{EV,t}$	EV battery SOC
$x_{BESS,t}^{Ch}$	binary variable for usage of battery for charging process
$x_{BESS,t}^{Disch}$	binary variable for usage of battery for discharging process
$x_{CHP,t}$	on/off status of CHP
$x_{EV,t}^{Ch}$	binary variable for charging EV
$x_{EV,t}^{Disch}$	binary variable for discharging EV
c	constant in $Fobj$ ($kgCO_2$/€)
$Fobj$	objective function of the multi-objective optimization problem
$SOC_{BESS,t}$	battery SOC
ω	weight value in $Fobj$

Parameters

$\overline{P_{BESS}^{Ch}}$	maximum charging power of battery (kW)
$\overline{P_{BESS}^{Disch}}$	maximum discharging power of battery (kW)
$\overline{P_{CHP}}$	maximum load of CHP (kW)
$\underline{P_{CHP}}$	minimum part load of CHP (kW)
$\overline{P_{EV}^{Ch}}$	maximum charging power for EV battery (kW)
$\overline{P_{EV}^{Disch}}$	maximum discharging power for EV battery (kW)
$\overline{SOC_{EV}}$	maximum EV SOC battery
A_{PV}	installed area of PV (m^2)

A_{ST}	installed area of ST (m²)
CI_{NG}	carbon intensity of natural gas (kgCO$_2$/Nm³)
CI_{PG}	carbon intensity of power grid (kgCO$_2$/kWh)
COP_{AC}	COP of the absorption chiller
COP_{HP}^{HM}	COP of heat pump in heating mode
Cap_{BESS}	capacity of battery (kWh)
Cap_{EV}	EV battery capacity (kWh)
DR_{CHP}	maximum ramp-down rate of CHP (kW)
$H_{dem,t}$	time-varying heat rate demand (kW)
I_t	hourly solar irradiance (kW/m²)
LHV_{NG}	lower heat value of natural gas (kWh/Nm³)
$P_{dem,t}$	time-varying power demand (kW)
Pr_{NG}	natural gas price (€/Nm³)
$Pr_{PG,t}$	time-varying unit price of grid power (€/kWh)
$\underline{\underline{SOC_{BESS}}}$	minimum SOC battery
$\overline{SOC_{BESS}}$	maximum SOC battery
$\underline{SOC_{EV}}$	minimum EV SOC battery
$SOC_{EV}^{Desired}$	desired EV battery SOC at departure time from charging station
$SOC_{EV}^{initial}$	EV SOC at arrival time at charging station
UR_{CHP}	maximum ramp-up rate of CHP (kW)
$t_{EV}^{arrival}$	arrival time of PEV to charging station (hr)
$t_{EV}^{departure}$	departure time of PEV from charging station (hr)
η_{BESS}^{Ch}	efficiency of charging process for battery
η_{BESS}^{Disch}	efficiency of discharging process for battery
η_{EV}^{Ch}	charging efficiency of EV battery
η_{EV}^{Disch}	discharging efficiency of EV battery
η_{Hsto}	efficiency of H$_2$ storage
$\eta_{e,CHP}$	electric efficiency of CHP
$\eta_{e,CHP\ FC}$	electric efficiency of CHP FC
$\eta_{e,EZ}$	electric efficiency of EZ
$\eta_{e,PV}$	electric efficiency of PV
$\eta_{th,GB}$	thermal efficiency of gas boiler
$\eta_{th,CHP}$	thermal efficiency of CHP
$\eta_{th,ST}$	thermal efficiency of ST
φ_{TES-Th}	TES storage loss fraction
Δt	length of the time interval (1 hr)
LHV_{H2}	lower heat value of hydrogen (kWh/m³)

Subscripts/Superscripts

HM	heating mode
SC	space cooling purposes
$self$	self-use in the MCES
$sell$	sold
Th	thermal purposes

NOTE

1   The cooling energy balance can be formulated in a similar way.

REFERENCES

[1] Intergovernmental Panel on Climate Change (IPCC)-Contribution of Working Group II to the Sixth Assessment Report of the Intergovernmental Panel on Climate Change, *Climate Change 2022 – Impacts, Adaptation and Vulnerability*. Cambridge University Press, 2023. doi: 10.1017/9781009325844.

[2] G. Comodi, G. Spinaci, M. Di Somma, and G. Graditi, "Transition potential of local energy communities," in *Technologies for Integrated Energy Systems and Networks*, Wiley, 2022, pp. 275–304. doi: 10.1002/9783527833634.ch11.

[3] EU Commission, "A European Green Deal Striving to be the first climate-neutral continent," https://ec.europa.eu/info/strategy/priorities-2019–2024/european-green-deal_en.

[4] IEA, "CCUS in clean energy transitions," Paris, Licence: CC BY 4.0, 2020. Accessed: June 10, 2024. [Online]. www.iea.org/reports/ccus-in-clean-energy-transitions.

[5] G. Graditi and Marialaura Di Somma, *Technologies for Integrated Energy Systems and Networks*. Wiley, 2022. doi: 10.1002/9783527833634.

[6] Md. B. Hossain, Md. R. Islam, K. M. Muttaqi, D. Sutanto, and A. P. Agalgaonkar, "Advancement of fuel cells and electrolyzers technologies and their applications to renewable-rich power grids," *Journal of Energy Storage*, vol. 62, p. 106842, June 2023. doi: 10.1016/j.est.2023.106842.

[7] M. Brenna, F. Foiadelli, D. Zaninelli, G. Graditi, and M. Di Somma, "The integration of electric vehicles in smart distribution grids with other distributed resources," in *Distributed Energy Resources in Local Integrated Energy Systems*. Elsevier, 2021, pp. 315–345. doi: 10.1016/B978-0-12-823899-8.00006-6.

[8] C. N. Papadimitriou, A. Anastasiadis, C. S. Psomopoulos, and G. Vokas, "Demand response schemes in energy hubs: A comparison study," in *Energy Procedia*, 2019. doi: 10.1016/j.egypro.2018.11.260.

[9] A. R. Dahiru, A. Vuokila, and M. Huuhtanen, "Recent development in Power-to-X: Part I—A review on techno-economic analysis," *Journal of Energy Storage*, vol. 56, p. 105861, 2022. doi: 10.1016/j.est.2022.105861.

[10] S. G. Simoes, J. Catarino, A. Picado, T. F. Lopes, S. Di Berardino, F. Amorim, F. Girio, C. M. Rangel, and T. Ponce de Leao, "Water availability and water usage solutions for electrolysis in hydrogen production," *Journal of Cleaner Production*, vol. 315, p. 128124, September 2021. doi: 10.1016/j.jclepro.2021.128124.

[11] M. Mohammadi, Y. Noorollahi, B. Mohammadi-ivatloo, M. Hosseinzadeh, H. Yousefi, and S. T. Khorasani, "Optimal management of energy hubs and smart energy hubs – A review," *Renewable and Sustainable Energy Reviews*, vol. 89, pp. 33–50, June 2018. doi: 10.1016/j.rser.2018.02.035.

[12] B. van der Holst, G. Verhoeven, M. Kazemi, C. Papadimitriou, M. Di Somma, and K. Kok, "Integrated flexibility solutions for effective congestion management in distribution grids," in *Integrated Local Energy Communities: From the Concept and Enabling Conditions to the Optimal Planning and Operation*, M. Di Somma, C. Papadimitriou, G. Graditi, and K. Kok, Eds., Wiley, 2024. doi: 10.1002/9783527843282.ch7

[13] I. Kountouris, L. Langer, R. Bramstoft, M. Münster, and D. Keles, "Power-to-X in energy hubs: A Danish case study of renewable fuel production," *Energy Policy*, vol. 175, p. 113439, April 2023. doi: 10.1016/j.enpol.2023.113439.

[14] A. Morch, H. Sæle, J. Merino, A. Cortés, M. Santos-Mugica, J. Fraile, Á. Gutiérrez, D. Jiménez, M. Di Somma, A. Buonanno, and M. Caliano, "eNeuron project: Local multi-vector energy systems within the European political and regulatory landscape: Scope and key priorities for the study," 2021.

[15] Directorate-General for Energy (European Commission), "The role and potential of Powerto-X in 2050," April 2019. Accessed: June 11, 2024. [Online]. https://op.europa.eu/en/publication-detail/-/publication/1e6b9012–6bbc-11e9–9f05–01aa75ed71a1/language-en/format-PDF/source96288622.

[16] IRENA, "Green hydrogen cost reduction: Scaling up electrolysers to meet the 1.5°C climate goal, International Renewable Energy Agency," Abu Dhabi, 2020.

[17] A. Morch, M. Di Somma, C. Papadimitriou, H. Sæle, V. Palladino, J. Fraile Ardanuy, G. Conti, M. Rossi, and G. Comodi, "Technologies enabling evolution of Integrated Local Energy Communities," in *2022 IEEE International Smart Cities Conference (ISC2)*, IEEE, September 2022, pp. 1–6. doi: 10.1109/ISC255366.2022.9922568.

[18] M. Qi, D. Nguyen Vo, H. Yu, C.-M. Shu, C. Cui, Y. Liu, J. Park, and I. Moon, "Strategies for flexible operation of power-to-X processes coupled with renewables," *Renewable and Sustainable Energy Reviews*, vol. 179, p. 113282, June 2023. doi: 10.1016/j.rser.2023.113282.

[19] H. Sale, A. Morch, A. Buonanno, M. Caliano, M. Di Somma, and C. Papadimitriou, "Development of energy communities in Europe," in *2022 18th International Conference on the European Energy Market (EEM)*, IEEE, September 2022, pp. 1–5. doi: 10.1109/EEM54602.2022.9921054.

[20] M. Di Somma, C. Papadimitriou, A. Morch, H. Sæle, P. Richardson, A. Coccia, and A. Buonanno, "An innovative toolbox for the optimal design and operation of integrated local energy communities," in *27th International Conference on Electricity Distribution (CIRED 2023)*, Institution of Engineering and Technology, 2023, pp. 676–680. doi: 10.1049/icp.2023.0460.

[21] N. Gray, S. McDonagh, R. O'Shea, B. Smyth, and J. D. Murphy, "Decarbonising ships, planes and trucks: An analysis of suitable low-carbon fuels for the maritime, aviation and haulage sectors," *Advances in Applied Energy*, vol. 1, p. 100008, February 2021. doi: 10.1016/j.adapen.2021.100008.

[22] EU Commission, "Clean energy for all Europeans package." Accessed: June 12, 2024. [Online]. https://energy.ec.europa.eu/topics/energy-strategy/clean-energy-all-europeans-package_en.

[23] N. Neyestani, "Modeling of multienergy carriers dependencies in smart local networks with distributed energy resources," in *Distributed Energy Resources in Local Integrated Energy Systems*, Elsevier, 2021, pp. 63–87. doi: 10.1016/B978-0-12-823899-8.00012-1.

[24] C. Papadimitriou, M. Di Somma, C. Charalambous, M. Caliano, V. Palladino, A. F. C. Borray, A. González-Garrido, N. Ruiz, and G. Graditi, "A comprehensive review of the design and operation optimization of energy hubs and their interaction with the markets and external networks," *Energies (Basel)*, vol. 16, no. 10, p. 4018, May 2023. doi: 10.3390/en16104018.

[25] J.-P. Jimenez-Navarro, K. Kavvadias, F. Filippidou, M. Pavičević, and S. Quoilin, "Coupling the heating and power sectors: The role of centralised combined heat and power plants and district heat in a European decarbonised power system," *Applied Energy*, vol. 270, p. 115134, July 2020. doi: 10.1016/j.apenergy.2020.115134.

[26] A. Wiedermann, M. Calderoni, O. Bernstrauch, C. Herce, T. Nowak, P. Jansohn, R. Pastor, D. Rutz, A. Iliceto, I. Ionel, and A. Ilo, "Coupling of heating/cooling and electricity sectors in a renewable energy-driven Europe, Publications Office of the European Union," 2022, Accessed: June 12, 2024. [Online]. https://op.europa.eu/en/publication-detail/-/publication/919a8405-6ed7-11ed-9887–01aa75ed71a1.

[27] D. Connolly, H. Lund, B. V. Mathiesen, S. Werner, B. Möller, U. Persson, T. Boermans, D. Trier, P. A. Østergaard, and S. Nielsen, "Heat roadmap Europe: Combining district heating with heat savings to decarbonise the EU energy system," *Energy Policy*, vol. 65, pp. 475–489, February 2014. doi: 10.1016/j.enpol.2013.10.035.

[28] EU, "Communication from the Commission to the European Parliament, The European Council, The Council, The European Economic and Social Committee and the Committee of the Regions REPowerEU Plan."

[29] G. Gahleitner, "Hydrogen from renewable electricity: An international review of power-to-gas pilot plants for stationary applications," *International Journal of Hydrogen Energy*, vol. 38, no. 5, pp. 2039–2061, February 2013. doi: 10.1016/j.ijhydene.2012.12.010.

[30] H. Q. Nguyen and B. Shabani, "Proton exchange membrane fuel cells heat recovery opportunities for combined heating/cooling and power applications," *Energy Conversion Management*, vol. 204, p. 112328, January 2020. doi: 10.1016/j.enconman.2019.112328.

[31] R. K. Pachauri and Y. K. Chauhan, "A study, analysis and power management schemes for fuel cells," *Renewable and Sustainable Energy Reviews*, vol. 43, pp. 1301 1319, March 2015. doi: 10.1016/j.rser.2014.11.098.

[32] Ö. Gönül, A. C. Duman, and Ö. Güler, "Electric vehicles and charging infrastructure in Turkey: An overview," *Renewable and Sustainable Energy Reviews*, vol. 143, p. 110913, June 2021. doi: 10.1016/j.rser.2021.110913.

[33] J. Martínez-Lao, F. G. Montoya, M. G. Montoya, and F. Manzano-Agugliaro, "Electric vehicles in Spain: An overview of charging systems," *Renewable and Sustainable Energy Reviews*, vol. 77, pp. 970–983, September 2017. doi: 10.1016/j.rser.2016.11.239.

[34] A. Monforti Ferrario, V. Cigolotti, A. M. Ruz, F. Gallardo, J. García, and G. Monteleone, "Role of hydrogen in low-carbon energy future," in *Technologies for Integrated Energy Systems and Networks*, Wiley, 2022, pp. 71–104. doi: 10.1002/9783527833634.CH4.

[35] H. Xing, C. Stuart, S. Spence, and H. Chen, "Fuel cell power systems for maritime applications: Progress and perspectives," *Sustainability*, vol. 13, no. 3, p. 1213, January 2021. doi: 10.3390/su13031213.

[36] S. Habib, M. M. Khan, F. Abbas, L. Sang, M. U. Shahid, and H. Tang, "A comprehensive study of implemented international standards, technical challenges, impacts and prospects for electric vehicles," *IEEE Access*, vol. 6, pp. 13866–13890, 2018. doi: 10.1109/ACCESS.2018.2812303.

[37] N. A. Diangelakis, C. Panos, and E. N. Pistikopoulos, "Design optimization of an internal combustion engine powered CHP system for residential scale application," *Computational Management Science*, vol. 11, no. 3, pp. 237–266, July 2014. doi: 10.1007/s10287-014-0212-z.

[38] H. I. Onovwiona, V. Ismet Ugursal, and A. S. Fung, "Modeling of internal combustion engine based cogeneration systems for residential applications," *Applied Thermal Engineering*, vol. 27, nos. 5–6, pp. 848–861, April 2007. doi: 10.1016/j.applthermaleng.2006.09.014.

[39] P. A. Pilavachi, "Mini- and micro-gas turbines for combined heat and power," *Applied Thermal Engineering*, vol. 22, no. 18, pp. 2003–2014, December 2002. doi: 10.1016/S1359-4311(02)00132-1.

[40] G.-P. Negreanu, I. Oprea, and V. Berbece, "Some design characteristics of micro steam turbines for agricultural biomass energy conversion," *E3S Web of Conferences*, vol. 180, p. 01017, July 2020. doi: 10.1051/e3sconf/202018001017.

[41] S. O. Sanjay and B. N. Prasad, "Thermodynamic evaluation of advanced combined cycle using latest gas turbine," in *Volume 3: Turbo Expo 2003*, ASMEDC, January 2003, pp. 95–101. doi: 10.1115/GT2003-38096.

[42] Y. Shi, M. Liu, and F. Fang, *Combined Cooling, Heating, and Power Systems*, Wiley, 2017. doi: 10.1002/9781119283362.

[43] M. Di Somma, L. Ciabattoni, G. Comodi, and G. Graditi, "Managing plug-in electric vehicles in eco-environmental operation optimization of local multi-energy systems," *Sustainable Energy, Grids and Networks*, vol. 23, p. 100376, September 2020. doi: 10.1016/j.segan.2020.100376.

[44] L. Jin, M. Rossi, L. Ciabattoni, M. Di Somma, G. Graditi, and G. Comodi, "Environmental constrained medium-term energy planning: The case study of an Italian university campus as a multi-carrier local energy community," *Energy Conversion Management*, vol. 278, p. 116701, February 2023. doi: 10.1016/j.enconman.2023.116701.

16 Power-to-X for Improved Flexibility in Modern Energy Landscapes

Mohammad Reza Sheibani, Abbas Marini,
Seyed Mohsen Hashemi, and Mehdi Zeraati

16.1 INTRODUCTION

A continuous increase in energy consumption and demand, population growth in the world, and as well as requests to decrease greenhouse gas (GHG) emissions, and resource reduction, are global concerns related to energy supply and security. These factors have motivated attaining sustainable development without relying on fossil fuel sources. Fossil fuels are the main reasons of emissions in the world, and energy produced from conventional resources affects the climate change. In 2018, fossil fuel–based power plants caused a 1.7% increase in global pollution [1–3].

In response to this challenge, world leaders are increasingly focused on evolving environmentally friendly approaches for generating and storing green energy. A significant step toward reaching these goals is the participation of several countries in the Paris Agreement in order to reduce greenhouse gas emissions [1–3].

In the roadmaps of 21 European countries, ambitious goals have been set to produce 20% of the total energy needed from renewable energy resources (RES). These new objectives aim to transition toward net-zero carbon energy production by 2050 in select countries. Within this context, wind and solar energies play pivotal roles. Notably, Germany leads as the largest wind energy producer, generating approximately 116 TWh of energy. Other prominent wind energy producers in Europe include England, Spain, and France. Globally, the installed wind capacity has surged from 17.4 GW in 2000 to 318.2 GW in 2013, and an impressive 1,021 GW in 2023. Meanwhile, China stands out as the world's largest wind energy producer, contributing around 651 TWh. Policy initiatives and market projections underscore the critical role of RES, particularly solar and wind energy, in shaping the energy landscape by 2030 and 2050 [3–5].

The intermittency nature of the RESs creates numerous challenges in providing the stability of the power grid. Solar energy and wind energy are constantly fluctuating due to changes in their primary energy sources (changes in sunlight and wind). Therefore, the amount of electricity supplied by them is constantly fluctuating. These fluctuations are normally offset by the electricity demand of the consumer [6–8].

The importance of reducing greenhouse gases and, on the other hand, the limitation of fossil fuels to provide the energy needed by different consumers have caused different energy systems based on different energy carriers to be modeled in an integrated way and for energy management to be done simultaneously for all of them. In fact, we will face modern energy systems, including different energy carriers. In modern energy systems, to cover the challenges caused by the presence of the RESs, the concept of flexibility is proposed to measure the ability of the systems to overcome the challenge of the changing and unpredictability of the output of the RESs.

Various elements can help provide flexibility in the system. Energy storage systems and demand response are among the solutions to improve flexibility, whose costs and technical limitations are among the upcoming challenges for their use [9–11]. Power-to-X (PtX) systems, due to their unique capabilities, are among the upcoming choices to improve the conditions for providing the flexibility of power systems, which is focused on in this chapter of the book.

The concept of PtX revolves around the conversion of power (electricity) into various substances (X), mostly chemical, which can be very diverse. In this chapter, PtX systems are studied with the aim of improving the flexibility of modern energy systems.

Modern power systems are networks that, by being equipped with modern equipment, have the ability to exchange energy with other energy systems (with different energy carriers other than electrical energy). In this way, the ability to simultaneously manage energy based on different energy carriers is created. For this purpose, the importance of flexibility in power systems is examined first. After that, different PtX systems will be studied as mediators of different energy systems. Then, PtX systems are compared with other tools for improving flexibility in power systems. Finally, how to operate modern power systems with the presence of PtX systems is studied, and how to plan for the development of these systems has been formulated.

16.2 SCOPE AND IMPORTANCE OF FLEXIBILITY IN ENERGY SYSTEMS

A *power system* is a vital infrastructure supporting other different infrastructural systems. So the appropriate performance of the power system is pivotal for the good performance of other infrastructures. In traditional power systems, electrical energy is produced in large power plants located in far distances from load centers and is transferred through large power lines. Fossil fuels, such as gas, gas oil, and coal, are the main primary energy carriers of large thermal power plants. Such power plants are one of the main generation resources of GHGs. The environmental constraints and the access limitation to fossil fuel resources in the future [12–14] have caused the transition to RESs as unlimited and clean resources. Although the operation cost of RESs is very low, their large investment costs at the beginning years of emerging were a significant obstacle for their extensive extension. However, by reducing their capital cost in recent years, economic obstacles have been removed, and they are introduced as a serious competitor for thermal power plants. Different countries are programming for net zero based on full penetration of RESs in their energy portfolio [15, 16]. However, there are some hard challenges for this condition. Remote areas may have high potentials for installing RESs; however, large power lines have to be installed to transfer huge amounts of power through large distances [17]. Other challenges are the system operation problems. Variable renewable energy (VRE) resources are a popular kind of RESs, including photovoltaic (PV) solar systems and wind turbines. Their power generation has fluctuations and depends on weather conditions. VREs cause challenges for system operators as they have to provide enough reserve capacity as backup of VREs. An important question is if it is possible to manage and control a bulk power system having high penetration of RESs? To consider such concerns, the flexibility concept has been introduced in the power system context. Different research studies have defined the flexibility concept in different ways. In general, *flexibility* is the capability to handle the variability and uncertainties of the power demand and generation in the most reliable and economic manner [18]. It may include different timescales, from some minutes to an hour. The need for flexibility to accommodate sudden and rapid events in power systems is not a new challenge. For instance, minute-to-minute or hourly load variations have long been addressed in conventional power grids. To mitigate this issue, certain power plants capable of quickly adjusting their output levels are typically utilized, although the issue of flexibility in modern power systems presents additional problems. In addition to uncertainties and variability of electricity demand, the power output from renewable energy sources also exhibits high levels of uncertainty and variability. Moreover, in power systems with significant penetration of renewable generation, the capacity of conventional thermal power plants may be insufficient to accommodate the aforementioned fluctuations. In other words, the sources of flexibility [19] are severely limited, and coping with sudden changes and uncertainties is one of

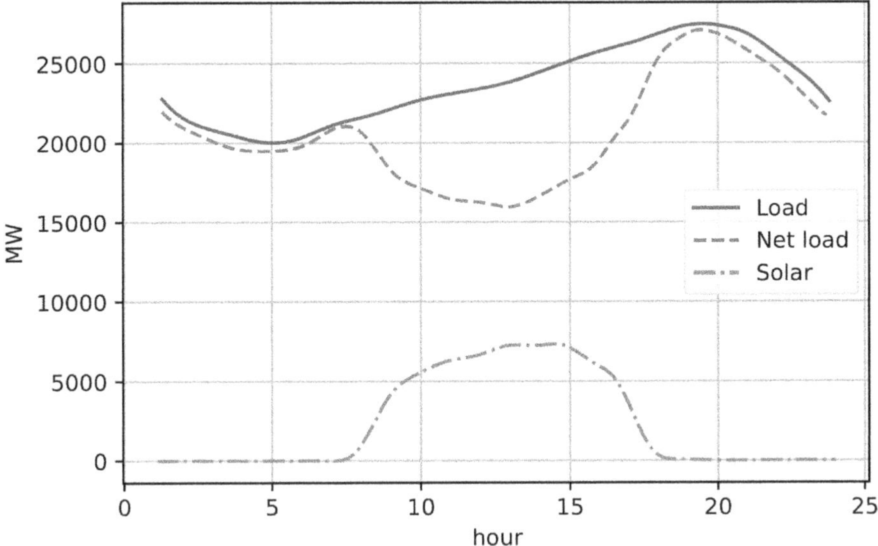

FIGURE 16.1 Duck curve related to California for October 22, 2016.
Source: [21].

the most critical challenges in operating these systems. Thus, in the last few years, the concept of flexibility has received the special attention of specialists in the field of power system studies, to the extent that it has become an important concern in the studies of operation and expansion planning of the power system.

16.2.1 Duck Curve

To illustrate the problem more clearly, Figure 16.1 depicts the net load curve in a power system, which looks like a duck [20]. *Net load* is the hourly electricity demand minus the power generation from renewable sources. In a scenario where a large-capacity solar power plant has been installed, the output covers a significant portion of electricity demand during the noon and afternoon hours (corresponding to peak solar radiation), and as expected, the net load during these hours is greatly reduced. As the day ends and the sun sets, the solar power output drops to zero within a short period, and the net load increases sharply. As shown in the figure, a steep load rise occurs in a brief time-frame, and the power system must be capable of supporting these changes. Several issues arise in this regard. First, is there sufficient power plant capacity to replace this volume of changes? Another question is, if such capacity exists, is it possible to execute this replacement quickly? An additional concern is that when the net load decreases drastically in the middle of the day, a substantial number of conventional thermal power plants need to be shut down. Is it feasible to bring them back online quickly to meet the late-day net load growth?

16.2.2 Reserve and Flexibility

The problem that was raised evaluated the need for flexibility in the operation and dispatching process in the power system. However, the widespread integration of RES has a significant impact on the dynamics of power systems. Many renewable generators inject their output into the grid without being synchronized with the power system, which means that in the event of sudden disturbances, such as a rapid load drop in a portion of the grid or an abrupt loss of generation, the network

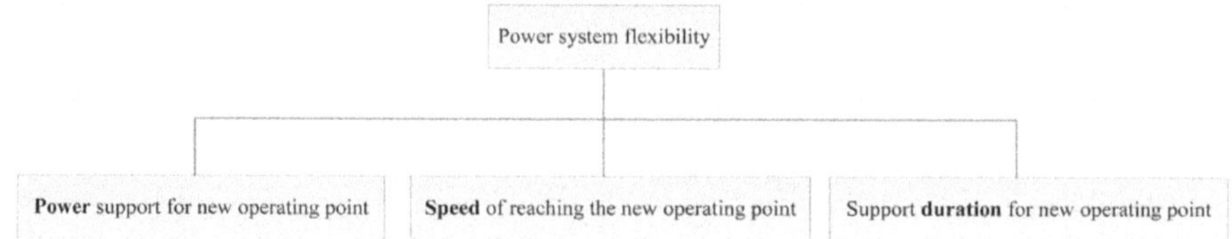

FIGURE 16.2   Different aspects of flexibility in power system studies.

experiences severe voltage and frequency fluctuations. This issue arises due to the reduction in system inertia caused by the decreased presence of conventional power plants [22].

A portion of the flexibility services is actually related to the provision of primary and secondary frequency control reserves that can effectively respond in such conditions and maintain operational variables within acceptable limits. These reserves are essential for managing the increased variability and uncertainty created by RES.

16.2.3   Flexibility in Energy Systems

In the previous section, the importance of flexibility in power systems was examined from different perspectives. This section delves into the details of the flexibility concept and introduces some of the most critical indices in the field of power system flexibility. Coping with sudden changes and uncertainties in load and generation, which formed the basis for the flexibility, can be studied in the following three categories depicted in Figure 16.2.

16.2.3.1   Power Support for New Operating Point

The power support for the new operating point can also be investigated in terms of reserve capacity. Naturally, when there is a high reserve capacity in the power system to support unexpected events [23], a higher level of flexibility is expected. This issue is related not only to power plant capacity but also the ability to utilize controllable loads [24], and storage systems can also improve system flexibility by providing a higher reserve capacity. In fact, with this capability, it is possible to achieve new operating points that arise as a result of changes in the production or consumption of electrical energy.

16.2.3.2   Speed of Reaching the New Operating Point

In many cases, the system's response speed is particularly crucial when facing sudden variations and uncertainties in load and generation [25, 26]. This aspect is more critical in power systems with high renewable penetration. The inertia in these systems is low; therefore, the occurrence of sudden events can quickly lead to system instability. For example, the presence of gas turbines or hydro-electric plants, compared to steam plants, enables faster response to sudden changes. Additionally, the use of storage systems and controllable loads significantly enhances the system's response speed to sudden events. The response speed to change the production and consumption power is usually evaluated as the ramp rate. In fact, the ramp rate depends on the duration of time needed to change power production or consumption; the shorter this period is, the more the speed of reaching the new operating point to balance production and power consumption will increase.

16.2.3.3   Support Duration for the New Operating Point

Apart from the unpredictability of the timing and magnitude of load and generation changes and uncertainties, the duration of these changes is also a determining factor [27, 28]. For instance, although storage systems have a high response speed, their production or consumption power has a time range proportional to their energy storage capacity. According to the explanations provided, this feature is relevant for energy storage elements.

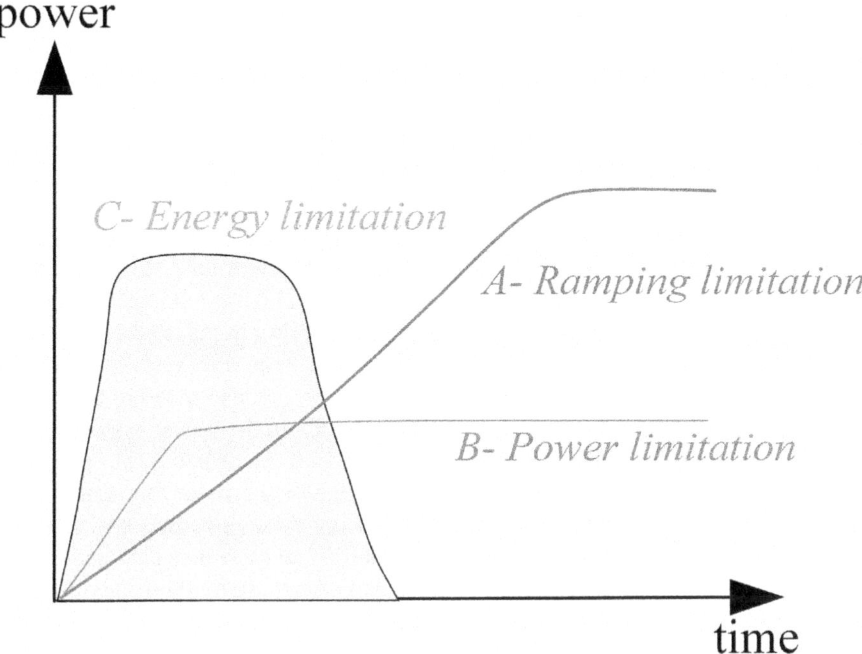

FIGURE 16.3 Different aspects of the system flexibility requirements.

For a better understanding of the aforementioned three categories, Figure 16.3 illustrates various aspects of flexibility. In Case A, the system can offer significant reserve power to address changes and uncertainties, yet it is constrained in terms of response speed. In Case B, although the system lacks high reserve capacity, it exhibits a rapid response speed. In Case C, the system can deliver a substantial reserve capacity in a short timeframe; however, sustaining these services over an extended period may not be feasible, potentially due to energy availability limitations.

16.2.4 Role of Flexibility in Addressing New Challenges

The modern power system with widespread renewable energy sources is accompanied by numerous challenges. Although using renewable energy sources as a sustainable and clean energy source enables the power system to respond to rapid and intense demand growth, these resources themselves pose fundamental challenges to the power system. In traditional power grids, the presence of thermal power plants provides confidence to grid operators that they can meet forecasted demand in the near future without significant concerns. However, this confidence cannot be expected in power systems with a high proportion of non-dispatchable renewable power plants. In traditional power systems, one of the most common strategies to deal with various uncertainties, such as demand forecast errors and sudden outages of large transmission lines and power plants, is to hold a reserve capacity in the different operating conditions of the system. Typically, the reserve capacity is determined based on the largest expected demand or the largest power plant in the system. In reality, this reserve capacity can handle the largest unexpected event in the system, which could be the sudden outage of the largest power plant. Although such an event is unlikely to occur, having a reserve capacity provides the ability to handle such situations. In power systems with a high penetration of renewable energy, the likelihood of high-intensity uncertainties is relatively high. For example, in situations where solar power plants have a high capacity in the system, the occurrence of cloudy skies and subsequent reduction in solar panel production is not considered a low-probability event. Therefore, using traditional methods for allocating reserve capacity cannot provide satisfactory results. In addition to sudden outages of large power plants, the uncertainty of renewable energy production must also be considered in determining reserve capacity. Considering the rapid expansion of renewable power plants, determining the reserve capacity of the network based on probabilistic criteria rather than deterministic criteria has become more important. For example, in PJM, a desired reliability level is determined, and ORDC[1] curves are used to determine the optimal reserve capacity for different operating periods.

In addition to determining reserve capacity as one of the dimensions of flexibility in the power system, utilizing various reserve sources or, in other words, flexibility supply sources is another issue that needs to be considered. Energy storage systems and controllable loads can be considered effective players in future power systems. These resources can be used alongside renewable energy sources as part of the solution for providing reserve and flexibility in the power system. Using these resources is also economically more beneficial, as the cost of providing reserve from these sources is significantly lower than from power plants. Given the high response speed of these resources and the resulting high flexibility due to their widespread presence in the power system, it is possible to handle the challenges arising from the widespread presence of renewable power plants.

In addition to the technical criteria required to determine the necessary flexibility in the system, such as reserve capacity, the economic mechanism for using these resources is also of great importance. These resources should be aligned with the flexibility services provided and the value added by them, which should be subject to revenue. In advanced electricity markets, flexible ramp services are traded, and the value of these services is determined based on different operating conditions of the system. Note that, in response to the increasing integration of renewable power plants, power markets are undergoing a paradigm shift toward novel structures designed to address the unique challenges associated with high penetration levels of intermittent renewable energy sources [29].

Beyond enhancing system flexibility through new elements and resources, many power systems have taken steps to update operating instructions or grid codes. For instance, in traditional power systems, the permitted frequency range is typically limited and defined. However, grid codes in power systems that host a large volume of renewable power plants indicate that standards for operation and the necessary requirements in this field are also evolving. For example, renewable power plants that are considered a threat to power system stability are now included in new grid codes as part of the solution to ensure system stability and must be able to remain connected to the network and participate in frequency control processes across a wide range of frequency deviations [30].

16.3 POWER-TO-X (PtX) SYSTEMS

The PtX technology represents a developing field that harnesses excess electricity generated by RESs. Through an electrochemical process, this surplus energy is converted into hydrogen, which subsequently reacts with carbon mixtures to produce product X. Among these PtX routes, the main considered routes can include the following:

- Power-to-gas (PtG)
- Power-to-liquids (PtL)
- Power-to-chemicals (PtC)

In addition to those mentioned, additional pathways, such as power-to-methane (PtM), power-to-food (PtF), power-to-heat (PtH), and power-to-hydrogen (PtH$_2$), have also been considered as potential PtX technologies. However, usually these routes are among the three main routes introduced.

Presently, there is growing interest, particularly in Europe, regarding extensive research and development in PtX technologies. This interest stems from the European Union's commitment to reducing GHG emissions and increasing the portion of RESs in energy production within Europe. As depicted in Figure 16.4, a PtX system utilizes surplus electricity produced by the RESs to achieve liquid or gaseous chemical products as energy carriers. In general, the end products of most

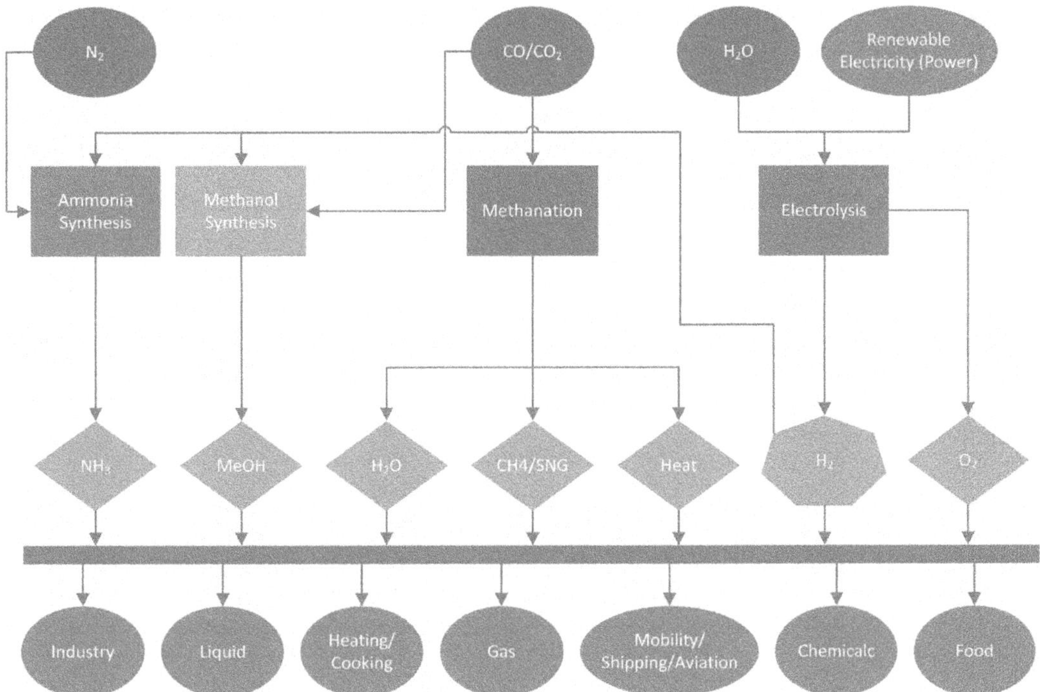

FIGURE 16.4 Schematic representation of the PtX conversion pathways.

PtX processes are hydrogen or methane, which are then further processed and converted into other materials. Due to the gaseous products produced in this process, PtX is closely related to PtG concepts, and typically, they may be used interchangeably. Considering that the main pathways of PtX include PtG, PtL, and PtC, these transformation processes will be described in this section. Before that, the water electrolysis (WE) and methanation concepts as the main process in most of the PtX systems will be described here.

The electrolysis of water to produce O_2 and H_2 is an electrochemical reaction which could be divided into two steps. At the cathode, with a negative charge, the reduction reaction occurs. Conversely, at the anode, with a positive charge, the oxidation reaction takes place in accordance. Depending on the employed technology, the charge carrier could be OH^-, H_3O^+, or O^{2-}.

Excess electrical energy generated from RES can be harnessed for hydrogen production through WE. The resulting H_2 can serve multiple purposes: it can be converted to methane when combined with an appropriate carbon source, directly injected into the gas grid, or utilized for fuel cell vehicles. Water electrolysis is a well-established technology, and advancements in fuel cell technology contribute to its continuous improvement. Within the PtG process, three primary electrolysis technologies are relevant: alkaline electrolysis (AEL), polymer electrolyte membrane (PEM) electrolysis, and solid oxide electrolysis (SOEC). The energy demand for water electrolysis can be met using either electrical or thermal energy. When RES serves as the energy source, the overall system achieves net-zero CO_2 emissions. The choice of a specific electrolysis technology depends on factors such as efficiency, flexibility, and lifetime.

Methanation refers to the process of converting carbon monoxide and carbon dioxide (CO_x) into methane (CH_4) through hydrogenation. The discovery of CO_x methanation reactions dates back to 1902, when Sabatier and Sanders first observed them. Since then, this reaction has found numerous practical applications. It serves as an effective method for removing CO from process gases and has also been explored as an alternative to preferential oxidation in fuel processors for mobile fuel cell applications. Additionally, methanation has been considered a viable approach for producing synthetic natural gas (SNG) since the 1970s. Recently, there has been interest in using PtG, which converts excess RES energy into methane, as a means of energy storage in conjunction with existing natural gas infrastructure.

In a PtG chain, SNG should exhibit properties akin to those of natural gas distributed in the gas network. While natural gas primarily consists of over 80% CH_4, it also contains significant amounts of higher hydrocarbons—such as ethane, propane, and butane—that enhance its calorific value compared to pure methane. Conversely, inert components like CO_2 or N_2, found in natural gas, reduce its overall calorific value. Notably, CO_2 methanation using Ni catalysts achieves nearly 100% selectivity. The absence of higher hydrocarbons in SNG could lead to a lower calorific value of SNG from CO_2 methanation than that of natural gas.

Methanation can be conducted using both biological and catalytic reactors. Various methanation technologies have been evaluated based on factors such as the quality of gas achievable from the product gas, the reactor volume necessary to achieve the desired gas quality and flow rate, and the complexity of the process setup.

16.3.1 POWER-TO-GAS (PTG)

The PtG process connects the electricity and gas networks by converting the surplus energy produced by the RESs into gases compatible with the gas grid. This connection takes place through a two-step process, in which, first, H_2 is produced through WE and then, by combining or methanating with CO or CO_2 (produced from an external source), CH_4 is produced. The produced H_2 can itself be used directly or be further converted into syngas, methanol, or LPG (Figure 16.5).

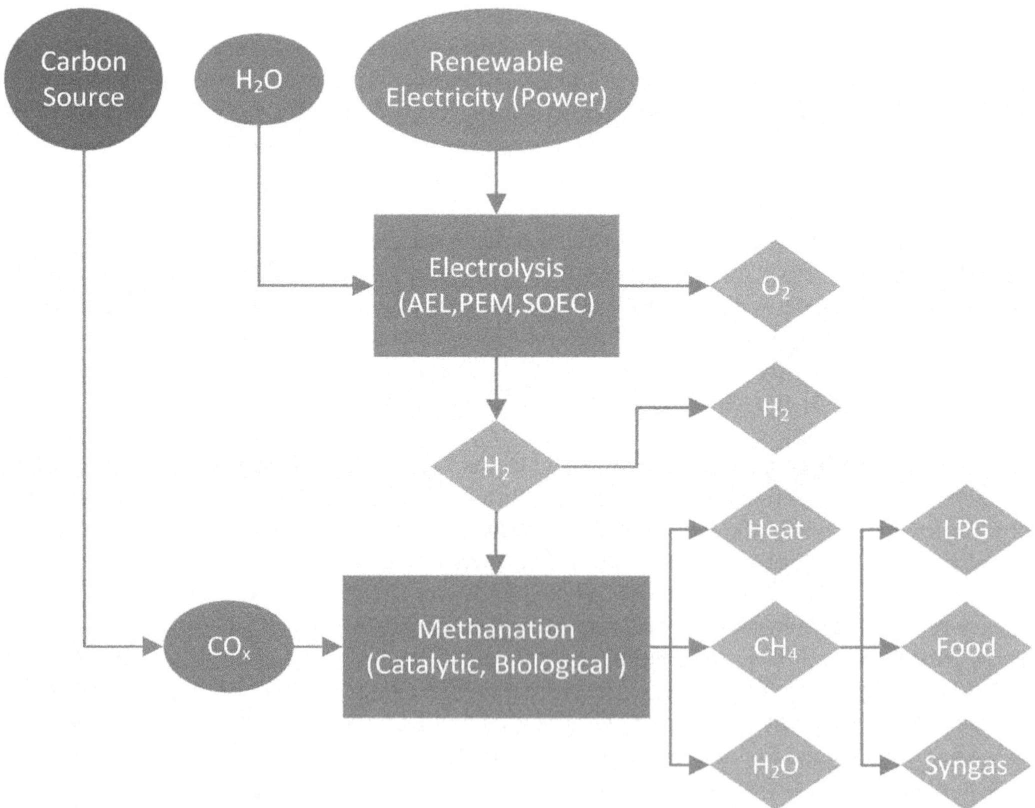

FIGURE 16.5   Schematic representation of PtG.

The CH_4 resulting from this two-step process, known as substitute natural gas or SNG, has several uses; this gas could be injected into the gas delivery network or existing gas storage, used as CNG engine fuel, or simply be used in other natural gas wells.

16.3.1.1 Power-to-Hydrogen (PtH$_2$)

Observations indicate that existing PtG systems initiate the process by utilizing electricity to perform WE. In contrast, in a PtH$_2$ system, the generated hydrogen is either injected into the natural gas grid or employed for transportation and industrial purposes, rather than being converted into a different gas. Hydrogen provides a distinct storable carrier that can be transported over time and space, enabling the integration of large volumes of RES. This issue is directly related to the goals of reducing GHG emissions and diseases caused by pollution. Compared to nuclear energy and fossil fuels, H_2 is a more reliable carrier and has less effects on the environment. Also, H_2 has created many more job opportunities than the large, limited state-owned nuclear and traditional power companies, which are constantly expanding.

When discussing hydrogen generated from RES, commonly referred to as "green hydrogen," it's essential to recognize the various "colors" associated with hydrogen production. Let's delve into these color distinctions:

- *Blue hydrogen.* Primarily derived from natural gas through a process known as steam reforming, blue hydrogen involves combining natural gas with heated water to produce steam. The end result is a mixture of H_2 and CO_2 as by-products. To mitigate environmental impact, carbon capture and storage (CCS) technology is crucial for capturing and safely storing the CO_2 emissions.
- *Gray hydrogen.* Produced using steam methane reforming from natural gas, gray hydrogen lacks the integration of GHG into the process. Unlike blue hydrogen, it doesn't utilize CCS technology.
- *Brown hydrogen.* By employing coal as a feedstock in the gasification process, brown hydrogen can be produced. However, similar to gray hydrogen, it also releases CO_2 during production.

Hydrogen finds application in various industrial processes, serves as rocket fuel, and plays a crucial role in fuel cells. Hydrogen has actively been explored as a supplement or replacement for natural gas in some natural gas power plants. H_2 also holds promise as an effective energy storage medium for electricity generation. In the United States, the majority of hydrogen consumption occurs within industries such as oil refining, metal refining, fertilizer production, chemical manufacturing, and food processing. Additionally, biofuel producers employ hydrogen to create hydrogenated refined vegetable oil (HVO), a renewable diesel substitute.

Hydrogen is converted into electricity through hydrogen fuel cells. These fuel cells generate electricity by combining hydrogen and oxygen. The products include electricity, water, and minimal heat. Currently, hydrogen fuel cells serve as energy sources for spacecraft electrical systems. Smaller fuel cells have been developed to power electronic devices like laptops and mobile phones. Furthermore, several car manufacturers have harnessed fuel cells to propel vehicles. In the future, fuel cells could potentially supply emergency power to buildings and remote locations not connected to the conventional power grid.

Hydrogen is also utilized as an alternative fuel in fuel cell vehicles due to its potential to power fuel cells in zero-emission vehicles. In fact, a fuel cell can be two to three times more efficient than an internal combustion engine fueled by gasoline. Several automakers have already introduced H_2 fuel cell vehicles in some regions of the United Sates. However, the high cost of fuel cells and the scarcity of hydrogen refueling infrastructure have constrained the widespread adoption of H_2-fueled vehicles. The production of these cars remains limited because consumers hesitate to purchase them without readily available refueling stations. Currently, the United States boasts approximately 56 hydrogen vehicle refueling stations, all of which are located in California.

16.3.1.2 Power-to-Methane (PtM)

The idea of power-to-methane (PtM) could be fulfilled by combining H_2 obtained from WE with the methanization procedure. At first, H_2 is produced using RES energies, and then CO_2 is hydrogenated to obtain methane.

Methane, a crucial energy carrier, serves as the backbone for heat generation, electricity production, and the synthesis of value-added chemicals. Its primary source lies in natural gas—a remarkably cost-effective fossil fuel. When produced sustainably, the output SNG holds immense potential for curbing GHG emissions. Unlike renewable hydrogen, SNG seamlessly integrates into the existing natural gas infrastructure. The emergence of PTM technology offers a promising avenue for sustainable methane production. Among the various PTM pathways, the catalytic hydrogenation of CO_2 (methanation) has undergone extensive scrutiny, with operational demonstration plants active across several countries. Additionally, recent research underscores the pathway's promising features, even though electrochemical CO_2 reduction remains at the laboratory validation stage.

Methane gas serves dual roles—it can act as a chemical feedstock or be converted back into electricity using conventional gas turbines. PTG technology enables the storage and transport of energy in the form of compressed gas, capitalizing on existing infrastructure designed for natural gas. Notably, Germany's natural gas network, which historically operated on coal gas (composed of 50–60% hydrogen), boasts an impressive storage capacity exceeding 200,000 GWh. This capacity far surpasses that of all German hydropower plants combined (approximately 40 GWh). Consequently, storing SNG within the gas network becomes an economically justifiable solution for seasonal renewable energy storage.

16.3.2 Power-to-Liquid (PtL) and Power-to-Chemicals (PtC)

The transport segment is an important contributor to global GHG emissions. For example, 25% of all energy-related CO_2 emissions have been reported from this sector in 2018. The worldwide demand for annual transportation is also continuously increasing. In this field, power-to-liquid (PtL) technologies have become relevant for liquid fuel production from RES energy. In PtL or power-to-liquid fuel, liquid fuel is produced directly from water, renewable electricity, and CO_2. The two main routes for the production of liquid fuels are MeOH synthesis and the Fischer–Tropsch (FT) (Figure 16.6). The PtL process produces synthetic fuels, for instance, hydrocarbons, MeOH, and dimethyl ether (DME).

In the FT reaction, at first, CO_2 is transformed to CO using the RWGS reaction, and H_2 is produced using WE. After that, H_2 and CO react in a conventional FT synthesis to produce valuable hydrocarbons, such as paraffins, olefins, and alcohols.

MeOH could be produced in two methods in PtL procedures. The first method is direct hydrogenation of CO_2, and the second one is electrochemical reduction of CO_2. The output MeOH could be transformed into liquid hydrocarbons in the next process. To reduce GHG emissions, numerous investigators have newly focused on the production of MeOH by CO_2 hydrogenation. Methanol is an essential chemical for the production of several industrial and consumer products. It also serves as a non-conventional fuel for transportation.

In methanol production, hydrogen and CO_2, in a formal ratio of 3/1 M, react to produce CH_3OH. Methanol is an important raw material in the chemical industry to produce olefins, DME, and liquid fuels, which has become an attractive alternative to fossil fuels. In addition, MeOH is one of the main raw materials produced in the world. It is a liquid energy carrier that is easier to transport than gases and solids.

Methanol has a great transportation and storing organization due to its significant position as an intermediate in chemical manufacturing. Until now, energy-efficient pipeline transportation has been the preferred mode of transportation between main manufacturers and customers nearby. Nonetheless, since methanol is predicted to have been an important future functionality due to its suitability as a proper energy carrier, transshipment of methanol by pipeline may be economically

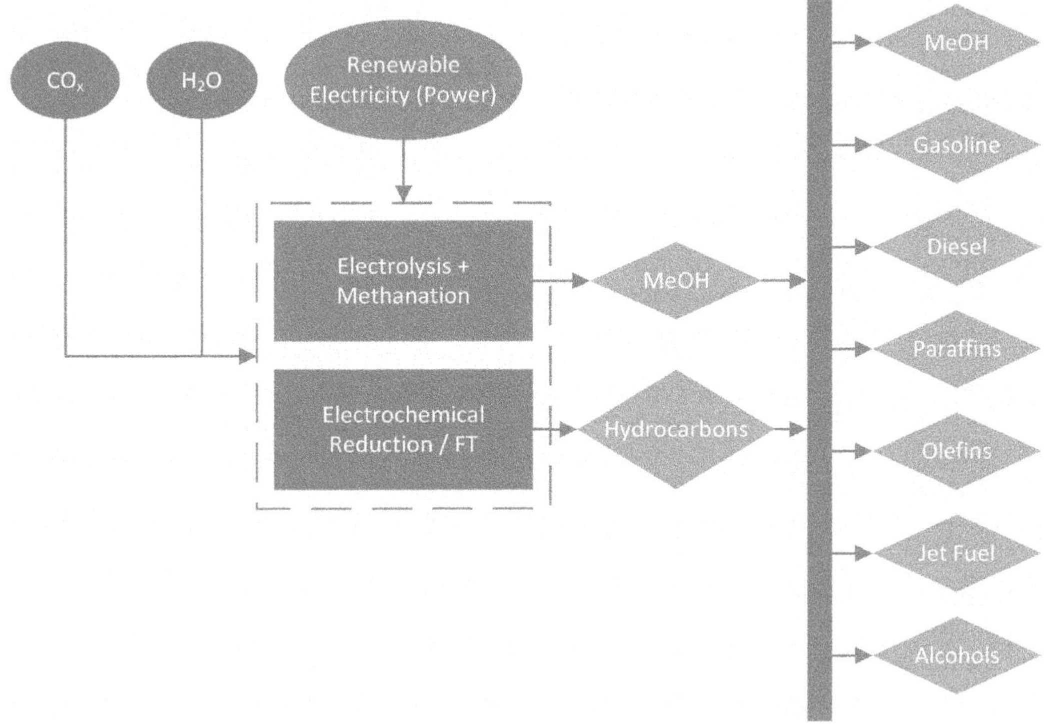

FIGURE 16.6 Schematic representation of the PtL conversion pathways.

feasible. A great volume of methanol is shipped by waterway via tankers or ships designed to carry oil, and tankers specifically intended to carry methanol.

Fuels produced from the FT process and MeOH are the main types of studied fuel in PtL processes. However, few investigations have been conducted on DME and liquid hydrocarbons. PEM electrolysis has been mainly used for PtL processes, and in addition, the use of carbon sources from biogenic, air, and industrial processes has also been investigated.

DME (dimethyl ether) serves as a transportation fuel additive, derived from feedstocks like biomass and fossil fuels. In industrial applications, DME can be obtained indirectly through synthesis of MeOH or straight fully from syngas.

Furthermore, dual PtX mixing—combining liquid and gas production—along with the recovery of unconverted feed gas from synthesis reactors, can reduce liquid fuel production costs. This approach minimizes the demand for operation of unit and costs of WE-produced hydrogen. Increasing the recycling ratio of unconverted feed gases in the PtL/PtG hybrid improves energy efficiency and reduces capital investment. However, as the non–feed gas recovery ratio increases, the hybrid process efficiency declines. Careful adjustment of the recycle ratio prevents the accumulation of inert gaseous hydrocarbons downstream.

Efforts to decarbonize the chemical industry necessitate alternative methods for synthesizing chemicals from alternative pollutant-free sources. Analogous to methanol (MeOH) production, two promising approaches are CO_2 hydrogenation and electrochemical reduction (ER) of CO_2. These methods enable the production of value-added chemicals through a two-step process known as power-to-chemical. In the initial step, hydrogen is generated via WE. Subsequently, hydrogen interacts with CO_2 or N_2 to yield chemicals. Researchers have taken interest in this flexible and rapidly kinetic process. In a typical ER process, CO_2 is reduced to chemical species within a temperature range of 20–80°C, depending on the desired end products. Notably, the ER of CO_2 can be conducted at temperature condition, allowing for straightforward adjustments of process parameters such as electrolyte composition, electrocatalyst, and redox potential.

In recent studies, the performance of methanol production through direct hydrogenation of CO_2 from biogas and fossil fuels, using ammonia (NH_3) as a catalyst, has been thoroughly evaluated. The findings indicate that connecting a PtL plant to biogas demonstrates superior performance compared to an NH_3-based power plant. Additionally, several factors significantly influence the overall MeOH production process. These include electricity generation costs, hydrogen production efficiency, and carbon taxation. Furthermore, when considering solar integration with coal and biomass gasified power plants, the estimated costs for MeOH production are approximately 300% to 500% higher than those associated with a conventional coal plant.

In the field of PtX, an intriguing development involves the production of ammonia (NH_3), which holds significant potential for decarbonizing the ammonia production industry. Unlike the conventional Haber–Bosch (H-B) process for NH_3 production, the power-to-ammonia ($PtNH_3$) process operates without emitting CO_2, as it requires no carbon input. Consequently, NH_3 emerges as a promising candidate for energy storage due to its high hydrogen content. Moreover, NH_3 can be liquefied under mild conditions (20°C, 8 bar) and can serve as a source for hydrogen production through hydrogenation.

16.4 PtX VS. OTHER SOLUTIONS FOR FLEXIBILITY

16.4.1 POWER SYSTEMS FLEXIBILITY RESOURCES

As mentioned before, flexibility as one of the vital requirements of energy systems, especially with the presence of renewable energy sources, can be provided in different ways. Of course, various technical and economic factors are influential in determining the method of providing flexibility. In order to compare between different solutions for flexibility challenge, various options for providing flexibility in power systems are classified and presented in Figure 16.7.

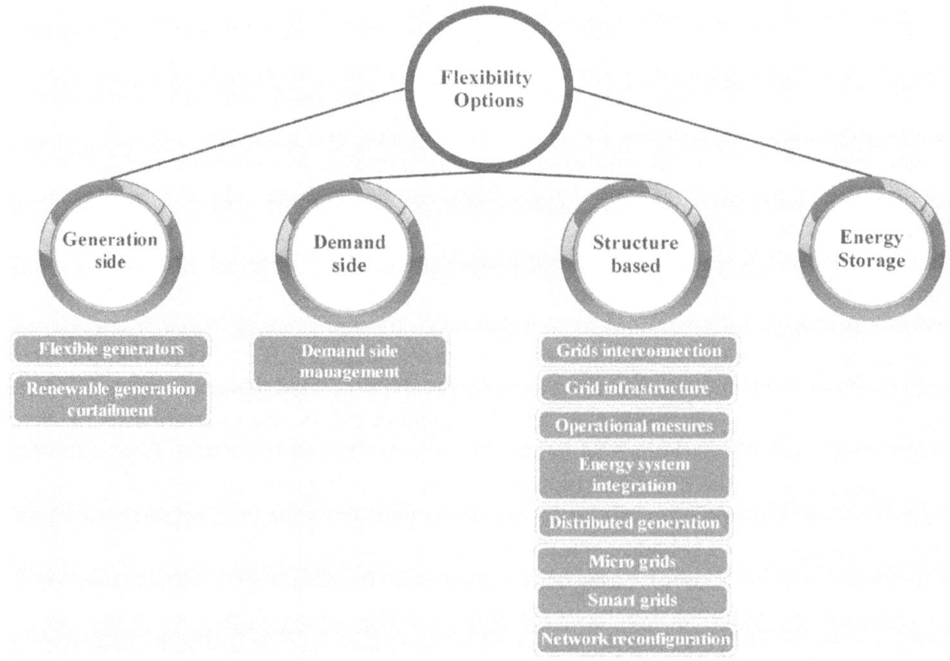

FIGURE 16.7 Flexibility options classification.

According to this figure, flexibility resources can be classified into the following general categories:

- Generation-side resources
- Demand-side resources
- Structure-based solutions for providing flexibility
- Energy storage systems

In fact, in the form of this division, it is determined in which part of the power system flexibility is provided. It is worth noting that flexibility is provided by the equipment available or by using different current processes in each side (such as management and planning activities). Of course, a number of options for providing flexibility, such as energy storage systems, can be used in different sides of the power system. In the following, different options for providing flexibility are examined.

16.4.1.1 Generation-Side Resources

These resources can be considered among the most important sources in providing flexibility. Two examples of these sources are explained in what follows.

16.4.1.1.1 Flexible Generators

Flexible generators are among the tools on the power generation side, which can compensate for the imbalance of production and power consumption, with the ability to change their output power in a short period of time, and thus increase the responsiveness of the power generation side. Gas turbines, CHPs, and run-of-river power plants are examples of flexible resources in power systems. The response speed in power plant units that use electromechanical energy conversion to produce electricity is limited due to the inertia of rotating equipment and technical issues, as well as destructive mechanical stresses caused by sudden changes in production. Storage systems such as batteries do not have such a structure and naturally do not have such limitations. Therefore, they have a very high response speed.

16.4.1.1.2 Renewable Generation Curtailment

Renewable generation curtailment occurs when the production of renewable sources exceeds the amount required to balance the supply and demand of electrical energy. In this condition, this capability can be one of the choices available to make the system flexible in times of excess production power.

16.4.1.2 Demand-Side Resources

Demand-side resources provide flexibility in the system through the participation of power consumers. For example, demand response programs (such as EDRP[2]) provide conditions that, in critical situations caused by network events, controllable loads can be interrupted and thereby contribute to improving the reliability and stability of the network. Demand response programs can be divided into two main categories: incentive-based and price-based programs [31]. In incentive-based programs, participants are rewarded for their cooperation with the network operator based on predetermined contracts. For example, in the EDRP program, consumers grant the network operator the authority to control consumption in emergency situations, including the possibility of power interruption. One of the most important features of these programs is the relative certainty of their occurrence and the subsequent guarantee of the required flexibility at the network level. However, providing complete monitoring and control facilities for each participating consumer is a challenge that may limit the use of these programs to large consumers and, naturally, make it difficult for small consumers (such as residential consumers) who form a wide range of electricity demand to benefit from them. On the other hand, some demand response programs are based on prices and cover a wide range of consumers. In some of these programs, such as CPP,[3] end users can monitor the real-time electricity prices and manage their energy consumption in response to sudden price increases during critical system conditions that may be caused by unexpected events in the network or by a shortage of generation due to the forced outage of power plants. This performance enables them to support the network and, in fact, improve flexibility.

16.4.1.3 Structure-Based Solutions for Providing Flexibility

Some solutions for providing flexibility are not directly related to generation or load; rather, they are related to network infrastructure issues and systems operation structures. Following, a number of flexibility improvement options related to this category are reviewed.

- **Grids interconnection.** Connecting power grids to each other can enable energy exchange between them, and as a result, increase flexibility. In this situation, the ability to balance local power is also increased for each of the power grids, and thus, the interconnected grids will become more flexible.
- **Grid infrastructure.** The use of infrastructures on the grid side, such as FACTS tools, provides the possibility of improving flexibility. In fact, with the use of these equipment, the possibility of easier energy transfer, and thus the possibility of using energy in different parts of the system in order to meet electrical loads, increases.
- **Operational measures.** Improving the operation procedure of different equipment on the generation and demand side can help increase the flexibility of the system. In fact, with the increasing importance of flexibility in the presence of renewable production sources, the improvement of flexibility indicators should also be included in the objective function or constraints of operations issues.
- **Energy system integration.** The integration of different energy systems, such as electricity, transportation, and heating systems, according to Figure 16.8, with multiple-energy carrier systems is another way to improve system flexibility [32]. Diversification of the electricity supply portfolio through the use of various types of energy carriers greatly increases the level of flexibility of electricity supply. Structures such as energy hubs provide the possibility of replacing the primary sources of electricity supply and thus provide more options to the network operators so that they can face uncertainties and unexpected changes.
- **Distributed generation.** The use of distributed generation units is one of these solutions. The expansion of DGs in the vicinity of electricity consumers and in small dimensions provides the conditions that allow the management of unexpected conditions caused by uncertainties in a distributed and more effective way.
- **Microgrids.** The development of microgrids within the electricity distribution network as a systematic method for applying DGs is among the other structural solutions to improve flexibility. These systems, by having a variety of distributed generation sources and storage systems, provide the possibility of local supply of the load, and in addition, they will have the possibility of using the distribution network structure as a backup. Furthermore, these systems are able to provide service to the distribution network in the conditions caused by the uncertainties of the distribution network, and thus, they can improve the flexibility of this network.
- **Network reconfiguration.** Among other structural solutions is the use of network reconfiguration methods in emergency conditions [33]. Based on this, changing the network configuration as a solution to face sudden conditions and changes in the form of network rearrangement plans can be considered as one of the most important structural solutions to improve the flexibility of the system.
- **Smart grids.** One of the most important requirements for improving system flexibility is the development of smart grids. In traditional systems, real-time operation was only possible at the level of transmission networks and large power plants, and the ability to send

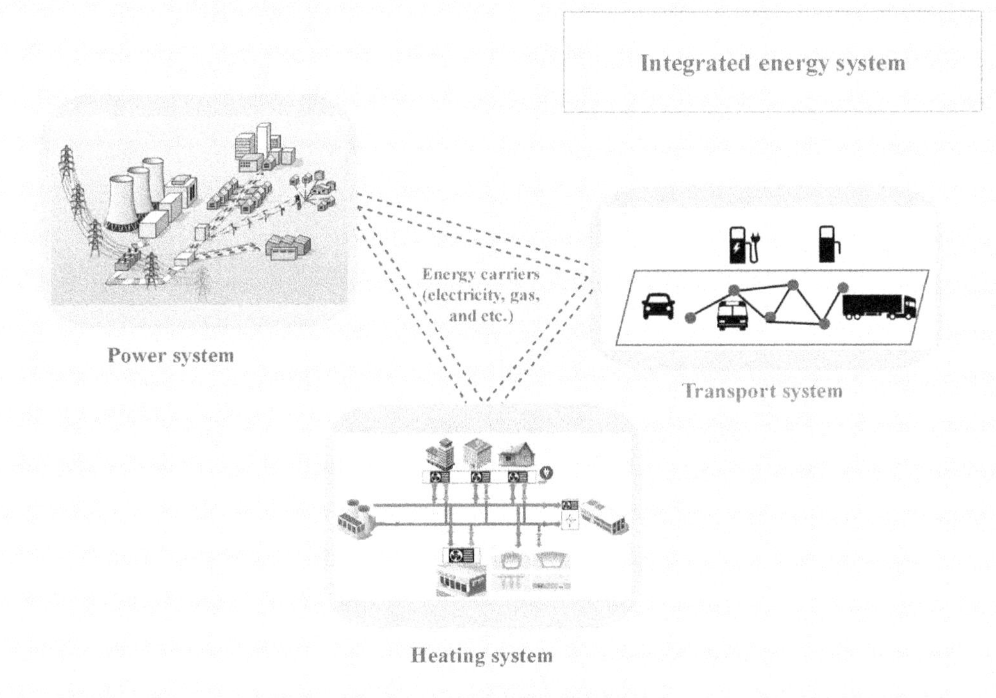

FIGURE 16.8 Integrated energy system schematic.

real-time dispatch instructions to end users at the distribution level was not available. The advent of smart grids and the gradual integration of its components into traditional distribution networks have enabled the development of advanced energy management systems (EMS) [34]. This integration has made concepts such as demand response a viable reality, allowing for more sophisticated and efficient management of energy distribution and consumption. One of the most important components of smart grids is advanced monitoring and control systems that operate based on information technology and enable bidirectional data exchange between power network end users and system operators. Implementing demand response programs as one of the tools for improving system flexibility is an important achievement of smart grids [35–37].

16.4.1.4 Energy Storage Systems

Energy storage systems can increase the flexibility of the system to a relatively high level due to the ability to receive and deliver energy as well as the relatively high response speed. Energy storage systems can be used in both stationary and mobile modes. The ability to carry portable energy storage such as batteries can increase flexibility. The ability to deliver and receive energy by energy storage systems provides the possibility of being placed on both sides of generation and demand, which also improves flexibility [38, 39].

16.4.2 PtX System: An Option for Improved Flexibility

In order to investigate the role of the PtX systems in improving the performance of modern energy systems, the integrated system shown in Figure 16.9 is considered. In the displayed system, it is possible to exchange energy between the systems based on different energy carriers by

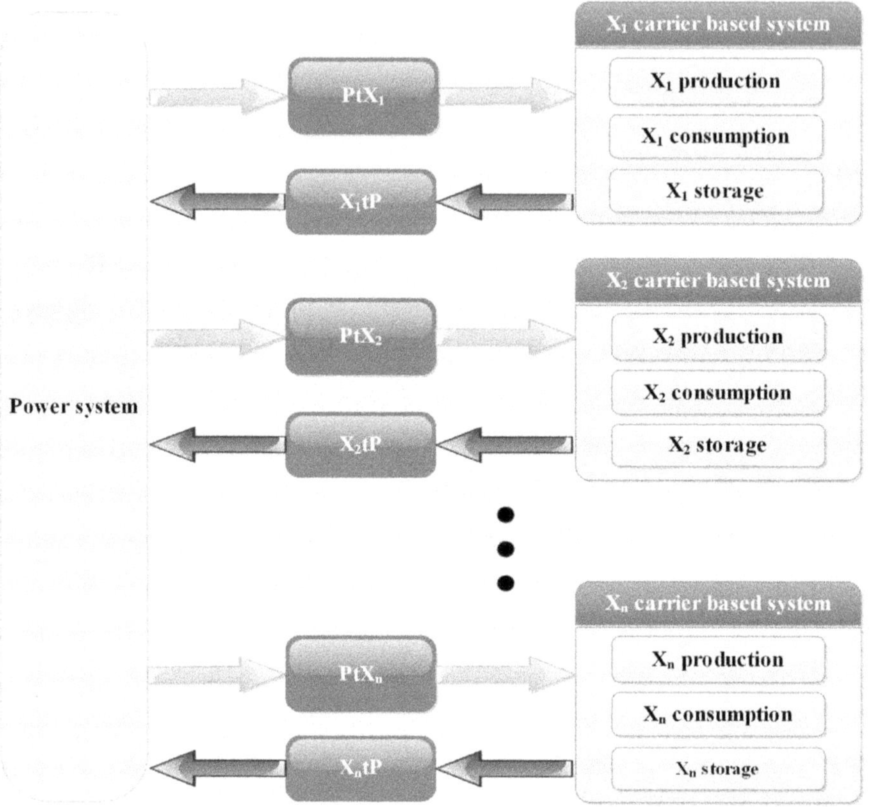

FIGURE 16.9 Generic schematic representation of an integrated energy system in the presence of multiple PtX systems.

intermediary systems. According to Figure 16.8, systems based on different energy carriers can include power, heating, and transportation systems. Intermediary systems actually make it possible to convert different energy carriers into each other. As mentioned before, PtX systems can be considered as a main intermediary in modern energy systems with several energy carriers that convert electric power to other carriers and act as one of the feeders of other energy systems. But in order to complete the energy cycle in modern energy systems, according to Figure 16.9, the XtP systems must also be used to enable the conversion of various energy carriers into electrical energy. According to this figure, in systems based on different energy carriers, energy exchange takes place between producers, consumers, and storage devices, and the existence of the PtX and XtP systems makes it possible to exchange energy between the desired systems and the power system. In the other words, the PtX system as an infrastructure according to Figure 16.10 can connect systems with different energy carriers and thus increase the opportunities to increase flexibility. In fact, the PtX system can be used as a supplement in energy systems and thus improve flexibility.

With the use of PtX systems, it will be possible to create an intermediate condition for the management of the energy (consumption or storage), which can be used to improve flexibility. For example, by using PtX systems, the electricity produced by renewable production sources can be converted into hydrogen, the hydrogen can be stored, and the stored hydrogen can be used to produce electricity during the time of electricity shortage (decrease in renewable production or sudden increase in load). This can be done using fuel cells. In this way, the flexibility of the system will increase during changes in the production and consumption of electrical energy.

As a real practical example to determine the role of PtX systems in improving the flexibility of the energy systems, we can refer to the RH$_2$-WKA project in Germany. In this project, alkaline electrolysis is used to produce hydrogen by electric energy produced by wind sources. In this way, the ability to store energy in the form of hydrogen will be created in different time periods, which has a significant effect on improving flexibility from the point of view of storage ability. On the other hand, the produced hydrogen is used to feed CHP units to produce electricity and heat. In fact, in this way, the flexibility of the system will also increase in terms of electrical energy production (generation side). Also, by producing heat by the CHP system as well as gas injected into the gas network, a part of the thermal load of the system can be provided. Therefore, the dependence of the thermal load on the electric energy produced by conventional production sources will be reduced, and the flexibility from the point of view of consumption will also be improved (demand side). On the other hand, with this PtX system, it is possible to connect different energy systems. Therefore, by examining this practical example, it can be seen that the investigated PtX system from all four aspects mentioned can help improve the flexibility of the energy system.

16.5 OPERATION AND EXPANSION PLANNING OF THE PTX SYSTEMS IN THE MODERN ENERGY SYSTEMS CONSIDERING THE FLEXIBILITY REQUIREMENTS

According to the explanation of the importance of the presence of PtX systems that was examined in the previous sections, one of the challenges facing researchers is to decide how to operate and plan the development of these systems in modern energy systems. In fact, with the presence of these equipment and their impact on the energy cycle process, how the integrated energy system is used and, in fact, the operating point of the equipment will change. In order to create a proper perspective, how to operate a modern energy system with the presence of PtX systems is described and formulated. In fact, in the form of the operation problem, the amount of production, consumption, and storage of energy in the different equipment is determined every hour. It should be noted that the problem is formulated from the point of view of the system operator, with the aim of minimizing the cost of energy supply for consumers.

The use of PtX systems, despite providing various capabilities, will also involve costs. Therefore, deciding for the presence of these systems in modern energy systems will require economic analysis and optimal decision-making. In fact, the cost-effectiveness of the costs applied to the system should be investigated with the presence of these systems. For this purpose, how to formulate the decision-making problem for the development of these systems is presented in what follows.

It should be noted that in order to consider flexibility in the studies of operation and expansion planning of PtX systems, this concept in modern energy systems is evaluated from the following aspects:

- *The ability to change the output power (decrease or increase).* From this point of view, the maximum rate of change of energy production or consumption is examined (flexibility due to ramping capability [FRC]).
- *Available power.* The maximum available power in the mode of reducing or increasing power is considered as one of the flexibility criteria (flexibility due to power delivering capability [FPDC]).
- *Available energy.* The ability to charge and discharge energy is another way to improve flexibility, based on which the mismatch between energy production and consumption can be covered (flexibility due to storing capability [FSC]).

16.5.1 FLEXIBILITY CONSIDERATION IN OPERATION OF THE PtX SYSTEMS

In order to formulate the problem of the operation of the modern energy systems in the presence of several energy carriers as well as PtX systems similar to Figure 16.9, the objective function is defined as the minimization of the cost of operation of various elements of the system according to the following relationship.

$$
\begin{aligned}
Of = \sum_{t=1}^{T}\Bigg(& \sum_{m=1}^{NE}OC_{mt}^{EG} + \sum_{g=1}^{NES}OC_{gt}^{ES} + \sum_{k=1}^{NEDR}OC_{kt}^{EDR} + \\
& \sum_{i=1}^{n}\left| OC_{it}^{PtX} + OC_{it}^{XtP} + \sum_{mi=1}^{NiE}OC_{mit}^{XiG} + \sum_{gi=1}^{NiES}OC_{git}^{XiS} + \sum_{ki=1}^{NiEDR}OC_{kit}^{XiDR} \right|
\end{aligned}
\tag{16.1}
$$

According to the preceding relationship, the operating cost of producers, responsive consumers, and energy storage systems for different energy carriers (electrical and non-electrical), as well as intermediates (PtX and XtP systems) between interconnected systems, are considered in the objective function.

The constraints of the operation problem are presented in the following [40–43].

- **Electrical generation units' constraints:**
  - Power generation capability:

$$
\underline{P_m^{EG}} \leq P_{mt}^{EG} \leq \overline{P_m^{EG}}
\tag{16.2}
$$

  - Ramping rate constraints:

$$
P_{mt}^{EG} - P_{mt-1}^{EG} \leq RUR_m^{EG}
\tag{16.3}
$$

$$
P_{mt-1}^{EG} - P_{mt}^{EG} \leq RDR_m^{EG}
\tag{16.4}
$$

  - Flexibility constraints:

The FPDC that can be provided by the mth electrical generation unit is determined as follows:

$$
FPDC_{mt}^{EG} = min\left\{ \overline{P_m^{EG}} - P_{mt}^{EG}, P_{mt}^{EG} - \underline{P_m^{EG}} \right\}
\tag{16.5}
$$

The FRC that can be provided by the mth electrical generation unit is also defined as follows:

$$
FRC_m^{EG} = min\left\{ RUR_m^{EG}, RDR_m^{EG} \right\}
\tag{16.6}
$$

- **Electrical storage units' constraints:**
  - Energy balance constraint:

$$
\sum_{t=1}^{T}\eta_g^{ESch} \times P_{gt}^{ESch} = \sum_{t=1}^{T}\frac{1}{\eta_g^{ESdis}} \times P_{gt}^{ESdis}
\tag{16.7}
$$

- State-of-charge calculation:

$$E_{gt}^{ES} = E_{gt-1}^{ES} + \eta_g^{ESch} \times p_{gt}^{ESch} - \frac{1}{\eta_g^{ESdis}} \times p_{gt}^{ESdis} \tag{16.8}$$

- Energy constraint:

$$\underline{E_g^{ES}} \leq E_{gt}^{ES} \leq \overline{E_g^{ES}} \tag{16.9}$$

- Power constraint:

$$0 \leq p_{gt}^{ESch}, p_{gt}^{ESdis} \leq \overline{P_g^{ES}} \tag{16.10}$$

- Flexibility constraint:

The FPDC that can be provided by the gth electrical storage unit is determined as follows:

$$FPDC_g^{ESG} = min\left\{\overline{P_g^{ES}} - \left(p_{gt}^{ESdis} - p_{gt}^{ESch}\right), \overline{P_g^{ES}} - \left(p_{gt}^{ESdis} - p_{gt}^{ESch}\right)\right\} \tag{16.11}$$

The FSC that can be provided by the gth electrical storage unit is also defined as follows:

$$FSC_g^{ESG} = min\left\{\overline{E_g^{ES}} - E_{gt}^{ES}, E_{gt}^{ES}\right\} \tag{16.12}$$

Also, the FRC that can be provided by the gth electrical storage unit is defined as follows:

$$FRC_g^{ESG} = min\left\{RR_g^{ES,charging}, RR_g^{ES,discharging}\right\} \tag{16.13}$$

- **Electrical responsible demands' constraints:**

Demand responsiveness is actually defined as the ability to decrease and increase the consumption load in the required time to increase flexibility.

- Responsibility constraint:

$$0 \leq PDR_{kt}^{EDR-}, PDR_{kt}^{EDR+} \leq \overline{PDR_k^{EDR}} \tag{16.14}$$

- Ramping constraint:

$$0 \leq PDR_{kt}^{EDR-}, PDR_{kt}^{EDR+} \leq RR_k^{ED} \tag{16.15}$$

- Flexibility constraint:

  The FPDC that can be provided by the kth responsible electrical load is determined as follows:

$$FPDC_k^{EDR} = min\left\{D_{kt}^{ED}, \overline{D_{kt}^{ED}} - D_{kt}^{ED}\right\} \tag{16.16}$$

  The FRC that can be provided by the kth responsible electrical load is also defined as follows:

$$FRC_k^{EDR} = RR_k^{ED} \tag{16.17}$$

- **PtX systems' constraints:**
  - Energy conversion equation:

$$C_{it}^{GX} = D_{it}^{EX} \times CC_i^{PtX} \tag{16.18}$$

  - Energy carrier constraint:

$$0 \leq C_{it}^{GX} \leq \overline{C_{it}^{GX}} \tag{16.19}$$

- **XtP systems' constraints:**
  - Energy conversion equation:

$$p_{it}^{EX} = C_{it}^{DX} \times CC_i^{XtP} \tag{16.20}$$

  - Power capacity constraint:

$$0 \leq p_{it}^{EX} \leq \overline{p_{it}^{EX}} \tag{16.21}$$

  - Ramping constraints:

$$p_{it}^{EX} - p_{it-1}^{EX} \leq RUR_i^{EX} \tag{16.22}$$

$$p_{it-1}^{EX} - p_{it}^{EX} \leq RDR_i^{EX} \tag{16.23}$$

  - Flexibility constraints:

  The FPDC that can be provided by the ith XtP system is defined as follows:

$$FPDC_i^{XtP} = min\left\{\overline{p_{it}^{EX}} - p_{it}^{EX}, p_{it}^{EX}\right\} \tag{16.24}$$

The FRC that can be provided by the ith XtP system is also determined as follows:

$$FRC_i^{XtP} = min\left\{RUR_i^{EX}, RDR_i^{EX}\right\} \tag{16.25}$$

- **Other energy systems' constraints (other than power system):**

It should be noted that the constraints related to the other energy carriers (production, storage, and consumption) are similar to the relationships expressed for the electrical system, and therefore, duplicate relationships have been avoided.

- **System constraints:**

  - Electrical power balancing constraint:

$$\sum_{m=1}^{NE} p_{mt}^{EG} + \sum_{g=1}^{NES}\left(\frac{1}{\eta_g^{ESdis}} \times p_{gt}^{ESdis} - \eta_g^{ESch} \times p_{gt}^{ESch}\right) + \sum_{i=1}^{n}\left(p_{it}^{EX}\right) +$$
$$\sum_{k=1}^{NEDR} PDR_{kt}^{EDR-} = \sum_{i=1}^{NEDR} PDR_{kt}^{EDR+} + Load_t^E + \sum_{i=1}^{n}\left(D_{it}^{EX}\right) \tag{16.26}$$

  - Requirement constraint of electrical flexibility due to the power delivering capability:

$$\sum_{m=1}^{NE} FPDC_m^{EG} + \sum_{g=1}^{NES} FPDC_g^{ESG} + \sum_{i=1}^{n} FPDC_i^{XtP} + \sum_{k=1}^{NEDR} FPDC_k^{EDR} \leq FPDC^{min} \tag{16.27}$$

  - Requirement of electrical flexibility due to the storing capability:

$$\sum_{m=1}^{NE} FSC_m^{EG} + \sum_{g=1}^{NES} FSC_g^{ESG} + \sum_{i=1}^{n} FSC_i^{XtP} + \sum_{k=1}^{NEDR} FSC_k^{EDR} \leq FSC^{min} \tag{16.28}$$

  - Requirement of electrical flexibility due to the ramping capability:

$$\sum_{m=1}^{NE} FRC_m^{EG} + \sum_{g=1}^{NES} FRC_g^{ESG} + \sum_{i=1}^{n} FRC_i^{XtP} + \sum_{k=1}^{NEDR} FRC_k^{EDR} \leq FRC^{min} \tag{16.29}$$

The preceding relationships show that the presence of PtX and XtP systems can contribute to various aspects of flexibility, including FPDC and FRC, in addition to adding the possibility of energy storage in the form of other energy carriers.

16.5.2 Flexibility Consideration in Expansion Planning of the PtX Systems

In order to create continuity between systems with different energy carriers, PtX systems can be developed. In order to make a correct decision for the development of these systems, all aspects related to different energy systems should be modeled as much as possible in the expansion planning problem. In this part, an example of the formulation of the development problem of these systems from the system operator's point of view is presented. For this purpose, the objective function of the development decision-making problem is the minimization of the following objective function.

$$
\begin{aligned}
&\sum_{i=1}^{n} (AIC_i^{PtX} + AOC_i^{PtX} + AOMC_i^{PtX} + AIC_i^{XtP} + AOC_i^{XtP} + \\
&AOMC_i^{XtP} + AOC_i^{EC} + AOMC_i^{EC}) + \sum_{m=1}^{NE} \left(AOC_m^{EG} + AOMC_m^{EG}\right) + \\
&\sum_{g=1}^{NES} \left(AOC_g^{ESG} + AOMC_g^{ESG}\right) + \sum_{k=1}^{NEDR} (AOC_k^{EDR} + AOMC_k^{EDR})
\end{aligned}
\tag{16.30}
$$

It is important to note that in the presented formulation, it is assumed that the decision is made only for the development of PtX and XtP systems and the capacity of other equipment is assumed to be fixed. However, if the capacity of other equipment is variable, the generality of the problem is not changed, and the cost of investment in this equipment is also considered in the objective function.

The constraints of the planning problem are similar to the limitations presented in the previous subsection, including the following:

- Operation constraints of the generation units
- Operation constraints of the storage units
- Operation constraints of the responsive demands
- Operation constraints of the PtX systems
- Operation constraints of the XtP systems

In addition to the preceding, development restrictions for the PtX and XtP systems can also be considered. It should be noted that the size of the PtX equipment is determined in the presented formulation. Flexibility aspects can be formulated and modeled for each of the equipment in the energy system, similar to the relationships presented in the previous subsection.

16.5.3 FLEXIBILITY IN THE PRESENCE OF THE PtX SYSTEMS

In order to examine the proposed method in this book chapter, the amount of flexibility available with and without the presence of PtX systems is examined for a case study. For this purpose, a case study with the following specifications is considered:

- Gas turbine (generation capacity: 10 MW; ramping rate: 1 MW/min).
- Battery energy storage (capacity: 1 MW/2 MWh; round-trip efficiency: 85%).
- Fuel cell (generation capacity: 2 MW; electrical efficiency: 60%; ramping rate: 2 MW/ min).
- Energy efficiency of the hydrogen process, including production, compression, and utilization as a fuel, is considered about 40%.
- Hydrogen energy storage (1MW/2MWh; round-trip efficiency: 40%).

The maximum available flexibility aspects with and without the presence of the PtX system for this case study are present in Figure 16.10. The results show a significant improvement in flexibility, and more importantly, with the PtX system, it will be possible to store for a longer period of time and thus provide flexibility at other times as well.

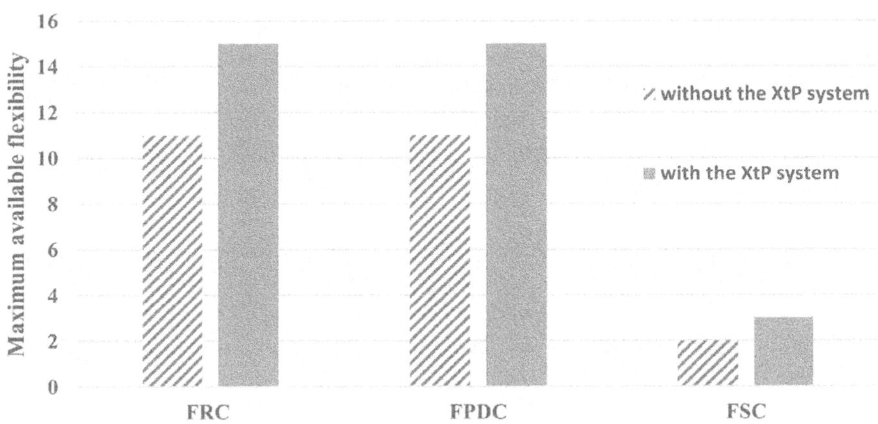

FIGURE 16.10 Maximum available flexibility for the considered case study.

16.6 REAL-WORLD APPLICATIONS OF PTX FOR FLEXIBILITY

In this section, we provide an introduction of the status of various projects worldwide, with a specific focus on the European Union. We delve into information related to global investments in different technologies and highlight trends that are expected to shape the coming years.

An analysis of the historical development of PtX projects commissioned in Europe reveals that 2018 witnessed the highest number of commissioned projects in Europe to date. Looking ahead, fewer projects are expected to be started, but installed capacity continues to grow quickly as well. The PtX development trajectory follows a pattern similar to a wave, with notable peaks occurring in the years 2015, 2018, and 2020. However, some projects initially scheduled for 2020 were delayed to 2021 due to technical issues, prolonged approval processes, and the unique circumstances arising from the COVID-19 pandemic circumstances. As an example, the year 2024 is anticipated to be significant, particularly in Germany. Regulatory sandboxes are expected to be commissioned in 2024. Additionally, the three HyPerformer projects are projected to commence in 2025, a conservative estimate, considering their funding notification in December 2019. Generally, commissioning tends to occur in the penultimate year of a project, according to insights from previous endeavors.

The studies show that Germany, with its substantial number of projects, has consistently held the main projected capacity during the previous five-year period, and this trend is expected to continue with significant growth in the next period. Notably, demonstration plans have also been successfully implemented in some additional countries. Looking ahead, it seems that these technologies would be adopted by fewer countries, but with greater strength. Germany, France, the Netherlands, Belgium, and Denmark emerge as key players in this landscape.

In WE, AEL has been employed in previous projects and is expected to continue playing a significant role in future endeavors, underscoring the ongoing advancement of this technology. PEM electrolyzers have captured a substantial market share since 2015, mainly due to dynamic performance and favorable partial load range. While there are some efforts to incorporate SOECs into PtX projects, it shows the least technology readiness level. An intriguing alternative is the alkaline solid polymer electrolyte electrolyzer, which combines features of both PEM and AEL electrolyzers. Although this technology has been utilized in some projects, the installed plants have been relatively modest, and upcoming demonstration schemes by means of this technology have not been published.

The cumulative installed capacity according to electrolyzer type is reported in [4]. The trend reveals a consistent increase, with a notably steeper rise from 2021 onward. By the end of 2020, a planned electrolyzer size about 93 MW is set to be connected. Among the notable projects, all boasting a 6 MW capacity, are the Audi e-gas project (commissioned in 2013, utilizing an alkaline electrolyzer) and the Energiepark Mainz (commissioned in 2015, employing a PEM electrolyzer), both located in Germany. Additionally, the new H_2Future project (commissioned in 2019, also using a PEM electrolyzer) is underway in Austria. The projected capacities for the upcoming years are shown, drawing from project announcements—many of them have protected funding. Looking ahead to 2025, the largest schemes are expected to reach a capacity of 100 MW.

Fields of application of PtX projects in Europe include the blending of the produced gas—mainly hydrogen or methane—into the national gas grids (about 50 projects); the production of fuels for mobility applications (about 80 projects), for example, hydrogen in fuel cell electric vehicles, methane, methanol, or Fischer–Tropsch fuels in internal combustion engines; and the use of the produced gases in CHP plants (about 40 projects) and the use of the gases in industry (about 35 projects), for example, refineries or steel plants. For some projects, no such purpose was detected.

The first PtM plant was established in 1996 by Japan, which constructed a prototype facility at Tohoku University. Since 2009, Europe has been at the forefront of major PtM projects, with Germany leading the way, followed by Switzerland, Denmark, and the Netherlands. To date, over 50 PtG initiatives have been developed, involving more than 180 public institutions and private companies. Electrolyzer sizes for PtM applications range from 1 kW to 12 MW, and the conversion of CO_2 into methane can occur through either biological or chemical pathways.

PtMe projects have gained global attention. The George Olah Renewable Methanol Plant, commissioned in 2011 in Svartsengi (Iceland), stands as the first industrial-scale facility for fuel production from CO_2. Since its full operational status in 2012, this groundbreaking plant captures carbon dioxide from an adjacent geothermal power facility. The captured CO_2 undergoes purification and compression before being combined with hydrogen generated by a 6 MW alkaline electrolyzer powered by Iceland's grid electricity. Following catalytic conversion, the resulting methanol is further purified, with water removal. Additionally, the largest PtMe plant is currently under construction in the Delfzijl industrial park in the Netherlands as part of the Djewels project. The primary goal is the working readiness of a 20 MW electrolyzer for renewable methanol production, advancing the technology from technology readiness level (TRL) 7 to TRL 8. This ambitious five-year project involves six European partners, although detailed information remains limited due to its ongoing development.

Unlike the PtM and PtMe projects, $PtNH_3$ initiatives focus on refurbishing existing ammonia production factories through a gradual replacement of fossil-sourced hydrogen with a WE-sourced one. The main manufacturing process for ammonia synthesis remains the Haber–Bosch technology. In 2018, the JGC Corporation and the National Institute of Advanced Industrial Science and Technology successfully demonstrated a $PtNH_3$ plant in Koriyama, Japan. This plant utilized renewable hydrogen and powered gas turbines fueled by synthesized ammonia. The project, which began in 2014, achieved operational status by 2018. Besides evaluating the performance of a novel ruthenium catalyst for efficient ammonia synthesis under low-temperature and low-pressure conditions, the consortium investigated ammonia production using hydrogen derived from solar panels, subjecting it to varying loading conditions. The plant achieved a daily ammonia production of 20 kg, and the ammonia-fueled gas turbine generated 47 kW of power.

The PtX market in non–Organisation for Economic Co-operation and Development (OECD) countries is still in its early stages. While more than 680 clean hydrogen projects have been announced recently, boasting a rated capacity of at least 1 MW and a total investment volume exceeding US$240 billion until 2030, only projects worth US$22 billion have progressed to financial investment decisions (FID) or are currently under construction or operational. The majority of these projects are concentrated in OECD countries. Notably, the number of projects in non-OECD countries with a rated electrolyzer capacity exceeding 10 MW and having reached FID is quite limited. According to the latest data from the IEA hydrogen project database, there are only 18 such projects, and out of these, merely 6 are located outside of China.

16.7 LESSONS LEARNED AND CONCLUDING REMARKS

In this book chapter, the PtX systems were studied as tools to improve the flexibility conditions of modern power systems. In modern power systems with a significant presence of RESs, the possibility of connecting the system with other energy systems (based on energy carriers other than electricity) is considered using the modern intermediary equipment. The concept of flexibility has also been studied as a tool to determine the level of readiness of the power system against load and production changes. It is worth noting that flexibility can be described for all forms of energy, but this chapter focuses on the electrical energy flexibility. In the following, the main findings obtained from the studies of this chapter are presented:

- The PtX equipment corresponding to that XtP as an intermediary makes it possible to implement an energy system as well as a modern power system. In fact, by using this equipment, it will be possible to create an integrated system with the presence of various energy carriers. In this way, it is possible to simultaneously manage energy for different energy systems.
- PtX and XtP systems can directly and indirectly affect the flexibility of the system. It should be noted that flexibility can be used for all types of energy forms. The focus of this chapter is on the flexibility of electrical energy, which can be analyzed for other forms of energy similar to the studies.
- Three capabilities to improve system flexibility include FRC, FPDC, and FSC.
- PtX systems provide the opportunity of storing the electrical energy in the form of other energy carriers, thus indirectly improving flexibility through the FSC. This opportunity can create the possibility of energy storage with high energy capacity as well as long storage time.
- XtP systems, with the possibility of providing electrical energy for the power system, provide the possibility of supplying the flexibility through FRC and FPDC directly.

Integrating PtX technologies into existing energy systems poses several challenges, which can be summarized as follows:

1. Resource requirements and availability

   - *Water supply.* PtX processes, particularly electrolysis, require significant amounts of ultra-pure water, which can be a challenge in regions with limited water resources.
   - *CO_2 supply.* For processes like power-to-methane or power-to-syngas, a reliable supply of CO_2 is necessary, which can be a bottleneck.

2. Grid stability and balancing

   - *Variability and unpredictability.* RESs used for PtX are often intermittent, making it challenging to ensure a stable power supply. Advanced forecasting models and smart energy management systems are needed to balance supply and demand.
   - *Grid infrastructure.* Integrating PtX into the grid requires modernization of existing infrastructure, which can be costly and complex. Distributed systems and smart grids can help mitigate these issues.

3. Technological and economic challenges

   - *Efficiency and cost.* The efficiency and cost-effectiveness of PtX technologies, such as electrolysis, are critical. Improving these aspects is essential for widespread adoption.

- *Scalability.* Scaling up PtX technologies to meet demand while maintaining efficiency and reducing costs is a significant challenge.

4. Sector coupling and integration

- *Sectoral decarbonization.* PtX aims to decarbonize sectors that are hard to electrify directly, such as transportation and industry. Ensuring a seamless integration of PtX products (like synthetic fuels) into these sectors is crucial.
- *Regulatory frameworks.* The lack of comprehensive regulatory frameworks can hinder the deployment of PtX technologies. Policymakers need to create supportive environments for these technologies.

5. Environmental impacts

- *Life cycle assessments.* The environmental impacts of PtX systems, including their production and end-of-life phases, need careful evaluation to ensure they contribute to overall sustainability.

6. Energy storage and conversion

- *Energy storage.* PtX technologies often involve energy storage, which must be efficient and reliable. Battery storage and other solutions can help manage the variability of renewable energy sources.

Addressing these challenges will be essential for the successful integration of PtX technologies into the energy mix, enabling a more sustainable and decarbonized energy future.

In order to complete the studies carried out in this book chapter, the following items are suggested:

- Studying the problem over a long period of time in order to demonstrate the ability of long-term storage
- Modeling of the energy carriers' networks
- Modeling the expansion planning problem from the investors' point of view

Nomenclature

t	index of the operation time period
m	index of the electrical generator
g	index of the electrical storage unit
k	index of the responsible electrical load
i	index of the energy carrier other than electricity
mi	index of the ith energy carrier generation unit
gi	index of the ith energy carrier storage unit
ki	index of the ith responsible energy carrier load
T	operation time period
NE	number of the electrical generation units
NES	number of the electrical storage units
$NEDR$	number of the responsible electrical loads
n	number of the energy carriers other than electricity
NiE	number of the ith energy carrier generating units
$NiES$	number of the ith energy carrier storage units
$NiEDR$	number of the ith responsible energy carrier loads

OC_{mt}^{EG}	operation cost of the mth electrical generation unit at time t
OC_{gt}^{ES}	operation cost of the gth electrical storage unit at time t
OC_{kt}^{EDR}	operation cost of the kth responsible electrical load at time t
OC_{it}^{PtX}	operation cost of the ith PtX unit at time t
OC_{it}^{XtP}	operation cost of the ith XtP unit at time t
OC_{mit}^{XiG}	operation cost of the mith generating unit that produces the ith energy carrier at time t
OC_{git}^{XiS}	operation cost of the gith storage unit that stores the ith energy carrier at time t
OC_{kit}^{XiDR}	operation cost of the gith responsible load that consumes the ith energy carrier at time t
p_{mt}^{EG}	power generation of the mth electrical generation unit at time t
$\overline{p_{m}^{EG}}$	maximum generation capability of the mth electrical generation unit
$\underline{p_{m}^{EG}}$	minimum generation capability of the mth electrical generation
$\overline{RUR_{m}^{EG}}$	maximum ramp-up rate of the mth electrical generation unit
$\overline{RDR_{m}^{EG}}$	maximum ramp-down rate of the mth electrical generation unit
η_{g}^{ESch}	charging efficiency of the gth electrical storage unit
η_{g}^{ESdis}	discharging efficiency of the gth electrical storage unit
p_{gt}^{ESch}	charging power of the gth electrical storage unit at time t
p_{gt}^{ESdis}	discharging power of the gth electrical storage unit at time t
E_{gt}^{ES}	electrical energy stored in the gth electrical storage unit at time t
$\underline{E_{g}^{ES}}$	minimum energy capacity of the gth electrical storage unit
$\overline{E_{g}^{ES}}$	maximum energy capacity of the gth electrical storage unit
$\overline{P_{g}^{ES}}$	maximum power capacity of the gth electrical storage unit
$RUR_{g}^{ES,charging}$	ramp rate of the electrical storage unit during charging mode
$RDR_{g}^{ES,discharging}$	ramp rate of the electrical storage unit during discharging mode
PDR_{kt}^{EDR-}	decremental demand response value of the kth electrical responsible demand at time t
PDR_{kt}^{EDR+}	incremental demand response value of the kth electrical responsible demand at time t
$\overline{PDR_{k}^{EDR}}$	maximum demand responsiveness of the kth electrical responsible demand
$\underline{D_{kt}^{EDR}}$	consumption of the kth responsible electrical load at time t
$\overline{D_{k}^{ED}}$	maximum consumption of the kth responsible electrical load
C_{it}^{GX}	amount of energy carrier produced by the ith PtX system at time t
D_{it}^{EX}	amount of electrical power consumed by the ith PtX system at time t
CC_{i}^{PtX}	conversion coefficient of the ith PtX system
RR_{k}^{ED}	ramp rate of the kth electrical load
p_{it}^{EX}	amount of electrical power produced by the ith XtP system at time t
C_{it}^{DX}	amount of energy carrier consumed by the ith XtP system at time t
CC_{i}^{XtP}	conversion coefficient of the ith XtP system
$\overline{p_{it}^{EX}}$	electrical power production capacity of the ith XtP system
$\overline{C_{it}^{GX}}$	energy carrier production capacity of the ith PtX system
RUR_{i}^{EX}	ramp-up rate of the ith XtP system
RDR_{i}^{EX}	ramp-down rate of the ith XtP system
$Load_{t}^{E}$	electrical demand at time t
$FPDC^{min}$	minimum required FPDC
FSC^{min}	minimum required FSC
FRC^{min}	minimum required FRC

AIC_i^{PtX}	annualized investment cost of the ith PtX system
AOC_i^{PtX}	annualized operation cost of the ith PtX system
$AOMC_i^{PtX}$	annualized maintenance cost of the ith PtX system
AIC_i^{XtP}	annualized investment cost of the ith XtP system
AOC_i^{XtP}	annualized operation cost of the ith XtP system
$AOMC_i^{XtP}$	annualized maintenance cost of the ith XtP system
AOC_i^{EC}	annualized operation cost of equipment in systems based on the ith energy carrier
$AOMC_i^{EC}$	annualized maintenance cost of equipment in systems based on the ith energy carrier
AOC_m^{EG}	annualized operation cost of the mth electrical generation unit
$AOMC_m^{EG}$	annualized maintenance cost of the mth electrical generation unit
AOC_g^{ESG}	annualized operation cost of the gth electrical storage unit
$AOMC_g^{ESG}$	annualized maintenance cost of the gth electrical generation unit
AOC_k^{EDR}	annualized operation cost of the kth responsible demand
$AOMC_k^{EDR}$	annualized maintenance cost of the kth responsible demand

NOTES

1 Operational reserve demand curve.
2 Emergency demand response programs.
3 Critical peak pricing.

REFERENCES

[1] M. K. Kane and S. Gil, "Green hydrogen: A key investment for the energy transition," June, 2022.
[2] M. J. Palys and P. Daoutidis, "Power-to-X: A review and perspective," Computers & Chemical Engineering, vol. 165, p. 107948, 2022.
[3] IEA, 2024. www.iea.org/.
[4] C. Wulf, P. Zapp, and A. Schreiber, "Review of power-to-X demonstration projects in Europe," Frontiers in Energy Research, vol. 8, p. 191, 2020.
[5] E. Bargiacchi, "Power-to-fuel existing plants and pilot projects," Power to Fuel, pp. 211–237, 2021.
[6] A. C. Ince, C. O. Colpan, A. Hagen, and M. F. Serincan, "Modeling and simulation of power-to-X systems: A review," Fuel, vol. 304, p. 121354, 2021.
[7] M. Ozturk and I. Dincer, "A comprehensive review on power-to-gas with hydrogen options for cleaner applications," International Journal of Hydrogen Energy, vol. 46, no. 62, pp. 31511–31522, 2021.
[8] A. R. Dahiru, A. Vuokila, and M. Huuhtanen, "Recent development in Power-to-X: Part I – A review on techno-economic analysis," Journal of Energy Storage, vol. 56, p. 105861, 2022.
[9] J. P. Shim, M. Warkentin, J. F. Courtney, D. J. Power, R. Sharda, and C. Carlsson, "Past, present, and future of decision support technology," Decision Support Systems, vol. 33, no. 2, pp. 111–126, 2002.
[10] M. Götz, J. Lefebvre, F. Mörs, A. M. Koch, F. Graf, S. Bajohr, R. Reimert, and T. Kolb, "Renewable power-to-gas: A technological and economic review," Renewable Energy, vol. 85, pp. 1371–1390, 2016.
[11] Z. Chehade, C. Mansilla, P. Lucchese, S. Hilliard, and J. Proost, "Review and analysis of demonstration projects on power-to-X pathways in the world," International Journal of Hydrogen Energy, vol. 44, no. 51, pp. 27637–27655, 2019.
[12] P. F. Borowski, "Mitigating climate change and the development of green energy versus a return to fossil fuels due to the energy crisis in 2022," Energies, vol. 15, no. 24, p. 9289, 2022.
[13] B. H. Kreps, "The rising costs of fossil-fuel extraction: An energy crisis that will not go away," The American Journal of Economics and Sociology, vol. 79, no. 3, pp. 695–717, 2020.
[14] S. M. Hashemi, M. Tabarzadi, F. Fallahi, M. R. N. Kalhori, D. Abdollahzadeh, and M. Qadrdan, "Water and emission constrained generation expansion planning for Iran power system," Energy, vol. 288, p. 129821, 2024.

[15] S. Fankhauser, S. M. Smith, M. Allen, K. Axelsson, T. Hale, C. Hepburn, J. M. Kendall, and M. Obersteiner, "The meaning of net zero and how to get it right," Nature Climate Change, vol. 12, no. 1, pp. 15–21, 2022.

[16] A. Ahmed, T. Ge, J. Peng, W.-C. Yan, B. T. Tee, and S. You, "Assessment of the renewable energy generation towards net-zero energy buildings: A review," Energy Building, vol. 256, p. 111755, 2022.

[17] S. M. Hashemi, B. Alizadeh, M. Sheibani, and F. Fallahi, "Assessing potential of renewable energy sources in Iran through practical and analytical data," in *2023 13th Smart Grid Conference (SGC)*, IEEE, 2023, pp. 1–8 (Accessed June 8, 2024). https://ieeexplore.ieee.org/abstract/document/10459316/.

[18] International Energy Agency, "Status of power system transformation 2018 advanced power plant flexibility," 2018 [Online]. www.oecd-ilibrary.org/energy/status-of-power-system-transformation-2018_9789264302006-en.

[19] M. Z. Degefa, I. B. Sperstad, and H. Sæle, "Comprehensive classifications and characterizations of power system flexibility resources," Electric Power Systems Research, vol. 194, p. 107022, 2021.

[20] I. Calero, C. A. Canizares, K. Bhattacharya, and R. Baldick, "Duck-curve mitigation in power grids with high penetration of PV generation," IEEE Transactions on Smart Grid, vol. 13, no. 1, pp. 314–329, 2021.

[21] *Duck curve* [Online]. https://en.wikipedia.org/wiki/Duck_curve.

[22] P. Hu, Y. Li, Y. Yu, and F. Blaabjerg, "Inertia estimation of renewable-energy-dominated power system," Renewable and Sustainable Energy Reviews, vol. 183, p. 113481, 2023.

[23] H. Nosair and F. Bouffard, "Reconstructing operating reserve: Flexibility for sustainable power systems," IEEE Transactions on Sustainable Energy, vol. 6, no. 4, pp. 1624–1637, 2015.

[24] S. M. Hashemi, V. Vahidinasab, M. S. Ghazizadeh, and J. Aghaei, "Security constrained operation of radial micro grids based on the loads' vulnerabilities and flexibilities," Tabriz Journal of Electrical Engineering, vol. 50, no. 3, Art. no. 3, 2020.

[25] Q. Wang and B.-M. Hodge, "Enhancing power system operational flexibility with flexible ramping products: A review," IEEE Transactions on Industrial Informatics, vol. 13, no. 4, pp. 1652–1664, 2016.

[26] S. Sreekumar, S. Yamujala, K. C. Sharma, R. Bhakar, S. P. Simon, and A. S. Rana, "Flexible ramp products: A solution to enhance power system flexibility," Renewable and Sustainable Energy Reviews, vol. 162, p. 112429, 2022.

[27] A. Ulbig and G. Andersson, "Analyzing operational flexibility of electric power systems," International Journal of Electrical Power & Energy Systems, vol. 72, pp. 155–164, 2015.

[28] S. M. Hashemi, H. Arasteh, M. Shafiekhani, M. Kia, and J. M. Guerrero, "Multi-objective operation of microgrids based on electrical and thermal flexibility metrics using the NNC and IGDT methods," International Journal of Electrical Power & Energy Systems, vol. 144, p. 108617, 2023.

[29] A. Marini, S. M. Hashemi, and M. Tabarzadi, "A SCOPF model for iran's hourly power market, a requirement to develop renewable power plants," in *2023 13th Smart Grid Conference (SGC)*, IEEE, 2023, pp. 1–8 (Accessed June 8, 2024). [Online]. https://ieeexplore.ieee.org/abstract/document/10459278/.

[30] IRENA, "Grid codes for renewable powered systems," 2022. [Online]. www.irena.org/publications/2022/Apr/Grid-codes-for-renewable-powered-systems.

[31] H. Arasteh, N. Moslemi, and S. M. Hashemi, "Demand response measurement and verification approaches: Analyses and guidelines," Industrial Demand Response Methods Best Practice Case Study Apple, p. 133, 2022.

[32] H. Arasteh *et al.*, "A System-of-Systems Planning Platform for Enabling Flexibility Provision at Distribution Level," in *Flexibility in Electric Power Distribution Networks*, CRC Press, pp. 41–65.

[33] S. P. Mirhoseini, S. M. Hashemi, B. Alizadeh, M. Tabarzadi, and F. Fallahi, "A reliability-based strategy for active distribution network expansion planning," in *2023 13th Smart Grid Conference (SGC)*, IEEE, 2023, pp. 1–8. [Online]. https://ieeexplore.ieee.org/abstract/document/10459273/ (Accessed June 8, 2024).

[34] S. M. Hashemi and V. Vahidinasab, "Energy management systems for microgrids," Microgrids: Advances in Operation, Control, and Protection, pp. 61–95, 2021.

[35] S. M. Hashemi, V. Vahidinasab, M. S. Ghazizadeh, and J. Aghaei, "Valuing consumer participation in security enhancement of microgrids," IET Generation, Transmission & Distribution, 2018.

[36] S. M. Hashemi, V. Vahidinasab, M. Ghazizadeh, and J. Aghaei, "Load control mechanism for operation of microgrids in contingency state," IET Generation, Transmission & Distribution, vol. 14, no. 23, Art. no. 23, 2020.

[37] S. M. Hashemi, V. Vahidinasab, M. S. Ghazizadeh, and J. Aghaei, "Reliability-oriented DG allocation in radial Microgrids equipped with smart consumer switching capability," in *2019 International Conference on Smart Energy Systems and Technologies (SEST)*, IEEE, 2019, pp. 1–6.

[38] M. R. Sheibani, G. R. Yousefi, M. A. Latify, and S. H. Dolatabadi, "Energy storage system expansion planning in power systems: A review," IET Renewable Power Generation, vol. 12, pp. 1203–1221, July 2018.

[39] M. R. Sheibani, and A. Moshari, "Operation planning of a microgrid considering the resiliency in the presence of energy storage systems," in *2020 10th Smart Grid Conference (SGC)*, 2020, pp. 1–6.

[40] M. R. Sheibani, G. R. Yousefi, and M. A. Latify, "Modeling the energy storage systems in the power system studies," in *Synergy Development in Renewables Assisted Multi-carrier Systems*, M. Amidpour, M. Ebadollahi, F. Jabari, M. R. Kolahi, and H. Ghaebi, Eds. Green Energy and Technology. Cham: Springer.

[41] M. R. Sheibani, G. R. Yousefi, H. Raoufi, and N. Moslemi, "Economics of energy storage options to support a conventional power plant: A stochastic approach for optimal energy storage sizing," The Journal of Energy Storage, vol. 23, January 2021.

[42] M. R. Sheibani, G. R. Yousefi, and M. A. Latify, "Stochastic price based coordinated operation planning of energy storage system and conventional power plant," Journal of Modern Power Systems and Clean Energy, vol. 7, pp. 1020–1032, July 2019.

[43] M. R. Sheibani, N. Moslemi, B. Alizadeh, and A. Marini, "Payback cycles: A new concept to decide for energy storage expansion planning," in *2024 9th International Conference on Technology and Energy Management (ICTEM)*, Behshar, Mazandaran, Iran, Islamic Republic of, 2024, pp. 1–5.

17 Modeling and Control of Grid-Connected Electrolyzer Plants

Chunjun Huang, Yi Zheng, and Shi You

17.1 INTRODUCTION

17.1.1 TRANSITION TO RENEWABLE-DOMINATED POWER SYSTEMS

Our modern society has been built upon the extensive use of fossil fuels, driving economic growth and technological advancements. However, the reliance on coal, oil, and natural gas has led to detrimental environmental consequences, including the alarming rise in global CO_2 emissions [1]. This upward trend directly correlates with the increase in global average temperatures, reaching nearly 1°C above pre-industrial levels by 2020 [2]. The urgency to address climate change has prompted a global shift toward harnessing renewable energy sources. Figure 17.1 illustrates the gradual increase in the utilization of renewables within global energy consumption [3]. While fossil fuels remain dominant, the upward trajectory of renewables signifies a transition for power systems toward renewable-dominated power systems (RPSs). This transition is particularly evident in the electricity sector, where low-carbon options like nuclear, hydropower, wind, and solar offer greater potential for replacing fossil fuels. As reported in [4], the share of low-carbon sources in the electricity mix surpassed 30% in 2019, exceeding that of the total energy mix. This highlights the strategic importance of integrating renewables into the electricity sector to accelerate decarbonization efforts.

The rise of renewable energy resources, particularly solar and wind power, introduces new complexities to the power grid due to the inherent variability and intermittency of these resources. Their generation depends on weather conditions, leading to fluctuations in power output that can disrupt the delicate balance between supply and demand within the power system. This intermittency poses a significant challenge to maintaining grid stability and reliability, potentially causing frequency and voltage fluctuations and, in extreme cases, even blackouts in power systems. Thus, the shift toward RPSs signifies a transition away from fossil fuel dependency and toward a more sustainable future, but it also necessitates addressing the technical challenges associated with integrating these variable resources into the grid.

One of the key challenges of RPSs is the reduction in system inertia due to the decreased use of traditional synchronous generators (SGs) and the rise of inertia-less renewable generators. System inertia plays a crucial role in maintaining grid stability by providing a buffer against sudden changes in frequency. The reduced inertia in RPS can lead to increased frequency deviations and rate of change of frequency (RoCoF), jeopardizing grid stability and potentially causing blackouts, as witnessed in the 2019 UK incident [5]. This incident highlighted the need for additional grid services, such as frequency containment reserve, to ensure the stability and reliability of power systems with high renewable penetration.

To address these challenges and facilitate the integration of larger shares of renewable energy, several solutions are being explored and implemented. Grid expansion and modernization are crucial for accommodating the geographically dispersed nature of renewables and improving grid flexibility. Demand-side management strategies can help shift electricity consumption patterns to better

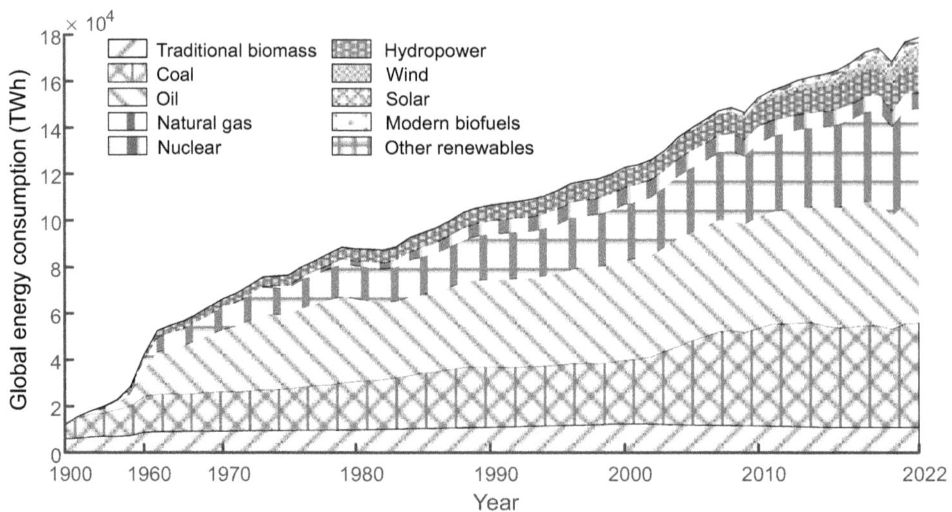

FIGURE 17.1 Different energy utilization within global energy consumption.

Source: [3].

match the availability of renewable energy. However, energy storage remains a critical need for balancing the variability of renewables and ensuring a consistent and reliable power supply. While various energy storage technologies exist, such as batteries and pumped hydro storage, they often face limitations in terms of duration, scalability, and environmental impact.

17.1.2 ROLE OF PtH$_2$ IN RENEWABLE-DOMINATED POWER SYSTEMS

Among the promising solutions to supporting the transition to RPSs, hydrogen-based solutions, particularly power-to-hydrogen (PtH$_2$) technology, have emerged as a key enabler due to its unique advantages and potential for renewable energy integration and deep decarbonization. The core value proposition of grid-connected PtH$_2$ lies in its ability to convert surplus renewable electricity into hydrogen, effectively storing excess energy that would otherwise be curtailed due to grid limitations or mismatches between supply and demand. This not only maximizes the utilization of renewable energy resources but also plays a crucial role in balancing the grid and enhancing its stability. By acting as a buffer and absorbing excess renewable generation, grid-connected PtH$_2$ plants help mitigate fluctuations in power output, thereby reducing the risk of frequency and voltage instability that can lead to blackouts.

Furthermore, hydrogen-based storage solutions, by integrating PtH$_2$ and hydrogen storage devices, show a unique ability to provide long-term and large-scale energy storage. Unlike batteries with limited storage durations, hydrogen can be stored indefinitely without significant energy loss, making PtH$_2$ ideal for addressing the seasonal variability of renewable energy sources and providing long-term energy security [6]. This capability enables the storage of surplus renewable energy generated during peak production periods for use during periods of low renewable generation, ensuring a reliable and consistent energy supply throughout the year.

Beyond its grid-balancing capabilities, PtH$_2$ facilitates sector coupling by connecting the electricity sector with other energy sectors, such as transportation and industry. Hydrogen produced from renewable electricity can be used as a clean fuel for transportation, powering fuel cell vehicles or blended with conventional fuels, and as a feedstock for industrial processes, reducing reliance on fossil fuels in these sectors [7, 8]. This cross-sectoral integration creates synergies and enables a holistic approach to decarbonization, extending the benefits of renewable energy beyond the electricity sector and contributing to a more comprehensive energy transition.

While PtH_2 technology offers significant potential, its widespread adoption is currently affected by crucial cost-related challenges that currently hinder its competitiveness against established energy solutions. A key economic hurdle for PtH_2 technology lies in the high levelized cost of hydrogen (LCOH) it produces. This cost is heavily influenced by several factors, particularly the significant electricity consumption during hydrogen production and the substantial capital expenditures required for electrolyzer systems and supporting infrastructure [9]. High electricity costs arise from both the current electricity price and PtH's limited operational efficiency, where a portion of the electricity input is inevitably wasted. Additionally, the initial investment costs for PtH_2 plants remain considerably higher compared to conventional energy alternatives due to the specialized equipment and infrastructure needs. These combined factors contribute to a high LCOH, hindering the widespread adoption of PtH_2.

However, these cost challenges can be mitigated through the development of advanced control and operation strategies for PtH_2 plants. Optimizing the control and operation of electrolyzers is crucial for reducing operational costs by minimizing system energy losses, thereby directly lowering electricity expenses. Furthermore, intelligent control strategies enable economic arbitrage in response to electricity price variations, further reducing operational costs. Additionally, by participating in energy markets and providing grid services, PtH_2 plants can generate revenue streams that offset operational costs and enhance economic viability. Effective planning and management of hydrogen infrastructure also play a crucial role in minimizing development and maintenance expenses. By addressing these cost challenges through innovative control and operation approaches, PtH_2 technology can achieve greater competitiveness and contribute significantly to a sustainable energy future.

Despite these challenges, the potential of PtH_2 in facilitating renewable energy integration and enabling a more sustainable and flexible energy system is undeniable. Ongoing research and development efforts are focused on improving the efficiency, reducing the cost, and optimizing the integration of PtH_2 systems within the power grid.

17.1.3 Development of PtH_2 Systems

The rising importance of PtH_2 in the context of renewables integration and decarbonization has spurred significant interest in the development and advancement of electrolyzer technologies. This trend is underscored by the number of large-scale PtH_2 and other hydrogen projects announced worldwide, exceeding 680 in 2022 and representing a staggering investment of \$240 billion [10]. These initiatives encompass a wide range of applications, from GW-level hydrogen production to integration within industrial processes, transportation systems, and essential infrastructure development. Europe stands out as a leader in this domain, with 314 projects demonstrating the region's commitment to leveraging hydrogen as a key driver for achieving decarbonization goals across various sectors.

Beyond its role in the energy sector, hydrogen also holds immense potential as a feedstock for producing valuable commodities, such as ammonia, methanol, and methane, through reactions with CO_2 or nitrogen. This broader concept, known as power-to-X (PtX), encompasses a diverse range of products derived from electricity, offering increased flexibility and cross-sectoral integration within the overall energy system. Consequently, PtX solutions are attracting growing attention and development efforts on a global scale.

As of 2019, a total of 192 PtX demonstration projects were already underway across 32 countries, with Europe leading the way with approximately 154 projects [11]. Denmark, with its abundant wind resources, serves as a prime example of a nation embracing PtX solutions. The country's ambitious national strategy aims to establish a substantial electrolysis capacity of 4–6 GW by 2030, focusing on green hydrogen production for challenging sectors, like heavy-duty transport and shipping [12]. This initiative is projected to significantly reduce CO_2 emissions, contributing to Denmark's ambitious climate goals. Figure 17.2 provides a comprehensive overview of Denmark's current and

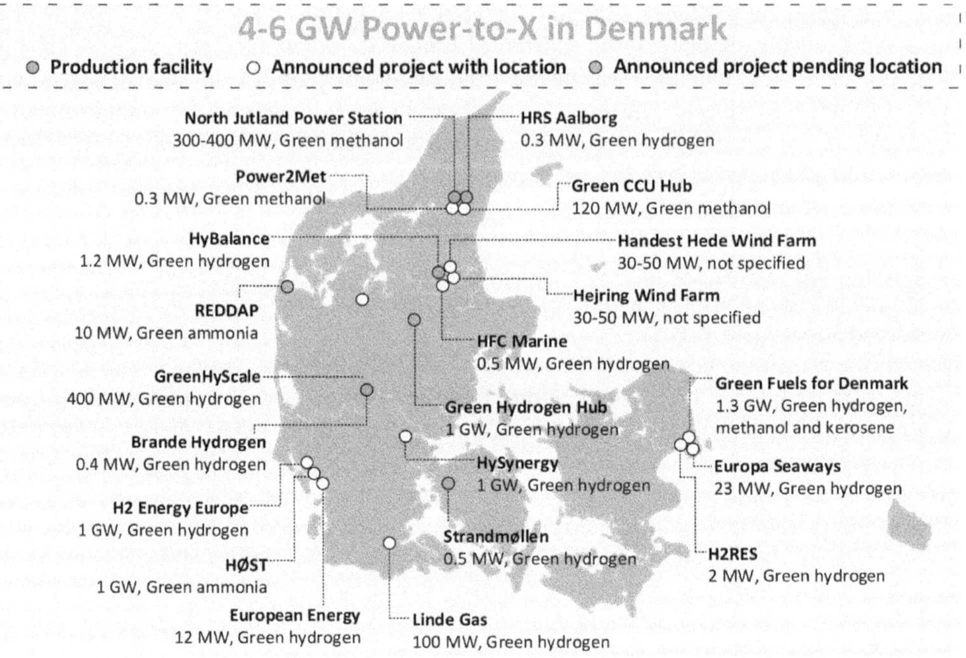

FIGURE 17.2   Overview of PtX projects in Denmark.
Source: [11].

planned PtX projects [13], illustrating the country's commitment to PtH$_2$ and the exploration of green ammonia and methanol production. The trend toward larger-scale PtX installations is evident, with plans for several GW-level green hydrogen and ammonia projects, further solidifying PtX's role in accelerating the transition toward a sustainable and carbon-neutral future.

17.1.4   SIGNIFICANCE OF MODELING AND CONTROL OF PtH$_2$

Currently, four main state-of-the-art PtH technologies exist, including alkaline electrolyzers (AEL), proton-exchange membrane electrolyzers (PEM), solid oxide electrolyzers (SOE), and anion exchange membrane electrolyzers (AEM). Among them, AEL stands out as the most mature and commercially available option, offering advantages such as low cost, long lifespan, and scalability to larger capacities [14, 15]. These characteristics have positioned AEL as a key technology for large-scale grid-connected applications and a focal point for research efforts.

Modeling of AEL systems plays a crucial role in understanding their fundamental principles, optimizing system design, and enabling effective control and operation strategies to improve their techno-economic performance. Current modeling efforts have primarily focused on capturing the complex, multi-physics nature of electrolyzers, covering electrical, electrochemical, thermal, and mass transfer domains [16]. These models range from mechanistic models based on physical laws to data-driven empirical and semi-empirical models [16–18]. These modeling approaches have proven effective in achieving high-accuracy performance prediction of electrolyzers through the development of precise and complex models. However, the inherent complexity of these models can hinder the development of efficient control strategies, highlighting the need for simplified yet accurate models that capture the essential dynamics of AEL systems while remaining computationally efficient for control purposes.

It is important to note that previous modeling and control efforts have primarily focused on understanding and optimizing electrolyzer behavior under a traditional operational mode, where stable power input conditions and maximum hydrogen production were the primary objectives. However, the increasing integration of renewable energy sources and the role of PtH_2 in grid support necessitate a shift in this paradigm. The economic viability of PtH_2 is also a critical factor influencing its widespread adoption.

The evolving operational context, characterized by fluctuating power inputs from renewable sources; the need for partial load operation to provide grid services; and the imperative to reduce operational costs, creates a critical need for new and improved modeling and control strategies for AEL plants. Existing models, while effective for traditional operation, may not adequately capture the dynamic behavior of electrolyzers under these variable conditions. Additionally, control strategies must be adapted to not only optimize hydrogen production but also enable grid support functions, enhance the overall system flexibility, and minimize operational costs.

Therefore, future research efforts are crucial to develop:

1. **Advanced models.** These models should accurately represent the dynamic behavior of AEL systems under variable power input and partial load conditions, considering the complex interplay of electrical, electrochemical, thermal, and mass transfer processes. This includes capturing the dynamic response of the electrolyzer to rapid changes in power input and the effects of operating at different load levels.
2. **Advanced control and operation strategies.** These strategies should enable AEL plants to operate effectively under partial load conditions in response to grid demands, thereby providing grid services like frequency regulation. These strategies also aim to optimize hydrogen production within the constraints of the electrolyzer's technical constraints, the grid, and renewable energy availability. This requires consideration of the dynamic interactions between the electrolyzer, the power grid, and renewable energy sources. Furthermore, these control strategies should aim to enable economic scheduling and operation for electrolyzers to minimize operational costs by optimally enabling economic arbitrage and providing grid services.

By addressing these modeling and control challenges, we can unlock the full potential of PtH_2 technology. Not only will this facilitate greater renewable energy integration and create a more sustainable and flexible energy system, but it will also improve the economic competitiveness of PtH_2, paving the way for its widespread adoption as a key component of the future energy landscape. Motivated by this, this chapter will mainly focus on providing a basic description of relevant issues related to the modeling and control of grid-connected electrolyzer plants.

17.2 MODELING AND CONTROL OF GRID-CONNECTED ELECTROLYZER PLANTS

The integration of electrolyzers into modern energy systems, particularly within the context of renewable energy sources and grid support services, necessitates a thorough understanding of their complex behavior and the development of effective control strategies. Modeling plays a crucial role in achieving these objectives, providing valuable insights into the dynamic interactions between various physical processes within the electrolyzer system. This section focuses on establishing a comprehensive modeling process for grid-connected AEL systems, to accurately represent the system's characteristics and enable efficient control and optimization.

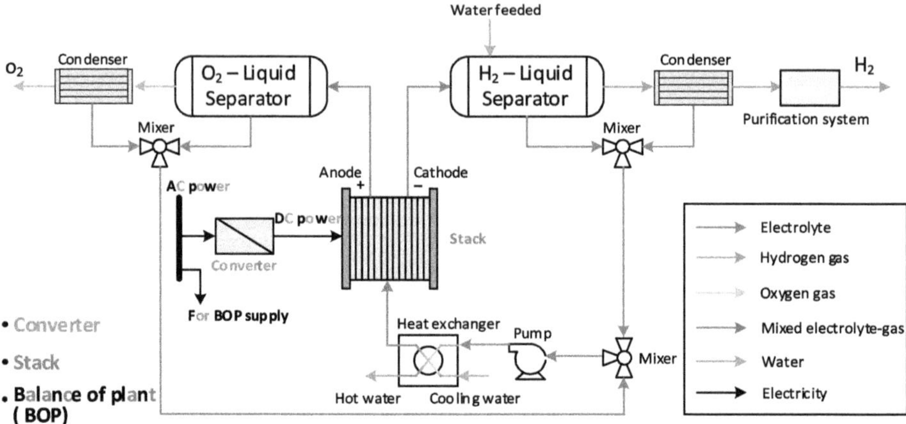

FIGURE 17.3 Schematic of an alkaline electrolyzer system.

17.2.1 Electrolyzer System Overview

Before delving into the intricacies of the AEL modeling, it is crucial to establish a solid understanding of the fundamental components and operational principles that govern an AEL system. Electrolyzers function as energy conversion devices, leveraging electrical energy to decompose water into its constituent elements, hydrogen and oxygen, through the process of electrolysis. The specific electrochemical reactions involved in AEL, the focus of this chapter, are outlined in equations (17.1)–(17.3) [19].

$$Anode : 2OH^-(aq) \rightarrow \frac{1}{2}O_2(g) + H_2O(l) + 2e^-$$

(17.1)

$$Cathode : 2H_2O(l) + 2e^- \rightarrow H_2(g) + 2OH^-(aq)$$

(17.2)

$$Overall : H_2O(l) \rightarrow H_2(g) + \frac{1}{2}O_2(g)$$

(17.3)

Figure 17.3 provides a generalized schematic representation of a complete AEL system, illustrating its key components and their interconnectedness. These components can be broadly categorized into three main sections:

1. **Stack.** The stack serves as the heart of the electrolyzer system, where the core process of electrolysis takes place. It consists of multiple electrolyzer cells, each containing an anode and a cathode separated by a diaphragm or membrane. These cells are typically connected in series or parallel to achieve the desired voltage and current levels for efficient hydrogen production.
2. **Balance of plant (BOP).** The BOP encompasses the supporting infrastructure and auxiliary systems that are essential for maintaining the proper operating conditions for electrolysis within the stack. These systems typically include heat exchangers, pump, gas–liquid separators, condenser, gas purification units, and so on.
3. **Converter.** The converter acts as the electrical interface between the electrolyzer system and the external power grid. It is responsible for converting the alternating current (AC) power supplied by the grid into stable direct current (DC) power, which is required for the

electrolysis process. Additionally, the converter regulates the voltage and current levels to meet the specific operational requirements of the electrolyzer, ensuring efficient and stable performance.

The operation of an electrolyzer system involves a complex interplay of various physical processes. Understanding these intricate interactions through comprehensive modeling is essential for developing effective control strategies and optimizing the system's performance in terms of hydrogen production, energy efficiency, and overall operational cost.

17.2.2 Modeling Development

To holistically characterize the complex multi-physics within AEL systems, especially the interaction among different physical domains, a comprehensive modeling process is presented. Particularly, various component models of AELs, each representing a specific aspect of the electrolyzer system's behavior, are detailed, including:

1. **UI model (polarization characteristic).** Describes the relationship between cell voltage and current density, capturing the electrochemical performance and efficiency of the electrolyzer cell.
2. **Electrolysis efficiency model.** Quantifies the efficiency of converting electrical energy into hydrogen, considering both voltage and Faraday efficiencies.
3. **Hydrogen production model.** Determines the rate of hydrogen production based on the electrical current and Faraday's law.
4. **Thermal model.** Analyzes the thermal dynamics and temperature variations within the stack, considering heat generation, heat loss, and cooling processes.
5. **Converter model.** Represents the electrical interface between the electrolyzer system and the power grid, including both AC–DC and DC–DC converters and their control systems.
6. **Scaling model.** Enables the scaling up of model parameters and variables from lower to higher scales, such as from cell to stack.

17.2.2.1 UI Model (Polarization Characteristic)

The polarization characteristic of an electrolyzer cell describes the relationship between cell voltage and current density during operation. This relationship, often depicted as a voltage–current (UI) curve, reveals how cell voltage increases with rising current density due to various electrochemical losses. The UI curve is a critical indicator for evaluating the efficiency and overpotential performance of the electrolyzer under different conditions. Based on the physical laws governing the electrolysis reaction, the UI curve can be analytically modeled as follows [16]:

$$U_{cell} = U_{rev} + U_{act} + U_{ohm} + U_{diff} \tag{17.4}$$

where U_{cell} is the cell voltage; U_{rev} is the reversible voltage; and U_{act}, U_{ohm}, and U_{diff} are the activation, ohmic, and diffusion overvoltages, respectively. These overvoltages arise from the kinetics of electronic charge transfer, mass transfer, and ohmic losses in the electrolyte.

The reversible voltage represents the theoretical minimum cell voltage required for the electrolysis reaction and varies with temperature and pressure conditions within the electrolyzer. It can be expressed as follows [16]:

$$U_{rev}(T,P) = \frac{R(T+273.15)}{zF} \ln\left(\frac{P_{H_2}\sqrt{P_{O_2}}}{\alpha_{H_2O}}\right) + U_{rev}^0(T) \tag{17.5}$$

where T is the operating temperature, P_{H_2} and P_{O_2} are the partial pressures of hydrogen and oxygen, R is the gas constant, F is the Faraday constant, z is the number of transferred electrons per reaction, α is the thermodynamic activity of water, and $U_{rev}^0(T)$ is the reversible voltage at standard conditions, which is temperature-dependent and can be calculated using the following equation [20]:

$$U_{rev}^0(T) = 1.5184 - 1.5421 \times 10^{-3}(T + 273.15) + 9.256 \times 10^{-5}(T + 273.15) \cdot \ln(T + 273.15)$$
$$+ 9.84 \times 10^{-8}(T + 273.15)^2 \tag{17.6}$$

The three overvoltage terms can be calculated using both mechanistic and empirical models. Due to their computational efficiency, low dimensionality, and reasonable accuracy, empirical models are commonly used in current research. In this chapter, we adopt the well-established Ulleberg model to calculate the overvoltage terms, leading to the following reformulated expression for cell voltage [21]:

$$U_{cell} = U_{rev} + \frac{(r_1 + r_2 \cdot T)}{A_e} I + s \cdot \ln\left(\left(t_1 + \frac{t_2}{T} + \frac{t_3}{T^2}\right) \cdot \frac{I}{A} + 1\right) \tag{17.7}$$

where I is the cell current; r_1, r_2, s, t_1, t_2, and t_3 are fitted parameters related to the calculation of overvoltages; and A_e is the electrode area.

17.2.2.2 Electrolysis Efficiency Model

Several indicators are used to assess the efficiency of an electrolyzer, which typically includes overall energy conversion efficiency, voltage efficiency, and Faraday efficiency. The overall energy conversion efficiency (η_{tot}) is defined as the ratio of consumed electrical energy to produced hydrogen energy and can be expressed as follows [15]:

$$\eta_{tot} = \frac{\dot{m}_{H_2}^c HHV}{U_{cell}I} \tag{17.8}$$

where $\dot{m}_{H_2}^c$ is the rate of hydrogen production within an electrolyzer cell and HHV is the higher heating value of hydrogen.

Another important efficiency metric is the voltage efficiency (η_v), defined as the ratio of the thermoneutral voltage U_{tn} to the cell voltage [15]:

$$\eta_v = \frac{U_{tn}}{U_{cell}} \tag{17.9}$$

The thermoneutral voltage represents the minimum voltage required for low-temperature electrolysis to occur without heat integration [15]. At this voltage, the electrolyzer cell neither generates nor absorbs heat. The value of U_{tn} depends on the operating temperature and pressure within the electrolyzer system and can be calculated for AELs using equations (17.10) to (17.16) [22].

$$U_{tn} = \frac{\Delta H}{zF} \approx U_{HHV} + \frac{\phi}{zF} \cdot Y \tag{17.10}$$

$$U_{HHV} = 1.4756 + 2.252 \times 10^{-4}T + 1.52 \times 10^{-8}T^2 \tag{17.11}$$

$$\phi = 1.5\frac{P_\omega}{P - P_\omega} \tag{17.12}$$

$$P_\omega = e^{\left(0.01621 - 0.138m + 0.1933\sqrt{m} + 1.024\ln P_\omega^*\right)} \tag{17.13}$$

$$\ln P_\omega^* = 37.04 - \frac{6276}{T + 273.15} - 3.416\ln(T + 273.15) \tag{17.14}$$

$$m = \frac{W_t\left(183.1221 - 0.56845(T + 273.15) + 984.5679e^{\frac{W_t}{115.96277}}\right)}{100 \times 56.105} \tag{17.15}$$

$$Y = 42.96 + 40.762T - 0.06682T^2 \tag{17.16}$$

where U_{HHV} is the high heating voltage; φ and Y are intermediate variables; P is the total pressure of the AEL; P_w is the aqueous vapor pressure of the alkaline solution within the AEL; $P*w$ is the vapor pressure of pure water; W_t is the concentration of the alkaline solution; m is the molarity of the aqueous alkaline solution.

Faraday efficiency η_F measures the ratio of actual hydrogen production rate to the theoretical rate and is typically near 100% under normal operating conditions. In this chapter, we assume $\eta_F = 1$, consistent with previous research [15, 16, 23]. The overall energy conversion efficiency of the electrolyzer system is related to both voltage efficiency and Faraday efficiency, as shown in equation (17.17). This relationship will be further elaborated upon in the hydrogen production model.

$$\eta_{tot} = \eta_v\eta_F \tag{17.17}$$

17.2.2.3 Hydrogen Production Model

According to Faraday's law, the rate of hydrogen production within an electrolyzer cell is directly proportional to the magnitude of the electric current:

$$\dot{m}_{H_2}^c = \eta_F \frac{I}{zF} M_{H_2} \times 3600 \tag{17.18}$$

where M_{H_2} is the molar mass of hydrogen.

By substituting equations (17.9), (17.10), and (17.18) into equation (17.8), and considering relation (17.19), we can derive the relationship between overall efficiency η_{tot}, voltage efficiency η_v, and Faraday efficiency η_F, as shown in equation (17.20).

$$\Delta H = 3600 M_{H_2} \cdot HHV \tag{17.19}$$

$$\eta_{tot} = \frac{\eta_F \dfrac{I}{zF} M_{H_2} \times 3600 \times HHV}{U_{cell}I} = \eta_F \frac{\dfrac{\Delta H}{zF}}{U_{cell}} = \eta_v\eta_F \tag{17.20}$$

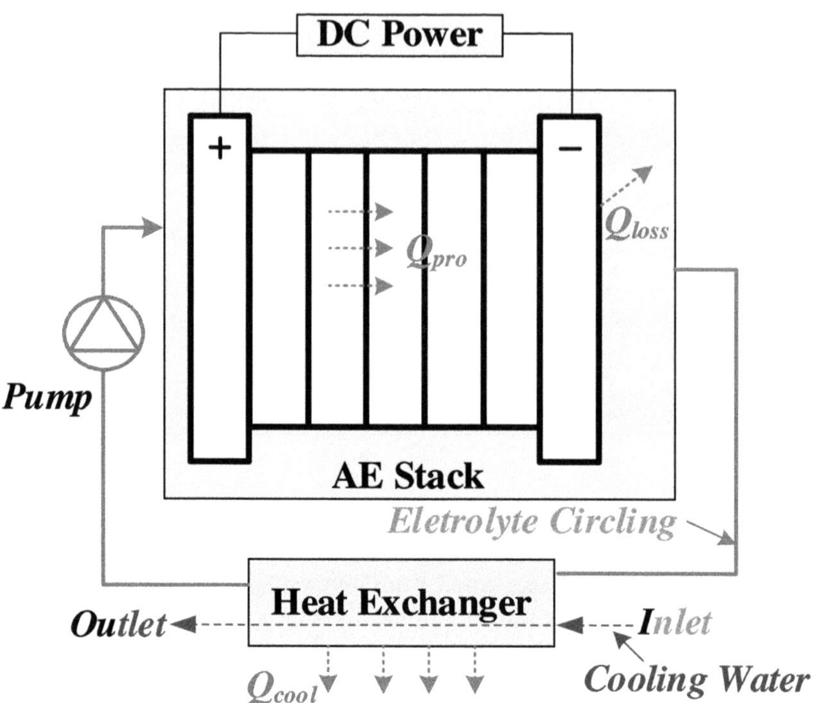

FIGURE 17.4   Schematic of thermal dynamics within an alkaline electrolyzer stack.

17.2.2.4 Thermal Model

The operating temperature of an electrolyzer significantly influences its performance and efficiency. To analyze the thermal dynamics and temperature variations within the stack, we employ a lumped thermal model that represents the stack as an equivalent thermal system with lumped capacitance, as illustrated in Figure 17.4. The thermal balance within the stack is determined by the heat generated during the electrolysis reaction ($\dot{Q}_{pro}$), heat loss to the environment ($\dot{Q}_{loss}$), and heat removed by the cooling water through a heat exchanger ($\dot{Q}_{cool}$). The lumped thermal model capturing the relationship between temperature evolution and thermal balance can be expressed as follows [21]:

$$C_{ely}\frac{dT}{dt} = \dot{Q}_{pro} - \dot{Q}_{loss} - \dot{Q}_{cool} \tag{17.21}$$

where C_{ely} is the lumped thermal capacitance of the electrolyzer stack.

Excess heat is generated during electrolyzer operation when the applied voltage exceeds the thermoneutral voltage. This heat, originating from the remaining electricity not used for hydrogen production, can be calculated as follows:

$$Q_{pro} = N_{cell}(U_{cell} - U_{tn})I = U_{stack}I(1 - \eta_{tot}) \tag{17.22}$$

The heat loss to the environment primarily occurs through heat convection and can be calculated using the following equation:

$$Q_{loss} = \frac{T - T_a}{R_{heat}} \tag{17.23}$$

where T_a is the ambient temperature and R_{heat} is the thermal resistance of the electrolyzer stack. To maintain the desired operating temperature, a cooling system is employed. As depicted in Figure 17.4, a parallel flow heat exchanger is used to dissipate the accumulated heat within the stack. The cooling power can be calculated based on the laws of thermodynamics and the characteristics of the heat exchanger:

$$\dot{Q}_{cool} = \dot{m}_{cool}c_{cw}\left(T - T_{cin}\right)\left[1 - e^{-\frac{UA}{\dot{m}_{cool}c_{cw}}}\right] \tag{17.24}$$

where $\dot{m}_{cool}$ is the mass flow rate of cooling water, c_{cw} is the specific heat capacity of cooling water, T_{cin} is the inlet temperature of cooling water, and U and A are the heat transfer coefficient and heat transfer area of the heat exchanger, respectively.

Equations (17.21) and (17.24) indicate that adjusting the mass flow rate of cooling water can regulate the cooling power $\dot{Q}_{cool}$, thereby controlling the temperature according to the set point. Therefore, a temperature-stabilizing controller can be developed to maintain the AEL's stack temperature at the desired set point. The detailed description of this controller will be presented in the following subsection.

17.2.2.5　Converter Model

The converter serves as the electrical interface between an electrolyzer module, which integrates multiple parallel stacks, and the AC power grid, as shown in Figure 17.5. A two-stage converter configuration is adopted here, which consists of an AC–DC converter followed by a DC–DC converter.

17.2.2.5.1　AC–DC Converter

A widely used pulse width modulation (PWM) rectifier is adopted for the AC–DC conversion due to its advantages in reducing harmonic distortion, enabling bidirectional power flow, and achieving a high power factor. Using dq rotational coordinates, the dynamic model of the AC–DC converter can be represented by the following voltage–current equations at the AC side:

$$u_{gd} = -L_g\frac{di_{gd}}{dt} - R_gi_{gd} + \omega_gL_gi_{gq} + e_{gd} \tag{17.25}$$

$$u_{gq} = -L_g\frac{di_{gq}}{dt} - R_gi_{gq} - \omega_gL_gi_{gd} + e_{gq} \tag{17.26}$$

where u_{gd} and u_{gq} are the rectified voltages, e_{gd} and e_{gq} are the grid voltages, i_{gd} and i_{gq} are the grid currents, L_g and R_g are the inductance and resistance of the filter, and ω_g is the grid frequency.

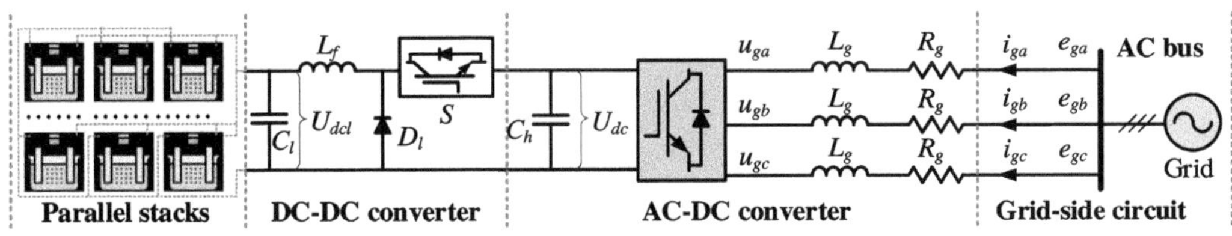

FIGURE 17.5　Schematic of a grid-connected electrolyzer module.

Then, the delivered active power P_g and reactive power Q_g at the AC side can be expressed as:

$$P_g = \frac{3}{2}(u_{gd}i_{gd} + u_{gq}i_{gq}) \tag{17.27}$$

$$Q_g = \frac{3}{2}(u_{gq}i_{gd} - u_{gd}i_{gq}) \tag{17.28}$$

Besides, the power transfer dynamics in the DC link are modeled as follows:

$$U_{dc}C_h \frac{dU_{dc}}{dt} = P_g - P_o \tag{17.29}$$

where U_{dc} is the DC link voltage, P_o is the power delivered to the DC–DC converter, and C_h is the capacitance of the DC link.

17.2.2.5.2   DC–DC Converter

A step-down buck converter is employed as the DC–DC converter to reduce the input voltage from the AC–DC converter to the desired level for the electrolyzer. The dynamic model for the buck circuit, considering different switching states (on or off) for, is given by:

$$\begin{cases} L_f \dfrac{dI_L}{dt} = U_{dc} - U_{dcl}, & \text{if } S = 1 \\[2mm] L_f \dfrac{dI_L}{dt} = -U_{dcl}, & \text{if } S = 0 \end{cases} \tag{17.30}$$

$$C_l \frac{dU_{dcl}}{dt} = I_L - I_{ely} \tag{17.31}$$

where L_f and C_l are the inductance and capacitance of the buck circuit, U_{dcl} is the output voltage, I_L is the inductor current, I_{ely} is the supplied current for the electrolyzer, and S represents the switching state (on or off).

Notably, the duty cycle (D) determines the durations of the on state and off state of the switch. At steady state, the relationship between input and output voltage can be expressed as equation (17.32). This equation highlights the control principle of the DC–DC converter, where the supplied voltage to the electrolyzer can be regulated by adjusting the duty cycle.

$$U_{dcl} = D \cdot U_{dc} \tag{17.32}$$

17.2.2.6   Scaling-Up Model

To scale up from the cell level to the stack level, we assume a bipolar design in which adjacent cells are connected electrically in series through a metal sheet, resulting in identical cell and stack currents. Assuming negligible differences among cells, the power consumption, voltage, and hydrogen production rate at the cell scale can be directly scaled up to the stack scale by multiplying by the number of cells in the stack:

$$U_{stack} = N_{cell}U_{cell}, \; I_{stack} = I, \; \dot{m}^s_{H_2} = N_{cell}\dot{m}^c_{H_2}, \; P_{stack} = N_{cell}U_{cell}I \tag{17.33}$$

Similar scaling-up methods and assumptions are also employed when it comes to scaling up from the stack level to the module, which integrates multiple stacks. That is, we assume negligible differences among stacks. Therefore, the power consumption, voltage, and hydrogen production rate at the module level can be obtained by multiplying the corresponding stack values by the total number of stacks within the system.

17.2.3 Control Design

Effective control strategies are crucial for guaranteeing safe operation and enabling the grid support capability of AEL systems. Three control strategies are developed for AEL systems to release their control flexibility on optimizing the operation performance at different aspects. They include:

1. **Temperature-stabilizing control.** This control strategy aims to maintain the temperature at the desired set point (e.g., 80°C). As the temperature has a significant impact on system efficiency and lifetime, this control is important to ensure the AEL system operates with an acceptable temperature condition, thereby enabling safe and efficient operation.
2. **Converter control.** This includes control strategies for both DC–DC converter and AC–DC converter, utilized for offering acceptable DC power for the stack. Moreover, a power control strategy is designed for the DC–DC converter, enabling the partial load operation of AEL systems at various load factors.
3. **Dynamic frequency control.** This control strategy enables the frequency regulation from AELs for power grids. It drives the AEL systems to implement demand response, that is, changing the consumed power, according to the grid frequency variation, thereby providing frequency support.

17.2.3.1 Temperature-Stabilizing Control

As presented in Section 17.2.2, the temperature can be controlled by adjusting the mass flow rate of cooling water. Building upon this, the idea of the temperature-stabilizing control is to calculate the error between the actual temperature and the reference temperature and to feed it into a proportional-integral (PI) regulator. The control signal from the PI controller is a scale factor (λ), as shown in (17.34), for the mass flow rate, which represents the extent of the opening of a valve that controls the flow of cooling water through the heat exchanger. Additionally, considering a first-order regulation delay, the mass flow rate of cooling water can thus be expressed as (17.35). Figure 17.6 shows the schematic of the proposed temperature-stabilizing control. Benefiting from the PI controller function of achieving zero error in steady state, the actual temperature can be finally controlled at its reference. This closed-loop control system ensures that the electrolyzer operates within the optimal temperature range, preventing overheating and ensuring stable and efficient hydrogen production.

$$\lambda = (T - T_{ref})\left(k_{pt} + \frac{k_{it}}{s}\right) \tag{17.34}$$

$$\dot{m}_{cool} = \rho_{cw}k_m\lambda\frac{1}{1+t_d s} \tag{17.35}$$

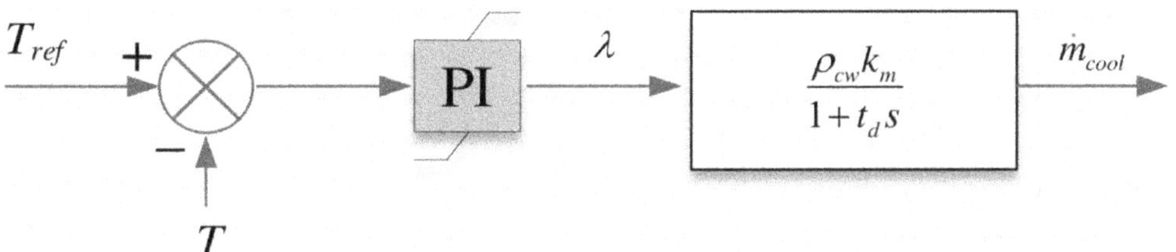

FIGURE 17.6 Schematic of the temperature-stabilizing control.

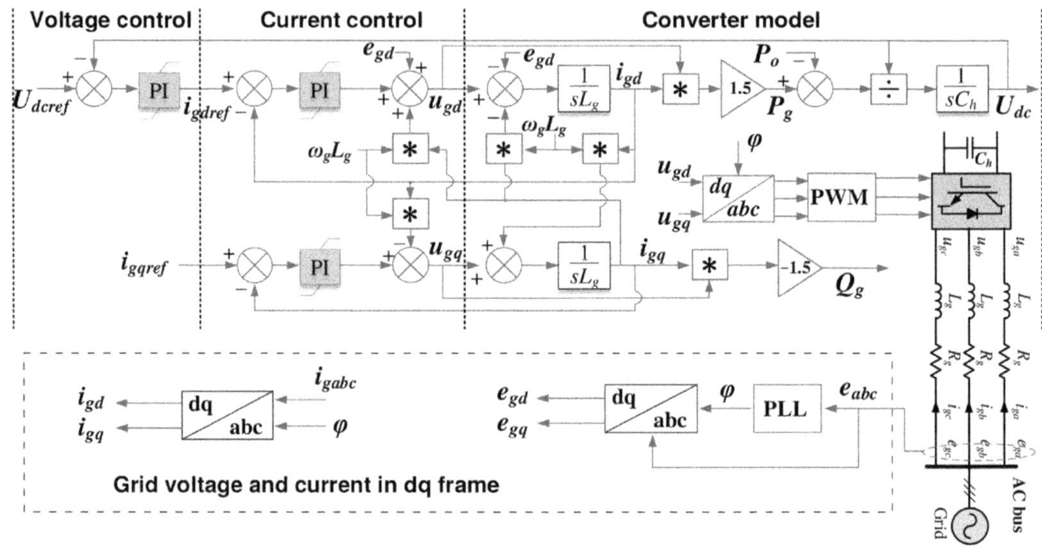

FIGURE 17.7　Control diagram of the AC–DC converter.

where T_{ref} is the temperature set point, k_{pt} and k_{it} are the PI controller parameters, ρ_{cw} is the water density, k_m is the maximum volume flow rate of cooling water, and t_d is the time constant of the valve adjustment process.

17.2.3.2　Converter Control

The converter plays a critical role in interfacing the electrolyzer system with the external power grid. It is responsible for converting AC power from the grid into stable DC power for the electrolyzer stack, while also regulating voltage and current levels to meet the specific operational requirements.

17.2.3.2.1　AC–DC Converter Controller

The AC–DC converter, typically a PWM rectifier, is controlled using a dual closed-loop structure, illustrated in Figure 17.7.

1. **Outer voltage control loop.** This loop regulates the DC link voltage (U_{dc}) by comparing it to a reference value (U_{dcref}) and adjusting the active current component (i_{gd}) accordingly. The PI controller in this loop generates a reference value for the d-axis current (i_{gdref}) based on the voltage error.
2. **Inner current control loop.** This loop controls the current component (i_{gd}, i_{gq}) according to their references to achieve the desired power factor. A PI controller in this loop generates a reference value for the d-axis and q-axis voltage u_{gd}, u_{gq}, based on the current error. An inductor current feedforward mechanism is incorporated to compensate for cross-coupling effects between the d-axis and q-axis current components, ensuring independent control and stability. The reference voltages (u_{gd}, u_{gq}) are then transformed into PWM signals to control the switching of the AC–DC converter.

   This dual closed-loop control structure ensures stable DC voltage output from the AC–DC converter and also enables reactive power control for optimal power factor and grid interaction.

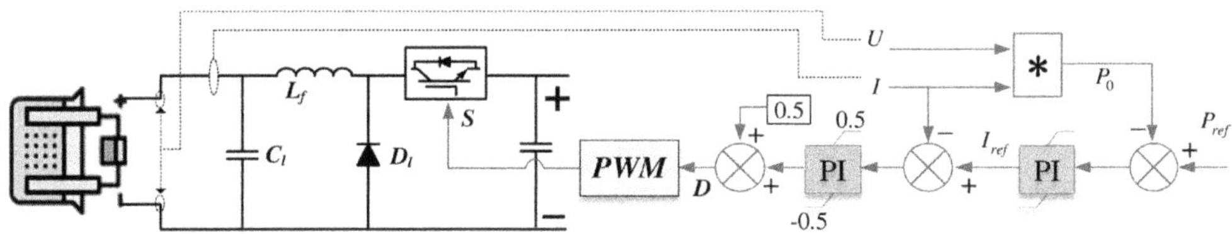

FIGURE 17.8 Control diagram of the DC–DC converter.

17.2.3.2.2 DC–DC Converter Controller

The DC–DC converter, typically a buck converter, is also controlled using a dual closed-loop structure, as shown in Figure 17.8.

1. **Outer power control loop.** This loop regulates the power consumed by the electrolyzer module (P_0) by comparing it to a reference value (P_{ref}) and adjusting the duty cycle of the converter accordingly. This allows for precise control of the power input to the electrolyzer, enabling partial load operation and flexible response to varying operational requirements. The PI controller in this loop generates a reference value for the consumed current of the electrolyzer (I_{ref}) based on the power error (P_{ref}-P_0).
2. **Inner current control loop.** This loop controls the consumed current of the electrolyzer (I) to follow the reference obtained from the outer loop. A PI controller in this loop compares the actual consumed current to the reference value and generates a duty cycle signal to control the switching of the DC–DC converter.

The dual closed-loop control structure for the DC–DC converter enables precise power control and facilitates efficient energy transfer from the DC link to the electrolyzer stack.

17.2.3.3 Dynamic Frequency Control (DFC)

The ability of AEL systems to operate at various load factors, enabled by the power control strategy of the DC–DC converter, opens up the possibility of providing frequency regulation services through demand response. This means that the AEL plant can adjust its power consumption in real time based on grid frequency deviations, effectively contributing to frequency stability within the power system. The dynamic frequency control strategy aims to leverage the operational flexibility of AEL plants to provide grid support functions, specifically frequency regulation, within renewables-dominated power systems.

17.2.3.3.1 Frequency Response in Wind-Dominated Power Systems

Understanding the frequency response characteristics of wind-dominated power systems is crucial for designing effective dynamic frequency control (DFC) strategies. The system frequency response (SFR) is typically modeled using the swing equation of the power system, which relates the rate of change of frequency (i.e., RoCoF) to the power imbalance within the system, as well as the system's inertia and damping properties:

$$\frac{d\Delta f^*}{dt} = \frac{1}{2(1-\rho)H_{sys}}\left(\Delta P_G^* - \Delta P_{L0}^* - D\Delta f^*\right) \tag{17.36}$$

where Δf^* represents the per-unit system frequency deviation from the nominal frequency; ΔP_G^* represents the change in per-unit power supplied by synchronous generators (SGs) due to frequency variation; ΔP_{L0}^{*} represents the initial per-unit power imbalance caused by sudden generator loss or load change; D is the system damping constant, representing the frequency dependency of loads; H_{sys} is the equivalent inertia constant of the power system, considering a lumped machine model for aggregated SGs; and ρ is the wind penetration rate in the system.

Equation (17.36) highlights the impact of wind penetration on system inertia. As the wind penetration rate (ρ) increases, the equivalent system inertia decreases proportionally by a factor of $1-\rho$, leading to larger frequency deviations and faster RoCoF following disturbances.

17.2.3.3.2 Dynamic Frequency Control Principle of Alkaline Electrolyzer

AEL plants can emulate the inertia and damping characteristics of conventional SGs by dynamically adjusting their power consumption based on frequency deviations. To achieve this, virtual inertia and droop constants (H_{AEL} and D_{AEL}) are introduced for the AEL, similar to the parameters in the swing equation. The active power–frequency characteristic of the AEL can then be expressed as:

$$\frac{d\Delta f^{*}}{dt} = \frac{1}{2H_{AEL}}\left(\Delta P_{AEL\_ref}^{*} - D_{AEL}\Delta f^{*}\right) \tag{17.37}$$

where $\Delta P_{AEL\_ref}^{*}$ represents the per-unit additional power consumption required by the AEL to participate in frequency regulation.

To determine the required power adjustment, the power reference of the AEL is designed as a proportional-derivative (PD) controller based on the frequency deviation:

$$\Delta P_{AEL\_ref}^{*}(s) = \left(D_{AEL} + 2H_{AEL}s\right)\Delta f^{*}(s) \tag{17.38}$$

It indicates that the DFC scheme consists of a virtual inertia control (VIC) and droop control. The integration of DFC into a wind-dominated power system modifies the overall system frequency response. The modified SFR model, incorporating the virtual inertia and droop control provided by the AEL plant, can be represented in the frequency domain, as illustrated in Figure 17.9. It can mathematically be expressed as follows:

$$\tilde{G}(s) = \frac{-\tilde{G}_{f}(s)}{1-\tilde{G}_{f}(s)G_{g}(s)(1-\rho)} \tag{17.39}$$

$$\tilde{G}_{f}(s) = \frac{1}{\left[2(1-\rho)H_{sys} + 2H_{AEL}G_{AEL}(s)\right]\cdot s + D + D_{AEL}G_{AEL}(s)} \tag{17.40}$$

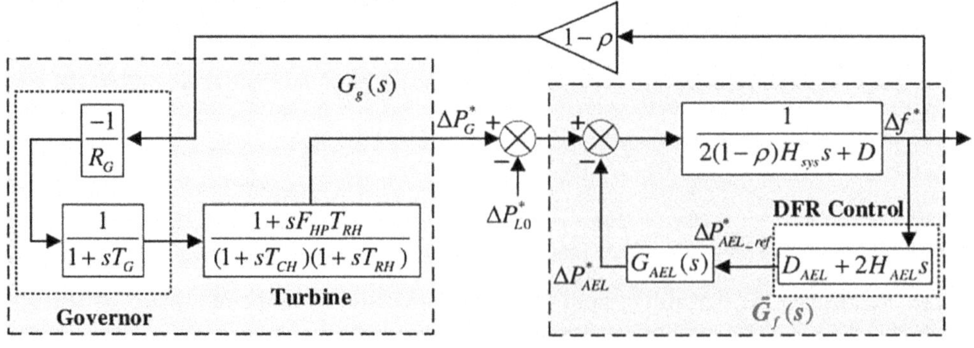

FIGURE 17.9 Modified system frequency response model integrated with the dynamic frequency control.

$$G_g(s) = \frac{-1}{R_G} \cdot \frac{1}{1+sT_G} \cdot \frac{1+sT_{HP}T_{RH}}{(1+sT_{CH})(1+sT_{RH})} \tag{17.41}$$

where $\tilde{G}(s)$ is the overall SFR of the wind-dominated electricity–hydrogen system (WEHS) with DFC; $\tilde{G}_f(s)$ is the frequency response of the WEHS considering the contribution of the AEL plant; and $G_g(s)$ is the primary frequency response of the conventional SGs in the system.

The term $2H_{AEL}G_{AEL}(s)$ in equation (17.40) represents the virtual inertia contribution of the AEL plant, effectively increasing the equivalent system inertia and reducing the RoCoF. Similarly, the term $D_{AEL}G_{AEL}(s)$ represents the virtual damping contribution, further enhancing frequency stability.

In addition, $G_{AEL}(s)$ characterizes the power-tracking response of the AEL plant to its power reference. It is typically modeled as a first-order system:

$$G_{AEL}(s) = \frac{K}{sT_d+1} = \frac{P_{AEL}^{rate}}{P_{grid}^{rate}} \cdot \frac{1}{sT_d+1} \tag{17.42}$$

where K is the ratio of the AEL plant's rated capacity (P_{AEL}^{rate}) to the total capacity of the power system (P_{grid}^{rate}), and T_d is the equivalent time constant representing the overall control delay of the AEL's power control system.

17.2.3.3.3 Analysis of DFC Performance

The effectiveness of DFC in improving frequency response can be evaluated using key metrics, such as RoCoF, frequency nadir, and steady-state frequency deviation. Equations (17.43) and (17.44) show how the steady-state frequency deviation (Δf_∞) and maximum RoCoF (RoCoF_{max}) are impacted by the virtual inertia and droop control provided by the AEL plant.

$$\Delta f_\infty = \frac{\Delta P_{L0}^* R_G f_0}{DR_G + D_{AEL}KR_G + 1 - \rho} \tag{17.43}$$

$$\mathrm{RoCoF}_{max} = \frac{-\Delta P_{L0}^*}{2(1-\rho)H_{sys} + 2H_{AEL}K} \tag{17.44}$$

These equations demonstrate that increasing the virtual inertia (H_{AEL}) and droop gain (D_{AEL}) of the AEL plant leads to a reduction in both the steady-state frequency deviation and the maximum RoCoF, thereby improving frequency stability.

As shown in (17.42), the AEL control delay (T_d) will impact the tracking performance between the power reference of the AEL and its actual value. Larger Td would slow down the response of the power regulation, thereby causing undesirable frequency support from the AEL. To clarify the impact of T_d, the frequency response of the power system under different T_d is examined, as shown in Figure 17.10. While the presence of DFC generally improves the frequency response compared to the scenario without DFC, a longer control delay can diminish the benefits, leading to larger frequency deviations and higher RoCoF.

Additionally, the stability of the DFC scheme can be analyzed using root locus techniques. The root locus method is a powerful tool for analyzing the stability of linear time-invariant systems. It involves plotting the trajectories of the closed-loop poles of the system as a specific parameter, such as the gain of a controller, is varied. By examining the locations of the poles in the complex s-plane, we can determine the stability and dynamic behavior of the system.

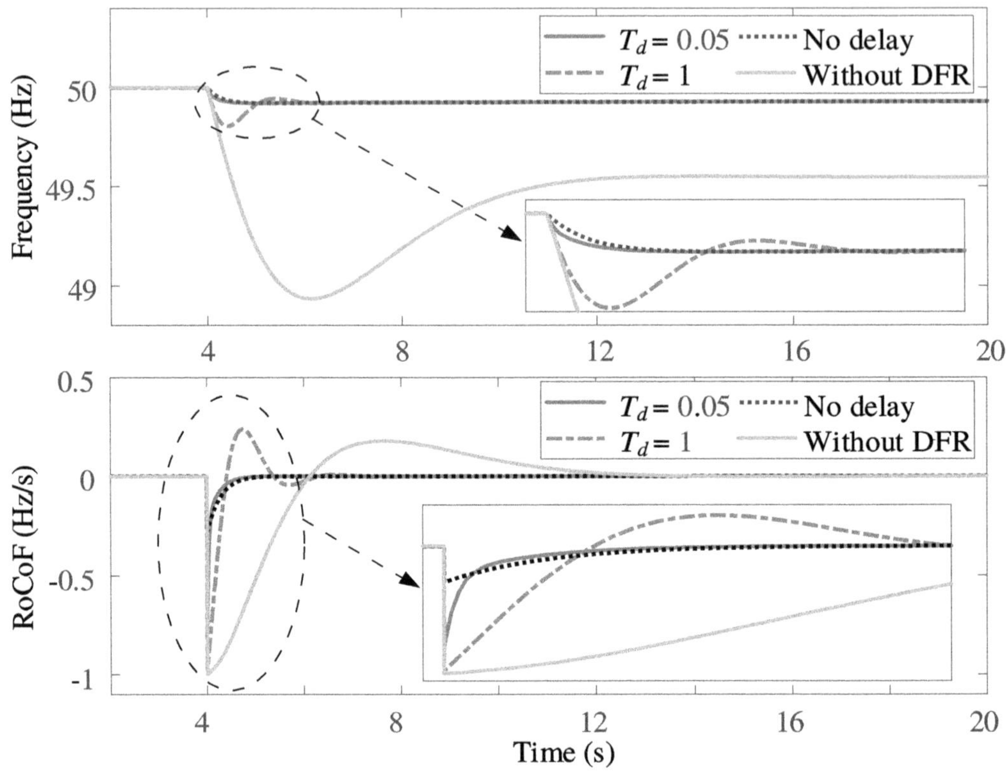

FIGURE 17.10 Frequency response under different control delays.

The root locus analysis can be applied to assess the impact of the virtual inertia (H_{AEL}) and droop gain (D_{AEL}) on the stability of the overall system frequency response. Based on the transfer function shown in (17.39), the equivalent open-loop transfer function for the system with DFC can be formulated as:

$$\tilde{G}_{eqo}(s) = H_{AEL} \cdot \frac{2G_{AEL}(s) \cdot s}{2(1-\rho)H_{sys}s + D + D_{AEL}G_{AEL}(s) - G_g(s)(1-\rho)} \qquad (17.45)$$

By plotting the root locus for this transfer function as H_{AEL} varies, while keeping D_{AEL} within a specified range (e.g., 0 to 500), we can observe the movement of the closed-loop poles. Figure 17.11 presents an example of such a root locus plot, demonstrating that the poles remain in the left-hand side of the s-plane (stable region) regardless of the virtual inertia value. Furthermore, as the droop gain D_{AEL} increases, the poles move further away from the imaginary axis, indicating improved damping and faster settling time of the system response. This analysis demonstrates that the DFC scheme maintains stability even with variations in virtual inertia and droop gain, highlighting its robustness to parameter uncertainties.

17.2.3.3.4   Control Framework of Dynamic Frequency Control

To implement DFC within a wind-dominated electricity–hydrogen system (WEHS), we consider the configuration depicted in Figure 17.12a. A grid-connected AEL plant, consisting of multiple AEL modules, is connected to a point of common coupling (PCC) within the power system. Each module is equipped with a DC–DC converter, followed by an AC–DC converter, as detailed in

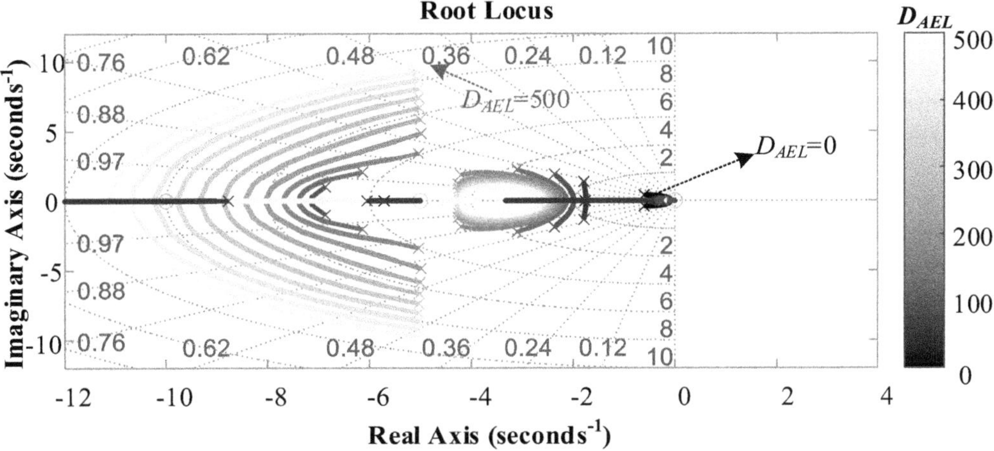

FIGURE 17.11   Root locus for the improved system frequency response model.

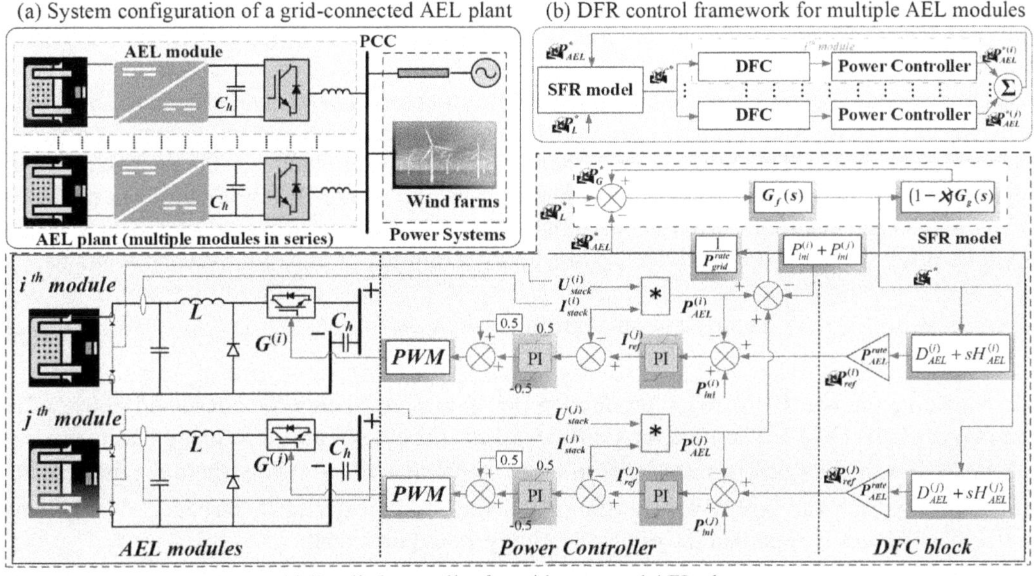

FIGURE 17.12   Configuration and detailed controller of a grid-connected electrolyzer plant for supporting system frequency.

Section 17.2.2. The DFC control functionality is integrated into the controllers of the DC–DC converters to enhance the frequency response of the external power system.

Figure 17.12b illustrates the general DFC-driven control framework that coordinates the operation of multiple AEL modules. Each module has its own dedicated controller, which includes a DFC block and a power control block for the DC–DC converter. The DFC block utilizes the measured system frequency (Δf^*) to generate a power reference signal based on the frequency deviation. The power control block then adjusts the actual power consumption of the module to match the reference signal, contributing to the overall frequency regulation effort.

For a clearer understanding of the control process, let's consider a specific case with two AEL modules, as shown in Figure 17.12c. The measured system frequency Δf^* is input to the DFC block of each module. This block, modeled by equation (17.38), generates additional power references ($\Delta P_{ref}^{(i)}$, $\Delta P_{ref}^{(j)}$) by multiplying the frequency deviation by the rated capacity of the AEL plant P_{AEL}^{rate}. These additional power references are then added to the pre-contingency operating power set points P_{ini}^i, P_{ini}^j to determine the final power reference for each module's DC–DC converter. Then, the power control strategy of the DC–DC converter, as described before, will adjust the actual consumed power (P_{AEL}^i, P_{AEL}^j) of each AEL module to meet the corresponding power reference. The actual consumed power of each module is calculated based on the stack's voltage and current. The sum of these actual power values is then compared to the sum of the pre-contingency operating powers to determine the overall power adjustment of the AEL plant in response to the frequency event. Particularly, it's important to note that each module's control process operates independently based on the measured system frequency, without direct communication between modules. This decentralized control approach minimizes communication delays and facilitates rapid frequency regulation.

17.3 OPERATION OF ELECTROLYZER PLANTS

17.3.1 Introduction

While electrolyzer control implies the process of adjusting the electrolyzer status to a pre-determined point, the operation of electrolyzer plants set the goals for control. The term "operation" means to find a feasible trajectory of electrolyzer states that meets certain requirements. In general, the owner may try to minimize the operational costs of producing a certain amount of hydrogen or achieve more profits. The owner of grid-connected electrolyzer plants has the possibility to participate in multiple power markets. In the simplest case, the owner is only involved in day-ahead market, purchasing electricity from the grid and selling hydrogen to relevant consumers. Given an electricity price λ_e and hydrogen price λ_H, the hourly marginal profits of the owner would be:

$$\text{Hourly marginal profits} = \dot{m}_H \lambda_h \Delta t - p_{ele} \lambda_e \Delta t \tag{17.46}$$

with $\dot{m}_H$ being the hourly hydrogen production (kg/h) and p_{ele} being the consumed power of the electrolyzer plants (MW). The time interval Δt takes 1 hr. To obtain the strategy that maximizes the marginal profits, let's try a simple electrolyzer model that $\dot{m}_H = \alpha p_{ele}$, where α represents the hydrogen conversion rate (kg/MWh). Assuming that the p_{ele} can arbitrarily be changed in the range $[0, \bar{p}]$, it can be easily seen that the optimal strategy would be trivial:

$$\begin{cases} p_{ele}^* = 0, \ if \ \alpha\lambda_H \leq \lambda_e \\ p_{ele}^* = \bar{p}, \ if \ \alpha\lambda_H > \lambda_e \end{cases} \tag{17.47}$$

The strategy would be all-or-nothing, depending on the price of electricity and hydrogen. Furthermore, the operation decision is usually made on a daily basis, and therefore, we use a subscript, t, to indicate the time-dependent nature of relevant parameters and variables. The owner seeks to maximize the daily marginal profits, which is:

$$\text{Daily marginal profits} = \sum_t (\dot{m}_{H,t} \lambda_{h,t} \Delta t - p_{ele,t} \lambda_{e,t} \Delta t) \tag{17.48}$$

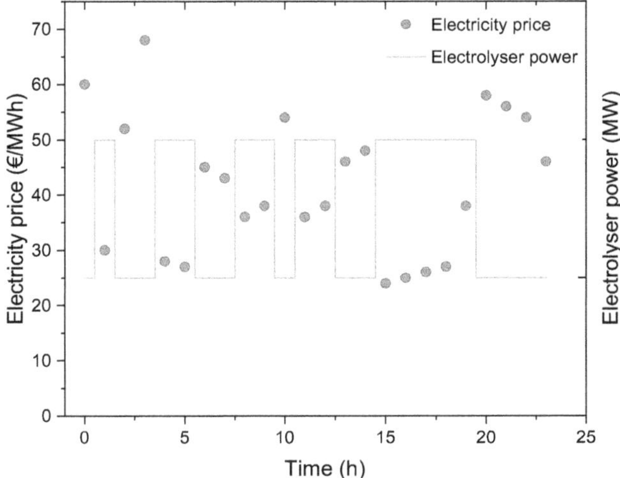

FIGURE 17.13 Trivial optimal trajectory of electrolyzer power given fluctuating electricity price. The marginal revenue is fixed in this case. The owner maximizes hydrogen production for positive marginal revenue and stops the production for negative revenue.

Providing that P_{ele,t_i} and P_{ele,t_j} are uncorrelated, the maximization of daily profits becomes the maximization of each hourly profits, resulting in a series of all-or-nothing decisions. Figure 17.13 shows the optimal trajectory of the electrolyzer power given fluctuating electricity price.

It is clearly shown that the electrolyzer owner tends to maximize the power consumption at lower electricity price and stop producing hydrogen if the price increases. Note that the hydrogen price is fixed in this case, which aligns with the reality.

17.3.2 Optimal Operation Problem: Mathematical Formulation

The stepwise power trajectory shown in Figure 17.13 is unlikely feasible. One of the main reasons is that we cannot adjust the electrolyzer without any technical limitations. Specifically, the process of adjusting the electrolyzer power is subject to the following constraints:

- Limited ramping-up/down rates
- Non-zero lower bound
- Zero power only reached when the electrolyzer is in standby or off state
- Hot/cold starts of electrolyzer time-consuming (*hot start*: transitions from standby to production state; *cold start*: transition from off to standby state)

The electrolyzer power cannot increase or decrease with an exceeding rate due to safety issues. Too fast current change will cause a delayed change of the concentration of "hydrogen in oxygen" and a delayed change of pressure. High concentration of hydrogen in oxygen may cause explosion risks, while pressure change may result in unbalanced liquid level in hydrogen and oxygen separators. This poses a limitation of $P_{ele,t}$ and $P_{ele,t+1}$. However, if the time interval is set as 1 hr, this limitation no longer works, because the electrolyzer can at least change its power by 20% nominal power per min, as reported. For shorter time intervals, the constraint must be considered.

Operating the electrolyzer plants at excessively low load levels will cause a mix of hydrogen and oxygen, leading to potential explosion risks. For example, the lower bound of alkaline electrolyzers is generally known as 20% of their nominal power. Keeping this in mind, the owner may choose to shut down the electrolyzer if the electricity price is high. To achieve this, the owner can either put the electrolyzer into a standby state or completely shut down the device. Operation in standby state consumes extra energy to maintain the temperature and pressure, but the electrolyzer can start up quickly. Putting the device into an off state means no energy cost but a long time to restart the device. Hot starts usually take tens of minutes, while cold starts take several hours. The choice between standby or off states depends on the anticipated duration of time when the electrolyzer has no power input, which is further influenced by electricity price trajectory.

Additionally, the modeling section of an electrolyzer indicates that the hydrogen conversion rate α relies on the load or power level, resulting in a nonlinear relationship between $\dot{m}_{H,t}$ and $p_{ele,t}$.

Keeping all the constraints in mind, the mathematical formulation of an optimal problem of the electrolyzers can be formulated. The operational objective can be reported from (17.49):

$$\max_{x} \sum_{t=0} (\dot{m}_{H,t} \lambda_{h,t} \Delta t - p_{ele,t} \lambda_{e,t} \Delta t) \tag{17.49}$$

where the decision vector x is composed of multiple time-dependent variables:

$$x = [p_{ele,t}, \dot{m}_{H,t}, I_t, p_t^b, s_t^b, i_t^b, w_{b,t}, s_t^p, s_t^l] \tag{17.50}$$

Besides the well-defined physical variables, including electrolyzer power $p_{ele,t}$, hourly hydrogen production $\dot{m}_{H,t}$, and electrolyzer current I_t, ancillary variables are introduced to describe the operational limitations, and their usage will be seen later. The optimization problems are subject to a series of technical constraints, given in the following.

17.3.2.1 Power Input and Hydrogen Output

The electrolyzer polarization curve reveals the voltage–current properties and, further, the relationship between electrolyzer power and current, as shown in Figure 17.14. It is clearly shown that the power–current relationship is nonlinear. Such a constraint will result in a hard-to-solve mathematical programming. Here, several points are chosen along the power–current curve, and note the use of a piecewise linear function to approximate the original nonlinear curve, as presented in Figure 17.14. By introducing ancillary variables $w_{b,t}$, the following constraints exist, quantifying the electrolyzer power–current relationship:

$$\begin{aligned} p_{ele,t} &= \sum_b w_{b,t} p_{ele,b} \\ I_t &= \sum_b w_{b,t} I_b \\ \sum_b w_{b,t} &= 1 \\ w_{b,t} &\in SOS_2 \end{aligned} \tag{17.51}$$

with parameter $p_{ele,b}$ denoting the reference power at breakpoint b, and I_b the corresponding current. $w_{b,t}$ represents the weight of breakpoint b at time t, belonging to a special ordered set of type 2 (SOS_2), where, at most, two adjacent components are non-zero. SOS_2 could be expressed by introducing binary variables [24]. So far, the power and current at time t are exclusively determined by a group of $w_{b,t}$.

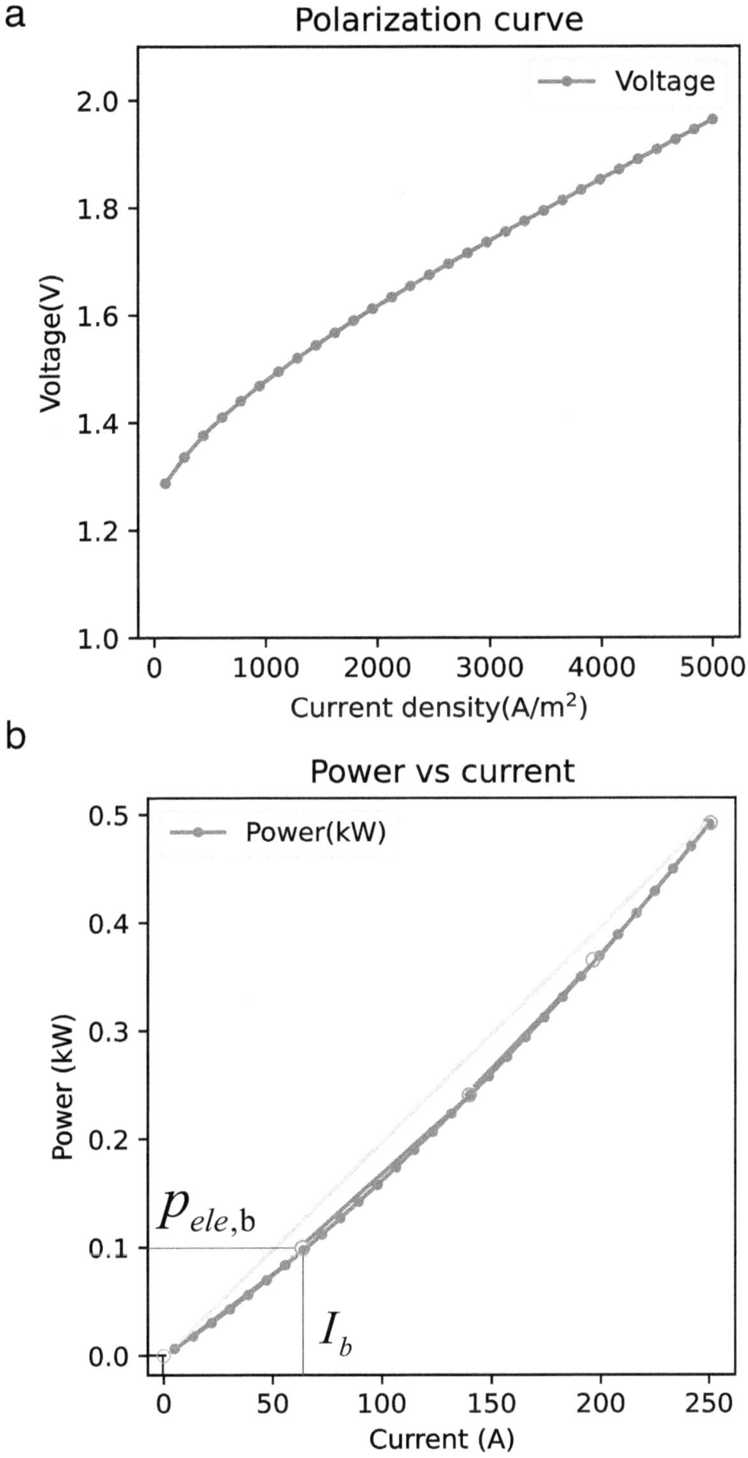

FIGURE 17.14 Typical (a) polarization curve and (b) power vs. current curve of electrolyzers.

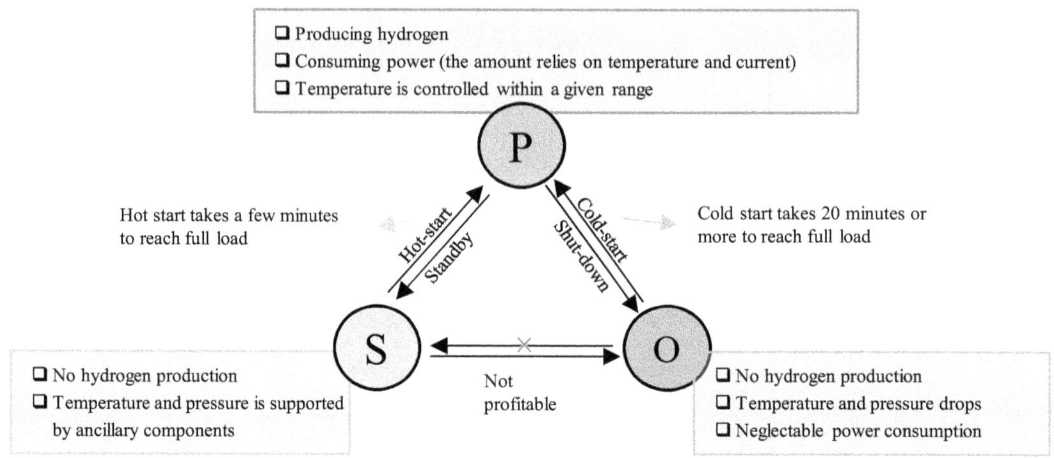

FIGURE 17.15 Production (P), standby (S), and off (O) states of alkaline electrolyzers.

Furthermore, Faraday's law links the electrolyzer current to the hydrogen mass production rate:

$$\dot{m}_{H,t} = \eta_F \frac{I_t}{zF} \tag{17.52}$$

where η_F is the Faraday efficiency, which approximates 100% if electrolyzer is working at a relatively high load factor. F represents the Faraday constant, and z is the mol of electrons per mol hydrogen produced, taking two in water splitting reaction.

17.3.2.2 State Transitions

Generally, three states exist during electrolyzer operation: production, standby, and off states, as shown in Figure 17.15. In production state, electrolyzers consume electricity and produce hydrogen. The electrolyzer power can be adjusted within a given range and ramping-up/down rates. On standby state, the power input and hydrogen output are both zero. However, extra energy is used to stabilize the electrolyzer temperature and pressure, which helps achieve fast transition from standby to production states (hot start). In the off state, the electrolyzer temperature and pressure will decrease, implying a longer time to put the device into the production state.

To express the limitations triggered by state transitions, three binary variables p_t^b, s_t^b, and i_t^b, are introduced to indicate whether the electrolyzer is in a certain state. The whole electrolyzer model involving state transition restrictions is as follows:

$$\begin{aligned}
P_{ele,t} &= \sum_b w_{b,t} P_{ele,b} + s_t^p \\
I_t &= \sum_b w_{b,t} I_b + s_t^I \\
-(s_t^b + i_t^b)M &\le s_t^p \le (s_t^b + i_t^b)M \\
-(s_t^b + i_t^b)M &\le s_t^I \le (s_t^b + i_t^b)M
\end{aligned} \tag{17.53}$$

The preceding four equations decouple the piecewise relationship between electrolyzer power and current on standby and off states by introducing slack variables s_t^p and s_t^I, which equals zero for production states but can be almost unlimited for standby and off states. Here, M is a significantly large number, ensuring that s_t^p is zero when both s_t^b and i_t^b are zero (when the electrolyzer is in production state). Meanwhile, s_t^p becomes unlimited if either s_t^b or i_t^b is 1.

$$p_t^b + s_t^b + i_t^b = 1$$
$$p_t^b \underline{I} \leq I_t \leq p_t^b \overline{I} \tag{17.54}$$
$$s_t^b P_s \leq P_{ele,t} \leq p_t^b \overline{P} + s_t^b P_s$$

The first equation ensures that the three states are incompatible, followed by the two equations that force the current and power to be located in a different range for distinct states. If $p_t^b = 1$, then the electrolyzer current is restricted within its safe range and electrolyzer power would be less than the nominal capacity. On standby state, the electrolyzer power would be P_s, that is, the power to maintain temperature and pressure. Both current and power would be zero on the off state.

To reflect the time required for hot/cold starts, we introduce two more binary variables that indicate the hot/cold start-up signals sent at time t.

$$Y_t^b = s_{t+(h-1)\Delta t}^b p_{t+h\Delta t}^b$$
$$Z_t^b = i_{t+(c-1)\Delta t}^b p_{t+c\Delta t}^b \tag{17.55}$$

Here, it is assumed that the hot start time is $h\Delta t$, and the cold start time $c\Delta t$. The equations basically mean that the start-up signal sent at time t will switch the electrolyzer to production state after a few time steps. During the start-up process, we have the following constraints:

$$\sum_{t}^{t+(h-1)\Delta t} s_t^b \geq h Y_t^b$$
$$\sum_{t}^{t+(c-1)\Delta t} i_t^b \geq c Z_t^b \tag{17.56}$$

The two inequalities ensure that the electrolyzer keeps standby/off state during the hot/cold start process. These equations can be simplified if h and c are relatively small. As operation optimization is usually performed at a time interval of 1 hr and electrolyzers have a fast start-up potential, it is reasonable to assume that $h = 1$ and $c = 2$, then we have:

$$Y_t^b = s_t^b p_{t+\Delta t}^b$$
$$Z_t^b = i_{t+\Delta t}^b p_{t+2\Delta t}^b \tag{17.57}$$
$$\sum_{t}^{t+\Delta t} i_t^b \geq 2 Z_t^b$$

17.3.2.3 Other Considerations

The aforementioned equations can serve as an effective modeling approach for electrolyzer operation optimization. However, there are more issues that should be pointed out for detailed electrolyzer modeling:

- Temperature variation
- Hydrogen in the oxygen
- Degradation caused by hot/cold starts

The electrolyzer temperature is another crucial operating variable affecting the overpotentials, diffusion rates, etc. An improved electrolyzer model should describe the temperature variation during operation, which typically involves the description of heat generation, heat losses, cooling flow, and control of electrolyte flow rate. Hydrogen in oxygen (HTO) plays a key role in electrolyzer safety, which is influenced by the electrolyzer temperature, pressure, and load factor. It is crucial to quantify HTO as a direct safety index during the decision-making of the electrolyzer operation.

Frequent hot/cold starts may cause extra degradation of electrolyzers and should be avoided during the operation. This could be achieved by introducing "degradation" costs in the objective function. However, the mechanism that causes hot/cold starts to influence degradation is still unclear.

17.3.3 UNCERTAINTY HANDLING

Even in the simplest scenario, where the electrolyzer owner only participates in the day-ahead market, the decision-maker faces uncertainties, that is, the uncertain price prediction errors. The owner has to make a decision based on imperfect information of price prediction. As a result, the operational strategy may not work well and may lead to poor economic performance. In this context, there arise multiple stochastic programming methods to cope with the uncertainty during the decision-making process. Here, we introduce a basic scenario-based stochastic optimization to examine how the uncertainty is controlled.

$$\max_{\mathbf{x}} \sum_{w} \gamma_w \sum_{t=0} (\dot{m}_{H,t} \lambda_{h,t} \Delta t - p_{ele,t} \lambda_{e,wt} \Delta t) \tag{17.58}$$

Here, the electricity price $\lambda_{e,wt}$ has a subscript, w, indicating the scenario w. The decision-maker tries to maximize the weighted sum of marginal profits over a certain amount of scenarios. This would be a plain way to cope with uncertainty. In such stochastic programming problems, there are factors worth mentioning, such as the generation and selection of scenarios, the computational burden, the value of the weights, and the risk management. For more details, please refer to [25].

17.4 EXAMPLES

In this section, a series of specific examples are conducted for analyzing the AEL performance under integration with the developed model, control, and operation strategies. Various case studies are implemented to examine the AEL performance in terms of different aspects.

17.4.1 EXAMPLE 1: AEL PLANT OPERATION IN VARYING POWER INPUT

The operation performance of an AEL plant is first examined under varying power input condition, to prove the effectiveness of the AEL modeling, as well as to test the effect of the developed temperature-stabilizing control and power control strategies.

To assess the effectiveness of the developed models, as presented in Section 17.2.2, they are implemented within the MATLAB/Simulink platform to simulate the operational performance for a grid-connected AEL plant. The validation process relies on experimental datasets from existing references, specifically drawing from two distinct electrolyzer systems: one with a 26.6 kW capacity, documented in [26], and another with a 3 MW capacity, as detailed in [27]. It's important to note that the fitted parameters within the models need to be adjusted for different electrolyzer systems to ensure an accurate representation of their specific characteristics. Table 17.1 summarizes the key parameters used in the simulations, with values obtained from existing references and empirical observations.

TABLE 17.1

Key Parameters for Electrolyzer Models

Parameters	Value
Fitted parameters r_1	$8.05 \times 10^{-3}\ \Omega \cdot m^2$
Fitted parameters r_2	$-2.5 \times 10^{-7}\ m^2{}^\circ C^{-1}$
Fitted parameters s	$0.185\ V$
Fitted parameters t_1	$1.002\ A^{-1}m^2$
Fitted parameters t_2	$8.424\ A^{-1}m^2{}^\circ C$
Fitted parameters t_3	$247.3\ A^{-1}m^2{}^\circ C^2$
Electrode area A_e	$0.25\ m^2$
Number of cells per stack N_{cell}	21
Number of stacks per module N_{stack}	6
Stack rated power	26.6 kW
Thermal capacitance C_{ely}	625,000 J/°C
Thermal resistance R_{heat}	0.167°C/W
Temperature set point T_{ref}	80°C
Ambient temperature T_a	20°C
Specific heat of the water c_{cw}	4,182 J/(kg°C)
Inlet temperature of cooling water T_{cin}	14.5°C
The product of heat transfer coefficient and heat transfer area UA	250 W/°C
Maximum mass flow of cooling water	5 m³/h
Time constant of valve adjustment t_d	30 s
PI controller parameters of the temperature-stabilizing control k_{pt}, k_{it}	0.9, 0.5
Grid voltage (line-to-line)	380 V
Grid frequency	50 Hz
DC link voltage reference of AC–DC converter	680 V

17.4.1.1　Validation of the UI Curve

The empirical UI model, represented by equation (17.7), was validated by comparing its predictions with experimental data from both kW-level and MW-level electrolyzer systems. Figure 17.16 illustrates this comparison, demonstrating the model's ability to accurately capture cell voltage behavior across a range of current densities and operating temperatures. This validates the model's effectiveness in representing the polarization characteristics of electrolyzer cells.

17.4.1.2　Validation of the Whole AEL System Model

The whole AEL system model, consisting of various component models and integrating with the proposed temperature-stabilizing control and converter control, was further evaluated by analyzing its ability to capture the dynamic behavior of a grid-connected AEL system.

Figure 17.17 presents the results of simulating an electrolyzer stack under varying input current conditions, especially the thermal dynamics. The results demonstrate the effectiveness of the proposed temperature-stabilizing control strategy in maintaining the stack temperature at the desired set point, even with fluctuating heat generation due to changes in electrical power consumption.

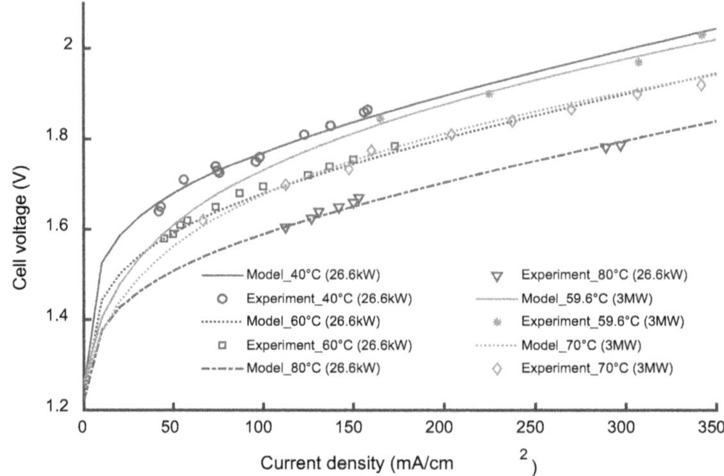

FIGURE 17.16 Comparisons between the empirical UI model and experimental data of two different electrolyzers.

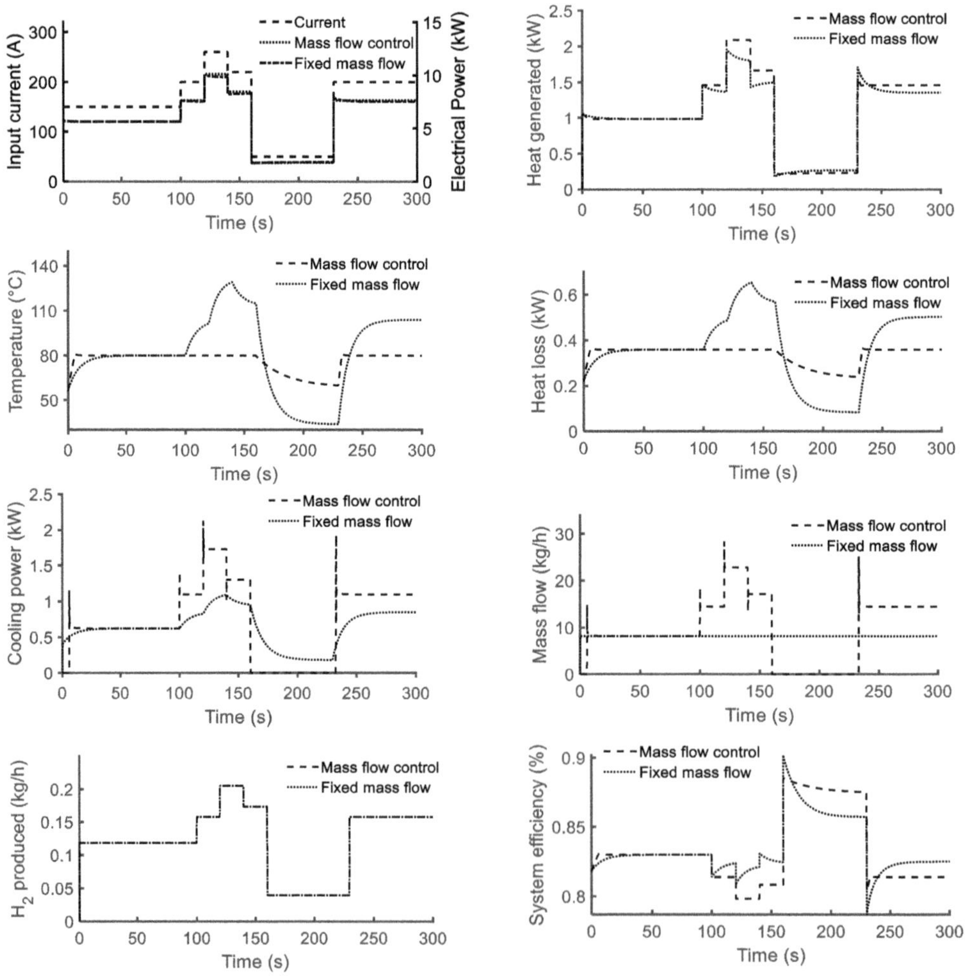

FIGURE 17.17 Operation performance of an electrolyzer stack.

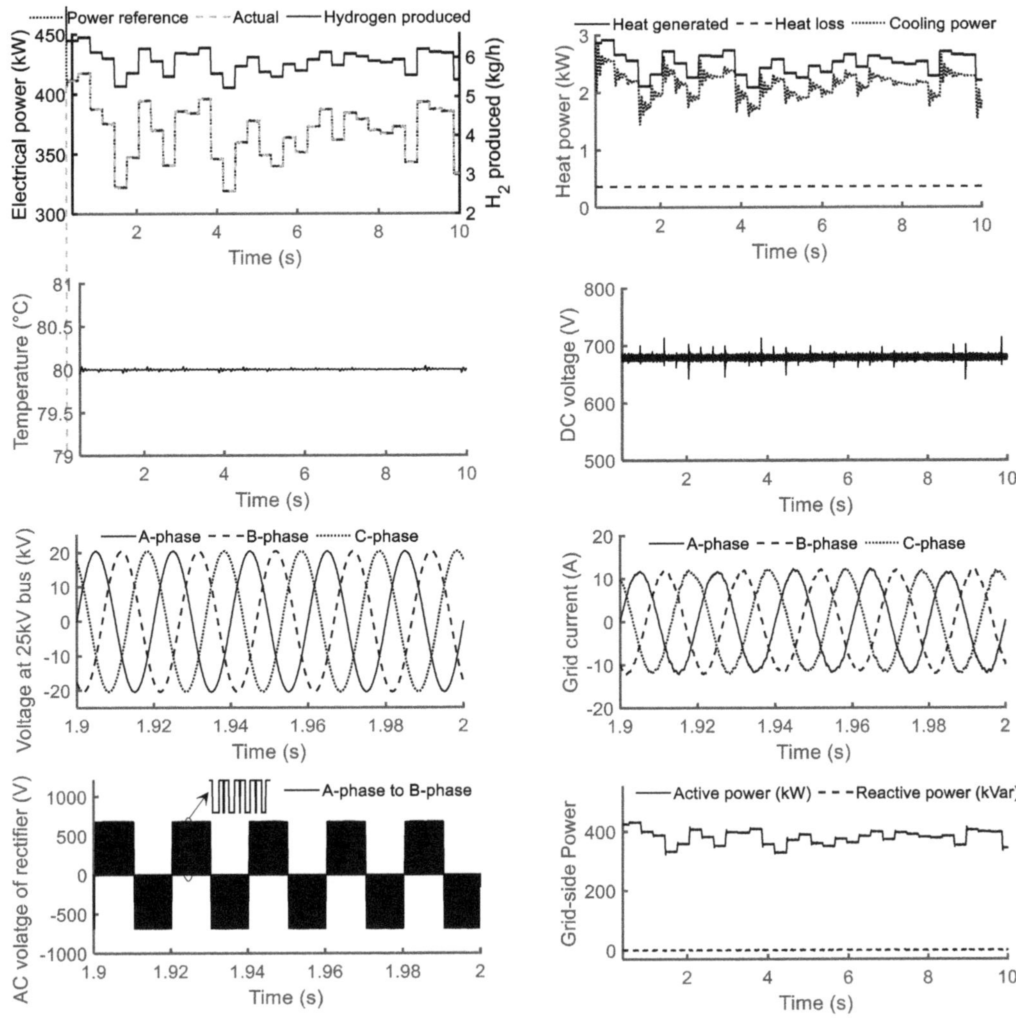

FIGURE 17.18   Partial-load operation performance of an electrolyzer module.

This shows the model's ability to accurately represent thermal behavior and the importance of temperature control for safe and efficient electrolyzer operation. Furthermore, the simulation results also showcase the model's capability to capture variations in hydrogen production and system efficiency with changes in electrical power and operating temperature, providing valuable insights for optimizing the system performance.

Additionally, the performance of an AEL system connected to an AC grid was evaluated, with the results presented in Figure 17.18. The simulations demonstrate the effectiveness of the developed converter control strategies in achieving accurate tracking of power references, ensuring fast partial load operation, and maintaining the desired DC link voltage and power factor while minimizing grid voltage and current distortion. This demonstrates the model's ability to represent the dynamic interactions between the electrolyzer and the power grid and highlights the importance of converter control for efficient and reliable operation.

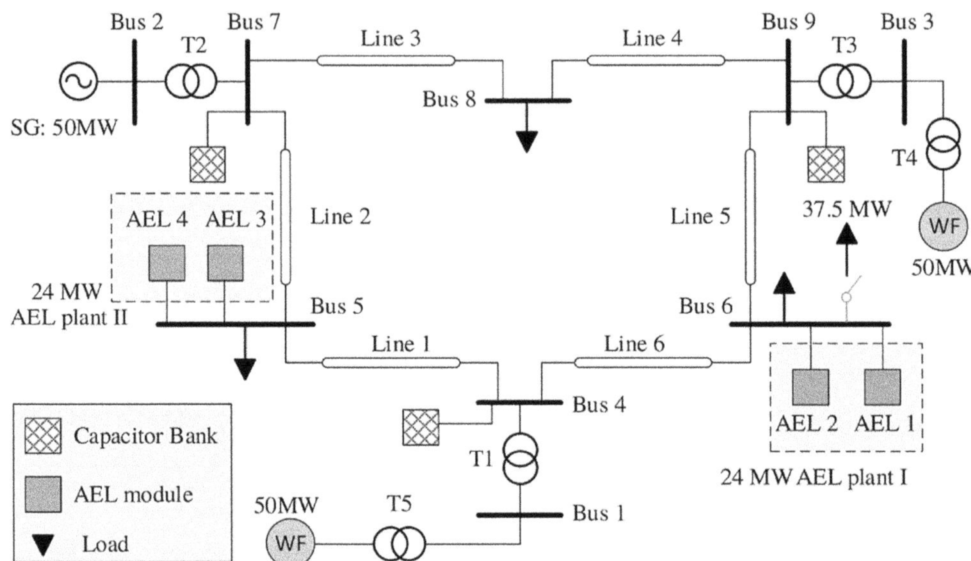

FIGURE 17.19   Modified IEEE 9-bus system integrated with two AEL plants and two wind farms.

17.4.2   EXAMPLE 2: FREQUENCY REGULATION SUPPORT FROM AEL PLANT

To demonstrate the effectiveness of the proposed DFC strategy in enhancing the frequency response, a case study is conducted using a modified IEEE 9-bus system to simulate a wind-dominated electricity–hydrogen system (WEHS).

The modified IEEE 9-bus system, as illustrated in Figure 17.19, includes a single 50 MW synchronous generator with rotational inertia at bus 2, two 50 MW wind farms without inertia at bus 1 and bus 3, and two 24 MW AEL plants at bus 5 and bus 6. Each AEL plant consists of two 12 MW AEL modules. The wind farms utilize grid-forming converters with droop control. Detailed simulation parameters can be found in [28]. Each AEL stack within each plant still adopts the parameters shown in Table 17.1, meanwhile assuming no difference between stacks. The AEL plant parameters can thus be estimated based on the AEL stack parameter and scaling-up model described in Section 17.2.2.

To evaluate the system's frequency response, a sudden load increase of 37.5 MW ΔP_{L0} is applied at bus 6, simulating a frequency contingency event. Four different cases are considered, each with varying levels of DFC implementation and a total frequency regulation capacity (FCR) provided by the AEL plants:

1. Case 1: No DFC. This serves as the baseline case without any frequency regulation from the AEL plants.
2. Case 2: DFC without VIC (i.e., virtual inertia constant H_{AEL} equals 0). In this case, the AEL plants provide frequency regulation using droop control only, with a total FCR capacity of 40 MW.
3. Case 3: DFC with VIC (FCR = 40 MW). This case includes both droop control and virtual inertia control (VIC), also with a total FCR capacity of 40 MW.
4. Case 4: DFC with VIC (FCR = 20 MW). Similar to Case 3, but with a reduced FCR capacity of 20 MW to evaluate the impact of limited regulation capability.

The frequency response of the system under the different case scenarios is analyzed using key metrics, such as RoCoF, frequency nadir Δf_{nadir}, and steady-state frequency deviation Δf_{∞}.

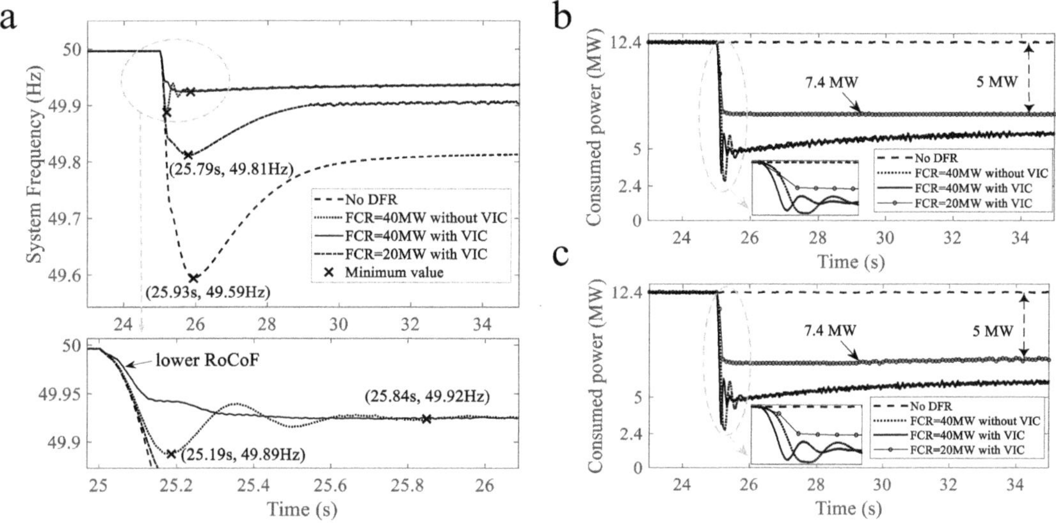

FIGURE 17.20 Simulation results: (a) system frequency response, (b) consumed power of AEL plant I, (c) consumed power of AEL plant II.

Figure 17.20a shows the system frequency response for each case following the load disturbance. Case 1, without DFC, exhibits poor frequency quality with a high negative RoCoF, significant frequency nadir, and a large steady-state deviation. The implementation of DFC in Cases 2, 3, and 4 significantly improves the frequency response by utilizing the FCR capacity of the AEL plants.

The results highlight the impact of FCR capacity on frequency regulation performance. Cases 2 and 3, with a higher FCR capacity of 40 MW, achieve better steady-state frequency deviation and lower frequency nadir compared to Case 4, which has a limited FCR capacity of 20 MW. This is because a higher FCR capacity corresponds to a larger virtual droop constant, leading to improved steady-state frequency regulation.

Furthermore, the inclusion of VIC in Cases 3 and 4 effectively reduces the initial RoCoF and frequency nadir, demonstrating the benefits of virtual inertia in enhancing the transient frequency response. However, as expected, VIC does not influence the steady-state frequency deviation.

Figure 17.20b and c illustrate the power response of the AEL plants during the frequency event. The figures show how the AEL plants reduce their power consumption to release the available FCR capacity and support the system frequency. The extent of power reduction is limited by the available FCR capacity, as seen in Case 4, where AEL 1 can only reduce its consumption to a certain level due to the lower FCR availability. A comparison between Cases 2 and 3 also highlights the faster power regulation achieved with VIC, leading to improved transient frequency response.

17.4.3 Example 3: Optimization of Electrolyzer Plant Participating in Day-Ahead Market

Accounting for the detailed operational constraints, an optimized strategy is examined for an electrolyzer owner participating in day-ahead market. The owner has the objective to maximize the overall marginal profits by performing cross-commodity arbitrage. The mathematical model introduced in Section 17.3.2 is utilized, and we assume the spot prices of the electricity are already known, which are listed in Table 17.2. It is also assumed that the produced hydrogen can be sold without any limitations so that the hydrogen storage is not involved.

TABLE 17.2
Spot Prices of the Electricity

Hour	1	2	3	4	5	6	7	8	9	10	11	12
Price (€/MWh)	60	30	52	68	28	27	45	43	36	38	54	36
Hour	13	14	15	16	17	18	19	20	21	22	23	24
Price (€/MWh)	38	46	48	24	25	26	27	38	58	56	54	46

TABLE 17.3
Electrolyzer Parameters

Hydrogen Conversion Rate @100% Load Factor	48.1 kWh/kg	Hydrogen Conversion Rate @50% Load Factor	45 kWh/kg	Hydrogen Conversion Rate @25% Load Factor	42.7 kWh/kg
Hot start time	20 min	Cold start time	1 h		

The electrolyzer owner needs to decide the electrolyzer power curve for the next 24 hr. Depending on the hydrogen price, the owner has distinct participating strategies selected individually, as presented in Figure 17.21.

With hydrogen price of €1.5/kg, the owner has little will to produce hydrogen using electricity. Only when the electricity price is considerably low will the owner put the electrolyzer into production state and sell hydrogen. From hour 6 to hour 15, the electrolyzer is completely shut down to save energy costs. When the hydrogen price rises to €2/kg, the owner chooses to produce hydrogen more frequently. It is clearly seen that the electrolyzer is operated at its lower bound, for example, from hour 6 to hour 8 and from hour 13 to hour 14. This implies that extra costs induced by high electricity price is less than the loss caused by shutting down the electrolyzer.

Hydrogen price of €2.5/kg results in a significantly higher load factor of the electrolyzer. In most time, the electrolyzer is operated at its upper bound to produce hydrogen as much as possible. Finally, if the hydrogen price reaches €3/kg, the electrolyzer is almost fully dispatched, except the fourth hour, when the electricity price skyrockets to nearly €70/MWh.

17.4.4 EXAMPLE 4: OPTIMIZATION OF ELECTROLYZER PLANT PROVIDING GRID SERVICES

Electrolyzers have shown excellent dynamic performance, enabling providing grid services, such as primary frequency containment reserve. Therefore, the electrolyzer owner has the motivation to participate in multiple power markets for revenue stacking. For example, the owner can adjust the electrolyzer power in real time to offer balancing energy in real-time market. Also, the reserving part of electrolyzer capacity for potential frequency regulation could be attractive to the owner due to the extra benefits from keeping reserve.

Figure 17.22 illustrates the investigated system and its behavior in multiple power markets. For more details on power markets, see [29]. The electrolyzer owner purchases energy from the utility and provides reserves. Part of the produced hydrogen is stored and further used to satisfy the local hydrogen loads, and the remainder is sold in the hydrogen market. Hydrogen purchase is not applicable in this case, implying that the electrolysis system needs to at least produce enough hydrogen for on-site usage.

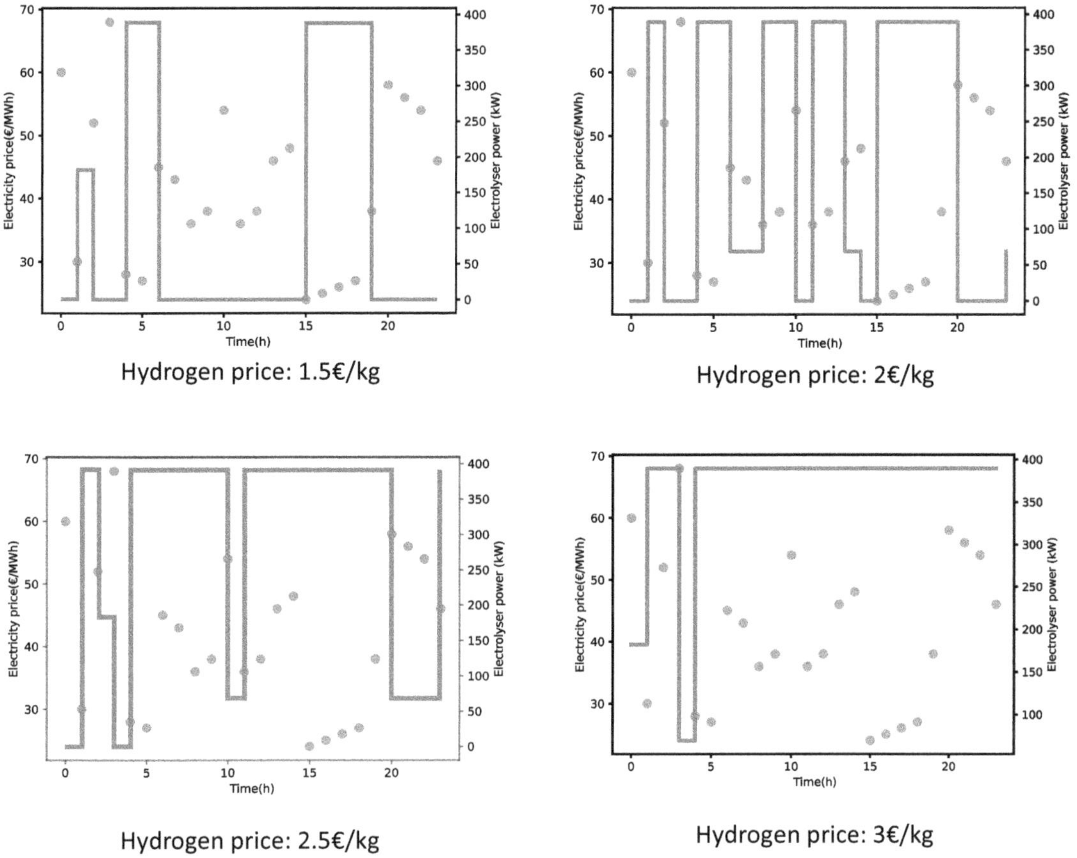

FIGURE 17.21 (a–d) Electrolyzer power trajectory in the cases of different hydrogen selling prices.

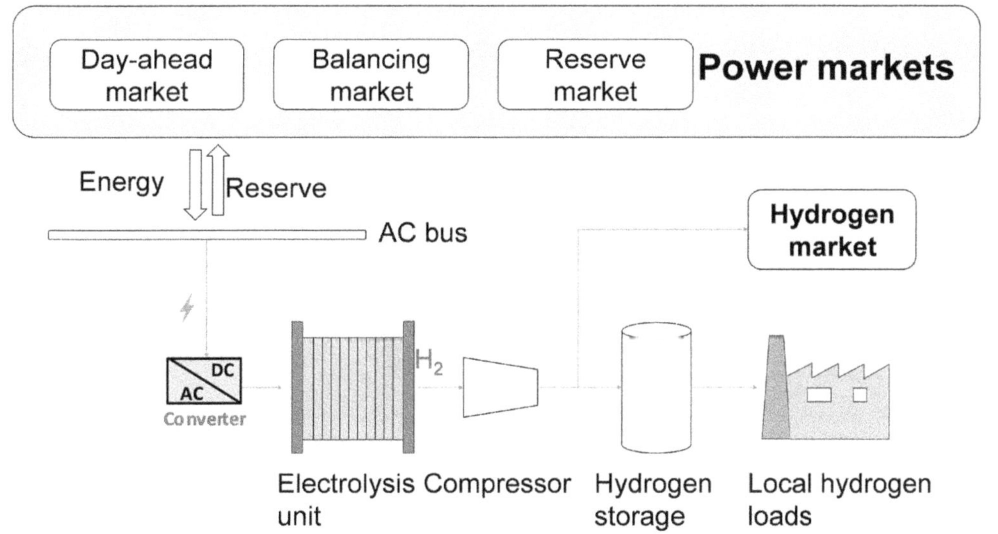

FIGURE 17.22 Schematics of an electrolysis system participating in multiple power markets.

Of the complexity of involving multiple markets, focus here is on describing market variables and using a simplified electrolyzer based on a linear assumption of the relationship between input power and hydrogen output. The objective function of the owner is to maximize the overall marginal profits, given as follows:

$$\max \ (\dot{m}_{H,t}^{exp} + D_{H,t})\lambda_{H,t}\Delta t + \lambda_{r,t}^{-}r_t^{-}\Delta t + \lambda_{r,t}^{+}r_t^{+}\Delta t - \lambda_t^{sp}p_t^{sp}\Delta t - \lambda_{p,t}^{+}\Delta p_t^{+}\Delta t + \lambda_{p,t}^{-}\Delta p_t^{-}\Delta t \qquad (17.59)$$

where $\dot{m}_{H,t}^{exp}$ is the hydrogen sold in the market and $D_{H,t}$ is the local hourly demand. $\lambda_{H,t}$ represents the hydrogen price, which is actually independent on time in the case. $\lambda_{r,t}^{-}$ and $\lambda_{r,t}^{+}$ are the price of up- and downregulation capacity, measured in €/MW/h. λ_t^{sp} denotes the spot price. r_t^{-} and r_t^{+} are the upward and downward reserve capacity for frequency regulation services. Note that the superscript "-" means reserving capacity, so that the electrolyzer can reduce its power consumption. From the perspective of the power system operator, such reserve helps increase the frequency, that is, an upward reserve. p_t^{sp} stands for the bidding power in day-ahead market. Δp_t^{+} and Δp_t^{-} are the downward- and upward-regulation power in the balancing market, respectively.

This mathematical programming is subject to a series of technical and market constraints. To get full access to the details, check out [30].

The annual revenue and cost of an electrolyzer system providing reserve service are examined. For each year, four cases, including providing no reserve service, frequency containment reserve (fcr), automatic frequency restoration reserve (afrr), and manual frequency restoration reseve (mfrr), are considered. The revenues come from providing real-time (RT) upregulation energy, up/down reserve capacity, and hydrogen sell, while the costs derived from real-time upregulation energy and energy cost in the spot market.

From Figure 17.23, the potential of an electrolyzer to stack revenues by providing real-time balancing services and different kinds of frequency regulation reserves can be seen. The system costs includes the cost from purchasing electricity in day-ahead market and performing down-regulation in real-time operation. While most of the revenues are from hydrogen selling, providing upward/downward reserve capacities also contributes to higher revenues, which are especially significant when offering frequency containment reserve and automatic frequency restoration reserve services.

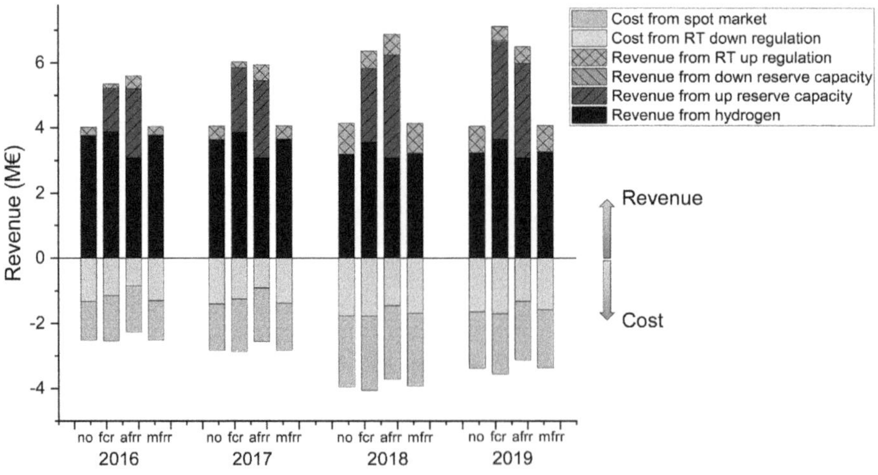

FIGURE 17.23   Revenue and cost breakdown of the electrolysis system during 2016 to 2019 in Denmark.

17.5 CONCLUSION AND OUTLOOK

This chapter delves into the modeling and control of alkaline electrolyzers, also including a brief introduction of electrolysis system operation. The roles of PtH_2 technologies in renewable-dominated power systems are firstly pointed out, namely, the grid-balancing service providers, long-term energy storage, and sector coupling enablers. The intermittent and fluctuating power output nature of renewable power generation highlights the importance of dynamic control and strategic operation of electrolyzers, which rely on detailed electrolyzer modeling.

The key modeling part focuses of electrolyzer modeling include polarization characteristics, electrolyzer efficiency, hydrogen production, thermal properties, converter models, and scaling-up methodologies. On the basis of the proposed models, temperature-stabilizing control, converter control, and dynamic frequency control of the electrolyzers are described, serving as the three dominated controlling scenarios. The chapter further introduces the operation problem of an electrolysis system, which is related to the determination of operating points within an extended timescale, such as in an hour. Electrolyzer models are prudently simplified in mathematical programming–oriented way. Key physical properties in the component level, such as the efficiency variation, are maintained to guarantee the technical feasibility of the resulting operational strategies.

The extensive utilization of PtH_2 technologies sees the emergence of more modeling, control, and operation issues:

- Except for the mentioned electrochemical and thermal modeling, the mass transfer, fluid transportation, and degradation of electrolyzers are attracting increasing attention. Furthermore, the interaction and coupling among multiple physical fields call for better electrolyzer modeling to support reliable control and operation. For instance, hydrogen-in-oxygen ratio is a key safety-related factor, influenced by electrolyzer current level, pressure, and temperature. Advanced models are required to describe the dynamic variation of hydrogen-in-oxygen ratio.
- Electrolyzer control is developing in a way that incorporates increasing consideration of multiple physical field and parameters. For example, coordinated control of temperature, pressure, and hydrogen in oxygen has been a new research focus. An increasing complexity of electrolyzer control also highlights the necessity of developing advance control methods, such as model predictive control, fuzzy control, and adaptive robust control. In parallel, electrolyzer control focuses are moving from a single device to multiple electrolyzers or large electrolysis group with distinctive electrolyzer types. An increase of spatial scale has raised up new challenges of group control methodologies, which becomes critical, especially in the context of large electrolysis projects.
- Electrolyzer operation is also challenged by the multi-physical field coupling within electrolysis devices. Therefore, future research on the operation will include the formulation of operational research problem considering more detailed device-level constraints. Also, new technologies that can cope with the uncertainties related to high-renewable and electrolysis penetration systems need to be further studied.

REFERENCES

[1] "CO$_2$ emissions in 2022," *CO$_2$ Emiss. 2022*, 2023. doi: 10.1787/12ad1e1a-en.

[2] "Average temperature anomaly, global," *Our World in Data*, 2023. https://ourworldindata.org/grapher/temperature-anomaly.

[3] "Global primary energy consumption by source," *Our World in Data*, 2023. https://ourworldindata.org/grapher/global-energy-substitution.

[4] "Low-carbon sources have a higher share in our electricity mix," *Our World in Data*, 2023. https://ourworldindata.org/energy-overview#low-carbon-sources-have-a-higher-share-in-our-electricity-mix.

[5] J. Bialek, "What does the GB power outage on 9 August 2019 tell us about the current state of decarbonised power systems ?," *Energy Policy*, vol. 146, August 2019, p. 111821, 2020. doi: 10.1016/j.enpol.2020.111821.

[6] A. Mayyas, M. Wei, and G. Levis, "Hydrogen as a long-term, large-scale energy storage solution when coupled with renewable energy sources or grids with dynamic electricity pricing schemes," *Int. J. Hydrogen Energy*, vol. 45, no. 33, pp. 16311–16325, 2020. doi: 10.1016/j.ijhydene.2020.04.163.

[7] S. Singh, S. Jain, P. S. Venkateswaran, A. K. Tiwari, M. R. Nouni, J. K. Pandey, and S. Goel, "Hydrogen: A sustainable fuel for future of the transport sector," *Renew. Sustain. Energy Rev.*, vol. 51, pp. 623–633, 2015. doi: 10.1016/j.rser.2015.06.040.

[8] S. Ehsan and M. A. Wahid, "Hydrogen production from renewable and sustainable energy resources: Promising green energy carrier for clean development," *Renew. Sustain. Energy Rev.*, vol. 57, pp. 850–866, 2016. doi: 10.1016/j.rser.2015.12.112.

[9] L. Viktorsson, J. T. Heinonen, J. B. Skulason, and R. Unnthorsson, "A step towards the hydrogen economy: A life cycle cost analysis of a hydrogen refueling station," *Energies*, vol. 10, no. 6, pp. 1–15, 2017. doi: 10.3390/en10060763.

[10] M. Hydrogen Council, "Hydrogen insights 2022: An updated perspective on hydrogen market development and actions required to unlock hydrogen at scale," 2022. [Online]. www.hydrogencouncil.com.

[11] Z. Chehade, C. Mansilla, P. Lucchese, S. Hilliard, and J. Proost, "Review and analysis of demonstration projects on power-to-X pathways in the world," *Int. J. Hydrogen Energy*, vol. 44, no. 51, pp. 27637–27655, 2019. doi: 10.1016/j.ijhydene.2019.08.260.

[12] K. E. og Forsyningsministeriet, "Regeringens strategi for Power-to-X," 2021. [Online]. https://ens.dk/sites/ens.dk/files/ptx/strategy_ptx.pdf.

[13] "New strategy accelerates Denmark's Power-to-X ambitions," *State of Green*, 2023. https://stateofgreen.com/en/news/new-strategy-kick-starts-denmark-production-of-green-hydrogen-and-e-fuels/.

[14] H. Lange, A. Klose, W. Lippmann, and L. Urbas, "Technical evaluation of the flexibility of water electrolysis systems to increase energy flexibility: A review," *Int. J. Hydrogen Energy*, vol. 48, no. 42, pp. 15771–15783, 2023. doi: 10.1016/j.ijhydene.2023.01.044.

[15] A. Buttler and H. Spliethoff, "Current status of water electrolysis for energy storage, grid balancing and sector coupling via power-to-gas and power-to-liquids: A review," *Renew. Sustain. Energy Rev.*, vol. 82, February 2017, pp. 2440–2454, 2018. doi: 10.1016/j.rser.2017.09.003.

[16] P. B. B. Pierre Oliviera and C. Bourasseau, "Low-temperature electrolysis system modelling: A review," *Renew. Sustain. Energy Rev.*, vol. 78, May, pp. 280–300, 2017. doi: 10.1016/j.rser.2017.03.099.

[17] C. Daoudi and T. Bounahmidi, "Overview of alkaline water electrolysis modeling," *Int. J. Hydrogen Energy*, vol. 49, pp. 646–667, 2024. doi: 10.1016/j.ijhydene.2023.08.345.

[18] S. Hu, B. Guo, S. Ding, F. Yang, J. Dang, B. Liu, J. Gu, J. Ma, and M. Ouyang, "A comprehensive review of alkaline water electrolysis mathematical modeling," *Appl. Energy*, vol. 327, September, p. 120099, 2022. doi: 10.1016/j.apenergy.2022.120099.

[19] Ø. Ulleberg, "Modeling of advanced alkaline electrolysers: A system simulation approach," *Int. J. Hydrogen Energy*, vol. 28, no. 1, pp. 21–33, 2003. doi: 10.1016/S0360-3199(02)00033-2.

[20] A. Ursúa and P. Sanchis, "Static-dynamic modelling of the electrical behaviour of a commercial advanced alkaline water electrolyser," *Int. J. Hydrogen Energy*, vol. 37, no. 24, pp. 18598–18614, 2012. doi: 10.1016/j.ijhydene.2012.09.125.

[21] Ø. Ulleberg, "Modeling of advanced alkaline electrolysers a system," *Hydrog. Energy*, vol. 28, pp. 21–33, 2003 [Online]. http://ac.els-cdn.com/S0360319902000332/1-s2.0-S0360319902000332-main.pdf?_tid=9de625b8-a749–11e2-af8d-00000aab0f6b&acdnat=1366194721_e2bd5a3ec87dd1ed6f-b5af4c9b55380e.

[22] M. Hammoudi, C. Henao, K. Agbossou, Y. Dubé, and M. L. Doumbia, "New multi-physics approach for modelling and design of alkaline electrolysers," *Int. J. Hydrogen Energy*, vol. 37, no. 19, pp. 13895–13913, 2012. doi: 10.1016/j.ijhydene.2012.07.015.

[23] D. Jang, W. Choi, H. S. Cho, W. C. Cho, C. H. Kim, and S. Kang, "Numerical modeling and analysis of the temperature effect on the performance of an alkaline water electrolysis system," *J. Power Sources*, vol. 506, May, p. 230106, 2021. doi: 10.1016/j.jpowsour.2021.230106.

[24] W. Wei and J. Wang, *Modeling and Optimization of Interdependent Energy Infrastructures*. Cham: Springer, 2020.

[25] F. L. John and R. Birge, *Introduction to Stochastic Programming*. Springer Science & Business Media, 2011.

[26] Ø. Ulleberg, Stand-alone power systems for the future: Optimal design, operation and control of solar-hydrogen energy systems. PhD thesis, Department of Thermal Energy and Hydropower, Norwegian University of Science and Technology, 1998.

[27] G. Sakas, A. Ibáñez-Rioja, V. Ruuskanen, A. Kosonen, J. Ahola, and O. Bergmann, "Dynamic energy and mass balance model for an industrial alkaline water electrolyser plant process," *Int. J. Hydrogen Energy*, vol. 47, no. 7, pp. 4328–4345, 2022. doi: 10.1016/j.ijhydene.2021.11.126.

[28] C. Huang, "Analytical modeling and control of grid-scale alkaline electrolyser plant for frequency support in wind-dominated electricity-hydrogen systems," *IEEE Trans. Sustain. Energy*, vol. 14, no. 1, pp. 217–232, 2023. doi: 10.1109/TSTE.2022.3208361.

[29] F. H. M. Jeremy Lin, *Electricity Markets: Theories and Applications.* Wiley-IEEE Press, 2017.

[30] Y. Zheng, C. Huang, S. You, and Y. Zong, "Economic evaluation of a power-to-hydrogen system providing frequency regulation reserves: A case study of Denmark," *Int. J. Hydrogen Energy*, vol. 48, no. 67, pp. 26046–26057, 2023. doi: 10.1016/j.ijhydene.2023.03.253.

18 The Power-to-X (PtX) Effect in Ancillary Service Markets

From Opportunities and Challenges toward Future Directions

*Mohammad Mohsen Hayati, Ali Aminlou, Mehdi Abapour,
Miadreza Shafie-khah, Hossein Shahinzadeh,
and Gevork B. Gharehpetian*

Abbreviations

ASMs	ancillary services markets
ASs	ancillary services
CEMs	competitive energy markets
CHP	combined heat and power
DERs	distributed energy resources
DG	distributed generation
EMs	energy markets
ESSs	energy storage systems
EU	European Union
GH_2	green hydrogen
LMPs	locational marginal prices
MCPs	market clearing prices
MGs	microgrids
NRAs	bational regulatory authorities
PtG	power-to-gas
PtX	power-to-X
RESs	renewable energy sources
SO	system operator
syngas	synthetic gas
TSOs	transmission system operators

18.1 INTRODUCTION

The energy generation landscape has undergone significant transformation in recent years, with a notable shift toward decentralized electrical systems heavily influenced by the integration of renewable and distributed energy resources (DERs) [1, 2]. These resources, which often exhibit high variability and rapid fluctuations, pose challenges to grid stability and must be effectively integrated into ancillary services markets (ASMs) to ensure balanced and secure system operations [3]. Ancillary services (ASs), which are essential for supporting the transmission of electric power from generators to consumers while ensuring the continuous balance of supply and demand, are now at

DOI: 10.1201/9781032719436-18

the forefront of these challenges. The traditional approaches to ASs are increasingly being tested by the fluctuating nature of renewable energy source (RES), necessitating innovative solutions to ensure a resilient and sustainable energy system [4]. The European Union (EU), through initiatives like the European Green Deal, aims to become the first climate-neutral continent by 2050 [5]. A crucial milestone in this journey was the target to achieve 20% of gross final energy consumption from renewable energy sources (RESs) by 2020, and 32% by 2030 [6]. As of 2019, RESs accounted for over 30% of the EU's electricity production, with variable sources such as wind and solar energy comprising 18% of total generation [7, 8].

Variable RESs, typically smaller in scale, are often connected to low- or medium-voltage networks and are classified as DERs. The penetration of DERs has increased significantly across the EU, driven by advancements in technology and a growing focus on decarbonization and energy efficiency [9]. These DERs predominantly consist of RES plants and small-scale, high-efficiency thermal plants, including combined heat and power (CHP) systems. For example, in countries like Denmark and Germany, DERs represent more than half of the total installed power, with wind energy and small-scale CHP plants playing a significant role in the energy mix [10, 11]. This evolving generation mix underscores the growing importance of managing RESs and DERs effectively, particularly as long-term decarbonization scenarios for 2050 emerge [12, 13]. Hence, ensuring the stability and security of the power system in this context requires innovative solutions to maintain the instantaneous balance between energy supply and demand. A key strategy in this effort is the production of hydrogen through water electrolysis using low-carbon electricity. This hydrogen can be utilized directly or converted into chemical compounds that can replace fossil fuels, offering a viable alternative for energy applications that are not easily electrified. The concept, known as power-to-X (PtX), is still in its developmental stages and is associated with considerable uncertainty [14].

Traditionally, ASs have been supplied predominantly by large-scale thermal and hydroelectric power plants. However, as the proportion of variable and intermittent renewable generation grows, maintaining grid security and reliability is likely to become increasingly difficult [15, 16]. The gradual decommissioning of conventional flexibility providers, such as large-scale gas and coal plants with rotating generators, exacerbates this challenge. This transition could lead to more frequent extreme events within power systems, such as significant frequency deviations, and result in higher costs associated with re-dispatching and system balancing [17–19].

In alignment with the European Climate and Energy Package, market design must evolve to incorporate flexibility services from emerging assets. Even as markets open up to new participants, the security of the electricity system must remain a top priority [20, 21]. Regular reviews of ASs and related products are essential to accommodate changes in the generation mix. Such reviews would enable all generators and consumers to contribute services they are technically capable of providing, from a regulatory standpoint. To minimize reliance on costly backup plants and avoid the need for extensive infrastructure upgrades, new flexibility providers—such as end users, energy storage systems (ESSs), innovative forms of flexible generation, and RESs—should have guaranteed access to the market. Integrating DERs fully into ASMs is crucial to meeting new demands effectively and economically, particularly since DERs are partly responsible for these new challenges [22, 23].

Amending ASMs can be viewed as a proactive strategy to maintain the reliability of the EU power system. Significant efforts have already been undertaken to upgrade distributed generation (DG) systems and prevent their premature disconnection [24, 25]. The next crucial step in bolstering network security involves integrating DG into electricity balancing and ASMs. Throughout the early 21st century, numerous initiatives have been undertaken to open up and redesign ASMs, aimed at removing regulatory barriers that hinder effective participation [26–28]. However, market reforms must be approached with caution by national regulatory authorities (NRAs), as each adjustment can involve complex regulatory trade-offs. Market stakeholders, particularly operators and service providers, often have differing perspectives. For example, new market participants, typically small-scale and distributed, can only submit small power bids. Although smaller bids

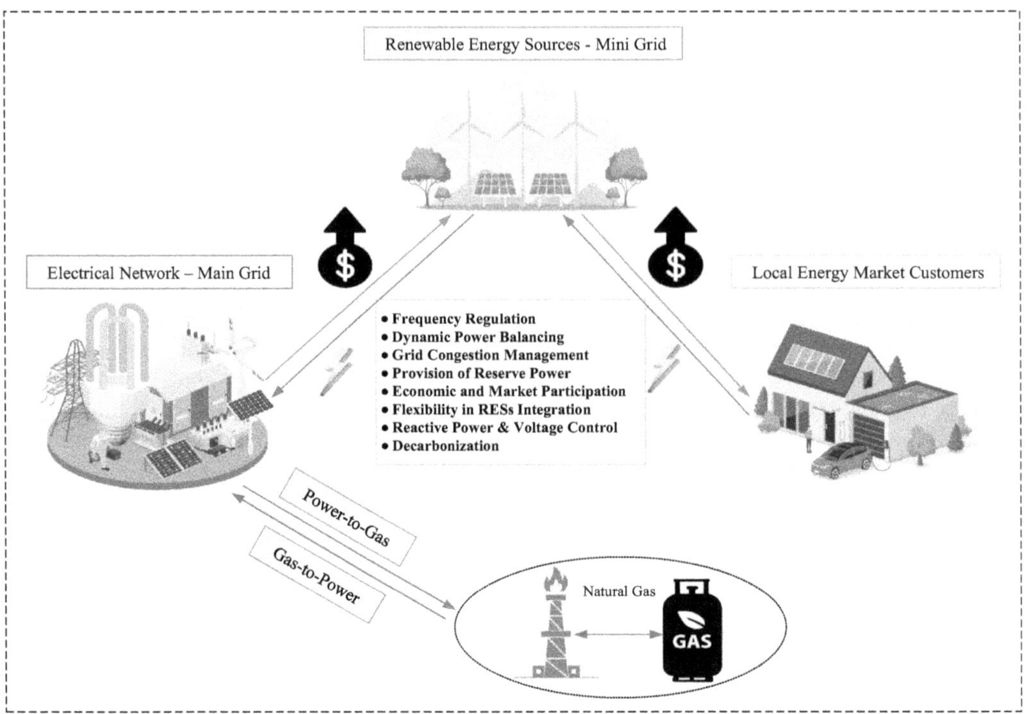

FIGURE 18.1   ASMs provided by interconnected renewable MGs in the presence of a PtX system.

increase the number of market transactions, they also demand more effort from system operators (SOs). Effective market reform requires a deep understanding of these underlying trade-offs, with an initial focus on identifying solutions that benefit all parties involved [29, 30].

PtX encompasses a broad spectrum of technologies that convert surplus electrical energy— often generated from renewable sources—into other forms of energy or valuable products, such as hydrogen, synthetic fuels, and chemicals (such as power-to-hydrogen), and synthetic natural gas (power-to-gas), liquid fuels (power-to-liquids), or heat. This conversion not only offers a viable solution for the storage and utilization of excess renewable energy but also presents new avenues for grid balancing, peak shaving, and the reduction of carbon emissions. This chapter explores emerging opportunities and innovative concepts that arise from the interplay between PtX and ASs. One such opportunity is the concept of dynamic energy pricing, which could optimize PtX operations by aligning energy production with demand patterns. Additionally, demand-side management strategies offer a means of enhancing grid resilience while maximizing the efficiency of PtX systems. The chapter further investigates the potential for cross-sector coupling, highlighting the synergies between the electricity, heating, and transportation sectors. This integration could lead to a more cohesive and sustainable energy system, where the benefits of PtX technologies extend beyond the electricity grid to other critical energy sectors. Figure 18.1 illustrates the role of interconnected microgrids (MGs) in providing ASs to the main electrical network and local energy markets in the presence of PTX technology cycle.

18.2   COORDINATED INVOLVEMENT IN ASMS

The procurement and management of ASs differ significantly across different countries, reflecting the unique regulatory frameworks and market structures in place within each region. These services are essential for maintaining the stability and reliability of power systems, and they are typically procured by transmission system operators (TSOs) to address issues such as frequency or voltage deviations [31, 32]. Key aspects, such as the definition of services offered, contracting methods, operational procedures, remuneration mechanisms, and prequalification criteria, are all tailored to the specific needs and conditions of each country. This variability poses challenges to conducting a comprehensive, comparative analysis of ancillary service procurement on a global scale. The differences in how countries approach these services underscore the complexity of harmonizing practices across international borders and highlight the need for region-specific strategies in ensuring the effective operation of power systems [33].

The restructuring of the power industry introduced significant shifts in power system operations, driven by the unbundling of generation, transmission, and distribution functions. This process allowed different entities to manage generation and distribution independently, while transmission generally remained a regulated monopoly due to its natural monopoly characteristics [34]. The primary goals of these reforms were to enhance service quality and lower costs by fostering competition within electricity markets [35, 36]. However, the deregulation of power markets also brought about new challenges. Key among these were the need to clearly define the roles and responsibilities of various network participants, set appropriate energy prices, and establish mechanisms for cost allocation. This complexity necessitated the development and refinement of ASs, which are essential for maintaining the reliability, stability, and security of power systems. AS typically includes services such as frequency regulation, voltage control, and operating reserves, which ensure that the grid operates smoothly and efficiently [37, 38].

The management, procurement, and restructuring process of ASs differ across global energy markets (EMs), reflecting the varied approaches to market design and regulation. In some markets, ASs are procured through competitive bidding processes, allowing market participants to offer these services under specified terms and conditions. In other cases, certain ASs remain under the direct control of the SO, who manages them to meet the operational needs of the power system. This divergence in management structures highlights the complexity and regional specificity of AS management in the context of restructured power systems [39, 40]. Generally, there is no universal standard for defining specific ASs applicable across all power systems, as the procurement of AS varies depending on the system in question. However, a simplified resource-based classification helps categorize AS into three main groups: voltage control services (through reactive power support), frequency control services (such as regulation, load following, and operating reserves), and emergency services (including black-start capabilities) [41, 42].

Over the past decades, global electricity consumption has surged by approximately 68%, driven by rising demand across various economic sectors [43, 44]. This escalating demand underscores the need for a more reliable and adaptable power system to support future global development. Furthermore, the increased frequency of natural disasters, exacerbated by climate change, necessitates power systems with enhanced resilience to handle abnormal and low-probability events. Renewable MGs have emerged as effective and reliable solutions to these challenges. Historically, most renewable MGs were implemented as off-grid systems, primarily serving remote rural areas disconnected from the main grid [45, 46]. However, their applications have expanded significantly, with renewable MGs now being deployed in grid-connected regions to boost supply reliability, reduce grid dependency, and lower electricity costs. These interconnected MGs can operate autonomously or in conjunction with the main grid, facilitating the integration of DERs and increasing the share of renewables in the energy mix [47, 48].

18.2.1 The Necessity of Integrated Microgrid Clusters in ASs Markets

The adoption of DG and MGs offers numerous benefits, including environmental sustainability, cost-effectiveness, improved power quality, and enhanced marketability [49]. The advantages of MGs and DG systems have been well-documented, with evidence supporting their effectiveness in delivering these benefits. Renewable energy–driven MGs, in particular, are seen as a promising solution for combating climate change, serving as a clean energy management framework [50, 51]. Authors of [52] highlight that DC MGs provide additional advantages by mitigating some of the challenges associated with renewable energy expansion, such as resource management and coordination among diverse energy providers.

The academic literature on MGs is extensive, covering various aspects including DG units, power electronics interface topologies, energy management strategies, and converter control schemes [53]. Additionally, there are comprehensive reviews on energy storage solutions, control methods in MGs, and the use of demand response in renewable energy MGs [54–56]. Critical analyses of decision-making strategies, solution methods for energy management systems, and optimization techniques for planning and operating islanded MGs have also been thoroughly explored [57–59]. Other studies focus on generation scheduling, peak load management, and the control of standalone MGs using renewables and battery storage to reduce reliance on fossil fuels, thereby enhancing reliability and resilience against fuel supply disruptions [60–62].

Transactive energy technologies have been increasingly integrated into multiple MGs within the emerging smart grid framework, with the goal of enhancing the reliability and sustainability of energy sharing and management. These technologies facilitate a more dynamic and efficient distribution of energy resources across interconnected MGs, thereby optimizing the overall performance of the energy system [63, 64]. Furthermore, several utility companies are progressively embracing networked MG technologies, leveraging them to deliver advanced ASs and actively contributing to grid operations. This approach not only supports grid stability but also enhances the flexibility and responsiveness of the power system [65, 66]. Recent studies, such as those by Han et al., have highlighted the growing focus on MG clusters, which aim to significantly increase the penetration of MGs within the energy infrastructure [67]. By organizing MGs into interconnected clusters, the resilience of power systems can be substantially improved. This networked approach allows for better management of energy flows and enhances the system's ability to withstand and recover from disruptions. The formation of these MG networks is seen as a critical strategy for bolstering the robustness and reliability of modern power systems, ensuring that they can adapt to and thrive in the face of evolving challenges and demands [68].

18.3 POWER-TO-X: CURRENT POSITION AND PERSPECTIVE

Over the past decade, the production of chemicals derived from RESs, particularly for use as fuels, has attracted significant attention within the energy sector. This growing interest is largely driven by the global push for decarbonization and the need to transition away from fossil fuels toward more sustainable alternatives. At the heart of this emerging field lies the concept of renewable hydrogen production, often referred to as power-to-hydrogen (PtH) [69–71]. Renewable hydrogen serves as a foundational element of the broader PtX framework, which seeks to convert surplus renewable electricity into various chemical forms that can be stored, transported, and utilized as fuel or feedstock [72]. Hence, PtX technology is rapidly emerging as a key component in the global energy transition [73].

Currently, PtX is in a developmental and early commercial stage, with numerous pilot projects and small-scale implementations around the world. Europe, particularly Germany and the Nordic countries, is leading in PtX development, supported by strong policy frameworks and investment in renewable energy. For instance, Germany's National Hydrogen Strategy emphasizes the role of green hydrogen (GH_2) produced through PtX as a cornerstone for achieving its climate goals. Similarly, countries like Japan and Australia are also investing heavily in PtX technologies, particularly in hydrogen production, to leverage their renewable resources and develop export markets [74, 75].

18.3.1 Power-to-X Technologies as a New Participant in ASMs

Historically, power system operators have focused on reducing the total cost of power generation by incorporating unit commitment into their operations. With the advent of power market liberalization, the focus has shifted toward optimizing social welfare for both generation companies and consumers [76, 77]. Growing concerns about the environmental impact of fossil fuel–based power plants and their contribution to global warming have spurred research into the increased use of RESs for both primary energy production and ASs worldwide. In a liberalized power market, a key responsibility of the SO is to facilitate power transactions while ensuring the demand–supply equilibrium is maintained [78]. This involves coordinating various market participants, such as generators, consumers, and service providers, to ensure that the electricity grid remains stable and efficient. One of the critical aspects of this task is the provision of ASs. These services are essential support services that help maintain the stability, reliability, and quality of the power system. These services include frequency regulation, voltage control, spinning reserve, and black start capability, among others. They are necessary for managing the variability and uncertainty inherent in power systems, especially with the increasing integration of RESs [79]. PtX technologies have emerged as a significant player in this landscape. These technologies offer flexibility to the power system by providing options for energy storage, demand response, and load management. For instance, during periods of excess electricity production, PtX technologies can convert the surplus power into storable forms of energy, which can then be used during periods of high demand or low generation [80].

The integration of PtX technologies into electricity markets is particularly important, as they can contribute to ASs by offering grid-balancing capabilities, participating in frequency regulation, and providing reserve power. As the capacity and efficiency of PtX technologies improve, they are expected to play a more significant role in electricity markets, supporting the SO in maintaining the equilibrium between supply and demand while enhancing the overall reliability and resilience of the power system [81].

The SOs, who are responsible for managing ASs, must navigate the challenges brought by market liberalization. They need to procure the necessary ASs from market participants to ensure that the power system operates within the required standards of quality and safety [82]. This involves balancing the grid in real time, congestion management, and addressing any disruptions that might arise due to the integration of PtX technologies. Furthermore, with the rise in transaction volumes and the increasing complexity of the energy mix, SOs face greater uncertainty and variability, necessitating more advanced tools and strategies for grid management. This issue itself includes leveraging advanced forecasting, machine learning, and other data-driven approaches to anticipate and mitigate potential issues, ensuring the continuous and secure operation of the power system.

Figure 18.2 represents a holistic approach to sustainable energy systems, particularly focusing on the integration of PtX technologies within ancillary service markets. The interconnected elements emphasize the crucial aspects required for a successful energy transition. This framework underscores the multidimensional approach required to harness PtX technologies effectively, addressing the challenges and opportunities in ancillary service markets.

18.3.2 Enhancing Energy Storage, Grid Integration, and Market Viability in PtX Systems

The PtX concept encompasses the conversion of electrical energy into a variety of fuels, chemicals, and heat, which are more easily stored and transported. A key application of PtX technology involves the production of hydrogen using RESs, as hydrogen (H_2), a sustainable molecule, does not naturally occur in its pure form and therefore requires synthesis [83]. In PtX systems, hydrogen serves as the fundamental output, with the potential to be used directly as an energy carrier or further converted into other valuable products, such as methane, synthesis gas, electricity, chemicals, and liquid fuels [84, 85]. The PtX concept offers several benefits, one of which is its capacity for long-term energy storage, referred to as vector-coupling storage [86]. This facilitates the use of hydrogen, methane, and other alternative fuels across various energy sectors, including electricity, heating, and transportation. This process also absorbs CO_2, aligning with decarbonization goals. Among the products

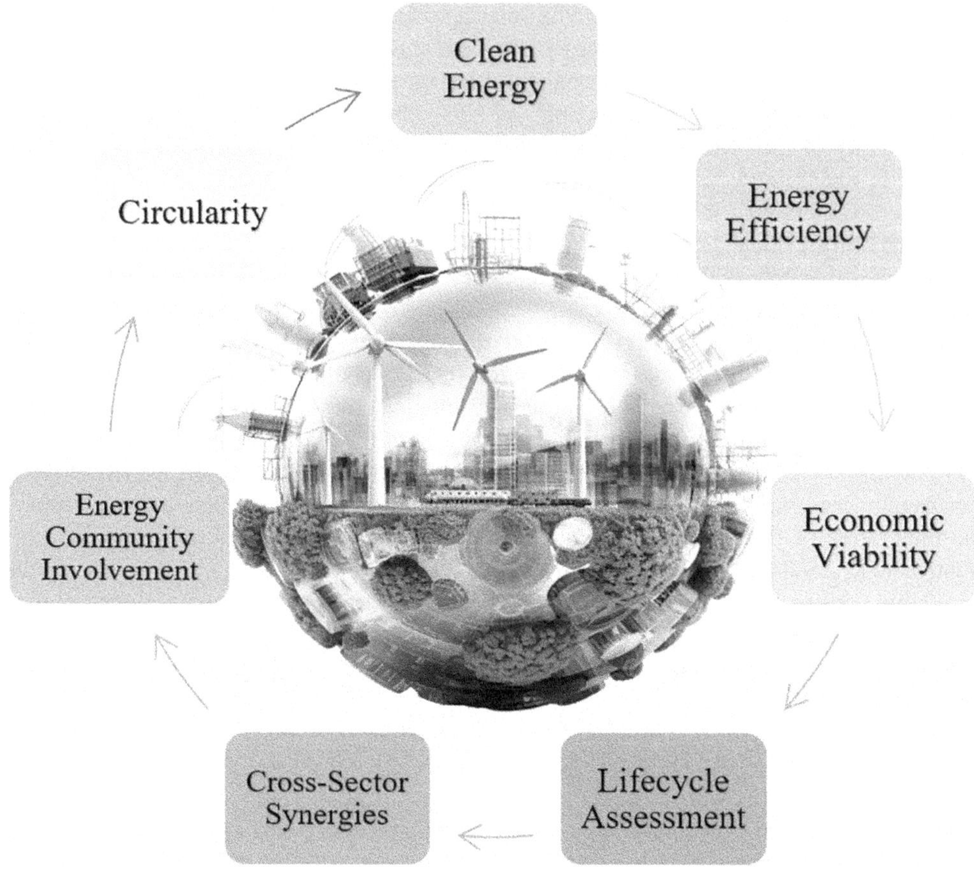

FIGURE 18.2 The integrated framework for sustainable energy transition in PtX systems.

of PtX technology, hydrogen stands out due to its flexible production sources (i.e., fossil fuels, thermal, and renewable), scalability (i.e., from watts to gigawatts), and diverse applications in industry, heating, and transportation. Hydrogen's high heating value and ability to convert between electricity and chemical energy make it a critical component of long-term energy storage and ASs [87, 88]. PtX systems are particularly valuable for storing renewable energy in different forms, such as hydrogen, which can be reused to balance the electrical grid during peak demand [89]. Additionally, systems such as power-to-gas (PtG) can convert electrical energy into natural gas via the methanation process, providing an alternative to large-scale electrical storage. Also, PtX technology can be used to demineralize rainwater, which can then be used in the electrolysis process to produce hydrogen. This hydrogen, generated from demineralized water, is already recognized as a zero-emission fuel for future transportation [90]. Hence, PtX systems and water electrolysis are expected to play a significant role in reducing greenhouse gas emissions and mitigating climate change. In California between 2017 and 2025, it is estimated that up to 7,800 GWh (gigawatt-hours) of renewable energy could be curtailed. To put this into perspective, 1 GWh is equivalent to 1 million kWh (kilowatt-hours), which is enough energy to power about 750 homes for a year. This excess energy could be converted into hydrogen or methane for storage or other uses. Of course, if synthetic methane were used for heating, it could meet the needs of around 370,000 homes. This difference in numbers arises because heating typically requires more energy than electrical appliances, but methane is a more efficient source of heating energy [91].

One of the major challenges in the widespread adoption of PtX technology is the high capital investment required for infrastructure development. To overcome this, governments could consider providing incentives and financial support. However, a more sustainable and competitive approach would involve developing markets for the sale of PtX services and products, thereby making these projects economically viable. Given that PtX technologies enhance the capture of renewable energy and enable the production of zero-emission fuels, allocating a portion of environmental protection incentives toward these technologies could further bolster their economic feasibility [92]. In addition to economic considerations, technical factors play a crucial role in determining the efficiency of PtX systems at an industrial scale. For instance, the efficiency of processes such as electrolysis and methanation directly impacts the overall efficiency of natural gas production. Current state-of-the-art PtG systems achieve a conversion efficiency of approximately 70%, which highlights the challenges of scaling up hydrogen production beyond laboratory settings [93]. Moreover, energy losses during conversion processes can undermine the efficiency of PtX systems. One potential solution to enhance efficiency is the co-location of PtX equipment with renewable energy plants. By integrating high-efficiency PtG systems with RESs, these setups could emerge as promising options for future decarbonized energy systems. The PtX concept, particularly the use of electrolysis to produce hydrogen from renewable energy, has the potential to serve as a central hub for various energy-related projects. However, much of the current research has focused on small-scale applications. Therefore, conducting pilot projects on a larger scale could provide valuable insights into the effects of scaling up and help identify optimal locations for PtX facilities, whether offshore or onshore [94, 95].

18.3.3 THE EFFECT OF PtG AS A KEY ENABLER FOR RESs INTEGRATION AND LONG-TERM ENERGY STORAGE

The integration of the power and gas sectors has gained significant attention in recent years, largely due to the flexibility offered by power-to-gas technologies and hydrogen storage. Water electrolysis is the key process that links electricity to the gas sector, enabling the conversion of excess electricity generated by renewable sources into GH_2 [96]. This approach allows for long-term energy storage by storing energy in molecular form rather than as electrons. Hydrogen can further undergo methanation to produce synthetic gas (syngas), which is primarily used for power generation in gas turbines and various industrial applications. Additionally, syngas can be safely injected into the natural gas grid, providing a method to manage fluctuations in power demand on both daily and seasonal timescales [97]. Power-to-gas (PtG) represents an emerging technology that offers a more efficient means of converting excess energy generated by RESs into gases that are compatible with existing gas infrastructure. PtG systems are capable of converting surplus energy from renewable or other sources into hydrogen via water electrolysis [98, 99]. Technologically, water electrolysis is a critical step within the PtG framework, as it serves as a necessary precursor for various energy conversion pathways. The hydrogen produced through this process can be utilized directly as an energy carrier or converted into methane, synthesis gas, and other compounds [100, 101]. PtG technology plays a crucial role in advancing toward a sustainable energy system, offering the key benefit of enabling long-term energy storage through the existing natural gas network [102]. By enhancing energy recovery from renewable sources, PtG systems improve efficiency and help reduce greenhouse gas emissions. Additionally, PtG technology can supply alternative fuels, such as hydrogen, methane, and synthesis gas for transportation and heating, while also providing medium-term storage capabilities through the natural gas network, known as vector-coupling storage. These versatile functions position PtG systems as essential components of future energy networks. Furthermore, developing markets for PtG products can aid investors in recouping significant capital investments in this technology [103, 104].

18.3.4 GLOBAL PtG INITIATIVES IN RESE INTEGRATION AND ASs

As said in previous sections, PtG projects have been increasingly implemented worldwide as a way to integrate RESs and provide flexibility in ancillary service markets. Following are some successful global PtG projects identified so far that have demonstrated the potential of this technology in supporting grid stability and Ass.

18.3.4.1 WindGas Falkenhagen (Germany)

This PtG project in Falkenhagen, launched in 2013 by Uniper, Germany, converts wind power into hydrogen through electrolysis. This 2 MW plant has been operational since May 2018, supplying hydrogen to a high-pressure natural gas network. The project provides balancing power by adjusting hydrogen production by about 360 m^3 of hydrogen per hour in response to grid demand, thus helping stabilize the grid. It also serves as a test bed for the integration of RESs into the gas network. In other words, the project has successfully demonstrated the technical feasibility and economic viability of PtG technology as a provider of grid services in a renewable energy–rich environment [105, 106].

18.3.4.2 Audi e-Gas Plant (Germany)

The Audi e-gas plant started operation in 2013 and is located in Werlte, Germany. It uses wind energy to produce hydrogen via electrolysis and then combines it with CO_2 to produce synthetic methane, which is fed into the natural gas grid. The plant participates in grid balancing by modulating its production based on grid needs, effectively using excess renewable energy, and providing a flexible load for grid operators. The plant has been a successful example of integrating PtG technology with the automotive sector, demonstrating the ability to stabilize the grid and create renewable fuels [107, 108]. **Figure 18.3** illustrates the functional principle of the "Audi e-gas plant" project. This project showcases how surplus renewable energy can be converted into synthetic methane, which is then injected into the natural gas network.

18.3.4.3 Green Industrial Hydrogen Project: "GrInHy" (Germany)

The GrInHy project in Salzgitter, Germany, started in 2016, focuses on producing hydrogen from renewable electricity for use in steel production. GrInHy has successfully demonstrated the potential of PtG to support energy-intensive industries while contributing to grid-balancing services by increasing or decreasing its power consumption based on grid demand. The GrInHy project focuses on producing green hydrogen through electrolysis powered by renewable electricity, while also offering grid management services as a reversible generator at the Salzgitter Flachstahl GmbH steel plant in Salzgitter [110, 111]. This project, shown in **Figure 18.4**, has received support funding from the Federal Ministry for Economic Affairs and Climate Action and the EU's Horizon 2020 research program.

18.3.4.4 AREVA PtG Project (France)

Located in France, this project by AREVA H$_2$Gen involves the production of hydrogen from renewable electricity through electrolysis. In this project, hydrogen is used for industrial purposes and injected into the natural gas grid. The project provides load balancing and grid stability services by adjusting hydrogen production in response to grid signals. Also, this project has contributed to the development of a market for GH$_2$ and demonstrated the role of PtG in integrating renewable energy into the broader energy system [113, 114]. The AREVA H$_2$Gen project has received support funding from the EU's Horizon 2020 research program.

18.3.4.5 Jupiter 1000 (France)

Jupiter 1000 is a French PtG project located in Fos-sur-Mer. It integrates renewable energy to produce hydrogen and synthetic methane, which are injected into the natural gas network. The project

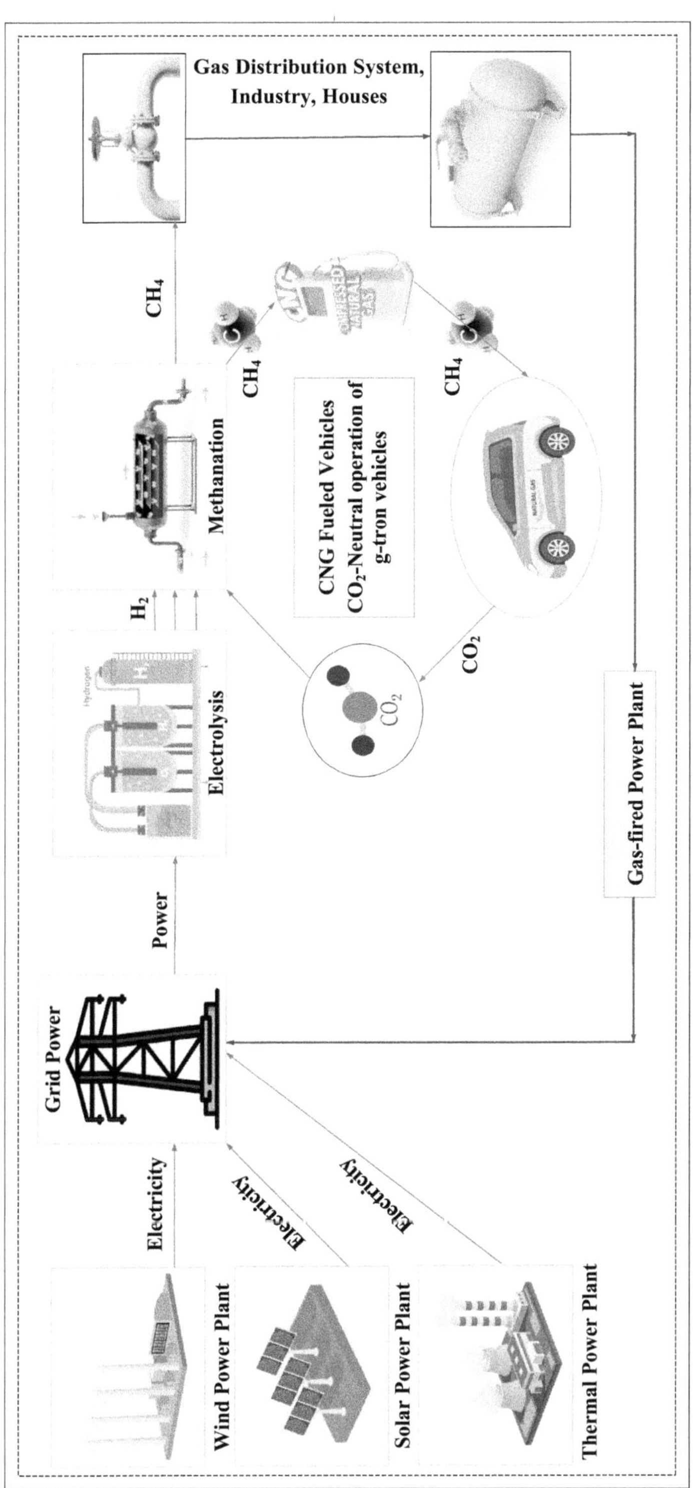

FIGURE 18.3 The functional principle of e-gas in "Audi e-gas plant" project.

Source: Adapted from [109].

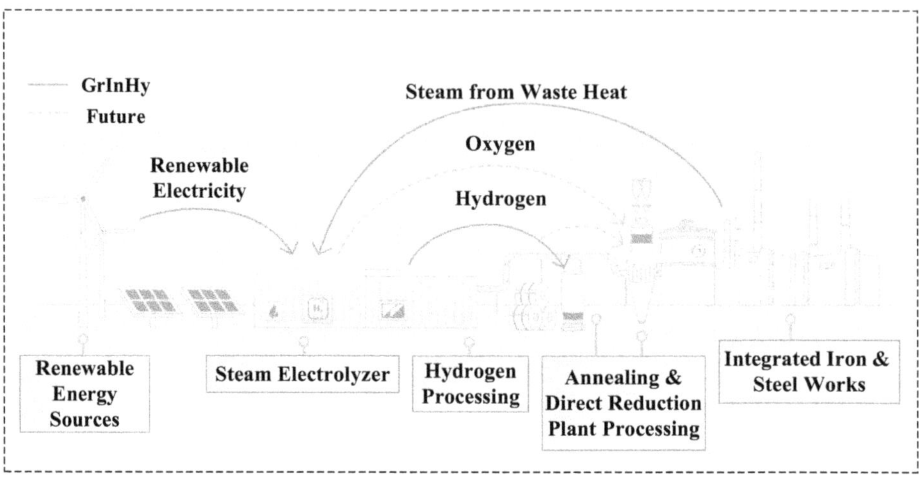

FIGURE 18.4 The schematic of the GrInHy project in Germany.

Source: Adapted from [112].

offers grid services such as frequency control by dynamically adjusting its operations based on grid requirements. Jupiter 1000 has been successful in demonstrating the potential of PtG for large-scale renewable energy storage and grid balancing, as well as its integration with existing gas infrastructure [115].

18.3.4.6 HyBalance (Denmark)

The HyBalance project in Hobro, Denmark, initiated in 2016, is a PtG facility that uses surplus wind energy to produce hydrogen. The HyBalance project is a collaborative effort involving seven European partners and is financially supported by the Fuel Cells and Hydrogen Joint Undertaking (FCH JU) as well as the Danish Energy Technology Development and Demonstration Program (EUDP). Hydrogen is utilized for industrial purposes and as a fuel for transportation. HyBalance contributes to grid stability by providing fast-responding balancing services and mitigating wind power fluctuations. The project has proven the scalability and integration potential of PtG in Denmark's electricity market, particularly in balancing the grid in real time [116].

18.3.4.7 STORE&GO Project (Europe)

STORE&GO is an EU-funded project that has implemented several PtG demonstration plants across Europe, including in Falkenhagen (Germany), Solothurn (Switzerland), and Troia (Italy). These plants focus on integrating PtG with the energy system. The project explores the role of PtG in providing balancing services, such as frequency regulation. Also, it enhances the flexibility of the European energy system and influences future policy and EM design in the EU [117].

18.3.5 Classification and Functions of Conventional ASs in PtG Systems

Conventional ASs can be categorized into two primary groups: (1) frequency-related AS, which focuses on maintaining the active power balance during normal operations and following contingencies, and (2) non-frequency-related AS, which involves services designed to ensure the balance of reactive power and provide reserves needed to restore the power system after major disruptions. These services maintain the active power balance in the grid during normal operation and after contingencies, such as a sudden loss of generation or load. PtG systems can contribute to both frequency-related and non-frequency-related ASs in the following ways.

18.3.5.1 Frequency-Related ASs in PtG Systems

These services maintain the active power balance in the grid during normal operation and after contingencies, such as a sudden loss of generation or load. PtG systems can contribute to this group in the following ways:

- **Frequency regulation.** PtG systems can help balance supply and demand by adjusting their power consumption based on grid frequency deviations. When there's an excess of electricity (e.g., during periods of high renewable generation and low demand), PtG systems can increase their power consumption to absorb the excess. Conversely, during a deficit, they can reduce their consumption or even supply power back to the grid if they have storage or cogeneration capabilities.
- **Load following/dispatchable load.** PtG systems can act as a flexible, dispatchable load that grid operators can control in real time to match generation with demand. This flexibility helps in stabilizing the grid, especially with the increasing share of intermittent RESs like wind and solar.
- **Spinning and non-spinning reserves.** Although traditionally provided by generators, PtG systems with integrated storage can also participate in reserve markets. They can rapidly reduce their electricity consumption to provide spinning reserves or, in some cases, be ready to increase consumption when there's excess generation (non-spinning reserves).

18.3.5.2 Non-Frequency-Related ASs in PtG Systems

These services ensure the reactive power balance and support system restoration after significant contingencies. PtG systems can contribute to this group as follows:

- **Voltage support/reactive power.** Non-frequency-related ASs are primarily focused on regulating reactive power imbalances by closely monitoring and responding to voltage fluctuations within the power system. PtG systems, especially those integrated with power electronics, can contribute to this function by compensating reactive power, which is crucial for maintaining voltage stability across the grid. The effectiveness of reactive power management depends heavily on the location within the grid. This is because reactive power losses increase with the distance from the generation source, making voltage control more challenging the farther the power needs to travel. Consequently, PtG systems situated in strategic locations can play a vital role in local voltage regulation, reducing the strain on transmission resources and enhancing overall grid stability. The provision of reactive power typically requires the deployment of generation and transmission resources. PtG systems can alleviate some of this demand by offering localized voltage support, thereby minimizing the need for long-distance transmission of reactive power and reducing associated losses.
- **Black start capability.** In the event of a significant grid failure, PtG systems could potentially provide black start capability. This involves restarting the electricity grid by providing the initial power required to kick-start other generating units, although this application is still largely theoretical for PtG systems and would require further technological development.
- **Restoration reserves.** PtG systems can store energy in the form of hydrogen or methane, which can be converted back into electricity when needed. This capability allows them to act as restoration reserves, supplying power to the grid during recovery phases after major contingencies, thereby aiding in the swift restoration of normal operations.

18.4 IMPACT OF PTX TECHNOLOGIES ON ENERGY MARKET PRICE

PtX technologies have several significant impacts on market prices within competitive energy markets (CEMs), particularly in terms of locational marginal prices (LMPs) and market clearing prices (MCPs) [118]. In the following, we have examined the impact of PtX on the EM on a case-by-case basis:

18.4.1 INCREASED SUPPLY AND DEMAND FLEXIBILITY

- **Supply-side.** PtX technologies convert surplus electricity into other forms of energy or chemical products (like hydrogen, synthetic fuels, or heat). This can reduce the curtailment of renewable energy, particularly during periods of high production and low demand. By effectively storing excess energy and converting it into useful products, PtX can increase the supply flexibility within the market, potentially lowering LMPs during times of high renewable output.
- **Demand-side.** PtX can also introduce new flexible demand into the system. For example, during periods of low electricity prices (due to high renewable generation), PtX facilities can ramp up their consumption to produce hydrogen or other products. This demand response can help balance supply and demand, preventing extreme dips in market prices and potentially stabilizing MCPs.

18.4.2 IMPACT ON MARGINAL COSTS AND MCPS

- **Lower marginal costs.** The integration of PtX with RESs can lead to an overall reduction in marginal costs across the system. Since renewables have very low or zero marginal costs, the introduction of PtX could reduce the reliance on higher-cost, fossil fuel–based generation during certain periods, pushing the MCP downward.
- **Negative prices.** In extreme cases, the integration of PtX might even exacerbate instances of negative pricing in electricity markets. When renewable generation exceeds demand and PtX cannot absorb all the excess, the market might still clear at negative prices. However, PtX technologies could mitigate the frequency and severity of these events by absorbing some of the excess energy.

18.4.3 TRANSMISSION AND LMPS

- **Reduction in transmission congestion.** PtX can be deployed strategically in locations with high renewable energy generation but limited transmission capacity. By converting excess electricity into other forms of energy locally, PtX could reduce transmission congestion and the associated costs, leading to lower LMPs in congested areas.
- **LMP smoothing.** By providing an outlet for surplus generation in specific locations, PtX technologies could smooth out extreme variations in LMPs between different nodes or zones. This could lead to a more balanced and equitable distribution of electricity prices across regions.

18.4.4 IMPACT ON ASMS

- **Increased participation.** PtX technologies could also participate in ASMs, such as frequency regulation and reserve markets, by quickly adjusting their electricity consumption or production. This additional flexibility can enhance system stability and potentially reduce the costs associated with ASs, indirectly influencing MCPs and LMPs.

18.4.5 Long-Term Market Dynamics

- **Investment signals.** The presence of PtX technologies may shift investment signals within CEMs. As PtX reduces the volatility of MCPs and LMPs, it could affect the profitability of traditional generation assets, particularly peaking plants that rely on periods of high prices. This could lead to a shift in investment toward more flexible and responsive technologies, including further development of PtX and storage solutions.

18.4.6 Decarbonization and Policy Impacts

- **Support for decarbonization.** By enabling the conversion of renewable electricity into carbon-neutral fuels and chemicals, PtX supports broader decarbonization goals. This can lead to policy shifts that further encourage the deployment of PtX technologies, influencing market dynamics and potentially leading to a reconfiguration of market price structures to accommodate the growing importance of PtX.

18.5 CHALLENGES, OPPORTUNITIES, AND PATH FORWARD IN INTEGRATING PTX INTO ASMS

Table 18.1 encapsulates the prominent challenges, opportunities, and suggested pathways for effectively integrating PtX technologies into ASMs, highlighting the need for a multifaceted approach to overcome barriers and leverage potential benefits.

18.6 CONCLUSION

The integration of RESs into the electricity grid presents both significant opportunities and formidable challenges, particularly in the context of ASs. This chapter has explored the transformative potential of PtX technologies, which facilitate the conversion of excess renewable energy into valuable products, such as hydrogen, synthetic fuels, and chemicals. These technologies not only contribute to grid balancing and peak shaving but also play a crucial role in reducing carbon emissions, thereby supporting the transition toward sustainable energy systems. The analysis indicates that PtX can enhance the reliability and stability of the grid by providing flexible and responsive solutions to the variable nature of renewable energy generation. However, the successful integration of PtX

TABLE 18.1

Navigating the Integration of PtX into ASMs: Challenges, Opportunities, and Future Directions

Challenges	Opportunities	Future Directions
Economic barriers related to investment and funding	Enhanced grid stability through flexible energy solutions	Developing supportive regulatory frameworks
Technical obstacles in integrating PtX systems	Creation of new markets for hydrogen and synthetic fuels	Developing collaboration among stakeholders
Regulatory barriers that hinder market participation	Potential for cross-sector coupling (electricity, heating, transportation)	Investing in research and development of technologies
Complexity in market design and operation	Optimization of energy production through dynamic pricing	Implementing pilot projects to test integration strategies
Need for technological advancements in PtX	Increased efficiency in energy utilization and storage	Incentivizing public–private partnerships for innovation

systems into existing EMs is not without challenges. Economic barriers, such as high initial investment costs and the need for supportive regulatory frameworks, must be addressed to unlock the full potential of these technologies. Additionally, technical obstacles related to the interoperability of PtX systems with current grid infrastructure require further research and development.

REFERENCES

[1] A. Aminlou, M. M. Hayati, H. Majidi-Garehnaz, H. Biabani, K. Zare, and M. Abapour, "Techno-economic analysis for decentralized GH_2 power systems," in *Green Hydrogen in Power Systems*, V. Vahidinasab, B. Mohammadi-Ivatloo, and J. Shiun Lim, Eds. Cham: Springer International Publishing, 2024, pp. 85–103.

[2] M. M. Hayati, A. Aminlou, K. Zare, and M. Abapour, "A two-stage stochastic optimization scheduling approach for integrating renewable energy sources and deferrable demand in the spinning reserve market," in *2023 8th International Conference on Technology and Energy Management (ICTEM)*, 2023, pp. 1–7.

[3] K. Zhang, S. Troitzsch, S. Hanif, and T. Hamacher, "Coordinated market design for peer-to-peer energy trade and ancillary services in distribution grids," *IEEE Transactions on Smart Grid*, vol. 11, no. 4, pp. 2929–2941, 2020.

[4] O. Stanojev, Y. Guo, P. Aristidou, and G. Hug, "Multiple ancillary services provision by distributed energy resources in active distribution networks," *arXiv Preprint arXiv:2202.09403*, 2022.

[5] C. Fetting, "The European green deal," in *ESDN Report*, December 2020, vol. 2.

[6] "Energy balance sheets: 2017 Data," Luxembourg: Publications Office of the European Union, Statistical Books, August 2019.

[7] G. Rancilio, A. Rossi, D. Falabretti, A. Galliani, and M. Merlo, "Ancillary services markets in europe: Evolution and regulatory trade-offs," *Renewable and Sustainable Energy Reviews*, vol. 154, p. 111850, 2022.

[8] L. J. Ossowska and D. A. Janiszewska, "Toward sustainable energy consumption in the European Union," *Polityka Energetyczna*, vol. 23, 2020.

[9] C. Eid, P. Codani, Y. Perez, J. Reneses, and R. Hakvoort, "Managing electric flexibility from distributed energy resources: A review of incentives for market design," *Renewable and Sustainable Energy Reviews*, vol. 64, pp. 237–247, 2016.

[10] K. L. Anaya and M. G. Pollitt, "Integrating distributed generation: Regulation and trends in three leading countries," *Energy Policy*, vol. 85, pp. 475–486, 2015.

[11] A. Tavakoli, A. Karimi, and M. Shafie-khah, "Optimal probabilistic operation of energy hub with various energy converters and electrical storage based on electricity, heat, natural gas, and biomass by proposing innovative uncertainty modeling methods," *Journal of Energy Storage*, vol. 51, p. 104344, 2022.

[12] R. Newell, D. Raimi, S. Villanueva, and B. Prest, "Global energy outlook 2021: Pathways from Paris," *Resources for the Future*, vol. 8, p. 39, 2021.

[13] "World Energy Outlook 2022, an updated roadmap to net zero emissions by 2050," in *Licence: CC BY 4.0 (report); CC BY NC SA 4.0 (Annex A)*, International Energy Agency, IEA, Paris, May 2021, www.iea.org/reports/world-energy-outlook-2022 (Accessed August 1, 2024).

[14] I. Sorrenti, T. B. H. Rasmussen, S. You, and Q. Wu, "The role of power-to-X in hybrid renewable energy systems: A comprehensive review," *Renewable and Sustainable Energy Reviews*, vol. 165, p. 112380, 2022.

[15] A. Banshwar, N. K. Sharma, Y. R. Sood, and R. Shrivastava, "Renewable energy sources as a new participant in ancillary service markets," *Energy Strategy Reviews*, vol. 18, pp. 106–120, 2017.

[16] K. Oureilidis, K.-N. Malamaki, K. Gallos, A. Tsitsimelis, C. Dikaiakos, S. Gkavanoudis, M. Cvetkovic, J. M. Mauricio, J. M. Maza Ortega, J. L. M. Ramos, and G. Papaioannou, "Ancillary services market design in distribution networks: Review and identification of barriers," *Energies*, vol. 13, no. 4, p. 917, 2020.

[17] S. Chondrogiannis, M. Poncela-Blanco, A. Marinopoulos, I. Marneris, A. Ntomaris, P. Biskas, and A. Bakirtzis, "Power system flexibility: A methodological analytical framework based on unit commitment and economic dispatch modelling," in *Mathematical Modelling of Contemporary Electricity Markets*. Elsevier, 2021, pp. 127–156.

[18] G. Papaefthymiou and K. Dragoon, "Towards 100% renewable energy systems: Uncapping power system flexibility," *Energy Policy*, vol. 92, pp. 69–82, 2016.

[19] E. M. Carlini, R. Schroeder, J. M. Birkebæk, and F. Massaro, "EU transition in power sector: How RES affects the design and operations of transmission power systems," *Electric Power Systems Research*, vol. 169, pp. 74–91, 2019.

[20] J. Lowitzsch, C. E. Hoicka, and F. J. van Tulder, "Renewable energy communities under the 2019 European Clean Energy Package–Governance model for the energy clusters of the future?" *Renewable and Sustainable Energy Reviews*, vol. 122, p. 109489, 2020.

[21] L. Ollier, F. Metz, A. Nuñez-Jimenez, L. Späth, and J. Lilliestam, "The European 2030 climate and energy package: Do domestic strategy adaptations precede EU policy change?" *Policy Sciences*, vol. 55, no. 1, pp. 161–184, 2022.

[22] O. Borne, K. Korte, Y. Perez, M. Petit, and A. Purkus, "Barriers to entry in frequency-regulation services markets: Review of the status quo and options for improvements," *Renewable and Sustainable Energy Reviews*, vol. 81, pp. 605–614, 2018.

[23] F. Gulotta, E. Daccò, A. Bosisio, and D. Falabretti, "Opening of ancillary service markets to distributed energy resources: A review," *Energies*, vol. 16, no. 6, p. 2814, 2023.

[24] "Dispersed generation impact on ce region security," in *Dynamic Study*, European Network of Transmission System Operators for Electricity, Brussels, Belgium, 2014, www.entsoe.eu/Documents/ Publications/SOC/Continental_Europe/141113_Dispersed_Generation_Impact_on_Continental_ Europe_Region_Security.pdf (Accessed August 1, 2024).

[25] "Commission Regulation (EU) 2016/631-Establishing a network code on requirements for grid connection of generators," Official Journal of the European Union, April 14, 2016, https://eur-lex.europa.eu/ legal-content/EN/TXT/PDF/?uri=CELEX:32016R0631&from=ES (Accessed August 1, 2024).

[26] C. Zhang, Y. Ding, N. C. Nordentoft, P. Pinson, and J. Østergaard, "FLECH: A Danish market solution for DSO congestion management through DER flexibility services," *Journal of Modern Power Systems and Clean Energy*, vol. 2, no. 2, pp. 126–133, 2014.

[27] M. Arosio and D. Falabretti, "DER participation in ancillary services market: An analysis of current trends and future opportunities," *Energies*, vol. 16, no. 5, p. 2443, 2023.

[28] K. De Vos, N. Stevens, O. Devolder, A. Papavasiliou, B. Hebb, and J. Matthys-Donnadieu, "Dynamic dimensioning approach for operating reserves: Proof of concept in Belgium," *Energy Policy*, vol. 124, pp. 272–285, 2019.

[29] R. A. Van der Veen and R. A. Hakvoort, "The electricity balancing market: Exploring the design challenge," *Utilities Policy*, vol. 43, pp. 186–194, 2016.

[30] T. Schittekatte, L. Meeus, T. Jamasb, and M. Llorca, "Regulatory experimentation in energy: Three pioneer countries and lessons for the green transition," *Energy Policy*, vol. 156, p. 112382, 2021.

[31] A. Banshwar, N. K. Sharma, Y. Sood, and R. Shrivastava, "An investigation of the various approaches for competitive procurement of Ancillary Services in deregulated power sector," in *2016 International Conference on Electrical Power and Energy Systems (ICEPES)*, 2016, pp. 492–497. IEEE.

[32] A. Zecchino, K. Knezović, and M. Marinelli, "Identification of conflicts between transmission and distribution system operators when acquiring ancillary services from electric vehicles," in *2017 IEEE PES Innovative Smart Grid Technologies Conference Europe (ISGT-Europe)*, 2017, pp. 1–6. IEEE.

[33] V. G. Suárez, J. L. R. Torres, B. W. Tuinema, A. P. Guerra, and M. A. M. M. v. d. Meijden, "Integration of power-to-gas conversion into Dutch electrical ancillary services markets," 2018.

[34] S. Borenstein and J. Bushnell, "Electricity restructuring: Deregulation or reregulation," *Regulation*, vol. 23, p. 46, 2000.

[35] V. Thomasi, J. C. M. Siluk, P. D. Rigo, and C. A. D. O. Pappis, "Challenges, improvements, and opportunities market with the liberalization of the residential electricity market," *Energy Policy*, vol. 192, p. 114253, 2024.

[36] R. Faia, F. Lezama, J. Soares, T. Pinto, and Z. Vale, "Local electricity markets: A review on benefits, barriers, current trends and future perspectives," *Renewable and Sustainable Energy Reviews*, vol. 190, p. 114006, 2024.

[37] D. Asija and R. Viral, "Renewable energy integration in modern deregulated power system: Challenges, driving forces, and lessons for future road map," in *Advances in Smart Grid Power System*. Elsevier, 2021, pp. 365–384.

[38] F. Billimoria, P. Mancarella, and R. Poudineh, "Market and regulatory frameworks for operational security in decarbonizing electricity systems: From physics to economics," *Oxford Open Energy*, vol. 1, p. oiac007, 2022.

[39] S. Panda, S. Mohanty, P. K. Rout, B. K. Sahu, S. M. Parida, I. S. Samanta, M. Bajaj, M. Piecha, V. Blazek, and L. Prokop, "A comprehensive review on demand side management and market design for renewable energy support and integration," *Energy Reports*, vol. 10, pp. 2228–2250, 2023.

[40] H. Khajeh, H. Firoozi, H. Laaksonen, and M. Shafie-Khah, "Business models for different future electricity market players," in *Deregulated Electricity Structures and Smart Grids*. CRC Press, 2022, pp. 63–81.

[41] D. Ribó-Pérez, L. Larrosa-López, D. Pecondón-Tricas, and M. Alcázar-Ortega, "A critical review of demand response products as resource for ancillary services: International experience and policy recommendations," *Energies*, vol. 14, no. 4, p. 846, 2021.

[42] P. Cappers, J. MacDonald, C. Goldman, and O. Ma, "An assessment of market and policy barriers for demand response providing ancillary services in US electricity markets," *Energy Policy*, vol. 62, pp. 1031–1039, 2013.

[43] "Net electricity consumption worldwide in select years from 1980 to 2022," in *Energy & Environment*, Statista, www.statista.com/statistics/280704/world-power-consumption/ (Accessed August 1, 2024).

[44] J. A. Villanueva-Rosario, F. Santos-García, M. E. Aybar-Mejía, P. Mendoza-Araya, and A. Molina-García, "Coordinated ancillary services, market participation and communication of multi-microgrids: A review," *Applied Energy*, vol. 308, p. 118332, 2022.

[45] L. Olmos and I. J. Pérez-Arriaga, "Regional markets," in *Regulation of the Power Sector*, I. J. Pérez-Arriaga, Ed. London: Springer London, 2013, pp. 501–538.

[46] *Renewable Energy Market Update-Outlook for 2023 and 2024*. International Energy Agency (IEA), June 2023, https://iea.blob.core.windows.net/assets/63c14514-6833-4cd8-ac53-f9918c2e4cd9/RenewableEnergyMarketUpdate_June2023.pdf (Accessed August 1, 2024).

[47] M. H. Andishgar, A. Fereidunian, and H. Lesani, "Healer reinforcement for smart grid using discrete event models of FLISR in distribution automation," *Journal of Intelligent & Fuzzy Systems*, vol. 30, pp. 2939–2951, 2016.

[48] M. Kiehbadroudinezhad, A. Merabet, A. G. Abo-Khalil, T. Salameh, and C. Ghenai, "Intelligent and optimized microgrids for future supply power from renewable energy resources: A review," *Energies*, vol. 15, no. 9, p. 3359, 2022.

[49] P. Rani, V. Parkash, and N. K. Sharma, "Technological aspects, utilization and impact on power system for distributed generation: A comprehensive survey," *Renewable and Sustainable Energy Reviews*, vol. 192, p. 114257, 2024.

[50] S. Dawn, A. Ramakrishna, M. Ramesh, S. S. Das, K. Dhananjay Rao, Md M. Islam, and T. Selim Ustun, "Integration of renewable energy in microgrids and smart grids in deregulated power systems: A comparative exploration," *Advanced Energy and Sustainability Research*, p. 2400088, 2024.

[51] M. Mohammadi and A. Mohammadi, "Empowering distributed solutions in renewable energy systems and grid optimization," in *Distributed Machine Learning and Computing: Theory and Applications*. Springer, 2024, pp. 141–155.

[52] S. Ullah, A. M. A. Haidar, P. Hoole, H. Zen, and T. Ahfock, "The current state of Distributed Renewable Generation, challenges of interconnection and opportunities for energy conversion based DC microgrids," *Journal of Cleaner Production*, vol. 273, p. 122777, 2020.

[53] H. O. Shami, A. Basem, A. H. Al-Rubaye, and K. Sabzevari, "A novel strategy to enhance power management in AC/DC hybrid microgrid using virtual synchronous generator based interlinking converters integrated with energy storage system," *Energy Reports*, vol. 12, pp. 75–94, 2024.

[54] S. Sharma and Y. R. Sood, "Microgrids: A review of status, technologies, software tools, and issues in indian power market," *IETE Technical Review*, vol. 39, pp. 411–432, 2020.

[55] R. Sepehrzad, A. Hedayatnia, M. Amohadi, J. Ghafourian, A. Al-Durra, and A. Anvari-Moghaddam, "Two-stage experimental intelligent dynamic energy management of microgrid in smart cities based on demand response programs and energy storage system participation," *International Journal of Electrical Power & Energy Systems*, vol. 155, p. 109613, 2024.

[56] M. R. Khan, Z. M. Haider, F. H. Malik, F. M. Almasoudi, K. S. S. Alatawi, and M. S. Bhutta, "A comprehensive review of microgrid energy management strategies considering electric vehicles, energy storage systems, and AI techniques," *Processes*, vol. 12, no. 2, p. 270, 2024.

[57] A. Meydani, H. Shahinzadeh, H. Nafisi, and G. B. Gharehpetian, "Optimizing microgrid energy management: Metaheuristic versus conventional techniques," in *2024 11th Iranian Conference on Renewable Energy and Distribution Generation (ICREDG)*, 2024, vol. 11, pp. 1–15. IEEE.

[58] H. R. Karimi and S. Jadid, "Optimal energy management for multi-microgrid considering demand response programs: A stochastic multi-objective framework," *Energy*, vol. 195, p. 116992, 2020.

[59] M. H. Imani, M. J. Ghadi, S. Ghavidel, and L. Li, "Demand response modeling in microgrid operation: A review and application for incentive-based and time-based programs," *Renewable and Sustainable Energy Reviews*, vol. 94, pp. 486–499, 2018.

[60] A. A. Khodadoost Arani, G. B. Gharehpetian, and M. Abedi, "Decentralised primary and secondary control strategies for islanded microgrids considering energy storage systems characteristics," *IET Generation, Transmission & Distribution*, vol. 13, pp. 2986–2992, 2019.

[61] M. Uddin, M. F. Romlie, M. F. Abdullah, C. Tan, G. M. Shafiullah, and A. H. A. Bakar, "A novel peak shaving algorithm for islanded microgrid using battery energy storage system," *Energy*, vol. 196, p. 117084, 2020.

[62] M. M. Hayati, B. Motallebi Azar, A. Aminlou, M. Abapour, and K. Zare, "Techno-economic analysis for centralized GH_2 power systems," in *Green Hydrogen in Power Systems*, V. Vahidinasab, B. Mohammadi-Ivatloo, and J. Shiun Lim, Eds. Cham: Springer International Publishing, 2024, pp. 59–83.

[63] A. Aminlou, M. M. Hayati, and K. Zare, "P2P energy trading in a community of individual consumers with the presence of central shared battery energy storage," in *Demand-Side Peer-to-Peer Energy Trading*, V. Vahidinasab and B. Mohammadi-Ivatloo, Eds. Cham: Springer International Publishing, 2023, pp. 143–159.

[64] M. Daneshvar, B. Mohammadi-ivatloo, M. Abapour, and S. Asadi, "Energy exchange control in multiple microgrids with transactive energy management," *Journal of Modern Power Systems and Clean Energy*, vol. 8, pp. 719–726, 2020.

[65] M. Daneshvar, B. Mohammadi-Ivatloo, and K. Zare, "An innovative transactive energy architecture for community microgrids in modern multi-carrier energy networks: A Chicago case study," *Scientific Reports*, vol. 13, no. 1, p. 1529, 2023.

[66] B. Chen, J. Wang, X. Lu, C. Chen, and S. Zhao, "Networked microgrids for grid resilience, robustness, and efficiency: A review," *IEEE Transactions on Smart Grid*, vol. 12, pp. 18–32, 2021.

[67] Y. Han, K. Zhang, H. Li, E. A. A. Coelho, and J. M. Guerrero, "MAS-based distributed coordinated control and optimization in microgrid and microgrid clusters: A comprehensive overview," *IEEE Transactions on Power Electronics*, vol. 33, pp. 6488–6508, 2018.

[68] L. Wu, J. Li, M. Erol-Kantarci, and B. Kantarci, "An integrated reconfigurable control and self-organizing communication framework for community resilience microgrids," *The Electricity Journal*, vol. 30, pp. 27–34, 2017.

[69] M. V. Bram, J. Liniger, S. S. Majidabad, H. R. Shabani, M. P. Teles, and X. Cui, "Challenges in power-to-X: A perspective of the configuration and control process for E-methanol production," *International Journal of Hydrogen Energy*, vol. 76, pp. 315–325, 2024.

[70] M. J. Palys and P. Daoutidis, "Power-to-X: A review and perspective," *Computers & Chemical Engineering*, vol. 165, p. 107948, 2022.

[71] V. Shahbazbegian, M. Shafie-khah, H. Laaksonen, G. Strbac, and H. Ameli, "Resilience-oriented operation of microgrids in the presence of power-to-hydrogen systems," *Applied Energy*, vol. 348, p. 121429, 2023.

[72] M. M. Hayati, A. Safari, M. Nazari-Heris, and A. Oshnoei, "Hydrogen-incorporated sector-coupled smart grids: A systematic review and future concepts," in *Green Hydrogen in Power Systems*, V. Vahidinasab, B. Mohammadi-Ivatloo, and J. Shiun Lim, Eds. Cham: Springer International Publishing, 2024, pp. 25–58.

[73] Report prepared for Food & Bio Cluster Denmark with financial support from the Danish Agency for Higher Education and Science. Aalborg University, AAU Energy. 2022, https://vbn.aau.dk/files/514146100/PtX_Report.pdf (Accessed December 15, 2024).

[74] M. Jensterle, J. Narita, R. Piria, S. Samadi, M. Prantner, K. Crone, S. Siegemund, S. Kan, T. Matsumoto, Y. Shibata, and J. Thesen, "The role of clean hydrogen in the future energy systems of Japan and Germany: An analysis of existing mid-century scenarios and an investigation of hydrogen supply chains," 2019.

[75] A.-K. Jannasch, H. Pihl, M. Persson, E. Svensson, S. Harvey, and H. Wiertzema, "Opportunities and barriers for implementation of Power-to-X (P2X) technologies in the West Sweden Chemicals and Materials Clusters process industries," Lund, 2020.

[76] L. Trevino, *Liberalization of the electricity market in Europe: An overview of the electricity technology and the market place*. Lausanne, Switzerland: College of Management of Technology, The École Polytechnique Fédérale de Lausanne (EPFL), 2008.

[77] P. L. Joskow, "Introduction to electricity sector liberalization: Lessons learned from cross-country studies," *Electricity Market Reform: An International Perspective*, vol. 1, pp. 1–32, 2006.

[78] D. Ngondya and J. Mwangoka, "Demand–supply equilibrium in deregulated electricity markets for future smartgrid," *Cogent Engineering*, vol. 4, no. 1, p. 1392410, 2017.

[79] B. Kirby, "Ancillary services: Technical and commercial insights," *Retrieved October*, vol. 4, p. 2012, 2007.

[80] F. Feijoo, A. Pfeifer, L. Herc, D. Groppi, and N. Duić, "A long-term capacity investment and operational energy planning model with power-to-X and flexibility technologies," *Renewable and Sustainable Energy Reviews*, vol. 167, p. 112781, 2022.

[81] P. C. Bhagwat, S. Bhagwat, and A. Parashar, "Expert survey on energy storage systems: Regulation and policy from an Indian power sector perspective," *The Electricity Journal*, vol. 34, no. 9, p. 107026, 2021.

[82] I. Marneris, A. V. Ntomaris, P. N. Biskas, C. G. Baslis, D. I. Chatzigiannis, C. S. Demoulias, K. O. Oureilidis, and A. G. Bakirtzis, "Optimal participation of RES aggregators in energy and ancillary services markets," *IEEE Transactions on Industry Applications*, vol. 59, no. 1, pp. 232–243, 2022.

[83] J. Salehi, A. Namvar, F. S. Gazijahani, M. Shafie-khah, and J. P. S. Catalão, "Effect of power-to-gas technology in energy hub optimal operation and gas network congestion reduction," *Energy*, vol. 240, p. 122835, 2022.

[84] M. Bajus, *Converting Power into Chemicals and Fuels: Power-to-X Technology for a Sustainable Future*. John Wiley & Sons, 2023.

[85] A. R. Dahiru, A. Vuokila, and M. Huuhtanen, "Recent development in power-to-X: Part I-A review on techno-economic analysis," *Journal of Energy Storage*, vol. 56, p. 105861, 2022.

[86] Z. M. Shoja, A. B. Oskouei, and M. Nazari-Heris, "Optimal scheduling of a community multi-energy system in energy and flexible ramp markets considering vector-coupling storage devices: A hybrid fuzzy-IGDT/stochastic/robust optimization framework," *Energy and Buildings*, vol. 318, p. 114465, 2024.

[87] J. Ramsebner, R. Haas, A. Ajanovic, and M. Wietschel, "The sector coupling concept: A critical review," *Wiley Interdisciplinary Reviews: Energy and Environment*, vol. 10, no. 4, p. e396, 2021.

[88] M. A. Mirzaei, M. Habibi, V. Vahidinasab, and B. Mohammadi-Ivatloo, *Power-to-Gas: Bridging the Electricity and Gas Networks*. Academic Press, 2023.

[89] M. Lappalainen, "Techno-economic feasibility of hydrogen production via polymer membrane electrolyte electrolysis for future power-to-X systems," 2019.

[90] J. Ogden, A. M. Jaffe, D. Scheitrum, Z. McDonald, and M. Miller, "Natural gas as a bridge to hydrogen transportation fuel: Insights from the literature," *Energy Policy*, vol. 115, pp. 317–329, 2018.

[91] "IRENA, innovation landscape for a renewable-powered future: Solutions to integrate variable renewables," 2019, www.irena.org/-/media/Files/IRENA/Agency/Publication/2019/Feb/IRENA_Innovation_Landscape_2019_report.pdf (Accessed August 2, 2024).

[92] L. Engstam, "Power-to-X-to-power in combined cycle power plants: A techno-economic feasibility study," Degree Project in Sustainable Energy Engineering, Master of Science Thesis, KTH, School of Industrial Engineering and Management (ITM), Stockholm, Sweden, 2021.

[93] E. Erdem, "Why P2X must be the part of the energy solution?," *Environmental Progress & Sustainable Energy*, vol. 40, no. 3, p. e13545, 2021.

[94] M. A. Mirzaei, M. Habibi, V. Vahidinasab, and B. Mohammadi-Ivatloo, "Chapter 1: Whole system approach to energy," in *Power-to-Gas*, M. A. Mirzaei, M. Habibi, V. Vahidinasab, and B. Mohammadi-Ivatloo, Eds. Academic Press, 2023, pp. 1–14.

[95] A. Risco-Bravo, C. Varela, J. Bartels, and E. Zondervan, "From green hydrogen to electricity: A review on recent advances, challenges, and opportunities on power-to-hydrogen-to-power systems," *Renewable and Sustainable Energy Reviews*, vol. 189, p. 113930, 2024.

[96] G. Fambri, C. Diaz-Londono, A. Mazza, M. Badami, T. Sihvonen, and R. Weiss, "Techno-economic analysis of power-to-gas plants in a gas and electricity distribution network system with high renewable energy penetration," *Applied Energy*, vol. 312, p. 118743, 2022.

[97] V. Garcia Suarez, "Exploitation of power-to-gas for ancillary services provision (within the context of synergy action TSO 2020)," Master Thesis, Faculty of Electrical Engineering, Mathematics and Computer Science, TU Delft, Holland, 2018.

[98] M. A. Mirzaei, K. Zare, B. Mohammadi-Ivatloo, M. Marzband, and A. Anvari-Moghaddam, "Robust network-constrained energy management of a multiple energy distribution company in the presence of multi-energy conversion and storage technologies," *Sustainable Cities and Society*, vol. 74, p. 103147, 2021.

[99] H. Khaloie, A. Abdollahi, M. Shafie-Khah, P. Siano, S. Nojavan, A. Anvari-Moghaddam, and J. P. S. Catalão, "Co-optimized bidding strategy of an integrated wind-thermal-photovoltaic system in deregulated electricity market under uncertainties," *Journal of Cleaner Production*, vol. 242, p. 118434, 2020.

[100] O. Sadeghian, A. M. Shotorbani, S. Ghassemzadeh, and B. Mohammadi-Ivatloo, "Energy management of hybrid fuel cell and renewable energy based systems: A review," *International Journal of Hydrogen Energy*, 2024.

[101] M. S. Ahmad, M. S. Ali, and N. Abd Rahim, "Hydrogen energy vision 2060: Hydrogen as energy carrier in Malaysian primary energy mix–developing P2G case," *Energy Strategy Reviews*, vol. 35, p. 100632, 2021.

[102] M. A. Lasemi and A. Arabkoohsar, "Participation of high-temperature heat and power storage system coupled with a wind farm in energy market," in *2020 International Conference on Smart Energy Systems and Technologies (SEST)*, 2020, pp. 1–5. IEEE.

[103] R. Zhang, T. Jiang, F. Li, G. Li, H. Chen, and X. Li, "Coordinated bidding strategy of wind farms and power-to-gas facilities using a cooperative game approach," *IEEE Transactions on Sustainable Energy*, vol. 11, no. 4, pp. 2545–2555, 2020.

[104] A. Maroufmashat and M. Fowler, "Transition of future energy system infrastructure; through power-to-gas pathways," *Energies*, vol. 10, no. 8, p. 1089, 2017.

[105] A. Task, S. Reuter, R.-R. Schmidt, N. Marx, and P. Ortmann, "IEA HPT ANNEX 57/Task 2: Potential future and conventional waste heat sources," 2022, https://heatpumpingtechnologies.org/annex57/wp-content/uploads/sites/69/2023/05/annex57-task-12-new-heat-pump-sources-v095.pdf (Accessed August 2, 2024).

[106] "The European Association for storage of energy-windgas for the energy transition with uniper's power-to-gas plant," 2018, https://ease-storage.eu/news/windgas-for-the-energy-transition-with-unipers-power-to-gas-plant-in-falkenhagen/ (Accessed August 2, 2024).

[107] F. Heymann, M. Rüdisüli, F. vom Scheidt, and A. S. Camanho, "Performance benchmarking of power-to-gas plants using composite indicators," *International Journal of Hydrogen Energy*, vol. 47, no. 58, pp. 24465–24480, 2022.

[108] B. Xiong, J. Predel, P. C. del Granado, and R. Egging-Bratseth, "Spatial flexibility in redispatch: Supporting low carbon energy systems with Power-to-Gas," *Applied Energy*, vol. 283, p. 116201, 2021.

[109] "The first industrial PtG plant – Audi e-gas as driver for the energy turnaround," in "Reinhard Otten|Sustainable Product Development|AUDI AG," Verona27th of May 2014, www.cedec.com/files/default/8-2014-05-27-cedec-gas-day-reinhard-otten-audi-ag.pdf (Accessed August 2, 2024).

[110] J. D. Ngando Ebba, M. B. Camara, M. L. Doumbia, B. Dakyo, and J. Song-Manguelle, "Large-scale hydrogen production systems using marine renewable energies: State-of-the-art," *Energies*, vol. 17, no. 1, p. 130, 2023.

[111] H. Elsheikh and V. Eveloy, "Assessment of variable solar-and grid electricity-driven power-to-hydrogen integration with direct iron ore reduction for low-carbon steel making," *Fuel*, vol. 324, p. 124758, 2022.

[112] "Green industrial hydrogen," https://salcos.salzgitter-ag.com/en/grinhy-20.html (Accessed August 2, 2024).

[113] C. S. Duncan, "Techno-economical modeling of a PtG plant for operational optimization in the context of gas grid injection in France," Master Thesis Report, KTH, School of Engineering Sciences in Chemistry, Biotechnology and Health (CBH), Chemical Engineering, Stockholm, Sweden, 2020.

[114] A. M. Abomazid, "Modeling and control of electrolysis based hydrogen production system," 2021.

[115] F. Brissaud, A. Chaise, K. Gault, and S. Soual, "Lessons learned from Jupiter 1000, an industrial demonstrator of Power-to-Gas," *International Journal of Hydrogen Energy*, vol. 49, pp. 925–932, 2024.

[116] M. Liao, C. Liu, and Z. Qing, "A recent overview of power-to-gas projects," in *2020 IEEE 4th Conference on Energy Internet and Energy System Integration (EI2)*, 2020, pp. 2282–2286. IEEE.

[117] R. Schlautmann, H. Böhm, A. Zauner, F. Mörs, R. Tichler, F. Graf, and T. Kolb, "Renewable Power-to-Gas: A technical and economic evaluation of three demo sites within the STORE&GO project," *Chemie Ingenieur Technik*, vol. 93, no. 4, pp. 568–579, 2021.

[118] S. Haghifam, H. Laaksonen, and M. Shafie-khah, "A market-based mechanism for local energy trading in integrated electricity-heat networks," in *Trading in Local Energy Markets and Energy Communities: Concepts, Structures and Technologies*, M. Shafie-khah and A. S. Gazafroudi, Eds. Cham: Springer International Publishing, 2023, pp. 241–261.

19 Strategies for Profit Maximization of Power-to-Gas Utilities within Carbon Emission Trading Market

Rufeng Zhang, Tao Jiang, Fangxing Li, Guoqing Li, Houhe Chen, and Xue Li

Nomenclature
Indices

i/j	index of buses
l	index of transmission lines
M/N	number of transmission lines/buses
T	number of time periods
t	index of time periods
w	index of wind farms
s	index of scenarios
S	number of scenarios
Θ_{P2G}/Θ_G	sets of buses with P2G facilities/thermal units

Parameters

ε	carbon emission permit factor (tCO$_2$/MW)
α_i/α_j	allocated factor of carbon emission permits of thermal units at buses i and j
α_{P2G}	carbon emission reduction factor of P2G (tCO$_2$/MW)
c_i/c_{wi}	bidding price of thermal unit/wind power at bus i
cp	carbon emission trading price ($/tCO$_2$)
$D_{i,t}$	load at bus i (MW)
e_i	carbon emission factor of generation unit i (tCO$_2$/MW)
Gmax i/Gmin i	upper/lower limits of generation (MW)
$GSF_{l\text{-}i}$	generation shift factor
$Limit_l$	transmission line flow limit of line l (MW)
Pmax P2G,i/Pmin P2G,i	hourly maximum/minimum operating power of P2G at bus i (MW)
Pmax w,i,t,s	maximum output of wind power units at bus i at period t of scenario s
pr gas	gas price of natural gas market at period t
η	energy conversion efficiency of P2G facility

DOI: 10.1201/9781032719436-19

Variables

Ψ	Lagrangian function
$\pi_{i,t,s}$	LMP of bus i at period t of scenario s
$CR_{P2G,i,t}$	reduced carbon emission of P2G facility of bus i at period t (tCO$_2$)
D^*i,t	total load, including power consumption, of P2G (MW)
$E_{i,t}$	difference between actual emission and permit of thermal unit
$Eto\ i,t$	actual emission of thermal unit of bus i at period t (tCO$_2$)
$F_{CD,i}$	carbon trading cost of thermal unit of bus i (\$)
$G_{i,t,s}$	power output of thermal unit of bus i at period t of scenario s (MW)
$PE_{i,t}$	carbon emission permit of thermal unit of bus i at period t (tCO$_2$)
PF^*P2G/PF_{P2G}	profit of P2G facility
$P_{P2G,i,t}$	power consumption of P2G facility of bus i at period t (MW)
$P_{w,i,t,s}$	wind power output of bus i at period t of scenario s (MW)
$S_{P2G,j,t}$	volume of SNG (kcf)

19.1 INTRODUCTION

19.1.1 Introduction to Profit Maximization Strategies for Power-to-Gas Utilities

P2G technology produces synthetic natural gas (SNG) consuming electric energy, which can enhance the flexibility of integrated electricity and natural gas systems [1]. In [2], the authors proposed a collaborative planning method for P2G and gas turbines to minimize the total operating cost of IES. In addition, an effective P2G bidding strategy can bring more benefits to IES. In order to allocate market interests, reference [3] proposed a bidding strategy based on cooperative game. A distributed supply coordination model for P2G facilities in energy grids to maximize the profit of P2G facilities was proposed in [4].

19.1.2 Understanding the Carbon Emission Trading Market and Its Impact on Profitability

In recent years, a carbon emission trading scheme (CETS) has been implemented in several countries [5], in which carbon emission permits can be traded at a price in the carbon market. The influence of carbon dioxide trading on generators' economic dispatch (ED) is analyzed in [6]. In [7], a unit commitment and dispatch model in microgrids considering demand response and CETS was developed, and the results demonstrated that low-emission permits and high-emission trading prices could reduce total carbon emission. In [8], the unit commitment problem considering the CETS was proposed in a smart grid environment. In [9], an approach to incorporate the effects of the CETS planning of distribution networks was proposed. In [10], the impact on allowance price and the shift of emissions across sectors were studied under a cap-and-trade system. At the same time, P2G facilities utilize carbon to produce SNG. If P2G facilities can participate in the carbon market and sell carbon permits by purchasing electricity power to produce SNG-consuming carbon, the profit of P2G facilities may increase.

19.1.3 The Role of Power-to-Gas Utilities in a Low-Carbon Economy

In the context of the global transition to a low-carbon economy, the development of the low-carbon power sector has important economic significance [11]. Hence, renewable energy penetration, including both wind and photovoltaic power generation, has been increasing rapidly in power systems. Wind power is one of the most important renewable generations in power systems, albeit volatile with significant uncertainties [12]. Due to the uncertainties and power network operating constraints, wind power curtailment occurs [13]. P2G technologies convert excess renewable energy into storable and transportable gases [14], which can then be used in heating and power generation. As the demand for renewable energy continues to grow, P2G utilities have the potential to play a significant role in balancing the intermittent nature of renewable energy sources and enhancing energy security.

19.1.4 KEY CHALLENGES AND OPPORTUNITIES FOR PROFIT MAXIMIZATION IN POWER-TO-GAS UTILITIES

Although P2G equipment consumes electrical energy and can reshape the load curve of the power system, the SNG generated by P2G facilities can be stored and utilized in the gas transmission network, making the two systems more closely coupled. However, the current operation economy of P2G facilities is not feasible, and very little research has focused on the economic operation of P2G facilities. The feasibility of P2G in electricity markets dominated by renewables has been analyzed in [15]. Day-ahead scheduling of P2G energy storage in wholesale electricity and gas markets was studied in [16]. However, the market clearing process and the impact of P2G technology on the low-carbon economy embedded operation of power systems have not been effectively considered in the aforementioned works. In [17], the combination of P2G facilities with natural gas generating units to utilize surplus renewables and reduce carbon emission has been presented as promising. At the same time, P2G facilities utilize carbon to produce SNG. If P2G facilities can participate in the carbon market and sell carbon permits by purchasing electricity power to produce SNG-consuming carbon, the profit of P2G facilities may increase.

19.1.5 FRAMEWORK FOR ENHANCING PROFITABILITY IN POWER-TO-GAS SYSTEMS

To enhance profitability in P2G systems, a framework for profit maximization is needed. This framework should consider the various factors that influence the profitability of P2G utilities, including market dynamics, regulatory frameworks, and technological advancements. By understanding these factors and developing strategies to address them, P2G utilities can maximize their profits and contribute effectively to the low-carbon transition.

19.1.6 OVERVIEW OF MARKET DYNAMICS AFFECTING POWER-TO-GAS PROFITABILITY

The profitability of P2G is affected by a series of market dynamics. In [18], the dynamic changes of renewable energy, natural gas prices, and carbon quota prices are considered, and a low-carbon economic dispatch model for integrated energy systems with P2G and carbon trading mechanisms to improve energy efficiency and minimize carbon emissions is established. In [19], the price of renewable energy can affect the cost of producing hydrogen and other gases through P2G technology, and the price of natural gas can also affect the demand for P2G production gas as an alternative fuel source. At the same time, the price of carbon quotas can provide financial incentives for emission reduction and the sale of excess quotas [20].

19.1.7 THE IMPORTANCE OF PROFIT MAXIMIZATION IN THE TRANSITION TO SUSTAINABLE ENERGY

With the deepening of energy transformation, the development of integrated energy with electricity as the core is an inevitable trend in the future [21], and profit maximization is crucial for P2G to effectively promote the transformation of traditional energy to sustainable energy. In reference [22], P2G devices attract investment and maximize profits by reducing their own operating costs and improving energy efficiency. In reference [23], due to the current intensification of the transition from traditional energy to sustainable energy, this, in turn, can make P2G technology more competitive and increase its adoption in the energy field.

19.1.8 Metrics for Profit Maximization in Power-to-Gas Utilities

To measure and optimize profitability in P2G utilities, a range of metrics can be used. These metrics can include financial indicators, such as revenue, profit margins, and return on investment, as well as operational indicators, such as energy efficiency, emissions reductions, and cost savings. By tracking and analyzing these metrics, P2G utilities can identify areas for improvement and develop strategies to maximize profits.

19.2 STRATEGIC BIDDING FOR POWER-TO-GAS UTILITIES IN ENERGY MARKETS

19.2.1 Concept Definition

A novel bi-level strategic bidding model for P2G in the electricity market considering the CETS-embedded LMP (CETS-LMP) is proposed here. The framework of the proposed bidding strategy is shown in Figure 19.1. Table 19.1 summarizes the state of the art of the problem addressed.

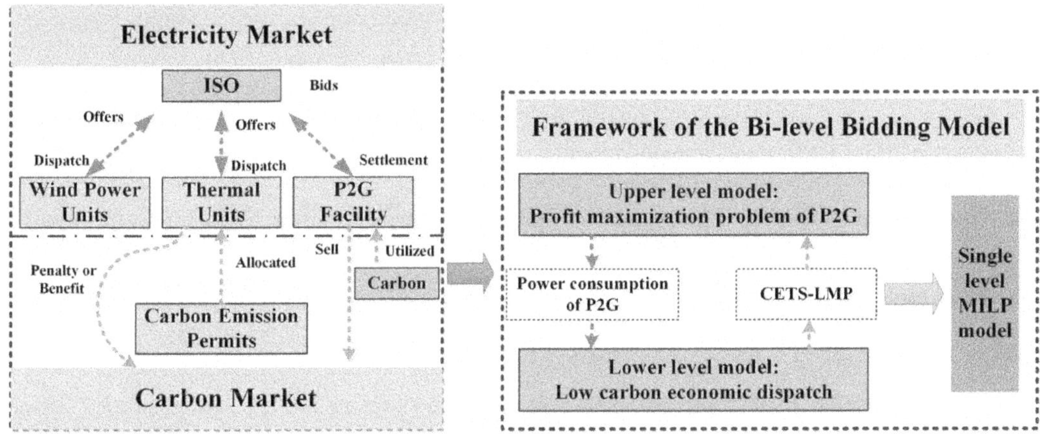

FIGURE 19.1 Framework of the proposed bidding strategy.

TABLE 19.1
Overview of the State of the Art

References	Optimal Operation of P2G Facilities	Bi-Level Model	Low-Carbon Economy	CETS	LMP Computation	Bidding of P2G Facilities
[11]			✓			
[22]	✓					✓
[1, 24, 15, 16, 4]	✓					
[3, 24]	✓					✓
[17]	✓		✓			
[5–10]				✓		
[25]		✓	✓	✓		
[26–28]		✓			✓	
Current Chapter	✓	✓	✓	✓	✓	✓

Thermal generation units and wind power units submit offers, and a P2G facility submits bids to an independent system operator (ISO). The ISO carries out the low-carbon ED considering the CETS and obtains dispatch decisions and the CETS-LMP. The carbon emission permits are allocated among thermal units, and P2G facilities are assumed to participate in the carbon market to sell carbon permits (i.e., the amount of utilized carbon). Then, the bi-level bidding model is formulated.

1. To increase the profit and motivate investment in P2G facilities, a novel bi-level optimization strategic bidding model for P2G considering CETS-LMP and wind power uncertainty is proposed, which is solved by the mathematical programming with equilibrium constraints (MPEC) technique.
2. Under the proposed strategy for a low-carbon economy, P2G facilities can strategically purchase electricity to maximize their profit according to the proposed CETS-LMP framework in the lower-level low-carbon ED model, which can help P2G facilities bid in the CETS-embedded electricity market. The benefits of utilizing a P2G facility include (a) purchasing electricity power at lower prices and then selling the produced SNG at natural gas prices, which can be profitable and help reduce wind power curtailments, and (b) reducing carbon emission directly while producing SNG and selling the resulting carbon emission permits for arbitrage.
3. The calculation method of the CETS-LMP based on DC optimal power flow is proposed and analyzed. A comparison of the CETS-LMP and the normal LMP is conducted. Note that the CETS-LMP may be negative when the wind power unit is the marginal unit, which can be beneficial to a P2G facility.
4. This work also analyzes the impact of a CETS, the values of carbon emission permits, and prices on the profit of P2G facilities. Thus, it discovers potential values of P2G facilities such that a more reasonable value proposition can be provided for decision-makers.

19.2.2 Mathematical Models for Addressing Strategic Behavior of Power-to-Gas Owners

P2G facilities utilize electricity energy to produce natural gas with a certain efficiency. The chemical process can be described as: $2H_2O \rightarrow 2H_2 + O_2$, $CO_2 + 4H_2 \rightarrow CH_4 + 2H_2O$. In the first step, gaseous hydrogen is formed by the process of electrolysis. In the second step, SNG is generated by utilizing carbon. Both processes can be achieved by consuming electric power, which achieves the transformation of electricity energy into a natural gas resource.

The relationship of volumetric quantity of produced SNG and consumed electric power is shown in (19.1).

$$S_{P2G,i,t} = \eta P_{P2G,i,t} \quad \forall t, \forall i \in \Theta_{P2G} \tag{19.1}$$

In the electricity market, for the owner of a P2G facility, the consumed electric power should be purchased at market prices, and the produced SNG can be stored and sold in the natural gas market at the gas price. If the gap between the electricity market price and the gas price is large enough, the owner of the P2G facility can realize arbitrage. The consumed electricity power of the P2G facility can be cleared at the LMP. In this chapter, we assume that the P2G facility can obtain LMP information from an ISO and make a decision on its power consumption. The payoff of a P2G facility without carbon emission trading can be shown as:

$$PF_{P2G}^* = \sum_{t}^{T} \sum_{i}^{N} \left(pr_{gas} S_{P2G,i,t} - \pi_{i,t} P_{P2G,i,t} \right) \quad \forall t, \forall i \in \Theta_{P2G} \tag{19.2}$$

The integration of P2G technology is regarded as an effective means of energy storage to aid renewable energy integration and reduce carbon emission. Hence, the benefits of P2G integration are twofold:

1. It reduces renewable power curtailments (e.g., wind power curtailments) by purchasing electricity power at lower prices and selling the produced SNG at the higher natural gas price. Renewable energy generation can replace fossil fuel–fired power generation with almost zero carbon emission. However, large-scale renewable energy generation facilities are usually installed far away from load centers, which causes renewable energy curtailments due to the operation constraints of electricity networks and generation units. A P2G facility can utilize low-priced surplus renewable energy to produce SNG to aid in the integration and utilization of renewable energy.

2. It reduces carbon emission directly. In the second stage of the chemical process of P2G, carbon is utilized to produce CH_4. The carbon utilized in the production process of SNG can be from the atmosphere, which is beneficial for carbon reduction. In [24], the impact of P2G on the reduction of carbon emission was quantified, and the CO_2 emission reduction from SNG production was measured at 180 kg/MWh. The relationship between the reduced carbon emission and the operating power consumption of P2G is shown as:

$$CR_{P2G,i,t} = \alpha_{P2G} P_{P2G,i,t} \quad \forall t, \forall i \in \Theta_{P2G} \tag{19.3}$$

In this chapter, a P2G facility is allowed to participate in the carbon market based on the CETS, and the carbon emission reduction of P2G can be paid (i.e., selling carbon emission permits) in the carbon market, which is an incentive for P2G facilities and will increase profit relative to the problem of an insufficient number of dispatching periods with low LMPs. Hence, the total profit of a P2G facility in the electricity market and the carbon market includes three parts: SNG sales revenue, electricity power purchasing cost, and revenue from the carbon market. The profit of P2G facilities can be shown as:

$$PF_{P2G} = \sum_{t}^{T} \sum_{i}^{N} \left(pr_{gas} S_{P2G,i,t} - \pi_{i,t} P_{P2G,i,t} + cp CR_{P2G,i,t} \right) \quad \forall t, \forall i \in \Theta_{P2G} \tag{19.4}$$

To strategically bid in the electricity market, the owner of the P2G facility would consider the LMPs in the electricity market, the natural gas price, and the carbon trading price in the carbon market to maximize its profit, as shown in (19.4).

19.2.3 Mathematical Models for Integration of Carbon Emission Trading Schemes

In the context of a CETS corresponding to a cap-and-trade carbon market, a compulsory cap on total carbon emission is set by the government [29]. A certain number of emission permits are freely allocated to generation units. The allocated carbon emission permits can be traded in the carbon market. A generation company can pay for excessive emission that exceeds the allocated permits and would be paid for surplus emission permits (i.e., if the number of permits is greater than the actual carbon emission). The carbon trading cost of a generation company considering emission permits can be expressed as:

$$F_{CD,i,t} = cp E_{i,t} = cp(E_{i,t}^{to} - PE_{i,t}) = cp(e_i G_{i,t} - PE_{i,t}) \quad \forall t, \forall i \in \Theta_G \tag{19.5}$$

The optimal allocation of carbon emission permits is of great importance and has been studied in [25, 30], which is not the focus of this chapter. In this chapter, we assume that the emission permits differ depending on time periods and generation units. The total number of carbon emission permits (i.e., the overall cap) of all the involved power units depends on the total demand and the regional preset emission permit factor ε, as shown in (19.6). Similar to the method in [8], the allocated carbon emission permits of each power unit (non-zero emission generation units) depend proportionally on its emission factor (i.e., amount of carbon emission for generating one unit of electricity power), that is, the larger the emission factor of a generation unit, the larger the percentage of emission permits allocated, as shown in (19.7). Note that the hourly carbon emission cost shown in (19.5) can be both positive and negative. When the emission of a generation unit exceeds the allocated permits, the cost is positive, and in contrast, when the number of allocated permits is greater than the emission of a generation unit, the cost would be negative.

$$\sum_{i \in \Theta_G}^{N} PE_{i,t} = \varepsilon \sum_{i=1}^{N} D_{i,t}^{*} \quad \forall t \tag{19.6}$$

$$\begin{cases} PE_{i,t} = \alpha_i \sum_{i}^{N} PE_{i,t} \\ \dfrac{PE_{i,t}}{PE_{j,t}} = \dfrac{\alpha_i}{\alpha_j} \end{cases} \quad (i,j) \in \Theta_G, \forall t \tag{19.7}$$

In this section, we formulate the bidding problem of P2G considering the CETS as a bi-level stochastic optimization model. The model is developed for one P2G company. The upper model is the profit maximization problem of P2G facilities, and the lower model is the low-carbon ED model considering wind power uncertainty.

Simplification and assumptions:

1. We assume the intraday natural gas price is constant.
2. The carbon emissions trading price is considered as a parameter for the lower-level problem.

19.2.3.1 Upper Model for P2G Facilities Profit Maximization

Given several P2G facilities belonging to a joint owner, the profit of P2G facilities participating in the carbon market considering the CETS in (19.4) should be maximized. The upper-level profit maximization problem of P2G facilities can be represented as follows:

$$\max \sum_{t}^{T} \sum_{i}^{N} \left(pr_{gas} S_{P2G,i,t} - \sum_{s=1}^{S} ps \times \pi_{i,t,s} P_{P2G,i,t} + cp CR_{P2G,i,t} \right) \quad \forall t, \forall i \in \Theta_{P2G} \tag{19.8}$$

s.t.

$$P_{P2G,i}^{min} \leq P_{P2G,i,t} \leq P_{P2G,i}^{max} \quad \forall t, \forall i \in \Theta_{P2G} \tag{19.9}$$

$$S_{P2G,i,t,gas} = \eta P_{P2G,i,t} \quad \forall t, \forall i \in \Theta_{P2G} \tag{19.10}$$

$$CR_{P2G,i,t} = \alpha_{P2G} P_{P2G,i,t} \quad \forall t, \forall i \in \Theta_{P2G} \tag{19.11}$$

The objective function (19.8) of the upper-level model is to maximize the expected profit of P2G facilities for a whole day. Constraint (19.9) indicates the maximum/minimum consuming power limits of a P2G facility. Constraints (19.10) and (19.11) denote the amount of volumetric quantity of produced SNG and reduced carbon emission. For the upper-level model, the decision variables are the consuming power of P2G facilities, and LMPs can be calculated by an ISO's ED.

19.2.3.2 Lower Model for Low-Carbon Economic Dispatch

ED is carried out by ISOs to realize market clearing and obtain generation dispatches. An ED model based on DC optimal power flow is utilized in [26]. In this chapter, the CETS and wind power uncertainty are considered. Wind power uncertainty is represented by stochastic scenarios. A modified low-carbon stochastic ED model based on the ED model in [26] is formulated as a linear programming (LP) problem and presented as follows:

$$\min \sum_{s=1}^{S} ps \times \sum_{t=1}^{T} \sum_{i=1}^{N} \left(c_i G_{i,t,s} + c_{wi} P_{w,i,t,s} + cp E_{i,t,s} \right) \tag{19.12}$$

$$PE_{i,t} = \alpha_i \times \varepsilon \sum_{i=1}^{N} D_{i,t}^* : \rho_{i,t} \tag{19.13}$$

$$\text{s.t. } D_{i,t}^* = D_{i,t} + P_{P2G,i,t}$$

$$\sum_{i=1}^{N} (G_{i,t,s} + P_{w,i,t,s}) = \sum_{i=1}^{N} D_{i,t}^* : \lambda_{t,s} \tag{19.14}$$

$$-\text{Limit}_l \leq \sum_{i=1}^{N} GSF_{l-i} (G_{i,t,s} + P_{w,i,t,s} - D_{i,t}^*) \leq \text{Limit}_l : \mu_{l,t,s}^{\min}, \mu_{l,t,s}^{\max} \tag{19.15}$$

$$G_i^{\min} \leq G_{i,t,s} \leq G_i^{\max} : \omega_{i,t,s}^{\min}, \omega_{i,t,s}^{\max} \tag{19.16}$$

$$0 \leq P_{w,i,t,s} \leq P_{w,i,t,s}^{\max} : \phi_{i,t,s}^{\min}, \phi_{i,t,s}^{\max} \tag{19.17}$$

The objective function (19.12) of the proposed low-carbon ED model is to minimize the total operation costs, including bidding costs (to cover the fuel costs of generators) of thermal units, bidding costs of wind power units, and carbon emission trading costs of thermal units. Constraint (19.13) is the carbon emission permit constraint. Constraints (19.14)–(19.17) denote the power balance, line flow limits, and upper and lower generation limits of thermal units and wind power units. $\lambda_{t,s}$, $\rho_{i,t}$, $\mu_{\min i,t,s}$, $\mu_{\max i,t,s}$, $\omega_{\min i,t,s}$, $\omega_{\max i,t,s}$, $\phi_{\min i,t,s}$, and $\phi_{\max i,t,s}$ are dual variables of objective function and constraints.

After obtaining the optimal solution of the low-carbon ED model, the CETS-LMP is calculated. The Lagrangian function of the ED model can be expressed as:

$$\psi = \sum_{t=1}^{T} \left\{ \begin{array}{l} \sum_{s=1}^{S} ps \times \sum_{i=1}^{N} [(c_i + cp \times e_i) G_{i,t,s} - cp PE_{i,t} + c_{wi} P_{w,i,t,s}] \\[2ex] -\lambda_{t,s} [\sum_{i=1}^{N} (G_{i,t,s} + P_{w,i,t,s}) - \sum_{i=1}^{N} D_{i,t}^*] \\[2ex] -\sum_{i \in \Theta_G}^{N} \rho_{i,t} (PE_{i,t} - \alpha_i \times \varepsilon \sum_{i=1}^{N} D_{i,t}^*) \\[2ex] -\sum_{l=1}^{M} \mu_{l,t,s}^{\min} [\text{Limit}_l + \sum_{i=1}^{N} GSF_{l-i} (G_{i,t,s} + P_{w,i,t,s} - D_{i,t}^*)] \\[2ex] -\sum_{l=1}^{M} \mu_{l,t,s}^{\max} [\text{Limit}_l - \sum_{i=1}^{N} GSF_{l-i} (G_{i,t,s} + P_{wi,t,s} - D_{i,t}^*)] \\[2ex] -\sum_{i=1}^{N} \omega_{i,t,s}^{\min} (G_{i,t,s} - G_i^{\min}) - \sum_{i=1}^{N} \omega_{i,t,s}^{\max} (G_i^{\max} - G_{i,t,s}) \\[2ex] -\sum_{i=1}^{N} \phi_{i,t,s}^{\min} (P_{w,i,t,s} - 0) - \sum_{i=1}^{N} \phi_{i,t,s}^{\max} (P_{w,i,s}^{\max} - P_{w,i,t,s}) \end{array} \right\} \tag{19.18}$$

Then, the CETS-LMP can be calculated by:

$$\pi_{i,t,s} = \frac{\partial \psi}{\partial D_{i,t}^{*}} = \lambda_{t,s} + \sum_{i \in \Theta_G}^{N} \rho_{i,t} \alpha_i \times \varepsilon + \sum_{l=1}^{M} G_{l-i} (\mu_{l,t,s}^{\min} - \mu_{l,t,s}^{\max})$$ (19.19)

The proposed bi-level model cannot be solved directly, and the consuming powers of both P2G and the LMPs are decision variables in the bidding process. Therefore, the objective function in (19.8) is nonlinear. The transformation and solution of the bi-level model are analyzed in Section 19.4.

In this section, we reformulate the proposed bi-level bidding model of P2G into a single-level mixed-integer linear programming (MILP) problem, which can be solved effectively by existing solvers. First, the bi-level model is re-formulated as an MPEC problem by replacing the lower-level model with complementary constraints based on Karush Kuhn Tucker (KKT) optimality conditions. Then, the MPEC model can be transformed into a MILP problem according to the strong duality theory. Finally, for the nonlinear term in the objective function, a linearization method is applied.

19.2.3.2.1 MPEC Formulation

The lower-level model is an LP problem, and the optimal solution should satisfy the KKT conditions. The lower-level model can then be inserted into the upper-level model as complementary constraints [26]. The reformulated MPEC model can be expressed as:

$$\max \quad \sum_{t}^{T} \sum_{i}^{N} \left(pr_{gas} S_{P2G,i,t} - \pi_{i,t} P_{P2G,i,t} + cpCR_{P2G,i,t} \right) \quad \forall t, \forall i \in \Theta_{P2G}$$ (19.20)

$$\text{s.t. Constraints } (19.9)-(19.11), (19.13)-(19.14)$$ (19.21)

$$\lambda_{t,s} + \sum_{l=1}^{M} GSF_{l-i} (\mu_{l,t,s}^{\min} - \mu_{l,t,s}^{\max}) + \omega_{i,t,s}^{\min} - \omega_{i,t,s}^{\max} = c_i + cp \times e_i$$ (19.22)

$$\lambda_{t,s} + \sum_{l=1}^{M} GSF_{l-i} \times (\mu_{l,t,s}^{\min} - \mu_{l,t,s}^{\max}) + \phi_{i,t,s}^{\min} - \phi_{i,t,s}^{\max} = c_{wi}$$ (19.23)

$$-\rho_{i,t} = cp$$ (19.24)

$$0 \leq \mu_{l,t,s}^{\min} \perp [Limit_l + \sum_{i=1}^{N} GSF_{l-i}(G_{i,t,s} + P_{w,i,t,s} - D_{i,t}^{*})] \geq 0$$ (19.25)

$$0 \leq \mu_{l,t,s}^{\max} \perp [Limit_l - \sum_{i=1}^{N} GSF_{l-i}(G_{i,t,s} + P_{w,i,t,s} - D_{i,t}^{*})] \geq 0$$ (19.26)

$$0 \leq \omega_{i,t,s}^{\min} \perp G_{i,t,s} - G_i^{\min} \geq 0$$ (19.27)

$$0 \leq \omega_{i,t,s}^{\max} \perp G_i^{\max} - G_{i,t,s} \geq 0$$ (19.28)

$$0 \leq \phi_{i,t,s}^{\min} \perp P_{w,i,t,s} \geq 0$$ (19.29)

$$0 \leq \phi_{i,t,s}^{\max} \perp P_{w,i}^{\max} - P_{w,i,t,s} \geq 0$$ (19.30)

19.2.3.2.2 MILP Formulation

The MPEC model of (19.20)–(19.30) is nonlinear and can be linearized based on the strong duality theory and the techniques in [26–28, 31]. According to the strong duality theory, the nonlinear term $\pi_{i,t,s}P_{P2G,i,t}$ can be replaced by a linear combination of variables. By introducing auxiliary binary variables and Bid-M constants, the constraints (19.25)–(19.30) can be linearized. The MILP formulation of the proposed bi-level bidding model for P2G facilities is given in (19.31)–(19.44).

$$
\max \sum_t^T \left[\begin{aligned}
& \sum_{i\in\Theta_{P2G}}^N (pr_{gas}S_{P2G,i,t} + cpCR_{P2G,i,t}) - \sum_{s=1}^S ps \times \\
& \left\{ \sum_{i=1}^N [(c_i + cp\times e_i)G_{i,t,s} - cpPE_{i,t} + c_{wi}P_{w,i,t,s}] \right. \\
& -\lambda_{t,s} \sum_{i\notin\Theta_{P2G}}^N D_{i,t} - \sum_{i\in\Theta_G}^N (\rho_{i,t}\alpha_i \times \varepsilon \sum_{i\notin\Theta_{P2G}}^N D_{i,t}) \\
& -\sum_{l=1}^M \mu_{l,t,s}^{\min}(-Limit_l + \sum_{i\notin\Theta_{P2G}}^N GSF_{l-i}D_{i,t}) \\
& -\sum_{l=1}^M \mu_{l,t,s}^{\max}(-Limit_l - \sum_{i\notin\Theta_{P2G}}^N GSF_{l-i}D_{i,t}) \\
& -\sum_{i=1}^N \omega_{i,t,s}^{\min}(G_i^{\min}) - \sum_{i=1}^N \omega_{i,t}^{\max}(-G_i^{\max}) \\
& \left. -\sum_{i=1}^N \phi_{i,t,s}^{\min}(0) - \sum_{i=1}^N \phi_{i,t}^{\max}(-P_{w,i,t,s}^{\max}) \right\}
\end{aligned} \right]
\tag{19.31}
$$

s.t.
$$
\text{Constraints (19.21)–(19.24)} \tag{19.32}
$$

$$
0 \le \mu_{l,t,s}^{\min} \le M_{\mu,t}^{\min}\nu_{\mu,l,t,s}^{\min} \tag{19.33}
$$

$$
0 \le \sum_{i=1}^N GSF_{l-i}(G_{i,t,s} + P_{w,i,t,s} - D_{i,t}^*) + Limit_l \le M_{\mu,t}^{\min}(1 - \nu_{\mu,l,t,s}^{\min}) \tag{19.34}
$$

$$
0 \le \mu_{l,t,s}^{\max} \le M_{\mu,t}^{\max}\nu_{\mu,l,t,s}^{\max} \tag{19.35}
$$

$$
0 \le -\sum_{i-1}^N GSF_{l-i}(G_{i,t,s} + P_{w,i,t,s} - D_{i,t}^*) + Limit_l \le M_{\mu,t}^{\max}(1 - \nu_{\mu,l,t,s}^{\max}) \tag{19.36}
$$

$$
0 \le \omega_{i,t,s}^{\min} \le M_{\omega,t}^{\min}\nu_{\omega,i,t,s}^{\min} \tag{19.37}
$$

$$
0 \le G_{i,t,s} - G_i^{\min} \le M_{\omega,t}^{\min}(1 - \nu_{\omega,i,t,s}^{\min}) \tag{19.38}
$$

$$
0 \le \omega_{i,t,s}^{\max} \le M_{\omega,t}^{\max}\nu_{\omega,i,t,s}^{\max} \tag{19.39}
$$

$$
0 \le G_i^{\max} - G_{i,t,s} \le M_{\omega,t}^{\max}(1 - \nu_{\omega,i,t,s}^{\max}) \tag{19.40}
$$

$$
0 \le \phi_{i,t,s}^{\min} \le M_{\phi,t}^{\min}\nu_{\phi,i,t,s}^{\min} \tag{19.41}
$$

$$
0 \le P_{w,i,t,s} \le M_{\phi,t}^{\min}(1 - \nu_{\phi,i,t,s}^{\min}) \tag{19.42}
$$

$$0 \leq \phi_{i,t,s}^{\max} \leq M_{\phi,t}^{\max} \nu_{\phi,i,t,s}^{\max} \tag{19.43}$$

$$0 \leq P_{\mathrm{w},i}^{\max} - P_{\mathrm{w},i,t,s} \leq M_{\phi,t}^{\max} (1 - \nu_{\phi,i,t,s}^{\max}) \tag{19.44}$$

where $M_{\min\,\mu}$, $M_{\max\,\mu}$, $M_{\min\,\omega}$, $M_{\max\,\omega}$, $M_{\varphi}^{\min}$, and $M_{\varphi}^{\max}$ are large enough constants. $\nu_{\min\,\mu,l,t,s}$, $\nu_{\mu,l,t,s}^{\max}$, $\nu_{\omega,i,t,s}^{min}$, $\nu_{\omega,i,t,s}^{max}$, $\nu_{\phi,i,t,s}^{min}$, and $\nu_{\phi,i,t,s}^{max}$ are auxiliary binary variables.

19.3 IMPLEMENTATION OF CETS-EMBEDDED LOCATIONAL MARGINAL PRICE

Note that the obtained CETS-LMPs are calculated with the carbon emission cost in the carbon market and are different from the values obtained without carbon emission cost, in which the term of the sum of $\rho i,t\alpha i \times \varepsilon$ is added when compared with the formulation in [26]. According to the strong duality theory, $\rho i,t$ is negative. αi and ε are positive constants. It seems that the CETS-LMP is smaller than the LMP without a carbon emission cost. However, the total bidding cost coefficient of thermal generation unit i increases by $cp \times ei$. Hence, the CETS-LMP may be larger than that of the normal LMP when a thermal unit is the marginal unit.

Remark 1: CETS-LMP may be negative. As it is known, LMP can represent the marginal price, and it depends on the energy price of a marginal unit. A 1 MWh increase on load would cause 1 MWh more output of the generation unit and bring in more carbon emission permits. If the marginal unit is a low-cost wind power unit, then thermal generation units would not cause any more actual emission and the increased carbon emission permits can be sold for arbitrage. Hence, the CETS-LMP is equal to the difference between the bidding price of a wind power unit and the revenue of selling surplus carbon emission permits, which may be negative. The negative CETS-LMPs can be beneficial to P2G facilities, which can increase load to reduce wind power curtailments.

19.4 CASE STUDIES

19.4.1 Introduction of Case Studies

In this section, the proposed bi-level bidding model for P2G facilities is performed on the modified PJM 5-bus system, and the IEEE 118-bus system is utilized to verify the applicability of the proposed model on large-scale systems. The MILP optimization model is solved using CPLEX in the general algebraic modeling system (GAMS) on a PC with Intel Core i7 3.00 GHz CPU and 8 GB RAM. The SCENRED tool provided in GAMS is utilized to perform the scenario reduction process. The probabilities for all scenarios before reduction are assumed to be the same, with a cumulative sum equal to one ($\Sigma P_{s} = 1$).

19.4.2 Implications for P2G Facilities under CETS-Embedded Locational Marginal Price

19.4.2.1 PJM 5-Bus System

The PJM 5-bus system is depicted in Figure 19.2, in which the generation capacities and bid prices are also included [32]. A wind farm is integrated at bus E, and the bid cost of wind power is set to $8. The load is equally distributed on three load buses. The installed capacity of the wind farm at bus E is 400 MW, and the hourly maximum operating power of the P2G facility at bus C is 200 MW. The emission factors of the five thermal units are assumed to be 0.65, 1.05, 1.05, 0.65, and 1.05 $tCO_2/$MWh. The carbon emission permit factor is set to be 0.7 $tCO_2/$MWh, and the carbon emission trading price is $23/t. The efficiency of P2G to produce SNG is 60%. The price of SNG is $16/MWh. The forecasted wind power output and load are shown in Figure 19.3. In this section, to verify the

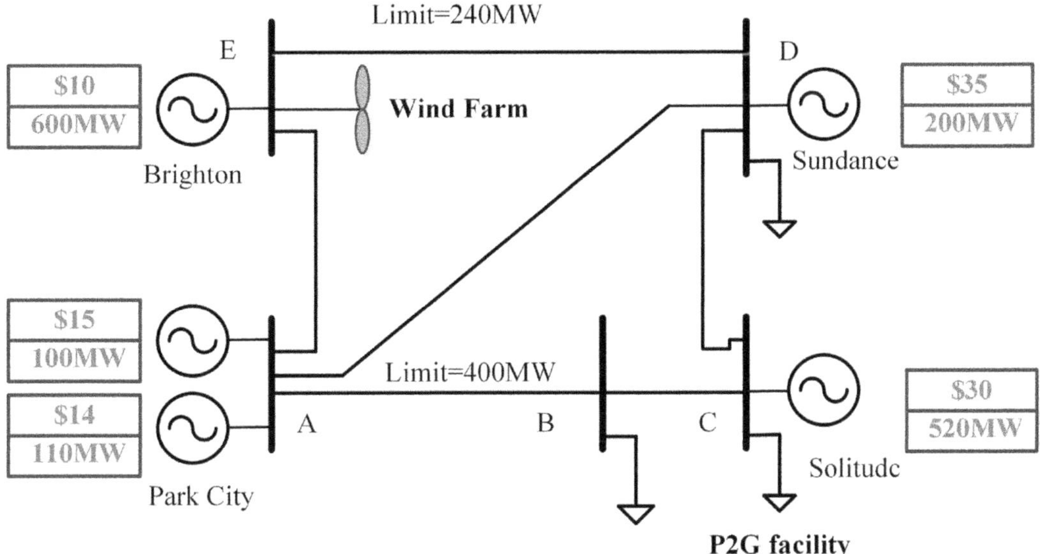

FIGURE 19.2   PJM 5-bus system.

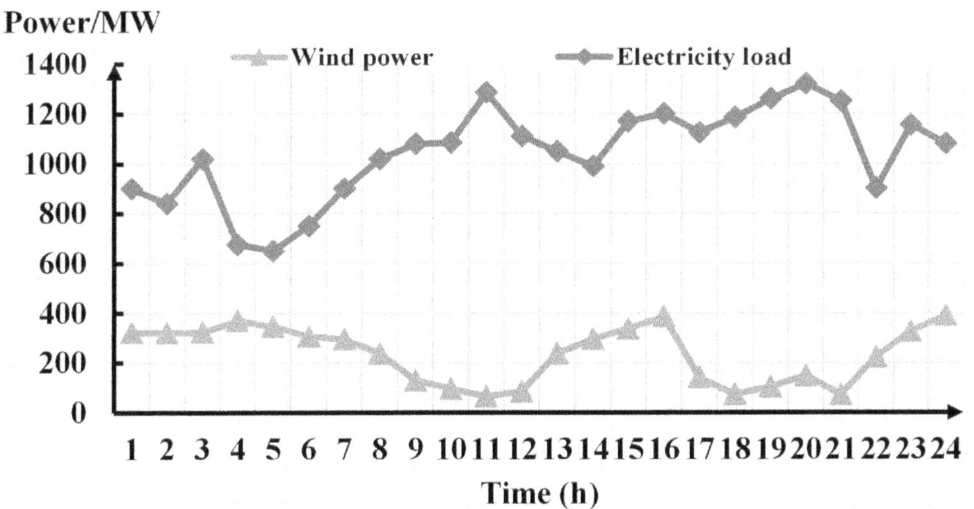

FIGURE 19.3   Forecasting output of wind power and electricity load.

impact of CETS on the profit of P2G facilities and LMPs, wind power uncertainty is not considered. The big-M constants are set as 10,000.

19.4.2.1.1   Impact of the CETS on Bidding of P2G Facility and LMPs

To verify the impact of the CETS on the bidding results of a P2G facility and LMPs obtained by an ISO's ED, the following two cases considering and not considering the CETS are carried out and compared.

Case 1: Bi-level bidding model considering the CETS.
Case 2: Bi-level bidding model without the CETS, setting the carbon emission trade price as 0.

TABLE 19.2

Comparison of Results of Cases 1 and 2

Cases	Case 1	Case 2
Total profit of P2G ($)	9,427.84	747.35
Total power consumption of P2G (MWh)	467.1	467.1
Wind power utilization rate	99.7%	99.7%
Total carbon emission of thermal units (tCO$_2$)	18,934.2	19,420.7

The comparison of the results of Cases 1 and 2 is listed in Table 19.2. It is important to note that the total profit of the P2G facility in Case 1 is increased to $9,427.84 from $747.35 in Case 2, a more than 10 times increase in profit. The profit of the P2G facility from the carbon market accounts for 12.3% of the total profit. The comparison demonstrates that the consideration of the CETS can bring in more profits to a P2G facility. The total power consumption of P2G facilities in the two cases is the same, which verifies that based on the proposed bidding method, a P2G facility can achieve more profit even with the same power consumption in the case without the CETS. The main reason for this is that when the CETS is considered, the CETS-LMPs can be lower than LMPs without the CETS during the periods when the wind power unit is the marginal unit. The wind power utilization rates in the two cases are the same. Hence, the CETS cannot change the scheduling results of wind power because the bidding price of wind power is the lowest under the utilized parameters in this case study. Moreover, the total carbon emission of thermal units in Case 1 shows a decrease of 486.5 tCO$_2$ when compared to the value of Case 2, which demonstrates that considering the CETS can reduce carbon emission from thermal units. Thus, considering the CETS can increase the profit of a P2G facility and reduce carbon emission from thermal units, which demonstrates the effectiveness of the proposed bidding model for P2G facilities.

The comparison of the CETS-LMP and the LMP results of bus C (i.e., the bus integrated with a P2G facility) in Cases 1 and 2 are shown in Figure 19.4a. It can be seen that in most periods, the values of the CETS-LMPs in Case 1 are larger than those of the LMPs in Case 2. However, in periods 4–6, the CETS-LMPs are much lower than normal LMPs. The reason for this is that when a thermal unit is a marginal unit (as in most periods), an increase of load leads to an increase in the output of the corresponding thermal unit, and the increase in carbon trading cost leads to an increase in LMP. When a wind power unit is the marginal unit, then an increase on load is balanced by wind power and increases the number of carbon emission permits from the carbon market. No emission is produced from wind power outputs, so the surplus emission permits can be sold. The CETS-LMPs in

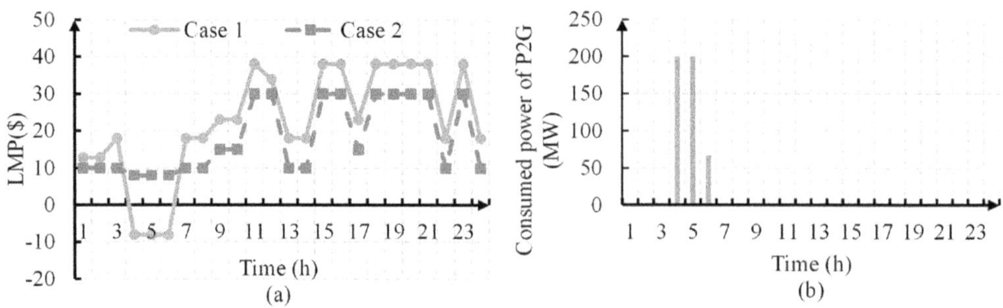

FIGURE 19.4   For (a) comparison of LMP of bus C in Cases 1 and 2, and for (b) consumed power results of P2G facility in Case 1.

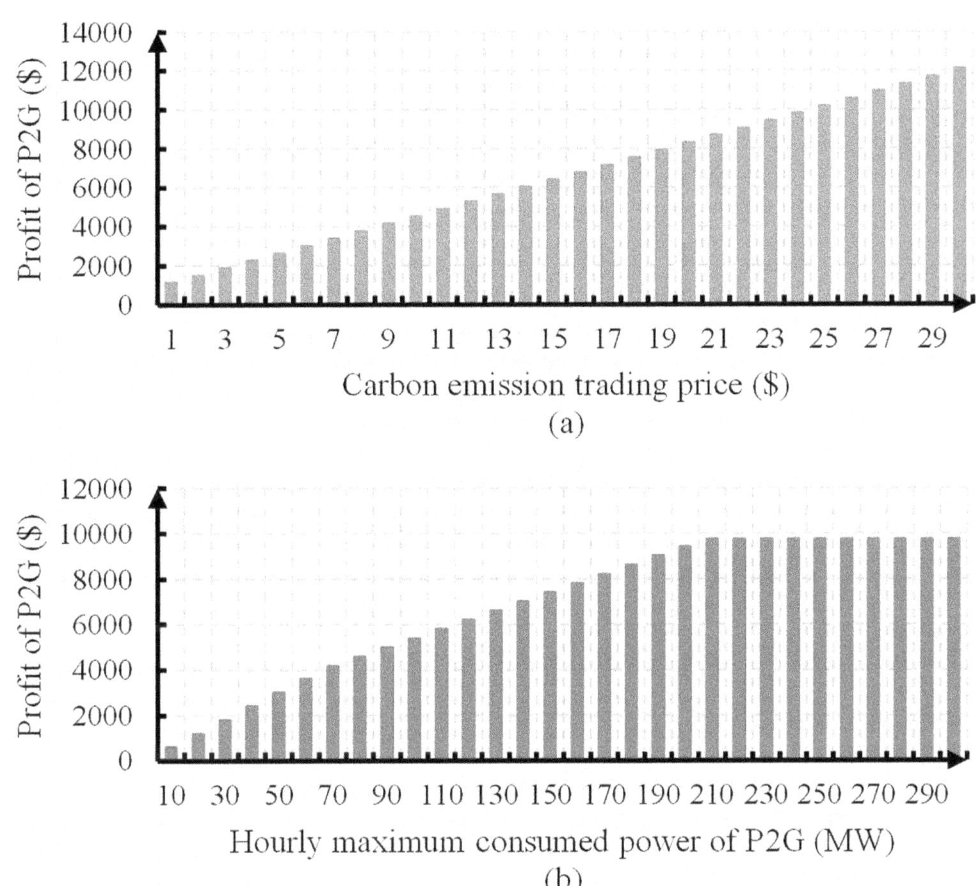

FIGURE 19.5 Profit results of P2G facility: (a) profit of P2G facility under different carbon emission trading prices; (b) profit of P2G facility under different hourly maximum consumed power.

these periods are equal to the difference between the wind power bidding price and the multiplication of the carbon trading price and the emission permit factor. Hence, the CETS-LMP is negative in periods 4–6. Figure 19.4b shows the consumed power of the P2G facility in Case 1. It is worth noting that the P2G facility consumes non-zero power in periods 4–6 with low CETS-LMPs; hence, the profit of the P2G facility in Case 1 is larger than in Case 2. The preceding results also demonstrate that the consideration of a CETS can increase the profit of the P2G facility.

19.4.2.1.2 Impact of Carbon Trading Price and Hourly Maximum Consumed Power on Profit of P2G Facility

Figure 19.5 shows the profit results of P2G under different carbon emission trading prices and at hourly maximum consumed power. As shown in Figure 19.5a, when the carbon trading price increases, the profit of P2G trends upward, mainly because a larger carbon trading price decreases the CETS-LMPs in the periods when the wind power unit is the marginal unit. The electricity power purchasing cost of the P2G facility decreases, and hence, the profit increases. It can be seen in Figure 19.5b that when the hourly maximum consumed power of the P2G facility gradually increases from 10 MW to 210 MW, the profit of the P2G facility increases from \$605.50 to \$9,478.40. When the hourly maximum consumed power of the P2G facility is larger than 210 MW, the profit of the P2G facility does not increase any more, because the lack of a wind power surplus corresponds to the low CETS-LMPs that can be purchased.

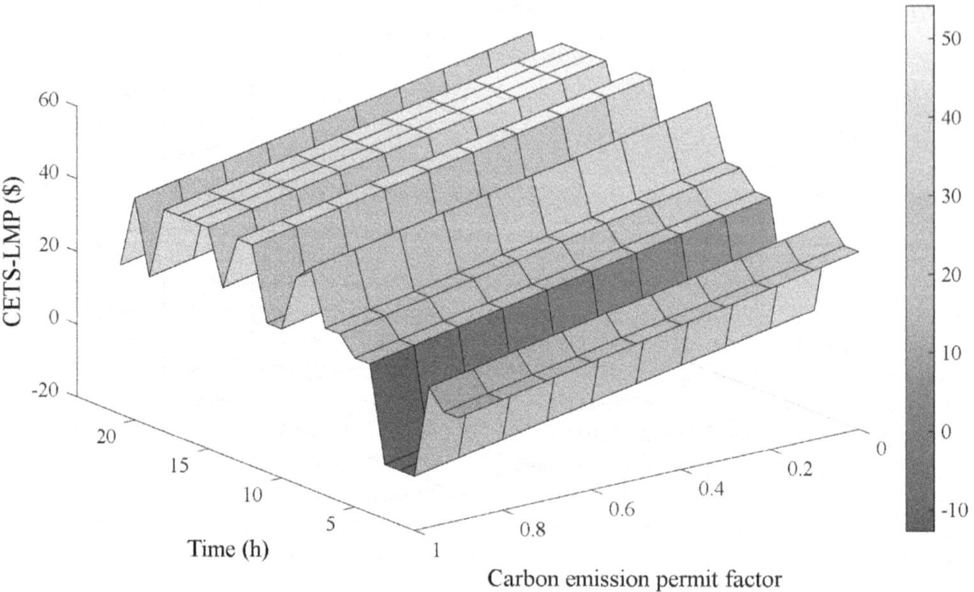

FIGURE 19.6 CETS-LMP results under various carbon emission permit factor.

19.4.2.1.3 Impact of a Carbon Trading Permit on the Profit of P2G Facility and LMP

A carbon trading permit affects the bidding cost of power units and then affects CETS-LMPs. The P2G facility determines the consumed power according to the CETS-LMPs. The hourly CETS-LMP results under different carbon emission permit factors are shown in Figure 19.6. When the carbon emission permit factor increases, the CETS-LMPs decrease correspondingly. The reason for this is that the larger the carbon emission permit factor, the more emission permits must be allocated to thermal units and the lower the emission cost.

The profit and total consumed power results of the P2G facility under different carbon emission permit factors are shown in Figure 19.7. Two conclusions can be obtained in Figure 19.7. First, the profit of the P2G facility increases as the carbon emission permit factor increases. The reason for this is that the larger the carbon emission permit factor, the lower the CETS-LMPs when the P2G facility purchases non-zero electricity power. Second, the total consumed power of the P2G facility does not change until the carbon emission factor reaches 0.8. The reason for this is that when the carbon emission factor reaches 0.8 or larger, the CETS-LMPs in certain periods are low enough for it to be beneficial for the P2G facility to purchase more electricity power.

19.4.2.2 IEEE 118-Bus System

To further verify the effectiveness of the proposed bi-level bidding model on large systems, the IEEE 118-bus system is utilized for case studies. In this section, wind power uncertainty is considered. The network data, bidding parameters of generation units, and load curve are similar to those in [22, 26]. The thermal limits of transmission lines are 175 MW and 500 MW. Two wind farms with the same capacities and forecasting output as in the PJM 5-bus system are integrated at buses 3 and 20. An emission factor of 0.65 tCO_2/MWh is assumed for 30 thermal units, and the other 24 thermal units are assumed to have the emission factor of 1.05 tCO_2/MWh. The carbon emission permit factor, SNG price, and CETS price are the same as in the PJM 5-bus system. A thousand initial scenarios are generated and reduced to five scenarios for computational convenience. The forecasting errors are all represented by a normal distribution function with mean equal to 0 and 15% standard deviation.

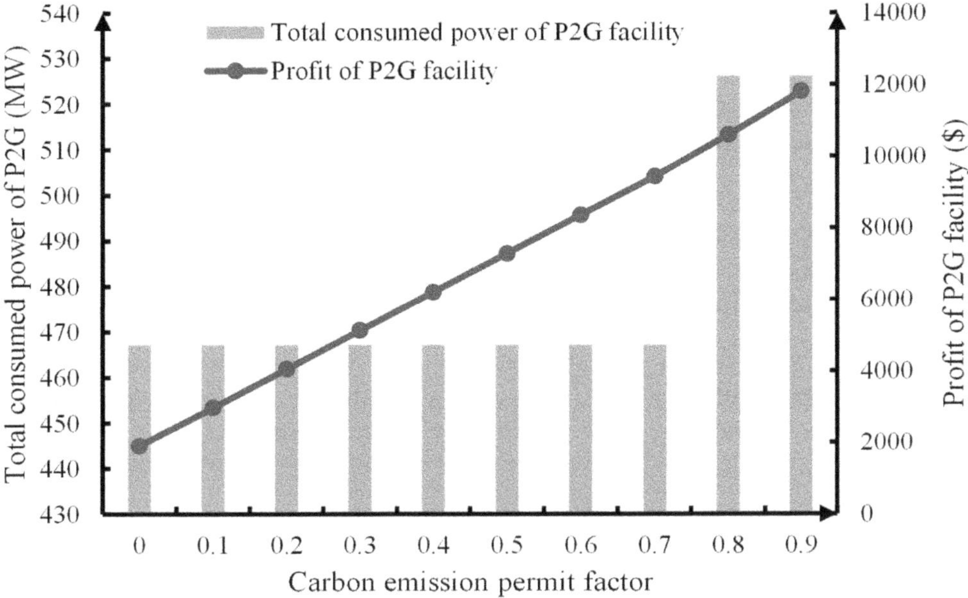

FIGURE 19.7 Profit and total consumed power results of P2G under various carbon emission permit factors.

TABLE 19.3
Comparison of Results of Different Locations of P2G

Locations	Buses 8 and 9	Buses 80 and 81
Expected profit of P2G ($)	1,380.8	316.2
Total power consumption of P2G (MWh)	291.8	482.2

Impact of location of P2G facilities. Two P2G facilities with 200 MW hourly maximum power consumption are integrated at different locations: (1) same area with wind farms at bus 8 and 9, and (2) different area with wind farms at bus 80 and 81. The comparison of bidding results with different locations of P2G facilities is listed in Table 19.3.

As can be observed from Table 19.3, the P2G facility's strategic behavior varies at different locations. When connected to wind farms at nearby buses (i.e., bus 8 and 9), P2G facilities purchase more power consumption compared to the case of P2G facilities connected at distant buses. The profit can be enormously increased from $316.2 to $1,380.8 because the CETS-LMPs are high when P2G facilities are connected at distant buses from wind farms due to transmission congestion.

The computation time of the stochastic bi-level model for the IEEE 118-bus system is 76.6 s. The preceding results further demonstrate the effectiveness of the proposed bi-level strategic bidding model for P2G facilities.

19.4.3 Future Directions in P2G Strategic Bidding

As an active market participant, a P2G facility can perform strategic bidding to maximize its profit. However, the revenue of purchasing low-price electricity in day-ahead energy market and selling natural gas to natural gas market cannot cover the investment cost. It's not limited by the bidding strategy. There are not enough low-price periods in the day-ahead electricity market. Hence, to increase the profit of P2G, more markets participation should be developed. On one hand, P2G can participate in auxiliary service markets by providing reserve, flexible ramping, and regulation services. Wind power and PV have great reserve and regulation requirement for their fluctuation and intermittence. P2G can provide auxiliary service to renewable power generation plants and purchase low-price renewable power, which can reduce the cost but increase the profit. On the other hand, P2G can consume carbon emission, which can bring profit from carbon market. In this chapter, we consider the CETS.

The future of P2G strategic bidding promises significant advancements through the integration of advanced technologies, policy and regulatory developments, market design innovations, decarbonization strategies, stakeholder collaboration, and consumer engagement.

> **Integration of advanced technologies.** By harnessing AI, machine learning, and big data analytics, P2G utilities can optimize operations, predicting demand patterns, enhancing maintenance schedules, and reducing downtime. These technologies can also facilitate real-time pricing strategies, enabling utilities to capitalize on market fluctuations and maximize profits.
>
> **Policy and regulatory developments.** Evolving carbon pricing mechanisms and regulations will incentivize P2G utilities to reduce emissions (utilizing carbon to produce SNG) and increase the use of renewable energy sources. Utilities can adapt by investing in carbon capture and storage technologies, diversifying their energy portfolios, and leveraging subsidies and incentives offered by governments.
>
> **Market design innovations.** Innovative market structures, such as virtual power plants and peer-to-peer energy trading platforms, can facilitate the integration of P2G utilities into broader energy systems. These designs can enable utilities to participate in multiple markets, diversify revenue streams, and maximize profits through flexible pricing and trading strategies.
>
> **Consumer engagement.** Enhancing consumer engagement and participation in carbon trading markets can positively influence the profitability of P2G utilities. By educating consumers about the benefits of P2G and offering incentives for adopting clean energy solutions, utilities can build demand, ultimately driving revenue growth and profitability.

REFERENCES

[1] Z. Zeng, T. Ding, Y. Xu, Y. Yang, and Z. Y. Dong, "Reliability evaluation for integrated power-gas systems with power-to-gas and gas storages," IEEE Trans. Power Syst., vol. 35, pp. 571–583.

[2] A. Zlotnik, L. Roald, S. Backhaus, M. Chertkov, and G. Andersson, "Coordinated scheduling for interdependent electric power and natural gas infrastructures," IEEE Trans. Power Syst., vol. 32, no. 1, pp. 600–610, Jan. 2017.

[3] Y. Li, W. Liu, M. Shahidehpour, F. Wen, K. Wang, and Y. Huang, "Optimal operation strategy for integrated natural gas generating unit and power-to-gas conversion facilities," IEEE Trans. Sustain. Energy, vol. 9, no. 4, pp. 1870–1879, Oct. 2018.

[4] D. Alkano and J. M. A. Scherpen, "Distributed supply coordination for Power-to-Gas facilities embedded in energy grids," IEEE Trans. Smart Grid, vol. 9, no. 2, pp. 1012–1022, Mar. 2018.

[5] R. Zhang, K. Yan, T. Jiang, G. Li, X. Li, and H. Chen, "Privacy-preserving decentralized power system economic dispatch considering carbon capture power plants and carbon emission trading scheme via over-relaxed ADMM," Int. J. Electr. Power Energy Syst. Energy, vol. 121, 106094, Oct. 2020.

[6] M. T. Tsai and C. W. Yen, "The influence of carbon dioxide trading scheme on economic dispatch of generators," Appl. Energy, vol. 88, no. 12, pp. 4811–4816, Dec. 2011.

[7] F. Li, J. Qin, and Y. Kang, "Closed-loop hierarchical operation for optimal unit commitment and dispatch in microgrids: A hybrid system approach," IEEE Trans. Power Syst., vol. 35, pp. 516–526, 2019.

[8] N. Zhang, Z. Hu, D. Dai, S. Dang, M. Yao, and Y. Zhou, "Unit commitment model in smart grid environment considering carbon emission trading," IEEE Trans. Smart Grid, vol. 7, no. 1, pp. 420–427, Jan. 2016.

[9] O. D. Melgar-Dominguez, M. Pourakbari-Kasmaei, M. Lehtonen, and J. R. Sanches Mantovani, "An economic-environmental asset planning in electric distribution networks considering carbon emission trading and demand response," Electr. Power Syst. Res., vol. 181, p. 106202, 2020.

[10] E. Delarue and K. Van den Bergh. "Carbon mitigation in the electric power sector under cap-and-trade and renewables policies," Energy Policy, vol. 92, pp. 34–44, 2016.

[11] R. Zhang, T. Jiang, L. Bai, G. Li, H. Chen, X. Li, and F. Li, "Adjustable robust power dispatch with combined wind-storage system and carbon capture power plants under low-carbon economy," Int. J. Electr. Power Energy Syst. Energy, vol. 113, pp. 772–781, Dec. 2019.

[12] Y. Sun, B. Zhang, L. Ge, D. Sidorov, J. Wang, and Z. Xu, "Day-ahead optimization schedule for gas-electric integrated energy system based on second-order cone programming," CSEE J. Power Energy Syst., vol. 6, no. 1, pp. 142–151, Mar. 2020.

[13] G. Li, G. Li, and M. Zhou, "Comprehensive evaluation model of wind power accommodation ability based on macroscopic and microscopic indicators," Protect. Control Mod. Power Syst., vol. 4, no. 4, pp. 215–226, 2019.

[14] Q. Hu, B. Zeng, Y. Zhang, H. Hu, and W. Liu, "Analysis of probabilistic energy flow for integrated electricity-gas energy system with P2G based on cumulant method," 2017 IEEE Conference on Energy Internet and Energy System Integration (EI2), Beijing, China, 2017, pp. 1–6.

[15] C. Leeuwen and M. Mulder, "Power-to-gas in electricity markets dominated by renewables," Appl. Energy, vol. 1232, pp. 258–272, Dec. 2018.

[16] H. Khani and H. E. Z. Farag, "Optimal day-ahead scheduling of power-to-gas energy storage and gas load management in wholesale electricity and gas markets," IEEE Trans. Sustain. Energy, vol. 9, no. 2, pp. 940–951, Apr. 2018.

[17] J. Yang, N. Zhang, Y. Cheng, C. Kang, and Q. Xia, "Modeling the operation mechanism of combined P2G and gas-fired plant with CO_2 recycling," IEEE Trans. Smart Grid, vol. 10, no. 1, pp. 1111–1121, Jan. 2019.

[18] Y. Li, J. Xu, M. Wu, L. Zeng, and N. Tong, "Low-carbon economic dispatch of integrated energy system considering P2G and carbon trading mechanism," 2022 IEEE 5th International Electrical and Energy Conference (CIEEC), Nanjing, China, 2022, pp. 519–524.

[19] N. Kopacak, H. C. Güldorum, and O. Erdinç, "Development of cost minimization-oriented optimization algorithm for green hydrogen production in a multi-energy system considering FCEV availability," 2023 14th International Conference on Electrical and Electronics Engineering (ELECO), Bursa, Turkiye, 2023, pp. 1–5.

[20] M. Zhang, H. Li, and Y. Zhu, "Optimal operation of a regional integrated energy system considering P2G and stepped carbon trading," 2023 5th Asia Energy and Electrical Engineering Symposium (AEEES), Chengdu, China, 2023, pp. 1796–1801.

[21] L. Hou, B. Chen, Y. Wang, A. Miao, K. Su, and Z. Zhang, "Optimal decision of electricity transaction for integrated energy service providers under renewable portfolio standards," 2019 IEEE Sustainable Power and Energy Conference (iSPEC), Beijing, China, 2019, pp. 1060–1065.

[22] R. Zhang, T. Jiang, F. F. Li, G. Li, H. Chen, and X. Li, "Coordinated bidding strategy of wind farms and power-to-gas facilities using a cooperative game approach," IEEE Trans. Sustain. Energy, vol. 11, no. 4, pp. 2545–2555, Oct. 2020.

[23] X. Qi, J. Zheng, F. Mei, J. Wu, and Y. Zhang, "Low carbon economic dispatch of integrated energy systems considering stepped carbon trading," 2023 7th International Conference on Green Energy and Applications (ICGEA), Singapore, 2023, pp. 66–71.

[24] S. Clegg and P. Mancarella, "Integrated modeling and assessment of the operational impact of power-to-gas (P2G) on electrical and gas transmission networks," IEEE Trans. Sustain. Energy, vol. 6, no. 4, pp. 1234–1244, Oct. 2015.

[25] Z. Feng, W. Tang, Z. Niu, and Q. Wu, "Bi-level allocation of carbon emission permits based on clustering analysis and weighted voting: A case study in China," Appl. Energy, vol. 228, pp. 1122–1135, Oct. 2011.

[26] X. Fang, Q. Hu, F. Li, and B. Wang, "Coupon-based demand response considering wind power uncertainty: A strategic bidding model for load serving entities." IEEE Trans. Power Syst., pp. 1–13, May 2015.

[27] M. Pourakbari-Kasmaei, M. Asensio, M. Lehtonen, and J. Contreras, "Trilateral planning model for integrated community energy systems and PV-based prosumers: A bilevel stochastic programming approach," IEEE Trans. Power Syst., vol. 35, no. 1, pp. 346–361, Jan. 2020.

[28] C. Ruiz and A. J. Conejo, "Pool strategy of a producer with endogenous formation of locational marginal prices," IEEE Trans. Power Syst., vol. 24, no. 4, pp. 1855–1866, Nov. 2009.

[29] P. Rocha, T. K. Das, V. Nanduri, and A. Botterud, "Impact of CO_2 cap-and-trade programs on restructured power markets with generation capacity investments," Int. J. Electr. Power Energy Syst., vol. 71, pp. 195–208, Oct. 2015.

[30] X. Zhou, G. James, A. Liebman, Z. Y. Dong, and C. Ziser, "Partial carbon permits allocation of potential emission trading scheme in Australian electricity market," IEEE Trans. Power Syst., vol. 25, no. 1, pp. 543–553, Feb. 2010.

[31] R. Zhang, T. Jiang, G. Li, X. Li, and H. Chen, "Stochastic optimal energy management and pricing for load serving entity with aggregated TCLs of smart buildings: A Stackelberg game approach," IEEE Trans. Ind. Inform. doi: 10.1109/TII.2020.2993112.

[32] R. Bo, F. Li, and K. Tomsovic, "Prediction of critical load levels for AC optimal power flow dispatch model," Int. J. Electr. Power Energy Syst., vol. 42, pp. 635–643, Nov. 2012.

20

Enhancing Energy Market Dynamics through Coordinated Wind–P2G Bidding Strategies

Rufeng Zhang, Tao Jiang, Fangxing Li, Guoqing Li, Houhe Chen, and Xue Li

Nomenclature
Indices

s/s'	index of scenarios
t	index of time periods
w	index of wind farms

Parameters

N_s/N_w	number of scenarios and wind farms
$P_{P2G,max}$	the maximum operating power of P2G in an hour
P_S	the probability of scenario s
T	number of time periods
η	energy conversion efficiency of P2G facility
$P_{ac,s\,w,t}$	actual power output of wind power at period t and scenario s
$P_{max\,w,im}$	maximum value of wind power imbalance
$P_{max\,w}$	installed capacity of the wind farm w
pr_{gas}	gas price of natural gas market at period t
$pr_{R,s\,t}$	reserve market price at period t and scenario s
$prs_{D,t}$	day-ahead market price at period t and scenario s
$prs_{r,t}$	real-time market price at period t and scenario s
$prs_{up,t/prs\,dn,t}$	positive and negative balancing prices at period t and scenario s

Variables

$\Delta P^s_{w,t,dn}/\Delta P^s_{w,t,up}$	positive and negative imbalance of the wind farm w at period t and scenario s
$\Delta^s_{w,t}$	total energy imbalance at period t and scenario s
$PF^s_{P2G,t}$	payoff of P2G facility at period t and scenario s
$PF^s_{P2Gr,t}$	payoff of P2G facility in reserve market at period t and scenario s
$F^s_{w,R,t}$	reserve cost of wind farm w at period t and scenario s
$F^s_{w,t}$	imbalance revenue at period t and scenario s
$PF^s_{w,t}$	revenue of wind power from day-ahead market at period t and scenario s
$Profit$	expected profit of the coalition
$P^s_{P2G,t}$	bidding power of P2G in day-ahead market at period t and scenario s

DOI: 10.1201/9781032719436-20

$P^s_{P2G,w,t}$	operating power of P2G provided by wind power at period t and scenario s
$P^s_{w,t}$	bid of wind farm w at period t and scenario s
$R^s_{P2Gr,t}/R^s_{P2G,w,t}$	reserve provided by the P2G facility in reserve market and to wind farm at period t and scenario s
$R^s_{P2G,t}$	total reserve provided by the P2G facility
$R^s_{w,t}$	reserve requirements in reserve market of wind power at period t and scenario s
$S^s_{P2G,t,gas}$	volume of SNG at period t and scenario s
$V(K)$	profit created by the alternation of the members of coalition
$v(K)$	profit of coalition K
ε	arbitrary small real number for profit
$\mu_{w,t,s}$	binary variable

20.1 INTRODUCTION TO COORDINATED BIDDING IN ENERGY MARKETS

20.1.1 BACKGROUND OF WIND–P2G SYNERGY

The penetration of renewable energy in power systems is increasing rapidly to cope with energy crises and greenhouse gas emissions. Wind power is one of the most important renewable generations in power systems; however, it is highly volatile [1]. The uncertainty of wind power results in power imbalances between forecasting outputs and the actual output values, which calls for additional reserve capacities from ancillary service markets to ensure real-time power balance.

Most deregulated electricity markets require producers to submit their energy offers to the day-ahead energy market for the next day. The selected power providers are paid at the obtained market clearing prices [2]. The power producers bid using strategies aimed at maximizing their profits. Wind producers offer hourly generation quantities based on forecasted wind speed and participate in electricity markets to maximize their profits [3–6]. However, due to the uncertainties of wind power forecasting, there are mismatches between the forecasted and the actual wind power. Such mismatches result in real-time market penalties. On the other hand, market prices are also uncertain, which further increases the financial risks of wind power producers [7]. Hence, several bidding strategies have been studied for wind power producers to mitigate wind power trading risks [2, 8–10]. An optimal day-ahead energy bidding strategy that maximizes the expectation of revenue has been derived in [8], in which the sensitivity of optimal expected profit to uncertainty is explored. Correlation between the wind power and imbalance prices is considered in the strategy proposed in [9]. Wind power producers can also participate in the reserve market to buy reserve capacities for the imbalances of wind power. In [10], a stochastic model to maximize a wind power producer's revenue is proposed, in which the wind power producer can purchase its own reserve capacity to offset the uncertainty of wind power. Optimal bidding strategies to maximize the total profits of wind and conventional power producers in both the energy market and bilateral reserve market are proposed in [11]. It is of great significance to develop effective bidding strategies to purchase reserve for wind power producers.

20.1.2 SIGNIFICANCE OF COORDINATED BIDDING STRATEGIES

Coordinated bidding of wind power with other resources is regarded as an effective way to mitigate wind power trading risks, such as energy storage [6, 12–15], thermal units [7], and hydropower or pumped storage [16, 17]. Power-to-gas (P2G) technology is regarded as a viable energy storage method to accommodate wind power in power system operations, which produces synthetic natural gas (SNG) by the consumption of electric energy [18]. P2G technology is considered a promising approach to realize high penetration of renewables and low-carbon emission [19]. The operational impacts of P2G on electricity and gas transmission networks are investigated in [20]. Under an electricity market framework, the power consumption of P2G corresponds to its operating cost, and the produced SNG can be traded in natural gas markets for revenue. The feasibility of P2G in electricity markets dominated by renewables is analyzed in [21]. An optimal day-ahead scheduling model of P2G energy storage and gas load management in electricity and gas markets is proposed in [22]. However, very little research has been carried out on bidding strategy development for P2G facilities to realize arbitrage. Moreover, based on its operating mechanism, P2G can be regulated to participate in an ancillary service market to increase its revenue [23]. In [23], a scenario-based stochastic model to determine the bidding strategy for integrated natural gas generating units (NGGs) and P2G facilities in energy and regulation markets is proposed. In fact, the regulation ability from P2G can be utilized to provide a reserve for wind power.

Cooperative game theory is widely used for making bidding strategies [24] and loss allocation [25]. In the proposed bidding strategy, wind farms and P2G facilities form a coalition and participate in day-ahead energy, real-time, and reserve markets. P2G facilities participate in the day-ahead electricity market and the reserve market for arbitrage. As an important renewable power source, wind power can bid with other resources to mitigate trading risks. Based on the proposed bidding strategy, the revenues of wind farms and P2G facilities can be increased. Hence, the operation economics of wind farms can be enhanced. Meanwhile, for P2G facilities, there are currently not enough time intervals with electricity prices low enough to realize arbitrage [21], and the investment costs of P2G facilities are difficult to be paid back if no additional revenue is explored. This chapter shows that the coordination between wind farms and P2G facilities can easily enhance the operational economics of P2G facilities. Cost–benefit analysis demonstrates that the payback period can be reduced based on the proposed coordinated bidding scheme. Thus, this motivates market players to invest or obtain P2G facilities. Hence, the proposed approach can enhance the financial portfolio of the integrated system with wind power and P2G facilities.

20.2 THEORETICAL FRAMEWORK OF COORDINATED BIDDING

20.2.1 COOPERATIVE GAME APPROACH FOR WIND FARMS AND P2G

20.2.1.1 Market Framework and Assumptions

In a pool-based electricity market, power suppliers and consumers submit day-ahead energy supply and demand bidding offers for each hour. The market operator aggregates the presented bidding curves to calculate hourly market clearing prices. In most US markets, the day-ahead energy and reserve markets are cleared at a given time of the current operating day. The real-time market is carried out to ensure the power balance [11]. In the real-time market, when the total supply of the system is less than the demand, the suppliers whose actual outputs are lower than their bidding outputs (i.e., negative imbalances) are penalized at real-time prices. If the imbalances are positive, the producers are paid at the real-time prices. For wind power producers, wind power uncertainty leads to imbalances, in which the negative imbalances are usually penalized and reduce the profits of wind power. The power imbalance can be mitigated by bidding in the reserve market to reduce the impact of imbalance penalties. In the bilateral reserve market mechanism proposed in [11], wind producers are allowed to buy energy from the bilateral reserve market to minimize the risk of penalties associated with power imbalances. In this chapter, we further propose a market mechanism in which wind power producers can purchase reserve capacities in advance, with the consideration of the potential negative imbalance during real-time market operation, which can hedge the risk of high negative imbalance penalty. Also, the amount of purchased reserve should be a trade-off between conservative over-purchase of reserve and the risk of high penalty due to negative imbalance during real-time operation. The purchased reserve bids of wind farms are submitted to reserve market and cleared in advance. However, real-time market imbalance penalty prices and reserve market prices are uncertain and different, and wind power producers can choose to pay for negative imbalances or reserve capacities according to price signals.

The two-price balancing market framework is considered in [10], and the mechanism for real-time and reserve market prices can be explained as follows:

1. If the bidding energy output of wind power is less than the actual output in the real-time market, the imbalance is positive and the wind power producer sells the extra energy in the real-time market at the positive imbalance price *prs up,t*, which is equal to the lower value between the day-ahead market price and real-time market price:

$$\begin{cases} pr_{up,t}^{s} = \min(pr_{D,t}^{s}, pr_{r,t}^{s}) & \forall t,s \\ pr_{dn,t}^{s} = pr_{r,t}^{s} & \forall t,s \end{cases} \quad (20.1)$$

2. If the bidding energy output of wind power is greater than the actual output in the real-time market, the imbalance is negative and the wind power producer pays for the imbalance in the real-time market at the negative imbalance price *prs dn,t*, which is equal to the higher value between the day-ahead market price and real-time market price:

$$\begin{cases} pr_{up,t}^{s} = pr_{r,t}^{s} & \forall t,s \\ pr_{dn,t}^{s} = \max(pr_{D,t}^{s}, pr_{r,t}^{s}) & \forall t,s \end{cases} \quad (20.2)$$

3. In this chapter, wind power producers only purchase reserve for negative imbalances (which cause a penalty). The reserve prices are also uncertain.

Several assumptions are included in this chapter: (1) Wind power producers are not market price-makers, and the operation cost of wind power generation is ignored. (2) In the introduced reserve market, the reserve is provided by the grid and P2G and consumed by wind power producers at reserve market prices. Wind farms cannot provide reserve capacity, and reserve market prices are uncertain.

20.2.1.2   Coordination Framework and Benefits

Because of the uncertainty of wind power prediction, wind power producers must pay high costs for negative imbalance penalties or reserve costs. Market clearing prices are uncertain as well. When the day-ahead market prices are low, the corresponding revenue of wind power producers decreases.

To increase the profits of wind power producers and realize arbitrage, P2G facilities, which can achieve bidirectional coupling of integrated electricity and natural gas systems, are introduced to bid in the electricity market coordinated with wind power. The P2G technology utilizes electricity power to produce SNG through the process of electrolysis. The chemical process can be described as: $2H_2O \rightarrow 2H_2 + O_2$, $CO_2 + 4H_2 \rightarrow CH_4 + 2H_2O$. In the first step, gaseous hydrogen is formed by the process of electrolysis with a certain efficiency. The generated hydrogen is injected into natural gas pipelines at a limited rate. In the second step, synthetic natural gas is produced. Both processes can be realized with the consumption of electric power as the input energy. More detailed models of P2G technology can be found in [18, 19]. The main advantage of P2G facility is that it can utilize electricity power during low-price periods and store the power in the form of hydrogen or SNG for long-term storage, which generally has a lower cost than battery-based energy storage technologies. The consumed power can come from surplus wind power, which may be otherwise curtailed. However, the investment costs of P2G facilities are high, and there are currently not enough periods with low enough electricity price to realize arbitrage [21]. Hence, the investment cost of P2G

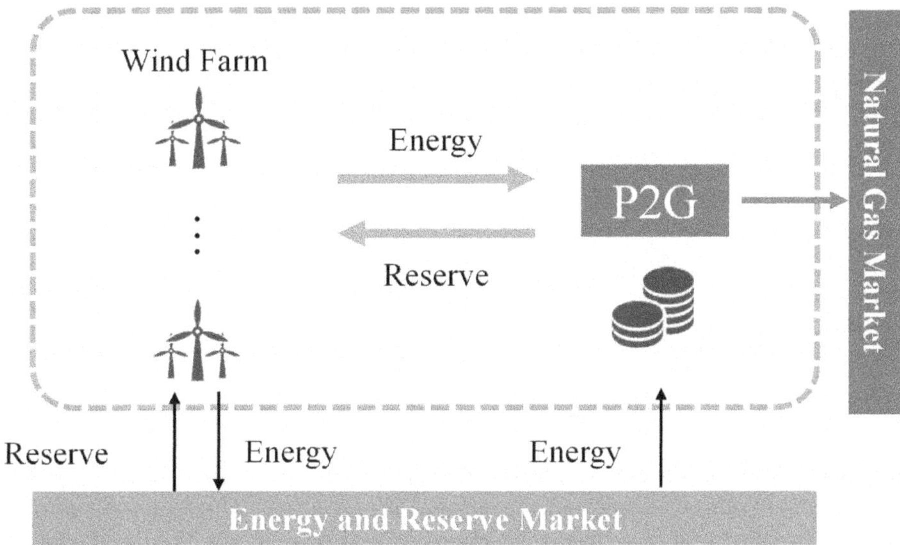

FIGURE 20.1 Schematic of wind farm and P2G with gas storage.

facilities is usually considered difficult to be paid back, while this chapter proposes bidding strategy for P2G facilities to gain additional benefits when operated in coalition with wind power.

The integration of P2G facilities enables their owners to utilize cheap electricity energy to realize arbitrage by selling the produced SNG in the natural gas market, which is beneficial when the price gap between electricity and natural gas is large enough [23]. Moreover, P2G facilities can participate in ancillary service markets by providing regulation or reserve services, which can be utilized by wind farms. A schematic of wind farms and P2G with natural gas storage is presented in Figure 20.1, and the bidding framework is discussed as follows.

Individual operation of a wind farm. For a wind farm w, the revenue of wind power from the day-ahead market is equal to the hourly bid, multiplied by the day-ahead market price, which can be expressed as:

$$PF_{w,t}^s = pr_{D,t}^s P_{w,t}^s \quad \forall t,s,w \tag{20.3}$$

The imbalance revenue of wind power is equal to the revenue from positive energy imbalances in the real-time market minus the penalty cost for negative energy imbalances in the real-time market, which can be expressed as:

$$F_{w,t}^s = pr_{up,t}^s \Delta P_{w,t,up}^s - pr_{dn,t}^s \Delta P_{w,t,dn}^s \quad \forall t,s,w \tag{20.4}$$

Remark 1: To avoid the high penalties of negative imbalances, wind power producers can choose to strategically purchase reserve in the reserve market when the reserve market price is lower than the real-time market negative imbalance price, and the cost of buying reserve is:

$$F_{w,R,t}^s = pr_t^R R_{w,t}^s \quad \forall t,s,w \tag{20.5}$$

Hence, the imbalance revenue of wind power is equal to the revenue from positive energy imbalances in the real-time market minus the penalty cost for negative energy imbalances, in which the penalized negative energy imbalances should be equal to the difference between total negative energy imbalances and purchased reserve in the reserve market:

$$F_{w,t}^s = pr_{up,t}^s \Delta P_{w,t,up}^s - pr_{dn,t}^s (\Delta P_{w,t,dn}^s - R_{w,t}^s) \quad \forall t, s \tag{20.6}$$

If the reserve bid exceeds the value of the wind power imbalances, it's not reasonable for wind power producers to pay for extra reserve. Hence, wind power producer w would bid reserve at a value less than the negative power imbalance:

$$R_{w,t}^s \leq \Delta P_{w,t,dn}^s \quad \forall t, s \tag{20.7}$$

Coordinated with P2G facility as a coalition. When wind farms and P2G facilities become a coalition, wind farms can provide energy to P2G facilities and get reserve capacities in return according to price signals through a bilateral deal. When the day-ahead market clearing price is low enough, a P2G facility can choose to purchase electricity energy to be transformed into SNG and sell SNG to the natural gas system for arbitrage. The utilized energy of P2G can be provided by wind farms. At the same time, the operating power of P2G can be reduced to provide reserve capacities when negative imbalances of wind power occur, which can reduce the reserve costs of the wind power producers. By reasonable profit allocation, profits of wind farms and P2G facilities can both be increased.

Hence, the coordinated bidding strategy of wind farms and P2G facilities as a coalition in the electricity market is required. The revenues and benefits are further explained as follows:

1. Electricity power can be transformed into SNG to be sold to the natural gas system at the natural gas price for arbitrage when the day-ahead electricity market prices are low enough. In the coalition, we assume wind farms provide low-priced energy to P2G facilities. The corresponding revenue for wind farms to provide low-priced energy to P2G facilities is equal to the cost of P2G. Hence, within the coalition, the energy provided by a wind farm to a P2G facility is free. When P2G is integrated, the revenue of P2G in the day-ahead market is equal to the revenue of selling SNG in the natural gas market minus the cost of buying electricity energy in the day-ahead electricity market:

$$PF_{P2G,t}^s = pr_{gas} S_{P2G,t,gas}^s - pr_{D,t}^s P_{P2G,t}^s \quad \forall t, s \tag{20.8}$$

The SNG total is generated by P2G facilities consuming energy from wind farms and the day-ahead market:

$$S_{P2G,t,gas}^s = \eta \left(P_{P2G,t}^s + \sum_{w}^{N_w} P_{P2G,w,t}^s \right) \quad \forall t, s \tag{20.9}$$

2. Providing reserve capacity for wind power negative imbalances to reduce imbalance penalties or reserve bid by wind power producers in the reserve market by reducing the operating power of P2G facilities, which can reduce the cost for wind power producers. Furthermore, there are only a few periods with adequate low electricity market prices for P2G to realize

arbitrage by selling SNG [21]. Participating in reserve markets can increase the revenue of P2G. Hence, the reserve provided by P2G includes two parts: reserve bids in the reserve market and low-priced reserve capacities provided to wind farms in the coalition:

$$R^s_{P2G,t} = R^s_{P2Gr,t} + \sum_w^{N_w} R^s_{P2G,w,t} \quad \forall t,s \tag{20.10}$$

The reserve provided by P2G is also free for the coalition. The reserve revenue of P2G in the reserve market is:

$$PF^s_{P2Gr,t} = pr^R_t R^s_{P2Gr,t} \quad \forall t,s \tag{20.11}$$

When considering reserve provided by P2G, the imbalance revenue of the wind farm w should be:

$$F^s_{w,t} = pr^s_{up,t} \Delta P^s_{w,t,up} - pr^s_{dn,t} (\Delta P^s_{w,t,dn} - R^s_{P2G,w,t} - R^s_{w,t}) \quad \forall t,s \tag{20.12}$$

Remark 2: For the joint profit of the coalition, the exchanged energy and reserve between the players are free. The reason is that the corresponding bilateral deals are within the coalition, and the buying cost equals the selling revenue.

Similarly, the total reserve for the wind farm should be less than the total imbalances for each time t under each scenario s:

$$R^s_{P2G,w,t} + R^s_{w,t} \leq \Delta P^s_{t,dn} \quad \forall t,s \tag{20.13}$$

20.2.2 Stochastic Programming in Bidding Strategies

The coordinated bidding problem of a coalition formed by wind farms and P2G facilities is formulated as a stochastic mixed-integer linear programming (MILP) model to trade energy in day-ahead, real-time, and reserve markets. Two bidding strategy models are proposed with day-ahead decision variables, depending on stochastic scenarios or not.

20.2.2.1 Bidding Strategy 1 (BS1): Day-Ahead Offers/Bids Depend on Scenarios

A stochastic optimization model with day-ahead decision variables depending on scenarios is proposed in this section for a coordinated bidding strategy of the coalition formed by wind farms and P2G facilities. The objective function of the coordinated bidding problem BS1 is to maximize the expected profit for the coalition. The optimal coordinated bidding strategy can be represented as follows:

$$\max \sum_s^{N_s} Ps \times \left\{ \sum_t^T \left[\left(\sum_w^{N_w} PF^s_{w,t} + F^s_{w,t} - F^s_{w,R,t} \right) + PF^s_{P2G,t} + PF^s_{P2Gr,t} \right] \right\} \quad \forall t,s \tag{20.14}$$

s.t.

$$\text{Constraints in (20.3), (20.5), (20.8)–(20.13)} \tag{20.15}$$

$$0 \leq P^s_{w,t} + P^s_{P2G,w,t} \leq P^{max}_w \quad \forall t,s \tag{20.16}$$

$$0 \leq P^s_{w,t} \leq P^{max}_w \quad \forall t,s \tag{20.17}$$

$$0 \leq P^s_{P2G,w,t} \leq P^{max}_w \quad \forall t,s \tag{20.18}$$

$$\Delta^s_{w,t} = P^{ac,s}_{w,t} - P^s_{w,t} - P^s_{P2G,w,t} \quad \forall t,s \tag{20.19}$$

$$\Delta^s_{w,t} = \Delta P^s_{w,t,up} - \Delta P^s_{w,t,dn} \quad \forall t,s \tag{20.20}$$

$$0 \leq \Delta P^s_{w,t,up} \leq \alpha_{w,t,s} P^{max}_{w,im} \quad \forall t,s \tag{20.21}$$

$$0 \leq \Delta P^s_{w,t,dn} \leq \left(1 - \alpha_{w,t,s}\right) P^{max}_{w,im} \quad \forall t,s \tag{20.22}$$

$$P^s_{w,t} = P^{s'}_{w,t} : pr^s_{D,t} = pr^{s'}_{D,t} \quad \forall t,s,w \tag{20.23}$$

$$(pr^s_{D,t} - pr^{s'}_{D,t})(P^s_{w,t} - P^{s'}_{w,t}) \geq 0 \quad \forall t,s,w \tag{20.24}$$

$$0 \leq P^s_{P2G,t} + \sum_{w}^{N_w} P^s_{P2G,w,t} \leq P_{P2G,max} \quad \forall t,s \tag{20.25}$$

$$0 \leq P^s_{P2G,t} \leq P_{P2G,max} \quad \forall t,s \tag{20.26}$$

$$0 \leq \sum_{w}^{N_w} P^s_{P2G,w,t} \leq P_{P2G,max} \quad \forall t,s \tag{20.27}$$

$$0 \leq R^s_{P2G,t} \leq P^s_{P2G,t} + \sum_{w}^{N_w} P^s_{P2G,w,t} \quad \forall t,s \tag{20.28}$$

$$0 \leq R^s_{P2Gr,t} \leq P^s_{P2G,t} + \sum_{w}^{N_w} P^s_{P2G,w,t} \quad \forall t,s \tag{20.29}$$

$$0 \leq R^s_{P2G,w,t} \leq P^s_{P2G,t} + \sum_{w}^{N_w} P^s_{P2G,w,t} \quad \forall t,s \tag{20.30}$$

The expected profit of the coalition consists of five terms, as shown in (20.14): (1) the revenue of wind production, (2) the imbalance revenue of wind power, (3) the revenue of SNG production, (4) reserve revenue of P2G, and (5) the reserve cost. The operation cost of wind power is negligible, and the cost of power consumption of P2G is equal to the operating power multiplied by the day-ahead market price. The natural gas price is assumed to be constant.

In the proposed model (20.14)–(20.30), $P^s_{w,t}$, $P^s_{P2G,t}$, $P^s_{P2G,w,t}$, $R^s_{w,t}$, $R^s_{P2Gr,t}$, and $R^s_{P2G,w,t}$ are day-ahead decision variables (i.e., all generation bids for energy and reserve). Further, $\Delta P_{w,t,dn}$, $\Delta P^s_{w,t,up}$, $\Delta^s_{w,t}$, $R^s_{P2G,t}$, and $S^s_{P2G,t,gas}$ are also unknown variables related to imbalance. All variables are listed in the "Nomenclature."

Constraints (20.16)–(20.18) indicate that the power bids of wind farms, including the hourly bid of wind power $P^s_{w,t}$ in the day-ahead market and the energy consumed by P2G, should be positive and cannot exceed the installed capacity. Constraints (20.19)–(20.22) determine the wind power

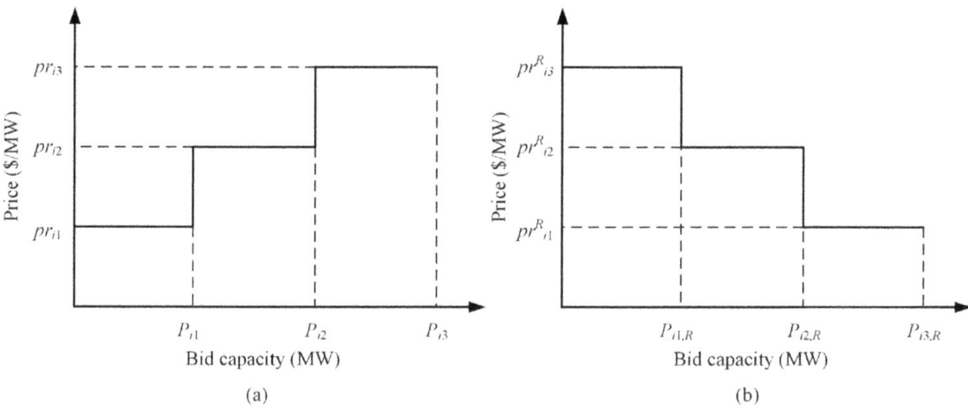

FIGURE 20.2 Scenario-based bidding curve for wind farm i (a) in energy market as a seller, (b) in reserve market as a buyer.

imbalances per period and scenario. Constraint (20.19) states that the total imbalance is calculated as the difference between the actual wind power generation and the hourly bid of wind power. The total imbalances consist of the positive and negative imbalances in (20.20). The positive and negative imbalances should be less than the allowable value $Pmax\,w,im$ in (20.21) and (20.22), and a binary variable, $\mu_{w,t,s}$, is introduced. The actual power of wind turbine is determined by uncertain scenarios, that is, the actual power of wind turbine equals the power in scenario s. To avoid unacceptable imbalances, we should add limitations on the amount of imbalances in (20.21) and (20.22). The allowable value $Pmax\,w,im$ limits wind power imbalances within a reasonable range. In most energy markets, hourly bidding offers should not decrease with respect to market prices [7]. Hence, constraints (20.23)–(20.24) are utilized to obtain non-decreasing offer curves. Constraints (20.25)–(20.27) denote the maximum operating power limits of P2G facilities. Constraints (20.28)–(20.30) enforce that the reserve provided by P2G cannot be more than its operating power.

Note that the bids of wind farms and P2G facilities in day-ahead and reserve market depend on scenarios, which would form an offer curve instead of a single value at each trading period. The day-ahead hourly bidding quantities of a wind farm obtained in BS1 correspond to market prices in different scenarios. Then, the obtained price and bid capacity pairs can be utilized to construct a piecewise non-decreasing bidding curve [26], with the consideration of constraints (20.23) and (20.24). In electricity markets, participants can submit the bidding curves as shown in Figure 20.2. In other words, if a specific scenario happens in real time, the actual bidding solution will always land on a point in the bidding curve in Figure 20.2.

20.2.2.2 Bidding Strategy 2 (BS2): Day-Ahead Offers/Bids Do Not Depend on Scenarios

In the general model discussion in previous section, the day-ahead offers of wind farms and bids of P2G facilities vary with scenarios and are calculated as offer/bid curves. In this subsection, we further propose a coordinated bidding strategy of the coalition formed by wind farms and P2G facilities with offers independent of scenarios, in which the day-ahead offer of each wind farm and the bid of P2G facilities are obtained as single values in each period. The optimization model of BS2 can be formulated based on BS1 by removing (20.23)–(20.24) and replacing $P^s_{w,t}$, $P^s_{P2G,t}$, $P_{sP2G,w,t}$, $R^s_{w,t}$, and $R^s_{P2Gr,t}$ by $P_{w,t}$, $P_{P2G,t}$, $P_{P2G,w,t}$, $R_{w,t}$, and $R_{P2Gr,t}$. The formulations in (20.23) and (20.24) are constraints utilized to obtain non-decreasing offer curves of wind power, so they should be removed from BS2. Note that the reserves provided by P2G to wind farms depend on scenarios to follow negative imbalances and should not be submitted to the reserve market.

20.2.2.3 Cooperative Game–Based Profit Allocation Method

The cooperative bidding strategy can increase the profit of the coalition, which is higher than the sum of the profits of the individual bidding strategies of wind farms and P2G facilities. The increased profit is allocated based on cooperative game theory–based allocation methods. *nucleolus theory* and *Shapley value* methods are utilized to allocate the profit of the coalition of wind farms and P2G facilities.

Nucleolus theory. The existence of a nucleolus is unique, and if the core of the game is non-empty, the nucleolus is always in the core [25, 27]. Suppose $\mathbf{X} = \{x_1, x_2, \ldots, x_n\}$ is the set of allocation profit of each participant (i.e., wind farm and P2G facility), $\mathbf{Y} = \{y_1, y_2, \ldots, y_n\}$ is the set of the profit allocation imputation, and $V(K)$ is the profit created by the alternation of the members of the coalition. By solving the following linear programming (LP) optimization model, the profit allocated to each wind farm and P2G facility can be determined fairly [25, 28]:

$$\min \quad \varepsilon \tag{20.31}$$

$$s.t. \quad V(K) = \sum_{i \in K_1} y_i \tag{20.32}$$

$$V(K) - \sum_{i \in K_2} y_i \leq \varepsilon \tag{20.33}$$

Finally, the allocated profit to each wind farm or P2G facility is calculated by the summation of the profit created by the coalition and the profit created by individual bidding.

$$x_i = y_i + v(i) \tag{20.34}$$

Shapley value. The Shapley value method is further adopted to allocate profit based on individual profits of the wind farm and the P2G facility. The Shapley value technique is a method used in cooperative game theory to allocate the total profit or payoff among the players of a coalition in a fair and equitable manner. It was introduced by Lloyd Shapley in 1953 and has since become a widely accepted and applied approach for solving profit allocation problems in various fields, including economics, finance, and operations research. In a cooperative game, players can form coalitions to achieve a common goal, such as maximizing profit. The Shapley value technique aims to distribute the total profit generated by the grand coalition (the coalition of all players) in a way that reflects each player's marginal contribution to the coalition. The Shapley value of a player is calculated by considering all possible coalitions that the player could join and the marginal contribution that the player makes to each of these coalitions. Specifically, the Shapley value of a player is the average of the player's marginal contributions across all possible permutations of the players in the grand coalition.

Profit allocation based on the Shapley value can be calculated by [23]:

$$x_i = \sum_{K \notin K_i} \frac{(n - |K|)!(|K| - 1)!}{n!} \left[v(K) - v(K - \{i\}) \right] \tag{20.35}$$

where $|K|$ is the number of participants in the coalition K.

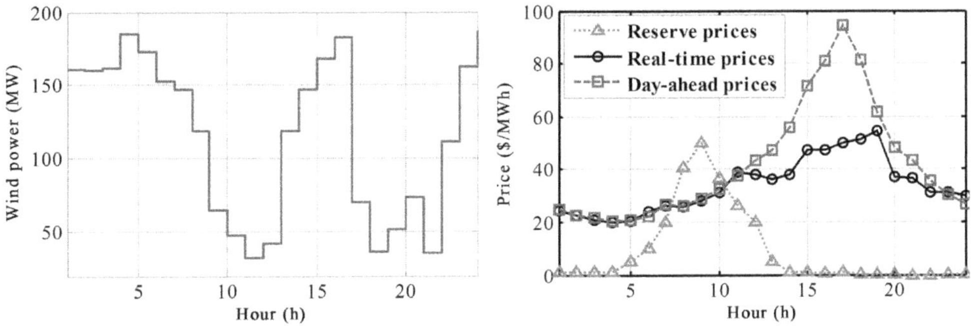

FIGURE 20.3 Expected wind power and market prices.

20.3 CASE STUDIES AND MODELING APPROACHES

20.3.1 Scenario Analysis in Electricity Markets

The proposed stochastic MILP optimization model is solved using CPLEX on a PC with Intel Core i7 3.00 GHz CPU and 8 GB RAM. We consider two wind farms with a P2G facility. The installed capacity of each wind farm is 200 MW with the same forecasting output, and the hourly maximum operating power of the P2G facility is 50 MW. In this chapter, the hourly wind power, electricity market clearing prices, and real-time market price forecasting errors are presented as a normal distribution function with mean equal to 0 and a preset standard deviation of the hourly forecasted value. By using the Monte Carlo simulation method [6], 1,000 scenarios are generated to simulate the uncertainties. To reduce the computation burden caused by the large number of scenarios, scenario reduction is applied. In this chapter, the SCENRED tool in the general algebraic modeling system (GAMS) is utilized for the scenario reduction process [29]. Probabilities for all scenarios before reduction are assumed to be the same, the sum of which is equal to 1 [30]. The expected day-ahead market clearing prices, reserve market prices, and real-time market prices are from the PJM historical market price data [31] and shown in Figure 20.3.

20.3.2 Evaluating Profit Maximization for Wind and P2G

20.3.2.1 Two-Player Game

In order to verify the effectiveness of the proposed coordinated bidding strategy, a two-player game of wind farm 1 and P2G, that is, $\Phi_2 = \{$wind farm 1 (WF1), P2G facility$\}$, is analyzed. The following three coordinating bidding strategies are carried out for comparison.

Case 1: The proposed coordinated bidding model.
Case 2: Coordinated bidding model, in which wind farm 1 and the P2G facility exchange no energy and reserve.
Case 3: Coordinated bidding model, in which wind farm 1 cannot purchase reserve for negative imbalances.

The comparison of results of the three bidding models based on BS1 is listed in Table 20.1. As can be observed in Table 20.1, the expected total profit of the coordinated bidding strategy of Case 1 based on cooperative game–based theory is higher than the values of Cases 2 and 3. The results verify that the coordination of the wind farm and P2G can bring in more profits. When the wind farm and P2G facility exchange energy and reserve in Case 1, the total operating power consumption of P2G is 460.41 MW to produce SNG and provide reserve for more revenue, which decreases

TABLE 20.1

Comparison of Results of Different Cases for Coordinated Bidding Strategies

Cases	Case 1	Case 2	Case 3
Expected profits ($10^5)	1.386	1.378	1.136
Expected total operating power of P2G (MWh)	460.41	322.45	330.66
Expected operating power of P2G in day-ahead market (MWh)	199.59	322.45	283.98

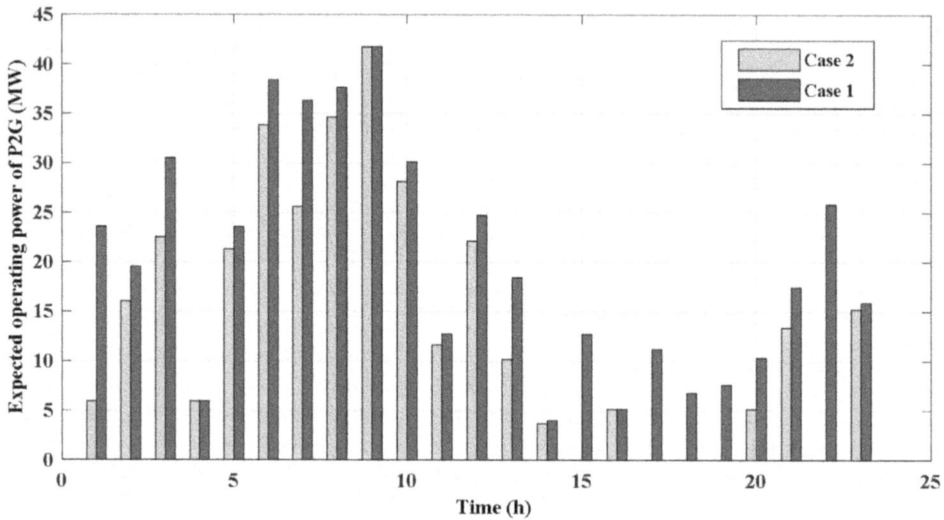

FIGURE 20.4   Expected operating power of P2G in Cases 1 and 2.

to 322.45 MW in Case 2. Note that the expected bid on operating power of P2G in the day-ahead market in Case 1 is 199.59 MWh, which shows a 38.1% reduction compared to the value in Case 2. For Case 3, the expected profit of the coalition sharply decreases compared to the values in Cases 1 and 2, which demonstrates that strategically purchasing reserve for negative imbalances can bring in more profit for the wind farm.

The comparison of the operating power of the P2G facility based on BS1 in Cases 1 and 2 is shown in Figure 20.4. It can be seen that in all periods, the expected operating power values of P2G in Case 1 are no less than those in Case 2. In hours 15, 17–19, and 22, no power is bid on by P2G in Case 2. However, when coordinated with the wind farm, low-priced energy can be provided by the wind farm in Case 1, and less-expensive energy from day-ahead market is bid on by P2G to avoid profit loss. The preceding results demonstrate that the coordination of the wind farm and P2G facility can increase the expected profit, demonstrating the effectiveness of the proposed coordinated bidding strategy of BS1.

To further verify the effectiveness of the proposed coordinated bidding strategy, the following two individual bidding cases for wind farm 1 and the P2G facility are proposed:

Case 4: Individual bidding model for wind farm 1 without P2G facility by setting P2G parameters as zeros.
Case 5: Individual bidding model for P2G facility without wind farm 1 by setting wind power parameters as zeros.

TABLE 20.2

Comparison of Results of Different Cases for Allocated Profits Based on BS1

Player	Uncoordinated Profits ($)	Using Nucleolus-Based Method		Using Shapley Value–Based Method	
		Profit ($)	*Surplus Profit ($)*	*Profit ($)*	*Surplus Profit ($)*
Wind farm 1	132,641 (Case 4)	133,162	520.9	133,162	520.9
P2G	4,904.5 (Case 5)	5,425.8	520.9	5,425.8	520.9

TABLE 20.3

Comparison of Results of Different Cases for Allocated Profits Based on BS2

Player	Uncoordinated Profits ($)	Using the Nucleolus-Based Method		Using the Shapley Value–Based Method	
		Profit ($)	*Surplus Profit ($)*	*Profit ($)*	*Surplus Profit ($)*
Wind farm 1	112,756 (Case 4)	113,103	347.40	113,103	347.40
P2G	2,961.30 (Case 5)	3,308.70	347.40	3,308.70	347.40

The comparison of results for coordinated bidding (Case 1) based on different profit allocation methods and uncoordinated bidding (Cases 4 and 5) based on BS1 is listed in Table 20.2. As seen in Table 20.2, the profits for the wind farm and P2G of coordinated bidding cases are all larger than those of uncoordinated bidding cases, showing a 0.4% increase in profit for wind farm 1 and a 9.6% increase in profit for the P2G facility. The total surplus profit can reach $1,041.8. Note that when only two players are included, the surplus profit is allocated equally, whether using the nucleolus-based method or the Shapley value–based method.

To verify the effectiveness of the proposed bidding strategy BS2, a comparison of results for coordinated bidding (Case 1) based on different profit allocation methods and uncoordinated bidding (Cases 4 and 5) based on BS2 is added into Table 20.3. As shown in the table, the profits of wind farm and P2G are both increased by $347.40 based on BS2 if compared with uncoordinated cases. This demonstrates that BS2 is also effective on increasing the profits of wind farm and P2G. If compared with the results in Table 20.2, the profit of the coalition based on BS2 is less than that of BS1, which demonstrates that submitting offer/bid curves tends to make more profits than submitting hourly offer/bid values under the test parameters and bidding methods. The reason is that BS1 gives a broader range of possible solution for wind farm and P2G facilities.

20.3.3 THREE-PLAYER GAME

To further verify the effectiveness of the proposed coordinated bidding strategy of BS1, a cooperative game for two wind farms and a P2G facility is analyzed in this section, that is, the coalition $\Theta_3 = \{$wind farm 1 (WF1), wind farm 2 (WF2), P2G facility$\}$. The parameters of market prices, wind farm 1, and the P2G facility are the same as the values in the two-player game.

The profits of coalitions in the coalition are listed in Table 20.4. Note that the profit increases with the enlargement of the scale of the coalition, which verifies again that coordination can bring in more profits. The results of the profit allocation using both the nucleolus-based and the Shapley value–based methods are listed in Table 20.5. As shown in Table 20.5, the surplus profits of the players are positive and larger than the values of surplus profits in the case of the two-player game,

TABLE 20.4
Profits of Coalitions in the Coalition

Coalition	Profits (10^5)
{WF1}	1.3264
{WF2}	1.2773
{P2G}	0.0490
{WF1, WF2}	2.6037
{WF1, P2G}	1.3859
{WF2, P2G}	1.3384
{WF1, WF2, P2G}	2.6968

TABLE 20.5
Comparison of Results of Allocated Profits Based on Different Methods

Player	Uncoordinated Profits (10^5)	Using the Nucleolus-Based Method		Using the Shapley Value–Based Method	
		Profit (10^5)	*Surplus Profit ($)*	*Profit (10^5)*	*Surplus Profit ($)*
WF1	1.3264	1.3411	1,470	1.3388	1,240
WF2	1.2773	1.2920	1,470	1.2905	1,320
P2G	0.0490	0.0637	1,470	0.0675	1,850
Total	2.6527	2.6968	4,410	2.6968	4,410

which demonstrates that the coordinated bidding strategy is effective for increasing profits of wind farms and P2G facilities, and enlarging the scale of coalition will bring in more profits. Note that, for a three-player game, results of allocated profits to each player are different for the nucleolus-based method and the Shapley value–based method. More profits are allocated to the P2G facility when using the Shapley value–based method.

20.3.4 DISTRIBUTION OF PROFITS IN A COOPERATIVE SETTING

In this section, the results of individual bidding of P2G facilities participating in the reserve market based on BS1 are compared to those participating in only the energy market and analyzed, and then the cost–benefit analysis of the P2G facility is carried out.

The comparison of hourly expected operating power of P2G facilities participating in the day-ahead energy market only and in both the day-ahead energy market and the reserve market based on BS1 is shown in Figure 20.5. The expected operating power of the P2G facility participating in both energy and reserve markets is more than the facility participating only in the energy market. The reason is that participating in the reserve market can bring in more revenue for P2G. The expected profits of P2G facilities participating in the energy market only and in both the day-ahead energy market and the reserve market are $196.7 and $4,904.5. The results demonstrate that participating in the reserve market can increase the revenue of the P2G facility.

The increase in profit incurred by P2G facilities participating in the reserve market and coordinating with wind farms should be analyzed against the investment costs of the P2G facility. In this section, we carry out a cost–benefit analysis of the P2G facility based on the individual bidding model and coordinated bidding models with various hourly maximum operating power (i.e.,

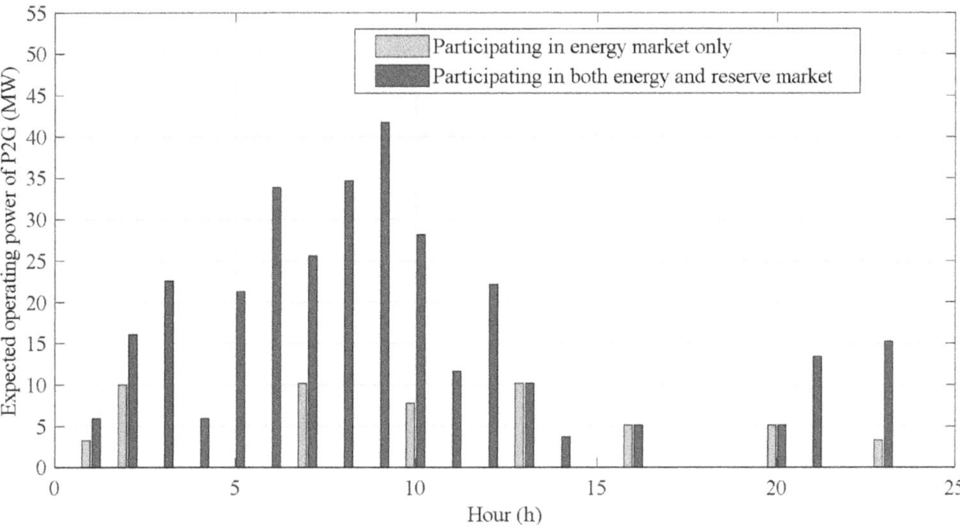

FIGURE 20.5 Comparison of expected operating power of P2G participating in the energy market only and in both the day-ahead energy market and the reserve market.

capacity) of P2G and natural gas prices based on BS1. Based on the cost parameters in [21] and [32], the investment cost of a P2G facility is assumed to be $1,137.6/kW. According to the results of sections A and B, coordinated bidding of P2G with wind farms will bring in more profit. The increase in profit of P2G can reach 10.62% in a two-player game, 30% using the nucleolus-based method, and 37.72% using the Shapley value–based method in a three-player game. The payback years of P2G with different capacities of P2G and natural gas prices based on different bidding strategies are shown in Figure 20.6.

Figure 20.6 shows the results of payback years with different capacities of P2G facilities and natural gas prices. It can be observed that when the capacity of a P2G facility is 70 MW and the natural gas price is $35/MWh, it takes 13 years to recover the investment costs with an individual bidding strategy. The increase on natural gas prices has limited influence on reducing the payback years, because the major revenue from the individual bidding of P2G facilities comes from participation in the reserve market. Hence, participating in an ancillary service market could increase the revenue of the P2G facility and reduce the payback years of the P2G facility. When coordinated with wind farms, the minimum payback years are as follows: 12 years for a two-player game, 10 years using the nucleolus-based method, and 9.5 years using the Shapley value–based method for a three-player game. The results demonstrate that the proposed coordinated bidding strategy can increase profits and reduce the number of payback years of P2G facilities.

To make the cost–benefit analysis more convincing, the payback periods are calculated based on a whole-year simulation. The simulation is carried out for individual bidding of P2G and a two-player game (i.e., a wind farm and a P2G facility with the same parameters in the previous subsection). Electricity market prices for one year are utilized and obtained from the PJM historical market price data [31]. Wind power forecasting output data are obtained from a wind farm in Inner Mongolia, China, as shown in Figure 20.7. Also, 1,000 uncertain scenarios are generated by Monte Carlo simulation method and then reduced to 5 representative scenarios for the convenience of calculation. Figure 20.8 shows the simulation results of payback years for individual and two-player bidding with different capacities of P2G facilities and natural gas prices. It can be seen that the payback years of the two-player bidding strategy are always less than those of individual bidding, and

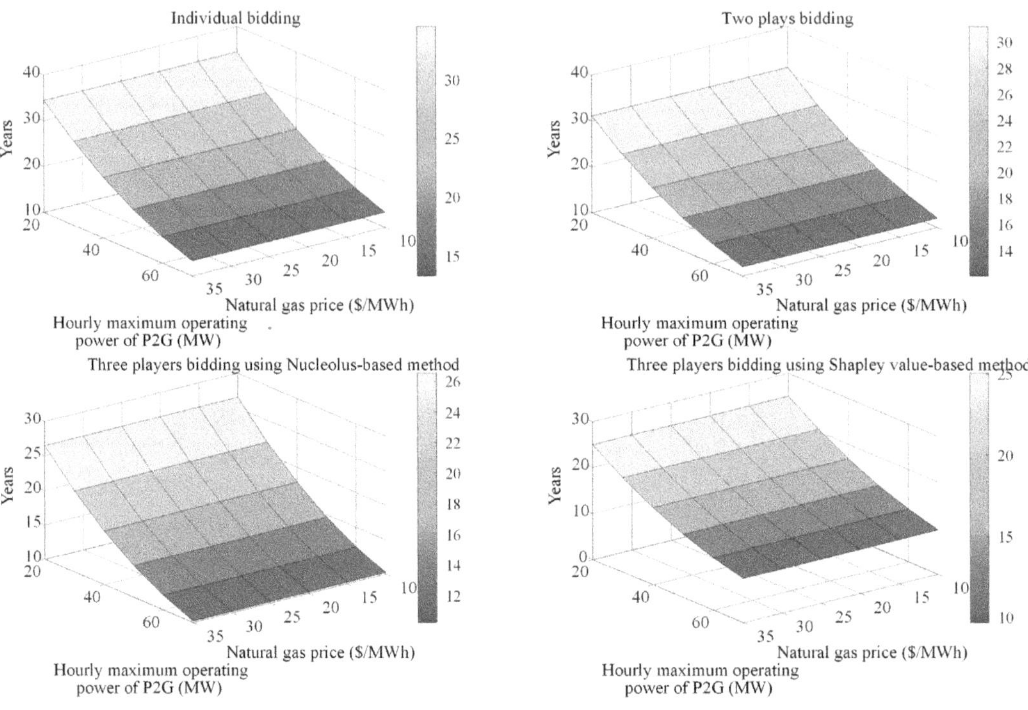

FIGURE 20.6 Payback years with different capacities of P2G and natural gas prices based on different bidding models.

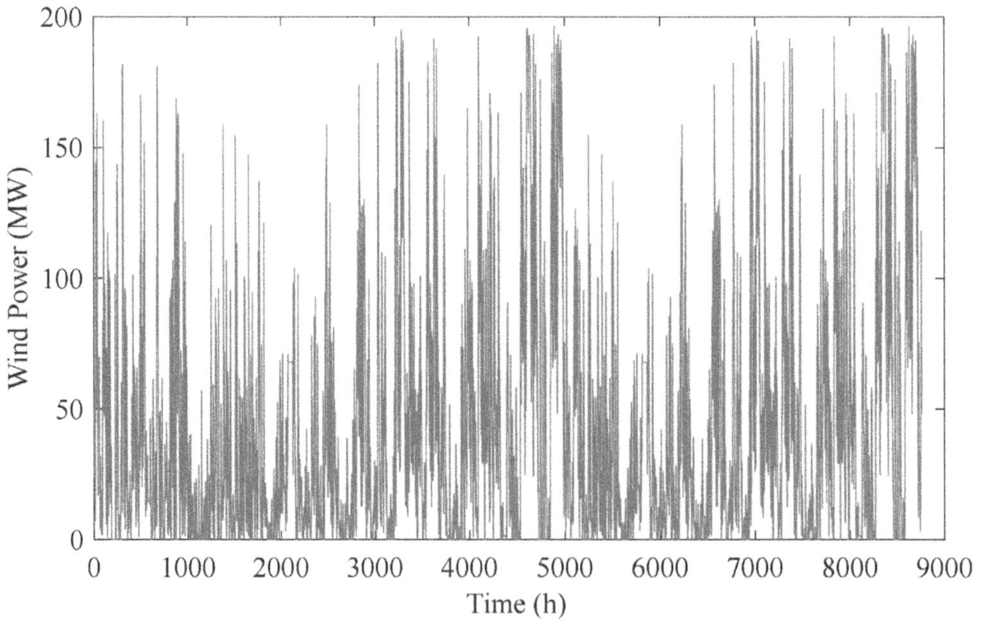

FIGURE 20.7 Annual wind power output.

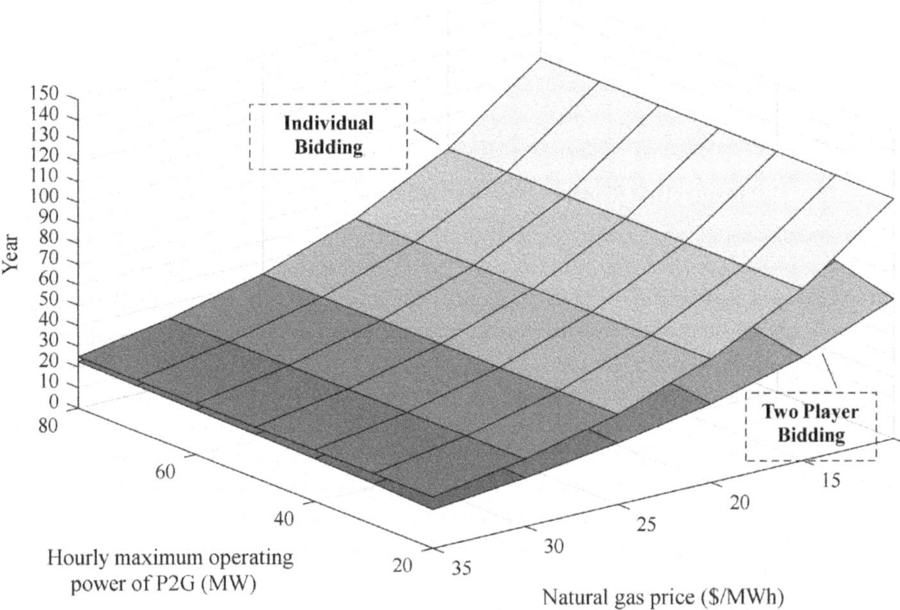

FIGURE 20.8 Payback years with different capacities of P2G and natural gas prices based on simulation for a year.

the minimum payback year of two-player bidding is about 18 years. The yearly simulation results also demonstrate that the proposed coordinated bidding strategy can reduce the number of payback years of P2G facilities.

As P2G technology continues to develop, the investment cost may decrease significantly [32], which would make the payback periods more acceptable.

20.4 IMPACTS ON ELECTRICITY MARKET AND ENERGY POLICY

20.4.1 IMPLICATIONS FOR MARKET DYNAMICS

In this chapter, a coordinated bidding strategy for wind farms and P2G facilities is proposed based on the cooperative game theory. The coalition of wind farms and P2G facilities exchanges energy and reserve to increase shared profit. The coordinated bidding strategy is modeled, the objective function of which is to maximize the expected profit of the coalition. Nucleolus- and Shapley value–based profit allocation methods are applied to fairly allocate the joint profits. Case studies based on real-world market prices are carried out to verify the effectiveness of the proposed bidding strategy. Key findings through case studies are:

1. This chapter provides a complete bidding strategy model using cooperative game with the consideration of wind farms and P2G facilities, which link the electricity and natural gas markets.
2. The key feature of the proposed algorithm is the utilization of the price gap between the two markets, gas and electricity, and thus further increase the profits of the coalition of wind farms and P2G facilities by exchanging energy and reserve capacities. For instance, strategic purchases of reserve for power imbalances according to price signals can increase the profits of a wind farm.

3. As demonstrated in the case studies, participation in the reserve market tends to provide a new means for P2G facilities to earn additional profit and thus increase the overall revenue and, moreover, reduce the payback years of the P2G facility.

20.4.2 POLICY RECOMMENDATIONS AND FUTURE DIRECTIONS

20.4.2.1 Policy Recommendations

Promotion of collaboration. Encourage collaboration between wind energy developers, PtG technology providers, utilities, and policymakers to identify optimal integration strategies. This includes joint research and development initiatives to improve technology efficiency and reduce costs.

Market mechanisms. Introduce market-based mechanisms, such as green certificates or carbon pricing schemes, that incentivize the use of renewable energy sources and PtG technologies. These mechanisms can provide a stable revenue stream for renewable energy producers and investors.

Grid infrastructure upgrades. Invest in grid infrastructure upgrades to accommodate the increased demand for renewable energy and ensure that PtG facilities can be connected to the grid efficiently. This includes upgrading transmission and distribution lines, installing smart grid technologies, and enhancing energy storage capabilities.

20.4.2.2 Future Directions

Advancements in PtG technology. Continued research and development of PtG technologies are crucial to improve their efficiency, reduce costs, and increase their scalability. Future directions should focus on developing more efficient electrolyzers, improving gas storage and transportation solutions, and integrating PtG with other renewable energy sources.

Integration with energy markets. Enhance the integration of PtG with energy markets by developing flexible bidding strategies that optimize the use of wind turbines and PtG facilities. This includes the development of advanced forecasting and scheduling algorithms that can predict energy demand and supply and optimize the use of renewable energy sources.

Cross-sector collaboration. Encourage cross-sector collaboration between the energy, transportation, and industrial sectors to leverage the benefits of PtG technology. For example, PtG can be used to produce hydrogen for transportation or as a feedstock for industrial processes, thereby reducing reliance on fossil fuels.

Sustainability and environmental considerations. As PtG technology is deployed at a larger scale, it is essential to consider its sustainability and environmental impacts. This includes evaluating the life cycle emissions of PtG systems, minimizing the use of non-renewable resources in their production, and ensuring that PtG facilities are constructed and operated in an environmentally responsible manner.

20.4.3 PROSPECTS FOR FUTURE RESEARCH AND APPLICATIONS

For future research, a more detailed cost–benefit analysis of P2G facilities would be performed, and risk constraints would be included.

REFERENCES

[1] H. Chen, F. Li, and Y. Wang, "Wind power forecasting based on outlier smooth transition autoregressive GARCH model," *Journal of Modern Power Systems and Clean Energy (Springer)*, vol. 6, no. 3, pp. 532–539, May 2018.

[2] V. Guerrero-Mestre, A. A. Sánchez de la Nieta, J. Contreras, and J. P. S. Catalão, "Optimal bidding of a group of wind farms in day-ahead markets through an external agent," *IEEE Transactions on Power Systems*, vol. 31, no. 4, pp. 2688–2700, July 2016.

[3] J. M. Morales, A. J. Conejo, and J. Perez-Ruiz, "Short-term trading for a wind power producer," *IEEE Transactions on Power Systems*, vol. 25, no. 1, pp. 554–564, February 2010.

[4] H. Huang and F. Li, "Bidding strategy for wind generation considering conventional generation and transmission constraints," *Journal of Modern Power System Clean Energy (Springer)*, vol. 3, no. 1, pp. 51–62, March 2015.

[5] P. Pinson, C. Chevallier, and G. N. Kariniotakis, "Trading wind generation from short-term probabilistic forecasts of wind power," *IEEE Transactions on Power Systems*, vol. 22, no. 3, pp. 1148–1156, August 2007.

[6] H. Wu, M. Shahidehpour, A. Alabdulwahab, and A. Abusorrah, "A game theoretic approach to risk-based optimal bidding strategies for electric vehicle aggregators in electricity markets with variable wind energy resources," *IEEE Transactions on Sustainable Energy*, vol. 7, no. 1, pp. 374–385, January 2016.

[7] A. T. Al-Awami and M. A. El-Sharkawi, "Coordinated trading of wind and thermal energy," *IEEE Transactions on Sustainable Energy*, vol. 2, no. 3, pp. 277–287, July 2011.

[8] E. Y. Bitar, R. Rajagopal, P. P. Khargonekar, K. Poolla, and P. Varaiya, "Bringing wind energy to market," *IEEE Transactions on Power Systems*, vol. 27, no. 3, pp. 1225–1235, August 2012.

[9] C. J. Dent, J. W. Bialek, and B. F. Hobbs, "Opportunity cost bidding by wind generators in forward markets: Analytical results," *IEEE Transactions on Power Systems*, vol. 26, no. 3, pp. 1600–1608, August 2011.

[10] E. Du, N. Zhang, C. Kang, B. Kroposki, H. Huang, M. Miao, and Q. Xia, "Managing wind power uncertainty through strategic reserve purchasing," *IEEE Transactions on Power Systems*, vol. 32, no. 4, pp. 2547–2559, July 2017.

[11] T. Dai and W. Qiao, "Trading wind power in a competitive electricity market using stochastic programing and game theory," *IEEE Transactions on Sustainable Energy*, vol. 4, no. 3, pp. 805–815, July 2013.

[12] H. Ding, P. Pinson, Z. Hu, and Y. Song, "Integrated bidding and operation strategies for wind-storage systems," *IEEE Transactions on Sustainable Energy*, vol. 7, no. 1, pp. 163–172, January 2016.

[13] H. Ding, P. Pinson, Z. Hu, and Y. Song, "Optimal offering and operating strategies for wind-storage systems with linear decision rules," *IEEE Transactions on Power Systems*, vol. 31, no. 6, pp. 4755–4764, November 2016.

[14] A. A. Thatte, L. Xie, D. E. Viassolo, and S. Singh, "Risk measure based robust bidding strategy for arbitrage using a wind farm and energy storage," *IEEE Transactions on Smart Grid*, vol. 4, no. 4, pp. 2191–2199, December 2013.

[15] T. Rodrigues, P. J. Ramírez, and G. Strbac, "Risk-averse bidding of energy and spinning reserve by wind farms with on-site energy storage," *IET Renewable Power Generation*, vol. 12, no. 2, pp. 165–173, February 2018.

[16] J. M. Angarita and J. G. Usoala, "Combined hydro-wind generation bids in a pool-based electricity market," *Electric Power Systems Research*, vol. 79, pp. 1038–1046, July 2009.

[17] M. Zima-Bockarjova, J. Matevosyan, M. Zima, and L. Soder, "Sharing of profit from coordinated operation planning and bidding of hydro and wind power," *IEEE Transactions on Power Systems*, vol. 25, no. 3, pp. 1663–1673, August 2010.

[18] G. Li, R. Zhang, T. Jiang, H. Chen, L. Bai, and X. Li, "Security-constrained bi-level economic dispatch model for integrated natural gas and electricity systems considering wind power and power-to-gas process," *Applied Energy*, vol. 194, pp. 696–704, May 2017.

[19] J. Yang, N. Zhang, Y. Cheng, C. Kang, and Q. Xia, "Modeling the operation mechanism of combined P2G and gas-fired plant with CO_2 recycling," *IEEE Transactions on Smart Grid*, vol. 10, no. 1, pp. 1111–1121, January 2019.

[20] S. Clegg and P. Mancarella, "Integrated modeling and assessment of the operational impact of power-to-gas (P2G) on electrical and gas transmission networks," *IEEE Transactions on Sustainable Energy*, vol. 6, no. 4, pp. 1234–1244, October 2015.

[21] C. Leeuwen and M. Mulder, "Power-to-gas in electricity markets dominated by renewables," *Applied Energy*, vol. 1232, pp. 258–272, December 2018.

[22] H. Khani and H. E. Z. Farag, "Optimal day-ahead scheduling of power-to-gas energy storage and gas load management in wholesale electricity and gas markets," *IEEE Transactions on Sustainable Energy*, vol. 9, no. 2, pp. 940–951, April 2018.

[23] Y. Li, W. Liu, M. Shahidehpour, F. Wen, K. Wang, and Y. Huang, "Optimal operation strategy for integrated natural gas generating unit and power-to-gas conversion facilities," *IEEE Transactions on Sustainable Energy*, vol. 9, no. 4, pp. 1870–1879, October 2018.

[24] D. J. Kang, B. H. Kim, and D. Hur, "Supplier bidding strategy based on non-cooperative game theory concepts in single auction power pools," *Electric Power Systems Research*, vol. 77, nos. 5–6, pp. 630–636, April 2007.

[25] S. H. Du, X. H. Zhou, L. Mo, and X. Hui, "A novel nucleolus-based loss allocation method in bilateral electricity markets," *IEEE Transactions on Power System*, vol. 21, no. 1, pp. 28–33, February 2006.

[26] T. Li, M. Shahidehpour, and Z. Li, "Risk-constrained bidding strategy with stochastic unit commitment," *IEEE Transactions on Power Systems*, vol. 22, no. 1, pp. 449–458, February 2007.

[27] Y. Du, Z. Wang, G. Liu, X. Chen, H. Yuan, Y. Wei, and F. Li, "A cooperative game approach for coordinating multi-microgrid operation within distribution systems," *Applied Energy*, vol. 222, pp. 383–395, 2018.

[28] E. A. Farsani, H. A. Abyanch, M. Abedi, and S. H. Hosseinian, "A novel policy for LMP calculation in distribution networks based on loss and emission reduction allocation using nucleolus theory," *IEEE Transactions on Power Systems*, vol. 31, no. 1, pp. 143–152, January 2016.

[29] GAMS/SCENRED documentation [Online]. http://www.gams.com/docs/document.htm.

[30] H. Chen, R. Zhang, L. Bai, T. Jiang, G. Li, H. Jia, and X. Li, "Stochastic scheduling of integrated energy systems considering wind power and multi-energy loads uncertainties," *Journal of Energy Engineering*, vol. 143, no. 5, October 2017.

[31] PJM—Markets & operations, data miner2 2018 [Online]. https://dataminer2.pjm.com.

[32] W. Liu, F. Wen, and Y. Xue, "Power-to-gas technology in energy systems: Current status and prospects of potential operation strategies," *Journal of Modern Power System and Clean Energy*, vol. 5, no. 3, pp. 439–450, May 2017.

21 PtX–P2P Intersection in Energy Markets

A Sustainable Solution to Future Energy Systems

*Mohammad Hasan Ghodusinejad, Hossein Yousefi,
Ahmad Nazari Gazik, and Khashayar Fardnia*

Abbreviations

AC	absorption chiller
CCHP	combined cooling, heating, and power
CHP	combined heat and power
CTES	cold thermal energy storage
EC	electrical chiller
EH	electrical heater
ES	electrical storage
EZ	electrolyzer
FC	fuel cell
GB	gas boiler
GS	gas storage
HP	heat pump
HRS	heat recovery system
HS	hydrogen storage
ICE	internal combustion engine
IS	ice storage
PV	photovoltaic
PVT	photovoltaic thermal
ST	solar thermal
TES	thermal energy storage
TS	thermal storage
WD	water desalination
WS	water storage
WT	wind turbine
WW	water well

21.1 RESTRUCTURING OF ENERGY MARKETS

The incorporation of peer-to-peer energy trading mechanisms signifies a fundamental transformation in the energy market, empowering individual energy producers to engage in the exchange of excess power within localized grids. This bottom-up approach to energy distribution is complemented

DOI: 10.1201/9781032719436-21

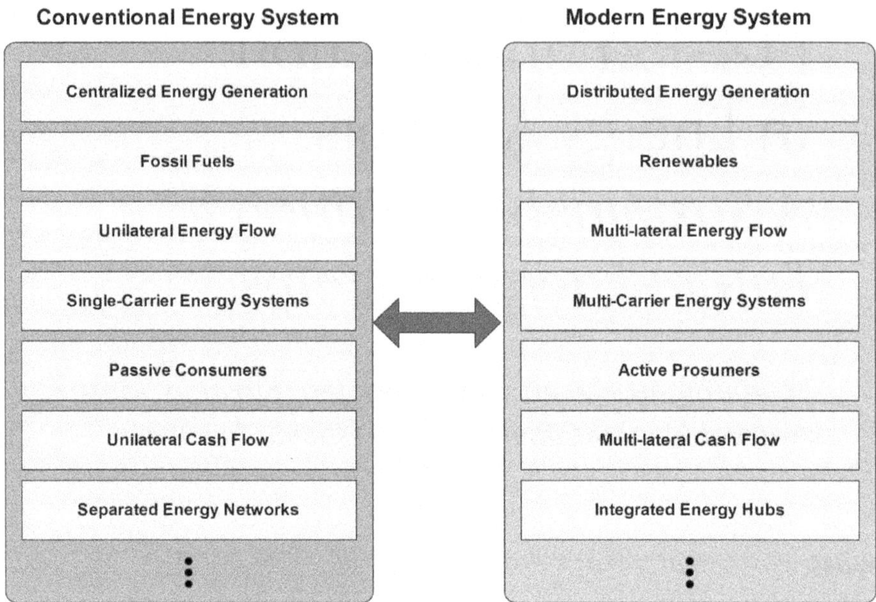

FIGURE 21.1   Comparative graph of conventional and modern energy systems.

by the power-to-X concept, which utilizes surplus renewable energy to produce environmentally friendly fuels and chemicals. These synthesized fuels play a crucial role in sectors where direct electrification is impractical, bridging the gap between renewable energy generation and its broader utilization. The ongoing technological progress in peer-to-peer trading and power-to-X processes is reshaping both regulatory and economic frameworks, necessitating adaptive policies to support these innovations. The proliferation of peer-to-peer trading requires updates to energy exchange regulations to ensure equitable and efficient transactions. Simultaneously, the power-to-X concept is spurring the creation of new market structures that facilitate the storage and reconversion of renewable energy, thereby enhancing the overall sustainability of the energy ecosystem. In summary, the transition in the energy market extends beyond a mere change in energy sources to encompass a comprehensive restructuring that incorporates peer-to-peer energy exchange and power-to-X technologies. These advancements are critical in realizing a decarbonized, efficient, and flexible energy system capable of meeting the modern society's demands while mitigating environmental impacts.

Figure 21.1 describes the transition regarding the most critical factors, comparing it with a perspective on conventional and modern energy systems in the past and the present.

The shift from centralized to distributed energy generation represents a significant change in the design of power systems. Traditionally, centralized energy generation has been controlled by large-scale power plants that distribute electricity through extensive transmission networks. On the other hand, contemporary systems are progressively embracing a distributed model, integrating small-scale renewable energy sources such as solar and wind into local grids. This decentralization improves energy efficiency, diminishes transmission losses, and bolsters the resilience of the energy infrastructure by enabling local production and consumption. Consequently, this reflects a more sustainable and flexible approach to energy management [1, 2].

The energy sector has experienced a significant shift, moving away from the predominance of traditional fossil fuels toward the increasing prominence of renewable energy sources. This change represents a deliberate reaction to the pressing concerns of climate change and sustainable practices, as renewable sources like wind, solar, and hydropower replace coal, oil, and natural gas [3, 4].

The transformation of energy systems from unilateral to bilateral energy flow signifies a substantial change in the distribution of power. Historic energy systems were marked by a single-directional transmission from centralized power plants to end users. In contrast, present-day systems demonstrate an intricate, multidirectional transmission, incorporating various energy origins and enabling both consumption and generation at the consumer level. This shift enables increased adaptability, effectiveness, and robustness within the contemporary energy framework [5–7].

Besides, the transition from relying on single-carrier to embracing multi-carrier energy systems represents a significant transformation within the energy sector. Conventional energy frameworks, which depend on a single energy carrier, such as electricity or gas, are being replaced by integrated systems that coordinate multiple carriers, encompassing heat, electricity, and gas, in a synergistic fashion. This integration fosters improved energy efficiency, resilience, and sustainability, embodying a comprehensive approach to addressing the varied energy needs of modern society [8].

The shift from passive to active consumers within energy systems represents a significant transformation in the role of the consumer. In traditional systems, consumers were confined to passive roles, serving solely as end users of centrally generated energy. However, contemporary systems acknowledge consumers as active participants who engage in the consumption, production, and management of energy. This active involvement contributes to grid stability and efficiency through the implementation of innovative practices, such as demand response and energy-sharing platforms [9].

In addition to energy flow, the financial structure of energy systems has transformed from a unilateral cash flow model, in which funds moved directly from consumers to centralized energy suppliers, to a multifaceted framework. In modern systems, financial dealings have become more intricate and mutually beneficial, mirroring the varied interactions among stakeholders in the energy value chain, encompassing prosumers [10, 11].

The shift from separated energy networks to integrated energy hubs represents a strategic advancement in the design of energy systems. Traditional systems function independently, with separate grids for individual energy sources. In contrast, contemporary systems are moving toward energy centers that harmoniously oversee various energy vectors, thereby improving efficiency and adaptability in catering to a wide range of energy needs [12].

As mentioned before, one of the main influential trends in energy transition that has received a lot of attention in recent years is energy market based on peer-to-peer (P2P) energy trading as well as the power-to-X energy systems. These two basic concepts are an essence of many concepts presented in Figure 21.1.

21.1.1 P2P Energy Trading

The transition toward decentralized energy systems is exemplified by the emergence of P2P energy trading in contemporary markets. This innovative model facilitates direct exchanges of energy between producers and consumers, known as "prosumers," bypassing traditional centralized intermediaries. P2P trading capitalizes on distributed energy resources, fostering a democratized energy landscape where individuals actively participate in the market. The fundamental concept underlying P2P energy trading revolves around prosumers, who represent the final links in the energy supply chain. Over the last two decades, the evolution of electricity markets has centered on increasing the participation of these end users in market management [13]. This transformative process began with demand response (DR) and demand-side management (DSM) programs and extended to peer-to-grid (P2G) energy exchange. Today, peer-to-peer energy trading, alongside other historical concepts, significantly enhances the intervention of end users in energy management. Market design plays a pivotal role in shaping P2P-based markets, which are categorized into three general types: centralized (community-based), decentralized (full P2P), and distributed (hybrid P2P), as shown in Figure 21.2. These categories reflect diverse approaches to integrating P2P trading within the energy market, emphasizing flexibility, sustainability, and active consumer engagement [14].

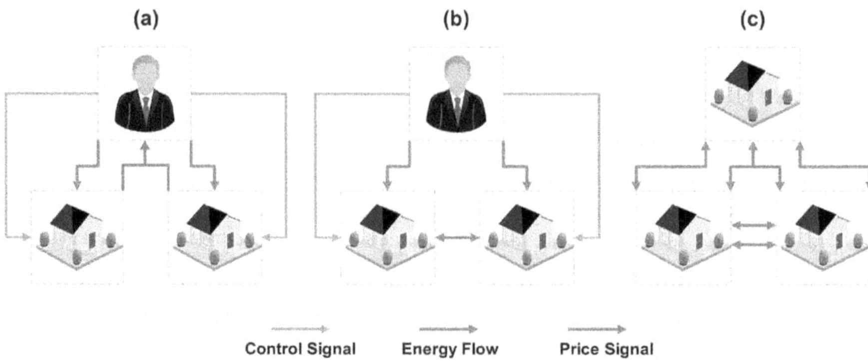

FIGURE 21.2 Three main P2P energy trading mechanisms: (a) centralized, (b) distributed, (c) decentralized.

In a centralized (community-based) market, a designated coordinator or community manager serves as the primary point of contact for all participating peers. The coordinator makes decisions regarding energy exchange based on information provided by the peers. A community is constituted by members who share common interests and objectives, such as the maximization of total income. The community manager allocates the income of the entire peer-to-peer community in accordance with pre-established regulations. Thus, a key advantage of a community-based market is its potential to enhance the overall social welfare of the market. However, drawbacks of these markets include the reduction of peers' autonomy and the potential compromise of market members' privacy.

In a decentralized (full peer-to-peer) market, energy trading is conducted through direct negotiation and exchange among peers, without the involvement of a market manager, in contrast to centralized markets. One of the key benefits of these markets is the robust protection of peers' privacy. Furthermore, decentralized markets offer improved scalability, allowing peers to seamlessly join or exit the market. However, the dynamic nature and absence of a central coordinator in decentralized P2P markets may lead to less-transparent and less-predictable outcomes within the network.

Distributed markets (hybrid P2P markets) represent a middle ground between centralized and decentralized markets in terms of their structure. Consequently, they are considered to harness the advantages of both previous types while mitigating their drawbacks, earning them the designation of hybrid P2P markets. In these hybrid markets, although a coordinator is present, its influence on the energy exchange among peers is indirect, often manifesting through the transmission of pricing signals. In comparison to decentralized markets, distributed markets exhibit a greater degree of control and coordination over peers' activities due to the involvement of a coordinator. Furthermore, contrasted with centralized markets, distributed markets typically afford peers a higher level of privacy and autonomy. Table 21.1 summarizes the advantages and disadvantages of the discussed energy markets.

21.1.2 PtX Concept

The concept of power-to-X (PtX) encompasses a suite of technologies that facilitate the conversion of surplus renewable electricity into a spectrum of energy carriers and chemical compounds. PtX serves as a pivotal strategy for decarbonizing sectors that are challenging to electrify, such as heavy transport and industrial processes. It also has to be noted that PtX is likely to involve electrolysis to produce hydrogen, which can then be transformed into synthetic natural gas, liquid fuels, or chemicals. This approach not only provides a pathway to store and utilize excess renewable energy but also contributes to the creation of a carbon-neutral society by offering alternatives to fossil-based fuels. PtX technologies are instrumental in coupling different sectors, such as electricity

TABLE 21.1

Advantages and Disadvantages of Different Types of Energy Markets

Energy Market	Advantages	Disadvantages
Centralized	• *Efficiency.* A central entity can optimize energy distribution, potentially leading to reduced energy waste. • *Control.* A centralized system offers greater control over energy flow, ensuring grid stability. • *Scalability.* It can handle large-scale energy trading efficiently.	• *Single point of failure.* If the central entity fails, the entire system is disrupted. • *Reduced flexibility.* Participants have limited control over their energy transactions. • *Potential for market manipulation.* The central entity could influence prices and create unfair advantages.
Distributed	• *Increased flexibility.* Participants have more control over their energy choices. • *Improved resilience.* The system is less vulnerable to single points of failure.	• *Complexity.* Managing energy transactions between multiple participants can be complex. • *Potential for market fragmentation.* Local energy markets might not be as efficient as a centralized one. • *Security risks.* Increased vulnerability to cyberattacks due to more interconnected systems.
Decentralized	• *Maximum flexibility.* Participants have complete control over their energy generation and consumption. • *Empowerment.* Promotes energy independence and community-driven energy solutions.	• *Coordination challenges.* Balancing energy supply and demand across a decentralized network is difficult. • *Potential for market volatility.* Prices can fluctuate significantly due to imbalances. • *Technical complexity.* Requires advanced technologies for efficient peer-to-peer transactions.

with heating, cooling, and mobility, thereby enhancing the overall efficiency of the energy system (Figure 21.3). The scalability of PtX solutions is prepared for future energy demands, positioning them as a key component in the global effort to limit climate change and achieve net-zero emissions. In essence, PtX represents a transformative shift in energy management, enabling the integration of renewable energy sources into a broader range of applications, fostering long-term sustainability, and supporting the transition to a greener future.

21.1.3 IMPACTS OF PtX ON MARKET RESTRUCTURING

The power-to-X technology plays a significant role in restructuring markets by facilitating the integration of renewable energy across various sectors, thereby promoting the transition toward decarbonized energy systems. PtX introduces demand-side flexibility to address the intermittency of renewable sources and enables energy storage and conversion, thereby offering a potential solution for uncertainty-related challenges. Moreover, by empowering local energy production and utilization, PtX technologies contribute to democratizing energy markets, thereby influencing energy justice. Furthermore, PtX's ability to synthesize sustainable fuels and chemicals accelerates the transition toward a net-zero-emission future, transforming energy markets and supply chains [15].

21.1.3.1 Energy Democratization

PtX technologies enable decentralized access to energy by effectively harnessing renewable energy sources. They offer alternatives for sectors that are challenging to electrify, thereby reducing

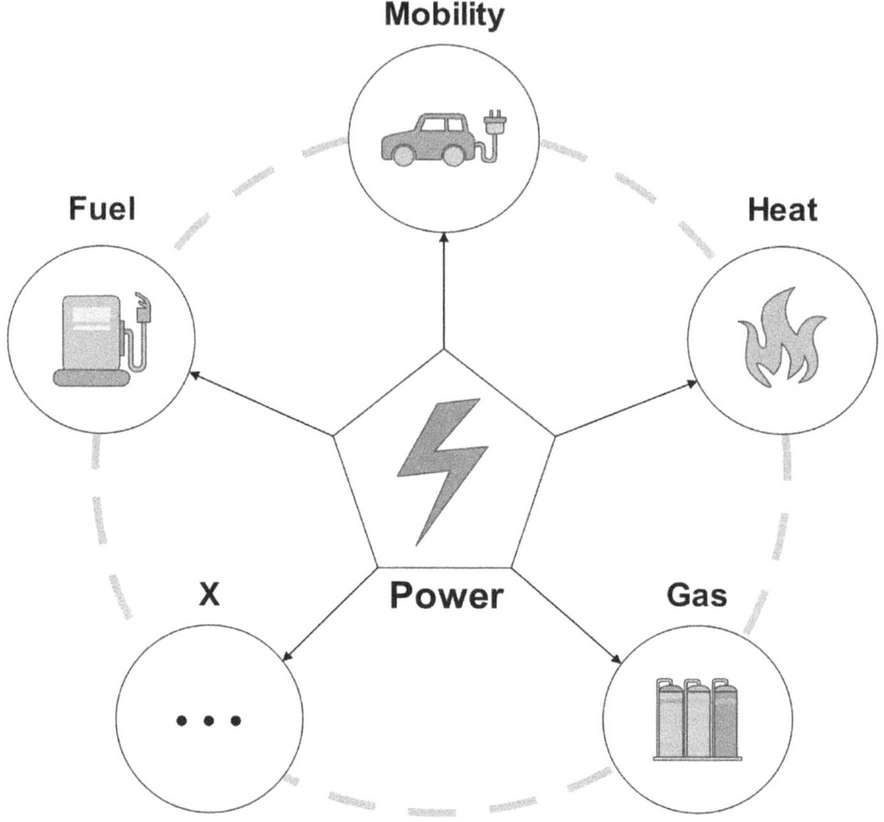

FIGURE 21.3   Power-to-X frameworks.

dependence on fossil fuels and contributing to a more equitable and sustainable energy framework. Also, by creating an integrated network of different energy carriers, in PtX systems, the effectiveness of prosumers is not limited to electricity and increases their prominent role in the energy network.

21.1.3.2   Uncertainty and Flexibility Issues

Considering the variability of renewable energy sources such as wind and solar, it is evident that there will be challenges related to their availability. Moreover, the high penetration of renewables amplifies the uncertainty of net load and diminishes the flexibility of power sources, underscoring the necessity for adaptable solutions. Nonetheless, PtX has the potential to bolster grid stability by converting surplus electricity into storable fuels or gases.

21.1.3.3   Energy Justice

PtX promotes justice by facilitating carbon-neutral alternatives in sectors that are resistant to electrification. It distributes advantages across various industries, effectively addressing environmental objectives. PtX serves as a link between renewable energy and sectors such as aviation or shipping.

21.1.3.4   Energy System Decarbonization

PtX converts electricity into synthetic fuels, supporting the integration of different sectors. It maximizes the utilization of biomass, aligning with sustainability standards. Furthermore, PtX plays a crucial role in achieving net-zero emissions.

In conclusion, PtX represents a versatile solution that will drive the transition toward cleaner, more accessible, and equitable energy systems.

21.2 PtX- AND P2P-BASED ENERGY SYSTEMS

In this section, the intersection of P2P and PtX concepts in energy systems is fully investigated. First, the interaction between P2P trading and PtX technologies and its benefits, such as reliability and sustainability, are discussed, and then technological requirements and trading technologies, including smart energy systems and digital technologies, are investigated. Finally, the last subsection is mainly focused on the economic implications of this integration on the energy markets and the impacts of this integration on the pricing and market accessibility of energy.

21.2.1 Synergies and Interactions between PtX and P2P Systems

Due to the availability of fundamental infrastructures for electricity trading worldwide, electricity is one of the main energy carriers in energy markets. Hence, most energy systems, such as peer-to-peer energy systems, are producing and trading energy in this form. As discussed in Section 21.2., this energy can be used in order to achieve secondary products, such as heat and fuels. However, in P2P energy systems, electricity is the only form of energy that is produced and traded between prosumers and consumers. In contrast, in PtX energy systems, each peer extracts and uses energy to produce water, heat, chemicals, and fuels, such as hydrogen. However, the energy and its final products are not tradable in PtX systems and energy hubs.

Overlaying the P2P trading and PtX technologies leads to conceptualizing a system in which each peer could extract energy and then trade both energy and its transforms, such as fuels and heat, with other peers, which could be a promising solution to the limitations of P2P and PtX energy systems. This newly developed concept, called multi-energy P2P systems or multi-carrier P2P energy systems, has come to solve this issue. In this concept, transactive energy systems and energy hubs are integrating with each other, leading to a P2P energy system in which other forms of energy, such as heat and water, are tradable. Figure 21.4 illustrates the main framework of this new concept.

In multi-energy P2P systems, on the one hand, energy hubs extract energy from different sources and produce energy. Then, using PtX technologies, the energy transforms into X (fuel, water, heat, etc.). On the other hand, the P2P trading systems are implicated to X in order to create an X trading market for the products. This provides a reliable market in which all the peers, whether they are prosumers or consumers, could exchange both energy and X at the same time based on their needs. This integration of P2P and PtX energy systems comes with promising benefits over the separated P2P and PtX energy systems, including profitability, reliability, and sustainability.

21.2.1.1 Profitability

P2P energy markets are suitable for peers to trade energy based on their demand. Adding fuels, water, heating, cooling, etc. into this P2P market assists prosumers in increasing their benefits by offering a wider variety of products to consumers. This win–win game helps consumers decrease costs by decreasing their capital costs for infrastructures, such as chillers or boilers, that provide heating and cooling. This means that consumers do not need to provide certain conventional energy-transforming utilities and infrastructures based on the peak of their demand; instead, they can use multi-energy PtX–P2P energy system market products in their time of need.

In addition, peers can focus on producing appropriate X based on their techno-environment-economical potential. For instance, the demand for hydrogen has increased in the past few years. This increase needs more hydrogen production, which could be accessed by consuming more energy. However, only a higher amount of energy extraction cannot generate hydrogen due to the need for the technology and utilizations that hydrogen production needs. In this case, peers with a

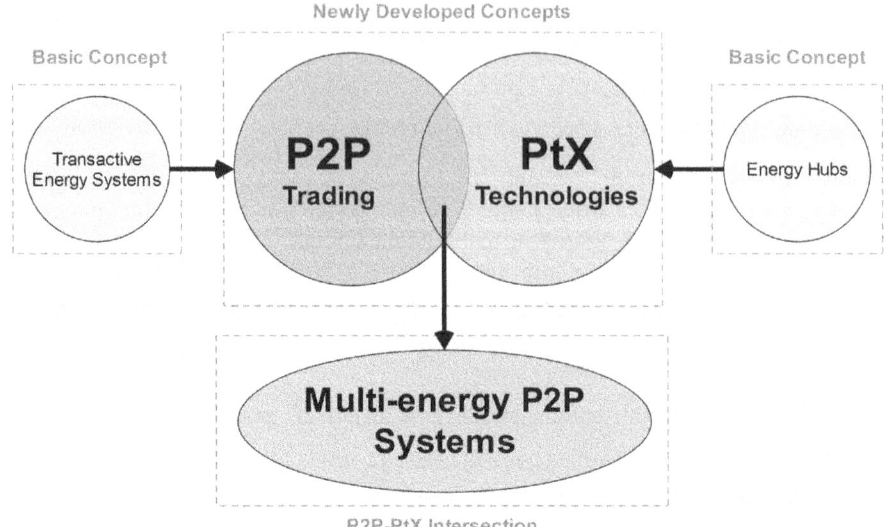

FIGURE 21.4 Conceptual framework for P2P energy trading and PtX technologies.

higher potential to produce hydrogen can use their potential to produce and sell hydrogen to other peers in case of need. This helps them increase their profit based on the market's demand.

21.2.1.2 Reliability

For a single peer, energy hubs are reliable sources for the extraction and transformation of energy. Transactive energy systems and decentralized energy systems are also highly reliable systems that provide electricity and increase reliability in order to reduce grid blackouts and peak shifting. Overlaying these systems significantly increases the reliability of the PtX–P2P systems. These P2P energy systems with multiple carriers could combine the reliability of both P2P and PtX energy systems. This means the system can produce electricity from dependable renewable sources and can also convert and exchange energy between peers with a strong emphasis on reliability.

21.2.1.3 Sustainability

PtX technologies convert renewable electricity into other forms of energy, like hydrogen or synthetic fuels. This allows for storing excess renewable energy and using it for applications where direct electrification isn't feasible, like transportation and industrial processes. This promotes a shift away from fossil fuels and reduces greenhouse gas emissions. Thus, the reliance on RES will increase.

On the other hand, regarding P2P energy trading platforms, prosumers are allowed to trade excess energy directly with each other, bypassing traditional utilities. This can incentivize renewable energy production and reduce reliance on centralized power plants. For example, considering skyscrapers in downtowns, due to space limitations, it is almost impossible for these buildings to extract enough energy to meet their demand from renewable energy resources, such as PV, and transform it into heating or cooling. In P2P systems, they must buy electricity from other peers and then use boilers or chillers to produce heating and cooling, which has a lot of CO_2 emissions and low efficiency. However, in multi-energy P2P systems, other peers, such as those with access to energy hubs, can meet the skyscraper's demands for heating and cooling with sustainable energy production methods such as PVT and heat transfer–based systems. This reduces the RES losses, which boosts the reduction of CO_2 emissions and, hence, increases sustainability.

In general, these multi-carrier PtX–P2P systems lead to more stability of energy networks in two basic axes:

1. *Reduced emissions.* By enabling a greater share of renewables and potentially lowering overall energy consumption, PtX and P2P systems can contribute to reducing greenhouse gas emissions and combating climate change.
2. *Grid resilience.* Increased local energy production and management can improve grid resilience by reducing peak demand and minimizing the impact of disruptions on the centralized system.

21.2.2 TECHNOLOGICAL INTEGRATION OF PtX AND P2P

PtX–P2P energy systems are not applicable using conventional technologies. Cash flow and trading ecosystems differ significantly from centralized energy systems, and the community of prosumers and consumers is not constant. Hence, these developing multi-carrier P2P systems require specific requirements and innovations to reach their full potential for restructuring the energy market. Figure 21.5 illustrates technological requirements and innovations which are necessary for PtX–P2P energy systems.

Data management and optimization are required to collect and organize peers' consumption and production data and safely share necessary information with other peers. It includes smart meters and sensors, machine learning, and blockchain technology, each of which has a specific role in the energy market.

Real-time data from smart meters on energy production (PtX) and consumption (P2P) allows for optimized trading decisions based on supply and demand fluctuations. In a peer-to-peer energy-sharing system, every prosumer has a smart meter that can determine if they should share their energy with other peers in the community based on available data about demand, generation, and market conditions. Smart meters can also communicate with each other using appropriate communication protocols. Machine learning algorithms can analyze energy usage patterns and predict future needs in P2P systems, facilitating efficient matching of buyers and sellers. Understanding the energy consumption patterns of various flexible loads in a building for P2P sharing is crucial.

Additionally, it is essential to comprehend how prosumers adjust their energy bids and prices in different market situations. AI techniques such as reinforcement learning, deep learning, and artificial neural networks have proven to be highly effective in capturing these learning objectives. Blockchain can provide a secure and transparent platform for PtX and P2P transactions, ensuring data integrity and facilitating participant trust. The P2P energy-sharing mechanism requires

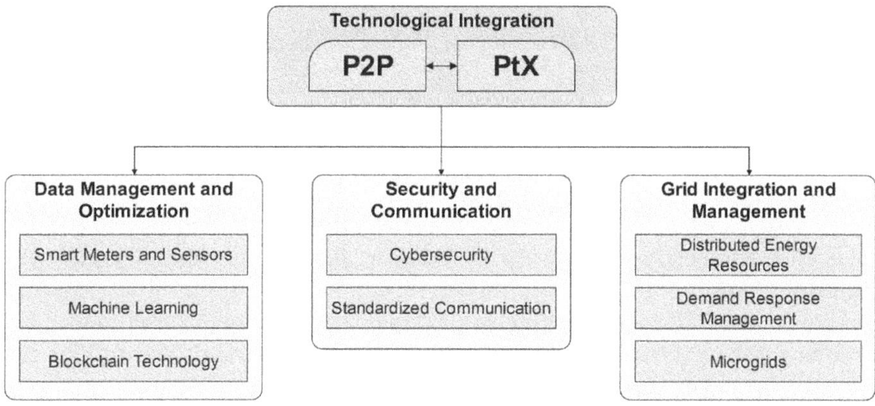

FIGURE 21.5 Overall framework for the technological integration of PtX and P2P.

high data security, data privacy, data integrity, and fast financial transactions between prosumers. Therefore, distributed ledger technologies like blockchain have demonstrated their effectiveness in resolving these issues by offering prosumers a secure platform for exchanging information related to energy and economic transactions without relying on trusted third parties.

Grid integration and management technologies, such as distributed energy resources, or DERs; demand response management; and microgrids, are another part of PtX–P2P technological integration. PtX systems, like those converting excess solar power to hydrogen, can act as DERs, feeding energy back into the grid and potentially reducing reliance on traditional power plants. DERs without coordination and control can cause problems, such as reverse power flows, voltage rise, and increased fault currents in the electricity network, potentially leading to a widespread catastrophic event, like a blackout, and may require network equipment reinforcement. To prevent this, significant recent efforts have been made to advance techniques for controlling DERs, such as PV inverters, battery storage, electric vehicles, and demand response assets (flexible loads). PV inverters can now autonomously adjust their active and reactive power injection to the electricity network based on the network condition [16]. Additionally, state-of-the-art algorithms and techniques are available for the opportune charging and discharging of battery storage to provide network services [17] and reduce costs [18] for the prosumers. The coordinated control of electric vehicle charging/discharging and the intelligent management of flexible loads now allow prosumers to enjoy the benefits of mobile storage [19] and participate in P2P trading using the energy stored in their vehicle batteries [20]. Furthermore, these DERs represent new points of control on distribution feeders that can be meaningfully aggregated to impact the bulk electricity system [21] through P2P trading. In addition, using demand response management, P2P platforms can incentivize energy consumers to adjust their usage during peak hours, alleviating stress on the grid (relevant for both PtX and P2P). Finally, PtX and P2P systems can be integrated into local microgrids, creating self-sufficient energy communities with increased resilience.

Security and communication technologies are completing the PtX–P2P technology puzzle. These technologies are essential for connecting peers and helping them communicate and interact securely. In this respect, some vital aspects, such as standardized communication and cybersecurity, are required.

Secure communication protocols are crucial for protecting sensitive data in both PtX and P2P systems. Cybersecurity technologies help multi-carrier P2P systems to achieve this goal. Standardized communication protocols allow different devices and platforms to interact seamlessly within PtX and P2P networks. A P2P energy-sharing system requires a communication infrastructure to identify prosumers and enable information sharing among them. The chosen communication infrastructure in a connected community must meet the necessary criteria for optimal system performance, such as latency, throughput, reliability, and security.

21.2.3 Economic and Market Dynamics of PtX–P2P Integration

21.2.3.1 Economic Aspects

Integrating PtX and P2P systems has many effects on the energy market. One of these main effects is the economic impact of PtX–P2P on the energy market. As discussed in Section 21.1, PtX–P2P integration restructures conventional energy markets, such as centralized energy systems. These modern energy systems strengthen peers' roles, especially those of prosumers, and add new revenue streams. This empowerment contributes to the balance and optimization of grids and increases market efficiency.

As stated previously, P2P energy trading is shaking up the traditional power market by allowing individuals to buy and sell electricity directly with each other. This shift toward P2P, often alongside the rise of PtX systems, like electric vehicles, hydrogen-related technologies, etc., is transforming the economic landscape of the energy sector. Figure 21.6 shows some of the economic aspects of PtX–P2P integration.

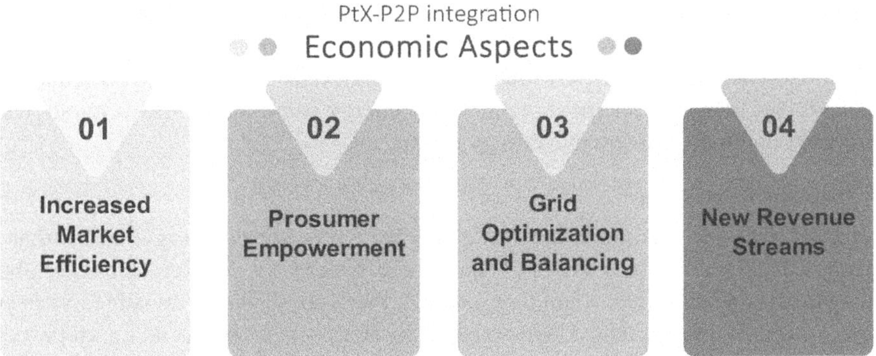

FIGURE 21.6 Economic aspects of PtX–P2P integration.

21.2.3.1.1 Increased Market Efficiency

P2P trading allows prosumers to sell excess power, potentially lowering overall energy costs. Besides, PtX allows for storing excess renewable energy and using it later, reducing reliance on traditional sources and potentially lowering overall energy costs.

P2P energy trading can boost market efficiency. This includes (1) matching local supply and demand, reducing transmission losses, and utilizing energy more effectively; (2) enabling dynamic pricing, reflecting real-time conditions, and incentivizing efficient consumption and production; and finally, (3) encouraging renewable energy adoption through financial incentives for prosumers. PtX technologies can also improve market efficiency by balancing supply and demand with renewable energy storage and conversion, boosting renewable integration by making renewables more reliable and optimizing prices through dynamic market participation.

21.2.3.1.2 Prosumer Empowerment

P2P trading lets individuals with solar panels or other generation capabilities sell their surplus energy directly to neighbors, as well as PtX facilities which integrate different carriers into the energy system, creating a more distributed and potentially cheaper energy market.

In an integrated PtX–P2P energy system, prosumers can monetize their excess energy more easily, resulting in gaining more benefits or lowering energy bills. The prosumers are also capable of boosting their self-sufficiency, utilizing technologies like battery storage or conversion to hydrogen or methane. This reduces the dependence on the grid, especially during peak demand or when renewable generation is low, and makes the prosumers more self-sufficient and less susceptible to price fluctuations. Moreover, by storing excess renewable energy and feeding it back into the grid during peak hours, prosumers can contribute to grid stability more effectively, making them an active participant, not just a passive consumer. Also, the flexibility awarded by PtX–P2P systems provides more options for how prosumers use their self-generated energy.

21.2.3.1.3 Grid Optimization and Balancing

By allowing for local energy exchange, P2P trading can reduce strain on the traditional grid, potentially delaying expensive upgrades. PtX technologies also allow for storing excess renewable energy (e.g., converting it to hydrogen) and injecting it back into the grid during peak demand, reducing reliance on traditional sources. This can promote demand response programs, such as peak shaving. All these in general can lead to the occurrence of positive economic effects.

21.2.3.1.4 New Revenue Streams

PtX facilities and P2P platforms create opportunities for renewable energy producers to sell power in other forms. These systems can create new business opportunities, like energy arbitrage (buying low and selling high) and providing grid balancing services. PtX–P2P integrated systems create new revenue streams by allowing prosumers to monetize their capabilities, such as renewable energy production, stored energy, PtX-derived fuels, etc. This not only benefits the prosumers financially but also incentivizes further investment in renewable energy sources and PtX technologies, contributing to a cleaner and more sustainable energy future.

21.2.3.2 Market Dynamics

As in other markets, there are numerous ever-changing forces within energy market that influence how it operates. In this context, a PtX–P2P integrated system is a marketplace where buyers and sellers interact, but instead of just a simple exchange, there are different invisible forces constantly tugging and pulling on the outcome. Understanding these dynamics is crucial for all parties, including regulators and policymakers, businesses, prosumers, etc. By staying informed about these ever-changing forces, all stakeholders can make informed decisions in the market. Figure 21.7 depicts the main market dynamics of the PtX–P2P integration.

By taking a close look at Figure 21.7 and analyzing it, the following can be said:

- *Disruptive potential.* P2P and PtX could challenge traditional utility monopolies, giving consumers more control and potentially lowering prices.

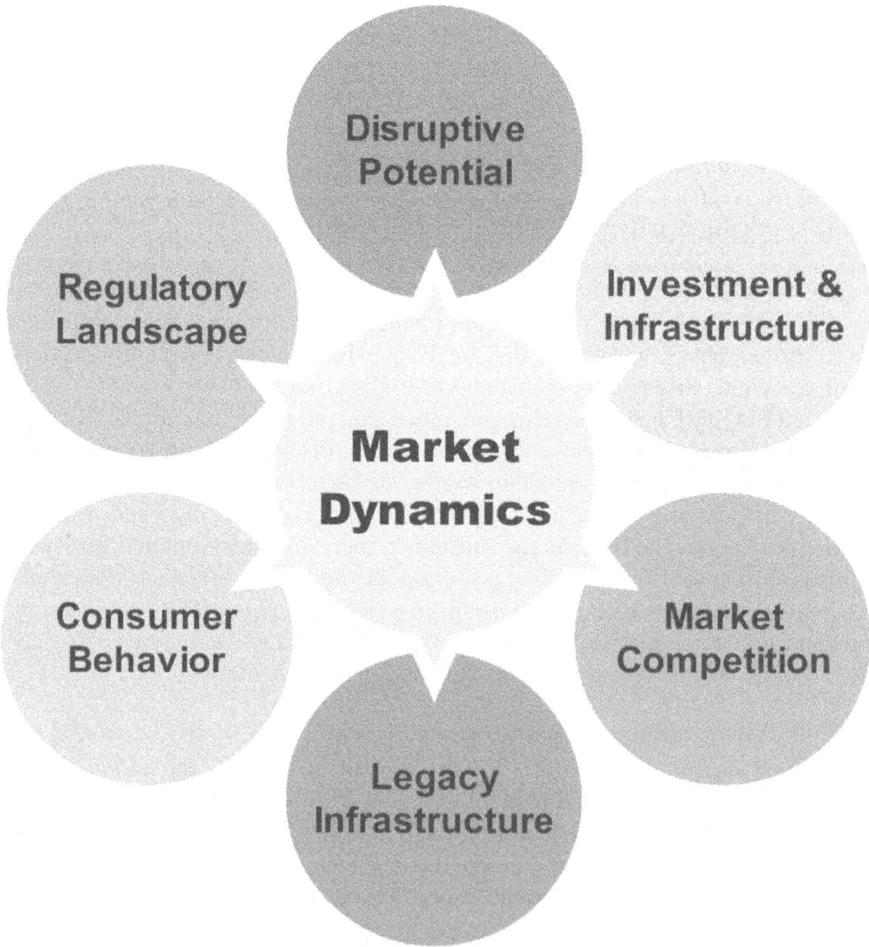

FIGURE 21.7 Market dynamics of the PtX–P2P integration.

- *Investment and infrastructure.* The development of PtX infrastructure like hydrogen conversion facilities, and its integration with P2P platforms, requires investment but can create new markets.
- *Market competition.* Competition among P2P platforms and PtX service providers will drive innovation and potentially lower costs.
- *Legacy infrastructure.* Integrating PtX and P2P with existing grids requires careful planning and investment.
- *Consumer behavior.* Consumer willingness to participate in P2P trading and to invest in PtX technologies will determine the market growth in the future.
- *Regulatory landscape.* Regulations need to adapt to facilitate P2P trading and PtX integration, ensuring fair market access and grid security. Governments need to develop regulations that incentivize PtX and P2P trading, while ensuring fair pricing and grid stability.

21.2.3.3 Market Accessibility and Challenges

PtX–P2P integration can specifically affect market accessibility. However, this can also come with some challenges. To make a brief clarification, PtX–P2P integration can increase the market accessibility as follows:

- *Prosumer participation.* P2P platforms allow individuals with renewable energy generation to become active participants in the market, increasing access for those who previously relied solely on traditional utilities. Also, the use of PtX equipment can make it easier for communities to access different energy carriers.
- *Localized markets.* PtX–P2P integrated energy systems can create local energy markets, potentially benefiting areas not well-served by the traditional grid infrastructure.
- *Investment opportunities.* The growth of PtX and P2P energy systems opens doors for new businesses and investment opportunities in these sectors.
- Despite all the aforementioned advantages, PtX- and P2P-based energy markets also face challenges. Not paying enough attention to these challenges can disrupt the future development of such markets. Some noteworthy challenges could be summarized as follows:
- *Price volatility.* Increased reliance on renewable energy sources and dynamic pricing can introduce more volatility into energy prices.
- *Grid integration costs.* Integrating PtX facilities and P2P systems with the existing grid might require infrastructure upgrades, impacting overall costs.
- *Regulatory hurdles.* Current regulations might not be fully adapted to P2P and PtX, potentially creating barriers to entry for new players.
- *The impact on low-income communities.* Ensuring equitable access to PtX and P2P benefits is crucial to avoid exacerbating energy poverty.

Overall, the impact of PtX and P2P on market accessibility and market challenges and costs is likely to be mixed. While increased competition and local markets could lead to lower prices and wider access, challenges like price volatility and grid integration costs need to be addressed. Hence, by carefully navigating the challenges and harnessing the opportunities, PtX and P2P integration can create a more dynamic, accessible, and affordable energy market.

21.2.4 The Impact of PtX–P2P Integrated Systems on Ancillary Service Markets

The integration of PtX technologies with P2P energy trading frameworks presents considerable potential to revolutionize ancillary service markets and generate novel revenue streams for stakeholders. As energy systems increasingly adopt decentralization and renewable energy sources

(RES) assume a more pivotal role, the demand for dependable ancillary services—such as frequency regulation, voltage stabilization, and reserve capacity—intensifies markedly. Integrated PtX–P2P systems are capable of addressing these requirements through innovative methodologies while simultaneously offering financial incentives to market participants.

21.2.4.1   Augmenting Flexibility and Responsiveness

A fundamental contribution of PtX–P2P systems to ancillary service markets resides in their capacity to enhance system flexibility and responsiveness. PtX technologies, which facilitate the conversion of surplus renewable electricity into alternative energy forms (such as hydrogen, methane, or thermal energy), can be strategically employed to mitigate grid imbalances. For instance, during intervals of surplus renewable generation, excess electricity can be converted and stored via PtX mechanisms, effectively diminishing curtailment and establishing a reservoir that can be utilized when demand surges or generation diminishes.

Within a P2P trading framework, participants are enabled to trade not only electricity but also the energy products that have been stored as a result of PtX processes. This decentralized paradigm empowers small-scale producers and consumers to bolster grid stability by providing ancillary services such as frequency regulation or balancing services. The responsiveness of P2P networks facilitates prompt reactions to grid necessities, thereby delivering a valuable service to grid operators and creating new revenue opportunities for participants capable of supplying these services.

21.2.4.2   Reduced Reliance on Centralized Ancillary Service Providers

The conventional framework for ancillary services is predominantly characterized by centralized entities, including large-scale power generation facilities and utility corporations. Nonetheless, the decentralized architecture of PtX–P2P systems contests this paradigm by facilitating a distributed methodology for service delivery. In this framework, each participant within the P2P network possesses the capacity to function as a service provider, thereby contributing negligible yet collectively substantial quantities of ancillary services.

This transformation has the potential to mitigate the grid's dependence on a limited number of substantial service providers while simultaneously fostering a more resilient and heterogeneous ancillary service marketplace. For example, residential or commercial edifices outfitted with PtX technology and intelligent metering could engage in demand response initiatives, rendering services such as load shedding or load shifting in return for financial compensation. This democratization of ancillary services not only enhances grid stability but also affords a continual revenue stream for participants who would otherwise be relegated to the status of mere energy consumers.

21.2.4.3   Creating New Market Mechanisms and Income Streams

The incorporation of P2P trading in conjunction with PtX systems mandates the formulation of novel market mechanisms to effectively harness the value of these services. Technologies such as blockchain and smart contracts can be utilized to automate the exchange of ancillary services within the P2P network, thereby ensuring both transparency and operational efficiency. Participants can receive real-time reward for the ancillary services they provide, with compensatory payments dynamically calibrated according to the prevailing requirements of the grid and the availability of services.

Additionally, the potential for ancillary services to be integrated with energy sales or rendered as an independent commodity gives rise to new business paradigms. For instance, a P2P platform could facilitate participants in selling frequency regulation services directly to grid operators or to other market stakeholders seeking to maintain their own grid stability. This innovation unveils a new revenue stream that is distinct from conventional energy sales, thereby enhancing the financial resilience of participants.

21.3 A LITERATURE REVIEW

The intersection of PtX and P2P energy systems is a new topic, and most previous literature has focused on transactive energy systems. However, several recent studies have investigated and optimized multi-carrier P2P energy systems. The objective functions for the optimization in these studies are different, varying between minimizing capital, operational, total costs, etc. and maximizing some other objective, such as net benefit or welfare.

Most recent studies are mainly focused on five energy carrier systems, including electricity, heating, cooling, water, hydrogen, and gas. Multi-objective optimizations were utilized in these studies, with a focus on minimizing costs, reducing emissions, and increasing water storage.

One of these studies, conducted by Zhang et al. [22], used renewable sources to produce a multi-energy P2P system based on electricity, heating, cooling, water, and hydrogen while minimizing operational and emission costs. This work proposes a hydrogen-based multi-carrier energy system containing electricity, renewable energy resources, and hydrogen markets as input energy carriers. The demands in this study include power, cooling, and heating. The benefits of hydrogen tanks are evident in the reduction of operation costs by 4.5%, and the significance of CCP in reducing operation costs underscores their importance. Jalilian et al. [23] conducted another research to reduce the overall expenses and the quantity of drinkable water drawn from water wells. Water extracted from water wells. Their research involves a comprehensive scheduling model to efficiently dispatch cooling, heating, power, gas, and water resources within an energy–water microgrid. In this microgrid, the operator engages in power, heat, and gas markets and employs energy conversion facilities to fulfill different requirements.

Most studies in the literature comprised three or four energy carriers, mainly including electricity, heating, cooling, gas, and/or hydrogen. In this aspect, a recent study by Jin et al. [24] proposed a new approach for planning medium-term energy strategies to achieve a 50% reduction in carbon emissions while considering financial investment factors. Their research focuses on a local energy community with various technologies, including photovoltaic, combined heat and power, gas-fired boilers, absorption, and electric chillers, to meet energy needs. The case studies demonstrate an economically optimized scenario for annual planning, ensuring a 50% decrease in carbon emissions. The findings emphasize the significance of leveraging synergies among energy carriers and the crucial role of renewable energy sources, batteries, and sector coupling technologies.

Many other studies have done the same optimizations on two and three energy carrier systems. However, the research conducted concerning multi-energy P2P systems can be examined from various aspects, such as different components, various energy carriers, equipment used in the system, objective function, etc. Hence, a brief overview of the literature has been provided in Table 21.2 in order to observe the details of recent papers in this field. The studies have been ordered based on the number of energy carriers.

21.4 CONCLUSION

P2P energy systems are restructuring energy markets and providing a beneficial trading system for energy trades. However, their limited ability to trade only one specific form of energy (electricity) is a disadvantage. PtX energy systems have the ability to transform electricity into other forms of energy carriers, such as heating, cooling, fuels, and hydrogen. Integrating these two types of systems with each other results in multi-carrier P2P energy systems. These systems have the advantages of both P2P trading system and PtX energy transformation potentials.

PtX–P2P integrated systems are able to extract, transform, and trade energy in different forms between prosumers and consumers. This leads to a revolutionary restructured energy trading system in which heating, cooling, fuels, water, electricity, and other forms of energy are tradable. Peers have prosumer and consumer roles based on the energy and utilities situation and are free to

TABLE 21.2

Literature Review of Some Recent Studies in Multi-Carrier P2P Energy Systems

Author	Carriers	System Components	Objective Function	Year
Zhang et al. [22]	Electricity/heating/cooling/water/hydrogen	WT/PV/PVT/FC/WS/WD/AC/HS	Minimize operational and emission costs	2023
Jalilian et al. [23]	Electricity/cooling/heating/gas/water	WT/EH/CHP/GB ES/IS/GS/AC/WD/WS/WW	Minimize the total cost and amount of potable water extracted from water wells	2022
Jin et al. [24]	Electricity/heating/cooling/hydrogen	PV/CHP/FC/HP/GB/AC/EC/ES/TES/CTES/EZ/HS	Maximize the emission reduction and minimize the capital and operational costs	2023
Ghodousinejad and Yousefi [25]	Electricity/heating/cooling/gas	PVT/ICE/GB/AC/EC	Minimize the operating costs	2023
Javadi et al. [26]	Electricity/cooling/heating/gas	PV/ES/EH/HP/CHP/GB/AC	Minimize the total cost	2022
Zhang et al. [27]	Electricity/cooling/heating/gas	PV/ES/CHP/HP	Maximize the net benefit	2022
Daneshvar et al. [28]	Electricity/cooling/heating/gas	WT/PV/CCHP/ES/TS	Minimize operating costs	2022
Davoudi and Moeini [29]	Electricity/cooling/heating/gas	PV/ES/CHP/EC/GB/TS/ST	Maximize welfare	2021
Davoudi et al. [30]	Electricity/cooling/heating/gas	PV/ES/CHP/EC/GB/TS/ST	Maximize the net benefit	2021
Khorasany et al. [31]	Electricity/heating/cooling/gas	WT/PV/GB/GE/ST/EC/ES/TS/AC/IS	Minimize operating costs	2021
Zhong et al. [32]	Electricity/heating/gas	WT/ES/CHP/HP/GB/TS	Minimize the social operational cost	2022
Zhang et al. [33]	Electricity/gas/water	WT/PV/ES/WS/WD/GE/GS/E2G	Minimize operating costs	2022
Daneshvar et al. [34]	Electricity/cooling/heating	WT/PV/TS/ES/EC/AC	Maximize the cost savings	2021
Li and Zhang [35]	Electricity/heating/gas	PV/CHP	Minimize operating costs	2021
Nguyen and Ishihara [36]	Electricity/heating/gas	FC/HE/TS/GB/HRS/CAS	Minimize energy costs	2021
Daneshvar et al. [37]	Electricity/cooling/heating	PV/GE/GB/AC/ES/TS	Minimize operating costs	2020
Liu et al. [38]	Electricity/heating/gas	CHP/PVT/HP	Minimize operating cost	2018
Wang et al. [39]	Electricity/heating	WT/PV/HP	Maximize the net benefit	2022
Liu et al. [40]	Electricity/hydrogen	WT/PV/ES//HS/HE	Minimize carbon emissions, maximize self-consumption, minimize utility bills	2021

decide whether they are prosumers or consumers in the market. This decentralized multi-carrier P2P energy trading system contributes to increasing energy democratization, grid resilience and flexibility issues, energy justice, and energy systems decarbonization.

The integration of PtX and P2P energy systems also increases the profitability and empowerment of prosumers and economic interactions between energy supply and demand contributors.

As the demand for a specific form of energy rises, the price of that energy form increases, leading to motivating prosumers to produce more energy. This significantly contributes to more reliable and sustainable energy systems. However, providing the essential technological infrastructures for such systems is not simple and requires large-scale implications of new technologies, such as smart meters, blockchain, and standardized secure communication.

REFERENCES

[1] C. Vezzoli, F. Ceschin, L. Osanjo, M. K. M'Rithaa, R. Moalosi, V. Nakazibwe, and J. C. Diehl, "Distributed/decentralised renewable energy systems," *Designing Sustainable Energy for All: Sustainable Product-Service System Design Applied to Distributed Renewable Energy*, 2018, pp. 23–39. doi: 10.1007/978-3-319-70223-0_2.

[2] M. Kopsakangas-Savolainen and R. Svento, "Modern energy markets: Real-time pricing, renewable resources and efficient distribution," *Green Energy and Technology*, vol. 105, 2012. doi: 10.1007/978-1-4471-2972-1.

[3] Chen Shi, ZHANG, Chongyu, Lu, Xi, "Energy Conversion from Fossil Fuel to Renewable Energy," in *Handbook of Air Quality and Climate Change*, H. Akimoto Hajime and Tanimoto, Ed., Singapore: Springer Nature Singapore, 2023, pp. 1–44. doi: 10.1007/978-981-15-2760-9_42.

[4] Scott Foster and David Elzinga, "The role of fossil fuels in a sustainable energy system|United Nations," *UN Chronicle, United Nations*, 2015.

[5] M. Zhu, C. Xu, S. Dong, K. Tang, and C. Gu, "An integrated multi-energy flow calculation method for electricity-gas-thermal integrated energy systems," *Protection and Control of Modern Power Systems*, vol. 6, no. 1, 2021. doi: 10.1186/s41601-021-00182-2.

[6] L. Kriechbaum, G. Scheiber, and T. Kienberger, "Grid-based multi-energy systems—modelling, assessment, open source modelling frameworks and challenges," *Energy, Sustainability and Society*, vol. 8, no. 1, pp. 1–19, Nov. 2018. doi: 10.1186/S13705-018-0176-X.

[7] J. Bushnell, E. T. Mansur, and K. Novan, "Review of the economics literature on US electricity restructuring," *Unpublished Manuscript, Department of Economics, University of California at Davis, Davis, CA*, 2017.

[8] O. Sadeghian, A. Oshnoei, B. Mohammadi-Ivatloo, and V. Vahidinasab, "Concept, definition, enabling technologies, and challenges of energy integration in whole energy systems to create integrated energy systems," *Power Systems*, 2022, pp. 1–21. doi: 10.1007/978-3-030-87653-1_1.

[9] E. Fox, C. Foulds, and R. Robison, *Energy & the active Consumer—A Social Sciences and Humanities Cross-Cutting Theme Report*, Cambridge: Shape Energy, Jun. 2017.

[10] IRENA, *Power System Flexibility for the Energy Transition, Part 1: Overview for Policy Makers*, International Renewable Energy Agency, Abu Dhabi, Dec. 2018.

[11] E. Shittu and J. Santos, "Electricity markets and power supply resilience: An incisive review," *Current Sustainable/Renewable Energy Reports*, vol. 8, August 2021. doi: 10.1007/s40518-021-00194-4.

[12] M. Mohammadi, Y. Noorollahi, and B. Mohammadi-Ivatloo, "An introduction to smart energy systems and definition of smart energy hubs," in *Operation, Planning, and Analysis of Energy Storage Systems in Smart Energy Hubs*, 2018. doi: 10.1007/978-3-319-75097-2_1.

[13] A. Das, S. D. Peu, Md. A. M. Akanda, and A. R. Md. T. Islam, "Peer-to-peer energy trading pricing mechanisms: Towards a comprehensive analysis of energy and network service pricing (NSP) mechanisms to get sustainable enviro-economical energy sector," *Energies (Basel)*, vol. 16, no. 5, 2023. doi: 10.3390/en16052198.

[14] Y. Zhou, J. Wu, and W. Gan, "P2P energy trading via public power networks: Practical challenges, emerging solutions, and the way forward," *Frontiers in Energy*, vol. 17, August 2023. doi: 10.1007/s11708-023-0873-9.

[15] R. Daiyan, I. Macgill, and R. Amal, "Opportunities and challenges for renewable power-to-X," *ACS Energy Letters*, vol. 5, pp. 3843–3847, August 2020. doi: 10.1021/acsenergylett.0c02249.

[16] S. Weckx and J. Driesen, "Optimal local reactive power control by PV inverters," *IEEE Transactions on Sustainable Energy*, vol. 7, no. 4, pp. 1624–1633, October 2016. doi: 10.1109/TSTE.2016.2572162.

[17] J. I. Chowdhury, N. Balta-Ozkan, P. Goglio, Y. Hu, L. Varga, and L. McCabe, "Techno-environmental analysis of battery storage for grid level energy services," *Renewable and Sustainable Energy Reviews*, vol. 131, October 2020. doi: 10.1016/J.RSER.2020.110018.

[18] T. Li and M. Dong, "Residential energy storage management with bidirectional energy control," *IEEE Transactions on Smart Grid*, vol. 10, no. 4, pp. 3596–3611, July 2019. doi: 10.1109/TSG.2018.2832621.

[19] H. Kikusato, K. Mori, S. Yoshizawa, Y. Fujimoto, H. Asano, Y. Hayashi, A. Kawashima, S. Inagaki, and T. Suzuki, "Electric vehicle charge-discharge management for utilization of photovoltaic by coordination between home and grid energy management systems," *IEEE Transactions on Smart Grid*, vol. 10, no. 3, pp. 3186–3197, May 2019. doi: 10.1109/TSG.2018.2820026.

[20] J. Kang, R. Yu, X. Huang, S. Maharjan, Y. Zhang, and E. Hossain, "Enabling localized peer-to-peer electricity trading among plug-in hybrid electric vehicles using consortium blockchains," *IEEE Transactions on Industrial Informatics*, vol. 13, no. 6, pp. 3154–3164, December 2017. doi: 10.1109/TII.2017.2709784.

[21] D. B. Arnold, M. D. Sankur, M. Negrete-Pincetic, and D. S. Callaway, "Model-free optimal coordination of distributed energy resources for provisioning transmission-level services," *IEEE Transactions on Power Systems*, vol. 33, no. 1, pp. 817–828, January 2018. doi: 10.1109/TPWRS.2017.2707405.

[22] H. Zhang, J. Wang, X. Zhao, J. Yang, and Z. A. Bu sinnah, "Modeling a hydrogen-based sustainable multi-carrier energy system using a multi-objective optimization considering embedded joint chance constraints," *Energy*, vol. 278, 2023. doi: 10.1016/j.energy.2023.127643.

[23] F. Jalilian, M. A. Mirzaei, K. Zare, B. Mohammadi-Ivatloo, M. Marzband, and A. Anvari-Moghaddam, "Multi-energy microgrids: An optimal despatch model for water-energy nexus," *Sustainable Cities and Society*, vol. 77, 2022. doi: 10.1016/j.scs.2021.103573.

[24] L. Jin, M. Rossi, L. Ciabattoni, M. Di Somma, G. Graditi, and G. Comodi, "Environmental constrained medium-term energy planning: The case study of an Italian university campus as a multi-carrier local energy community," *Energy Conversion Management*, vol. 278, 2023. doi: 10.1016/j.enconman.2023.116701.

[25] M. H. Ghodusinejad and H. Yousefi, "Peer-to-Peer Energy Trading in Multi-Carrier Energy Systems," in *Demand-Side Peer-to-Peer Energy Trading,* Springer, 2023. doi: 10.1007/978-3-031-35233-1_9.

[26] M. S. Javadi, A. Esmaeel Nezhad, A. R. Jordehi, M. Gough, S. F. Santos, and J. P. S. Catalão, "Transactive energy framework in multi-carrier energy hubs: A fully decentralized model," *Energy*, vol. 238, 2022. doi: 10.1016/j.energy.2021.121717.

[27] H. Zhang, S. Zhang, X. Hu, H. Cheng, Q. Gu, and M. Du, "Parametric optimization-based peer-to-peer energy trading among commercial buildings considering multiple energy conversion," *Applied Energy*, vol. 306, 2022. doi: 10.1016/j.apenergy.2021.118040.

[28] M. Daneshvar, B. Mohammadi-Ivatloo, and K. Zare, "A novel transactive energy trading model for modernizing energy hubs in the coupled heat and electricity network," *Journal of Cleaner Production*, vol. 344, 2022. doi: 10.1016/j.jclepro.2022.131024.

[29] M. Davoudi and M. Moeini-Aghtaie, "Local energy markets design for integrated distribution energy systems based on the concept of transactive peer-to-peer market," *IET Generation, Transmission and Distribution*, vol. 16, no. 1, 2022. doi: 10.1049/gtd2.12274.

[30] M. Davoudi and M. Moeini-Aghtaie, "Local energy markets design for integrated distribution energy systems based on the concept of transactive peer-to-peer market," *IET Generation, Transmission and Distribution*, vol. 16, no. 1, 2022. doi: 10.1049/gtd2.12274.

[31] M. Khorasany, A. Najafi-Ghalelou, R. Razzaghi, and B. Mohammadi-Ivatloo, "Transactive energy framework for optimal energy management of multi-carrier energy hubs under local electrical, thermal, and cooling market constraints," *International Journal of Electrical Power and Energy Systems*, vol. 129, 2021. doi: 10.1016/j.ijepes.2021.106803.

[32] X. Zhong, W. Zhong, Y. Liu, C. Yang, and S. Xie, "Optimal energy management for multi-energy multi-microgrid networks considering carbon emission limitations," *Energy*, vol. 246, 2022. doi: 10.1016/j.energy.2022.123428.

[33] G. Zhang, W. Hu, D. Cao, Z. Zhang, Q. Huang, Z. Chen, and F. Blaabjerg, "A multi-agent deep reinforcement learning approach enabled distributed energy management schedule for the coordinate control of multi-energy hub with gas, electricity, and freshwater," *Energy Conversion Management*, vol. 255, 2022. doi: 10.1016/j.enconman.2022.115340.

[34] M. Daneshvar, B. Mohammadi-Ivatloo, K. Zare, S. Asadi, and A. Anvari-Moghaddam, "A novel operational model for interconnected microgrids participation in transactive energy market: A hybrid IGDT/stochastic approach," *IEEE Transactions on Industrial Informatics*, vol. 17, no. 6, 2021. doi: 10.1109/TII.2020.3012446.

[35] L. Li and S. Zhang, "Peer-to-peer multi-energy sharing for home microgrids: An integration of data-driven and model-driven approaches," *International Journal of Electrical Power and Energy Systems*, vol. 133, 2021. doi: 10.1016/j.ijepes.2021.107243.

[36] D. H. Nguyen and T. Ishihara, "Distributed peer-to-peer energy trading for residential fuel cell combined heat and power systems," *International Journal of Electrical Power and Energy Systems*, vol. 125, 2021. doi: 10.1016/j.ijepes.2020.106533.

[37] M. Daneshvar, B. Mohammadi-Ivatloo, S. Asadi, A. Anvari-Moghaddam, M. Rasouli, M. Abapour, and G. B. Gharehpetian, "Chance-constrained models for transactive energy management of interconnected microgrid clusters," *Journal of Cleaner Production*, vol. 271, 2020. doi: 10.1016/j.jclepro.2020.122177.

[38] N. Liu, B. Guo, Z. Liu, and Y. Wang, "Distributed energy sharing for PVT-HP prosumers in community energy internet: A consensus approach," *Energies (Basel)*, vol. 11, no. 7, 2018. doi: 10.3390/en11071891.

[39] N. Wang, Z. Liu, P. Heijnen, and M. Warnier, "A peer-to-peer market mechanism incorporating multi-energy coupling and cooperative behaviors," *Applied Energy*, vol. 311, 2022. doi: 10.1016/j.apenergy.2022.118572.

[40] J. Liu, H. Yang, and Y. Zhou, "Peer-to-peer trading optimizations on net-zero energy communities with energy storage of hydrogen and battery vehicles," *Applied Energy*, vol. 302, 2021. doi: 10.1016/j.apenergy.2021.117578.

Index

A

Alkaline Electrolyser, 55, 432
artificial intelligence (AI), 133, 252, 275

C

carbon dioxide capture, 102, 104–105, 250
carbon-neutral fuels, 2, 223, 447
cross-sectoral synergy, 107, 112

D

decentralized energy systems, 226, 495, 500
district heating system, 41, 62, 133, 157, 179, 340
dynamic energy pricing, 436
dynamic frequency control, 409, 411–412, 414, 431

E

e-fuels, 71, 104, 171–172, 209, 219–220, 242, 245, 432
electricity network, 168, 340, 459
electrolysis technologies, 104–105, 221–222, 235, 373
energy conversions, 31, **32**, 33, **174**, 189, 223
energy economics, 126
energy resilience, 111, 129, 173, 223
energy transformation, 31, 52, 456, 507
excess energy, 2, 75–76, 78, 79, **134**, 158–159, 196, 206–207, 228, 316, 398, 440–441, 446, 500, 503

F

false data injection attack, 286, 296–297

G

game theory, 42, 56, 161, 164, 180, 475, 482, 489, 491–492
gas storage, 91, 93, 142, 146, **147, 150**, 316, 318, 324, **325**, 330–332, 336, 338, 340, 343, 375, 470, 477, 490, 493
grid integration, 50, 57, 73, 113, 189, 344, 346, 439, 502, 505

H

heat sector, 245, 250
high efficiency, 22, 30, 33, 46–47, **49**, 50–51, 90, 98–99, 109, 137–138, 328, 331, 435, 441
hybrid energy, 56, 80, 83, 85, 159, 161, 167, 180, 225, 251, 270
hydrogen market, 128, 170, 173, 183, 428, 432

I

initiative, 3, 27, 67, 74, 117, 130, 399

L

local energy community, 366, 507, 510
local energy markets, 169, 173, 436, 453, **497**, 505, 510
locational marginal price, 198, 434, 446, 471

M

mathematical programming, 170, 418, 430–431, 458
multi-agent system, 302–303, *304*
multi-carrier energy systems, 316, 339–340, 343, 495, 510
multi-objective optimization, 84, 161, 179, 356, 361–362, 507, 510

N

non-dominated solutions, 290

O

optimization model, 155, 162, 186, 241, 249, 340, 353, 357, 359, 361, 460, 464, 479, 481–483

P

P2P energy trading, 130, **171**, 451, 495, *496*, *500*, 502–503, 505, 508, 509
photovoltaics, **135**, 159, 162, 167
power-to-gas (P2G), 79, 335, 474

R

ramp rate, 350, 352, 370, 393
renewable energy source, 238, 435
reserve market, 445–446, 475, 479, 486

S

smart grid technologies, 185, 265, 449, 490
stochastic optimization, 155, 163, 324, 340, 422, 448, 460, 479
strategic planning, 113, 142, 219, 253
sustainable development, 47, 83, 105, 111, 125–127, 129, 133, 216–217, 222, 225, 258, 341–342, 367
synthetic fuels
system integration, 81, 126–127, 203–204, 219, 296, 380, 453, 471

T

techno-economic analysis, 27, 53, 55, 84, 125–126, 179, 182, 227, 253, 269–270, 364, 394, 451–452
temperature control, 101, 157, 256, 425
transport sector, **68**, 71, 80, 239, **240**, 242, 245, 432

V

value storage, 64–65, 73–74
vehicle-to-grid, 141, 164, 181, 286, 341, 355
virtual power plants, 169, 190, 271, 280, 470
voltage regulation, 141, 301, 309, 312, 445

For Product Safety Concerns and Information please contact our EU
representative GPSR@taylorandfrancis.com
Taylor & Francis Verlag GmbH, Kaufingerstraße 24, 80331 München, Germany